Social-Ecologically More Sustainable Agricultural Production

Social-Ecologically More Sustainable Agricultural Production

Editors

Moritz von Cossel
Joaquín Castro-Montoya
Yasir Iqbal

Basel • Beijing • Wuhan • Barcelona • Belgrade • Novi Sad • Cluj • Manchester

Editors
Moritz von Cossel
University of Hohenheim
Stuttgart
Germany

Joaquín Castro-Montoya
University of El Salvador
San Salvador
El Salvador

Yasir Iqbal
Hunan Agricultural University
Changsha
China

Editorial Office
MDPI
St. Alban-Anlage 66
4052 Basel, Switzerland

This is a reprint of articles from the Special Issue published online in the open access journal
Agronomy (ISSN 2073-4395) (available at: https://www.mdpi.com/journal/agronomy/special_issues/
Social-Ecologically).

For citation purposes, cite each article independently as indicated on the article page online and as
indicated below:

Lastname, A.A.; Lastname, B.B. Article Title. *Journal Name* **Year**, *Volume Number*, Page Range.

ISBN 978-3-7258-0797-0 (Hbk)
ISBN 978-3-7258-0798-7 (PDF)
doi.org/10.3390/books978-3-7258-0798-7

Cover image courtesy of Moritz von Cossel

Contents

About the Editors

Moritz von Cossel

Driven by a passion for social-ecologically more sustainable agricultural production, Moritz von Cossel dedicates his research in crop science to developing more resilient biomass cropping systems, contributing to a future world which is worth living in. In view of current global challenges such as climate change adaptation, biodiversity conservation, groundwater protection, and many others, he is investigating various strategies: perennial nectar-producing wild plants for bioenergy and biobased products, species enrichment of intensively used grassland, and strip-intercropping of non-edible industrial crops for more eco-friendly biomass provision. These innovative approaches pave the way for novel biomass utilization pathways, ultimately fortifying agricultural resilience in a thriving bioeconomy. His enthusiasm for advancing scientific progress shines through over 252 verified peer reviews and 43 editor entries (Web of Science ResearcherID: I-6596-2019, Web of Science, Clarivate, 2024). As a dedicated contributor to the scientific community, he shapes future generations by lecturing in national and international study programs and guiding both experimental and theoretical research through thesis supervision.

Joaquín Castro-Montoya

Joaquín Castro-Montoya graduated with a degree in Agricultural Sciences from the University of El Salvador, an MSc degree in Tropical and Subtropical Agriculture from the University of Hohenheim, Germany, and a Ph.D. in Applied Biological Sciences from the University of Gent, Belgium. Joaquín is a senior lecturer and researcher in the Faculty of Agricultural Sciences of the University of El Salvador and is the Director of the Research Institute for Agrifood and Environmental Sciences. The main research projects of Joaquín focus on the efficient use of resources in agricultural systems, particularly in animal production, with a later emphasis on data-driven decision making and the integration of soil–environment–people in the evaluation of production systems and interventions on them. He has authored more than 40 scientific articles and is a consultant for diverse projects on environment-driven changes in agriculture and other industries.

Yasir Iqbal

Yasir Iqbal is a crop scientist passionate about developing sustainable cropping systems. His research focuses on maximizing biomass production through the establishment of perennial energy crops on marginal and contaminated lands. He delves into the potential of these resource-efficient systems to enhance ecosystem services, particularly soil health improvements. His work goes beyond providing sustainably produced biomass feedstock. He also investigates the quality of biomass for various end-uses, including combustion and bioethanol production, ensuring the optimal utilization of resources.

MDPI

Editorial

Social-Ecologically More Sustainable Agricultural Production

Moritz von Cossel [1,*], Joaquín Castro-Montoya [2] and Yasir Iqbal [3]

[1] Biobased Resources in the Bioeconomy (340b), Institute of Crop Science, University of Hohenheim, 70599 Stuttgart, Germany
[2] Institute of Agri-Food and Environmental Sciences, Faculty of Agricultural Sciences, University of El Salvador, San Salvador 01101, El Salvador
[3] College of Bioscience and Biotechnology, Hunan Agricultural University, Changsha 410128, China
* Correspondence: moritz.cossel@uni-hohenheim.de

Citation: von Cossel, M.; Castro-Montoya, J.; Iqbal, Y. Social-Ecologically More Sustainable Agricultural Production. *Agronomy* **2023**, *13*, 2818. https://doi.org/10.3390/agronomy13112818

Received: 1 November 2023
Accepted: 13 November 2023
Published: 15 November 2023

Planet Earth is facing numerous imminent challenges, from climate change to ecological dysfunction, which are largely attributed to anthropogenic activities. In the long term, this puts humans as a species under threat. It is understandable that humanity's survival depends primarily on the provision of food, drinking water, and a safe habitable environment [1,2]. However, to ensure this, the current production methods employed in leading sectors such as agriculture must adopt a more holistic approach rather than focusing only on food production. On top of that, a large part (in the European Union, a whole 50%) of the plant and animal kingdom is directly or indirectly linked with agricultural systems, making agrobiodiversity a fundamental component of basic agricultural productivity [3]. It would therefore be foolish not to consider the full repertoire of ecosystem services within safe and just Earth system boundaries when developing social-ecologically more sustainable agricultural systems, or at least to strive towards achieving a more holistic view [1,4,5].

Social-ecologically more sustainable agricultural production intends to (i) meet the increasing food demand, (ii) reduce environmental degradation, and (iii) improve a number of other ecosystem services such as the provision of medicinal resources, climate regulation, erosion mitigation, groundwater protection, disturbance modulation, nutrient cycling, habitat functioning, and aesthetic information. Given the importance and relevance of these aspects, this Special Issue has been established to bring together the latest findings from current research. Within the aforementioned overarching theme, this Special Issue received a total of 21 contributions in forms of research articles, review articles, and communications. To facilitate reading, these contributions are briefly presented below.

The first contribution to this Special Issue, a study by Von Cossel et al. [6], reported on the potential trade-off between biomass provision and biodiversity support when species-rich polycultures of perennial flowering wild plant species are cultivated instead of maize (*Zea mays* L.) monocultures [6]. The biomasses of perennial flower-rich wild plants mugwort (*Artemisia vulgaris* L.), brown knapweed (*Centaurea nigra* L.), and common tansy (*Tanacetum vulgare* L.) were found to produce only 72 to 74% of methane compared to maize. This knowledge can help biogas plant operators better implement these types of more biodiversity-friendly biogas substrates to their biogas production value web. Future research should look at the process-relevant biochemical and physical effects caused by the admixture of wild plants as a co-substrate during anaerobic fermentation in the biogas plant [7,8].

In terms of bioenergy crops for combustion, woody species such as aspen (*Populus tremula* L.), Siberian elm (*Ulmus pumila*), and willow (*Salix* spp.) are among the commonly used species [9,10], followed by perennial herbaceous crops such as miscanthus (e.g., *Miscanthus × giganteus* Greef et Deuter) [11–13] and Sida (*Sida hermaphrodita* L. var. Rusby) [14]. All of these perennial bioenergy crops have in common that they are potentially suitable to grow on certain types of marginal land, that is, land that is only marginally suitable for food crop cultivation [9,10,13–16]. Thus, the cultivation of perennial bioenergy crops on unused marginal agricultural land could expedite the development of the bioeconomy and facilitate transition towards a fossil-free future without impeding food security.

For optimal cultivation of bioenergy crops, it is advisable to ensure that genetic material is always adapted to the growing economic and environmental challenges [2]. In this context, Liu et al. [11] applied a sampling strategy on a miscanthus primary core collection to evaluate its role in reducing the size of the initial collection whilst retaining genetic diversity in the collection. This approach was found to improve the range of the coincidence rate without affecting the mean difference percentage. Overall, these findings could contribute to a social-ecologically more sustainable agriculture by reducing the trade-off between biomass provisioning and other ecosystem services through the efficient development of novel miscanthus genotypes in the face of increasing environmental challenges such as climate change-related impacts on agricultural production [2].

However, not only climate change but also soil-related marginality constraints might affect future designs of social-ecologically more sustainable agriculture [15]. A particularly important aspect in this context is the contamination of soil with heavy metals [13]. Liao et al. [13] found that in regions with high concentrations of heavy metals in the soil, the above-ground increment of miscanthus accumulates large amounts of heavy metals each year. According to the authors, this means that if the miscanthus biomass is not used, the heavy metals organically bound in the miscanthus can spread further and impair the balance of natural nutrient cycles in surrounding ecosystems. From Liao et al.'s findings, it can be concluded that, in terms of lower environmental impacts, it would, therefore, make more sense to harvest the annual miscanthus biomass grown in such regions, use it as a bioresource (e.g., for bioenergy purposes), and dispose of the heavy metals contained in the remaining biomass residues in a controlled manner or use them elsewhere. The latter would be a win–win scenario for the bioeconomy approach of growing miscanthus for bioenergy or biobased products on marginal agricultural land.

In addition to nutrient cycling, biomass production systems on marginal land can also affect plant diversity, according to Zuševica et al. [17]. This research group from Latvia found that the establishment of woody crops on organic soils (from former peat extraction) can positively affect plant diversity, whereby the application of ash-based fertilizers and the distances to drainage ditches require special consideration [17]. Furthermore, a better understanding of plant–root bacterial interactions may help to improve the nutrient use efficiency in biomass production on low-yielding (poor) soils. This was found by Wu et al. [18] using the example of ramie (*Boehmeria nivea* L.), which offers great breeding potential for the more efficient use of soil nitrogen and phosphorus, subsequently increasing and stabilizing long-term biomass yields. In addition, Kitzcak et al. [14] succeeded in determining both the minimum organic fertilizer amounts and the optimal seeding rates for the economically feasible cultivation of *Sida* on light (sandy) soils in Poland.

Beyond maintaining agricultural productivity, there are also many chances for its recovery, for instance, through ameliorating contaminated or poor soils with the help of dedicated crops, as was recently reported by Testa et al. [19] and Wu et al. [18]. Given the ever-continuing degradation of agricultural soils worldwide [20], it would therefore be of existential importance to further intensify research on the challenges in modeling bioenergy crop performance, as highlighted by Haberzettl et al. [21], in order to adequately plan and implement bioenergy cropping systems at the interface of provisioning and regulating ecosystem services. Many of the problems for realistic and meaningful modeling approaches lie in the fact that there are not yet sufficient empirical data on the long-term performance of biomass crops on marginal land [21]. Neither for shallow soils [10] nor for soils with adverse soil texture [22] is there sufficient information available to derive biomass production projections for different site conditions worldwide, especially considering the uncertain impacts of climate change on agricultural systems.

A very fundamental effect of climate change is the shift in the water supply and available agricultural land, as reported by Li et al. [23] in their study on spatiotemporal changes in the geographic imbalances between crop production and farmland–water resources in China from 1990 to 2015. From their study, it can be concluded that the cultivation areas of the most important staple foods of rice, wheat, and maize in China will have to be shifted significantly due to climate change-induced fluctuations in precipitation distribution patterns in order to ensure a secure food supply in the future.

Looking at the cropping concept level, Zimmermann et al. [24] elaborated on an option beyond organic and conventional farming that certainly deserves more attention to further optimize the long-term sustainability of crop production. The approach is called mineral–ecological cropping and it aims to increase the benefits of agricultural production for agrobiodiversity whilst maintaining productivity. It builds on the many potential synergies between traditional and modern agricultural practices in a cropping system that exclude the use of synthetic chemical pesticides but allow the use of mineral fertilizers [24]. In this way, Zimmermann et al. suggest that ecosystem services can be increased without reducing productivity. However, further optimization is still needed to implement this new cultivation concept, as it currently appears very difficult to maintain both food crop quality and yield while dispensing with the usual synthetic chemical crop protection agents. In contrast, a new farming concept reported by Arunrat and Sereenonchai [25] seems to be more successful. It is a mixed farming system of rice and fish coculture, which is already used on many farms in Thailand [25]. As the study reveals, the holistic ecosystem services of rice and fish coculture can be increased by 14% in monetary terms compared to the monoculture of rice [25]. Unlike the outdoor farming concepts addressed by Zimmermann et al. and Arunrat and Sereenonchai, indoor farming concepts seem to be more focused on the provision of biomass because they are much less interlinked with the nutrient- and lifecycles of the natural environment. Here, Cichocki et al. [26] provided valuable insights on the opportunities and challenges of providing food directly in and for office buildings [26].

A more holistic recognition of the ecosystem services provided by agricultural value webs could take the form of a true cost–benefit assessment, as Wagner et al. [16] have shown using the example of growing miscanthus. Such insights into the true dimensions of agricultural value webs that have so far been rather neglected could then ideally be incorporated into the design of social-ecologically more sustainable certificates for food, fodder, and other agricultural products in the long term. This would enable a fairer compensation for any opportunity costs on the part of farmers and other involved stakeholders. This is already being sought, for example, for viticulture and wine production worldwide, according to Marques and Teixeira [27] and Wagner et al. [28].

Nevertheless, fairer remuneration must be preceded by the application of more sustainable cultivation practices, and here the views and perceptions of the decision makers directly or indirectly involved also play decisive roles, as Sereenonchai and Arunrat [29] and Huang et al. [30] report. In terms of opportunities for farmer influence, Sereenonchai and Arunrat [29] found that more sustainable cropping systems are usually implemented only when farmers are also aware of the ecosystem benefits. Presenting non-burning uses of rice straw and rice stubble as examples, Sereenonchai and Arunrat found that appropriate communication strategies are needed to ensure that more sustainable farming practices are implemented in a meaningful way in the long term [29]. A similar situation applies to the management strategies of companies that have an indirect link to agricultural production, according to Huang et al. [30]. In their communication article, based on a hierarchical linear modeling approach, Huang et al. suggest to promote the implementation of more sustainable environmental strategies through targeted increases in social responsibility [30].

However, all efforts to encourage farmers or gardeners (in urban areas) to implement social-ecologically more sustainable agricultural production will fail unless the community, as well as political decision makers, endorse it. In this area of research, Wu et al. [31] have made great strides using the example of urban community gardens. Wu et al. [31] have outlined new ways to create more clarity in communities about the potential advantages and disadvantages of such social-ecologically more sustainable urban land use systems. As also highlighted in the studies by Sereenonchai and Arunrat [29] and Huang et al. [30], an appropriate communication strategy about the pros and cons seems to be the key to success in implementing urban community gardens [31]. A trivial solution at first glance, but its justification requires elaborate research adapted to local socio-political as well as geophysical conditions [31].

In summary, this Special Issue offers a wide range of insights into problems, solutions, and next steps towards social-ecologically more sustainable agricultural production. The articles of this Special Issue cover almost at all levels of agricultural production and thus make an important contribution to the agricultural systems of tomorrow.

Author Contributions: Conceptualization: M.v.C.; writing and original draft: M.v.C.; visualization: M.v.C.; manuscript review and editing: J.C.-M., M.v.C. and Y.I. All authors have read and agreed to the published version of the manuscript.

Acknowledgments: The Guest Editors would like to thank first and foremost all the authors who contributed to this Special Issue. Special thanks to Nicolai David Jablonowski for providing constructive feedback on this Editorial. Finally, we would like to thank all Editors and technical staff of the journal Agronomy who were involved in the supervision of this Special Issue for the smooth processing of the submitted manuscripts and for the website support.

References

1. Morizet-Davis, J.; Marting Vidaurre, N.A.; Reinmuth, E.; Rezaei-Chiyaneh, E.; Schlecht, V.; Schmidt, S.; Singh, K.; Vargas-Carpintero, R.; Wagner, M.; Von Cossel, M. Ecosystem Services at the Farm Level—Overview, Synergies, Trade-Offs and Stakeholder Analysis. *Glob. Chall.* **2023**, *7*, 2200225. [CrossRef]
2. Pörtner, H.-O.; Roberts, D.C.; Tignor, M.; Poloczanska, E.S.; Mintenbeck, K.; Alegría, A.; Craig, M.; Langsdorf, S. *IPCC 2022: Climate Change 2022: Impacts, Adaptation and Vulnerability. Contribution of Working Group II to the Sixth Assessment Report of the Intergovernmental Panel on Climate Change 2022*; Cambridge University Press: Cambridge, UK, 2001; Available online: https://www.ipcc.ch/report/ar6/wg2/downloads/report/IPCC_AR6_WGII_FullReport.pdf (accessed on 20 October 2023).
3. European Commission. Enhancing Agricultural Biodiversity. Available online: https://agriculture.ec.europa.eu/sustainability/environmental-sustainability/biodiversity_en (accessed on 20 October 2023).
4. Rockström, J.; Gupta, J.; Qin, D.; Lade, S.J.; Abrams, J.F.; Andersen, L.S.; Armstrong McKay, D.I.; Bai, X.; Bala, G.; Bunn, S.E.; et al. Safe and Just Earth System Boundaries. *Nature* **2023**, *619*, 102–111. [CrossRef]
5. de Groot, R.; Brander, L.; van der Ploeg, S.; Costanza, R.; Bernard, F.; Braat, L.; Christie, M.; Crossman, N.; Ghermandi, A.; Hein, L.; et al. Global Estimates of the Value of Ecosystems and Their Services in Monetary Units. *Ecosyst. Serv.* **2012**, *1*, 50–61. [CrossRef]
6. Von Cossel, M.; Pereira, L.A.; Lewandowski, I. Deciphering Substrate-Specific Methane Yields of Perennial Herbaceous Wild Plant Species. *Agronomy* **2021**, *11*, 451. [CrossRef]
7. Hahn, J.; Westerman, P.R.; de Mol, F.; Heiermann, M.; Gerowitt, B. Viability of Wildflower Seeds After Mesophilic Anaerobic Digestion in Lab-Scale Biogas Reactors. *Front. Plant Sci.* **2022**, *13*, 942346. [CrossRef] [PubMed]
8. Müller, J.; Hahn, J. Ensilability of Biomass from Effloresced Flower Strips as Co-Substrate in Bioenergy Production. *Front. Bioeng. Biotechnol.* **2020**, *8*, 14. [CrossRef] [PubMed]
9. Celma, S.; Sanz, M.; Ciria, P.; Maliarenko, O.; Prysiazhniuk, O.; Daugaviete, M.; Lazdina, D.; von Cossel, M. Yield Performance of Woody Crops on Marginal Agricultural Land in Latvia, Spain and Ukraine. *Agronomy* **2022**, *12*, 908. [CrossRef]
10. Reinhardt, J.; Hilgert, P.; von Cossel, M. Biomass Yield of Selected Herbaceous and Woody Industrial Crops across Marginal Agricultural Sites with Shallow Soil. *Agronomy* **2021**, *11*, 1296. [CrossRef]
11. Liu, S.; Zheng, C.; Xiang, W.; Yi, Z.; Xiao, L. A Sampling Strategy to Develop a Primary Core Collection of Miscanthus Spp. in China Based on Phenotypic Traits. *Agronomy* **2022**, *12*, 678. [CrossRef]
12. Clifton-Brown, J.; Hastings, A.; von Cossel, M.; Murphy-Bokern, D.; McCalmont, J.; Whitaker, J.; Alexopoulou, E.; Amaducci, S.; Andronic, L.; Ashman, C.; et al. Perennial Biomass Cropping and Use: Shaping the Policy Ecosystem in European Countries. *GCB Bioenergy* **2023**, *15*, 538–558. [CrossRef]
13. Liao, X.; Wu, Y.; Fu, T.; Iqbal, Y.; Yang, S.; Li, M.; Yi, Z.; Xue, S. Biomass Quality Variations over Different Harvesting Regimes and Dynamics of Heavy Metal Change in Miscanthus Lutarioriparius around Dongting Lake. *Agronomy* **2022**, *12*, 1188. [CrossRef]
14. Kitczak, T.; Jarnuszewski, G.; Łazar, E.; Malinowski, R. Sida Hermaphrodita Cultivation on Light Soil—A Closer Look at Fertilization and Sowing Density. *Agronomy* **2022**, *12*, 2715. [CrossRef]
15. Von Cossel, M.; Lewandowski, I.; Elbersen, B.; Staritsky, I.; Van Eupen, M.; Iqbal, Y.; Mantel, S.; Scordia, D.; Testa, G.; Cosentino, S.L.; et al. Marginal Agricultural Land Low-Input Systems for Biomass Production. *Energies* **2019**, *12*, 3123. [CrossRef]
16. Wagner, M.; Winkler, B.; Lask, J.; Weik, J.; Kiesel, A.; Koch, M.; Clifton-Brown, J.; von Cossel, M. The True Costs and Benefits of Miscanthus Cultivation. *Agronomy* **2022**, *12*, 3071. [CrossRef]

17. Zuševica, A.; Celma, S.; Neimane, S.; von Cossel, M.; Lazdina, D. Wood-Ash Fertiliser and Distance from Drainage Ditch Affect the Succession and Biodiversity of Vascular Plant Species in Tree Plantings on Marginal Organic Soil. *Agronomy* **2022**, *12*, 421. [CrossRef]

18. Wu, S.; Jie, H.; Jie, Y. Role of Rhizosphere Soil Microbes in Adapting Ramie (*Boehmeria nivea* L.) Plants to Poor Soil Conditions through N-Fixing and P-Solubilization. *Agronomy* **2021**, *11*, 2096. [CrossRef]

19. Testa, G.; Corinzia, S.A.; Cosentino, S.L.; Ciaramella, B.R. Phytoremediation of Cadmium-, Lead-, and Nickel-Polluted Soils by Industrial Hemp. *Agronomy* **2023**, *13*, 995. [CrossRef]

20. AbdelRahman, M.A.E. An Overview of Land Degradation, Desertification and Sustainable Land Management Using GIS and Remote Sensing Applications. *Rend. Fis. Acc. Lincei* **2023**, *34*, 767–808. [CrossRef]

21. Haberzettl, J.; Hilgert, P.; von Cossel, M. A Critical Review on Lignocellulosic Biomass Yield Modeling and the Bioenergy Potential from Marginal Land. *Agronomy* **2021**, *11*, 2397. [CrossRef]

22. Reinhardt, J.; Hilgert, P.; Von Cossel, M. A Review of Industrial Crop Yield Performances on Unfavorable Soil Types. *Agronomy* **2021**, *11*, 2382. [CrossRef]

23. Li, D.; Zhang, H.; Xu, E. Spatiotemporal Changes in the Geographic Imbalances between Crop Production and Farmland-Water Resources in China. *Agronomy* **2022**, *12*, 1111. [CrossRef]

24. Zimmermann, B.; Claß-Mahler, I.; von Cossel, M.; Lewandowski, I.; Weik, J.; Spiller, A.; Nitzko, S.; Lippert, C.; Krimly, T.; Pergner, I.; et al. Mineral-Ecological Cropping Systems—A New Approach to Improve Ecosystem Services by Farming without Chemical Synthetic Plant Protection. *Agronomy* **2021**, *11*, 1710. [CrossRef]

25. Shah, T.M.; Tasawwar, S.; Bhat, M.A.; Otterpohl, R. Intercropping in Rice Farming under the System of Rice Intensification—An Agroecological Strategy for Weed Control, Better Yield, Increased Returns, and Social–Ecological Sustainability. *Agronomy* **2021**, *11*, 1010. [CrossRef]

26. Cichocki, J.; von Cossel, M.; Winkler, B. Techno-Economic Assessment of an Office-Based Indoor Farming Unit. *Agronomy* **2022**, *12*, 3182. [CrossRef]

27. Marques, A.; Teixeira, C.A. Vine and Wine Sustainability in a Cooperative Ecosystem—A Review. *Agronomy* **2023**, *13*, 2644. [CrossRef]

28. Wagner, M.; Stanbury, P.; Dietrich, T.; Döring, J.; Ewert, J.; Foerster, C.; Freund, M.; Friedel, M.; Kammann, C.; Koch, M.; et al. Developing a Sustainability Vision for the Global Wine Industry. *Sustainability* **2023**, *15*, 10487. [CrossRef]

29. Sereenonchai, S.; Arunrat, N. Farmers' Perceptions, Insight Behavior and Communication Strategies for Rice Straw and Stubble Management in Thailand. *Agronomy* **2022**, *12*, 200. [CrossRef]

30. Huang, S.Y.B.; Lee, S.-C.; Lee, Y.-S. Why Can Green Social Responsibility Drive Agricultural Technology Manufacturing Company to Do Good Things? A Novel Adoption Model of Environmental Strategy. *Agronomy* **2021**, *11*, 1673. [CrossRef]

31. Wu, C.; Li, X.; Tian, Y.; Deng, Z.; Yu, X.; Wu, S.; Shu, D.; Peng, Y.; Sheng, F.; Gan, D. Chinese Residents' Perceived Ecosystem Services and Disservices Impacts Behavioral Intention for Urban Community Garden: An Extension of the Theory of Planned Behavior. *Agronomy* **2022**, *12*, 193. [CrossRef]

 *agronomy*

Article

Deciphering Substrate-Specific Methane Yields of Perennial Herbaceous Wild Plant Species

Moritz von Cossel [1,*], Lorena Agra Pereira [2] and Iris Lewandowski [1]

[1] Biobased Resources in the Bioeconomy (340b), Institute of Crop Science, University of Hohenheim, Fruwirthstr. 23, 70599 Stuttgart, Germany; iris_lewandowski@uni-hohenheim.de
[2] Soil Science, Luiz de Queiroz College of Agriculture, University of São Paulo, Av. Pádua Dias, 11, 13418-900 Piracicaba, Brazil; lorena.agra.pereira@alumni.usp.br
* Correspondence: moritz.cossel@uni-hohenheim.de; Tel.: +49-711-459-23557

Abstract: The global demand for plant biomass to provide bioenergy and heat is continuously increasing because of a growing interest among many industrialized and developing countries towards climate sound and renewable energy supply. The exacerbation of land-use conflicts proliferates social-ecological demands on future bioenergy cropping systems. Perennial herbaceous wild plant mixtures (WPMs) represent an approach to providing social-ecologically more sustainably produced biogas substrate that has gained increasing public and political interest only in recent years. The focus of this study lies on three perennial wild plant species (WPS) that usually dominate the biomass yield performance of WPM cultivation. These WPS were compared with established biogas crops in terms of their substrate-specific methane yield (SMY) and lignocellulosic composition. The plant samples were investigated in a small-scale mesophilic discontinuous biogas batch test for determining the SMY. All WPS were found to have significantly lower SMY (241.5–248.5 l_N kgVS^{-1}) than maize (337.5 l_N kgVS^{-1}). This was attributed to higher contents of lignin (9.7–12.8% of dry matter) as well as lower contents of hemicellulose (9.9–11.5% of dry matter) in the WPS. Only minor, non-significant differences to cup plant and Virginia mallow were observed. Thus, when planning WPS as a diversification measure in biogas cropping systems, their lower SMY should be considered.

Keywords: anaerobic digestion; *Artemisia vulgaris* L.; biodiversity; biogas production; brown knapweed; *Centaurea nigra* L.; common tansy; mugwort; perennial crops; *Tanacetum vulgare* L.

Citation: von Cossel, M.; Pereira, L.A.; Lewandowski, I. Deciphering Substrate-Specific Methane Yields of Perennial Herbaceous Wild Plant Species. *Agronomy* **2021**, *11*, 451. https://doi.org/10.3390/agronomy11030451

Academic Editor: Steven R. Larson

Received: 2 February 2021
Accepted: 24 February 2021
Published: 28 February 2021

Publisher's Note: MDPI stays neutral with regard to jurisdictional claims in published maps and institutional affiliations.

1. Introduction

Supplying "clean" energy is a major component of the growing bioeconomy, the core goal of which is the complete replacement of fossil and nuclear resources with renewable energy and bioenergy [1]. The full extent of this challenge can be seen in the fact that the share of renewable energy in total global energy consumption seems to have stalled at between 12 and 14% over the last 20 years, despite various efforts and scientific progress. While the amount of renewable energy has increased from 54.4 to 82.7 EJ, the amount of fossil fuels, such as coal, oil, and gas, has also increased significantly over the same period, from 337.7 EJ to 486 EJ [1]. Apart from the end use sectors heat and transport, bioenergy makes up only a small share of 2.4% of total renewable energy production [1]. However, bioenergy cropping systems are assumed to have a promising future for two important reasons:

1. By growing bioenergy crops, unused land can be returned to agricultural production and, if necessary, even protected from further degradation by adhering to best management practices.
2. Bioenergy production enables a stable basis for the reliable provision of electricity and heat compared to wind and solar energy, which are subject to strong fluctuations.

Many other ecosystem functions besides the provision of biomass are currently only being discovered bit by bit or investigated in connection with bioenergy cropping systems. The additional ecosystem services resulting from these ecosystem functions could be a turning point in the history of bioenergy cropping systems, as monetization of them could increase land conversion many times over. For example, the monetary value of all ecosystem services of growing Miscanthus (*Miscanthus* ANDERSSON), a very well-known perennial bioenergy crop [2–4], in a case study region in Germany varies between 1200 and 4183 € per hectare and year [5]. Several other perennial second generation lignocellulosic crops such as switchgrass (*Panicum virgatum* spp.) [6,7], willow (*Salix* spp.) [8–11], cup plant (*Silphium perfoliatum* L.) [12–16] and Virginia mallow (*Sida hermaphrodita* L. Rusby) [13,17] have been intensively researched worldwide for decades [18]. All these bioenergy cropping systems have one thing in common: they are monocultures. Therefore, it is to be expected that agricultural biodiversity could be better promoted by a more diverse bioenergy cropping system. In the search for more diverse bioenergy cropping systems, the first reports were published during the last nine years on how species-rich flowering mixtures of annual, biennial, and perennial wild plants can significantly enhance many nursery services compared with the abovementioned mono-perennials [19–22]. These so-called "perennial wild plant mixtures" (WPM) were investigated by several German institutions over the past decade for their use as second generation co-substrates in anaerobic digestion [19–21,23–30]. Whether WPMs are also suitable for other bioenergy production pathways such as combustion, pyrolysis or bioethanol production has not yet been explored [22].

It was found that WPM cultivation for anaerobic digestion, under the best circumstances, provides both a notable farm productivity, as indicated by a five-year average annual dry matter yield (DMY) of 12.5 Mg ha^{-1} at an annual nitrogen fertilization of 50 kg ha^{-1} [28,31,32], and an improvement of various social-ecological services [20,25,27,33–35]. However, the successful cultivation of WPMs strongly depends on several factors such as the seed-bed preparation, the sowing procedure, the weather conditions, the soil heterogeneity and the weed pressure [22,23,31,36]. After successful establishment, WPM cultivation provides high biomass yields each year accompanied by a dynamic change in the WPM species composition over the years [31]. Annual species dominate the plant stand in the first year of cultivation, biennial species in the second year, and perennial species from the third year onwards [25,31,36]. Therefore, perennial wild plant species (WPS) such as common tansy (*Tanacetum vulgare* L.), common knapweed (*Centaurea nigra* L.) and mugwort (*Artemisia vulgaris* L.) have the highest impact on the overall yield performance of the WPM in the long-term [22]. This is because the WPM can grow up to 5 years and even longer [22,25,33–35], and the perennial WPS have the highest share of total accumulated DMY [22,31,36].

Despite the fact that the DMY is the main determinant for the methane yield per hectare (MYH) of biogas crops [37–39], the substrate-specific methane yield (SMY) also plays a vital role in biogas plant management, with regard to (i) the organic loading of the fermenter (the higher the SMY the better the organic loading efficiency), (ii) the retention time of the co-substrate in the fermenter (the higher the SMY, the shorter the retention time in the biogas plant), and (iii) the secondary effects on the digestibility of the other fermentation substrate components, for example through the provision of essential trace elements [22,25,40–42]. However, little is known about the substrate-specific methane yield (SMY) of perennial WPS [19,43,44]. In most of the few studies on the methane yield potential of WPM, the mixtures are considered as a whole (plant stand level) and not examined for individual plant performance [21,23,36,45,46]. In addition, there are large differences within the limited data available. For example, SMY values from 287.5 [19] to 362.0 l$_N$ kgVS^{-1} [47] are reported for common and brown knapweed, respectively. For the other promising WPS, only single values are available, accounting for 233 l$_N$ kgVS^{-1} (common tansy) and 346 l$_N$ kgVS^{-1} (mugwort) [19]. Therefore, this study aims at investigating the potential SMY of relevant perennial WPS and compare them with relevant annual and perennial alternative biogas co-substrates. The results are expected to help better understanding the relevance of the WPM species composition dynamics [31] towards the development of social-ecologically more sustainable bioenergy cropping systems.

2. Materials and Methods

2.1. Origin and Harvest of Plant Material

The investigations in this study are based on above-ground biomass harvested from common tansy, brown knapweed, mugwort, cup plant, Virginia mallow, and maize (*Zea mays* L.) (Table 1). Cup plant, maize and Virginia mallow served as reference crops. All biomass samples were taken from the same field trial in Hohenheim, southwest Germany (407 m AMSL, N 48°42′57.024″, O 9°12′52.956″) (Figure 1).

Table 1. Overview of the crops (sorted alphabetically) used in this study.

Trivial Name	Botanical Name	Life Cycle	Origin
Common knapweed	*Centaurea nigra* L.	Perennial	Temperate Europe
Common tansy	*Tanacetum vulgare* L.	Perennial	Temperate Europe and Asia
Cup plant	*Silphium perfoliatum* L.	Perennial	Northern America
Maize	*Zea mays* L.	Annual	Central America
Mugwort	*Artemisia vulgaris* L.	Perennial	Temperate Europe, Alaska, Northern Africa and Asia
Virginia mallow	*Sida hermaphrodita* L. Rusby	Perennial	Northern America

Figure 1. Overview of the crop species investigated in this study: (**a**) common knapweed (**b**) common tansy (**c**) cup plant (**d**) maize (**e**) mugwort and (**f**) Virginia mallow.

This field trial was established in a randomized block design with three (maize, Virginia mallow, cup plant) and five replicates (WPM), respectively, in 2014. The plots were of square shape and their gross area was 36 m². The distance between the plots was 1.5 m, and the distance between the blocks was 5 m. The site is characterized by homogeneous favorable abiotic growth conditions, such as (i) clayey loam (Luvisol) [36], (ii) an average annual air temperature of 10.1 °C in 2016 (Figure 2), 1.4 °C higher compared with long-term data, and (iii) an annual precipitation of 595 mm in 2016 (Figure 2), which was 103 mm less compared with long-term data. The harvest dates of the biomass samples for this study varied according to the crop-specific demands. The WPS (common tansy, common knapweed and mugwort) and Virginia mallow were harvested in August 2016. Cup plant and maize were harvested in October 2016. Only fully developed individual plants from the WPM plots were selected for harvest of the WPS, with three plots each found for common tansy and common knapweed, but only one plot for mugwort. For cup plant, only plant samples of two randomly selected representative plots of the three existing plots were chosen due to technical reasons. For all crops, harvesting was done by hand using a pruning shear.

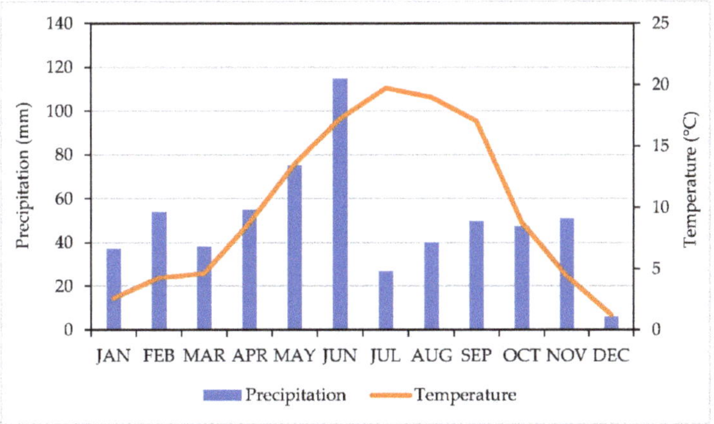

Figure 2. Overview of monthly precipitation and monthly average temperature conditions at the field trial site (407 m AMSL, N 48°42′57.024″, O 9°12′52.956″) in the year of harvest (2016).

2.2. Determination of C- and N-Content, Fibre Analyses

After harvesting and drying to constant weight (at 58 °C), the samples were milled using a cutting mill (SM 200, Retsch, Haan, Germany) with a 1 mm sieve for further analysis (including the biogas batch test). For the following analyses, the plant sample material was not pre-treated, e.g., through enzymatic hydrolysis. To measure nutrient detergent fiber content (NDF), acid detergent fiber (ADF), lignin (ADL), total carbon (C_T) and total nitrogen (N_T) all samples were prepared as follows: The ash content of plant samples was estimated according to Kiesel and Lewandowski [48], by drying a 1 g subsample at 105 °C in a cabinet dryer (to determine residual moisture) and burning at 550 °C in a muffle furnace to constant weight. After that, the contents of NDF, ADF and ADL were analyzed according to VDLUFA Method Book III, methods 6.5.1, 6.5.2 and 6.5.3 [49]. The contents of cellulose (CL) and hemicellulose (HC) were calculated using the following Equations:

$$CL = ADF - ADL \tag{1}$$

$$HC = NDF - ADF. \tag{2}$$

The contents of N_T and C_T were measured according to DIN ISO 5725 using the elemental analyzer 'Vario Max CNS' (Elementar Analysensysteme GmbH, Langenselbold, Germany).

2.3. Biogas Batch Test

The biogas batch test was conducted according to Von Cossel et al. [50]. The test commenced on 8 April 2019 and ended on 13 May 2019 with the duration of the experiment fixed in the implementation protocol of the biogas batch test. For the biogas batch test (wet fermentation), 200 mg of organic dry matter of the plant samples was mixed with 30.0 ± 0.1 g inoculum (4% DMC, origins from a biogas plant, degassed under the conditions intended for the biogas batch test) in 100 mL air-tight bottles and kept at 39 °C for 35 days, a standard procedure according to VDI guideline 4630 [48,51,52]. The substrate to inoculum ratio accounted for 1:3 on a volatile solids (VS) basis. The actual plant material per batch flask ranged from 229.2 mg DM (Virginia mallow) to 234.5 mg DM (cup plant) due to differences in ash content. Therefore, the DMC in the test bottles was about 4.7%. Each field replicate of the plant samples was repeated four times within the biogas batch test, and gas was collected a total of four times. After each gas collection, each bottle was emptied with a hollow needle. A hand-held pressure gauge for external pressure sensors (HND-P pressure gauge, Kobold Messring GmbH, Hofheim, Germany) was used to measure the pressure rise in order to calculate the gas production, taking into account the respective ambient air pressure. At the beginning of the biogas batch test, measurements were taken daily, while towards the end measurements were taken every three days due to decreasing gas production. The pressure increase was measured 19 times during the batch test and converted into standardized values (standard conditions: 0 °C and 1013 hPa). The control (inoculum without plant material) and ambient atmospheric pressure was required to calculate the accumulated substrate-specific net biogas yield (SBY). This is because biogas production still occurs even when the inoculum is starved, and its volume must be subtracted from the total volume per plant sample. A thermal conductivity detector (gas chromatograph GC-2014, Shimadzu, Kyoto) was used to determine the methane content (MC) of the collected biogas at a detection temperature of 120 °C. Under an oven temperature of 50 °C and the carrier gas argon, two columns (Haye-Sep and Molsieve column) were used [48]. All gas samples were injected with a Combi-xt PAL autosampler (CTC Analytics AG, Zwingen, Switzerland) [48]. The substrate-specific methane yield (SMY) was calculated following Equation (3):

$$\text{SMY} = \text{SBY} \times \text{MC}. \tag{3}$$

2.4. Statistical Analysis

Data curation was conducted using MS Excel. The biogas batch test was analyzed in accordance with [50]. The F-tests for the effects of the different crops on SMY and the biochemical constituents were conducted as adapted from according to [50] following Equation (4):

$$y_i = \mu + \tau_i + e_i \tag{4}$$

where μ is the intercept and e_i is the error of observation y_i with crop-specific variance. τ_i is the fixed effect for the ith crop species.

If differences were found, a multiple *t*-test was performed to create a letter display [53]. The assumptions of normality and homogeneous error variance were checked graphically. The Akaike information criterion (AIC) [54] was used to selected the best model. All analysis run using the PROC MIXED procedure of the SAS® Proprietary Software 9.4 TS level 1M5 (SAS Institute Inc., Cary, NC, USA). For the correlation matrix and SMY prediction, PROC CORR and PROC REG (SAS® Proprietary Software 9.4 TS level 1M5, see above) were used. Both degrees of freedom and standard errors were approximated using the Kenward-Roger method [55].

3. Results and Discussion

Both the lignocellulose composition studies, and the biogas batch tests showed significant differences between the WPS and the reference crop species. Only results from one crop year are available here, which means that there is not yet any information on the possibility of an interaction between crop type and climatic variations with respect to SMY. This could be assumed, since seasonal climatic conditions usually have a large influence on crop-specific biomass yield and quality [51]. However, no information is yet available on this with regard to WPS and it was not possible to investigate this in this study. Therefore, the use of plant samples from two or more seasons would be appropriate in future studies to examine the year effects on both specific biomass yield and quality of different biogas crops or biogas cropping systems. In the following, the results of the two categories lignocellulose and biogas batch test are presented and discussed separately.

3.1. Lignocellulosic Composition

The analyses of lignocellulosic composition revealed a large variation across plant species in contents of DM of lignin (3.2–12.6%), cellulose (25.8–48.8%) and hemicellulose (5.0–27.4%) (Table 2).

Table 2. Lignocellulosic composition of the biogas crops (sorted alphabetically) investigated in this study. Additionally, the standard error is provided. The color scaling indicates per parameter the meaning of the value for the use of biomass as biogas substrate from good (dark green) to bad (deep red).

Crop	NDF (% of DM)	ADF (% of DM)	ADL (% of DM)	Cellulose (% of DM)	Hemicellulose (% of DM)
Common knapweed	57.6 + 1.9 ab	47.3 + 1.9 a	9.7 + 0.7 b	37.6 + 1.3 a	10.3 + 0.6 b
Common tansy	62.4 + 1.9 a	50.9 + 1.9 a	12.8 + 0.7 a	38.1 + 1.3 a	11.5 + 0.6 b
Cup plant	52.0 + 2.4 b	44.6 + 2.3 a	6.7 + 0.9 c	37.9 + 1.6 a	7.4 + 0.7 c
Maize	52.7 + 1.9 b	29.0 + 1.9 b	3.3 + 0.7 d	25.8 + 1.3 b	23.7 + 0.6 a
Mugwort	61.9 + 3.4 a	52.0 + 3.3 a	12.6 + 1.3 ab	39.4 + 2.3 a	9.9 + 1.0 bc
Virginia	58.7 + 1.9 ab	47.8 + 1.9 a	7.0 + 0.7 c	40.8 + 1.3 a	10.9 + 0.6 b

NDF = neutral detergent fiber, ADF = acid detergent fiber, ADL = acid detergent lignin, DM = dry matter, n = number of field replicates. Different lower case letters denote for significant ($p < 0.05$) differences between crops within parameter.

The C:N ratio was highest for mugwort (127.1) and lowest for maize (55.2) (Table 3). Considering that a C:N ratio of 15–30:1 is required for a stable anaerobic digestion process in the biogas plant [56], all crops show too high a C:N ratio (Table 4). While there are no data in the literature for mugwort, common tansy and common knapweed that could be used for comparison, the values for maize compare well with those in the literature [57], although they appear somewhat too high (>36.2:1). This may be due to the difference in sample preparation, as the values in the literature are based on maize silage [57], whereas in this study dried maize samples were available that had not been ensiled beforehand. In any case, it can be seen that with an increasing share of WPS in the biogas crop rotation [58], attention should be paid to appropriate N supply to the fermenter in the biogas production process, which can usually be realized by adding residues from animal husbandry (slurry, manure). The C:N ratio of mugwort was thus much higher than that of straw, which is 69.5:1. But still, the SMY of mugwort was notable higher than that of straw, which is about 189 l_N kgVS^{-1} [59]. This could be due to the low ash content and mediocre hemicellulose content of mugwort (Tables 2 and 3) compared to the other crops studied. However, the C:N-ratio alone does not allow an evaluation for or against one of these wild plant species in comparison with maize.

The ash content of dry matter was highest for cup plant (9.7%) and intermediate in wild plant species (5.2–6.4%) indicating the highest ash dry matter content (Table 3).

Table 3. Contents of nitrogen, carbon, C_T:N_T ratio, ash and dry matter content (right before entering the biogas batch test) within the plant material (sorted alphabetically). Additionally, the standard error is provided. The color scaling indicates per parameter the meaning of the value for the use of biomass as biogas substrate from good (dark green) to bad (deep red).

Crop	N_T (% of DM)	C_T (% of DM)	C_T:N_T Ratio	Ash (% of DM)	DMC_{DS} (%)
Common knapweed	0.7 + 0.1 bc	46.1 + 0.3 bc	68.3 + 4.2 bc	6.4 + 0.3 b	93.6 + 0.3 c
Common tansy	0.6 + 0.1 bd	47.3 + 0.3 a	75.5 + 4.2 b	6.1 + 0.3 bc	93.9 + 0.3 bc
Cup plant	0.6 + 0.1 cd	44.0 + 0.3 d	77.9 + 5.2 b	9.2 + 0.3 a	90.8 + 0.3 d
Maize	0.8 + 0.1 b	45.4 + 0.3 c	57.2 + 4.2 c	4.1 + 0.3 d	95.9 + 0.3 a
Mugwort	0.4 + 0.1 d	46.8 + 0.3 ab	127.1 + 7.3 a	5.2 + 0.4 cd	94.8 + 0.4 ab
Virginia	1.2 + 0.1 a	45.7 + 0.3 bc	38.0 + 4.2 d	6.7 + 0.3 b	93.3 + 0.3 c

N_T = total nitrogen content, DM = dry matter, C_T = total carbon content, DMC_{DS} = dry matter of the dried plant substrate right before entering the biogas batch test. Different lower case letters denote for significant ($p < 0.05$) differences between crops within parameter.

Table 4. Methane content and substrate-specific methane yield of the crops (sorted alphabetically). Additionally, the standard error is provided. The color scaling indicates per parameter the meaning of the value for the use of biomass as biogas substrate from good (dark green) to bad (deep red).

Crop	CH_4 (%)	SMY (l_N kgVS^{-1})
Common knapweed	53.7 + 0.2 ab	248.5 + 4.1 c
Common tansy	54.2 + 0.2 a	243.2 + 4.1 c
Cup plant	53.3 + 0.3 bc	264.7 + 5.0 b
Maize	52.9 + 0.2 c	337.5 + 4.1 a
Mugwort	53.5 + 0.4 ac	241.5 + 7.0 c
Virginia	54.1 + 0.2 ab	267.2 + 4.1 b

N = norm conditions, CH_4 = methane content, SMY = substrate-specific methane yield, vs. = volatile solids. Different lower case letters denote for significant ($p < 0.05$) differences between crops within parameter.

3.2. Methane Content and Substrate-Specific Methane Yield

The methane content of the substrate-specific biogas was highest for common tansy (54.2%) and lowest for maize (52.9%) (Table 4). The SMY ranged from 241.5 l_N kgVS^{-1} (mugwort) to 337.5 l_N kgVS^{-1} (maize). The net velocity of biogas production was lowest for the WPS compared with maize, Virginia mallow and cup plant (Figure 3). This resulted in a lower slope of the accumulated substrate-specific net biogas production of the WPS (Figure 4). For all crops however, the duration of the biogas batch test appears to have been long enough to reach the maximum specific biogas yield potential because no significant biogas production was observed after the 34th day of the biogas batch test (Figure 4).

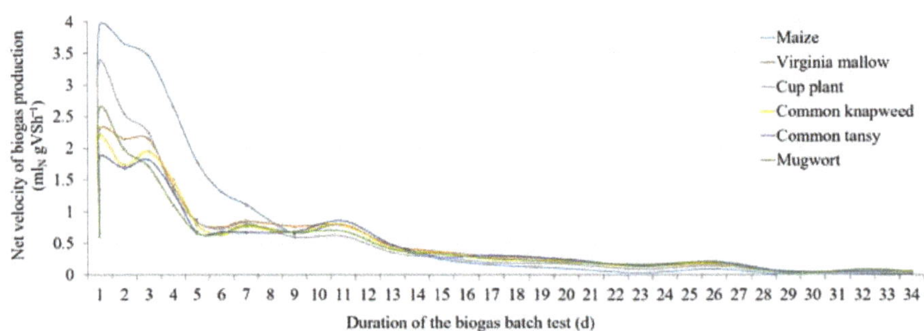

Figure 3. Net velocity of biogas production per gram volatile solids from the crops tested in this study. For each measurement and for each crop except mugwort, the error bars indicate the standard deviation for the replicates of the crop species in the field trial.

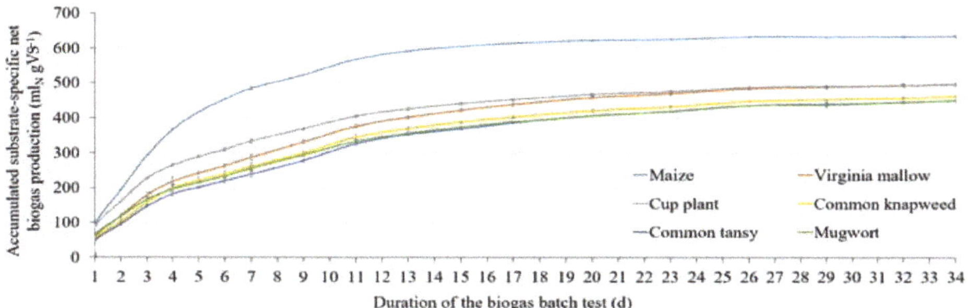

Figure 4. Accumulated substrate-specific net biogas production of the crops investigated in this study. For each measurement and for each crop except mugwort, the error bars indicate the standard deviation for the replicates of the crop species in the field trial.

Both, methane content and SMY are slightly lower than reported by [60,61]. This is likely because of variations in pre-treatment; the plant samples were ensiled before biogas batch test by [60]; whereas in our study, the plant samples were not ensiled. Ensilaging is known to increase SMY to some extend [60,62,63]. However, the results of biogas batch tests are generally not directly comparable due to large variations of methodological settings and conditions [60]. Against this backdrop, it also makes sense to compare the ratios between plant species within the studies. In [60] for example, the SMY of maize was about 1.6 times higher than for cup plant. In this study, the SMY of maize was also notably (1.3 times) higher compared with cup plant (Table 4). In [60], this was drawn back to differences in biochemical composition. This also applies to the results in this study, because maize has (i) significantly lower contents of lignin, which negatively correlates with the SMY (0.92, $p < 0.001$), and (ii) higher contents of N, which positively correlates with the SMY (0.54, $p < 0.05$) (Table 5).

Table 5. Pearson's correlation coefficients matrix of substrate-specific biochemical compositions and the key parameters of the biogas batch test. The levels of significance are indicated by asterisks. Significant Pearson's correlation coefficients were colorized to emphasize negative (dark red) and positive values (dark green).

	NDF	ADF	ADL	CEL	HC	Ash	N_T	C_T	CNR	SMY
ADF	0.78 **									
ADL	0.83 ***	0.87 ***								
CEL	0.65 *	n.r.	n.r.							
HC	n.s.	n.r.	−0.61 *	−0.89 ***						
Ash	n.s.	n.s.	n.s.	n.s.	−0.80 **					
N_T	n.s.	−0.15 *	−0.41 *	0.03 *	n.s.	n.s.				
C_T	0.67 **	n.s.	0.70 **	n.s.	n.s.	n.s.	n.s.			
CNR	n.s.	0.33 *	0.54 *	n.s.	n.s.	n.s.	n.r.	n.s.		
SMY	−0.66 *	−0.96 ***	−0.88 ***	−0.89 ***	0.90 ***	n.s.	0.26 *	n.s.	−0.39 *	
CH_4	n.s.	0.69 **	0.59 *	0.66 **	−0.51 **	0.32 *	n.s.	n.s.	n.s.	−0.64 **

N_T = total nitrogen content, C_T = total carbon content, CNR = C_T:N_T ratio, NDF = neutral detergent fiber, ADF = Acid detergent fiber, ADL = acid detergent lignin, CEL = cellulose, HC = hemicellulose, CH_4 = methane content, SMY = specific methane yield, * = $p < 0.05$, ** = $p < 0.01$, *** = $p < 0.0001$, n.s. = not significant, n.r. = not relevant.

The results from the lignocellulosic analyses (Tables 2 and 3) helped to interpret the results of the biogas batch test. Across plant species, lignin content had the strongest (negative) effect on SMY. This is in line with literature [43,64,65] (Table 5). Other correlations between SMY and biochemical constituents of the crops were either weak or not significant (Table 5). Regression analyses revealed a well-fitting ($R^2 = 0.9825$, $p < 0.0001$) prediction model shown in Equation (5):

$$SMY = 305.15579 + 2.94265 \times NDF - 3.79094 \times ADF - 4.20099 \times ADL, \tag{5}$$

with NDF = neutral detergent fiber, ADF = acid detergent fiber, ADL = acid detergent lignin. As expected, the strong negative influence of lignin on SMY also has a great significance in this SMY prediction equation.

However, following the findings of [43], the high accuracy of this prediction model is very likely due to the large variation of biochemical composition between the crops (Tables 2 and 3). Overall, lignin was found to be most relevant for SMY prediction (Table 5). But this is mostly the case for so-called "across-crop" prediction models [43,64–66]. Such models may be useful for the prediction of the SMY of mixtures whose species compositions of are known, for instance regarding crop rotation planning or national biomass potential analyses [67]. But for selecting the best genotypes within individual crop species such as WPS, species-specific prediction models would be required [43]. However, lignin content is an important parameter for the SMY of WPS [43,64–66]. Therefore, it is necessary to learn more about how to reduce the lignin content of WPS through advanced agronomic practices, e.g., harvest determination and planting geometry, in the future. Breeding could probably also help further improving WPS, which is currently being investigated in a German research project that focuses on common tansy [68].

As Table 5 further shows, the SMY correlates strong positively (R = 0.90) and highly significantly ($p < 0.0001$) with hemicellulose. Since hemicellulose is relatively low in WPS, this is also another reason for the low slope of the accumulated substrate-specific net biogas production of the WPS (Figure 4). This is also in line with expectations, since hemicellulose is easily digestible in anaerobic digestion [43,64–66]. Thus, it seems reasonable to pay attention to increasing the hemicellulose content for improving the biogas substrate quality of WPS. Furthermore, lignin and hemicellulose were found to be significantly ($p < 0.05$) moderately (R = |0.4| − |0.7|) correlated with methane content. For lignin, the correlation was positive, and for hemicellulose, the correlation was negative. Therefore, it would be expected that a decrease in lignin content combined with an increase in hemicellulosic content could result in a reduction in methane content of the biogas produce. However, as shown by the low methane content of maize (Table 4), this should not be a hindrance to increasing the overall SMY of WPS.

If only relatively small areas, such as field margins, are to be managed with WPS in a biogas scenario, only relatively small amounts of WPS silage would be available for biogas production. These could then be mixed in the biogas plant with more fermentable biomass from other biogas crops or manure. In this case, WPS would provide an opportunity to promote agrobiodiversity in the biogas crop rotation, at least on a small scale, without causing significant net income losses. If these small quantities were to be used in the alternative utilization pathway of combustion, additional investments might be required (e.g., for pellet production), which would not be worthwhile for small substrate quantities. However, the currently still lower specific methane yield of WPS compared to maize should be carefully considered for biogas plant management. It remains to be seen how the development of new seed mixtures [58,69] or breeding of new genotypes [68] will help reduce these qualitative differences between WPS and the more established biogas crops.

4. Conclusions

In this study, those WPS which most strongly contribute to the accumulated biomass yield of WPM over the whole multi-annual growth period (five years and longer) were analyzed for their specific biogas yield. All of them yield less biogas than the comparison plant species: conventional annual (maize), or perennial (cup plant, Virginia mallow). This is mostly due to the unfavorable ratio of lignin (too high) and hemicellulose (too low) in the biomass of those perennial WPS. Therefore, other energetic end uses, such as combustion, may be more appropriate. For combustion high lignin contents are desirable and therefore the crops are harvested later and stay longer in the field [8,70,71]. This brings additional positive effects in terms of other ecosystem services, such as (i) extended protection for animals from the weather and from predators (nursery services), and (ii) extended feed provision.

Author Contributions: Conceptualization, M.v.C. and L.A.P.; Data curation, M.v.C. and L.A.P.; Formal analysis, M.v.C.; Investigation, M.v.C. and L.A.P.; Methodology, M.v.C.; Project administration, M.v.C. and I.L.; Resources, M.v.C.; Supervision, M.v.C. and I.L.; Validation, M.v.C.; Visualization, M.v.C.; Writing—original draft, M.v.C., L.A.P. and I.L.; Writing—review & editing, M.v.C. All authors have read and agreed to the published version of the manuscript.

Funding: This research received funding from the European Union's Horizon 2020 research and innovation program under grant agreement No 727698, and the University of Hohenheim. The APC was funded by the European Union's Horizon 2020 research and innovation program under grant agreement No 727698.

Data Availability Statement: Data is contained within the article.

Acknowledgments: The authors would like to thank the staff of the Department of Biobased Resources in the Bioeconomy involved in the laboratory work. Special thanks go to Eva Lewin for improving the language quality of the manuscript.

Conflicts of Interest: The authors declare no conflict of interest. The funders had no role in the design of the study; in the collection, analyses, or interpretation of data; in the writing of the manuscript, or in the decision to publish the results.

References

1. WBA. *Global Bioenergy Statistics 2020*; World Bioenergy Association: Stockholm, Sweden, 2020.
2. Beale, C.V.; Long, S.P. Seasonal Dynamics of Nutrient Accumulation and Partitioning in the Perennial C4-Grasses Miscanthus x Giganteus and Spartina Cynosuroides. *Biomass Bioenergy* **1997**, *12*, 419–428. [CrossRef]
3. Heaton, E.A.; Dohleman, F.G.; Long, S.P. Meeting US Biofuel Goals with Less Land: The Potential of Miscanthus. *Glob. Chang. Biol.* **2008**, *14*, 2000–2014. [CrossRef]
4. Lewandowski, I.; Scurlock, J.M.; Lindvall, E.; Christou, M. The Development and Current Status of Perennial Rhizomatous Grasses as Energy Crops in the US and Europe. *Biomass Bioenergy* **2003**, *25*, 335–361. [CrossRef]
5. Von Cossel, M.; Winkler, B.; Mangold, A.; Lask, J.; Wagner, M.; Lewandowski, I.; Elbersen, B.; Eupen, M.; Mantel, S.; Kiesel, A. Bridging the Gap Between Biofuels and Biodiversity Through Monetizing Environmental Services of *Miscanthus* Cultivation. *Earth's Future* **2020**, *8*. [CrossRef]
6. Alexopoulou, E.; Zanetti, F.; Papazoglou, E.G.; Christou, M.; Papatheohari, Y.; Tsiotas, K.; Papamichael, I. Long-Term Studies on Switchgrass Grown on a Marginal Area in Greece under Different Varieties and Nitrogen Fertilization Rates. *Ind. Crop. Prod.* **2017**, *107*, 446–452. [CrossRef]
7. David, K.; Ragauskas, A.J. Switchgrass as an Energy Crop for Biofuel Production: A Review of Its Ligno-Cellulosic Chemical Properties. *Energy Environ. Sci.* **2010**, *3*, 1182–1190. [CrossRef]
8. Krzyżaniak, M.; Stolarski, M.J.; Szczukowski, S.; Tworkowski, J.; Bieniek, A.; Mleczek, M. Willow Biomass Obtained from Different Soils as a Feedstock for Energy. *Ind. Crop. Prod.* **2015**, *75*, 114–121. [CrossRef]
9. McElroy, G.H.; Dawson, W.M. Biomass from Short-Rotation Coppice Willow on Marginal Land. *Biomass* **1986**, *10*, 225–240. [CrossRef]
10. Stolarski, M.J.; Niksa, D.; Krzyżaniak, M.; Tworkowski, J.; Szczukowski, S. Willow Productivity from Small-and Large-Scale Experimental Plantations in Poland from 2000 to 2017. *Renew. Sustain. Energy Rev.* **2019**, *101*, 461–475. [CrossRef]
11. Stolarski, M.J.; Szczukowski, S.; Tworkowski, J.; Klasa, A. Willow Biomass Production under Conditions of Low-Input Agriculture on Marginal Soils. *For. Ecol. Manag.* **2011**, *262*, 1558–1566. [CrossRef]
12. Bufe, C.; Korevaar, H. *Evaluation of Additional Crops for Dutch List of Ecological Focus Area: Evaluation of Miscanthus, Silphium Perfoliatum, Fallow Sown in with Melliferous Plants and Sunflowers in Seed Mixtures for Catch Crops*; Wageningen Research Foundation (WR) Business Unit Agrosystems Research: Lelystad, The Netherlands, 2018.
13. Franzaring, J.; Holz, I.; Kauf, Z.; Fangmeier, A. Responses of the Novel Bioenergy Plant Species *Sida Hermaphrodita* (L.) Rusby and *Silphium Perfoliatum* L. to CO_2 Fertilization at Different Temperatures and Water Supply. *Biomass Bioenergy* **2015**, *81*, 574–583. [CrossRef]
14. Gansberger, M.; Montgomery, L.F.R.; Liebhard, P. Botanical Characteristics, Crop Management and Potential of *Silphium Perfoliatum* L. as a Renewable Resource for Biogas Production: A Review. *Ind. Crop. Prod.* **2015**, *63*, 362–372. [CrossRef]
15. Hartmann, A.; Lunenberg, T. Yield Potential of Cup Plant under Bavarian Cultivation Conditions. *J. Fur Kult.* **2016**, *68*, 385–388. [CrossRef]
16. Šiaudinis, G.; Jasinskas, A.; Šarauskis, E.; Steponavičius, D.; Karčauskienė, D.; Liaudanskienė, I. The Assessment of Virginia Mallow (Sida Hermaphrodita Rusby) and Cup Plant (*Silphium Perfoliatum* L.) Productivity, Physico–Mechanical Properties and Energy Expenses. *Energy* **2015**, *93 Pt 1*, 606–612. [CrossRef]
17. Jablonowski, N.D.; Kollmann, T.; Meiller, M.; Dohrn, M.; Müller, M.; Nabel, M.; Zapp, P.; Schonhoff, A.; Schrey, S.D. Full Assessment of Sida (Sida Hermaphrodita) Biomass as a Solid Fuel. *Gcb Bioenergy* **2020**, *12*, 618–635. [CrossRef]

18. Von Cossel, M.; Lewandowski, I.; Elbersen, B.; Staritsky, I.; Van Eupen, M.; Iqbal, Y.; Mantel, S.; Scordia, D.; Testa, G.; Cosentino, S.L.; et al. Marginal Agricultural Land Low-Input Systems for Biomass Production. *Energies* **2019**, *12*, 3123. [CrossRef]
19. Vollrath, B.; Werner, A.; Degenbeck, M.; Illies, I.; Zeller, J.; Marzini, K. *Energetische Verwertung von Kräuterreichen Ansaaten in der Agrarlandschaft und im Siedlungsbereich—eine Ökologische und Wirtschaftliche Alternative bei der Biogasproduktion*; Energie aus Wildpflanzen; Bayerische Landesanstalt für Weinbau und Gartenbau: Veitshöchheim, France, 2012; p. 207.
20. Emmerling, C. Impact of Land-Use Change towards Perennial Energy Crops on Earthworm Population. *Appl. Soil Ecol.* **2014**, *84*, 12–15. [CrossRef]
21. Schmidt, A.; Lemaigre, S.; Delfosse, P.; von Francken-Welz, H.; Emmerling, C. Biochemical Methane Potential (BMP) of Six Perennial Energy Crops Cultivated at Three Different Locations in W-Germany. *Biomass Conv. Bioref.* **2018**, *8*, 873–888. [CrossRef]
22. Von Cossel, M. Renewable Energy from Wildflowers—Perennial Wild Plant Mixtures as a Social-Ecologically Sustainable Biomass Supply System. *Adv. Sustain. Syst.* **2020**, *4*, 2000037. [CrossRef]
23. Brauckmann, H.; Broll, G. *Biogaserzeugung-Upscaling Der FuE-Ergebnisse Zu Neuen Kulturen Und Deren Implementierung*; Universität Osnabrück: Osnabrück, Germany, 2016.
24. Zuercher, A.; Stolzenburg, K.; Messner, J.; Wurth, W.; Löffler, C. Was Leisten Alternative Kulturen im Vergleich zu Energiemais? Landinfo. 9 January 2013, pp. 45–50. Available online: https://www.lfl.bayern.de/mam/cms07/ipz/dateien/aggf_2015_wurth_et_al.pdf (accessed on 2 February 2021).
25. Vollrath, B.; Werner, A.; Degenbeck, M.; Marzini, K. *Energetische Verwertung von Kräuterreichen Ansaaten in der Agrarlandschaft—eine Ökologische und Wirtschaftliche Alternative bei der Biogasproduktion (Phase II)*; Energie aus Wildpflanzen; Bayerische Landesanstalt für Weinbau und Gartenbau: Veitshöchheim, Germany, 2016; p. 241.
26. Ruf, T.; Makselon, J.; Udelhoven, T.; Emmerling, C. Soil Quality Indicator Response to Land-Use Change from Annual to Perennial Bioenergy Cropping Systems in Germany. *Gcb Bioenergy* **2018**, *10*, 444–459. [CrossRef]
27. Emmerling, C.; Schmidt, A.; Ruf, T.; von Francken-Welz, H.; Thielen, S. Impact of Newly Introduced Perennial Bioenergy Crops on Soil Quality Parameters at Three Different Locations in W-Germany. *J. Plant Nutr. Soil Sci.* **2017**, *180*, 759–767. [CrossRef]
28. Friedrichs, J.C. Wirtschaftlichkeit des Anbaus von Wildpflanzenmischungen zur Energiegewinnung—Kalkulation der Erforderlichen Förderung zur Etablierung von Wildpflanzenmischungen. 2013. Available online: http://lebensraum-brache.de/wp-content/uploads/2014/04/Gutachten-32-13b-Wildpflanzenmischungen-zur-Energieerzeugung_Netzwerk-Lebensraum-Feldflur.pdf (accessed on 28 February 2021).
29. Hahn, J.; De Mol, F.; Müller, J.; Knipping, M.; Minderlen, R.; Gerowitt, B. *Wildpflanzen-Samen in Der Biogas-Prozesskette—Eintrags- Und Überlebensrisiko Unter Dem Einfluss von Prozessparametern*; Universität Rostock: Rostock, Germany, 2018.
30. Heiermann, M.; Plogsties, V. *Wildpflanzen-Samen in Der Biogas-Prozesskette—Eintrags- Und Überlebensrisiko Unter Dem Einfluss von Prozessparametern—Teilprojekt 2*; Leibniz-Institut für Agrartechnik und Bioökonomie e.V.: Potsdam, Germany, 2018.
31. Von Cossel, M.; Lewandowski, I. Perennial Wild Plant Mixtures for Biomass Production: Impact of Species Composition Dynamics on Yield Performance over a Five-Year Cultivation Period in Southwest Germany. *Eur. J. Agron.* **2016**, *79*, 74–89. [CrossRef]
32. Baum, G. Betriebswirtschaftliche Betrachtung der Wildpflanzennutzung für Biogasbetriebe 2019. Available online: https://baden-wuerttemberg.nabu.de/imperia/md/content/badenwuerttemberg/vortraege/baum_betriebswirtschaftl_wildpflanzen_f__r_biogas_ver__ffentlichung.pdf (accessed on 28 February 2021).
33. Kuhn, W.; Zeller, J.; Bretschneider-Herrmann, N.; Drenckhahn, K. Energy from Wild Plants—Practical Tips for the Cultivation of Wild Plants to Create Biomass for Biogas Generation Plants. In *The Field Habitat Network*; Deutscher Jagdverband e.V. (DJV): Berlin, Germany, 2014; Volume 1, ISBN 978-3-936802-16-0.
34. Kuhn, W. Interview Kuhn 2020. Available online: https://www.youtube.com/watch?v=K12eOkYxb_U (accessed on 28 February 2021).
35. Frick, M.; Pfender, G. AG Wildpflanzen-Biogas Kißlegg 2019. Available online: https://baden-wuerttemberg.nabu.de/imperia/md/content/badenwuerttemberg/vortraege/frick_pr__sentation_hohenheim_12.03.2019_power_point.pdf (accessed on 28 February 2021).
36. Von Cossel, M.; Steberl, K.; Hartung, J.; Agra Pereira, L.; Kiesel, A.; Lewandowski, I. Methane Yield and Species Diversity Dynamics of Perennial Wild Plant Mixtures Established Alone, under Cover Crop Maize (*Zea Mays* L.) and after Spring Barley (*Hordeum Vulgare* L.). *Gcb Bioenergy* **2019**, *11*, 1376–1391. [CrossRef]
37. Mast, B.; Lemmer, A.; Oechsner, H.; Reinhardt-Hanisch, A.; Claupein, W.; Graeff-Hönninger, S. Methane Yield Potential of Novel Perennial Biogas Crops Influenced by Harvest Date. *Ind. Crop. Prod.* **2014**, *58*, 194–203. [CrossRef]
38. Theuerl, S.; Herrmann, C.; Heiermann, M.; Grundmann, P.; Landwehr, N.; Kreidenweis, U.; Prochnow, A. The Future Agricultural Biogas Plant in Germany: A Vision. *Energies* **2019**, *12*, 396. [CrossRef]
39. Weiland, P. Biogas Production: Current State and Perspectives. *Appl. Microbiol. Biotechnol.* **2010**, *85*, 849–860. [CrossRef]
40. Choong, Y.Y.; Norli, I.; Abdullah, A.Z.; Yhaya, M.F. Impacts of Trace Element Supplementation on the Performance of Anaerobic Digestion Process: A Critical Review. *Bioresour. Technol.* **2016**, *209*, 369–379. [CrossRef]
41. Sauer, B.; Ruppert, H. *Spurenelemente in Biogasanlagen: Eine Ausreichende Versorgung Durch Zufuhr Unterschiedlicher Energiepflanzenmischungen Oder Gülle Ist Möglich*; Biogaskongress; Interdisziplinäres Zentrum für Nachhaltige Entwicklung der Universität Göttingen & Geowissenschaftliches Zentrum der Georg-August-Universität Göttingen: Göttingen, Germany, 2011.
42. Schattauer, A.; Abdoun, E.; Weiland, P.; Plöchl, M.; Heiermann, M. Abundance of Trace Elements in Demonstration Biogas Plants. *Biosyst. Eng.* **2011**, *108*, 57–65. [CrossRef]

43. Von Cossel, M.; Möhring, J.; Kiesel, A.; Lewandowski, I. Optimization of Specific Methane Yield Prediction Models for Biogas Crops Based on Lignocellulosic Components Using Non-Linear and Crop-Specific Configurations. *Ind. Crop. Prod.* **2018**, *120*, 330–342. [CrossRef]
44. Siaudinis, G.; Jasinskas, A.; Slepetiene, A.; Karcauskiene, D. The Evaluation of Biomass and Energy Productivity of Common Mugwort (*Artemisia Vulgaris* L.) and Cup Plant (*Silphium Perfoliatum* L.) in Albeluvisol. *Žemdirbystè (Agric.)* **2012**, *99*, 357–362.
45. Stolzenburg, K. Anbauerfahrungen und Erträge aus einem Dauerkulturen-Projekt des Landes BW 2019. Available online: https://baden-wuerttemberg.nabu.de/natur-und-landschaft/landwirtschaft/biogas/index.html (accessed on 28 February 2021).
46. Zürcher, A. Permanent Crops as an Alternative to Maize—Wild Plant Mixtures, Jerusalem Artichoke, Cup Plant, Virginia Mallow and Szarvasi 2014. Available online: https://docplayer.org/53901435-Dauerkulturen-als-alternativen-zu-mais-wildartenmischungen-topinambur-durchwachsene-silphie-virginiamalve-und-riesenweizengras.html (accessed on 28 February 2021).
47. Seppälä, M.; Laine, A.; Rintala, J. Screening Novel Plants for Biogas Production in Northern Conditions. *Bioresour. Technol.* **2013**, *139*, 355–362. [CrossRef]
48. Kiesel, A.; Lewandowski, I. Miscanthus as Biogas Substrate—Cutting Tolerance and Potential for Anaerobic Digestion. *Gcb Bioenergy* **2017**, *9*, 153–167. [CrossRef]
49. Naumann, C.; Bassler, R. *VDLUFA Methodenbuch: Die Chemische Untersuchung von Futtermitteln*; VDLUFA-Verlag: Darmstadt, Germany, 2006.
50. von Cossel, M.; Mangold, A.; Iqbal, Y.; Lewandowski, I. Methane Yield Potential of Miscanthus (Miscanthus × Giganteus (Greef et Deuter)) Established under Maize (*Zea Mays* L.). *Energies* **2019**, *12*, 4680. [CrossRef]
51. Von Cossel, M.; Möhring, J.; Kiesel, A.; Lewandowski, I. Methane Yield Performance of Amaranth (*Amaranthus Hypochondriacus* L.) and Its Suitability for Legume Intercropping in Comparison to Maize (*Zea Mays* L.). *Ind. Crop. Prod.* **2017**, *103*, 107–121. [CrossRef]
52. VDI. *VDI 4630: Fermentation of Organic Materials—Characterization of the Substrate, Sampling, Collection of Material Data, Fermentation Tests*; Verein Deutscher Ingenieure e.V.—Gesellschaft Energie und Umwelt: Düsseldorf, Germany, 2016. Available online: https://infostore.saiglobal.com/en-us/Standards/VDI-4630-2016-1115305_SAIG_VDI_VDI_2590568/ (accessed on 28 February 2021).
53. Piepho, H.-P. An Algorithm for a Letter-Based Representation of All-Pairwise Comparisons. *J. Comput. Graph. Stat.* **2004**, *13*, 456–466. [CrossRef]
54. Wolfinger, R. Covariance Structure Selection in General Mixed Models. *Commun. Stat.-Simul. Comput.* **1993**, *22*, 1079–1106. [CrossRef]
55. Kenward, M.G.; Roger, J.H. Small Sample Inference for Fixed Effects from Restricted Maximum Likelihood. *Biometrics* **1997**, *53*, 983–997. [CrossRef]
56. Koch, K.; Post, M.; Auer, M.; Lehuhn, M. *Einsatzstoffspezifische Besonderheiten in Der Prozessführung*; Biogas Forum Bayern; Arbeitsgemeinschaft Landtechnik und landwirtschaftliches Bauwesen in Bayern e.V.: Freising, Germany, 2017.
57. Ohly, N. Verfahrenstechnische Untersuchungen Zur Optimierung Der Biogas-Gewinnung Aus Nachwachsenden Rohstoffen. Ph.D. Thesis, Technische Universität Bergakademie Freiberg, Freiberg, Germany, 2006.
58. Krimmer, E.; Marzini, K.; Heidinger, I. Wild Plant Mixtures for Biogas: Promoting Biodiversity in a Production-Integrated Manner—Practical Trials for Ecological Enhancement of the Landscape. *Nat. Und Landsch.* **2021**, *53*, 12–21. [CrossRef]
59. Amon, T.; Amon, B.; Kryvoruchko, V.; Machmüller, A.; Hopfner-Sixt, K.; Bodiroza, V.; Hrbek, R.; Friedel, J.; Pötsch, E.; Wagentristl, H.; et al. Methane Production through Anaerobic Digestion of Various Energy Crops Grown in Sustainable Crop Rotations. *Bioresour. Technol.* **2007**, *98*, 3204–3212. [CrossRef]
60. Herrmann, C.; Idler, C.; Heiermann, M. Biogas Crops Grown in Energy Crop Rotations: Linking Chemical Composition and Methane Production Characteristics. *Bioresour. Technol.* **2016**, *206*, 23–35. [CrossRef] [PubMed]
61. Herrmann, C.; Plogsties, V.; Willms, M.; Hengelhaupt, F.; Eberl, V.; Eckner, J.; Strauß, C.; Idler, C.; Heiermann, M. Methane Production Potential of Various Crop Species Grown in Energy Crop Rotations. *Landtechnik* **2016**, *71*, 194–209.
62. Ohl, S.; Hartung, E. Comparative Assessment of Different Methods to Determine the Biogas Yield. In Proceedings of the International Conference on Agricultural Engineering-AgEng 2010: Towards Environmental Technologies, Clermont-Ferrand, France, 6–8 September 2010.
63. Mangold, A.; Lewandowski, I.; Hartung, J.; Kiesel, A. Miscanthus for Biogas Production: Influence of Harvest Date and Ensiling on Digestibility and Methane Hectare Yield. *Gcb Bioenergy* **2019**, *11*, 50–62. [CrossRef]
64. Dandikas, V.; Heuwinkel, H.; Lichti, F.; Drewes, J.E.; Koch, K. Correlation between Biogas Yield and Chemical Composition of Energy Crops. *Bioresour. Technol.* **2014**, *174*, 316–320. [CrossRef]
65. Triolo, J.M.; Sommer, S.G.; Møller, H.B.; Weisbjerg, M.R.; Jiang, X.Y. A New Algorithm to Characterize Biodegradability of Biomass during Anaerobic Digestion: Influence of Lignin Concentration on Methane Production Potential. *Bioresour. Technol.* **2011**, *102*, 9395–9402. [CrossRef] [PubMed]
66. Thomsen, S.T.; Spliid, H.; Østergård, H. Statistical Prediction of Biomethane Potentials Based on the Composition of Lignocellulosic Biomass. *Bioresour. Technol.* **2014**, *154*, 80–86. [CrossRef] [PubMed]
67. Niu, W.; Han, L.; Liu, X.; Huang, G.; Chen, L.; Xiao, W.; Yang, Z. Twenty-Two Compositional Characterizations and Theoretical Energy Potentials of Extensively Diversified China's Crop Residues. *Energy* **2016**, *100*, 238–250. [CrossRef]

68. FNR. Erhöhung Des Ertragspotentials Heimischer Wildpflanzenmischungen Unter Berücksichtigung von Biodiversität Und Wasserschutz. Available online: https://www.fnr.de/index.php?id=11150&fkz=2219NR215 (accessed on 21 January 2021).
69. Knapkon. *Energie Für Die Biogasanlagen*; Knapkon: Frickenhausen, Germany, 2020.
70. Iqbal, Y.; Kiesel, A.; Wagner, M.; Nunn, C.; Kalinina, O.; Hastings, A.F.S.J.; Clifton-Brown, J.C.; Lewandowski, I. Harvest Time Optimization for Combustion Quality of Different Miscanthus Genotypes across Europe. *Front. Plant Sci.* **2017**, *8*, 727. [CrossRef] [PubMed]
71. Christian, D.G.; Riche, A.B.; Yates, N.E. Growth, Yield and Mineral Content of Miscanthus x Giganteus Grown as a Biofuel for 14 Successive Harvests. *Ind. Crop. Prod.* **2008**, *28*, 320–327. [CrossRef]

Article

Yield Performance of Woody Crops on Marginal Agricultural Land in Latvia, Spain and Ukraine

Santa Celma [1,*], Marina Sanz [2], Pilar Ciria [2], Oksana Maliarenko [3], Oleh Prysiazhniuk [3], Mudrite Daugaviete [1], Dagnija Lazdina [1] and Moritz von Cossel [4,*]

[1]   Latvian State Forest Research Institute SILAVA, LV 2169 Salaspils, Latvia; mudrite.daugaviete@silava.lv (M.D.); dagnija.lazdina@silava.lv (D.L.)
[2]   Centre for the Development of Renewable Energies, CEDER-CIEMAT, Autovía de Navarra A-15, Salida 56, 42290 Lubia, Spain; marina.sanz@ciemat.es (M.S.); pilar.ciria@ciemat.es (P.C.)
[3]   Institute of Bioenergy Crops and Sugar Beet NAAS, 03110 Kyiv, Ukraine; o.malyarenko@ukr.net (O.M.); olpris@mail.ru (O.P.)
[4]   Biobased Resources in the Bioeconomy (340b), Institute of Crop Science, University of Hohenheim, Fruwirthstr. 23, 70599 Stuttgart, Germany
*   Correspondence: santa.celma@silava.lv (S.C.); mvcossel@gmx.de (M.v.C.)

Abstract: Agricultural land abandonment due to biophysical and socioeconomic constraints is increasing across Europe. Meanwhile there is also an increase in bioenergy demand. This study assessed woody crop performance on several relevant types of marginal agricultural land in Europe, based on field experiments in Latvia, Spain and Ukraine. In Latvia, hybrid aspen was more productive than birch and alder species, and after eight years produced 4.8 Mg ha^{-1} y^{-1} on stony soil with sandy loam texture, when best clone and treatment combination was selected. In Spain, Siberian elm produced up to 7.1 Mg ha^{-1} y^{-1} on stony, sandy soil with low organic carbon content after three triennial rotations. In Ukraine, willow plantations produced a maximum of 10.8 Mg ha^{-1} y^{-1} on a soil with low soil organic carbon after second triennial rotation. The productivity was higher when management practices were optimized specifically to address the limiting factors of a site. Longer rotations and lower biomass yields compared to high-value land can be expected when woody crops are grown on similar marginal agricultural land shown in this study. Future studies should start here and investigate to what extent woody crops can contribute to rural development under these conditions.

Keywords: abandoned agricultural land; bioeconomy; bioenergy; biophysical constraints; birch; black alder; hybrid aspen; short-rotation forestry; Siberian elm; willow

Citation: Celma, S.; Sanz, M.; Ciria, P.; Maliarenko, O.; Prysiazhniuk, O.; Daugaviete, M.; Lazdina, D.; von Cossel, M. Yield Performance of Woody Crops on Marginal Agricultural Land in Latvia, Spain and Ukraine. *Agronomy* **2022**, *12*, 908. https://doi.org/10.3390/agronomy12040908

Academic Editor: Mark P. Widrlechner

Received: 21 February 2022
Accepted: 7 April 2022
Published: 9 April 2022

Publisher's Note: MDPI stays neutral with regard to jurisdictional claims in published maps and institutional affiliations.

1. Introduction

An increase in abandoned and marginal agricultural land area can be observed in most parts of Europe [1,2]. In a large portion of Eastern Europe, land abandonment is driven by socioeconomic factors, where landowners are often absent or uninterested in pursuing conventional agronomical practices [3]. However, biophysical constraints and inappropriate land management leading to degradation of the land are the main reasons for land abandonment [4]. Such land is often referred to as marginal. Passive restoration processes and natural succession happens on the abandoned land, if it is left unmanaged. The succession and ecological value of the land can be very diverse depending on a wide range of site conditions [5]. In cases where the natural vegetation cover development is impeded by biophysical constraints or if there is a high risk of colonization by invasive species, active restoration may be more suitable [6,7]. Such abandoned and marginal areas could be purposefully utilized for tree plantations or woody crops and contribute to meeting the bioenergy demand that is increasing throughout Europe [8]. Despite demand for bioenergy being expected to rise and that solid biomass already makes up about half of renewable energy sources, energy crops take up only a small percentage of European land [9–11]. While short- and medium-rotation tree plantations on agricultural land have become common in some countries, there is a lack of knowledge regarding the yields that can be expected from plantations established on unfavorable or marginal land. Knowledge on associations between specific biophysical constraints, species and biomass accumulation is also lacking. This complicates evidence-based decision-making for landowners. When planting woody crops on agricultural land, stakeholders primarily turn to short-rotation poplar and willow plantations, as these are well known for their rapid growth rate and are easy to propagate via cuttings. However *Salicaceae* species are rather water demanding and in areas with arid climatic conditions appropriate species for the region should be favored to avoid irrigation costs [12–14]. Some such species with fast biomass accumulation rates are Siberian Elm (*Ulmus pumila*), Black locust (*Robinia pseudoacacia*) and *Eucalyptus* spp. [15,16]. In addition, species can be selected with intention to alleviate a particular land constraint or to improve a specific ecological function of the land [17–20], thus, (unless established on high nature-value land) ensuring ecosystem services of higher quality than abandoned marginal land or marginal land that is under high input management [21–27].

While there is still ongoing discussion about a common definition of marginality and lack of united marginality factor classification [28–35], the limiting factors are similar across Europe; however, appropriate management practices are specific to each geographic location. The aim of this study was to obtain yield data from field studies carried out on marginal land in Latvia, Ukraine and Spain that represent the three environmental zones of Europe–boreonemoral, Atlantic and Continental. The objectives were to evaluate the survival of plantations on marginal land, and to summarize the yield results from these case study sites in the context of other research carried out on marginal land across Europe.

2. Materials and Methods

2.1. Case Study Sites and Data Collection

Marginal land area was determined using MAGIC-Maps [36]. Local marginality factors were assessed in accordance with Elbersen et al. [37] marginality factor classification thresholds.

Leading marginality factors in the case study countries are mainly associated with adverse climate and low soil fertility and limitations in rooting (Table 1). Adverse climate in Latvia refers to cold winters and short vegetation period (length of growing period $\leq$ 180 days; or degree days $\leq$ 1500 days). In Spain and Ukraine it is associated with the lack of precipitation in some areas (annual precipitation/potential evapotranspiration $\leq$ 0.5).

Table 1. Leading marginality factors affecting countries where case studies were carried out (according to MAGIC-Maps [36,37]).

Country	Leading Marginality Factors	Affected Area (km^2)	Affected Area of Total Utilized Agricultural Area (%)
Latvia	Adverse climate	8980	30
	Excessive wetness	3602	12
	Limitations in rooting	1475	5
	Total:	12,161	41
Spain	Adverse climate	77,490	23
	Limitations in rooting	76,179	22
	Low soil fertility	33,166	10
	Total:	148,496	44
Ukraine	Low soil fertility	37,000	9
	Adverse climate	30,000	7
	Limitations in rooting	29,100	7
	Total:	133,920	31

2.1.1. Latvia

The field experiment was conducted in Skrīveri municipality, Latvia (coordinates can be found in Table 2). Five fast-growing tree species were planted in the spring of 2011: hybrid aspen (*Populus tremula* L. × *P. tremuloides* Michx) clone 4 and clone 28, gray alder (*Alnus incana* (L.) Moench.), black alder (*Alnus glutinosa* (L.) Gaertn.), hybrid alder (*A. incana* × *A. glutinosa*) and birch (*Betula pendula* Roth) grown in different nursery containers—type 1 (Lannen Plantek 35F) and type 2 (Rootrainers Sherwood). Hybrid aspen clones were grown in three densities, with 2 × 2 m, 3 × 3 m and 2.5 × 5 m distance between trees (planting density of 2500, 1273 and 1227 trees ha^{-1} respectively); alder and birch were grown in plots with 2.5 × 2.5 m distance between trees (1636 trees ha^{-1}). Hybrid aspen was grown under four fertilization treatments applied prior to planting—control, wood ash (6 Mg$_{DW}$ ha^{-1}, total N 2.6, total P 65, total K 190 kg ha^{-1}), sewage sludge (10 Mg$_{DW}$ ha^{-1}, total N 259, total P 163, total K 22 kg ha^{-1}) and digestate (30 Mg ha^{-1}, total N 65, total P 12, total K 100 kg ha^{-1}) in four replications, with plot size of 240 m^2 for spacing 2 × 2 and 3 × 3 m, and plot size 360 m^2 for 2.5 × 5 m spacing.

Table 2. Summary of case study design considered in this article.

Country	Location	Establishment Year	Marginality	Density, Plants ha^{-1}	Species	Treatment
Latvia	56.69 N, 25.14 E	2011	Poor rooting conditions—unfavorable soil texture and stoniness	2500; 1273; 1227;	Hybrid aspen	Control; Wood ash; Sewage sludge; Digestate
				1636	Gray alder	Control; Wood ash; Sewage sludge
				1636	Black alder	
				1636	Hybrid alder	
				1636	Birch	
Spain	41.36 N, 2.30 W	2009	Unfavorable soil texture, stoniness and soil organic carbon < 1%	6666	Siberian elm	Rain-fed; Irrigated
Ukraine	48.99 N, 27.46 E	2016	Soil organic carbon < 1%, soil pH < 5	20,000	Willow	Fertilizer (N60)
		2013	Soil organic carbon < 1%			Control
		2016; 2011;	Clay soil (clay content > 50%), soil organic carbon < 1%			
		2011; 2013; 2016;	None			

Birch and alder species were grown under three fertilization treatments—control, wood ash (6 Mg$_{DW}$ ha^{-1}, total N 2.6, total P 65, total K 190 kg ha^{-1}), sewage sludge (10 Mg$_{DW}$ ha^{-1}, total N 259, total P 163, total K 22 kg ha^{-1}) in four replicates with plot size of 240 m^2.

Marginality of the site includes poor rooting conditions—unfavorable soil texture and stoniness (Table 2). The type of soil was classified as Luvic Stagnic Phaeozem (Hypoalbic) or Mollic Stagnosol (Ruptic, Calcaric, Endosiltic) according to the FAO [38] with the dominant loam (at 0–20 cm depth) and sandy loam (at 0–20 cm and 20–80 cm depth) soil texture. The climatic conditions of the site during the study period can be seen in Table A1 and Figure A1.

Tree height and diameter at breast height (DBH) were measured for all species after eight growing seasons. Biomass was calculated according to a methodology Liepiņš J. [39] specifically developed for young tree stands in local conditions.

2.1.2. Spain

The experimental fields are located in the north-central part of Spain, in Cubo de la Solana municipality at an altitude of about 1100 m above sea level (coordinates can be found in Table 2). Fields with a total area of 2500 m^2 were established manually, using Siberian elm (*Ulmus pumila* L.) rooted plants in 2009. The site was divided in to two plots, rain-fed and under irrigation conditions. The average water supplied in the irrigated plot was 1500 m^3 ha^{-1} year^{-1} during the summer months (from June to September). Irrigation was applied every year using a drip system. Density was 6666 trees per hectare. The experimental duration was nine years. Siberian elm was harvested every 3 years. Therefore, three harvests of the crop have been obtained.

The marginality factors of the planting site are unfavorable soil texture, stoniness and low soil organic carbon (Table 2). The soil analysis was performed on samples collected at depths of 0–30 cm. The soil has a sandy texture (sand 86%, lime and clay < 10%), about 28% coarse elements with good drainage. Moreover, it has pH of about 6, content in oxidable organic matter (0.4%), nitrogen content (0.03%) and its cation exchange capacity (CEC) is 3 cmol kg^{-1}, field capacity 6.6%, water utility 3.9% and wilting point 2.7%. The climatic conditions of the site during the study period can be seen in Table A2 and Figure A2.

Siberian elm trees were harvested by hand using a chainsaw. Each plant was cut down to 10–15 cm above the ground level. The number of tree samples were 15 per treatment and harvesting cycle. The fresh weight of each tree over the studied period was determined by weighing whole plants immediately after harvesting at the field. The representative biomass samples were taken to the laboratory to determine the dry matter content by drying it in an oven at 60 °C. Dry biomass yield per hectare and mean annual increment (MAI) was estimated from the harvest data of each plot.

2.1.3. Ukraine

Experiments were carried out at the Yaltushkiv Experimental Breeding Station, Chereshneve, Vinnytsia region, Bar district (coordinates can be found in Table 2). The fields are located in the forest steppe zone of sufficient moisture, which covers 33% of the territory of Ukraine. The climatic conditions of the site during the study period can be seen in Table A3 and Figure A3.

Willow (*Salix viminalis*) variety Zbruch was used in the experiments. Willow planting density was 20,000 plants ha^{-1}. Experimental plots have various marginality factors and were established in different years. One willow plot was established in 2011 on clay soil with low soil organic carbon. In 2013, another willow plot was laid out on soil with low soil organic carbon. In 2016, one more plot was laid out on clay soil with low soil organic carbon and unfavorable soil texture and another on soil with low organic carbon and high soil acidity (Table 2). In addition, in each establishment year, one plot was also established on land with no marginality factors serving as a control. Weed control was carried out, and fertilization with N rate of 60 kg ha^{-1} was done in plots established in 2016 on soil with low organic carbon content and low pH. Willow was harvested triennially. Only the data of latest harvest biomass yield was further assessed (first rotation data of plots established in 2016, second rotation data of plots established in 2013 and third rotation data of plots established in 2011).

2.2. Statistical Analysis

Data analysis and visualization was done using R version 4.0.5 [40]. The Shapiro−Wilk test was used to test normality of data and Levene's test was used to test homogeneity of variances assumptions. Data was not normally distributed; therefore, the Mann−Whitney U test was used to compare the two groups for the case study in Spain and Kruskal−Wallis and Dunnett's multiple comparison tests were used to compare groups for the case study in Latvia.

3. Results

3.1. Latvia

The MAI of studied species is shown in Figure 1. In the studied conditions, fastest biomass accumulation was achieved by the hybrid aspen clone 4 under digestate treatment and in the densest planting density (2 × 2 m).

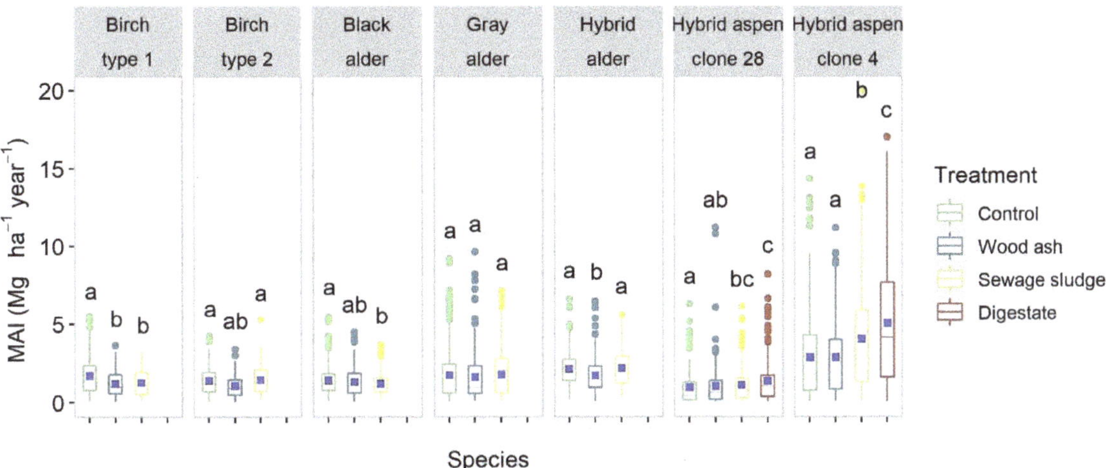

Figure 1. Mean annual increments (MAI) of dry above-ground biomass of eight-year-old fast-growing tree species stands depending on fertilization treatment (boxes represent interquartile range; median is shown as center horizontal line in the box; whiskers show minimum and maximum observed values; dots show outliers; mean values are represented by the blue squares; different letters represent significant ($p < 0.05$) differences among treatments within each species).

The hybrid aspen clone 4 showed significantly ($p < 0.05$) more rapid growth compared to clone 28 across all treatments and had a better survival rate (94% and 89% respectively). Across all treatments, the MAIs of hybrid aspen clones 4 and 28 were 3.7 and 1.1 Mg ha^{-1} y^{-1} after eighth growing season. Both clones responded in a similar pattern to stand density. In this study, the highest stand density yielded the most biomass (Figure 2). Tree height and mean breast height diameter followed the same pattern and were also the largest in plots planted in a grid of 2 × 2 m (2500 trees ha^{-1}).

Application of digestate had a positive effect on hybrid aspen yield, except in the cases of the 3 × 3 m plots for both clones and the 2 × 2 m plot for clone 28. Plots where wood ash was applied performed the worst in terms of biomass accumulation across all densities and regardless of clone. Sewage sludge did not have a positive effect in most cases, compared to the control—a positive effect on yield was observed only in planting density of 2.5 × 5 m for both clones; however, this was more likely due to soil differences across fields rather than interaction between density and fertilization treatment.

For birch, there were no significant differences between container types when initial planting material height differences were taken into account ($p = 0.09$). Birch stands with type 1 planting material produced an average of 1.5 Mg ha^{-1} y^{-1} and type 2 planting material produced 1.3 Mg ha^{-1} y^{-1} after the eight growing season. Compared to control, there was no evidence of a significantly positive effect of any of the fertilization treatments, either on birch type 1 or birch type 2.

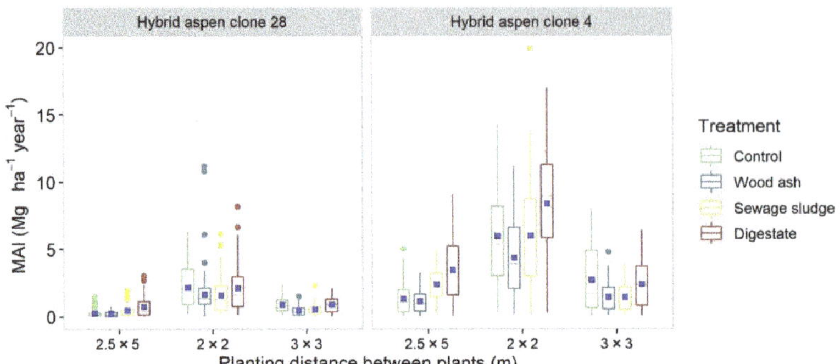

Figure 2. Mean annual increments (MAIs) of above-ground biomass dry matter of eight-year-old hybrid aspen stands depending on planting distance (boxes represent interquartile range; median is shown as center horizontal line in the box; whiskers show minimum and maximum observed values; dots show outliers; mean values are represented by the blue squares).

Similarly, neither of the fertilization treatments had a positive effect on any of the alder species' biomass accumulation rate. Hybrid alder, black alder and gray alder after the eight growing season produced 2.0, 1.7 and 1.3 Mg ha^{-1} y^{-1}, respectively, (mean of all treatments).

From the studied species, hybrid aspens', hybrid alders', black alders' and gray alders' overall survival rate was higher than 88%. Birch had the lowest survival rate—73% and 76% (type 1 and type 2, respectively).

3.2. Spain

The Siberian elm plantation exhibited a 100% survival rate during the study period. Irrigation had a statistically significant positive effect on biomass accumulation in Siberian elm in the first and third rotation ($p = 0.004$ and $p = 0.02$, respectively) but not in the second rotation ($p > 0.05$). The yield in irrigated plots was double of that in rain-fed plots. MAI increased with every rotation; however, the difference between rotations was not statistically significant. The increase from first to third rotation was from 1.79 to 3.66 and from 4.54 to 7.05 Mg ha^{-1} y^{-1} in rain-fed and irrigated plots, respectively (Figure 3).

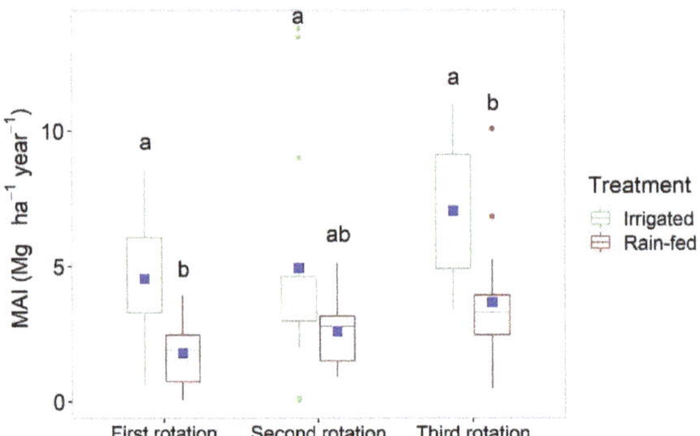

Figure 3. Mean annual increment (MAI) of Siberian elm above-ground biomass dry matter after first, second and third triennial rotation depending on treatment (boxes represent interquartile range; median is shown as center horizontal line in the box; whiskers show minimum and maximum observed values; mean values are represented by the blue squares; different letters represent significant ($p < 0.05$) differences between all treatments and rotations).

The acquired MAI corresponds to total yield per area of 5.37 and 13.63 Mg ha^{-1} after first rotation, 7.76 and 14.83 Mg ha^{-1} after second rotation and 10.97 and 21.14 Mg ha^{-1} after third rotation in rain-fed and irrigated plots respectively.

3.3. Ukraine

Willow plots established in 2016 yielded 8.35 and 8.63 Mg ha^{-1} y^{-1} during the first three-year rotation in unfertilized plots of soil with low soil organic matter and unfavorable soil texture and fertilized plots of soil with low soil organic matter and low pH, respectively (Figure 4). In the plots established in 2013 on land with low soil organic matter, MAI was 10.82 Mg ha^{-1} y^{-1} in the second three-year rotation. The plot established in 2011 on soil with low organic matter and unfavorable soil texture yielded slightly less—9.54 Mg ha^{-1} y^{-1} in the third three-year rotation. Survival rate was above 88% in all stands regardless of stand age or treatment. In plots with no known marginality factors, survival rate was above 94%. The yield was also higher in all plots on non-marginal land—12.66, 14.83 and 14.04 Mg ha^{-1} y^{-1} after first, second and third rotation, respectively.

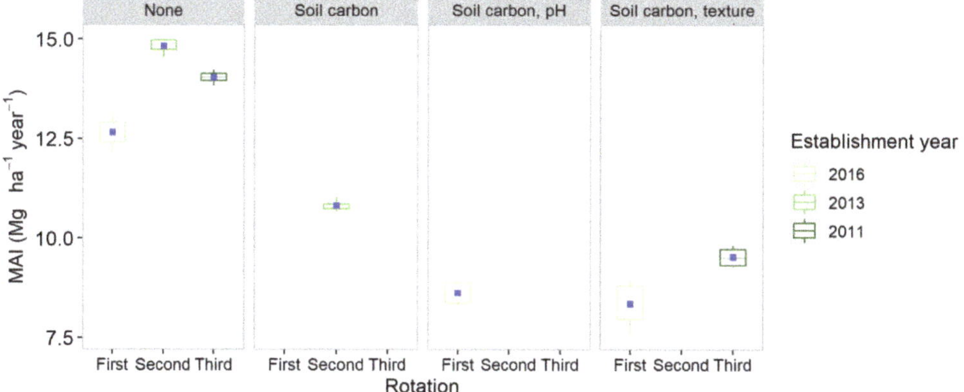

Figure 4. MAI (mean annual increment) of willow above-ground biomass dry matter in Ukraine after first, second and third rotation, under different marginality factors—no marginality (none), low soil organic matter content (soil carbon), low soil pH (pH), unfavorable soil texture (texture) (boxes represent interquartile range; median is shown as center horizontal line in the box; whiskers show minimum and maximum observed values; mean values are represented by the blue squares).

Both total biomass per hectare and the weight of individual plant was the highest (32.5 Mg^{-1} ha^{-1} and 1.8 Kg plant^{-1}, respectively) after the second rotation, in the plots that had only one marginality factor at play—low soil organic carbon—compared to plots with two marginality factors.

4. Discussion

4.1. Yield Performance

Depending on the intended application, biomass yield and crop performance can be measured in various ways. To alleviate the comparison between the case studies and available literature, we focus on MAI expressed as Mg$_{DW}$ ha^{-1} y^{-1} as it is one of the prevailing measurements used in other studies regarding woody biomass.

Similarly to results of other studies (Table A4) on both marginal land and on land with no known marginality, high variation in yield within a plantation was observed in the case studies [41–45]. Yield results from the case study in Latvia show that at the age of eight years, productivity of all planted species was low, compared to yields acquired in other studies. This can mainly be attributed to relatively low initial stand density, as the trees were grown for their trunks. Longer rotations (10 years and above) would be recommended in such a case, as is also recommended by other authors [46]. In similar density stands, similar growth results have been obtained [44,47]. Planting density is related to the target produce—lower stand density allows for thicker trunks and more dry matter per tree, but denser stands typically output more biomass per ha [48,49]. Hybrid aspen offers better financial returns, if grown for log production, according to Tullus et al. [50]. Thus, lower stand densities and longer rotations are favored in the current state of the market. Density effect on total biomass yield is more evident in the early age of the stand, but later on in-group competition causes natural thinning and suppresses tree growth, thus, in older stands density affects tree dimensions more than total yield per area [51–53]. However, in relatively low-density hybrid aspen stands in Latvia, the densest stand design (planted in 2×2 m) showed a slightly better survival rate; furthermore, no in-group competition was observed, as both the height and the DBH was bigger in the densest stand. This suggests that denser stands ($\geq$2500 trees ha^{-1}) of hybrid aspen can be established without compromising wood quality. Survival of Siberian elm in a short-rotation plantation (6666 plants ha^{-1}) was 100% in trials based in Spain, and survival did not decrease with stand age. Similarly, no effect of in-group competition on survival was observed in high-density stands in Ukraine, where willow survived equally well in all established plots (survival rate 88–90%), when planted in density of 20,000 plants ha^{-1}. In coppice systems, willow (or poplar) is typically grown in density of 10,000–20,000 plants per hectare (with 10,000–15,000 plants per hectare being recognized as the highest yielding [54]) for up to a total of 25 years with typical rotation length being 3–5 years [55,56]. Rotation of 3–5 years is considered optimal for willow short-rotation coppice (SRC) plantations even on marginal land [52]. MAI typically increases with stand density and stand age (up to certain point), thus providing a basis to favor longer rotation periods for non-coppice woody crops. However, in coppicing systems, the opposite trend can be observed, where MAI is increasing during first rotations, but in the long term it is often negatively correlated to the number of harvests, with some clones showing a decline in yield sooner than others [57–60]. Due to different establishment years and results from only three rotations, it is complicated to assess the rotation count effect on yield in the case study based in Ukraine. In this study, MAI was initially low in younger plots, established in 2016 (8.5 Mg ha^{-1} y^{-1}) compared to plots established earlier, in 2013 and 2011 (10.8 and 9.5 Mg ha^{-1} y^{-1}). However, the marginality and treatment of these plots also varied and therefore the differences in yield cannot be clearly attributed to plot age, especially since the establishment year was not the same for all plots. Regardless of the stand age, all plots established on marginal land produced around 30% less biomass per year compared to plots established on non-marginal land. On marginal land, total yield per hectare was the highest after the second harvest in plots established in 2013, possibly due to being affected only by one marginality factor—low soil organic matter—whereas other plots had two constraining factors—low soil organic matter combined with low pH or unfavorable (clay) soil texture. The obtained yields are within the range found in other studies on marginal lands (typically a wide range from 3 up to 12 Mg ha^{-1} y^{-1}) and, possibly due to the small scale of the experiment, even exceed the estimated bioenergy crop yields that can be achieved at a production scale (around 6 to 7 Mg ha^{-1} y^{-1}) (Table A4) [61,62]. Vande Walle et al. [63] found lower yields under similar soil conditions—on sandy soil with low organic matter and high acidity (3.4 Mg ha^{-1} y^{-1} after four growing seasons and 20,000 plants ha^{-1} density).

The biomass yield of Siberian elm trees depended on age and even more so water regimen. Siberian elm produced twice as much biomass under irrigated conditions compared to rain-fed, and the biomass accumulation increased with each rotation (even though not significantly). Therefore, the regrowth capacity of Siberian elm after harvesting can be considered as good. However, the biomass yields were lower than in the studies carried out in Madrid and Teruel under rain-fed conditions. In Madrid with the same planting density, the average biomass yield when elms finished the second cycle was estimated at 5.2 Mg ha^{-1} y^{-1} and after 3 years at 13.2 Mg ha^{-1} y^{-1} [64], while in Villarquemado (Teruel) the biomass yield was 5.1 Mg ha^{-1} y^{-1} with a density of 3333 trees ha^{-1} [65], although the biomass yield was similar to that in Oropesa (Toledo) 1.86 Mg ha^{-1} y^{-1} in second cycle [66]. In eastern Kansas, elm biomass yield varied from 4.7 to 9.8 Mg DM ha^{-1} y^{-1} with planting density of 1400–7000 trees ha^{-1}, respectively, harvested 7 years after planting in rain-fed conditions [67]. However, the biomass yield after 3 years ranged from 0.7 Mg ha^{-1} y^{-1} to 5.2 Mg ha^{-1} y^{-1} in different plots distributed throughout the state of Kansas [68] and between 4.5 and 16.9 Mg ha^{-1} when elms were cut annually for 6 years using a spacing of 0.3 × 0.3 m^2 in the same North American state [15].

According to available research (Table A4), on agricultural land that is simply classified as abandoned or fallow, yields are higher compared to those on land with known and defined land constraints. Comparison of marginality factor effect on yield is complicated as there are multiple factors at play, and the marginality itself can be of various degrees. Woody crop productivity is generally still good on sites with low soil organic matter. Spoil heaps and extracted mining sites, on the other hand, are especially limiting for growth, as these often include a combination of unfavorable soil qualities—adverse chemical conditions, limitations in rooting, low soil fertility and also adverse terrain conditions [69–71].

Some research suggest that yields presented based on small-scale experimental plantation sites are overly optimistic (due to increased edge effect, intense tending, limited pest damage, etc.) [72,73]. However, it is not clear if the same is expected on marginal land, but if the management practices are kept the same when upscaling the cultivation, results will most likely be similar to what has been obtained in smaller-scale experiments.

4.2. Species Suitability

Due to numerous possible species and marginality factor combinations, there is still lack of knowledge regarding each species' performance under unfavorable conditions. Attempts to narrow the knowledge gap can be made by compiling existing research on marginal land and knowledge on growth requirements of particular species. However, the intra-species variation can be high [52,74] as can also be seen from hybrid aspen clone 4 and clone 28 results from the case study in Latvia. Even more so for some clones, high inter-replicate variation can be observed [75]. Besides yield, characteristics of each genotype should be considered in context of the site. Depending on current and future risks, and intended target produce-stress tolerance, chemical composition, physical properties and disease susceptibility may be more important in planting material selection than tree growth rate [76]. Mixed genotype stand composition can be expected to increase the overall stand stability and resilience.

4.2.1. Willow and Aspen

Willow and hybrid aspen both are *Salicaceae* family species. The performance of willow is more studied compared to hybrid aspen; however, in terms of growth requirements and recommended management practices hybrid aspen is similar to another widely grown species from this family, poplar [47]. Compared to poplar, aspen is at higher animal browsing damage risk, but can better withstand colder temperatures and poorer site conditions [46]. It has been found that willow uptakes more nutrients from the soil compared to poplar, but poplar uptakes more than eucalyptus or paulownia, which can lead to faster soil depletion, and could be especially problematic if the land is initially low in nutrients [77,78]. In general, these species are not suited for highly acidic soils, soils that are poor in nutrients and sandy soils with low water availability [26,55,63,79,80]. Being water demanding, these species can withstand moderate flooding, with some clones being more tolerant than others [46,81]. The growth difference depending on clone was also evident in the case study in Latvia; however, the response of both clones to fertilization treatment and stand density followed the same pattern. The *Salicaceae* species can also be grown on contaminated soils, but due to their phythoremediating properties (especially of willow), accumulation of heavy metals in biomass can occur and compromise the quality and safety of obtained feedstock [82]. Willow is better adapted to colder climates, where poplar can suffer frost damage [62]. Overall, higher willow and poplar yields can be expected in milder climates—British Isles, Central and west Europe—but lower yields can be expected in Northern and Eastern Europe due to a colder climate and in Southern Europe due to limited precipitation, and thus, water availability [83].

4.2.2. Siberian Elm

Under arid climatic conditions, species such as Siberian elm, black locust and eucalyptus can be grown on marginal land. There are limited data on Siberian elm cultivation for biomass in Europe. In Spain, elm started to be studied as an energy crop in short-rotation coppices (SRCs) around 2000 [65]. Siberian elm is a hard wooded and a fast-growing tree that features greater resistance to Dutch elm disease than other species in genus *Ulmus*. Its drought tolerance, adaptation to different environments and sprouting capacity determine that this species grown as a SRC can produce high-biomass lignocellulosic yield under low input management [84,85]. However, this species can have an invasive nature, as seen in North America [86] and Serbia [87] and has a potential of hybridization with native species [88]. So far there is limited data on its invasiveness in the conditions of Europe, and in Spain it is not considered invasive at the moment [89]. Caution should be taken when planning Siberian elm plantations. Siberian elm plantations can successfully be established on unfavorably textured soil, as was also evident from the study case results. In the case study, the survival rate was equal to that in trials in Madrid, 100%, and higher than that at Villarquemado (Teruel), 96.5% [65], and Casale Moferrato (Alessandria), between 68–87% (Pérez et al., 2012). It is suited to conditions where other species fail to thrive, especially stony and coarse soil, as it is typically found occurring naturally in such soil [90], making it a very suitable species for marginal lands. In addition, SRC plantation can enhance soil carbon content. In a study carried out in Spain, the capacity to sequester C in the uppermost layer of the soil (0–30 cm) of black locust, Siberian elm and Euroamerican poplar was $0.36–0.83$ Mg ha^{-1} y^{-1} of C [91]. Alternative species for warmer and dryer climatic conditions are black locust and eucalyptus. Unfortunately, similar to Siberian elm, these two species also possess the potential to become invasive in some areas of Europe. Eucalyptus species vary in tolerance to different constraints. Most are heat and drought tolerant and some can withstand saline soils, flooding relatively well [16]. The yield of black locust is negatively affected by dryness during the initial planting and growth period; under such conditions low yield has been found in Spain—0.91 Mg ha^{-1} y^{-1}—ten times higher biomass has also been obtained on well managed (weed control, fertilizer and irrigation) sites with sandy soil and low organic matter—9.20 Mg ha^{-1} y^{-1} [92,93]. Black locust is often found on well aerated, relatively dry and stony soil, but is not suited for areas with compact or shallow soil and stagnant water. It can tolerate a broad range of soil reactions [94]. As a benefit to soil, black locust is a nitrogen-fixing species and can grow on nitrogen-poor soils [71,95].

4.2.3. Birch

Birch (*B. pubescence* and *B. pendula*) is another typically planted species across Europe and, if left abandoned, natural afforestation of agricultural land in a large portion of Europe happens with birch as a pioneer species. The high natural regeneration capacity of this species suggest its suitability for growing under a wide range of site conditions. Due to birches' ability to effectively propagate via seeds, dense naturally afforested birch stands can be used for biomass harvesting, thus avoiding the initial planting costs [96]. Compared to the *Salicaceae* family, birch is better suited to acidic soils and lower moisture levels, but due to its slower growth rate, should be grown in longer rotations if intended for bioenergy production. In the Baltic region with assumed stand density of 2000 trees per ha, birch stand MAI on marginal land can be expected to be 1.7 and 3.9 Mg ha^{-1} y^{-1} at the age of 8 and 15 years, respectively, but on non-marginal land 2.9 and 4.7 Mg ha^{-1} y^{-1} (based on Daugaviete et al., 2017). In the case study plot, an 8-year-old birch stand produced 1.4 Mg ha^{-1} y^{-1} due to a relatively low initial planting density and low survival rate of around 73%. Even lower survival rates have been observed in 15-year-old birch stands in

Latvia (both marginal and non-marginal) and in 4-year-old stands on a reclaimed oil shale mining area in Estonia [70,97]. The survival of birch can be significantly affected by the lack of sunlight, as it is light-demanding species [98].

4.2.4. Black and Gray Alder

Other pioneer species typical to Europe are black and gray alder. These species are known to be tolerant of extended periods of flooding and can grow in a relatively broad soil pH range [99]. Low soil pH and excess moisture is typical to the northern part of Europe. Alder is suited to coppice systems and is fit for short-rotation forestry, since it reaches half its mature height at around 25 years [48,81,100]. On marginal land biological rotation can be expected to be reached later than on non-marginal land. The most productive period is also reached later on marginal land, and longer rotation periods are advisable [52]. Due to nitrogen fixing bacteria, alder has shown to be beneficial to nitrogen-poor soil, and thus, to growth of admixed woody species [101–104]. Based on performance data from stands established some 15–20 years ago in Latvia, on marginal land black alder produced 2.1 Mg ha^{-1} y^{-1} at the age of 8 and 7.0 Mg ha^{-1} y^{-1} at the age of 15 at assumed stand density of 2000 trees per hectare (based on Daugaviete et al. [97]). MAI had almost tripled from 8 to 15 years, thus confirming that longer rotations are better suited to low-density forest species stands. The results showed high variance both within a stand and between the stands. In the experimental plot studied in this research, trees were planted at a lower density (1636 trees ha^{-1}) and the plantations of black alder, gray alder and hybrid alder yielded 1.3, 1.7 and 2.0 Mg ha^{-1} y^{-1}, respectively. In studies conducted on abandoned agricultural land, but no defined constraints, gray alder produced 15.86 Mg ha^{-1} (3.2 Mg ha^{-1} y^{-1}) in Estonia after 5 growing seasons. On average, the performance of plantations established on marginal land is around 60–80% of that on non-marginal land, based on research done in Latvia [97]. However, due to high variation, pests, lack of management or initially unidentified site constraints, some less productive plantations on non-marginal land are comparable in terms of yield to promising plantations on marginal land.

4.3. Treatment

Management practices determine the environmental impacts as well as the financial feasibility of forestry. It has been shown that economically viable woody crop plantations can be achieved by selection of appropriate planting material and management practices [16,46,105]. To secure profitability and more importantly, survival, on marginal land, treatment, such as soil preparation, weed control, fertilization, liming, irrigation or animal damage prevention is often necessary. Based on other research (Table A4), mechanical or chemical weed control is most commonly applied treatment in the early stage of plantation establishment [69,106–108]. Removal of competing vegetation has been shown to increase the biomass production twofold in small-scale field trials in Latvia. Weed control by covering the surrounding area with plastic film had an adverse effect on tree survival, when used for shallow-rooted trees [97].

Treatment is expected to be the most effective when it is selected to counteract the main limiting factors. However, species vary in their sensitivity to site conditions [109]. A particular marginality factor can have little effect on some species, but can be determining to other. Treatment can have indirect effects as well. For example, in this study irrigation had a significant positive effect on yield of Siberian elm planted in experimental plots established in Spain, where soil texture was recognized as the main marginality factor. Evidently, Siberian elm was also significantly affected by the water regimen of the site. Siberian elm withstands high temperature periods by increasing its transpiration rate [90]. Therefore, despite it being drought tolerant, sufficient water supply is needed to support the transpiration demand and to allow better nutrient uptake and resource allocation to biomass accumulation instead of defense mechanism processes [110–113]. Irrigation has also show a positive effect on willow plantations in more temperate climate [114,115].

With unfavorable soil texture and stoniness at the experimental site in Latvia, the fertilization effect was species-specific, with hybrid aspen being the only species showing a clear positive response to fertilization with digestate. Even some negative growth response to sewage sludge and wood ash was observed. This could be induced by changes in soil that affect either nutrient availability by raising the pH or mycorrhiza of these species [116]. Other studies have also found that response to fertilization is clone-specific [69]. It is important to consider the necessity of fertilizer application, as it does increase management costs and pose environmental risks (surplus nutrient leeching), but does not always lead to increased yield [57,69]. The effects of fertilization have shown to be more prominent on land, where plant growth is directly limited by nutrient availability [117,118]. Even in such cases, the lowest effective dose should be applied because increasing the dose typically does not provide significant additional effect on yield. Multiple applications of low-dose fertilizer are preferred over a single application of high-dose fertilizer in terms of environmental safety. In the Ukraine-based case study of a willow plantation with low soil organic matter, nitrogen fertilizer had minimal positive effect on yield. Just like in the case study carried out in Latvia, fertilization did not address the main limiting factors. Furthermore, high soil acidity was also present and could have an immobilizing effect on the added nutrients. Thus, in case of acidic soils (for example peat soils in Northern Europe) liming should be combined with fertilization. Liming agents often possess absorption and adsorption properties—soil treatment with lime and bisphosphonates, as well as biochar has shown a positive effect on willow growth on contaminated soils, most likely due to sorbent properties [108,119]. If the soil is already alkaline, different sorbents should be used. In addition, by promoting biomass accumulation, treatment can be used to improve phytoremediation of such sites.

Treatment is crucial when trying to establish a tree plantation or woody crop on especially challenging land, such as post-mining sites, spoil heaps and highly acidic soils (Table A4) [71,120,121]. While acidity can be mitigated by application of liming material that often also promotes nutrient availability, mining sites and spoil sites are more complicated to recultivate.

There are also some unconventional tools to improve site conditions for growth of woody crops, for example, utilization of other species as nurse plants that provide shade, improve soil structure and water-holding capacity or nitrogen fixation in soil [101,122–124].

5. Conclusions

It was found that tree plantations and woody crops can be successfully established in terms of survival on marginal land across Latvia, Spain and Ukraine. While the marginality factors addressed in this article are similar across the study sites and countries, the management and species vary depending on specific soil and climatic conditions of each site. In the more northern region, Latvia, hybrid aspen performed better than the indigenous pioneer species birch and alder on a site with loam and sandy loam texture, but there was significant difference between the hybrid aspen clones. Thus, the specific genetic material might be even more determining than the species. In the warmer and dryer climate of Spain, Siberian elm proved to be suitable for cultivation on stony soil with sandy texture and low organic carbon content, where most other crops would fail to thrive. In the continental agro-ecological zone of Ukraine, high-density stands of willow proved to be tolerant to low soil organic carbon content and produced yields that can compete with forest residues in terms of financially feasible biomass supply. This suggests high-density SRC woody crops are more productive for cultivation on marginal land compared to tree plantations. However, the feasibility strongly depends on the current state of legislation and socioeconomic factors that are a subject to constant change. Future studies should investigate the potential of growing woody crops and tree plantations on marginal land to contribute to rural development, biodiversity conservation, environmental protection and climate change adaptation.

Author Contributions: Conceptualization, all authors; methodology, D.L., M.D., P.C., M.S., O.P. and O.M.; formal analysis, S.C.; investigation, M.D., M.S., P.C., O.P. and O.M.; data curation, S.C.; writing—original draft preparation, S.C., M.S. and O.M.; writing—review and editing, M.v.C., M.S., D.L., P.C. and O.M.; visualization, S.C.; supervision, D.L., P.C. and M.v.C.; project administration, D.L.; funding acquisition, D.L. All authors have read and agreed to the published version of the manuscript.

Funding: This research received funding from the European Union's Horizon 2020 research and innovation program under grant agreement No 727698 (project: Marginal Lands for Growing Industrial Crops: Turning a burden into an opportunity) and the European Regional Development Fund project Elaboration of innovative white willow-perennial grass agroforestry systems on marginal mineral soils improved by wood ash and less demanded peat fractions amendments (1.1.1.1/19/A/112). The article's processing charge was funded by the European Regional Development Fund project Elaboration of innovative white willow-perennial grass agroforestry systems on marginal mineral soils improved by wood ash and less demanded peat fractions amendments (1.1.1.1/19/A/112).

Data Availability Statement: Data available on request made to the corresponding author S.C.

Conflicts of Interest: The authors declare no conflict of interest.

Appendix A

Table A1. Average climatic conditions of the study site located in Latvia during the study period.

Yearz	Precipitation (mm)	Mean Temperature (°C)	Maximum Temperature (°C)	Minimum Temperature (°C)
2011	692.7	7.2	31.1	−24.7
2012	935.4	6.0	31.8	−29.6
2013	652.5	6.9	31.5	−21.1
2014	855.5	7.3	32.3	−18.8
2015	687.4	7.6	31.7	−18.2
2016	894.4	6.9	31.6	−24.4
2017	874.7	6.7	30.6	−28.1
2018	363.9	7.4	32.9	−23.8
Annual average	744.6	7.0	31.7	−23.6

Table A2. Average climatic conditions of the study site located in Spain during the study period.

Year	Precipitation (mm)	Mean Temperature (°C)	Maximum Temperature (°C)	Minimum Temperature (°C)	Solar Irradiance (kWh/m^2)
2009	107.4	4.8	18.9	−12.8	98.0
2010	598.5	9.6	33.4	−11.0	1508.0
2011	379.6	11.1	35.7	−11.9	1629.2
2012	344.4	10.6	37.0	−10.3	1659.6
2013	594.5	9.5	33.8	−8.8	1547.8
2014	595.5	10.7	33.3	−6.7	1599.6
2015	488.1	11.0	36.0	−9.5	1605.8
2016	540.2	10.7	34.7	−7.5	1404.5
2017	314.8	11.9	35.3	−11.7	1654.6
2018	668.7	10.4	35.3	−8.3	1294.5
Annual average	463.2	10.0	33.3	−9.8	1400.2

Table A3. Average climatic conditions of the study site located in Ukraine during the study period.

Year	Precipitation (mm)	Mean Temperature (°C)	Maximum Temperature (°C)	Minimum Temperature (°C)
2011	446.0	8.3	30.8	−17.2
2012	497.4	8.4	36.8	−28.2
2013	618.2	8.6	29.7	−19.0
2014	549.0	8.5	32.9	−23.5
2015	372.0	9.8	35.2	−18.5
2016	466.5	9.0	33.1	−22.2
2017	538.1	9.0	33.4	−21.6
2018	566.1	8.8	30.0	−22.3
2019	535.8	9.9	33.1	−12.5
Annual average	509.9	8.9	32.8	−20.6

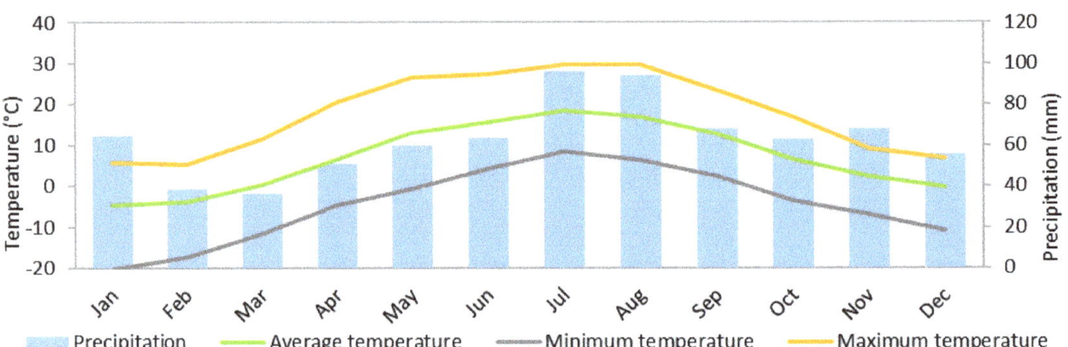

Figure A1. Average monthly climatic conditions of the study site located in Latvia during the study period 2011–2018.

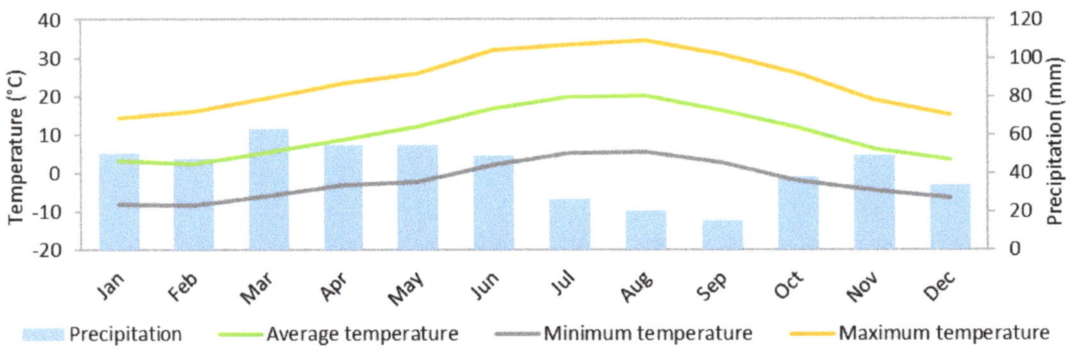

Figure A2. Average monthly climatic conditions of the study site located in Spain during the study period 2009–2018.

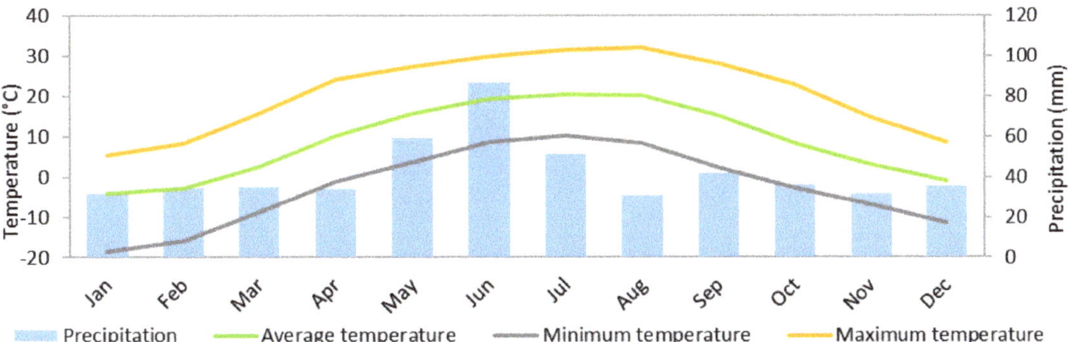

Figure A3. Average monthly climatic conditions of the study site located in Ukraine during the study period 2011–2019.

Table A4. Results of woody crop above-ground biomass yield on marginal land based on research done in Europe.

Species	Site Condition	Planting Density, Plants ha^{-1}	Rotation Length, Years	* Above-Ground Biomass Yield, Dry Weight, Mg ha^{-1} y^{-1}	** Treatment	Location	Source
Hybrid aspen (2 clones)	Poor rooting conditions— unfavorable soil texture and stoniness	1261–2500	8	0.9–4.8 depending on clone and treatment	Fertilizer, weed control, animal prevention	Latvia	
Hybrid alder				2.0			
Black alder		1636;		1.3			
Gray alder				1.7			
Birch				1.4			
Siberian elm	Sandy soil with unfavorable soil texture, stoniness and low soil organic carbon (<1%)	6666	3	1.8 rain-fed and 4.5 irrigated (first rotation); 2.6 rain-fed and 4.9 irrigated (second rotation); 3.7 rain-fed and 7.1 irrigated (third rotation);	Rain-fed and irrigation	Spain	
Willow	Clay soil with low soil organic carbon	20,000	3;	8.4 (first rotation); 9.5 (third rotation);		Ukraine	
Willow	Soil with low soil organic carbon	20,000	3;	10.8 (second rotation)		Ukraine	
Willow	Soil with low soil organic carbon and high soil acidity	20,000	3;	8.6 (first rotation)	Fertilizer	Ukraine	
Poplar (12 clones)	Former agricultural land with sandy soil and limited drainage	8000	2	1.5–7.2 (3.0–14.4 Mg ha^{-1}) (first rotation) and 7.4–16.2 (14.8–32.4 Mg ha^{-1}) (second rotation) depending on clone	Weed control	Belgium	[45,76]
Poplar (17 clones)	Former waste disposal site covered with a 2 m thick layer of sand, clay and rubble	10,000	4	2.2–11.4 depending on clone	Weed control	Belgium	[75]
Birch	Former agricultural land, sandy soil with soil organic matter <1% and pH$_{KCl}$ 4.5	6667 (birch, maple); 20,000 (poplar, willow)	4	2.6		Belgium	[63]
Maple				1.2			
Poplar				3.5			
Willow				3.4			

Table A4. *Cont.*

Species	Site Condition	Planting Density, Plants ha^{-1}	Rotation Length, Years	* Above-Ground Biomass Yield, Dry Weight, Mg ha^{-1} y^{-1}	** Treatment	Location	Source
Willow	Contaminated, dry, nutrient poor, sandy soils	18,000	3	4.2–6.6	Weed control	Belgium	[26]
Poplar				1.1–1.5			
Willow	Former agricultural land	14,800; 17,800;	5	3.1 (15.4 Mg ha^{-1}) control and 4.9–5.3 (24.7–26.3 Mg ha^{-1}) irrigated	Irrigated	Estonia	[114]
Birch	Naturally afforested abandoned agricultural land (and 1 planted site)	36,200	8	2.9 (22.8 Mg ha^{-1})		Estonia	[53]
		13,900		2.8 (22.0 Mg ha^{-1})			
		28,260		1.3 (10.2 Mg ha^{-1})			
		3060		0.8 (6.0 Mg ha^{-1})			
		4400 (planted)		1.7 (13.3 Mg ha^{-1}) (planted)			
Birch	Leveled quarry spoil	1017	7	0.02 (0.2 Mg ha^{-1})		Estonia	[121]
Alder		2100		0.36 (2.6 Mg ha^{-1})			
Pine		3042		0.27 (1.9 Mg ha^{-1})			
Willow	Restored landfill	1000–10,500	3	10.5; 18.8–22.6 (irrigated);	Irrigation	Finland	[115]
Birch and willow	Naturally afforested cut-away peatland	12,800	14	2.7–4.4	Fertilizer	Finland	[117]
Hybrid aspen	Fallow agricultural land	900	1	5.2;	Fertilizer (in second season)	Finland	[125]
			2	8.7 (17.4 Mg ha^{-1}) control and 9.95 (19.9 Mg ha^{-1}) fertilized			
			3	7.9 (23.9 Mg ha^{-1}) control and 9.5 (28.9 Mg ha^{-1}) fertilized			
Birch	Organic soils—cutaway peatlands—naturally afforested		10–27	3–4		Finland	[126]
Poplar (14 clones)	Trace element contaminated site		7	3.1–8.5		France	[82]
Poplar	Abandoned agricultural land	7272	2	1.9 (3.7 Mg ha^{-1}) (first rotation) 4.3 (8.6 Mg ha^{-1}) (second rotation)	Weed control	France	[77]
Willow	Abandoned agricultural land	9697	2	2.07 (4.1 Mg ha^{-1}) (first rotation) 11.0 (21.9 Mg ha^{-1}) (second rotation)	Weed control	France	[77]
Poplar (8 clones)	Disturbed, marginally fertile post-mine site	8333	8	0.4–6.0 (3.5–46.7 Mg ha^{-1})	Fertilizer	Germany	[120]
Black locust	Post-mine site with substrate from overburden sediments dumped during opencast lignite mining and low nitrogen content	6579;	14	2.7	Fertilizer	Germany	[71]
		10,929;	3	1.9, 2.5 and 1.8 (first, second and third rotation)			
		9200;	4	0.5			
		8736;	4	-			
Willow, poplar and black locust	Land with high sand content	6700	2	4.3, 7.7 and 9.2 (first, second and third rotation)		Italy	[57]
Willow, poplar and black locust	Land with low soil organic matter	6700	2	3.3, 12.9 and 12.2 (first, second and third rotation)		Italy	[57]
Birch; Pine;	Unfavorable soil texture, limited drainage	3300 (birch); 5000 (pine)	8	0.7 (birch) and 0.3 (pine)	Weed control, animal prevention	Latvia	[97]
			15	1.9 (birch) and 3.8 (pine)			

Table A4. *Cont.*

Species	Site Condition	Planting Density, Plants ha^{-1}	Rotation Length, Years	* Above-Ground Biomass Yield, Dry Weight, Mg ha^{-1} y^{-1}	** Treatment	Location	Source
Aspen	Limited soil drainage, periodic flooding, low temperatures	3300	8	0.5	Weed control, animal prevention	Latvia	[97]
			15	4.5			
Spruce	Acidic soil	3300	8	0.5–1.5 (depending on site)	Weed control, animal prevention	Latvia	[97]
			15	2.8–8.4 (depending on site)			
Black alder	Acidic soil	3300	8	1.3–3.3 (depending on site)	Weed control, animal prevention	Latvia	[97]
			15	2.5–15.9 (depending on site)			
Birch	Acidic soil	2000–3300	8	0.7–4.0 (depending on site)	Weed control, animal prevention	Latvia	[97]
			15	2.4–7.4 (depending on site)			
Birch; Spruce;	Acidic soil, excess moisture, low P and N content, low temperatures	3300 (spruce)	8	2.1 (birch) and 1.0 (spruce)	Weed control, fertilizer, animal prevention	Latvia	[97]
			15	4.9 (birch) and 4.5 (spruce)			
Willow	Marginal gley soils	20,000	3	12–15	Weed control	Northern Ireland	[56]
Siberian elm	Heavy black soil with a heavy clay granulometric composition	3448–51,282	7	5.2 (first rotation)	Rain-fed	Poland	[127]
Willow	Poor agricultural soils (loose, sandy soil with periodical dryness)	11,000	4	5.1–10.3	Weed control, fertilizer	Poland	[118]
Poplar				5.5–10.5			
Black locust				1.6–3.7			
Siberian elm	Sandy soil with low organic matter content (0.92%), low nitrogen (0.03%), many gravels (39.9%) and pH 5.90	3333	3	1.18 rain-fed and 2.43 irrigated (first rotation)	Rain-fed and irrigation (4167 m^3 ha^{-1}y^{-1})	Spain	[110]
Siberian elm	Sandy soil with low organic matter content (0.92%), low nitrogen (0.03%), many gravels (39.9%) and pH 5.90	6666	3	1.63, 5.19 rain-fed and 4.93 irrigated (first rotation)	Rain-fed and irrigation (2250 m^3 ha^{-1}y^{-1} and 4167 m^3 ha^{-1}y^{-1})	Spain	[110]
Siberian elm	Sandy soil with unfavorable soil texture, stoniness 28% and low soil organic matter content (0.4%)	3333	4	2.6 rain-fed and 6.0 irrigated (first rotation)	Rain-fed and irrigation (3400 m^3 ha^{-1} y^{-1})	Spain	[84]
		6666		2.5 rain-fed and 6.5 irrigated (first rotation)			
Siberian elm	Sandy clay loamy texture, pH 8.30, organic matter 4.0%, total nitrogen 0.35%, 27 ppm P (Olsen) and extreme climate	3333	3	5.1 (first rotation)	Rain-fed	Spain	[65]
Siberian elm	Basic soil with an excess of calcium. Entisol orden and Xerofluvent greatgroup.	6666	2	5.2 (first rotation)	Rain-fed	Spain	[64]
			3	13.2 (first rotation)			
Siberian elm	Sandy loam texture, low organic matter content (0.75%), nitrogen 0.08% and pH 5.87	6666	2.5	3.46 (first rotation)	Rain-fed	Spain	[66]
Poplar	Sandy soil with low organic matter content in semi-arid climatic conditions	10,000	2	1.9 (first rotation) 12	Fertilizer	Spain	[58]
Willow			3 (for 9 years)	9			
Black locust				7			
Sycamore				3			

Table A4. *Cont.*

Species	Site Condition	Planting Density, Plants ha^{-1}	Rotation Length, Years	* Above-Ground Biomass Yield, Dry Weight, Mg ha^{-1} y^{-1}	** Treatment	Location	Source
Willow (3 clones)	Former mining area	9876; 14,815;	5	0.3, 0.7, 1.7 (1,.3, 3.6, and 8.6 Mg ha^{-1}) depending on clone, 0.2, 1.1, 1.3 (1.1, 5.4, 6.6 Mg ha^{-1}) depending on treatment, 0.8, 1.0 (4.0 and 5.2 Mg ha^{-1}) depending on density	Fertilizer, weed control	Spain	[69]
Poplar	Degraded soils	5000	3	12.3–17.9 (36.9–53.8 Mg ha^{-1})	Fertilizer, irrigation	Spain	[78]
Eucalyptus	Degraded soils	5000	3	14.7–18.3 (44.2–55.0 Mg ha^{-1})	Fertilizer, irrigation	Spain	[78]
Paulownia	Degraded, acidic soils	5000	3	1.1–1.7 (3.3–5.1 Mg ha^{-1})	Fertilizer, irrigation	Spain	[78]
Eucalyptus	Degraded, acidic soils	5000	3	13.5–19.7 (40.4–59.2 Mg ha^{-1})	Fertilizer, irrigation	Spain	[78]

*- If yield is measured as Mg ha^{-1} in the source, the values are given in parentheses. **- Treatment for all or part of the experimental site (control plots are also present in most cases).

References

1. Ustaoglu, E.; Collier, M.J. Farmland abandonment in Europe: An overview of drivers, consequences, and assessment of the sustainability implications. *Environ. Rev.* **2018**, *26*, 396–416. [CrossRef]
2. Perpiña Castillo, C.; Kavalov, B.; Diogo, V.; Jacobs-Crisioni, C.; Batista e Silva, F.; Lavalle, C. Agricultural Land Abandonment in the EU within 2015–2030. In *JRC Report 113718*; European Commission: Ispra, Italy, 2018.
3. Abolina, E.; Luzadis, V.A. Abandoned agricultural land and its potential for short rotation woody crops in Latvia. *Land use policy* **2015**, *49*, 435–445. [CrossRef]
4. Shengfa, L.; Xiubin, L. Global understanding of farmland abandonment: A review and prospects. *J. Geogr. Sci.* **2017**, *27*, 1123–1150. [CrossRef]
5. Ruskule, A.; Nikodemus, O.; Kasparinska, Z.; Kasparinskis, R.; Brūmelis, G. Patterns of afforestation on abandoned agriculture land in Latvia. *Agrofor. Syst.* **2012**, *85*, 215–231. [CrossRef]
6. Rey Benayas, J.M. Restoring forests after land abandonment. In *Forest Restoration in Landscapes*; Springer: New York, NY, USA, 2005; pp. 356–360. [CrossRef]
7. Munroe, D.K.; van Berkel, D.B.; Verburg, P.H.; Olson, J.L. Alternative trajectories of land abandonment: Causes, consequences and research challenges. *Curr. Opin. Environ. Sustain.* **2013**, *5*, 471–476. [CrossRef]
8. Lange, L.; Connor, K.O.; Arason, S.; Bundgård-Jørgensen, U.; Canalis, A.; Carrez, D.; Gallagher, J.; Gøtke, N.; Huyghe, C.; Jarry, B.; et al. Developing a Sustainable and Circular Bio-Based Economy in EU: By Partnering Across Sectors, Upscaling and Using New Knowledge Faster, and For the Benefit of Climate, Environment & Biodiversity, and People & Business. *Front. Bioeng. Biotechnol.* **2021**, *8*, 619066. [CrossRef]
9. Lindegaard, K.N.; Adams, P.W.R.; Holley, M.; Lamley, A.; Henriksson, A.; Larsson, S.; von Engelbrechten, H.G.; Esteban Lopez, G.; Pisarek, M. Short rotation plantations policy history in Europe: Lessons from the past and recommendations for the future. *Food Energy Secur.* **2016**, *5*, 125–152. [CrossRef]
10. Mandley, S.J.; Daioglou, V.; Junginger, H.M.; van Vuuren, D.P.; Wicke, B. EU bioenergy development to 2050. *Renew. Sustain. Energy Rev.* **2020**, *127*, 109858. [CrossRef]
11. United Nations Environment Programme. *Renewables 2020 Global Status Report*; United Nations Environment Programme: Nairobi, Republic of Kenya, 2020; ISBN 978-3-948393-00-7.
12. Guidi, W.; Piccioni, E.; Bonari, E. Evapotranspiration and crop coefficient of poplar and willow short-rotation coppice used as vegetation filter. *Bioresour. Technol.* **2008**, *99*, 4832–4840. [CrossRef]
13. Bloemen, J.; Fichot, R.; Horemans, J.A.; Broeckx, L.S.; Verlinden, M.S.; Zenone, T.; Ceulemans, R. Water use of a multigenotype poplar short-rotation coppice from tree to stand scale. *GCB Bioenergy* **2017**, *9*, 370–384. [CrossRef]
14. Xi, B.; Clothier, B.; Coleman, M.; Duan, J.; Hu, W.; Li, D.; Di, N.; Liu, Y.; Fu, J.; Li, J.; et al. Irrigation management in poplar (Populus spp.) plantations: A review. *For. Ecol. Manag.* **2021**, *494*, 119330. [CrossRef]
15. Geyer, W.A. Biomass production in the Central Great Plains USA under various coppice regimes. *Biomass Bioenergy* **2006**, *30*, 778–783. [CrossRef]
16. Quinn, L.D.; Straker, K.C.; Guo, J.; Kim, S.; Thapa, S.; Kling, G.; Lee, D.K.; Voigt, T.B. Stress-Tolerant Feedstocks for Sustainable Bioenergy Production on Marginal Land. *Bioenergy Res.* **2015**, *8*, 1081–1100. [CrossRef]

17. Nocentini, A.; Monti, A. Land-use change from poplar to switchgrass and giant reed increases soil organic carbon. *Agron. Sustain. Dev.* **2017**, *37*, 23. [CrossRef]

18. Mafa-Attoye, T.G.; Thevathasan, N.V.; Dunfield, K.E. Indications of shifting microbial communities associated with growing biomass crops on marginal lands in Southern Ontario. *Agrofor. Syst.* **2020**, *94*, 735–746. [CrossRef]

19. Kahle, P.; Janssen, M. Impact of short-rotation coppice with poplar and willow on soil physical properties. *J. Plant Nutr. Soil Sci.* **2020**, *183*, 119–128. [CrossRef]

20. Georgiadis, P.; Vesterdal, L.; Stupak, I.; Raulund-Rasmussen, K. Accumulation of soil organic carbon after cropland conversion to short-rotation willow and poplar. *GCB Bioenergy* **2017**, *9*, 1390–1401. [CrossRef]

21. Bardos, R.P.; Bone, B.; Andersson-Sköld, Y.; Suer, P.; Track, T.; Wagelmans, M. Crop-based systems for sustainable risk-based land management for economically marginal damaged land. *Remediation* **2011**, *21*, 11–33. [CrossRef]

22. Finzi, A.C.; Canham, C.D.; Van Breeman, N. Erratum: Canopy tree-soil interactions within temperate forests: Species effects on pH and cations (Ecological Applications 8 (447–454)). *Ecol. Appl.* **1998**, *8*, 905. [CrossRef]

23. Kang, S.; Post, W.M.; Nichols, J.A.; Wang, D.; West, T.O.; Bandaru, V.; Izaurralde, R.C. Marginal Lands: Concept, Assessment and Management. *J. Agric. Sci.* **2013**, *5*, 129–139. [CrossRef]

24. McKay, H. (Ed.) *Short Rotation Forestry: Review of Growth and Environmental Impacts*; Forest Research: Surrey, UK, 2011; Volume 2, ISBN 978-0-85538-827-0.

25. Mertens, J.; Van Nevel, L.; De Schrijver, A.; Piesschaert, F.; Oosterbaan, A.; Tack, F.M.G.; Verheyen, K. Tree species effect on the redistribution of soil metals. *Environ. Pollut.* **2007**, *149*, 173–181. [CrossRef] [PubMed]

26. Ruttens, A.; Boulet, J.; Weyens, N.; Smeets, K.; Adriaensen, K.; Meers, E.; van Slycken, S.; Tack, F.; Meiresonne, L.; Thewys, T.; et al. Short rotation coppice culture of willows and poplars as energy crops on metal contaminated agricultural soils. *Int. J. Phytoremediation* **2011**, *13*, 194–207. [CrossRef] [PubMed]

27. Thiry, Y.; Colle, C.; Yoschenko, V.; Levchuk, S.; Van Hees, M.; Hurtevent, P.; Kashparov, V. Impact of Scots pine (*Pinus sylvestris* L.) plantings on long term 137Cs and 90Sr recycling from a waste burial site in the Chernobyl Red Forest. *J. Environ. Radioact.* **2009**, *100*, 1062–1068. [CrossRef] [PubMed]

28. Elbersen, B.; van Eupen, M.; Mantel, S.; Alexopoulou, E.; Zanghou, B.; Boogaard, H.; Carrasco, J.; Ceccarelli, T.; Ciria, C.S.; Ciria, P.; et al. Mapping Marginal land potentially available for industrial crops in Europe. In Proceedings of the European Biomass Conference and Exhibition Proceedings 26thEUBCE, Copenagen, Denmark, 14–17 May 2018; p. 71.

29. Emery, I.; Mueller, S.; Qin, Z.; Dunn, J.B. Evaluating the Potential of Marginal Land for Cellulosic Feedstock Production and Carbon Sequestration in the United States. *Environ. Sci. Technol.* **2017**, *51*, 733–741. [CrossRef] [PubMed]

30. Gerwin, W.; Repmann, F.; Galatsidas, S.; Vlachaki, D.; Gounaris, N.; Baumgarten, W.; Volkmann, C.; Keramitzis, D.; Kiourtsis, F.; Freese, D. Assessment and quantification of marginal lands for biomass production in Europe using soil-quality indicators. *Soil* **2018**, *4*, 267–290. [CrossRef]

31. Kang, S.; Post, W.; Wang, D.; Nichols, J.; Bandaru, V.; West, T. Hierarchical marginal land assessment for land use planning. *Land Use Policy* **2013**, *30*, 106–113. [CrossRef]

32. Lewis, S.M.; Kelly, M. Mapping the potential for biofuel production on marginal lands: Differences in definitions, data and models across scales. *ISPRS Int. J. Geo-Information* **2014**, *3*, 430–459. [CrossRef]

33. Mellor, P.; Lord, R.A.; João, E.; Thomas, R.; Hursthouse, A. Identifying non-agricultural marginal lands as a route to sustainable bioenergy provision—A review and holistic definition. *Renew. Sustain. Energy Rev.* **2021**, *135*, 110220. [CrossRef]

34. Shortall, O.K. 'Marginal land' for energy crops: Exploring definitions and embedded assumptions. *Energy Policy* **2013**, *62*, 19–27. [CrossRef]

35. Turley, D.; Taylor, M.; Laybourn, R.; Hughes, J.; Kilpatrick, J.; Procter, C.; Wilson, L.; Edgington, P. Assessment of the availability of 'marginal' and 'idle' land for bioenergy crop production in England and Wales. *Res. Proj. Final Rep.* **2010**, 86.

36. MAGIC-MAPS. Available online: https://iiasa-spatial.maps.arcgis.com/apps/webappviewer/index.html?id=270aa7d778c2 45228fe82dc826cbd703 (accessed on 10 February 2022).

37. Elbersen, B.; van Eupen, M.; Verzandvoort, S.; Boogaard, H.; Mucher, S.; Cicarreli, T.; Elbersen, W.; Mantel, S.; Bai, Z.; Iqbal, Y.; et al. *Methodological Approaches to Identify and Map Marginal Land Suitable for Industrial Crops in Europe*; EU Horiz. 2020; MAGIC; GA-No. 727698; Wageningen University & Research: Wageningen, The Netherlands, 2018.

38. FAO. *The State of Food and Agriculture: Food Aid for Food Security?* FAO: Rome, Italy, 2006.

39. Liepiņš, J. *Latvijas Kokaudžu Biomasas Un Oglekļa Uzkrājuma Novērtēšanas Metodes Forest Stand Biomass and Carbon Stock Estimates in Latvia*; Latvijas Lauksaimniecības Universitāte: Jelgava, Latvia, 2020.

40. R Core Team. *R: A Language and Environment*; R Foundation for Statistical Computing: Vienna, Austria, 2021.

41. Johansson, T. Biomass production of hybrid aspen growing on former farm land in Sweden. *J. For. Res.* **2013**, *24*, 237–246. [CrossRef]

42. Keoleian, G.A.; Volk, T.A. Renewable energy from willow biomass crops: Life cycle energy, environmental and economic performance. *CRC. Crit. Rev. Plant Sci.* **2005**, *24*, 385–406. [CrossRef]

43. Rosso, L.; Facciotto, G.; Bergante, S.; Vietto, L.; Nervo, G. Selection and testing of Populus alba and Salix spp. as bioenergy feedstock: Preliminary results. *Appl. Energy* **2013**, *102*, 87–92. [CrossRef]

44. Tullus, A.; Tullus, H.; Vares, A.; Kanal, A. Early growth of hybrid aspen (Populus × wettsteinii Hämet-Ahti) plantations on former agricultural lands in Estonia. *For. Ecol. Manag.* **2007**, *245*, 118–129. [CrossRef]

45. Verlinden, M.S.; Broeckx, L.S.; Van den Bulcke, J.; Van Acker, J.; Ceulemans, R. Comparative study of biomass determinants of 12 poplar (Populus) genotypes in a high-density short-rotation culture. *For. Ecol. Manag.* **2013**, *307*, 101–111. [CrossRef]
46. Kauter, D.; Lewandowski, I.; Claupein, W. Quantity and quality of harvestable biomass from Populus short rotation coppice for solid fuel use—A review of the physiological basis and management influences. *Biomass Bioenergy* **2003**, *24*, 411–427. [CrossRef]
47. Tullus, A.; Rytter, L.; Tullus, T.; Weih, M.; Tullus, H. Short-rotation forestry with hybrid aspen (*Populus tremula* L. × *P. tremuloides* Michx.) in Northern Europe. *Scand. J. For. Res.* **2012**, *27*, 10–29. [CrossRef]
48. Daugaviete, M.; Bārdulis, A.; Daugavietis, U.; Lazdiņa, D.; Bārdule, A. Potential of Producing Wood Biomass in Short-Rotation Grey Alder (Alnus Incana Moench) Plantations on Agricultural Lands. In Proceedings of the 25th NJF Congress, Riga, Latvia, 16–18 June 2015; pp. 394–399.
49. Liesebach, M.; Von Wuehlisch, G.; Muhs, H.J. Aspen for short-rotation coppice plantations on agricultural sites in Germany: Effects of spacing and rotation time on growth and biomass production of aspen progenies. *For. Ecol. Manag.* **1999**, *121*, 25–39. [CrossRef]
50. Tullus, A.; Lukason, O.; Vares, A.; Padari, A.; Lutter, R.; Tullus, T.; Karoles, K.; Tullus, H. Economics of Hybrid Aspen (*Populus tremula* L. × *P. tremuloides* Michx.) and Silver Birch (*Betula pendula* Roth.) Plantations on Abandoned Agricultural Lands in Estonia. *Balt. For.* **2012**, *18*, 288–298.
51. Johansson, T. Biomass equations for determining fractions of European aspen growing on abandoned farmland and some practical implications. *Biomass Bioenergy* **1999**, *17*, 471–480. [CrossRef]
52. Oliveira, N.; Pérez-Cruzado, C.; Cañellas, I.; Rodríguez-Soalleiro, R.; Sixto, H. Poplar short rotation coppice plantations under mediterranean conditions: The case of Spain. *Forests* **2020**, *11*, 1352. [CrossRef]
53. Uri, V.; Vares, A.; Tullus, H.; Kanal, A. Above-ground biomass production and nutrient accumulation in young stands of silver birch on abandoned agricultural land. *Biomass Bioenergy* **2007**, *31*, 195–204. [CrossRef]
54. Bergkvist, P.; Ledin, S. Stem biomass yields at different planting designs and spacings in willow coppice systems. *Biomass Bioenergy* **1998**, *14*, 149–156. [CrossRef]
55. Kellomäki, S.; Kilpeläinen, A.; Alam, A. Forest bioenergy production: Management, carbon sequestration and adaptation. In *Forest BioEnergy Production: Management, Carbon Sequestration and Adaptation*; Springer: New York, NY, USA, 2013; p. 268. [CrossRef]
56. McElroy, G.; Dawson, W.M. Biomass from Short-rotation Coppice Willow on Marginal Land. *Biomass* **1986**, *10*, 225–240. [CrossRef]
57. Amaducci, S.; Facciotto, G.; Bergante, S.; Perego, A.; Serra, P.; Ferrarini, A.; Chimento, C. Biomass production and energy balance of herbaceous and woody crops on marginal soils in the Po Valley. *GCB Bioenergy* **2017**, *9*, 31–45. [CrossRef]
58. Fernández, M.J.; Barro, R.; Pérez, J.; Ciria, P. Production and composition of biomass from short rotation coppice in marginal land: A 9-year study. *Biomass Bioenergy* **2020**, *134*, 105478. [CrossRef]
59. Kopp, R.F.; Abrahamson, L.P.; White, E.H.; Volk, T.A.; Nowak, C.A.; Fillhart, R.C. Willow biomass production during ten successive annual harvests. *Biomass Bioenergy* **2001**, *20*, 1–7. [CrossRef]
60. Di Nasso, N.O.; Guidi, W.; Ragaglini, G.; Tozzini, C.; Bonari, E. Biomass production and energy balance of a 12-year-old short-rotation coppice poplar stand under different cutting cycles. *GCB Bioenergy* **2010**, *2*, 89–97. [CrossRef]
61. Don, A.; Osborne, B.; Hastings, A.; Skiba, U.; Carter, M.S.; Drewer, J.; Flessa, H.; Freibauer, A.; Hyvönen, N.; Jones, M.B.; et al. Land-use change to bioenergy production in Europe: Implications for the greenhouse gas balance and soil carbon. *GCB Bioenergy* **2012**, *4*, 372–391. [CrossRef]
62. European Environmental Agency (EEA). *Estimating the Environmentally Compatible Bioenergy Potential from Agriculture*; European Environmental Agency (EEA): Copenhagen, Denmark, 2007.
63. Vande Walle, I.; Van Camp, N.; Van de Casteele, L.; Verheyen, K.; Lemeur, R. Short-rotation forestry of birch, maple, poplar and willow in Flanders (Belgium) I-Biomass production after 4 years of tree growth. *Biomass Bioenergy* **2007**, *31*, 267–275. [CrossRef]
64. Sanz, M.; Curt, M.D.; Plaza, A.; García-Müller, M.; Fernández, J. Assessment of siberian elm coppicing cycle. In Proceedings of the 19th European Biomass Conference, Berlin, Germany, 6–10 June 2011; pp. 601–605.
65. Fernández, J.; Iriarte, L.; Sanz, M.; Curt, M.D. Preliminary study of Siberian elm (*Ulmus Pumila* L.) as an energy crop in a continental-Mediterranean Climate. In Proceedings of the 17th European Biomass Conference & Exhibition—From Research to Industry and Markets, Hamburg, Germany, 29 June–3 July 2009; pp. 148–153.
66. Sanchéz, J.; Sanz, M.; Curt, M.D.; Fernández, J.; Mosquera, F. Influence of planting season on Siberian elm yield and economic prospects. In Proceedings of the 23rd European Biomass Conference and Exhibition, Vienna, Austria, 1–4 June 2015; pp. 198–204.
67. Geyer, W.A.; Argent, R.M.; Walawender, W.P. Biomass properties and gasification behavior of 7 year-old Siberian elm. *Wood Fiber Sci.* **1987**, *19*, 176–182.
68. Geyer, W.A. Influence of environmental factors on woody biomass productivity in the Central Great Plains, USA. *Biomass Bioenergy* **1993**, *4*, 333–337. [CrossRef]
69. Castaño-Díaz, M.; Barrio-Anta, M.; Afif-Khouri, E.; Cámara-Obregón, A. Willow short rotation coppice trial in a former mining area in Northern Spain: Effects of clone, fertilization and planting density on yield after five years. *Forests* **2018**, *9*, 154. [CrossRef]
70. Kuznetsova, T.; Rosenvald, K.; Ostonen, I.; Helmisaari, H.S.; Mandre, M.; Lõhmus, K. Survival of black alder (*Alnus glutinosa* L.), silver birch (*Betula pendula* Roth.) and Scots pine (*Pinus sylvestris* L.) seedlings in a reclaimed oil shale mining area. *Ecol. Eng.* **2010**, *36*, 495–502. [CrossRef]
71. Grünewald, H.; Böhm, C.; Quinkenstein, A.; Grundmann, P.; Eberts, J.; von Wühlisch, G. *Robinia pseudoacacia* L.: A lesser known tree species for biomass production. *Bioenergy Res.* **2009**, *2*, 123–133. [CrossRef]

72. Hansen, E.A. Poplar woody biomass yields: A look to the future. *Biomass Bioenergy* **1991**, *1*, 1–7. [CrossRef]

73. Stolarski, M.J.; Niksa, D.; Krzyżaniak, M.; Tworkowski, J.; Szczukowski, S. Willow productivity from small- and large-scale experimental plantations in Poland from 2000 to 2017. *Renew. Sustain. Energy Rev.* **2019**, *101*, 461–475. [CrossRef]

74. Fang, S.; Liu, Y.; Yue, J.; Tian, Y.; Xu, X. Assessments of growth performance, crown structure, stem form and wood property of introduced poplar clones: Results from a long-term field experiment at a lowland site. *For. Ecol. Manag.* **2021**, *479*, 118586. [CrossRef]

75. Laureysens, I.; Bogaert, J.; Blust, R.; Ceulemans, R. Biomass production of 17 poplar clones in a short-rotation coppice culture on a waste disposal site and its relation to soil characteristics. *For. Ecol. Manag.* **2004**, *187*, 295–309. [CrossRef]

76. Verlinden, M.S.; Broeckx, L.S.; Ceulemans, R. First vs. second rotation of a poplar short rotation coppice: Above-ground biomass productivity and shoot dynamics. *Biomass Bioenergy* **2015**, *73*, 174–185. [CrossRef]

77. Guénon, R.; Bastien, J.C.; Thiébeau, P.; Bodineau, G.; Bertrand, I. Carbon and nutrient dynamics in short-rotation coppice of poplar and willow in a converted marginal land, a case study in central France. *Nutr. Cycl. Agroecosyst.* **2016**, *106*, 293–309. [CrossRef]

78. Madejón, P.; Alaejos, J.; García-Álbala, J.; Fernández, M.; Madejón, E. Three-year study of fast-growing trees in degraded soils amended with composts: Effects on soil fertility and productivity. *J. Environ. Manag.* **2016**, *169*, 18–26. [CrossRef] [PubMed]

79. Pulford, I.D.; Watson, C. Phytoremediation of heavy metal-contaminated land by trees—A review. *Environ. Int.* **2003**, *29*, 529–540. [CrossRef]

80. Jensen, J.K.; Holm, P.E.; Nejrup, J.; Larsen, M.B.; Borggaard, O.K. The potential of willow for remediation of heavy metal polluted calcareous urban soils. *Environ. Pollut.* **2009**, *157*, 931–937. [CrossRef]

81. Glenz, C.; Schlaepfer, R.; Iorgulescu, I.; Kienast, F. Flooding tolerance of Central European tree and shrub species. *For. Ecol. Manag.* **2006**, *235*, 1–13. [CrossRef]

82. Chalot, M.; Girardclos, O.; Ciadamidaro, L.; Zappelini, C.; Yung, L.; Durand, A.; Pfendler, S.; Lamy, I.; Driget, V.; Blaudez, D. Poplar rotation coppice at a trace element-contaminated phytomanagement site: A 10-year study revealing biomass production, element export and impact on extractable elements. *Sci. Total Environ.* **2020**, *699*, 134260. [CrossRef]

83. Ericsson, K.; Rosenqvist, H.; Nilsson, L.J. Energy crop production costs in the EU. *Biomass Bioenergy* **2009**, *33*, 1577–1586. [CrossRef]

84. Sanz, M.; Pérez, J.; Carrasco, J.E.; Ciria, P. Biomass yield of Siberian elm under different crop conditions on marginal agricultural land. In Proceedings of the 28th European Biomass Conference & Exhibition, Marseille, France, 6–9 July 2020; pp. 238–241.

85. Martynov, V.V.; Nikulina, T.V. Population surge of zigzag elm sawfly (Aproceros leucopoda (Takeuchi, 1939): Hymenoptera: Argidae) in the Northern Cis-Azov Region. *Russ. J. Biol. Invasions* **2017**, *8*, 135–142. [CrossRef]

86. Oswalt, C.M.; Oswalt, S.N. Chapter 8: Invasive Plants on Forest Land in the United States. In *General Technical Report SRS 207*; USDA-Forest Service, Southern Research Station: Asheville, NC, USA, 2013; pp. 123–134.

87. Lakicevic, M.; Reynolds, K.M.; Orlovic, S.; Kolarov, R. Measuring dendrofloristic diversity in urban parks in Novi Sad (Serbia). *Trees, For. People* **2022**, *8*, 100239. [CrossRef]

88. Brunet, J.; Zalapa, J.E.; Pecori, F.; Santini, A. Hybridization and introgression between the exotic Siberian elm, *Ulmus pumila*, and the native Field elm, U. minor, in Italy. *Biol. Invasions* **2013**, *15*, 2717–2730. [CrossRef]

89. BOE. *Real Decreto 1628/2011, de 14 de Noviembre, por el que se Regula el Listado y Catálogo Español de Especies Exóticas Invasoras*. 2011. Available online: https://www.boe.es/buscar/doc.php?id=BOE-A-2011-19398 (accessed on 31 March 2022).

90. Dulamsuren, C.; Hauck, M.; Nyambayar, S.; Bader, M.; Osokhjargal, D.; Oyungerel, S.; Leuschner, C. Performance of Siberian elm (*Ulmus pumila*) on steppe slopes of the northern Mongolian mountain taiga: Drought stress and herbivory in mature trees. *Environ. Exp. Bot.* **2009**, *66*, 18–24. [CrossRef]

91. Alesso, S.P.; Tapias, R.; Alaejos, J.; Fernández, M. Biomass Yield and Economic, Energy and Carbon Balances of *Ulmus pumila* L., *Robinia pseudoacacia* L. and Populus × euroamericana (Dode) Guinier Short-Rotation Coppices on Degraded Lands under Mediterranean Climate. *Forests* **2021**, *12*, 1337. [CrossRef]

92. Monedero, E.; Hernández, J.J.; Collado, R. Combustion-related properties of poplar, willow and black locust to be used as fuels in power plants. *Energies* **2017**, *10*, 997. [CrossRef]

93. Sixto, H.; Cañellas, I.; van Arendonk, J.; Ciria, P.; Camps, F.; Sánchez, M.; Sánchez-González, M. Growth potential of different species and genotypes for biomass production in short rotation in Mediterranean environments. *For. Ecol. Manag.* **2015**, *354*, 291–299. [CrossRef]

94. Vítková, M.; Tonika, J.; Müllerová, J. Black locust-Successful invader of a wide range of soil conditions. *Sci. Total Environ.* **2015**, *505*, 315–328. [CrossRef]

95. De Marco, A.; Arena, C.; Giordano, M.; Virzo De Santo, A. Impact of the invasive tree black locust on soil properties of Mediterranean stone pine-holm oak forests. *Plant Soil* **2013**, *372*, 473–486. [CrossRef]

96. Hytönen, J.; Aro, L.; Jylhä, P. Biomass production and carbon sequestration of dense downy birch stands on cutaway peatlands. *Scand. J. For. Res.* **2018**, *33*, 764–771. [CrossRef]

97. Daugaviete, M.; Bambe, B.; Lazdiņš, A.; Lazdiņa, D. *Plantāciju Mežu Augšanas Gaita, Produktivitāte un Ietekme uz Vidi*; LVMI Silava: Salaspils, Latvia, 2017; p. 470.

98. Dubois, H.; Verkasalo, E.; Claessens, H. Potential of birch (betula pendula roth and b. pubescens ehrh.) for forestry and forest-based industry sector within the changing climatic and socio-economic context ofwestern Europe. *Forests* **2020**, *11*, 336. [CrossRef]

99. Christersson, L.; Sennerby-Forsse, L.; Zsuffa, L. The role and significance of woody biomass plantations in Swedish agriculture. *For. Chron.* **1993**, *69*, 687–693. [CrossRef]

100. Claessens, H.; Oosterbaan, A.; Savill, P.; Rondeux, J. A review of the characteristics of black alder (*Alnus glutinosa* (L.) Gaertn.) and their implications for silvicultural practices. *Forestry* **2010**, *83*, 163–175. [CrossRef]

101. Vurdu, H. *Anatomical Characteristics of Stem, Branch and Root Wood in European Black Alder (Alnus glutinosa L. Gaertn.)*; Iowa State University: Ames, Iowa, 1977.

102. Giardina, C.P.; Huffman, S.; Binkley, D.; Caldwell, B.A. Alders increase soil phosphorus availability in a Douglas-fir plantation. *Can. J. For. Res.* **1995**, *25*, 1652–1657. [CrossRef]

103. Myrold, D.D.; Huss-Danell, K. Alder and lupine enhance nitrogen cycling in a degraded forest soil in Northern Sweden. *Plant Soil* **2003**, *254*, 47–56. [CrossRef]

104. Uri, V.; Tullus, H.; Lo, K. Biomass production and nutrien accumulation in short-rotation grey alder. *For. Ecol. Manag.* **2002**, *161*, 169–179. [CrossRef]

105. Yrjälä, K.; Zheng, H. Renewable energy from woody biomass of poplar and willow src coupled to biochar production. *Handb. Environ. Chem.* **2021**, *99*, 133–150. [CrossRef]

106. Sevel, L.; Ingerslev, M.; Nord-Larsen, T.; Jørgensen, U.; Holm, P.E.; Schelde, K.; Raulund-Rasmussen, K. Fertilization of SRC Willow, II: Leaching and Element Balances. *Bioenergy Res.* **2014**, *7*, 338–352. [CrossRef]

107. Böhlenius, H.; Nilsson, U.; Salk, C. Liming increases early growth of poplars on forest sites with low soil pH. *Biomass Bioenergy* **2020**, *138*, 105572. [CrossRef]

108. Lebrun, M.; Miard, F.; Nandillon, R.; Hattab-Hambli, N.; Scippa, G.S.; Bourgerie, S.; Morabito, D. Eco-restoration of a mine technosol according to biochar particle size and dose application: Study of soil physico-chemical properties and phytostabilization capacities of Salix viminalis. *J. Soils Sediments* **2018**, *18*, 2188–2202. [CrossRef]

109. Callesen, I.; Raulund-Rasmussen, K.; Jrgensen, B.B.; Kvist-Johannsen, V. Growth of beech, oak, and four conifer species along a soil fertility gradient. *Balt. For.* **2006**, *12*, 14–23.

110. Pérez, I.; Pérez, J.; Carrasco, J.; Ciria, P. Siberian elm responses to different culture conditions under short rotation forestry in Mediterranean areas. *Turkish J. Agric. For.* **2014**, *38*, 652–662. [CrossRef]

111. Hasanuzzaman, M.; Nahar, K.; Alam, M.M.; Roychowdhury, R.; Fujita, M. Physiological, biochemical, and molecular mechanisms of heat stress tolerance in plants. *Int. J. Mol. Sci.* **2013**, *14*, 9643–9684. [CrossRef]

112. Park, G.E.; Lee, D.K.; Kim, K.W.; Batkhuu, N.O.; Tsogtbaatar, J.; Zhu, J.J.; Jin, Y.; Park, P.S.; Hyun, J.O.; Kim, H.S. Morphological characteristics and water-use efficiency of siberian elm trees (*Ulmus pumila* L.) within arid regions of northeast asia. *Forests* **2016**, *7*, 280. [CrossRef]

113. Bista, D.R.; Heckathorn, S.A.; Jayawardena, D.M.; Mishra, S.; Boldt, J.K. Effects of drought on nutrient uptake and the levels of nutrient-uptake proteins in roots of drought-sensitive and -tolerant grasses. *Plants* **2018**, *7*, 28. [CrossRef] [PubMed]

114. Holm, B.; Heinsoo, K. Municipal wastewater application to Short Rotation Coppice of willows—Treatment efficiency and clone response in Estonian case study. *Biomass Bioenergy* **2013**, *57*, 126–135. [CrossRef]

115. Ettala, M.O. Short-rotation tree plantations at sanitary landfills. *Top. Catal.* **1988**, *6*, 291–302. [CrossRef]

116. Kjøller, R.; Cruz-paredes, C.; Clemmensen, K.E. *Soil Biological Communities and Ecosystem Resilience*; Springer: Berlin/Heidelberg, Germany, 2017; pp. 223–252. [CrossRef]

117. Hytönen, J.; Kaunisto, S. Effect of fertilization on the biomass production of coppiced mixed birch and willow stands on a cut-away peatland. *Biomass Bioenergy* **1999**, *17*, 455–469. [CrossRef]

118. Stolarski, M.J.; Krzyżaniak, M.; Szczukowski, S.; Tworkowski, J.; Załuski, D.; Bieniek, A.; Gołaszewski, J. Effect of Increased Soil Fertility on the Yield and Energy Value of Short-Rotation Woody Crops. *Bioenergy Res.* **2015**, *8*, 1136–1147. [CrossRef]

119. Salam, M.M.A.; Mohsin, M.; Pulkkinen, P.; Pelkonen, P.; Pappinen, A. Effects of soil amendments on the growth response and phytoextraction capability of a willow variety (S. viminalis × S. schwerinii × S. dasyclados) grown in contaminated soils. *Ecotoxicol. Environ. Saf.* **2019**, *171*, 753–770. [CrossRef]

120. Bungart, R.; Hüttl, R.F. Growth dynamics and biomass accumulation of 8-year-old hybrid poplar clones in a short-rotation plantation on a clayey-sandy mining substrate with respect to plant nutrition and water budget. *Eur. J. For. Res.* **2004**, *123*, 105–115. [CrossRef]

121. Kuznetsova, T.; Lukjanova, A.; Mandre, M.; Lõhmus, K. Aboveground biomass and nutrient accumulation dynamics in young black alder, silver birch and Scots pine plantations on reclaimed oil shale mining areas in Estonia. *For. Ecol. Manag.* **2011**, *262*, 56–64. [CrossRef]

122. Löf, M.; Bolte, A.; Jacobs, D.F.; Jensen, A.M. Nurse Trees as a Forest Restoration Tool for Mixed Plantations: Effects on Competing Vegetation and Performance in Target Tree Species. *Restor. Ecol.* **2014**, *22*, 758–765. [CrossRef]

123. Joffre, R.; Rambal, S. Linked references are available on JSTOR for this article: How tree cover influences the water balance of mediterranean rangelands. *Ecology* **1993**, *74*, 570–582. [CrossRef]

124. Al-Namazi, A.A.; Bonser, S.P. Plant strategies in extremely stressful environments: Are the effects of nurse plants positive on all understory species? *J. Plant Interact.* **2020**, *15*, 233–240. [CrossRef]

125. Hytönen, J. Biomass, nutrient content and energy yield of short-rotation hybrid aspen (P. tremula x P. tremuloides) coppice. *For. Ecol. Manag.* **2018**, *413*, 21–31. [CrossRef]

126. Jylhä, P.; Hytönen, J.; Ahtikoski, A. Profitability of short-rotation biomass production on downy birch stands on cut-away peatlands in northern Finland. *Biomass Bioenergy* **2015**, *75*, 272–281. [CrossRef]
127. Berbeć, A.K.; Matyka, M. Planting density effects on grow rate, biometric parameters, and biomass calorific value of selected trees cultivated as src. *Agriculture* **2020**, *10*, 583. [CrossRef]

 *agronomy*

Review

Biomass Yield of Selected Herbaceous and Woody Industrial Crops across Marginal Agricultural Sites with Shallow Soil

Jana Reinhardt [†], Pia Hilgert [†] and Moritz von Cossel *

Biobased Resources in the Bioeconomy (340b), Institute of Crop Science, University of Hohenheim, 70599 Stuttgart, Germany; jana.reinhardt93@gmail.com (J.R.); pia.hilgert@uni-hohenheim.de (P.H.)
* Correspondence: moritz.cossel@uni-hohenheim.de
† Authors contributed equally to the work.

Abstract: Agricultural land in Europe is affected by low rooting depth (LRD) on 27.9 Mha. This marginal agricultural land can potentially be used to grow industrial crops without directly threatening food security or biodiversity conservation. However, little is known about the yield performance of industrial crops at LRD conditions. This study therefore compiles and discusses the meaningful data available in scientific literature. Twelve relevant industrial crops were identified for Europe. Currently, robust information on good growth suitability for LRD conditions is available for only one industrial crop, namely reed canary grass (RCG). Because this information was taken from field trial results from a single site, it remains unclear what role other growing conditions such as soil quality and climate play on both the yield level and the biomass quality of RCG under LRD conditions. These uncertainties about the quantitative as well as qualitative performance of industrial crop cultivation on marginal agricultural land characterized by LRD represent a major agronomic knowledge gap. Here, more knowledge needs to be compiled through both expanded crop science activities and improved international information exchange to make more optimal use of the large LRD areas available for the transition to a bioeconomy.

Keywords: annual crops; low rooting depth; marginal land; perennial crops; shallow rooting depth; unfavorable growth conditions

Citation: Reinhardt, J.; Hilgert, P.; von Cossel, M. Biomass Yield of Selected Herbaceous and Woody Industrial Crops across Marginal Agricultural Sites with Shallow Soil. *Agronomy* **2021**, *11*, 1296. https://doi.org/10.3390/agronomy11071296

Academic Editor: Jerome H. Cherney

Received: 2 June 2021
Accepted: 24 June 2021
Published: 26 June 2021

Publisher's Note: MDPI stays neutral with regard to jurisdictional claims in published maps and institutional affiliations.

1. Introduction

More than 47% of agricultural land in Europe is characterized by low rooting depth (LRD) [1]. This so called marginal agricultural land is deemed available for the cultivation of herbaceous and woody industrial crops for various biomass utilization pathways. Availability is based on the assumption that the cultivation of herbaceous and woody industrial crops on marginal agricultural land does not conflict with food production or nature conservation and can thus potentially contribute to a social-ecologically more sustainable utilization of these areas [2,3]. However, methods for regional assessment of the actual utilization status of European marginal land affected by LRD and other constraints are still under development [4]. This is necessary because not all of this land is actually abandoned and not used for agricultural production. In exceptional cases, such unproductive agricultural areas are in fact used for the cultivation of food crops through irrigation, which can be determined, for example, through satellite imagery [4]. On such land, there would therefore be a certain conflict of use between the cultivation of industrial and food crops, which must be taken into account when determining the actual available area of marginal agricultural land with LRD conditions. Nevertheless, the utilization of marginal agricultural land affected by LRD represents a tremendous potential for the production of biomass [5] that is urgently required to successfully and promptly manage the transition to a bioeconomy [6,7].

A well-developed and deep root system, however, is a key factor for crops, including herbaceous and woody industrial crops, to adapt to environmental conditions and for good growth and high yield levels [8]. It facilitates access to nutrients and water and increases the ability of the plant to compete with other species [9]. Roots in the upper soil layers, in particular, serve for the efficient use of mineral nutrients. Deep roots, on the other hand, contribute to a better water supply, especially under drought stress [10]. Where there are no barriers to root growth, most crop roots extend to depths of 60 to 120 cm [11]. The maximum root depth of winter wheat in a trial in Austria was 150–160 cm [10]. The maximum root depth of wheat is generally between 80–180 cm [12]. Therefore, a LRD means a limited availability of nutrients and water and thus an impairment of plant growth [8,13]. This also restricts soil cultivation [13]. Under LRD conditions, the rooting depth is generally limited by coherent hard rock or a dense soil layer. Following Rossiter et al. [11], sites are described as marginal due to their low LRD with a soil depth of ≤35 cm. This limit value is chosen very low, because a rooting depth of 35 cm means a very strong restriction of plant growth [11]. Hence, for this work the limit value of Mueller et al. [14] will be used; it gives a limit of <50 cm for flat and very flat soils. If a site is described as having shallow soil, it can also be classified as marginal due to LRD.

Consequently, among many socioeconomic challenges [15–19], the success of cultivating herbaceous and woody industrial crops on marginal agricultural land with LRD conditions is critical, with biomass yield being one of the most important components [20,21] that is currently being investigated, for example, at University of Hohenheim (Figure 1).

Figure 1. Impression of an ongoing field trial with herbaceous industrial crops hemp, camelina, and calendula in comparison with common food/feed crops maize and winter wheat under low rooting depth conditions (0.2–0.4 m rootable top soil above hard rock) at an experimental station of University of Hohenheim in southwest Germany. The photo was taken in summer 2020.

As little as is known about the performance of herbaceous and woody industrial crops on marginal agricultural land characterized by LRD [1,22], it can already be seen that there are very different specialists among the known herbaceous and woody industrial crops [23], which means that suitable biomass production systems could be developed for almost any type of marginal agricultural land [2]. The relevance of agricultural marginality constraints can be expressed by the size of agricultural land affected. The total area of available European marginal agricultural land with shallow soil accounts for about 27.9 Mha [1]. Since there is still much uncertainty about the link between LRD on marginal agricultural land and biomass yield of herbaceous and woody industrial crops [1,24], Gerwin et al. [1] claim that this must be investigated more in detail in the future. Therefore, this study focuses on the following main research question: How do herbaceous and woody industrial crops perform in terms of biomass yield on marginal agricultural land with shallow soil?

2. Material and Methods

In this study, the biomass yield of herbaceous and woody industrial crops on favorable agricultural land (favorable climate and fertile soils) across the northern hemisphere [25,26] was used as reference for the herbaceous and woody industrial crops' yield performance on marginal agricultural land characterized by LRD. This was intended to enable first insights into the future potential of biomass production on that specific type of marginal agricultural land.

2.1. Identification of Most Relevant Herbaceous and Woody Industrial Crops in Europe

Only those herbaceous and woody industrial crops were selected that were involved at least four times in one of the EU projects that started or ended in the period 1 January 2014 to 6 December 2019. Using the Community Research and Development Information Service (CORDIS) [27], 24 EU projects were found and considered for this purpose: Becool, BIO4A, COMETHA, COSMOS, DENDROMASS 4EUROPE, EUPOBIOREF, FIBRA, FIRST2RUN, FORBIO, GRACE, GRASSMARGINS, LIBBIO, AGIC, Mediopuntia, MULTIBIOPRO, Multihemp, OPTIMA, OPTIMISC, PANACEA, PHYSIO-POP, SEEMLA, SUNLIBB, SWEET fuel, and WATBIO (Table A1). In this way, twelve herbaceous and woody industrial crops were found relevant (Table A2), which is mostly in line with the crop selections in other studies [2,23]. These crops were examined in more detail in the remainder of this study, i.e., the literature review described in the following sections.

2.2. Literature Search

To obtain information about the yigeld performance of the most relevant herbaceous and woody industrial crops in Europe under LRD conditions, an extensive literature search in the database Scopus® (Elsevier BV, Amsterdam, The Netherlands) was carried out. To make the search queries in Scopus® as precise as possible, the advanced search function was used which enables a complex search order by means of field codes and Boolean operators [28].

First, the database was searched for papers in which the plant name appears in the title, abstract, or keywords. Keywords were selected for each plant as follows:

- camelina: *Camelina*;
- cardoon: *Cynara cardunculus*, artichoke thistle,
- cardoon; crambe: crambe;
- cup plant: *Silphium perfoliatum*, cup plant;
- giant reed: *Arundo donax*, giant reed;
- hemp: Cannabis sativa, hemp;
- Miscanthus: Miscanthus;
- poplar: *Populus*, poplar;
- reed canary grass: *Phalaris arundinacea*, reed canary grass;
- sorghum: Sorghum bicolor;
- switchgrass: *Panicum virgatum*, switchgrass;
- willow: *Salix*, willow.

Then, articles with terms such as "model", "gis", or "gene" were removed as they would most likely not be original articles with field trial results. Afterward, the results were screened for the marginality constraint LRD. The results were then further selected for those articles including terms such as "biomass", "yield", or "harvest". Of the studies found, those that met certain requirements were then selected, such as the existence of field trials, and information on yield, time, and location of the experiment. Weather conditions and soil properties also had to be adequately specified. Studies were excluded if their sites had high salinity, heavy metal contamination, or acidic soils. The climate of the sites found in the studies needed to be similar to one of the main European agroecological zones (Mediterranean, Atlantic, Continental/Boreal) and located in the northern hemisphere, if the study was not from Europe. Special selection criteria were also chosen for willow and sorghum—only field trials with short rotation coppice for willow and no grain use varieties for sorghum were taken into account.

2.3. Suitability Ranking of the Identified Crops

Based on the yield data and discussion in further literature, a classification of the cultivation suitability according to a classification of Ramirez-Almeyda et al. [5] was conducted.

If no sites were found that met (or properly describe) the threshold for LRD (soil depth of ≤50 cm), the classification was based only on further literature. The suitability values are then given in brackets. The suitability values denote as follows: 4 = "very good", much higher yields than average yields on favored sites; 3 = "good", yields approximately equivalent to the average yields on favored sites; 2 = "average", lower yield compared to average yields on favored sites; and 1 = "low" much lower yields than average yields on favorable sites. If the plant is classified as "0 = unsuitable", it cannot grow on sites with LRD.

3. Results and Discussion

According to the methodology described above, seven sites were found in seven papers that comply with the threshold values for LRD as defined by Rossiter et al. [11] (Figure 2, Table 1). Further details on the field trials of the identified references are provided in the annex (Tables A3 and A4). Some sites and their respective tests, which also meet the criteria for LRD, could not be identified using the methodology described. The reason for this was that in order to classify a site into marginal, certain information about the site was required. However, this information is often missing in papers.

Figure 2. Overview of the field trial locations of the studies identified in the literature review. Colors indicate the respective industrial crops investigated in the field trials. The numbers (1–7) denote for the site numbers as presented in Table 1. Map created using GPSvisualizer (www.gpsvisualizer.com).

Table 1. Overview of studies and locations identified in the literature search.

Site Number	Crop	Country	City	m a.s.l.	Constraints Other Than LRD		Reference
					UST	Steep Slope	
1	Poplar	Germany	Göttingen	206			[29]
2	Poplar	Czech Republic	Domanínek	578		1	[30]
3	*Miscanthus*	Germany	Durmersheim	118	1 S		[31]
4	*Miscanthus*	Great Britain	Aberystwyth	39	1 St		[32]
5	Hemp	Italy	Udine	109			[33]
6	Giant reed	Turkey	Sakarya	244			[34]
7	Reed canary grass	Ireland	Carlow	15	1S St		[35]

LRD = low rooting depth, UST = unfavorable soil type (S = coarse sand, St = stoniness).

3.1. Field Trial Observations

In the following sections, the cultivation suitability of the twelve identified herbaceous and woody industrial crops under LRD conditions is discussed. Each crop is only considered individually, which means that no further statements can yet be made on their implementability in existing agricultural systems. As already mentioned in the introduction, this is an overview of crop-specific suitability for LRD conditions. This is an important first step for the extensive effort needed to implement a bio-based economy that requires further follow-up research, such as the integration to existing agricultural systems, which is justified in more detail at the end of the discussion. Here, all yield figures are listed as yield per year (biomass, stalk, oil, grain, and ethanol yields, respectively), and only the aboveground biomass of the crops is considered. All crops for which field trial data were found in the literature were cultivated in single cropping. This means that no intercropping approaches were considered here, which is not to say that this is fundamentally irrelevant. Intercropping often helps to make better use of limited nutrient supply of cultivation-limited sites, such as LRD dominated marginal agricultural land [36]. However, so far only information on food and forage crops can be found, not on industrial herbaceous and woody crops.

3.1.1. Giant Reed

About 50% of all dry matter (DM) yield data of giant reed are in the range of 25–40 Mg ha^{-1} [37]. Under favorable growing conditions, giant reed shows very high DM yields of 33.8 Mg ha^{-1} [38] to 37.7 Mg ha^{-1} [39]. This is in line with Pilu et al. [40] who report DM yields of 36–55 Mg ha^{-1}.

One experiment was found for giant reed cultivation on marginal agricultural land with LRD (Table 2). This experiment was carried out on a site that is marginal due to its eroded flat soil profile. It is located in Sakarya, Turkey. In Ozdemir et al. [34], unfortunately, no information is given on the exact rooting depth of the Sakarya site. However, the experiment shows that giant reed produces a yield of 8.3 Mg ha^{-1} DM on marginal agricultural land with this factor without fertilization and irrigation. Compared to favorable sites, this is much lower [39,40]. In a further test variant on this site, fertilization with poultry slaughter sludge was applied, which led to a higher yield (16.5 Mg ha^{-1} DM in the second–third year of cultivation at the highest fertilization level 200 N kg ha^{-1}). In comparison to a fertile location, however, this yield is also lower. To develop a good rhizome system a deep soil is required. On a soil with a low root depth the establishment of giant reed is therefore limited, as is the agricultural cultivation of giant reed [13]. A location with a root depth of <35 cm is challenging and according to Parenti et al. [13] unsuitable for the cultivation of giant reed, while Von Cossel et al. [23] consider this kind of location slightly suitable for cultivation. A location with a root depth of 35–80 cm has a medium suitability for cultivation [23]. The test results of Ozdemir et al. [34] confirm the low suitability of giant reed for cultivation on sites with LRD.

3.1.2. Hemp

On favorable sites, fiber hemp achieves a stem yield of 10 Mg ha^{-1} DM [41]. Struik et al. [42] investigated five different varieties of hemp grown for fiber on three favored sites in Europe (Italy, United Kingdom, and the Netherlands) over three years with different fertilization and plant density variants. A maximum aboveground biomass yield of 22.5 Mg ha^{-1} DM and a stalk yield of 18.5 Mg ha^{-1} DM were achieved. The average aboveground biomass DM yield over all locations, years, varieties, and trial variants was 14 Mg ha^{-1} and the stem DM yield 11 Mg ha^{-1} [42]. In addition, on a sandy loamy soil in Foulum, Denmark, an aboveground biomass DM yield of 13–15 Mg ha^{-1} was achieved [43].

One site was found that meets the criteria for this agricultural marginality constraint (Tables 1 and 2). The site in Udine (Italy) has a rooting depth of 50 cm and achieved a grain yield of 0.55 Mg ha^{-1} and an oil yield of 0.11 Mg ha^{-1} [33]. Eight different monoecious varieties were grown there. The average yield of 6.1 Mg ha^{-1} DM (Table 2) was lower than the average yield of 11 Mg ha^{-1} DM on sites described by Struik et al. [42]. The highest stalk yield in Udine was achieved with the Futura variety; it was 8.3 Mg ha^{-1} DM. This was also below the yield of Futura reported by Struik et al. [42]. In addition, the thousand-grain mass on Udine was 7.6 g, which is significantly less than the stated thousand-grain mass on favored sites (17–23 g) [41]. The plants reached an average height of 208 cm, which is in the optimal range for industrial processing of the fibers [41]. In a field test of Amaducci et al. [44] in Cadriano, Italy, 50% of the root biomass was found in the upper 50 cm in a deep soil. The root length density was highest in the upper 10 cm of the soil. Hemp can root to a depth of 2–2.5 m, especially when it grows in dry environments [41,44]. It is likely that hemp yields are reduced on flat soils because it has a poorly developed root system compared to other economically important plants [41]. This is in line with Von Cossel et al. [23], who found that hemp has a low suitability for cultivation in soils with a depth of 35–80 cm and is unsuitable for soils with a depth less than 35 cm. However, the observations from Udine indicate that hemp is moderately suitable for cultivation when the soil depth is 50 cm, but the choice of the most suitable variety is crucial for this.

Table 2. Annual stalk and biomass yields of monoecious giant reed (*Arundo donax* L.), hemp (*Cannabis sativa* L.), *Miscanthus* (*Miscanthus* x *giganteus* Greef et Deuter), poplar (*Populus* L.), and reed canary grass (*Phalaris arundinacea* L.) on marginal agricultural land characterized by low rooting depth (LRD).

Site Number	Crop	Year of Cultivation (Poplar: Age of Trunk)	Dry Matter Stem Yield [Mg ha^{-1}]	Dry Matter Yield [Mg ha^{-1}]	Source
1	Poplar	3		0.1	[29]
2	Poplar	7		11.7	[30]
3	*Miscanthus*	4		8.9	[31]
4	*Miscanthus*	4		16	[32]
5	Hemp		6.1		[33]
6	Giant reed	2–3		8.3	[34]
7	Reed canary grass	2–3		13	[35]

3.1.3. *Miscanthus*

It was found that 50% of all yield data of *Miscanthus* on favored sites are in the range of 13–28 Mg ha^{-1} DM [37]. With irrigation, yields of over 30 Mg ha^{-1} DM can be achieved on sites in Southern Europe (average temperatures of 15.4 °C) [45]. In Central and Northern Europe (from Austria to Denmark), where global irradiation and average temperatures are lower (7.3–8.0 °C), yields without irrigation are typically 10–25 Mg ha^{-1} DM [45]. In a long-term test of Angelini et al. [39] without irrigation on a favored site near Pisa, the average yield over the 2nd–12th year of cultivation was 28.7 Mg ha^{-1} DM. As with Amaducci et al. [46], the yield of the first year of cultivation of this experiment was much lower than in the following years. From the first to the second year the yield increases, in the 3rd–8th year the highest yields are achieved and from the 9th–12th year the yield decreases [39]. The complete establishment of a *Miscanthus* population takes three to five years [45]. Under optimal location conditions, the full yield potential is achieved after three years, under suboptimal conditions only after five years [47].

Two sites were found that fulfill the criteria of the agricultural marginality constraint LRD (Table 2). The Durmersheim site also fulfilled the criterion [11] for unfavorably high soil content of sand (UST_S) and at the Aberystwyth site, the criterion [11] for stoniness (UST_St) in the topsoil was fulfilled. *Miscanthus* achieved a yield of 8.9 Mg ha^{-1} DM in the fourth year of cultivation in Durmersheim. This is less than on favorable sites in Central and Northern Europe [45]. The summer of the fourth year of cultivation on this site was exceptionally dry. In addition, the two agricultural marginality constraints, UST-S and LRD, have a negative influence on each other [48]. Sandy soils are generally characterized by low fertility, high nutrient leaching, and low water retention capacity. A low soil depth increases these effects, as the volume of water storage in the soil is limited [48]. This could explain the positive effect of fertilization on this site; the stated yield is from a trial variant with 50 kg N ha^{-1}, another variant with 100 kg N ha^{-1} yielded 13.7 Mg ha^{-1} DM, which is in the lower stated yield range in Central and Northern Europe [31]. The Aberystwyth site has a rooting depth of 30–50 cm. In the fourth year of cultivation the yield of *Miscanthus* in Aberystwyth was 16 Mg ha^{-1} DM. This yield corresponds to the yield range in Northern Europe [49]. This site has high rock content (35%) in addition to the shallow depth. These two agricultural marginality constraints create a negative synergy, since a high stone content further reduces the effective soil volume for rooting, water storage, and nutrient supply in a shallow soil [48]. Following Von Cossel et al. [23], the cultivation of *Miscanthus* is unsuitable on sites with soil depths below 35 cm, mediocre on soils that are 35–80 cm deep, and good at 80–120 cm. Findings in Ramirez-Almeyda et al. [5] slightly deviate from this classification; at a soil depth of 40–80 cm the suitability for cultivation of *Miscanthus* is low, at 80–120 cm it is average, and similar to Von Cossel et al. [23], it is unsuitable for a depth of less than 40 cm. According to Ferrarini et al. [50], *Miscanthus* has a deep fine root system on profound soils. Although *Miscanthus* produces lower yields on shallow soils than on deep soils [51], these are still comparable with yields on favored sites (rather lower half of the yield range), provided there is sufficient water supply and fertilization. For this reason, a medium suitability for cultivation on sites with LRD can be assumed for *Miscanthus*.

3.1.4. Poplar

In a systematic literature search and meta-analysis on poplar yields [37], it was found that 50% of all yield data are in the range 7–10 Mg ha^{-1} DM. Berendonk et al. [52] indicate that a yield of 8–12 Mg ha^{-1} DM can be expected for short rotation plantations. On a favorable location in Mira in Northern Italy, a yield of 20 Mg ha^{-1} DM in the second rotation and 15 Mg ha^{-1} DM in the first rotation was achieved [53].

For poplar cultivation under LRD conditions, two studies from Göttingen and Domanínek were found (Table 2). The soil depth of the Göttingen site is 20–50 cm. The poplar hybrids Max 1 (*Populus nigra* × P. *maximowiczii*) and H275 (*Populus trichocarpa* × P. *maximowiczii*) were cultivated on this site, which achieved a very low average yield of 0.1 Mg ha^{-1} DM. Furthermore, a survival rate of 25% for H275 and 60% for Max 1 was found. The experiment shows that soils with LRD can strongly restrict the growth of the poplar hybrids studied [29]. At the Domanínek site, *Populus nigra* × P. *maximowiczii* achieved a yield of 11.7 Mg ha^{-1} DM in the second rotation (which lasted seven years), which is much higher than at the Göttingen site, and also higher than the yields stated in Berendonk et al. [52]. This site has a soil depth of 30–50 cm and additionally a strongly sandy loam. A survival rate of 70% was determined [30].

The division into the classification system proposed by Ramirez-Almeyda [5] states that the cultivation of poplar is unsuitable on sites with soil depths of <40 cm, low at 40–80 cm, and medium at 80–120 cm. Von Cossel et al. [23] also classify soils with a depth of <35 cm as unsuitable, while soils with a depth of 35–80 cm are classified as medium, and soils with a depth of 80–120 cm as well suited for cultivation. The experiment in Moffat and Houston [54] shows that the survival rate of poplar plants is proportional to the soil depth, with a minimum soil depth of 20 cm. Overall, the suitability for cultivation of poplar on soils with low rooting depth is expected to be mediocre in comparison to favored locations.

3.1.5. Reed Canary Grass

Reed canary grass achieves dry matter biomass yields of 5–13 Mg ha^{-1} under favorable conditions [49,55,56]. For example, reed canary grass achieved an average dry matter yield of 8 Mg ha^{-1} (yields of 3.9–13.8 Mg ha^{-1} were measured) on four different favored sites in the Czech Republic from the second to the sixth year of cultivation [57].

One study from Carlow (Ireland) was found, where reed canary grass was grown under LRD conditions (Tables 1 and 2). This site has a shallow soil with a high proportion of gravel and sand with poor water retention capacity and therefore fulfils the requirements of LRD, UST-S, and UST_ST [35]. The location factors UST-S and UST_ST each have a negative synergy with LRD [48]. A high stone content also reduces the effective rooting volume. Sand increases the effect of the low water storage capacity of a soil with LRD [48]. The yield of the Carlow site is comparable to that of the favored sites [49]. Therefore, reed canary grass has a good suitability for cultivation on sites with LRD. Due to the negative synergies described above at this location, it can be assumed that higher yields can be achieved at locations with good soil and exclusively LRD than at the Carlow site. This contradicts Von Cossel et al. [23] and Ramirez-Almeyda et al. [5], who indicated a low suitability of reed canary grass for cultivation on sites with a soil depth of 35–80 cm or 40–80 cm.

3.2. Presumptions on Crops Where No Studies Were Found

For seven of the preselected herbaceous and woody industrial crops, no literature on biomass yield performance on shallow soil was found. The following sections provide a brief overview of the expected suitability of these herbaceous and woody industrial crops for LRD conditions based on topic-related literature.

3.2.1. Camelina

In a field study (winter camelina in Morris, Minnesota, on a clay soil), it was found that camelina has a rather compact root system: 82% of the root density was located in the upper 30 cm of the soil, 12% in 30–60 cm, and only about 6% in 60–100 cm soil depth [58]. This supports the statement in Von Cossel et al. [23] that camelina can be grown well on shallow soil below 35 cm and very well in soil over 35 cm depth. Although camelina roots are mainly found in the upper soil layers, they can grow in dry areas at depths of up to 140 cm to reach water [59]. However, whether or not camelina can be grown well under low rooting depth conditions is not yet proven by field trial results.

3.2.2. Cardoon

Following Von Cossel et al. [23], cardoon is unsuitable for cultivation under low rooting depth conditions and is only slightly suitable for soils with a depth of 35–80 cm. This matches the specification in Gominho et al. [60] that a soil depth of 50–150 cm is required for the cultivation of cardoon. Cardoon builds up a deep root system, which can reach a depth of more than 5 m [61]. This suggests that soil depth is an important factor influencing the growth and thus the yield of cardoon [62]. Cardoon is therefore deemed unsuitable for cultivation on sites with LRD.

3.2.3. Crambe

For the agricultural marginality constraint LRD, no literature was found that meets the threshold values (<50 cm for flat and very flat soils) described. After a classification by Von Cossel et al. [23], crambe has good suitability for cultivation on sites with a root depth of less than 35 cm and a very good cultivation suitability on soils with a rooting depth greater than 35 cm. However, this is not yet proven by field trial results. In a three-year trial with minirhizotrons, an average depth of 58 cm and a maximum depth of 118 cm of the crambe roots were determined [63,64].

3.2.4. Cup Plant

Cup plant has a deep and extensive root system [47,65]. At the Braunschweig site, the maximum root depth was 80–240 cm [66,67]. These data suggest that cup plant cannot grow well on sites with LRD and therefore has a low suitability for cultivation on shallow soils.

3.2.5. Sorghum

After a classification by Von Cossel et al. [23] sorghum is unsuitable for growing in locations with a root depth of less than 35 cm, medium at a depth of 35–80 cm, and good at 80–120 cm. Sorghum forms a deep root system [68,69], preferably in soils with a depth of at least 1 m [68]. This suggests that high yields cannot be achieved on sites with LRD and that sorghum therefore has a low suitability for cultivation.

3.2.6. Switchgrass

Switchgrass is unsuitable for cultivation on sites with a soil depth of less than 35 cm, medium for soils with a depth of 35–80 cm, and good for soils with a depth of 80–120 cm. Ramirez-Almeyda et al. [5] classify the suitability of switchgrass for cultivation at different soil depths as follows: unsuitable below 40 cm, 40–80 cm low, and 80–120 cm medium. Switchgrass has a deep root system that can reach a soil depth of 3 m [70,71]. At the Mason site, switchgrass rooted to an average depth of 81 cm [72]. The information provided here indicates that switchgrass cannot be grown on sites with a LRD. In addition, Parenti et al. [13] assessed sites with these characteristics as unsuitable for the cultivation of switchgrass.

3.2.7. Willow

The experiments conducted by Moffat and Houston [54] show that more willows survive at increased soil depths and that none can grow at a soil depth of only 20 cm. Von Cossel et al. [23] classify the suitability of willow for cultivation as follows: low for soils with a depth of less than 35 cm, good with 35–80 cm depth, and very suitable with 80–120 cm depth. According to Ramirez-Almeyda [5], soils with a depth of less than 40 cm are classified as unsuitable for growing willow, soils with a depth of 40–80 cm as shallow, and soils with a depth of 80–120 cm as moderately suitable. Since willow has a deep fine root system [50] it probably has only a limited suitability for cultivation in locations with LRD.

3.3. Recommendations for Cultivation of Herbaceous and Woody Industrial Crops under LRD Conditions

Following the approach by Ramirez-Almeyda et al. [5], herbaceous and woody industrial crops whose suitability for cultivation on sites with a certain marginality is rated as good or very good may be recommended for cultivation. Good and very good suitability for cultivation means that the biomass yields are comparable or even better than under favorable growth conditions. After a thorough literature search, it was found that this seems to apply only to a few herbaceous and woody industrial crops, such as reed canary grass, camelina, and crambe (Figure 3). This assumption could also only be confirmed for reed canary grass by field trial data, where an expected yield level of 13–15 Mg ha^{-1} was found. For camelina and crambe, it remains open whether or not the assumptions derived from the related literature can be confirmed by future field trial data. For all other herbaceous and woody industrial crops, either no field trial data were available or the suitability value was less than three (Figure 3).

This review article has thus shown that there are large gaps in knowledge about the best possible use of regions where agricultural land is marginal for LRD. Much more research is needed on the suitability of known herbaceous and woody industrial crops for LRD conditions in order to make the best use of such sites for biomass production in terms of more social-ecologically sustainable agriculture [73,74]. Certainly, this need does not exist if a site is allowed to lose arable status, because reforestation with cold-insensitive woody crops will then be the best solution. If arable status is to remain, however, there is a need for adapted herbaceous and woody industrial crops and the associated resource-conserving and environmentally friendly cultivation methods.

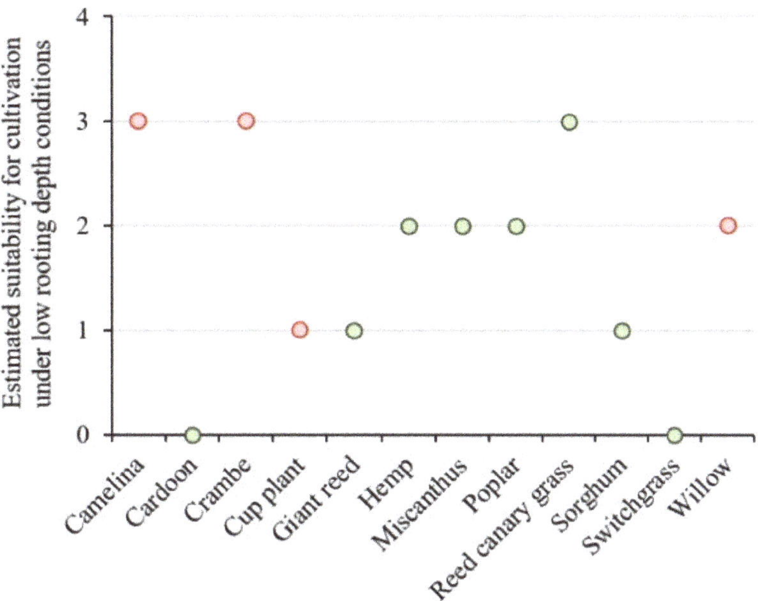

Figure 3. Overview of estimated suitability for cultivation of the preselected herbaceous and woody industrial crops under low rooting depth conditions ($\leq$35 cm soil depth). Green dots denote suitability values based on field trial data; red dots denote for suitability values based on associated literature and assumptions by the authors, because no field trial data were available.

Moreover, in light of climate change, regions where drought conditions are expected to increase, for example, from Spain to Central Europe [74,75], could further expand LRD limitation due to negative combined effects [48]. These dynamics should also be considered in field research on the use of LRD sites by herbaceous and woody industrial crops, which will be a major crop science challenge. A network of experimental stations, such as the ECOFE network, could be helpful, as they can detect changes in genotype-environment interaction with a very high breeding efficiency [76]. However, in addition to the further development of cultivable herbaceous and woody industrial crops, the ongoing adaptation of logistics and the processing industry is likely to be a far greater challenge if it is not yet even known which crop species will be grown in a particular region in ten or twenty years' time.

Author Contributions: Conceptualization, J.R., P.H., and M.v.C.; methodology, J.R. and M.v.C.; investigation, J.R.; resources, J.R.; data curation, J.R.; writing—original draft preparation, J.R., P.H., and M.v.C.; writing—review and editing, P.H. and M.v.C.; visualization, M.v.C.; supervision, M.v.C.; project administration, M.v.C.; funding acquisition, M.v.C. All authors have read and agreed to the published version of the manuscript.

Funding: This research received funding from the European Union's Horizon 2020 research and innovation program under grant agreement No. 727698, the Federal Ministry of Education and Research (01PL16003), and University of Hohenheim. The APC was funded by the European Union's Horizon 2020 research and innovation program under grant agreement No. 727698.

Institutional Review Board Statement: Not applicable.

Informed Consent Statement: Not applicable.

Data Availability Statement: Data sets for this research are included in the references.

Acknowledgments: The authors are very grateful to Iris Lewandowski for making this work possible.

Conflicts of Interest: The authors declare no conflict of interest. The funders had no role in the design of the study; in the collection, analyses, or interpretation of data; in the writing of the manuscript, or in the decision to publish the results.

Appendix A

Table A1. Industrial crops investigated in EU projects that were started or finished in the period from 1 January 2014 to 12 June 2019: a = MAGIC, b = PANACEA, c = SEEMLA, d = GRACE, e = Multihemp, f = FIBRA, g = COSMOS, h = MULTIBIOPRO, i = LIBBIO, j = GRASSMARGINS, k = OPTIMA, l = OPTIMISC, m = Becool, n = WATBIO, o = PHYSIO-POP, p = COMETHA, q = DENDROMASS 4EUROPE, r = SUNLIBB, s = FIRST2RUN, t = EUPOBIOREF, u = FORBIO, v = SWEET fuel, w = BIO4A, x = Mediopuntia.

Industrial Crop (Common Name)	Sum	a	b	c	d	e	f	g	h	i	j	k	l	m	n	o	p	q	r	s	t	u	v	w	x
Miscanthus	12	X	X	X	X						X	X	X	X	X				X		X	X			
Giant reed	9	X	X	X								X		X	X		X				X	X			
Poplar	7	X	X	X					X						X	X		X							
Hemp	6	X	X		X	X	X							X											
Switchgrass	6	X	X	X								X		X							X				
Camelina	5	X	X					X													X			X	
Cardoon	5	X	X	X								X								X					
Sorghum	5	X	X											X								X	X		
Willow	5	X	X	X																	X	X			
Crampe	4	X	X					X													X				
Reed canary grass	4	X	X	X							X														
Black locust	3	X	X	X																					
Castor	3	X	X																		X				
Eucalyptus	3		X	X										X											
Kenaf	3		X				X							X											
Lupin	3	X	X							X															
Safflower	3	X	X																		X				
Cup plant	2			X																		X			
Ethiopian mustard	2	X	X																						
Flax	2		X				X																		
Sunflower	2		X																					X	
Alfa-alfa	1																				X				
Cocksfoot	1										X														
Columbus grass	1			X																					
Common reed	1			X																					
Cuphea	1																				X				
Festulolium	1										X														
Giant Knotweed	1			X																					
Guayule	1		X																						
Indian fig opuntia	1																								X
Jatropha	1																				X				
Jerusalem artichoke	1			X																					
Lavender	1		X																						
Lesquella	1																				X				
Paulownia	1			X																					
Pennycress	1	X																							
Peppermint	1		X																						
Pine	1			X																					
Rapeseed (HEAR)	1		X																						
Rosemary	1		X																						
Russian dandelion	1		X																						
Siberian elm	1	X																							
Sugarbeet	1		X																						
Sugarcane	1																			X					
Tall fescue	1										X														
Tall wheat grass	1	X																							
Tobacco	1						X																		
Tree of heaven	1			X																					
Triticale	1		X																						
Virginia mallow	1			X																					
Wild sugarcane	1	X																							

Table A2. Overview of the selected industrial crops, sorted alphabetically by common name (adapted from [2]).

Botanical Name	Trivial Name	Life Cycle	Photosynthetic Pathway	Use	Projects	Frequency of Studies
Arundo donax L.	Giant reed	Perennial	C3	L	9	8
Camelina sativa (L.) Crantz	Camelina	Annual	C3	O	5	6
Cannabis sativa L.	Hemp	Annual	C3	L/O	6	10
Crambe abyssinica Hochst ex R.E.Fr.	Crambe	Annual	C3	O	4	3
Cynara cardunculus L.	Cardoon	Perennial	C3	L/O	5	5
Miscanthus Andersson	Miscanthus	Perennial	C4	L	12	13
Panicum virgatum L.	Switchgrass	Perennial	C4	L	6	10
Phalaris arundinacea L.	Phalaris	Perennial	C3	L	4	6
Populus L.	Poplar	Perennial	C3	W	7	11
Salix L.	Willow	Perennial	C3	W	5	13
Silphium perfoliatum L. [a]	Cup plant	Perennial	C3	L/C	2	4
Sorghum bicolor L. Moench	Sorghum	Annual	C4	L/C	5	7

[a] Cup plant was investigated in only two projects, but considered relevant because the European Parliament included it in the list of ecological priority areas in 2017 [77]; L = lignocellulose, O = oil, C = carbohydrates, W = wood.

Table A3. Overview of key agronomical data of the field trials conducted in the references identified in the literature search.

Reference Number	Field Trial Period				Variety	Precrop	Sowing or Planting Density [Seeds or Pants m^{-2}]	Irrigation [mm]	N Fertilization Per Year [kg ha^{-1}]
	Begin	End	Sowing	Harvest					
1	2011	2014	March 2011	Heights and basal diameter were measured bevor the growing season in 2014; harvest were carried out when the terminal buds of main shoots of poplars had been formed	Max 1 (P. *nigra* × P. *maximowiczii*) and H275 (P. *trichocarpa* × P. *maximowiczii*)	unmanaged grassland	13,333	Watered just after planting and in August 2011	0
2	2009	2015	April 2001	2015 (first harvest later in autumn 2008)	*Populus nigra* × P. *maximowiczii*	potatoes and cereals	9316	0	0
3	1991	1995	May 1991	February 1995				First two vegetation periods	0
4	2012	2015	May 2012	February to April	*Miscanthus* × *giganteus* Greef et Deu.	grassland (low quality)	1.96	0	60
5	2016	2017	April 2016		8 varieties	2016: wheat; 2017: Oat	130	45	80
6	2015	2017					1	0	0
7	2012	2014	May 2012	(4–7 June, 29–31 July, 24–23 September) 2013, (27–29 June, 29–31 July, 18 September) 2014 (end of 2012 growing season pruning took place).	Bamse, Cheifton	long term cereal cultivation	30	0	325

Table A4. Overview of yield relevant data from the field trials of the studies identified in the literature search.

Reference Number	Number of Vegetation Periods or Number of Cuts	Year of Cultivation	Age of the Tree	Aboveground Biomass Dry Matter Yield [Mg ha^{-1}]	Grain Yield [Mg ha^{-1}]	Stem Dry Matter [Mg ha^{-1}]	Oil Yield [L ha^{-1}]	Thousand Kernel Weight [g]
1	1		3	0.1		0.17		
2	1		7	11.7				
3		4		8.9				
4		4		16.0				
5	2				0.55	6.1	110	7.6
6	2–3			8.3				
7	2–3			13				

References

1. Gerwin, W.; Repmann, F.; Galatsidas, S.; Vlachaki, D.; Gounaris, N.; Baumgarten, W.; Volkmann, C.; Keramitzis, D.; Kiourtsis, F.; Freese, D. Assessment and Quantification of Marginal Lands for Biomass Production in Europe Using Soil-Quality Indicators. *Soil* **2018**, *4*, 267–290. [CrossRef]
2. von Cossel, M.; Lewandowski, I.; Elbersen, B.; Staritsky, I.; Van Eupen, M.; Iqbal, Y.; Mantel, S.; Scordia, D.; Testa, G.; Cosentino, S.L.; et al. Marginal Agricultural Land Low-Input Systems for Biomass Production. *Energies* **2019**, *12*, 3123. [CrossRef]
3. Elbersen, B.; Van Eupen, M.; Alexopoulou, E.; Bai, Z.; Boogaard, H.; Carrasco, J.E.; Ceccarelli, T.; Ciria Ramos, C.; Ciria, P.; Cosentino, S.L.; et al. Mapping Marginal Land Potentially Available for Industrial Crops in Europe. In Proceedings of the European Biomass Conference and Exhibition Proceedings 26thEUBCE, Copenhagehn, Denmark, 14–18 May 2018. [CrossRef]
4. Elbersen, B.; Hart, K.; Koper, M.; van Eupen, M.; Keenleyside, C.; Verzandvoort, S.; Kort, K.; Cormont, A.; Giadrossi, A.; Baldock, D. *Analysis of Actual Land Availability in the EU Trends in Unused, Abandoned and Degraded (Non-)Agricultural Land and Use for Energy and Other Non-Food Crops*; Wageningen Environmental Research: Wageningen, The Netherlands, 2020.
5. Ramirez-Almeyda, J.; Elbersen, B.; Monti, A.; Staritsky, I.; Panoutsou, C.; Alexopoulou, E.; Schrijver, R.; Elbersen, W. Assessing the Potentials for Nonfood Crops. In *Modeling and Optimization of Biomass Supply Chains*; Panoutsou, C., Ed.; Academic Press: Cambridge, MA, USA, 2017; Chapter 9; pp. 219–251. ISBN 978-0-12-812303-4.
6. Dahmen, N.; Lewandowski, I.; Zibek, S.; Weidtmann, A. Integrated Lignocellulosic Value Chains in a Growing Bioeconomy: Status Quo and Perspectives. *GCB Bioenergy* **2019**, *11*, 107–117. [CrossRef]
7. Lange, L.; Connor, K.O.; Arason, S.; Bundgård-Jørgensen, U.; Canalis, A.; Carrez, D.; Gallagher, J.; Gøtke, N.; Huyghe, C.; Jarry, B.; et al. Developing a Sustainable and Circular Bio-Based Economy in EU: By Partnering Across Sectors, Upscaling and Using New Knowledge Faster, and For the Benefit of Climate, Environment & Biodiversity, and People & Business. *Front. Bioeng. Biotechnol.* **2021**, *8*. [CrossRef]
8. Clark, L.J.; Whalley, W.R.; Barraclough, P.B. How Do Roots Penetrate Strong Soil? *Plant Soil* **2003**, *255*, 93–104. [CrossRef]
9. Place, G.; Bowman, D.; Burton, M.; Rufty, T. Root Penetration through a High Bulk Density Soil Layer: Differential Response of a Crop and Weed Species. *Plant Soil* **2008**, *307*, 179–190. [CrossRef]
10. Schweiger, P.; Petrasek, R.; Ableidinger, C.; Hartl, W. Tiefenverteilung von Wurzeln bei Winterweizen. In *Beiträge zur 10. Wissenschaftstagung Ökologischer Landbau: Zürich, 11–13 Februar 2009; Werte–Wege–Wirkungen: Biolandbau im Spannungsfeld zwischen Ernährungssicherung, Markt und Klimawandel*; Mayer, J., Alföldi, T., Leiber, F., Dubois, D., Fried, P., Heckendorn, F., Hillmann, E., Klocke, P., Lüscher, A., Riedel, S., et al., Eds.; Köster: Berlin, Germany, 2009; ISBN 978-3-03736-033-0.
11. Rossiter, D.; Schulte, R.; van Velthuizen, H.; Le-Bas, C.; Nachtergaele, F.; Jones, R.; van Orshoven, J. *Updated Common Bio-Physical Criteria to Define Natural Constraints for Agriculture in Europe*; Terres, J., Toth, T., Van Orshoven, J., Eds.; EUR, Scientific and Technical Research Series; Publications Office: Luxembourg, 2014; Volume 26638, ISBN 92-79-38190-3.
12. Kirkegaard, J.A.; Lilley, J.M. Root Penetration Rate—A Benchmark to Identify Soil and Plant Limitations to Rooting Depth in Wheat. *Aust. J. Exp. Agric.* **2007**, *47*, 590. [CrossRef]
13. Parenti, A.; Lambertini, C.; Monti, A. Areas with Natural Constraints to Agriculture: Possibilities and Limitations for The Cultivation of Switchgrass (Panicum Virgatum L.) and Giant Reed (Arundo Donax L.) in Europe. In *Land Allocation for Biomass Crops: Challenges and Opportunities with Changing Land Use*; Li, R., Monti, A., Eds.; Springer International Publishing: Cham, Switzerland, 2018; pp. 39–63. ISBN 978-3-319-74536-7.
14. Mueller, L.; Schindler, U.; Behrendt, A.; Eulenstein, F.; Dannowski, R.; Schlindwein, S.L.; Shepherd, T.G.; Smolentseva, E.; Rogasik, J. *The Muencheberg Soil Quality Rating (SQR): Field Manual for Detecting and Assessing Properties and Limitations of Soils for Cropping and Grazing*; Leibniz-Centre for Agricultural Landscape Research (ZALF) e. V.: Muencheberg, Germany, 2007.

15. Panoutsou, C.; Singh, A. A Value Chain Approach to Improve Biomass Policy Formation. *GCB Bioenergy* **2020**, *12*, 464–475. [CrossRef]

16. Fernando, A.L.; Rettenmaier, N.; Soldatos, P.; Panoutsou, C. Sustainability of Perennial Crops Production for Bioenergy and Bioproducts. In *Perennial Grasses for Bioenergy and Bioproducts*; Alexopoulou, E., Ed.; Academic Press: Cambridge, MA, USA, 2018; pp. 245–283. ISBN 978-0-12-812900-5.

17. Ferreira, J.A.; Brancoli, P.; Agnihotri, S.; Bolton, K.; Taherzadeh, M.J. A Review of Integration Strategies of Lignocelluloses and Other Wastes in 1st Generation Bioethanol Processes. *Process Biochem.* **2018**, *75*, 173–186. [CrossRef]

18. Ubando, A.T.; Felix, C.B.; Chen, W.-H. Biorefineries in Circular Bioeconomy: A Comprehensive Review. *Bioresour. Technol.* **2020**, *299*, 122585. [CrossRef]

19. Vogelpohl, T. Transnational Sustainability Certification for the Bioeconomy? Patterns and Discourse Coalitions of Resistance and Alternatives in Biomass Exporting Regions. *Energy Sustain. Soc.* **2021**, *11*, 3. [CrossRef]

20. Baum, G. Betriebswirtschaftliche Betrachtung der Wildpflanzennutzung für Biogasbetriebe 2019. Available online: https://baden-wuerttemberg.nabu.de/natur-und-landschaft/landwirtschaft/biogas/index.html (accessed on 25 June 2021).

21. Winkler, B.; Mangold, A.; Von Cossel, M.; Clifton-Brown, J.; Pogrzeba, M.; Lewandowski, I.; Iqbal, Y.; Kiesel, A. Implementing Miscanthus into Farming Systems: A Review of Agronomic Practices, Capital and Labour Demand. *Renew. Sustain. Energy Rev.* **2020**, *132*, 110053. [CrossRef]

22. Hassan, S.S.; Williams, G.A.; Jaiswal, A.K. Moving towards the Second Generation of Lignocellulosic Biorefineries in the EU: Drivers, Challenges, and Opportunities. *Renew. Sustain. Energy Rev.* **2019**, *101*, 590–599. [CrossRef]

23. Von Cossel, M.; Iqbal, Y.; Scordia, D.; Cosentino, S.L.; Elbersen, B.; Staritsky, I.; Van Eupen, M.; Mantel, S.; Prysiazhniuk, O.; Maliarenko, O.; et al. *Low-Input Agricultural Practices for Industrial Crops on Marginal Land*; University of Hohenheim: Stuttgart, Germany, 2018.

24. Haberzettl, J.; Hilgert, P.; Von Cossel, M. A Critical Assessment of Lignocellulosic Biomass Yield Modeling and the Bioenergy Potential from Marginal Land–A Review. *Agron. Sustain. Dev.* (under review).

25. BMEL. *Agrarexporte Verstehen-Fakten Und Hintergründe*; BMEL: Berlin, Germany, 2018.

26. UBA, (Umweltbundesamt). Toward Ecofriendly Farming. Available online: https://www.umweltbundesamt.de/en/topics/soil-agriculture/toward-ecofriendly-farming (accessed on 20 August 2020).

27. European Commission CORDIS EU Research Results. Available online: https://cordis.europa.eu/projects/en (accessed on 25 June 2021).

28. Elsevier How Can I Best Use the Advanced Search? Available online: https://service.elsevier.com/app/answers/detail/a_id/11365/c/10546/supporthub/scopus/ (accessed on 20 August 2020).

29. Euring, D.; Ayegbeni, S.; Jansen, M.; Tu, J.; Gomes Da Silva, C.; Polle, A. Growth Performance and Nitrogen Use Efficiency of Two Populus Hybrid Clones (P. Nigra × P. Maximowiczii and P. Trichocarpa × P. Maximowiczii) in Relation to Soil Depth in a Young Plantation. *iForest Biogeosci. For.* **2016**, *9*, 847. [CrossRef]

30. Orság, M.; Fischer, M.; Tripathi, A.M.; Žalud, Z.; Trnka, M. Sensitivity of Short Rotation Poplar Coppice Biomass Productivity to the Throughfall Reduction–Estimating Future Drought Impacts. *Biomass Bioenergy* **2018**, *109*, 182–189. [CrossRef]

31. Lewandowski, I.; Kicherer, A. Combustion Quality of Biomass: Practical Relevance and Experiments to Modify the Biomass Quality of Miscanthus x Giganteus. *Eur. J. Agron.* **1997**, *6*, 163–177. [CrossRef]

32. Kalinina, O.; Nunn, C.; Sanderson, R.; Hastings, A.F.S.; van der Weijde, T.; Özgüven, M.; Tarakanov, I.; Schüle, H.; Trindade, L.M.; Dolstra, O.; et al. Extending Miscanthus Cultivation with Novel Germplasm at Six Contrasting Sites. *Front. Plant Sci.* **2017**, *8*. [CrossRef] [PubMed]

33. Baldini, M.; Ferfuia, C.; Piani, B.; Sepulcri, A.; Dorigo, G.; Zuliani, F.; Danuso, F.; Cattivello, C. The Performance and Potentiality of Monoecious Hemp (Cannabis Sativa L.) Cultivars as a Multipurpose Crop. *Agronomy* **2018**, *8*, 162. [CrossRef]

34. Ozdemir, S.; Yetilmezsoy, K.; Nuhoglu, N.N.; Dede, O.H.; Turp, S.M. Effects of Poultry Abattoir Sludge Amendment on Feedstock Composition, Energy Content, and Combustion Emissions of Giant Reed (Arundo Donax L.). *J. King Saud Univ. Sci.* **2020**, *32*, 149–155. [CrossRef]

35. Meehan, P.; Burke, B.; Doyle, D.; Barth, S.; Finnan, J. Exploring the Potential of Grass Feedstock from Marginal Land in Ireland: Does Marginal Mean Lower Yield? *Biomass Bioenergy* **2017**, *107*, 361–369. [CrossRef]

36. Altieri, M.A.; Nicholls, C.I.; Montalba, R. Technological Approaches to Sustainable Agriculture at a Crossroads: An Agroecological Perspective. *Sustainability* **2017**, *9*, 349. [CrossRef]

37. Laurent, A.; Pelzer, E.; Loyce, C.; Makowski, D. Ranking Yields of Energy Crops: A Meta-Analysis Using Direct and Indirect Comparisons. *Renew. Sustain. Energy Rev.* **2015**, *46*, 41–50. [CrossRef]

38. Scordia, D.; Testa, G.; Cosentino, S.L. Perennial Grasses as Lignocellulosic Feedstock for Second-Generation Bioethanol Production in Mediterranean Environment. *Ital. J. Agronomy* **2014**, *9*, 84. [CrossRef]

39. Angelini, L.G.; Ceccarini, L.; Di Nassi o Nasso, N.; Bonari, E. Comparison of Arundo Donax L. and Miscanthus x Giganteus in a Long-Term Field Experiment in Central Italy: Analysis of Productive Characteristics and Energy Balance. *Biomass Bioenergy* **2009**, *33*, 635–643. [CrossRef]

40. Pilu, R.; Manca, A.; Landoni, M. Arundo Donax as an Energy Crop: Pros and Cons of the Utilization of This Perennial Plant. *Maydica* **2013**, *58*, 54–59.

41. Bócsa, I.; Karus, M.; Lohmeyer, D. *Der Hanfanbau*; Vollst. überarb. und erg. 2. Aufl.; Landwirtschaftsverl.: Münster-Hiltrup, Germany, 2000; ISBN 3-7843-3066-5.
42. Struik, P.C.; Amaducci, S.; Bullard, M.J.; Stutterheim, N.C.; Venturi, G.; Cromack, H.T.H. Agronomy of Fibre Hemp (Cannabis Sativa L.) in Europe. *Ind. Crop. Prod.* **2000**, *11*, 107–118. [CrossRef]
43. Manevski, K.; Lærke, P.E.; Jiao, X.; Santhome, S.; Jørgensen, U. Biomass Productivity and Radiation Utilisation of Innovative Cropping Systems for Biorefinery. *Agric. For. Meteorol.* **2017**, *233*, 250–264. [CrossRef]
44. Amaducci, S.; Zatta, A.; Raffanini, M.; Venturi, G. Characterisation of Hemp (Cannabis Sativa L.) Roots under Different Growing Conditions. *Plant Soil* **2008**, *313*, 227. [CrossRef]
45. Lewandowski, I.; CLIFTON-BROWN, J.C.; Scurlock, J.M.O.; Huisman, W. Miscanthus: European Experience with a Novel Energy Crop. *Biomass Bioenergy* **2000**, *19*, 209–227. [CrossRef]
46. Amaducci, S.; Facciotto, G.; Bergante, S.; Perego, A.; Serra, P.; Ferrarini, A.; Chimento, C. Biomass Production and Energy Balance of Herbaceous and Woody Crops on Marginal Soils in the Po Valley. *GCB Bioenergy* **2017**, *9*, 31–45. [CrossRef]
47. Bufe, C.; Korevaar, H. *Evaluation of Additional Crops for Dutch List of Ecological Focus Area*; Report/WPR; Wageningen Research Foundation (WR) Business Unit Agrosystems Research: Lelystad, The Netherlands, 2018; Volume 793.
48. Confalonieri, R.; Jones, B.; Van Diepen, K.; Van Orshoven, J. *Scientific Contribution on Combining Biophysical Criteria Underpinning the Delineation of Agricultural Areas Affected by Specific Constraints: Methodology and Factsheets for Plausible Criteria Combinations*; Terres, J., Hagyo, A., Wania, A., Eds.; Publications Office: Luxembourg, 2014.
49. Lewandowski, I.; Scurlock, J.M.O.; Lindvall, E.; Christou, M. The Development and Current Status of Perennial Rhizomatous Grasses as Energy Crops in the US and Europe. *Biomass Bioenergy* **2003**, *25*, 335–361. [CrossRef]
50. Ferrarini, A.; Fornasier, F.; Serra, P.; Ferrari, F.; Trevisan, M.; Amaducci, S. Impacts of Willow and Miscanthus Bioenergy Buffers on Biogeochemical N Removal Processes along the Soil–Groundwater Continuum. *GCB Bioenergy* **2017**, *9*, 246–261. [CrossRef]
51. Panacea. *Scientific Papers and Training Materials–Panacea. Summary Factsheet Miscanthus*; Agricultural University of Athens: Athens, Greece, 2020.
52. Berendonk, C.; Dahlhoff, A.; Dickeduisberg, M.; Dissemond, A.; Erhardt, N.; Gruber, W.; Hartmann, H.-B.; Holz, J.; Günter, J.; Kasten, P.; et al. *Nachwachsende Rohstoffe Vom Acker*; Landwirtschaftskammer Nordrhein-Westfalen: Münster, Germnay, 2012.
53. Paris, P.; Mareschi, L.; Sabatti, M.; Pisanelli, A.; Ecosse, A.; Nardin, F.; Scarascia-Mugnozza, G. Comparing Hybrid Populus Clones for SRF across Northern Italy after Two Biennial Rotations: Survival, Growth and Yield. *Biomass Bioenergy* **2011**, *35*, 1524–1532. [CrossRef]
54. Moffat, A.; Houston, T. Tree Establishment and Growth at Pitsea Landfill Site, Essex, U.K. *Waste Manag. Res.* **1991**, *9*, 35–46. [CrossRef]
55. Tilvikiene, V.; Kadziuliene, Z.; Dabkevicius, Z.; Venslauskas, K.; Navickas, K. Feasibility of Tall Fescue, Cocksfoot and Reed Canary Grass for Anaerobic Digestion: Analysis of Productivity and Energy Potential. *Ind. Crop. Prod.* **2016**, *84*, 87–96. [CrossRef]
56. Usťak, S.; Šinko, J.; Muňoz, J. Reed Canary Grass (Phalaris Arundinacea L.) as a Promising Energy Crop. *J. Cent. Eur. Agric.* **2019**, *20*, 1143–1168. [CrossRef]
57. Strašil, Z. Evaluation of Reed Canary Grass (Phalaris Arundinacea L.) Grown for Energy Use. *Res. Agr. Eng.* **2012**, *58*, 119–130. [CrossRef]
58. Gesch, R.W.; Johnson, J.M.-F. Water Use in Camelina–Soybean Dual Cropping Systems. *Agron. J.* **2015**, *107*, 1098–1104. [CrossRef]
59. Hunsaker, D.J.; French, A.N.; Clarke, T.R.; El-Shikha, D.M. Water Use, Crop Coefficients, and Irrigation Management Criteria for Camelina Production in Arid Regions. *Irrig. Sci.* **2011**, *29*, 27–43. [CrossRef]
60. Gominho, J.; Curt, M.D.; Lourenço, A.; Fernández, J.; Pereira, H. Cynara Cardunculus L. as a Biomass and Multi-Purpose Crop: A Review of 30 Years of Research. *Biomass Bioenergy* **2018**, *109*, 257–275. [CrossRef]
61. Archontoulis, S.V.; Struik, P.C.; Vos, J.; Danalatos, N.G. Phenological Growth Stages of Cynara Cardunculus: Codification and Description According to the BBCH Scale. *Ann. Appl. Biol.* **2010**, *156*, 253–270. [CrossRef]
62. Curt, M.D.; Mosquera, F.; Sanz, M.; Sánchez, J.; Sánchez, G.; Esteban, B.; Fernández, J. Effect of Land Slope on Biomass Production of Cynara Cardunculus L. In Proceedings of the 20th EU Biomass Conference and Exhibition, Milan, Italy, 18–22 June 2012; pp. 186–190.
63. Merrill, S.D.; Tanaka, D.L.; Hanson, J.D. Root Length Growth of Eight Crop Species in Haplustoll Soils. *Soil Sci. Soc. Am. J.* **2002**, *66*, 913. [CrossRef]
64. Merrill, S.D.; Tanaka, D.L.; Hanson, J.D. Comparison of Fixed-Wall and Pressurized-Wall Minirhizotrons for Fine Root Growth Measurements in Eight Crop Species. *Agron. J.* **2005**, *97*, 1367–1373. [CrossRef]
65. Favorite, J. *Cup Plant*; USDA NRCS National Plant Data Center: Washington, DC, USA, 2003.
66. Schittenhelm, S.; Schoo, B.; Schroetter, S. Yield physiology of biogas crops: Comparison of cup plant, maize, and lucerne-grass. *J. Kult.* **2016**, *68*, 378–384.
67. Schoo, B.; Schroetter, S.; Kage, H.; Schittenhelm, S. Root Traits of Cup Plant, Maize and Lucerne Grass Grown under Different Soil and Soil Moisture Conditions. *J. Agron. Crop Sci.* **2017**, *203*, 345–359. [CrossRef]
68. Panoutsou, C.; Singh, A. *Training Materials for Agronomists and Students*; Imperial College London: London, UK, 2019.
69. Zeller, F.J. Sorghum (Sorghum Bicolor L. Moench): Utilization, Genetics, Breeding. *Bodenkultur* **2000**, *51*, 71–85.

70. Alexopoulou, E.; Monti, A.; Elbersen, H.W.; Zegada-Lizarazu, W.; Millioni, D.; Scordia, D.; Zanetti, F.; Papazoglou, E.G.; Christou, M. Switchgrass: From Production to End Use. In *Perennial Grasses for Bioenergy and Bioproducts*; Alexopoulou, E., Ed.; Academic Press: Cambridge, MA, USA, 2018; pp. 61–105. ISBN 978-0-12-812900-5.
71. Parrish, D.J.; Fike, J.H. The Biology and Agronomy of Switchgrass for Biofuels. *Crit. Rev. Plant Sci.* **2005**, *24*, 423–459. [CrossRef]
72. Porensky, L.M.; Davison, J.; Leger, E.A.; Miller, W.W.; Goergen, E.M.; Espeland, E.K.; Carroll-Moore, E.M. Grasses for Biofuels: A Low Water-Use Alternative for Cold Desert Agriculture? *Biomass Bioenergy* **2014**, *66*, 133–142. [CrossRef]
73. Moore, K.J.; Kling, C.L.; Raman, D.R. A Midwest USA Perspective on Von Cossel et Al.'s Prospects of Bioenergy Cropping Systems for a More Social-Ecologically Sound Bioeconomy. *Agronomy* **2020**, *10*, 1658. [CrossRef]
74. Von Cossel, M.; Wagner, M.; Lask, J.; Magenau, E.; Bauerle, A.; Von Cossel, V.; Warrach-Sagi, K.; Elbersen, B.; Staritsky, I.; Van Eupen, M.; et al. Prospects of Bioenergy Cropping Systems for a More Social-Ecologically Sound Bioeconomy. *Agronomy* **2019**, *9*, 605. [CrossRef]
75. Teuling, A.J. A Hot Future for European Droughts. *Nat. Clim. Chang.* **2018**, *8*, 364. [CrossRef]
76. Stützel, H.; Brüggemann, N.; Inzé, D. The Future of Field Trials in Europe: Establishing a Network Beyond Boundaries. *Trends Plant Sci.* **2016**, *21*, 92–95. [CrossRef] [PubMed]
77. European Commission. (EC) COMMISSION DELEGATED REGULATION (EU) 2018/1784–of 9 July 2018–Amending Delegated Regulation (EU) No 639/2014 as Regards Certain Provisions on the Greening Practices Established by Regulation (EU) No 1307/2013 of the European Parliament and of the Council. 2018, p. 4. Available online: https://eur-lex.europa.eu/legal-content/en/TXT/?uri=CELEX:32018R1784 (accessed on 25 June 2021).

Article

A Sampling Strategy to Develop a Primary Core Collection of *Miscanthus* spp. in China Based on Phenotypic Traits

Shuling Liu [1,†], Cheng Zheng [2,†], Wei Xiang [3,†], Zili Yi [1,4,*] and Liang Xiao [1,4,*]

[1] College of Bioscience and Biotechnology, Hunan Agricultural University, Changsha 410128, China; yayaan@163.com

[2] College of Agronomy, Hunan Agricultural University, Changsha 410128, China; chengzheng@stu.hunau.edu.cn

[3] Crop Research Institute, Hunan Academy of Agricultural Science, Changsha 410128, China; xiangwei@hunaas.cn

[4] Hunan Engineering Laboratory for Ecological Application of Miscanthus Resources, Hunan Agricultural University, Changsha 410128, China

[*] Correspondence: yizili@hunau.net (Z.Y.); xiaoliang@hunau.edu.cn (L.X.)

[†] These authors contributed equally to this work.

Abstract: Core collections can act as a genetic germplasm resource for biologists and breeders. Thirty-seven phenotypic traits from 471 Miscanthus accessions in China were used to design 203 sampling schemes to screen the genetic variations in different sampling strategies. The sampling was analyzed using the unweighted pair group method with arithmetic mean (*UPGMA*) and the Euclidean distance (*Euclid*). Several parameters including the variance of phenotypic value (*VPV*), Shannon–Weaver diversity index (*H*), coefficient of variation (*CV*), variance of phenotypic frequency (*VPF*), ratio of phenotype retained (*RPR*), the mean difference percentage (*MD%*) and the variance difference percentage of traits (*VD%*), the range coincidence rate (*CR%*) and the variable rate of quantitative traits (*VR%*) were used to evaluate the level of representation of the primary core collections developed by the different sampling schemes. Based on the optimal sampling strategies of prior selecting accessions, a primary core collection was constructed that maintained > 99.5% of the *VPV* and a *CR%* of 100%. This study indicates that the optimal sampling scheme consisted of prior and deviation sampling methods (PD) combined with a logarithmic proportional sampling strategy (LG) of 37.4% of the actual sampling ratio. Sampling before clustering can improve several parameters including the *H*, *CV*, *RPR*, *VPF*, and *CR%*. Sampling strategies including the genetic diversity index (G), logarithmic proportional (LG) and the square root proportional strategy (SG) can improve the *H*, whilst the constant strategy (C) can improve the *RPR* and *VPF* when the sampling scale was >30%. Furthermore, the proportional strategy (P) can improve the *VPV*.

Keywords: core collection; flora distribution; *Miscanthus* phenotypic trait

1. Introduction

Environmental concerns including the greenhouse effect and increasing demands for fossil fuels have stimulated research into renewable energy resources [1]. Indigenous materials have the potential to supply energy with lower emission of greenhouse gases and are more environmentally favorable [2]. One promising plant material that can be used to produce efficiently and economically biofuels with a lower land requirement is *Miscanthus* [3–5]. *Miscanthus* is a raw material candidate of lignocellulosic biomass [6] that is a perennial C_4 tall grass of the Gramineae Family, *Miscanthus* spp. It belongs to *subtribe Saccharinae Griseb., tribe Andropogoneae Dumort., Subfam. Panicoideae A. Braun of Poaceae* [7]. *Miscanthus* Andersson grows widely in Eastern and Southeastern Asia, the Pacific Islands, and Africa [8,9]. Fourteen different species of *Miscanthus* Andersson are found around the world of which seven different species are native to various provinces in China [8]. China

Citation: Liu, S.; Zheng, C.; Xiang, W.; Yi, Z.; Xiao, L. A Sampling Strategy to Develop a Primary Core Collection of *Miscanthus* spp. in China Based on Phenotypic Traits. *Agronomy* **2022**, *12*, 678. https://doi.org/10.3390/agronomy12030678

Academic Editors: Moritz Von Cossel, Joaquín Castro-Montoya and Yasir Iqbal

Received: 17 January 2022
Accepted: 9 March 2022
Published: 11 March 2022

Publisher's Note: MDPI stays neutral with regard to jurisdictional claims in published maps and institutional affiliations.

is the distribution center of the genus *Miscanthus*, and *M. lutorioriparius* Andersson is the endemic species of China [10].

Miscanthus is used in various industries including papermaking, animal feeds, and soil and water conservation [11]. Studies by the *Miscanthus* Research Institute of Hunan Agricultural University (HUNAU) have identified >3000 wild *Miscanthus* populations in >800 counties of 30 provinces in China since 2006. There are more than 1000 representative accessions for the seven native species that have been collected and grown in the *Miscanthus* germplasm garden in HUNAU. However, there is a need for a germplasm collection that can be used to improve the utilization and management of plant germplasm resources. Core collections can conserve germplasm collections and inform optimized plant breeding strategies. Whilst there have been extensive core collection efforts in species including wheat, rice, soybean, maize, sesame, and barley [12–18], there have been no reported studies on the core collection methods of Miscanthus.

Core collections can improve the conservation, evaluation and utilization of germplasm. Core collections selected as subsets can represent the maximum genetic diversity of the initial collection with the minimum redundancies [19]. The development of core collections includes the collection and analysis of data obtained from fields or greenhouses, implementing the principle of stratified sampling by dividing the accessions into different groups, determining the sampling proportion of each group within the core collection, the selection of samples at random or based on representative criteria and evaluating the diversity and representativeness of the core collection. Furthermore, for studies conducted using core collections, the most important procedure is the development of a robust sampling strategy including sampling scale, stratified principle, sampling proportion and the sampling method. In this study, we report on the sampling strategy of a *Miscanthus* primary core collection and its role in reducing the size of the initial collection whilst retaining genetic diversity in the collection.

2. Materials and Methods

2.1. Plant Materials

The research materials used in this study were a subset of Miscanthus collected by the *Miscanthus* Research Institute of HUNAU from different areas across China from 2006 to 2008. The collection consisted of >1000 accessions including *M. sinensis*, *M. floridulus*, *M. sacchariflorus*, and *M. lutarioriparius*. The materials were planted in red soil at the *Miscanthus* Germplasm Garden of HUNAU, Changsha, China (Lat. 28°11′ N, Long. 113°4′ E). Each accession was grown on a 2 m × 2 m plot. More than 60 quantitative traits relating to different developmental phases and various uses of the plant were measured annually. Four hundred and seventy-one accessions were used for sampling of the primary core collection that were originally located in 3 Kingdoms, 4 Sub Kingdoms, and 11 regions of China according to the Floristics of Seed Plants [20]. The numbers of accessions in each flora region are presented in Table S1.

2.2. Evaluation of Phenotypic Traits

Thirty-seven phenotypic traits were studied including 14 qualitative traits and 23 quantitative traits. The biological and agronomic trait data (Traits 1–25: Table 1) were collected during the reproductive stage. Yield and energy-related quality data (Traits 26–37: Table 1) were collected during the harvest season in December. The quantitative traits were used to calculate the mean (X) values and standard deviation (σ) to quantify the observation values (X_i) into categories. Each category represented one specific phenotype. The subdividing range of quantitative traits ranged from the category where $X_i > X - n{\cdot}\sigma$ to $X_i < X + m{\cdot}\sigma$ ($m > n$), with the interval between the two neighbor categories being 0.5 σ (Table 2).

Table 1. The description and classes of the *Miscanthus* Phenotypic traits.

	Traits	Abbreviation	Description and Classes
1	Date of bud emergence	DBE	Emergence date of second leaf
2	Date of beginning flowers	DBF	Flowering date of first flower
3	Days to beginning flowering	DsBF	Days from bud emergence to any plant produces flower
4	Plant height	PH	Height of largest over-ground complete plant
5	Stem length	SL	Length of over-ground complete stem
6	First internode length	FIL	First complete internode length of above-ground stem
7	stem axis long diameter of FIL	SALD	stem axis long diameter of FIL's middle
8	Node number of per stem	NS	Node number of per over-ground complete stem
9	Largest leaf length	LL	Length of visual largest leaf
10	Largest leaf width	LW	Width of visual largest leaves
11	Fresh weight of per stem	FWS	Fresh weight of stem after reproductive stage
12	Dry weight of per stem	DWS	Weighted after fresh stem was dried three days at 45 °C
13	Node hairiness	NH	Does node have hairiness? (No = "0", Yes = "1")
14	Leaf back hairiness	LBH	Does leaf back have hairiness? (No = "0", Yes = "1")
15	Sheath hairiness	ShH	Does sheath have hairiness? (No = "0", Yes = "1")
16	Sheath mouth hairiness	ShMH	Does sheath mouth have hairiness? (No = "0", Yes = "1")
17	Internode waxiness	IWa	Does internode have waxiness? (No = "0", Yes = "1")
18	Node waxiness	NWa	Does node have waxiness? (No = "0", Yes = "1")
19	Leaf waxiness	LWa	Does leaf have waxiness? (No = "0", Yes = "1")
20	Sheath waxiness	ShWa	Does sheath have waxiness? (No = "0", Yes = "1")
21	Stem color	StC	0 = Yellow, 1 = Light green, 3 = Green, 5 = Dark green, 7 = lilac or pale-purple speckles interspersed; 9 = purple-red speckles interspersed
22	Leaf color	LC	1 = Light green, 3 = Green, 5 = Dark green
23	Sheath color	ShC	0 = Yellow, 1 = Light green, 3 = Green, 5 = Dark green, 7 = lilac or pale-purple speckles interspersed; 9 = purple-red speckles interspersed
24	Axillary bud on culm	ABC	0 = No, 1 = Yes Does node have waxiness?
25	Angle of Stem	AS	1 = Erect or $\theta \geq 80°$, 3 = $80° > \theta \geq 60°$, 5 = $60° > \theta \geq 40°$, 7 = $40° > \theta \geq 20°$, 9 = $\theta < 20°$ or Prostrate (Angle between plant outside stem and ground)
26	Tillers number per plot	TNP	Total number of tillers to plant on one plot
27	Dry matter content	DM	Dry matter content after fresh stem was dried to constant weight at 45 °C and at 105 °C
28	Neutral detergent fiber content	NDF	Determined with detergent fiber analysis
29	Acid detergent fiber content	ADF	Determined with detergent fiber analysis
30	Hemi-fibre content	HF	Determined with detergent fiber analysis
31	Fibre content	FC	Determined with detergent fiber analysis
32	Acid dissoluble lignin content	ADL	Determined with detergent fiber analysis
33	Acid insoluble ash content	AIA	Determined with detergent fiber analysis
34	Total ash content	TA	Ash content of matter incinerated in muffle furnace at 550 °C three hours
35	Total moisture content	TM	Total water content after fresh matter was dried to constant weight at 45 °C and at 105 °C
36	Total biomass per plot	TMP	Total biomass production of plants in one plot
37	Withered state	WiS	0 = No, 1 = Yes (Have the plants begun to wither?)

2.3. Sampling Strategy

The primary core collection was constructed using several methods based on grouping and ungrouping strategies (Figure 1). The ungroup-based strategy randomly selected three replicates from the initial collections. The primary core collection sampled using the random strategy was labeled as the non-group random sampling group (NGR). The group-based strategy involved a hierarchical two-level grouping approach in which each type of variety was grouped by flora after being grouped by species. In the hierarchical two-level grouping strategy, the accessions were divided into 23 hierarchical groups (Table S1).

Hierarchical two-level grouping methods and different sampling strategies were combined in this study. The primary core collections were selected from each group based on the given number of different sampling strategies. The clustering sampling methods were based on a stepwise clustering sampling method. A prior strategy of selecting accessions with the traits expressing maximum or minimum values as the primary core collections before clustering was used. The following sampling methods were used:

(1) A non-group random sampling method (NGR): In this method, the primary core collection was randomly selected from every subgroup with two germplasms at the lowest standard of categorizing. When one germplasm was in the subgroup, it was immediately selected for the cluster analysis. The procedures for the clustered and selected germplasms were repeated until the group scale was reduced to a given number.

(2) Deviation sampling (D): In this method, the degree of deviation degree of two germplasms were contrasted in each subgroup at the lowest standard of categorizing. The germplasm with the higher degree of deviation was selected for the following cluster analysis. When one germplasm was present in the subgroup, it was immediately selected for cluster analysis. The subsequent germplasms were processed similarly to the preceding step. The other procedures were similar to the stepwise clustering method.

Table 2. The number of Category for each trait.

Quantitative Trait	Number of Category	Qualitative Trait	Number of Category
DBE	7	NH	2
DBF	26	LBH	2
DsBF	29	ShH	2
PH	10	ShMH	2
SL	15	IWa	2
FIL	13	NWa	2
SALD	12	LWa	2
NS	14	ShWa	2
LL	10	StC	6
LW	13	LC	3
FWS	12	ShC	6
DWS	12	ABC	2
TNP	13	AS	5
DM	14	WiS	2
NDF	11	-	-
ADF	12	-	-
HF	14	-	-
FC	10	-	-
ADL	12	-	-
AIA	11	-	-
TA	14	-	-
TM	13	-	-
TMP	12	-	-

The degree of deviation of each quantitative trait was confirmed by the equation:

$$S_i^2 = \sum_{j=1}^{m} \frac{g_{ij}^2}{\sigma_j^2} \quad i = 1, 2, \ldots, n, \ j = 1, 2, \ldots, m \tag{1}$$

where g_{ij} represents the ith value of the jth trait, and σ_j^2 represents the variance of the jth trait [21].

(3) Prior sampling (PR): Germplasms with the traits expressing the maximum or minimum values were chosen as core collections before clustering. The residual germplasms were processed using a method similar to the random clustering method.

(4) Prior and Deviation sampling (PD): This strategy was based on the prior sampling method. Germplasms were processed in a similar way to the deviation sampling method after the germplasms with the traits expressing the maximum or minimum values were selected as the core collections.

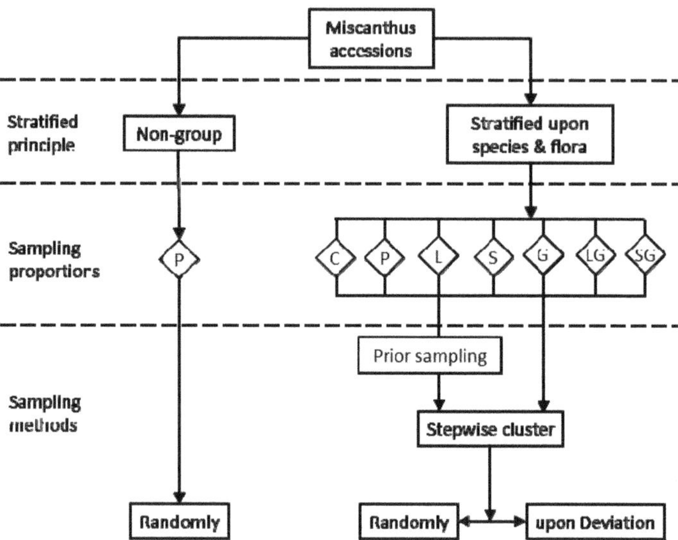

Figure 1. Sampling schemes of developing *Miscanthus* primary core collection. Constant strategy (C), Proportional strategy (P), Logarithm strategy (L), Square root strategy (S), Genetic diversity index strategy (G), Genetic diversity index adjusted with logarithmic proportional strategy (LG), Genetic diversity index adjusted with square root proportional strategy (SG). (1) Constant strategy (C)—the number of selected accessions sampled from each group was an equal number of accessions randomly; (2) Proportional strategy (P)—the number of selected accessions sampled from each group was proportional to the group size in the basic collection; (3) Logarithm strategy (L)—the number of selected accessions sampled from each group was proportional to the logarithmic group size in the basic collection; (4) Square root strategy (S)—sampling core collection from each group was proportional to the square root group size in the basic collection; (5) Genetic diversity index strategy (G)—sampling core collection from each group with proportional to genetic diversity index of the group in basic collection; (6) Genetic diversity index adjusted with logarithmic proportional strategy (LG)—sampling core collection from each group with the proportional to Shannon–Weaver diversity index was adjusted with logarithmic proportion; (7) Genetic diversity index adjusted with square root proportional strategy (SG)—sampling core collection from each group with the proportional to Shannon–Weaver diversity index was adjusted with square root proportion.

In the four clustering sampling methods, to reserve important biological types, it was decided that the groups including only one accession were selected as the primary core collection, e.g., two accessions of *M. floridulus* from the flora region of IIID12 and IVG22, one accession of *M. sinensis* from the flora region of IIIE14. Three accessions were selected as the primary core collections prior to clustering. Other groups were sampled using four clustering sampling methods. For comparison, the NGR was used to select a candidate primary core collection. Finally, 203 different sampling schemes were designed to develop a primary core collection of the *Miscanthus* in China.

To determine the optimal scale of a primary core collection, 20–50% ratios from the initial collections were considered as the ideal proportions for sampling. The actual numbers of selected accessions were calculated using different sampling proportions and combined with different sampling strategies and methods (Table 3).

2.4. Evaluating the Parameters for the Core Collection

Five parameters including *H, CV, VPV, VPF,* and the *RPR* were used to evaluate 203 sampling schemes [16].

Table 3. The sampling number of the primary core collections within different sampling strategies.

Sampling Strategy	Ideal		Actual			
	Number	Ratio (%)	Non-prior	Ratio (%)	Prior	Ratio (%)
C	92	20	81	17.2	134	28.5
	115	25	98	20.8	141	29.9
	138	30	115	24.4	150	31.8
	161	35	131	27.8	159	33.8
	184	40	146	31.0	167	35.5
	207	45	161	34.2	177	37.6
	253	50	187	39.7	197	41.8
G	96	20	99	21.0	138	29.3
	119	25	120	25.5	150	31.8
	142	30	140	29.7	161	34.2
	165	35	160	34.0	174	36.9
	186	40	178	37.8	189	40.1
	212	45	198	42.0	203	43.1
	234	50	213	45.2	216	45.9
L	95	20	98	20.8	134	28.5
	116	25	119	25.3	145	30.8
	144	30	147	31.2	163	34.6
	163	35	166	35.2	177	37.6
	189	40	189	40.1	194	41.2
	211	45	210	44.6	210	44.6
	237	50	231	49.0	231	49.0
LG	93	20	96	20.4	130	27.6
	118	25	121	25.7	144	30.6
	142	30	145	30.8	159	33.8
	166	35	169	35.9	176	37.4
	188	40	191	40.6	194	41.2
	214	45	217	46.1	217	46.1
	235	50	237	50.3	237	50.3
P	95	20	98	20.8	127	27.0
	118	25	121	25.7	138	29.3
	142	30	146	31.0	157	33.3
	164	35	167	35.5	177	37.6
	187	40	190	40.3	198	42.0
	211	45	214	45.4	220	46.7
	244	50	244	51.8	248	52.7
S	97	20	97	20.6	131	27.8
	116	25	116	24.6	140	29.7
	143	30	140	29.7	155	32.9
	165	35	162	34.4	170	36.1
	186	40	181	38.4	185	39.3
	212	45	206	43.7	208	44.2
	236	50	227	48.2	228	48.4
SG	95	20	98	20.8	129	27.4
	120	25	123	26.1	142	30.1
	143	30	146	31.0	157	33.3
	164	35	167	35.5	174	36.9
	188	40	191	40.6	194	41.2
	213	45	216	45.9	218	46.3
	234	50	237	50.3	238	50.5

Note: Constant strategy (C), Proportional strategy (P), Logarithm strategy (L), Square root strategy (S), Genetic diversity index strategy (G), Genetic diversity index adjusted with logarithmic. Proportional strategy (LG), Genetic diversity index adjusted with square root proportional strategy (SG).

The mean percentage difference (*MD%*), variance percentage difference (*VD%*), range coincidence rate (*CR%*) and variable rate (*VR%*) of the quantitative traits were compared by assessing the optimal sampling strategy [22]:

$$CR\% = \frac{1}{m} \sum_{j=1}^{m} \frac{R_C}{R_I} \times 100$$

$$VR\% = \frac{1}{m} \sum_{j=1}^{m} \frac{CV_C}{CV_I} \times 100 \tag{2}$$

where M_C = the mean of the core collection, M_I = the mean of the initial collection, R_C = the average scope of the quantitative traits of core collections, R_I = the average range of the quantitative traits of the initial collections, CV_C = the coefficient of variation of traits for the core collections, CV_I = the coefficient of variation of traits for the initial collection, m = the number of the quantitative traits.

Core collections are required to meet two criteria to accurately represent the genetic diversity of the initial collection. Specifically, core collections should include ≤20% of the traits possessed by diverse means (α = 0.05) between the core and initial collections, and the core collection should retain a range coincidence rate (*CR%*) ≥80% of the traits [23,24].

In developing the primary core collection, the four sampling methods (i.e., R, D, PR, and PD), seven sampling strategies (i.e., C, P, L, S, G, LG, and SG), and seven different sampling proportions (20%, 25%, 30%, 35%, 40%, 45%, and 50%) were applied. Then, 203 potential primary core collections were constructed and denoted as R-C20, D-P25, PR-L30, PD-LG35, etc. In contrast, seven non-group primary core collections were constructed using a combined proportional strategy (P) with different sampling proportions that were denoted as NGR20 to NGR50.

3. Results

3.1. The Tendency of Parameters for Sampling Methods in Different Sampling Scales

The variation tendency of the five parameters obtained using five sampling methods at seven sampling scales was processed (Figure 2). The group-based strategy was shown to be superior to the non-group strategy, and the methods of sampling before the clustering methods (PD and PR) were superior to the other clustering methods. The prior sampling strategy was potentially optimal for sampling. The variances of phenotypic value (*VPV*) of the primary core collections increased when reducing the sampling scale. The primary core collections constructed by the group strategy had similar *VPV*. Furthermore, the *VPVs* were all higher compared to the primary core collections constructed by NGR (Figure 2a). The *H* of the PD and PR methods increased with reducing sampling scale, yet the *H* of the method of deviation sampling (D) and random clustering (R) decreased with reducing sampling scale. The *VPV* of the NGR method showed no obvious regularity (Figure 2b). The coefficient of variation (*CV*) of the primary core collections constructed using various sampling methods showed undulating changes at a high sampling scale that then declined at a lower scale (Figure 3c). The ratio of phenotype retained (*RPR*) of the PD and PR methods was similar to the different sampling scales. The *RPR* of other methods decreased with reducing sampling scale (Figure 2d). The tendency of the ratio of variance of the phenotypic frequency (*VPF*) increased with a reducing sampling scale (Figure 2e). The *H*, *CV* and *RPR* of the PD and PR methods were similar or higher than the parameters of the other methods. The *H*, *CV* and *RPR* of the D and R methods were similar and higher than the NGR method. The *VPFs* of PD and PR methods were almost the same and lower than methods D and R. The group strategy was superior to the NGR, and the *VPV* of the prior strategy was superior to those of other strategies. In conclusion, the clustering methods of the P and D sampling methods were optimal.

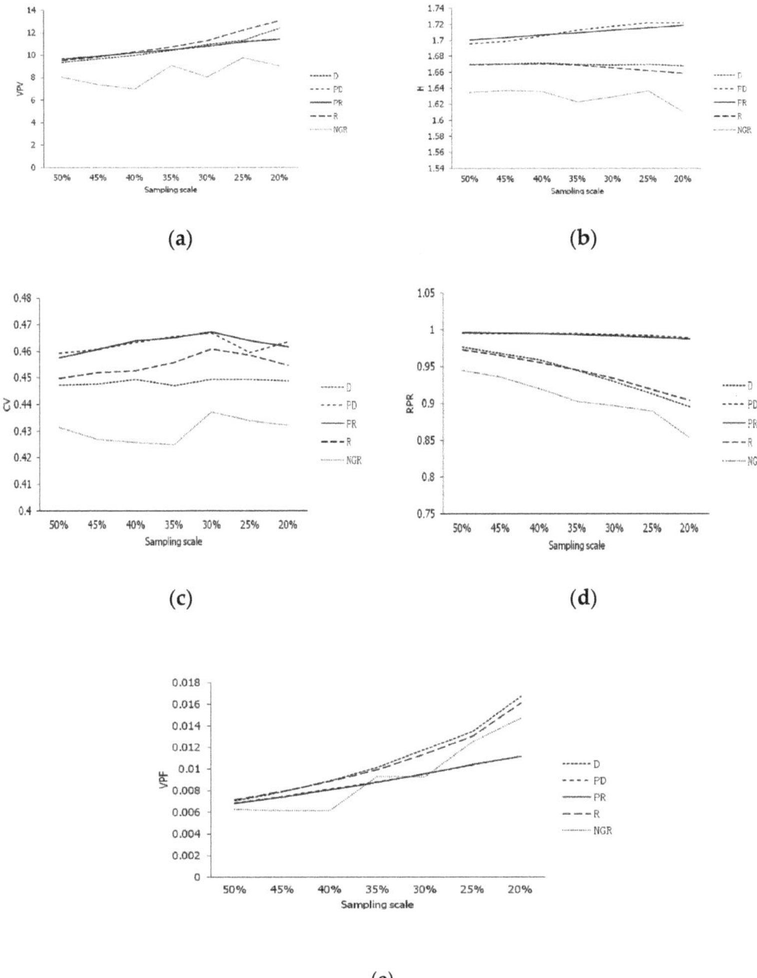

Figure 2. Tendency of parameters for sampling methods in different sampling scales. (**a**) Tendency of *VPV*; (**b**) Tendency of *H*; (**c**) Tendency of *CV*; (**d**) Tendency of *RPR*; (**e**) Tendency of *VPF*. *H*: Shannon–Weaver diversity index, *CV*: coefficient of variation, *VPV*: variance of phenotypic value, *VPF*: variance of phenotypic frequency, *RPR*: ratio of phenotype retained. D, PD, PR, R, and NGR stand for deviation sampling (D), prior and deviation sampling (PD), prior sampling (PR), random clustering (R), and non-group random sampling method (NGR), respectively.

3.2. The Tendency of the Parameters for Sampling Strategies in Different Sampling Scales

The five parameters obtained from the seven sampling strategies at the seven sampling scales were compared (Figure 3). The *VPV* for various core collections increased with a reduced sampling scale. The *VPV* of the core collections was highest when constructed using the constant strategy (C) and lowest when using the proportional strategy (P) (Figure 3a). The general tendency of *H* increased with reduced sampling scales. The value of *H* fluctuated when the sampling scale was <30% (Figure 3b). The tendency of the *CV* had no obvious regularity and mostly increased at a high sampling scale and decreased at a lower scale (Figure 3c). The *RPR* of all methods was similar and decreased with reducing sampling scales (Figure 3d). The *VPF* increased with reduced sampling scales and the C strategy was inferior to other strategies (Figure 3e). The *RPR* and *VPF* of the C strategy and the *VPV* and *H* of the P strategy performed the worst. These data indicated that the two sampling strategies were not applicable. Sampling strategies G, LG, SG, L, and S could potentially be used.

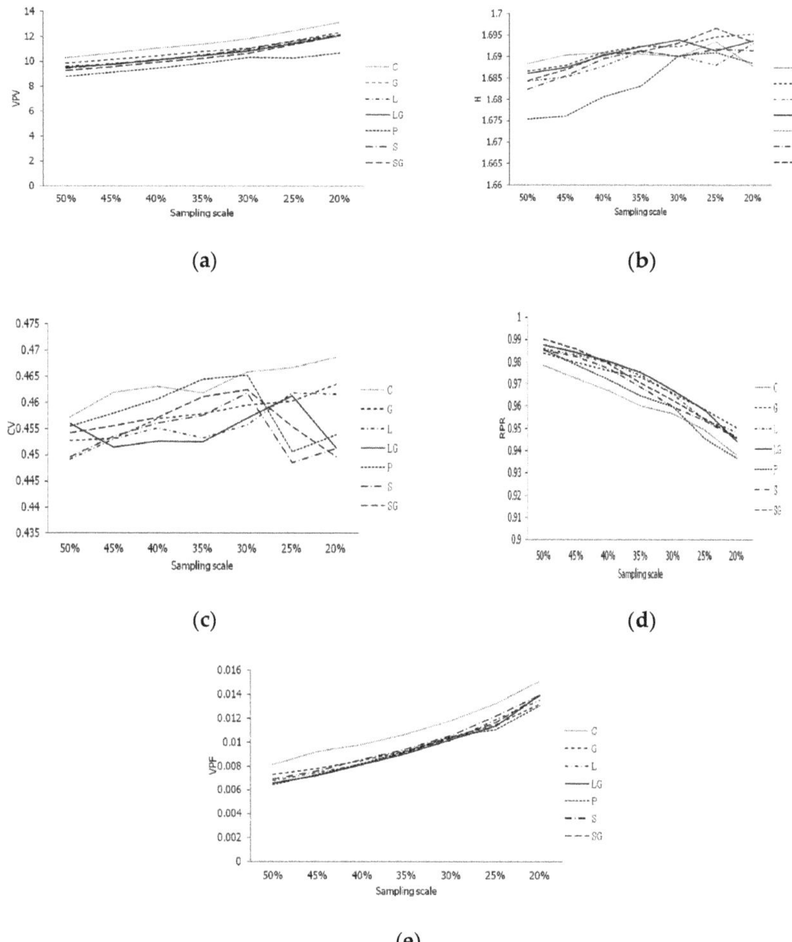

Figure 3. Tendency of parameters for sampling strategies in different sampling scales. (**a**) Tendency of *VPV*; (**b**) Tendency of *H*; (**c**) Tendency of *CV*; (**d**) Tendency of *RPR*; (**e**) Tendency of *VPF*. Shannon–Weaver diversity index (*H*), coefficient of variation (*CV*), variance of phenotypic value (*VPV*), variance of phenotypic frequency (*VPF*), and ratio of phenotype retained (*RPR*). C, G, L, LG, P, S, and SG stand for Constant strategy (C), Genetic diversity index strategy (G), Logarithm strategy (L), Genetic diversity index adjusted with logarithmic proportional strategy (LG), Proportional strategy (P), Square root strategy (S), Genetic diversity index adjusted with square root proportional strategy (SG), respectively.

3.3. The Relationship of the Parameters between Different Sampling Strategies and Methods

The five parameters obtained from the four clustering methods used in the different sampling strategies were compared at different sampling scales (Figure 4). From the results, the prior sampling strategy methods led to improved effectiveness in *H*, *CV*, *RPR*, and *VPF* amongst the different sampling strategies. The *VPV* values calculated for the four sampling methods were similar (Figure 4a). The *H*, *CV*, and *RPR* calculated from the primary core collections using the prior sampling strategy were higher than those for the other sampling strategies. The *VPF* calculated from the core collections using the prior sampling strategy was lower than for the other sampling strategies (Figure 4b–e). There were no significant differences in the five parameters between the PD and PR clustering methods. Prior sampling before clustering resulted in higher *H*, *CV* and *RPR* but a lower *VPF* of the primary core collections compared to the other two sampling methods. These data indicate that the methods of prior sampling before clustering were superior to directly clustering.

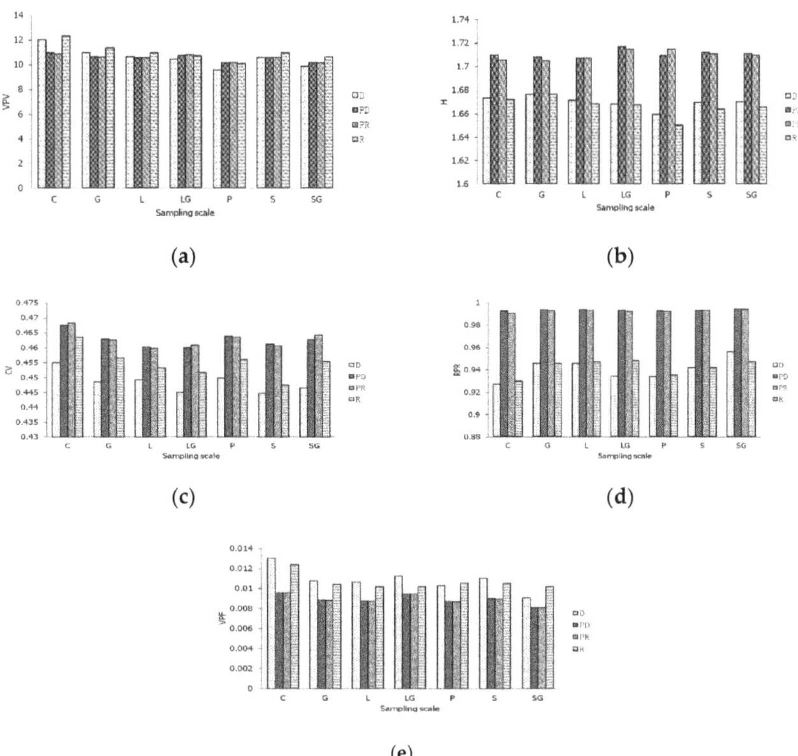

Figure 4. Difference of parameters for sampling strategies in different sampling scale. (**a**) Difference of *VPV*; (**b**) Difference of *H*; (**c**) Difference of *CV*; (**d**) Difference of *RPR*; (**e**) Difference of *VPF*. Shannon–Weaver diversity index (*H*), coefficient of variation (*CV*), variance of phenotypic value (*VPV*), variance of phenotypic frequency (*VPF*), and ratio of phenotype retained (*RPR*). D, PD, PR, R stand for deviation sampling (D), prior and deviation sampling (PD), prior sampling (PR), random clustering (R), respectively.

3.4. Comparison of the Sampling Strategies and Methods

The five parameters of the different sampling proportions, strategies and methods within the group were compared using Duncan's multiple range tests. The results are presented in Table 4 where the same ranking score implies that the data were not significantly different. The different ranking scores indicate superior to inferior assets.

Seven types of sampling strategies were compared across the groups using hierarchical cluster sampling. The results indicated that sampling according to the genetic diversity index strategy (G) was optimal, followed by the genetic diversity index adjusted with logarithmic proportional strategy (LG) and the genetic diversity index adjusted with a square root proportional strategy (SG). The square root strategy (S) gave the worst results. The ranking of the sampling schemes strategies from superior to inferior was G > LG > SG > C > L > P > S. In the same table, five sampling methods were compared. The results indicated that the hierarchical cluster methods were superior to the NGR methods.

The non-group-based strategy was performed worst. The prior sampling strategies, PD and PR, performed better than the non-prior sampling strategies (D, R and NGR). The superior-to-inferior order of the sampling schemes was PD > PR > D > R > NGR.

Table 4. The rank of sampling strategies, sampling methods and sampling scales within group in 203 primary core collections.

Parameter	Sampling Strategy							Sampling Method				
	C	G	L	LG	P	S	SG	PR	PD	R	D	NGR
VPV	1	2	4	5	7	3	6	4	3	2	1	5
H	4	1	6	3	7	5	2	2	1	3	4	5
VPF	1	3	5	6	2	7	4	1	2	4	3	5
RPR	7	5	4	2	1	6	3	1	2	5	4	3
CV	7	4	2	1	6	5	3	2	1	4	3	5
Sum of rank	20	15	21	17	23	26	18	10	9	18	15	23

Note: Shannon–Weaver diversity Index (*H*), coefficient of variation (*CV*), variance of phenotypic value (*VPV*), variance of phenotypic frequency (*VPF*), and ratio of phenotype retained (*RPR*), Constant strategy (C), Proportional strategy (P), Logarithm strategy (L), Square root strategy (S), Genetic diversity index strategy (G), Genetic diversity index adjusted with logarithmic proportional strategy (LG), Genetic diversity index adjusted with square root proportional strategy (SG). PR, PD, R, D, NGR stand for prior sampling, prior and deviation sampling, random clustering, deviation sampling, and non-group random sampling method (NGR), respectively.

The averages of the ranking scores of the five parameters of all 203 sampling schemes combining the different sampling strategies with different sampling methods are summarized in Table 5. When comparing the 203 sampling schemes based on sampling strategies, we found that the L and LG sampling methods of PD had the highest scores. The sampling scheme of PD-LG resulted in the highest score among all schemes.

Table 5. Comparison of the average ranking scores of the five parameters for sampling strategies with methods.

Sampling Strategies	Sampling Methods					Average
	D	R	PD	PR	NGR	
C	19.0	16.2	8.8	10.2	-	13.55
G	17.6	15.6	8.4	11.2	-	13.20
L	19.6	16.2	8.2	8.2	-	13.05
LG	17.6	17.4	7.8	9.4	-	13.05
P	25.0	25.6	11.8	10.0	25.8	19.64
S	22.8	19.0	9.2	10.8	-	15.45
SG	23.6	18.6	10.2	10.6	-	15.75
Average	20.74	18.37	9.20	10.06	25.80	14.81

Note: Random clustering (R), Deviation sampling (D), Prior sampling (PR), Prior and Deviation sampling (PD), non-group random sampling method (NGR). Constant strategy (C), Proportional strategy (P), Logarithm strategy (L), Square root strategy (S), Genetic diversity index strategy (G), Genetic diversity index adjusted with logarithmic proportional strategy (LG), Genetic diversity index adjusted with square root proportional strategy (SG).

3.5. Comparison of the Sampling Scale of the Core Collection

Comparison between the seven sampling scales showed that scales of 25%, 30% and 35% performed significantly better than the other sampling scales and followed the order of 30% > 25% = 35% > 40% > 20% = 45% > 50% (Table 6). The *CR*% increased with increasing sampling scale except for the *CR*% from 20% and 25% sampling proportions combined with sampling methods (Table 7). Furthermore, the sampling strategies did not influence the *CR*% results. The *CR*% values reached 100% when the sampling scales were >35%.

3.6. Assessment of the Core Collections with 21 Quantitative Traits

The results from different sampling schemes are summarized in Table S2. Of these, 176 primary core collections had 100% *VD*%. The *MD*% of these accessions was significantly different (*MD*% ≥ 33.3%) from the initial collections. All the *CR*% values were >80% and 96 of those reached 100% indicating a high range of variation of the traits. Prior sampling before clustering gave the largest *CR*% values. The *VD*% of deviation sampling strategies combined with prior sampling were lower than the random sampling strategies. The

VR% of the grouped sampling core collections were >100% and 53 *VR*% of the primary core collections had >110%. These data may be caused by the increased variation of traits after removing redundant germplasms by sampling germplasms with the traits expressing maximum or minimum values prior. Twenty (PD-LG35, PD-S35, PR-LG30, PD-P30, PD-SG30, PR-P30, PR-SG25, PD-P25, PR-P25, PD-L20, PR-C20, PR-L20, PD-S20, PR-S20, PD-LG20, PR-LG20, PD-SG20, PR-SG20, PD-P20, PR-P20) core collections had the highest *VD*% and *CR*% values, the lowest *MD*%, and the higher *VR*% in which PD-LG35 had the largest number of accessions.

Table 6. Sum of the rank of sampling scales within groups in 203 primary core collections.

Parameter	Sampling Scale						
	20%	25%	30%	35%	40%	45%	50%
VPV	1	2	3	4	5	6	7
H	2	1	3	4	5	6	7
VPF	5	3	1	2	4	6	7
RPR	7	6	5	4	3	2	1
CV	7	6	5	4	3	2	1
Sum of rank	22	18	17	18	20	22	23

Note: Shannon–Weaver diversity index (*H*), coefficient of variation (*CV*), variance of phenotypic value (*VPV*), variance of phenotypic frequency (*VPF*), and ratio of phenotype retained (*RPR*).

Table 7. Comparison of the range coincidence rates (CR%) of the primary core collections.

		Sampling Scale						
		20%	25%	30%	35%	40%	45%	50%
Sampling Method	D	90.24	96.40	91.18	96.28	100.00	100.00	100.00
	R	92.35	97.53	92.46	97.32	100.00	100.00	100.00
	PD	94.38	98.01	93.85	100.00	100.00	100.00	100.00
	PR	95.67	89.76	95.30	100.00	100.00	100.00	100.00
Sampling Strategy	C	94.09	95.14	96.18	96.37	97.22	98.13	98.40
	G	95.45	96.28	96.69	98.13	98.28	98.40	98.95
	LG	94.91	95.98	96.35	97.81	97.96	98.76	98.95
	SG	95.54	96.29	96.87	97.10	98.04	98.91	99.03
	L	95.18	95.97	97.32	98.15	98.39	98.82	99.01
	P	94.91	95.56	96.34	96.74	97.25	97.48	98.60
	S	94.91	95.98	97.21	97.36	98.32	98.68	98.89

Note: Random clustering (R), Deviation sampling (D), Prior sampling (PR), Prior and deviation sampling (PD), Constant strategy (C), Proportional strategy (P), Logarithm strategy (L), Square root strategy (S), Genetic diversity index strategy (G), Genetic diversity index adjusted with logarithmic proportional strategy (LG), Genetic diversity index adjusted with square root proportional strategy (SG).

3.7. Determination of the Sampling Scheme of the Core Collection

The *H*, *CV* and *RPR* of the primary core collections developed according to the combined PD and PR and G and LG strategies within all the sampling proportions are compared in Table 8. From Table S3, the *RPR* of all the candidate core collections were reduced by reducing the proportion of sampling, whilst the *H* and *CV* increased by reducing the sampling proportion. The *RPRs* were about 98.8%, 99.2%, 99.4%, 99.5% and 99.6%, respectively, two of which have reached 99.6%. No significant difference between those of all candidate primary core collections. The *H* and *CV* were larger compared to the initial collections.

The results of the sampling schemes were grouped using hierarchical clustering methods of the PD and PR and G and LG sampling strategies as summarized in Table S3. The rank of *VPV*, *H*, *CV*, *VPF* and *RPR* of all the sampling ratios from the whole collections indicated that the PD sampling method combined with the LG sampling strategy performed best at a sampling proportion of 35%. This sampling scheme developed a core collection with 176 accessions in which the actual sampling ratio is 37.4% (Table S3).

Table 8. Comparison of the sampling rations in candidate primary core collections.

Parameter	Sampling Scheme	Sampling Scale						
		20%	25%	30%	35%	40%	45%	50%
H	PD-G	1.719	1.716	1.710	1.708	1.705	1.700	1.698
	PD-LG	1.724	1.725	1.721	1.716	1.706	1.698	1.699
	PR-G	1.713	1.709	1.702	1.704	1.702	1.704	1.701
	PR-LG	1.722	1.717	1.714	1.713	1.709	1.705	1.702
CV	PD-G	46.764	46.513	46.566	46.267	46.186	45.938	45.802
	PD-LG	46.717	46.409	46.087	46.066	45.961	45.745	45.193
	PR-G	46.726	46.637	46.793	46.122	46.179	45.869	45.511
	PR-LG	46.932	46.359	46.369	45.858	45.775	45.489	45.147
RPR	PD-G	98.900	99.300	99.400	99.400	99.400	99.500	99.500
	PD-LG	98.900	99.300	99.400	99.500	99.500	99.500	99.500
	PR-G	98.800	98.800	99.300	99.400	99.400	99.500	99.600
	PR-LG	98.700	99.100	99.300	99.500	99.500	99.500	99.600

Note: Shannon–Weaver diversity index (*H*), coefficient of variation (*CV*), and ratio of phenotype retained (*RPR*). PD-G, PD-LG, PR-G, PR-LG stand for prior and deviation sampling method combined with genetic diversity index strategy, prior and deviation sampling method combined with logarithmic proportional strategy, prior sampling method combined with genetic diversity index strategy, prior sampling method combined with logarithmic proportional strategy, respectively.

4. Discussion

4.1. Phenotype Data Construction of a Primary Core Collection

The aim of developing a core collection is to build a population with minimal samples whilst maintaining maximum genetic diversity. Many core collections of crop germplasms have been successfully constructed including rice, wheat, soybean, and other commercial crops [14–30]. Currently, several types of data are used to construct core germplasm collections including habitat, phenotypic, and genomic data [31]. The distribution information and biological and agronomic traits were used in this study. It is difficult to establish core collections by assessing the genetic diversity of a whole germplasm resource using phenotype traits. Although molecular markers have been used for evaluating genetic diversity at the DNA level in crop germplasm resources [32], the application of such approaches to entire collections is laborious and costly. The development of primary core collections based on phenotype traits could reduce the scale of entire collections along with labor intensity and costs.

Phenotypic data has been previously used to build core collections in Miscanthus [33]. This approach showed that the grouping method based on the original geography data was the best strategy compared with the other grouping methods such as single phenotypic, random, administrative province, and non-grouping methods. In this study, we used phenotype data to establish core collections using different strategies in Miscanthus. We used five parameters including *H*, *CV*, *VPV*, *VPF*, and *RPR* to screen 203 candidate core collections. Our results showed that the PD-LG35 sampling strategy (prior and deviation sampling method, genetic diversity index adjusted with logarithmic proportional strategy, and 35% sampling ratio) was used to develop a core collection with 176 accessions, had the highest genetic diversity and optimum number of samples. Considering the data collected from the same observation station named Miscanthus germplasm garden built in 2006 in Hunan agricultural university [34], theoretically believe that all germplasm growth was in the same environment, therefore, the difference of phenotype traits able to stand for the genetic variation among individuals.

4.2. The Method to Establish the Primary Core Collection

Sampling strategies are a key factor in establishing a primary core collection. Studies have used different approaches to construct primary core collections such as the proportional strategy (P) in the apricot germplasm in China [28] and the genetic diversity index strategy (G) in safflower germplasm [35]. The scale of the sampling ratio is also an important factor that impacts the efficiency of primary core collections. Moreover, the scale of the primary core collection to the whole collection should be determined according to the size of the initial collection group. The sampling proportions may vary depending on the size of the initial collections. In spite of previous studies not suggesting any referable ration or any appropriate size for the primary core collection of *Miscanthus*, the ratio of the core collection to the whole collection for core collections established worldwide for different species is around 5–30% [26,36,37]. According to our preliminary study of the sampling strategy of *Miscanthus* in China, core collections of sampling before selecting core collections strategies retain a higher proportion of the phenotype characteristics (*RPR* > 98.8%). The PD-LG at 35% sampling proportion had the highest *H* and *CV* in the schemes compared to the PD-LD at 40%, 45%, and 50% which had the same or larger *RPR*.

Our data show that the group-based strategy was superior to the non-group strategy in different sampling scales or sampling strategies. Germplasm materials with similar heredity characteristics can be classified as one group using the group-based strategy. The methods of prior sampling before clustering methods were superior to the other clustering methods in different sampling scales because the germplasms with greater research value and special traits were not excluded. The *VPV* calculated based on the four sampling methods on different sampling scales or using sampling strategies were very similar and may be attributed to the rich genetic diversity of *Miscanthus* caused by intraspecific crossing. The constant strategy performed the worst at different sampling scales and may be attributed to the nonuniform genetic diversity of the intra-group *Miscanthus* as well as the proportional strategy (P). The sampling according to G, LG, and SG gave better results probably due to the affirmation of sampling ratio according to genetic diversity.

5. Conclusions

The PD-LG35 sampling strategy was used to develop a primary core collection with 176 accessions that had the best performance in this study. The actual sampling ratio was 37.4% suggesting that this was the optimal sampling scheme for selecting core collections. With such a moderate number of *Miscanthus* in China, PD methods combined with the LG at 37.4% of the actual sampling ratio was the optimum strategy. Furthermore, prior sampling before clustering could improve *H, CV, RPR* and *VPF*, with little impact on *VPV*. This sampling strategy also could improve the range of the *CR*% without affecting on the *MD*%. The sampling strategies using G, LG, and SG could improve *H*. Meanwhile, the C had the disadvantage of improving the *RPR* and *VPF* when the sampling scale was more than 30%, whilst the P had the disadvantage of improving the *VPV*.

Supplementary Materials: The following supporting information can be downloaded at: https://www.mdpi.com/article/10.3390/agronomy12030678/s1, Table S1: The number of accessions in each flora and species; Table S2: Comparison of the percentages for the differences between the primary core collections and the initial collections; Table S3: Rank of the integrative score of 5 parameters from candidate core collections. Shannon–Weaver diversity index (*H*), coefficient of variation (*CV*), variance of phenotypic value (*VPV*), variance of phenotypic frequency (*VPF*), and ratio of phenotype retained *(RPR)*.

Author Contributions: Conceptualization, Z.Y.; methodology, S.L.; software, S.L. and C.Z.; validation, W.X. and C.Z.; investigation, L.X.; writing—original draft preparation, S.L. and W.X.; writing—review and editing, L.X.; funding acquisition, Z.Y. All authors have read and agreed to the published version of the manuscript.

Funding: This study was financially supported by the Foundation for the Construction of the Innovative Hunan Province (grant number: 2019NK2011).

Institutional Review Board Statement: Not applicable.

Informed Consent Statement: Not applicable.

Data Availability Statement: The data used to support the findings of this study are available from the corresponding author upon request.

Acknowledgments: We thanks to these M.D candidate students during 2009 and 2012 in Hunan Agricultural University, who collected data in the field, included Cong lin, De Xue, Bin hu, Zuhong Wang, Yueyue Zhou, Yu Wang, and Guote Deng.

Conflicts of Interest: The authors declare no conflict of interest.

References

1. Londoño-Pulgarin, D.; Cardona-Montoya, G.; Restrepo, J.C.; Muñoz-Leiva, F. Fossil or bioenergy? Global fuel market trends. *Renew. Sustain. Energ. Rev.* **2021**, *143*, 110905. [CrossRef]
2. Demirbas, A. Political, economic and environmental impacts of biofuels: A review. *Appl. Energy* **2019**, *86*, 108–117. [CrossRef]
3. Agostini, A.; Serra, P.; Giuntoli, J.; Martani, E.; Ferrarini, A.; Amaducci, S. Biofuels from perennial energy crops on buffer strips: A win-win strategy. *J. Clean. Prod.* **2021**, *297*, 126703. [CrossRef]
4. Dubis, B.; Jankowski, K.J.; Załuski, D.; Bórawski, P.; Szempliński, W. Biomass production and energy balance of Miscanthus over a period of 11 years: A case study in a large-scale farm in Poland. *GCB Bioenergy* **2019**, *11*, 1187–1201. [CrossRef]
5. Shumny, V.K.; Veprev, S.G.; Nechiporenko, N.N.; Goryachkovskaya, T.N.; Slynko, N.M.; Kolchanov, N.A.; Peltek, S.E. A new form of Miscanthus (Chinese silver grass, Miscanthus sinensis—Andersson) as a promising source of cellulosic biomass. *Adv. Biosci. Biotechnol.* **2015**, *1*, 167–170. [CrossRef]
6. Clifton-Brown, J.C.; Lewandowski, I.; Andersson, B.; Basch, G.; Teixeira, F. Performance of 15 miscanthus genotypes at five sites in Europe. *Agron. J.* **2008**, *93*, 1013–1019. [CrossRef]
7. Qian, S.; Lin, Q.I.; Zi-Li, Y.I.; Yang, Z.R.; Zhou, F.S. A taxonomic revision of miscanthus s.l. (Poaceae) from China. *Bot. J. Linn. Soc.* **2010**, *164*, 178–220.
8. Chen, S.L.; Renvoize, S.A. Miscanthus Andersson. In *Flora of China*; Beijing Science Press: Beijing, China, 2006.
9. Hastings, A.; Clifton-Brown, J.; Wattenbach, M.; Stampfl, P.; Mitchell, C.P.; Smith, P. Potential of Miscanthus grasses to provide energy and hence reduce greenhouse gas emissions. *Agron. Sustain. Dev.* **2008**, *28*, 465–472. [CrossRef]
10. Liu, L.; Zhu, M.; Zhu, T.P. Exploitation and utilization of Miscanthus & Triarrhena. *J. Nat. Resour.* **2001**, *16*, 562–563. (In Chinese)
11. Yi, Z.L. Exploitation and utilization of Miscanthus as energy plant. *J. Hunan Agric. Univ.* **2012**, *38*, 455–463. (In Chinese) [CrossRef]
12. Hao, C.Y.; Zhang, X.Y.; Wang, L.F.; Dong, Y.S.; Shang, X.W.; Jia, J.Z. Genetic diversity and core collection evaluations in common wheat germplasm from the nothwestern spring wheat region in China. *Mol. Breeding.* **2006**, *17*, 69–77. [CrossRef]
13. Hao, C.Y.; Dong, Y.C.; Wang, L.F.; You, G.X.; Zhang, H.N.; Ge, H.M.; Jia, J.Z.; Zhang, X.Y. Genetic diversity and construction of core collection in Chinese wheat genetic resources. *Chin. Sci. Bull.* **2008**, *53*, 1518–1526. (In Chinese)
14. Zhang, H.L.; Zhang, D.L.; Wang, M.X.; Sun, J.L.; Qi, Y.W.; Li, J.J.; Wei, X.H.; Han, L.Z.; Qiu, Z.E.; Tang, S.X.; et al. A core collection and mini core collection of Oryza sativa L. in China. *Theor. Appl. Genet.* **2011**, *122*, 49–61. [CrossRef] [PubMed]
15. Zhao, L.M.; Dong, Y.S.; Liu, B.; Hao, S.; Wang, K.J.; Li, X.H. Establishment of a core collection for the Chinese annual wild soybean (Glycine Soja). *Chin. Sci. Bull.* **2015**, *50*, 989–996. (In Chinese) [CrossRef]
16. Li, Z.C.; Zhang, H.L.; Zeng, Y.W.; Yang, Z.Y.; Shen, S.Q.; Sun, C.Q.; Wang, X.K. Studies on sampling schemes for the establishment of core collection of rice landraces in Yunnan, China. *Genet. Resour. Crop. Evol.* **2002**, *49*, 67–74. [CrossRef]
17. Zhang, Y.X.; Zhang, X.R.; Hua, W.; Wang, L.H.; Che, Z. Analysis of genetic diversity among indigenous landraces from sesame (*Sesamum indicum* L.) core collection in China as revealed by SRAP and SSR markers. *Genes. Genom.* **2010**, *32*, 207–215. [CrossRef]
18. Hintum, T.L.; Morales, E.A.V. (Eds.) *Core Collections of Plant Genetic Resources*; John Wiley and Sons: Chichester, UK, 1995; pp. 127–146.
19. Malosetti, M.; Avadie, T. Sampling strategy to develop a core collection of Uruguayan maize landraces based on morphological traits. *Genet. Resour. Crop. Evol.* **2001**, *48*, 381–390. [CrossRef]
20. Wu, Z.Y. *Floristics of Seed Plants from China*; Science Publisher: Beijing, China, 2010; pp. 52–56.
21. Hu, J.; Zhu, J.; Xu, H.M. Methods of constructing core collections by stepwise clustering with three sampling strategies based on the genotypic values of crops. *Theor. Appl. Genet.* **2000**, *101*, 264–268. [CrossRef]
22. Li, C.T.; Shi, C.H.; Wu, J.G.; Xu, H.M.; Zhang, H.Z.; Ren, Y.L. Methods of developing core collections based on the predicted genotypic value of rice (*Oryza sativa* L.). *Theor. Appl. Genet.* **2004**, *108*, 1172–1176. [CrossRef]
23. Balfourier, F.; Roussel, V.; Strelchenko, P.; Exbrayat-Vinson, F.; Sourdille, P.; Boutet, G.; Koenig, J.; Ravel, C.; Mitrofanova, O.; Beckert, M.; et al. A worldwide bread wheat core collection arrayed in a 384-well plate. *Theor. Appl. Genet.* **2007**, *114*, 1265–1275. [CrossRef] [PubMed]
24. Oliveira, M.F.; Nelson, R.L.; Geraldi, I.O.; Cruz, C.D.; Toledo, J.F.F.D. Establishing a soybean germplasm core collection. *Field Crops Res.* **2010**, *119*, 277–289. [CrossRef]
25. Chandra, S.; Huaman, Z.; Krishna, S.H.; Ortiz, R. Optimal sampling strategy and core collection size of Andean tetraploid potato based on isozyme data—A simulation study. *Theor. Appl. Genet.* **2002**, *104*, 1325–1334. [CrossRef] [PubMed]
26. Li, Y.X.; Li, T.H.; Zhang, H.L.; Qi, Y.W. Sampling Strategy for a Primary Core Collection of Peach (*Prunus persica* (L.) Batsch.) Germplasm. *Eur. J. Hortic. Sci.* **2007**, *72*, 268–274.

27. Zhang, J.; Wang, Y.; Zhang, X.Z.; Li, T.Z.; Wang, K.; Xu, X.F.; Han, Z.H. Sampling strategy to develop a primary core collection of apple cultivars based on fruit traits. *Afr. J. Biotechnol.* **2010**, *9*, 123–127.
28. Wang, Y.Z.; Zhang, J.H.; Sun, H.Y.; Ning, N.; Yang, L. Construction and evaluation of a primary core collection of apricot germplasm in China. *Sci. Hortic.* **2011**, *128*, 311–319. (In Chinese) [CrossRef]
29. Zhang, Y.; Zhang, Q.; Yang, Y.; Luo, Z. Revision of a Primary Core Collection of Persimmon (Diospyros kaki Thunb.) Originated in China. *Acta Hortic.* **2013**, *996*, 213–217. (In Chinese) [CrossRef]
30. Martínez, I.B.; Cruz, M.V.; Nelson, M.R.; Bertin, P. Establishment of a Core Collection of Traditional Cuban Theobroma cacao Plants for Conservation and Utilization Purposes. *Plant Mol. Biol. Report.* **2016**, *35*, 47–60. [CrossRef]
31. Liu, X.L.; Cai, Q.; Ma, L.; Wu, C.W.; Lu, X.; Ying, X.M.; Fan, Y.H. Strategy of sampling for pre-core collection of sugarcane hybrid. *Acta Agron. Sin.* **2009**, *35*, 1209–1216. (In Chinese) [CrossRef]
32. Lerceteau, E.; Robert, T.; Petiard, V.; Crouzillat, D. Evaluation of the extent of genetic variability among Theobroma cacao accessions using RAPD and RFLP markers. *Theor. Appl. Genet.* **1997**, *95*, 10–19. [CrossRef]
33. Xiao, L.; Yi, Z.L.; Jiang, J.X.; Liu, S.L.; Qin, J.P.; Yi, J.Q.; Yang, S. Establishment of China's Miscanthus sinensis Primary Core Collection. *J. Plant Genet. Resour.* **2014**, *15*, 1196–1201. (In Chinese)
34. Xiang, W.; Xue, S.; Liu, F.; Qin, S.; Xiao, L.; Yi, Z. MGDB: A database for evaluating *Miscanthus* spp. to screen elite germplasm. *Biomass Bioenergy* **2020**, *138*, 105599. [CrossRef]
35. Kumar, S.; Ambreen, H.; Variath, M.T.; Rao, A.R.; Agarwal, M.; Kumar, A.; Goel, S.; Jagannath, A. Utilization of Molecular, Phenotypic, and Geographical Diversity to Develop Compact Composite Core Collection in the Oilseed Crop, Safflower (*Carthamus tinctorius* L.) through Maximization Strategy. Front. *Plant Sci.* **2016**, *7*, 1554. [CrossRef] [PubMed]
36. Sun, Y.; Dong, S.; Liu, Q.; Chen, J.; Pan, J.; Zhang, J. Selection of a core collection of Prunus sibirica L. germplasm by a stepwise clustering method using simple sequence repeat markers. *PLoS ONE* **2021**, *16*, e0260097. [CrossRef] [PubMed]
37. Xu, J.; Wang, L.; Wang, H.; Mao, C.; Kong, D.; Chen, S.; Zhang, H.; Shen, Y. Development of a Core Collection of Six-Rowed Hulless Barley from the Qinghai-Tibetan Plateau. *Plant Mol. Biol. Rep.* **2020**, *38*, 305–313. [CrossRef]

Article

Biomass Quality Variations over Different Harvesting Regimes and Dynamics of Heavy Metal Change in *Miscanthus lutarioriparius* around Dongting Lake

Xionghui Liao [1,2,†], Yini Wu [1,†], Tongcheng Fu [1,3], Yasir Iqbal [1,3], Sai Yang [4], Meng Li [1,3], Zili Yi [1,3] and Shuai Xue [1,3,*]

[1] Hunan Provincial Key Laboratory of Crop Germplasm Innovation and Utilization, College of Bioscience & Biotechnology, Hunan Agricultural University, Changsha 410128, China; hnnydxlxh@stu.hunau.edu.cn (X.L.); wuyouyi@stu.hunau.edu.cn (Y.W.); futongcheng@hotmail.com (T.F.); yasir.iqbal1986@googlemail.com (Y.I.); mengli@hunau.edu.cn (M.L.); yizili@hunau.net (Z.Y.)
[2] College of Resources & Environment, Hunan Agricultural University, Changsha 410128, China
[3] Hunan Branch, National Energy R & D Center for Non-Food Biomass, Hunan Agricultural University, Changsha 410128, China
[4] Orient Science & Technology College of Hunan Agricultural University, Changsha 410128, China; yangsai1116@163.com
* Correspondence: xue_shuai@hunau.edu.cn
† These authors contributed equally to this paper.

Abstract: *Miscanthus lutarioriparius* has a growing area of 100,000 ha and an annual biomass production of 1 Mt around Dongting Lake. However, due to serious soil pollution, there is a concern that the *M. lutarioriparius* biomass could have high heavy metal (HM) concentrations. This necessitates investigation of biomass quality to find the appropriate end use. Thus, this study aims to investigate the dynamics of HM elements in the *M. lutarioriparius* biomass and their impact on biomass quality across different growing areas and harvest times. We analyzed the HM concentrations in soil and biomass from 11 sites under different harvesting times (April, August and December). Results showed that Cd in soil samples was 9.43-fold higher than the national standards. The heavily polluted soil caused a high HM concentration in the biomass and the accumulation increased with the delayed harvest. The fresh young shoots in April met the food limitation for Cd and Cr, whereas Pb concentration was slightly higher than the threshold limit. The mature biomass from the southern part had higher Mn, Cd and Pb, but lower Cu, Zn and Cr concentrations than that from the eastern part. These results can provide guidance for guaranteeing the consistent quality of the *M. lutarioriparius* biomass for bio-based industry.

Keywords: biomass quality; eco-industrial crop; heavy metal; *Miscanthus*; phytoremediation; wetland plant

Citation: Liao, X.; Wu, Y.; Fu, T.; Iqbal, Y.; Yang, S.; Li, M.; Yi, Z.; Xue, S. Biomass Quality Variations over Different Harvesting Regimes and Dynamics of Heavy Metal Change in *Miscanthus lutarioriparius* around Dongting Lake. *Agronomy* **2022**, *12*, 1188. https://doi.org/10.3390/agronomy12051188

Academic Editors: Stefano Amaducci and Carmelo Maucieri

Received: 27 March 2022
Accepted: 13 May 2022
Published: 15 May 2022

Publisher's Note: MDPI stays neutral with regard to jurisdictional claims in published maps and institutional affiliations.

1. Introduction

Dongting Lake, the second largest freshwater lake in China, plays a vital role in maintaining the ecosystem functioning of the Yangtze River Basin. This is mainly achieved through supporting the wild *Miscanthus lutarioriparius* community with an area of approximately 100,000 ha [1]. *M. lutarioriparius* is a perennial herbaceous plant characterized by tall stems, high biomass yield and extensive belowground systems [2]. The developed rhizome network can reduce soil erosion and simultaneously adsorb and accumulate contaminants from soil or water [3,4]. In addition, its high biomass can sequester a great amount of atmospheric CO_2 with a potential of approximately 100 t CO_2 ha^{-1} annually [5]. The tall stems provide an ideal shield for wild animals, indicating a biodiversity-increasing potential. Moreover, the *M. lutarioriparius* community also supports the biomass-based industry in the Dongting Lake region through production of 1 Mt cellulose-rich biomass.

Since the 1950s, this biomass feedstock has been used to produce pulp [6,7]. However, due to the serious water pollution concerns associated with the paper mills, almost all the paper-making plants have been closed by the Hunan provincial government since 2019. At present, there are still no appropriate alternate pathways identified for the commercial utilization of the *Miscanthus* biomass. With the closure of paper mills and lack of industrial demand, management practices and harvesting of the *M. lutarioriparius* biomass has been almost abandoned. Thus, there is an increasing concern that the unharvested plants will cause a series of ecological problems such as water eutrophication caused by the leaching of nutrients and an increasing risk of wildfire causing negative impacts on biodiversity. The aforementioned issues indicate that harvesting and commercial utilization of the biomass are the key management measures to preserve the *M. lutarioriparius* community and the Dongting Lake wetland ecosystem [4].

Compared to the most prevalent herbaceous plant species, the biomass of *M. lutarioriparius* is characterized with high contents of lignin (132 g kg^{-1}) and cellulous (620 g kg^{-1}) and a low content of ash (84 g kg^{-1}) [8]. These biomass components can be utilized to produce a wide range of biobased products such as biofuel [9], light-weight concrete [10], biochar and mushroom growth substrate [1], etc. Of all these utilization possibilities, biochar production and use as growth substrate for mushroom and bio-fertilizer are mainly recommended because of the availability of well-established methods and matured technology. In the short term, these utilization pathways can immediately accommodate huge quantities of available unharvested *Miscanthus* biomass and address the aforementioned imminent ecological challenge. Furthermore, the young shoots of *M. lutarioriparius* harvested in early spring can be used to produce pickles. The production of pickled shoots has reached approximately 20,000 t with production values of about CNY0.5 billion [2], which can also contribute toward overcoming the ecological problem and maintaining the ecosystem functioning.

The main water sources of the Dongting Lake are the Xiangjiang River, Yuanshui River, Zishui River and Lishui River [11], which flow through the main mining area of Hunan. Consequently, the water and soil around Dongting Lake are contaminated by heavy metal (HM) elements such as Cd, Cu, Mn, Zn, Cr and Pb [3,11]. The miscanthus species, especially *Miscanthus floridulus*, is known for their high HM adsorption and accumulation capabilities and is considered as a promising candidate plant for phytoremediation [12]. From this, it can be deducted that the *M. lutarioriparius* biomass from Dongting Lake is contaminated by the HM elements. Currently, *M. lutarioriparius*-based bioproducts (e.g., biochar-based fertilizer, pickles and mushroom growth substrate) are mainly used in agriculture and food production. Therefore, there is a growing concern about the safety of these bioproducts, especially their impact on food safety. This is also the main barrier in the development and expansion of bio-based industries.

M. lutarioriparius is a perennial herbaceous plant with three main growth phases, including seedling, mature and senescence. There are differences in the ability to bioaccumulate HM elements by the plant organs [3]. However, it is still unclear how they vary and what determines the uptake of HM elements in *M. lutarioriparius* over different growing phases. In addition, *M. lutarioriparius* is a plant with a nutrients re-translocation ability at the end of the growth [13]. It is still not known if HM elements in the leaves and stems also return back into the underground organs or not. The aim of this study is to investigate the dynamics of HM elements in the *M. lutarioriparius* biomass and their impact on biomass quality across different growing areas and harvest times. Based on the outcomes of this study, recommendations will be formulated to ensure the efficient and safe use of the *M. lutarioriparius* biomass around the Dongting Lake area.

2. Materials and Methods

2.1. Study Area and Sampling Strategy

The total water-flooded area in the Dongting Lake region (28°38′–29°45′ N, 111°40′–113°10′ E) is approximately 20,000 km^2 during the wet season (from April to September), whereas during the dry season (from November to next March) it shrinks to 16,400 km^2. This region is characterized by a subtropical monsoon climate with a mean annual temperature of 17 °C and precipitation of 1200–1400 mm. The study area was comprised of 11 sampling sites evenly distributed across the whole Dongting Lake region as shown in Figure 1. There are six sampling sites locating in the southern Dongting Lake (S1–S6) and five in the eastern Dongting Lake (S7–S11). Each sampling site was comprised of evenly distributed areas of more than 1 ha of *M. lutarioriparius*.

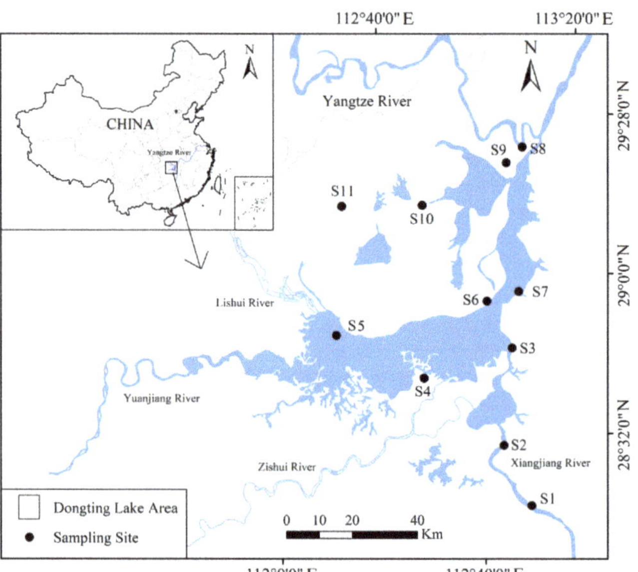

Figure 1. Location of the 11 in situ sampling sites within the Dongting Lake area.

Plant and soil sampling were conducted in 2018 at three different harvest times (3–8 April, 20–25 August and 15–20 December). Prior to the sampling, three 1×1 m quadrats were randomly selected at each site. The distance between each sampling quadrat was set to be more than 100 m. For each sampling, all the aboveground plants within the quadrat were harvested and weighed; then stems and leaves were separated; the belowground rhizomes and roots within the depth of 0–30 cm were collected, washed and weighed. About 500 g of fresh stems, leaves and underground organs (rhizomes and roots together) were collected and then taken to the laboratory for biomass yield determination and quality analysis. Afterward, all the plant samples were oven-dried at 80 °C until constant weight [8] and then milled to a powder (100 mesh) for the HM concentration determination. For soil sampling, five points were randomly selected within each sampling quadrat by following the "S"-shape principle. Five soil cores (0–30 cm) within the quadrat were mixed thoroughly to prepare a composite sample (approximately 500 g) and taken to the laboratory. The soil samples were air dried and then ground to pass through a 100-mesh sieve for the soil chemical analysis and HM concentration determination.

2.2. Chemical Analysis of Soils and Plants

Soil pH and electrical conductivity were determined using a portable pH meter (Bante-221, BANTE, Shanghai, China) and a conductivity meter (Bante-950-UK, BANTE, Shanghai, China), respectively. The soil suspension was prepared at soil: distilled water (w/v) of 1:2.5. Soil organic matter (SOM) was determined by the $K_2Cr_2O_7$–H_2SO_4 oxidation method. Soil total nitrogen (TN) was determined using the Kjedahl method. Both total phosphorous (TP) and available phosphorous (AP) of the soil were determined using the molybdenum antimony colorimeter method after samples were digested in $HClO_4$–H_2SO_4 (1:10, v/v) and extracted in $NaHCO_3$ (0.5 mol L^{-1}), respectively. Soil total potassium (TK) and available potassium (AK) were determined using the flame photometry method after soil samples were digested in $HClO_4$–H_2SO_4 (1:10, v/v) and extracted in NH_4OAc (1.0 mol L^{-1}). Plant and soil samples were digested by $HClO_4/HNO_3$ (1:4, v/v) and HCl/HNO_3 (1:3, v/v), respectively. The concentrations of Cu, Mn and Zn were determined by a flame atomic absorption spectrometer (TAS-A3, PERSEE, Beijing, China) with air-acetylene, while concentrations of Cd, Cr and Pb were determined by an inductively coupled plasma spectrometer (ICAP-7200, Thermo Fisher Scientific, Waltham, MA, USA). The procedures of soil and plant chemical analysis followed in this study are described in detail by Bao [14].

2.3. Data Analysis

The bioaccumulation factor (BF) is used to evaluate capability to absorb a specific HM element from soils by a plant [15,16], while the translocation factor (TF) is a measure of the internal mobility of a given HM element across plant organs [17]. They are calculated by the following equations:

$$BF = C_{underground}/C_{soil} \tag{1}$$

$$TF_{stem/underground} = C_{stem}/C_{underground} \tag{2}$$

$$TF_{leaf/underground} = C_{leaf}/C_{underground} \tag{3}$$

$$TF_{leaf/stem} = C_{leaf}/C_{stem} \tag{4}$$

where C_{soil}, C_{stem}, C_{leaf} and $C_{underground}$ are the concentrations (mg kg^{-1} DW) of a given HM element in soils, stems, leaves and underground organs (rhizomes and roots), respectively.

For each HM element, its total amount of element bioaccumulation (TAB) by each plant part (stem, leaf, rhizome and root) was calculated based on the following Equation (5):

$$TAB_{HM} = \sum^{organ=3} BW_{organ} \times C_{organ} \tag{5}$$

where BW_{organ} is the dry weight of the stems, leaves and underground organs (rhizomes and roots), which was calculated based on fresh weight and their corresponding water content; C_{organ} is the HM concentrations in the corresponding organs of stems, leaves and underground organs.

The statistical analysis was based on log-transformed data. A one-way ANOVA followed by pairwise t-tests was used to assess the effects of sampling location on soil chemical properties and HM concentrations. The significances of variance in terms of HM concentration in biomass among different sampling months and plant organs were determined using a one-way ANOVA followed by Duncan's post hoc test. The significances of variation in terms of soil chemical properties, HM concentrations in soils and biomass among different sampling months, sites and plant organs were evaluated using a two-way ANOVA. Pearson's coefficient analysis was used to reveal relationships between soil chemical properties and HM absorption-translocation capability and changes of HM concentrations in plant organs over growing seasons. Statistical analysis was performed using the statistical software IBM SPSS Version 22.0 (IBM, Armonk, NY, USA).

3. Results

3.1. Variation in Soil Chemical Properties and HM Concentrations across 11 Sampling Sites within the Dongting Lake Region

Results presented in Table 1 showed the variation in soil chemical properties and HM concentrations among the 11 sampling sites. The data supported that there was less variation in soil chemical properties than HM concentrations. The average variation coefficient of the eight soil chemical properties was 38.93% (10.68–69.82%), while it was 43.41% (22.18–95.17%) of the HM concentrations. Of the eight soil chemical properties, different sites had significant differences in terms of TN (p = 0.001), TP (p = 0.047), AP (p = 0.026), AK (p = 0.023) and SOM (p = 0.016). There was a general trend that the sites in the southern Dongting were characterized to have higher TN (0.10 vs. 0.06%), TP (0.07 vs. 0.06%), AP (26.67 vs. 16.99 mg kg^{-1}), AK (172.55 vs. 112.17 mg kg^{-1}) and SOM (2.72 vs. 2.11%) than that in the eastern Dongting. This trend was stable across different sampling months as indicated by the nonsignificant (p > 0.05) effect of 'Month' and the interaction of 'Month × Site'. Also, the southern Dongting sites generally had more serious HM pollution problems than the eastern Dongting sites. For example, the average Mn concentration of the southern Dongting sites (368.38 mg kg^{-1}) was 1.63-fold higher (p = 0.001) than that of the eastern Dongting sites (226.65 mg kg^{-1}). The significant differences between sampling sites in terms of Zn, Cd, Cr and Pb concentrations were recorded as indicated by their p values of <0.001. The variation coefficient of soil Cd concentration was the highest (95.17%), followed by soil Mn (37.89%), Cu (36.58%), Pb (34.47%), Cr (23.56%) and Zn (22.18%). Soil from Dongting Lake had 9.43-fold higher concentration of Cd than the national standards of China (GB15618-2018) for HM contamination. Although soil Cu concentration was still high in the southern Dongting sites, these differences were not significant (p = 0.346). The above results analysis suggest that the southern Dongting area is characterized by more fertile and seriously polluted soil than the eastern Dongting area.

Table 1. Soil chemical properties and heavy metal concentrations in the study area.

	Parameter [a]	Location [b]		Statistic		Sources of Variance in ANOVA (p-Value)		
		Southern Dongting	Eastern Dongting	Mean	Variation Coefficient (%)	Month	Site	Month × Site
Chemical properties	pH	7.34	7.62	7.47	10.68	0.106	0.350	0.131
	EC (μs cm^{-1})	425.83	367.89	399.49	36.63	0.961	0.487	0.994
	TN (%)	0.10A	0.06B	0.08	37.70	0.984	0.001	0.997
	TP (%)	0.07a	0.06b	0.06	28.45	0.668	0.047	0.378
	TK (%)	1.12	1.05	1.09	34.24	0.397	0.435	0.826
	AP (mg kg^{-1})	26.67a	16.99b	22.27	69.82	0.993	0.026	0.680
	AK (mg kg^{-1})	172.55a	112.17b	145.11	61.03	0.341	0.023	0.340
	SOM (%)	2.72a	2.11b	2.44	32.85	0.339	0.016	0.094
Heavy metal concentration (mg kg^{-1})	Cu	32.14	29.94	31.14	36.58	0.995	0.346	0.995
	Mn	368.38A	226.65B	303.96	37.89	0.998	0.001	0.997
	Zn	190.77A	141.44B	168.35	22.18	0.882	<0.001	0.867
	Cd	4.06A	1.37B	2.83	95.17	0.715	<0.001	0.903
	Cr	78.00A	57.89B	68.86	23.56	0.997	<0.001	0.999
	Pb	78.87A	51.21B	66.30	45.10	0.245	<0.001	0.730

[a] pH—soil pH; EC—soil electrical conductivity; TN—soil total nitrogen content; TP—soil total phosphorous content; TK—soil total potassium content; AP—soil available phosphorous content; AK—soil available potassium content; SOM—soil organic matter content. [b] Values are means of three sampling times (April, August and December) in 2018. Different lowercase and uppercase letters within the same row indicate significant differences at p < 0.05 and at p < 0.01 for the same indicators, respectively.

3.2. Seasonal Changes in HM Concentrations in Different Plant Organs

The pattern of the HM concentrations in the plant organs over the growing season was complex and varied for each HM element (Figure 2). Except for Zn and Pb, there was a general trend that HM concentrations in the underground organs decreased toward the end of the growing season. This was particularly true for Mn (Figure 2d) with significant changes in concentrations in the underground organs. However, a general opposite trend of dynamics through the growing seasons was observed for the HM concentrations in leaves and stems. This was especially true for Mn (Figure 2d) and Zn (Figure 2f) with a significant decrease in stems and a significant increase in leaves over the growing season. For example, the Mn concentrations in stems in April (49.66 mg kg^{-1}) were 33.7% and 49.7%, higher than that in August (32.91 mg kg^{-1}) and December (25.00 mg kg^{-1}), respectively. The Mn concentrations in leaves increased from 34.24 mg kg^{-1} in April to 39.74 mg kg^{-1} in August and then finally reached to 75.97 mg kg^{-1} in December. At the end of growth

(December), there was a significant increase in the HM concentrations in leaves, especially for Cd (Figure 2b), Mn (Figure 2d), Pb (Figure 2e) and Zn (Figure 2f).

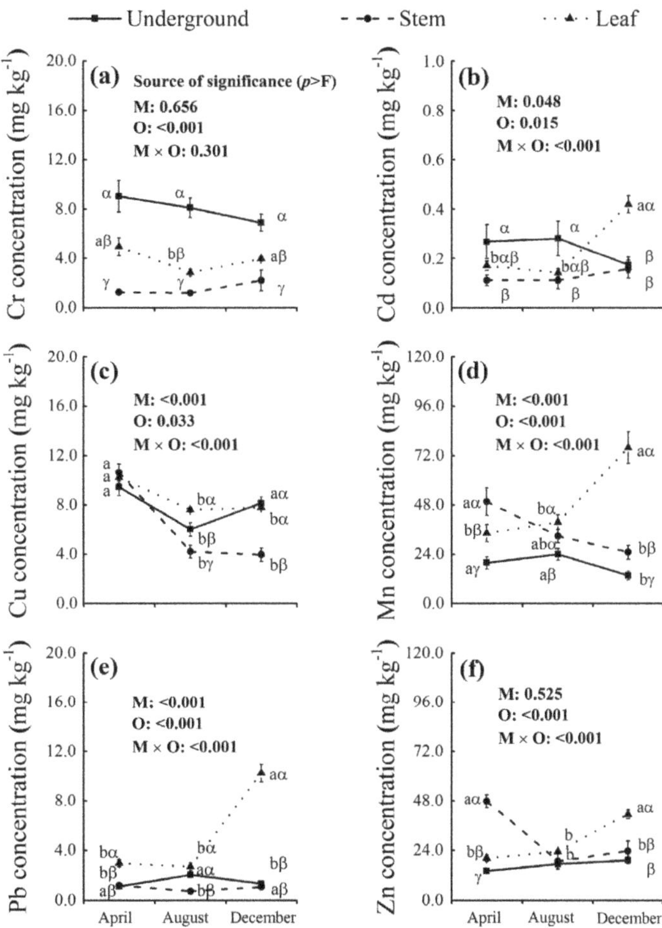

Figure 2. Seasonal changes of heavy metal concentrations in the underground organs, stems and leaves. M and O and represent sampling months and plant organs. Bars indicate standard errors of means. Different English letters indicate significant differences among the three sampling months at $p < 0.05$. Different Greek letters indicate significant differences among the underground organs, stems and leaves at $p < 0.05$.

April is the harvest time of aboveground young shoots for pickle production and December is the harvest time for mature biomass (Table 2). Based on dry weight, young shoots from only two sites (S5 = 0.05 mg kg^{-1}, S9 = 0.04 mg kg^{-1}) met the food safety standard (<0.05 mg kg^{-1}) of China (GB 2762–2017) for Cd concentration (Table 2). Furthermore, the young shoots in all sites had two- to fourfold higher Cr concentrations (0.84–1.78 mg kg^{-1}) and nine- to nineteenfold higher Pb concentrations (0.86–1.85 mg kg^{-1}) than the standard limitation (0.5 mg kg^{-1} for Cr, 0.1 mg kg^{-1} for Pb) (Table 2). The young shoots harvested during April are generally used to produce pickles or eaten freshly. Therefore, it is more practical to assess the quality of the April-harvested biomass based on the fresh weight instead of dry weight. Based on the fresh weight (converted by 85% water content), the young shoots from all the sampling sites met the food safety standards for Cd and Cr, but not for Pb. The fresh young shoots had varied Pb concentrations (0.13–0.28 mg kg^{-1}), which was still one- to threefold higher than the standard limits. For the mature biomass harvested in December, the concentrations of different HM elements were shown in a descending order as Mn (36.32 mg kg^{-1}), Zn (27.89 mg kg^{-1}), Cu (4.92 mg kg^{-1}), Pb (3.18 mg kg^{-1}), Cr (2.74 mg kg^{-1}) and Cd (0.22 mg kg^{-1}) (Table 2). In comparison with eastern Dongting Lake, the mature biomass in southern Dongting Lake was generally characterized to have higher Mn, Cd and Pb concentrations (Table 2). For example, the average Mn concentration in the harvested biomass was 40.73 mg kg^{-1} in southern Dongting Lake, which was 21.7% higher than that of eastern Dongting Lake (31.91 mg kg^{-1}). In addition, the produced biomass in southern Dongting Lake had a higher Cd concentration than that of eastern Dongting Lake by 9.1% (0.22 vs. 0.2 mg kg^{-1}) (Table 2). The average Pb concentration (3.21 mg kg^{-1}) in the harvested biomass from southern Dongting Lake was slightly higher (1.6%) than that in eastern Dongting Lake (3.16 mg kg^{-1}) (Table 2). A different trend was found in terms of Cu, Zn and Cr concentrations. In particular, biomass in eastern Dongting Lake had higher Cu, Zn and Cr concentrations than that in southern Dongting Lake by 28.1% (5.73 vs. 4.12 mg kg^{-1}), 6.9% (28.89 vs. 26.89 mg kg^{-1}) and 35.2% (3.32 vs. 2.15 mg kg^{-1}), respectively (Table 2).

Table 2. Heavy metal concentration of the April-harvested stems and December-harvested above-ground biomass (stems and leaves together) across the 11 sampling sites.

Sampling Sites	April (mg kg^{-1})						December (mg kg^{-1})					
	Cu	Mn	Zn	Cd	Cr	Pb	Cu	Mn	Zn	Cd	Cr	Pb
S1	7.64	28.36	37.09	0.13	1.20	1.13	4.02	38.15	33.58	0.26	1.45	3.34
S2	7.82	92.72	45.80	0.12	0.84	0.95	3.55	56.29	31.87	0.17	1.53	3.43
S3	10.96	76.75	62.50	0.31	1.46	1.02	3.80	42.13	26.72	0.36	4.77	2.93
S4	11.13	42.29	46.74	0.07	0.91	0.91	4.62	13.13	15.33	0.12	1.71	2.48
S5	10.84	29.28	49.88	0.05	1.14	0.86	3.87	43.66	19.66	0.12	1.52	2.49
S6	10.95	42.70	50.36	0.09	1.49	1.20	4.85	51.02	34.17	0.33	1.94	4.58
S7	n/a	n/a	n/a	n/a	n/a	n/a	4.59	35.33	17.35	n/a	n/a	n/a
S8	12.25	50.12	54.58	0.09	1.29	1.85	5.15	46.10	33.10	0.12	1.73	2.89
S9	7.69	60.44	26.37	0.04	1.25	1.38	5.05	22.07	14.99	0.14	7.85	2.81
S10	13.06	22.85	44.61	0.13	1.78	1.65	5.06	31.15	20.95	0.19	1.95	3.55
S11	14.13	51.08	60.86	0.08	1.17	1.03	8.78	24.92	58.08	0.38	1.74	3.38
Southern Dongting	9.89	52.02	48.73	0.13	1.17	1.01	4.12	40.73	26.89	0.22	2.15	3.21
Eastern Dongting	11.78	46.12	46.61	0.09	1.37	1.48	5.73	31.91	28.89	0.20	3.32	3.16
Overall mean	10.65	49.66	47.88	0.11	1.25	1.20	4.92	36.32	27.89	0.21	2.74	3.18

Note: n/a represents the data unavailable because of the missing samplings. Southern Dongting includes sampling sites S1–S6 and eastern Dongting includes sampling sites S7–S11.

3.3. Adsorption and Translocation Characteristics of HM Elements in Different Organs

The average TF values of all HM elements were higher than that of BF during the whole growth season (Figure 3). Among the three sampling dates, the BF$_{underground/soil}$ value of Cu was higher than that of the other HM elements (Figure 3a). It is particularly true for the BF$_{underground/soil}$ value (0.32) of Cu in April, suggesting a strong Cu absorption ability of the underground organs during this time period. The BF$_{underground/soil}$ value of Pb (0.016–0.037) was always the lowest among all the HM elements. For all the HM elements except Cu, no variations were observed between different sampling dates in terms of the BF$_{underground/soil}$ value. More HM elements absorbed by the underground organs were transformed to leaves than stems. In April, a higher TF$_{leaf/underground}$ value (Figure 3c) was observed for the most HM elements than the TF$_{stem/underground}$ value (Figure 3b). For example, the TF$_{leaf/underground}$ values for Pb, Cu, Cd and Cr in April reached to 5.46, 1.16, 1.04 and 0.76, respectively, whereas values for TF$_{stem/underground}$ were 1.82, 1.15, 0.56 and 0.17, respectively. Toward the end of the growing season, more HM elements were transformed to leaves than stems. The average TF$_{leaf/underground}$ value for all the HM elements was 2.36, whereas for the TF$_{stem/underground}$ it was only 1.13. At the end of the growing season (December), the highest TF$_{leaf/stem}$ value (10.16) for Pb was recorded.

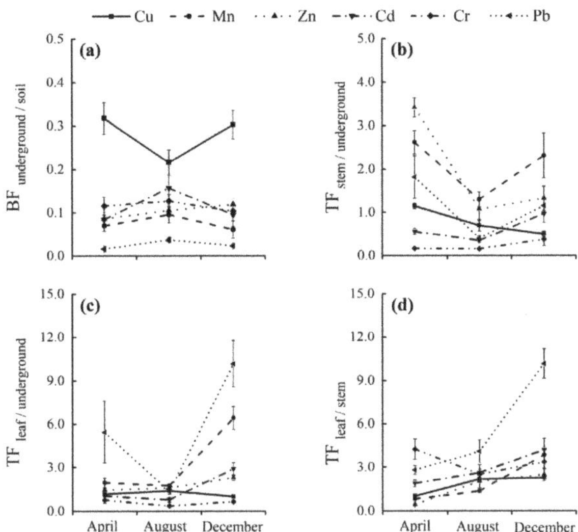

Figure 3. The bioaccumulation factor (BF) and translocation factor (TF) in the different parts (underground organs, stems and leaves) of *Miscanthus lutarioriparius* for Cu, Mn, Zn, Cd, Cr and Pb across different sampling time. Bars indicate standard errors of means.

3.4. Factors Contributing to the Variation in HM Absorption and Translocation Capabilities of M. lutarioriparius

Pearson's analysis results presented the contributions of soil chemical properties and HM absorption-translocation capability on the changes in HM concentrations in different plant organs over different harvest times (Figure 4). For the effects of the initial soil HM concentrations, the Cd concentrations in the underground organs were positively correlated with the soil Zn (r = 0.62, $p < 0.01$), Cd (r = 0.72, $p < 0.001$) and Cr (r = 0.50, $p < 0.05$) concentrations. The Mn and Pb concentrations in the underground organs positively (r = 0.56–0.60, $p < 0.05$) correlated to soil Cd concentration. The Mn, Zn, Cr and Pb concentrations in the underground organs were also found to be positively (r = 0.52–0.88, $p < 0.05$) associated with their bioaccumulation factors. However, to assess the contribution of the translocation factor, the Mn, Cd, Cr and Pb concentrations in the underground organs were negatively (r = 0.65–0.73, $p < 0.01$) correlated with the $TF_{leaf/underground}$ of the corresponding HM elements. The Cu, Mn, Zn, Cr and Pb concentrations in stems were positively (r = 0.53–0.91, $p < 0.05$) associated with the $TF_{stem/underground}$ of the corresponding HM elements. The concentrations of the HM elements except Pb in stems showed negative (r = 0.70–0.81, $p < 0.001$) relationships with the $TF_{leaf/stem}$ of the corresponding HM elements. The Mn, Zn, Cr and Pb concentrations in leaves were positively correlated with the translocation capability of the corresponding HM elements from the underground organs (r = 0.60–0.78, $p < 0.01$) and stems (r = 0.53–0.87, $p < 0.05$) to leaves. Furthermore, the Zn and Cd concentrations in leaves were positively correlated with the translocation capability of Mn (r = 0.62–0.71, $p < 0.01$) and Pb (r = 0.73–0.79, $p < 0.001$) from stems to leaves.

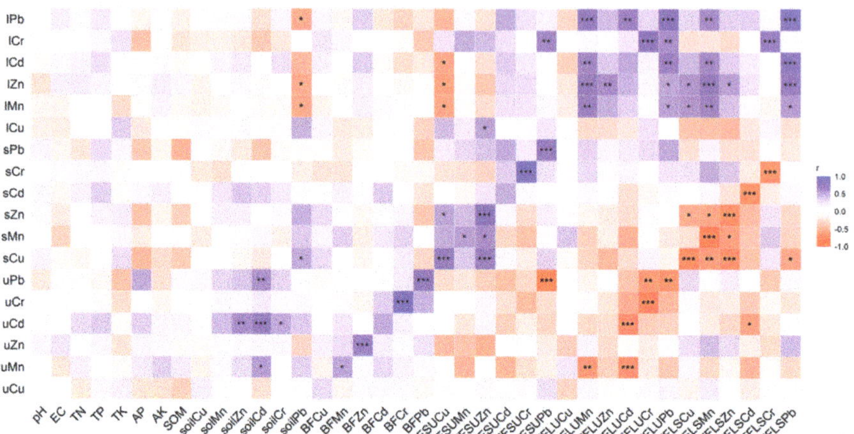

Figure 4. Pearson's coefficient analysis of heavy metal concentrations in different plant organs, soil chemical properties and heavy metal absorption-translocation capability. The first letters of the vertical axis indicate: 'l' represent plant leaves, 's' indicates stems and 'u' for underground organs followed by the chemical symbol of each heavy metal. In the horizontal axis, pH—soil pH; EC—soil electrical conductivity; TN—soil total nitrogen; TP—soil total phosphorous; TK—soil total potassium; AP—soil available phosphorous; AK—soil available potassium; SOM—soil organic matter. Moreover, the indicators of "BF + heavy metal chemical symbol" represent the bioaccumulation factor of the corresponding heavy metal. TFSUCu, TFSUMn, TFSUZn, TFSUCd, TFSUCr and TFSUPb represent translocation factor for Cu, Mn, Zn, Cd, Cr and Pb from underground organs to stems, respectively. It is same for TFLUCu, TFLUMn, TFLUZn, TFLUCd, TFLUCr, TFLUPb, TFLSCu, TFLSMn, TFLSZn, TFLSCd, TFLSCr and TFLSPb, where LU and LS represent translocation from the underground organs to leaves, and stems to leaves, respectively. Asterisks indicate significance: * $p < 0.05$, ** $p < 0.01$, *** $p < 0.001$.

4. Discussion

4.1. Factors Affecting the Seasonal Dynamics of HM Concentrations in M. lutarioriparius Biomass

Miscanthus, a rhizomatous perennial herbal species, is characterized by the translocation of nutrients back to rhizomes before harvest to be used in next growth season [2]. Cu, Mn and Zn, as micro-nutrients, participate in plant growth and metabolism [17]. This can explain that why more Cu, Mn and Zn were transferred from the underground organs to stems and leaves at the seedling stage (Figure 2). For a compartmentalization tolerance strategy [16], nonessential phytotoxic elements Cd and Cr were mainly stored in the underground organs (rhizomes and roots) (Figure 2) and enabled *M. lutarioriparius* to tolerate a high level of HM contamination. During the growth season, substantial amounts of essential nutrients are consistently taken up by the plant for photosynthesis, growth and reproduction [13], which lead to a co-transfer of HM elements to aboveground organs [18]. At the end of the growth season, unlike the macro-nutrients of N, P and K, the HM elements (especially for Zn, Mn, Cd and Pb) in the aboveground were not re-transferred to the underground but accumulated in leaves of *M. lutarioriparius* (Figure 2). These HM elements in leaves may be stored in the cell wall structure and retained by secondary metabolites [19].

The outcomes of correlation analysis between soil HM concentration and plant HM tolerance ability indicate that there is no consistent trend. A few studies have reported that soil HM concentration is positively correlated with the HM uptake amount by miscanthus plants [20,21]. The difference in genetic make-up could be the leading factor in terms of defining the ability to absorb, accumulate and tolerate. In the present study, the Cd concentrations in the underground organs were positively correlated with the soil Cd concentration (Figure 4). Previous studies indicated that HM elements might be co-taken with other essential elements [18,22]. In contrast with this, our results showed that HM uptake by *M. lutarioriparius* had no significant correlations with soil nutrients, but significantly correlated to other soil HM elements (Figure 4). For example, the Mn and Pb concentrations in the underground organs positively correlated to the soil Cd concentration and the Cd concentrations in the underground organs positively correlated with the soil Zn and Cr concentrations. Overall, the uptake of a specific HM element by the underground organs of *M. lutarioriparius* can be promoted by other HM elements. More HM elements are transferred and stored in the leaves during the whole growth season.

4.2. Recommendations for the Utilization of M. lutarioriparius Biomass Based on the Seasonal Dynamics of HM Concentrations

In our study, the sampling sites suffer different extents of HM contamination. Most sampling sites have high soil HM concentrations by exceeding the sediment background values, which is consistent with most previous studies [23–25]. High translocation potential from the underground to the aboveground organs would pose serious food safety risk and also negatively affect other industrial applications. At the seedling phase (in April), there were relatively high stem–underground translocation potentials for Cd, Cr and Pb (Figure 3), which would increase their concentrations in plant shoots (Figure 2 and Table 2) and consequently influence pickle quality. For pickled shoot manufacturing, the Pb concentrations in fresh young stems across the sampling Dongting Lake areas slightly exceeded the limitation set by the food safety standard of China (GB 2762–2017). The young shoots are generally comprised of sheath and unelongated stems. For pickle production, only the unelongated stems are used. It is evident from the literature that a high HM concentration is observed in sheaths [26]. This indicates that the young shoots with slightly excessive Pb can still be used for pickle processing. The processed sheath and substandard young shoots can be used as superior raw materials for silage production, according to the China Hygienical Standard for Feed GB13078-2017 (1, 30 and 5 mg kg^{-1} set limitations for Cd, Pb and Cr, respectively). The mature *M. lutarioriparius* biomass produced in Dongting Lake can be used for biogas production because of its high cellulose content [7] and the biogas residues with HM elements can be further processed safely as biofertilizers [15]. According to the current standards for fertilizers NY/T 3618-2020, permissible values for Cd, Pb and Cr are 3, 50 and 150 mg kg^{-1}, respectively. The *M. lutarioriparius* biomass produced within the Dongting Lake area is safe to be exploited to produce biofertilizers [6,27]. However, one must be careful not to exploit such contaminated biomass as substrate for mushroom cultivation because edible fungi have a high ability to absorb HM elements, which can consequently pose serious health risks [28]. As no HM elements in the aboveground organs were transferred back to the underground organs of *M. lutarioriparius* at the end of growth (Figure 2), this suggests that a delayed harvest will not improve the biomass quality in terms of HM contents. In addition, the delayed harvest will reduce the biomass yield because of the foliage falling [13]. For these reasons, the optimal harvest time for *M. lutarioriparius* in Dongting Lake is in August because of the high biomass yield and relatively low HM contents.

4.3. Potential Ecological Risks Posed by the Unharvested M. lutarioriparius Biomass around the Dongting Lake

Although *M. lutarioriparius* is not a hyperaccumulator, a great amount of HM elements is removed annually by the aerial shoots with the harvested biomass. It is estimated that $2.02–2.64 \times 10^4$ kg Cu, $1.54–1.77 \times 10^5$ kg Mn, $1.02–1.19 \times 10^5$ kg Zn, $0.56–0.84 \times 10^3$ kg Cd, $0.79–1.05 \times 10^4$ kg Cr and $0.62–1.24 \times 10^4$ kg Pb can be absorbed by the plants annually and removed from the soil (Table S1). These potentials are achieved by harvesting the *M. lutarioriparius* biomass and using it to produce bio-products (mainly making paper during last decades). However, since 2019, all the paper mills around Dongting Lake were closed because of the concerns that they are obstacles to improving the water quality of Dongting Lake. With the shutdown of the paper mills and a lack of cost-effective and environmentally friendly techniques, *M. lutarioriparius* biomass utilization is impeded. In addition, during recent years, a lack of management practices and harvesting of the *M. lutarioriparius* biomass have posed serious ecological threats, such as water eutrophication risks [29]. The unharvested biomass would submerge into the water and become a source of soil and water contamination from decomposition and the leaching of biomass constituents. In addition, carbon sequestered by the *M. lutarioriparius* biomass would be released into the atmosphere in the form of greenhouse gases (CO_2 and CH_4) because of microbial decomposition under aerobic and anaerobic conditions, consequently reducing the carbon sequestration potential of the Dongting Lake wetland ecosystem [30,31]. Dead plant biomass releases large amounts of dissolved organic resources into waters and affects microbial communities' diversity [32]. From this it can be concluded that the unharvested *M. lutarioriparius* biomass will induce a series of ecological issues in the Dongting Lake wetland. Thus, cost-effective and environmentally friendly techniques are needed to expand the applications of the miscanthus biomass.

5. Conclusions

Across the Dongting Lake area, the soil is facing a severe Cd contamination problem. The heavily polluted soil finally results in a high HM concentration of the *M. lutarioriparius* biomass. The HM concentrations in the aboveground organs, except Mn and Zn in stems, do not reduce but keep increasing with the ongoing growing season. This is particularly true for the HM concentrations in leaves. The young shoots harvested in April can meet the food safety standard limitation of Cd and Cr, but not Pb. For the mature biomass harvested in December, the concentrations of different kinds of HM elements show in a descending order as Mn, Zn, Cu, Pb, Cr and Cd. For differences between southern and eastern Dongting Lake, the mature biomass from the southern part generally has higher Mn, Cd and Pb, but lower Cu, Zn and Cr concentrations than that from the eastern part.

Supplementary Materials: The following are available online at https://www.mdpi.com/article/10.3390/agronomy12051188/s1, Table S1: Average amounts of heavy metal bioaccumulated by *M. lutarioriparius* at different harvest time.

Author Contributions: X.L.: investigation, formal analysis, writing—original draft preparation; Y.W.: formal analysis, visualization, writing—original draft preparation; T.F.: writing—review and editing; Y.I.: writing—review and editing; S.Y.: investigation; M.L.: data curation; Z.Y.: methodology, supervision; S.X.: conceptualization, supervision, project administration, writing—review and editing. All authors have read and agreed to the published version of the manuscript.

Funding: This study was financially supported by the National Natural Science Foundation of China (32000259), the Open Foundation of Hunan Provincial Key Laboratory for Crop Germplasm Innovation and Utilization (19KFXM04) and the Foundation for the Construction of Innovative Hunan (2019RS1051 and 2019NK2021).

Data Availability Statement: Not applicable.

Acknowledgments: We appreciate Mengqi Guo for his help in collecting the plant samples.

Conflicts of Interest: The authors declare no conflict of interest.

References

1. Xu, Y.; Wu, S.; Huang, F.; Huang, H.; Yi, Z.; Xue, S. Biomodification of feedstock for quality-improved biochar: A green method to enhance the Cd sorption capacity of *Miscanthus lutarioriparius*-derived biochar. *J. Clean. Prod.* **2022**, *350*, 131241. [CrossRef]
2. Xue, S.; Kalinina, O.; Lewandowski, I. Present and future options for *Miscanthus* propagation and establishment. *Renew. Sustain. Energy Rev.* **2015**, *49*, 1233–1246. [CrossRef]
3. Yao, X.; Niu, Y.; Li, Y.; Zou, D.; Ding, X.; Bian, H. Heavy metal bioaccumulation by *Miscanthus sacchariflorus* and its potential for removing metals from the Dongting Lake wetlands, China. *Environ. Sci. Pollut. Res.* **2018**, *25*, 20003–20011. [CrossRef]
4. Xue, S.; Yi, Z.; Huang, H.; Yang, S. Development of the Dongting Lake grown *Miscanthus lutarioriparius* derived bio-industries: Problems and solutions. *Environ. Ecol.* **2022**, *4*, 33–38.
5. Xue, S.; Lewandowski, I.; Wang, X.; Yi, Z. Assessment of the production potentials of *Miscanthus* on marginal land in China. *Renew. Sustain. Energy. Rev.* **2016**, *54*, 932–943. [CrossRef]

6.   Liao, X.; Long, Q.; Wang, H.; Yi, Z.; Xue, S. Interaction effects of *Miscanthus lutarioriparius*-derived biochar and cadmium passivators on rice cadmium content and yield. *J. Agro-Environ. Sci.* **2018**, *37*, 1818–1826.

7.   Xue, S.; Guo, M.; Iqbal, Y.; Liao, J.; Yang, S.; Xiao, L.; Yi, Z. Mapping current distribution and genetic diversity of the native *Miscanthus lutarioriparius* across China. *Renew. Sustain. Energy. Rev.* **2020**, *134*, 110386. [CrossRef]

8.   Zheng, C.; Iqbal, Y.; Labonte, N.; Sun, G.; Feng, H.; Yi, Z.; Xiao, L. Performance of switchgrass and *Miscanthus* genotypes on marginal land in the Yellow River Delta. *Ind. Crops Prod.* **2019**, *141*, 111773. [CrossRef]

9.   Pidlisnyuk, V.; Stefanovska, T.; Lewis, E.E.; Erickson, L.E.; Davis, L.C. *Miscanthus* as a productive biofuel crop for phytoremediation. *Crit. Rev. Plant Sci.* **2014**, *33*, 1–19. [CrossRef]

10.   Alexopoulou, E. *Perennial Grasses for Bioenergy and Bioproducts: Production, Uses, Sustainability and Markets for Giant Reed, Miscanthus, Switchgrass, Reed Canary Grass and Bamboo*; Academic Press: Cambridge, MA, USA, 2018.

11.   Liu, J.; Liang, J.; Yuan, X.; Zeng, G.; Yuan, Y.; Wu, H.; Huang, X.; Liu, J.; Hua, S.; Li, F.; et al. An integrated model for assessing heavy metal exposure risk to migratory birds in wetland ecosystem: A case study in Dongting Lake Wetland, China. *Chemosphere* **2015**, *135*, 14–19. [CrossRef]

12.   Wang, C.; Kong, Y.; Hu, R.; Zhou, G. *Miscanthus*: A fast-growing crop for environmental remediation and biofuel production. *GCB Bioenergy* **2021**, *13*, 58–69. [CrossRef]

13.   Yang, S. Research on Genetic Diversity and Cultivation Techniques for High Biomass of *Miscanthus lutarioriparius*. Doctoral Dissertation, Hunan Agricultural University: Changsha, China, 2019.

14.   Bao, S.D. *Soil Agrochemical Analysis*, 3rd ed.; China Agricultural Press: Beijing, China, 2000.

15.   Ali, H.; Khan, E.; Sajad, M.A. Phytoremediation of heavy metals—Concepts and applications. *Chemosphere* **2013**, *91*, 869–881. [CrossRef] [PubMed]

16.   Bonanno, G.; Vymazal, J.; Cirelli, G.L. Translocation, accumulation and bioindication of trace elements in wetland plants. *Sci. Total Environ.* **2018**, *631–632*, 252–261. [CrossRef] [PubMed]

17.   Bonanno, G.; Cirelli, G.L. Comparative analysis of element concentrations and translocation in three wetland congener plants: *Typha domingensis*, *Typha latifolia* and *Typha angustifolia*. *Ecotoxicol. Environ. Saf.* **2017**, *143*, 92–101. [CrossRef] [PubMed]

18.   Ghori, Z.; Iftikhar, H.; Bhatti, M.F.; Nasarum, M.; Sharma, I.; Kazi, A.G.; Ahmad, P. Chapter 15—Phytoextraction: The Use of Plants to Remove Heavy Metals from Soil. In *Plant Metal Interaction*; Ahmad, P., Ed.; Elsevier: Amsterdam, The Netherlands, 2016; pp. 385–409.

19.   Krzesłowska, M. The cell wall in plant cell response to trace metals: Polysaccharide remodeling and its role in defense strategy. *Acta Physiol. Plant.* **2011**, *33*, 35–51. [CrossRef]

20.   Adamczyk-Szabela, D.; Markiewicz, J.; Wolf, W.M. Heavy metal uptake by herbs. IV. influence of soil pH on the content of heavy metals in *Valeriana officinalis* L. *Water Air Soil Pollut.* **2015**, *226*, 106. [CrossRef]

21.   Antonkiewicz, J.; Kołodziej, B.; Bielińska, E.J. The use of reed canary grass and giant miscanthus in the phytoremediation of municipal sewage sludge. *Environ. Sci. Pollut. Res.* **2016**, *23*, 9505–9517. [CrossRef]

22.   van der Ent, A.; Baker, A.J.M.; Reeves, R.D.; Pollard, A.J.; Schat, H. Hyperaccumulators of metal and metalloid trace elements: Facts and fiction. *Plant Soil* **2013**, *362*, 319–334. [CrossRef]

23.   Li, F.; Huang, J.; Zeng, G.; Yuan, X.; Li, X.; Liang, J.; Wang, X.; Tang, X.; Bai, B. Spatial risk assessment and sources identification of heavy metals in surface sediments from the Dongting Lake, Middle China. *J. Geochem. Explor.* **2013**, *132*, 75–83. [CrossRef]

24.   Liang, J.; Liu, J.; Yuan, X.; Zeng, G.; Lai, X.; Li, X.; Wu, H.; Yuan, Y.; Li, F. Spatial and temporal variation of heavy metal risk and source in sediments of Dongting Lake wetland, mid-south China. *J. Environ. Sci. Health A* **2015**, *50*, 100–108. [CrossRef]

25.   Peng, D.; Liu, Z.; Su, X.; Xiao, Y.; Wang, Y.; Middleton, B.A.; Lei, T. Spatial distribution of heavy metals in the West Dongting Lake floodplain, China. *Environ. Sci.-Proc. Impacts* **2020**, *22*, 1256–1265. [CrossRef] [PubMed]

26.   Xiao, J. Study on Flavonoids and Analyse on Ingredients of *Triarrhena lutarioriparia* L. Liu in the Dongting Lake. Master's Thesis, Hunan Agricultural University, Changsha, China, 2016.

27.   Liao, X.; Zhou, X.; Cai, D.; Wang, H.; Yi, Z.; Xue, S. Effects of application of *Miscanthus lutarioriparius*-derived biochar based-soil conditioner on photosynthetic characteristics and yield of rice (*Oryza sativa* L.). *J. Agric. Sci. Technol.* **2019**, *21*, 132–139.

28.   Ab Rhaman, S.M.; Naher, L.; Siddiquee, S. Mushroom quality related with various substrates' bioaccumulation and translocation of heavy metals. *J. Fungi* **2022**, *8*, 42. [CrossRef] [PubMed]

29.   Geng, M.; Wang, K.; Yang, N.; Li, F.; Zou, Y.; Chen, X.; Deng, Z.; Xie, Y. Evaluation and variation trends analysis of water quality in response to water regime changes in a typical river-connected lake (Dongting Lake), China. *Environ. Pollut.* **2021**, *268*, 115761. [CrossRef]

30.   Kayranli, B.; Scholz, M.; Mustafa, A.; Hedmark, Å. Carbon storage and fluxes within freshwater wetlands: A critical review. *Wetlands* **2010**, *30*, 111–124. [CrossRef]

31.   Carmichael, M.J.; Helton, A.M.; White, J.C.; Smith, W.K. Standing dead trees are a conduit for the atmospheric flux of CH_4 and CO_2 from wetlands. *Wetlands* **2018**, *38*, 133–143. [CrossRef]

32.   Geng, M.; Zhang, W.; Hu, T.; Wang, R.; Cheng, X.; Wang, J. Eutrophication causes microbial community homogenization via modulating generalist species. *Water Res.* **2022**, *210*, 118003. [CrossRef]

agronomy

MDPI

Article

Sida hermaphrodita Cultivation on Light Soil—A Closer Look at Fertilization and Sowing Density

Teodor Kitczak, Grzegorz Jarnuszewski *, Elżbieta Łazar and Ryszard Malinowski

Department of Environmental Management, Faculty of Environmental Management and Agriculture, West Pomeranian University of Technology in Szczecin, 70-310 Szczecin, Poland
* Correspondence: gjarnuszewski@zut.edu.pl; Tel.: +48-91-449-64-10

Abstract: *Sida hermaphrodita* (L.) Rusby is a promising perennial biomass crop to provide sustainable bioenergy via combustion. This study investigated cultivation practices for *Sida hermaphrodita* (L.) Rusby on light soils in temperate climates. Therefore, two cultivation factors were varied over 8 years in a field trial: (i) fertilization with compost from urban green spaces (0, 10 and 20 t ha^{-1}), and (ii) seeding amount (1, 2 and 3 kg ha^{-1}). Compost fertilization and high seeding amount contributed to an increase in the number and height of *Sida* shoots while their thickness decreased. The applied compost fertilization increased the dry matter yield (DMY) of the plants by 24.9% and 50.7%, respectively, in all experimental years compared to the control. Compared to the lowest seeding rate, increasing the seeding rate to 2 and 3 kg ha^{-1} increased the DMY by 35.0% and 71.6%, respectively. Thus, the highest energy value of DMY of *Sida hermaphrodita* plants per unit area was also obtained for combining the highest organic compost fertilization and seeding strength. From this, it can be deduced that on light soils, it does not seem reasonable to choose a compost fertilizer rate below 20 kg ha^{-1} and a seeding amount below 3 kg ha^{-1}.

Keywords: bioenergy; biomass yield; low-input cultivation; perennial crop; sustainable agricultural intensification

Citation: Kitczak, T.; Jarnuszewski, G.; Łazar, E.; Malinowski, R. *Sida hermaphrodita* Cultivation on Light Soil—A Closer Look at Fertilization and Sowing Density. *Agronomy* **2022**, *12*, 2715. https://doi.org/10.3390/agronomy12112715

Academic Editors: Moritz Von Cossel, Joaquín Castro-Montoya and Yasir Iqbal

Received: 5 October 2022
Accepted: 29 October 2022
Published: 2 November 2022

Publisher's Note: MDPI stays neutral with regard to jurisdictional claims in published maps and institutional affiliations.

1. Introduction

The growing demand for energy and the need to protect the environment and the independence of man from the effects of limited resources of fossil fuels means that in the energy development strategies of many countries, especially the European Union, despite numerous discussions on economic profitability, more and more emphasis is placed on obtaining energy from renewable sources [1–3]. Obtaining energy in this way, according to many authors [4–9], results in increasing the energy security of a region (especially areas with underdeveloped energy infrastructure), the reduction of greenhouse gas emissions, gradual replacement of conventional fuels with renewable energy sources, revitalisation of the rural economy and economic recovery by creating new jobs and new technologies.

In Poland, due to the favourable climate and good geographical location and habitat conditions, biomass is considered as one of the essential components of future supplies for the bioeconomy [9–11]. In Poland, biomass for energy originates mainly from forest residues and the wood industry [12]. The continuous increases in energy demand and the implementation of EU commitments necessitate obtaining it from specially cultivated plants [12,13].

So far, the most popular among growers are perennial plants such as willow (*Salix viminalis*), poplar (*Populus* L.), sugar miscanthus (*Miscanthus sacchariflorus*) and giant miscanthus (*Miscanthus × giganteus*), as well as species containing sufficient amounts of oil and carbohydrates necessary to obtain energy carriers [14,15]. Little attention is paid to the cultivation of Virginia fanpetals (hereafter referred to as *Sida*) which can be established by sowing (most species introduced into Poland must be planted from prepared cuttings) and the use of compost from a waste of green areas for cultivating them as fertilizer. Currently,

Sida is grown in Central Europe, mainly in experiments in research centres and universities, and small areas for production purpose are found in Austria, Romania, Lithuania, and Hungary, as well as in Poland (approximately 300 ha) and Germany—approximately 100–150 ha [14,16].

Sida develops an extensive root system allowing access to water and nutrients and grows well on light soils (marginal lands) [8,10,17,18]. *Sida*'s biomass dry matter yields (DMYs) vary between 9 t ha^{-1} on light soils and 25 t ha^{-1} on rich soils [16,17]. Energy crops should be located on poor-quality soils to avoid competition with food production. In such conditions, biomass DMYs range from 10 t ha^{-1} to 12 t ha^{-1} [16,19–21] after the first two years of growth in various mineral fertilization variants. Generally, plants respond well to nitrogen fertilization, which significantly increases biomass yield, while phosphorus fertilization promotes better stem formation and yield, especially in poor soils [22,23]. Biomass dry matter production of *Sida* depends on several factors: soil quality, climatic conditions, cultivation, fertilization, plant density and establishment method (seeds, seedlings, root cuttings) [16,22]. Sowing seeds is the most cost-efficient method for growing *Sida* but the germination rate can be very low (even 5–15%) and slow. This unpredictable germination rate can cause weed growth and low biomass yield [16,21,22]. A useful number of seeds for *Sida* plantation establishment is considered to be approximately 200,000–300,000 seeds per hectare or 1 kg ha^{-1} [21,22]. As Cumplido-Marin [22] indicated on the basis of her research review, the number of seeds used ranged from 1.5 to even 9 kg ha^{-1}, and usually, a higher seedling amount results in higher yields. Seed preparation methods such as extension of the storage period, scarification of seeds, or pretreatment in hot water is used to increase the germination rate up to 90% [16,21,22]. In the case of energy crops grown in marginal soils (light soils), fertilization can substantially increase biomass DMY [19]. Research indicates that *Sida is* efficiently applicable using the nutrients from alternative sources such as sewage sludge [24], sludge compost [25], and digestates from anaerobic digestion [17,23,26,27]. Research conducted in Germany by Veste et al. [28] and in Poland by Ociepa-Kubicka and Pachura [29] on the influence of differentiated fertilization on the DMY of *Sida* indicates the possibility of using a combination of different doses of compost and mineral fertilization, which significantly influences the increase in DMY. However, few studies report a period longer than four years of cultivation of *Sida*, which is vital for plants with the highest DMY after the second year of use and alternative sources of nutrients in the fertilization of poor-quality soils intended for energy crops.

The work aimed to determine the effect of organic fertilization with urban green compost and mineral NPK fertilization on the DMY of *Sida* cultivated under varying sowing densities for energy purposes in light soil conditions. The underlying research question was whether the establishment via sowing is suitable for cultivating *Sida* on light soils.

2. Materials and Methods

2.1. Study Site and Experimental Design

A *Sida* field trial was established in the West Pomeranian region of Poland, in Lipnik, near Stargard (N 53°20′35.8″, E 14°98′10.8″), in 2009, using the random sub-block (split-plot) method in triplicate with a single plot area of 12 m^2. *Sida* was sown on 5 May 2009 with a row seeder with an inter-row spacing of 0.5 m in the amount assumed in the second study factor. The seeds came from another plantation of energy crops in Poland (West Pomeranian voivodeship), and the seed germination capacity before sowing was 49%.

The soil in the experiment was made of loamy sand, and, according to the WRB, represented Haplic Luvisols (Humic) [30].The content of total macroelement forms in the soil before the experiment was as follows: C—8.10 g kg^{-1} DM, N—0.92 g kg^{-1} DM, S—0.02 g kg^{-1} DM, P—0.45 g kg^{-1} DM, K—0.66 g kg^{-1} DM, Mg—0.91 g kg^{-1} DM. The soil reaction was acid (pH$_{KCl}$—5.2). The observation period ranged from years 2009 to 2016. The experiment was established in a post after oats were harvested for seeds. Before planting, the following agrotechnical operations were carried out: post-harvest tillage and medium ploughing. In the spring, before sowing, compost fertilization mixed with the soil was applied in the amount assumed in the first factor of the study. The compost was made from waste from the care of urban greenery in Szczecin, which included: a green mass of cut plants, leaves (of trees and shrubs), cones and other plant waste. Immediately before sowing the plants in the first year of the experiment, mineral fertilization with phosphorus and potassium in 80 kg ha^{-1} and 100 kg ha^{-1} was applied to all the study objects, and in the full years of cultivation, fertilization with phosphorus and potassium was applied before starting the plant vegetation Phosphorus was used as 19% superphosphate (phosphorus in the form of P_2O_5) and potassium as 60% potassium salt (potassium in the form of K_2O). Nitrogen fertilization in the form of ammonium nitrate was used in the amount of 100 kg ha^{-1} (every year of cultivation) in the year of establishment in two equal doses (the first month after sowing the plants, while the second one was used six weeks later), and in the years of full use, once in spring, before starting the plant vegetation. The tested factors included: factor I—compost doses: 0, 10 and 20 t ha^{-1} of dry matter and factor II—seeding amount: 1, 2 and 3 kg ha^{-1}.

2.2. Physicochemical Properties of the Compost

Basic physicochemical parameters were determined in the compost. The compost reaction (pH in 1M KCl and in H_2O) was determined potentiometrically, and the specific electrical conductivity was performed conductometrically in a water suspension. Total carbon, nitrogen and sulphur content were determined by means of elementary analyser COSTECH ECS 4010. The content of macroelements (P, K, Mg, Na) and microelements (Fe, Mn, Cr, Zn, Cd), soluble in the mixture of concentrated acids $HNO_3 + HClO_4$, was determined by compost mineralisation in this mixture, using atomic absorption spectrophotometer Unicam Solaar 929. Phosphorus was obtained colorimetrically. From the analysis of the chemical composition of the compost, it should be stated that the compost reaction was neutral, the total carbon, total nitrogen and magnesium content was low, and phosphorus, potassium, sulphur, calcium, and sodium were high (Table 1). The contents of trace elements in the compost (Cu, Fe, Mn, Cr, and Zn) did not exceed their permissible amounts adopted in the industry standard (BN–89/9103–09).

Table 1. Physicochemical properties of the compost used in this study and total doses of minerals brought in with compost fertilization.

Parameter	pH in 1M KCl	pH in H_2O	EC [1] μS cm^{-1}	Ctot [2] g kg^{-1}	Ntot [3]	C/N [4]	P	K	Ca	Mg	Na	S	Cu	Fe	Mn	Cr	Zn	Cd
									g kg^{-1}						mg kg^{-1}			
Value	6.78	7.08	624.10	142.01	9.52	14.93	1.95	3.54	34.973	2.91	0.34	0.74	26.39	8698.30	312.18	12.13	172.50	1.26

Total doses of minerals brought in with the compost in kg ha^{-1}

	N	P	K	Ca	Mg	Na	S	Cu	Fe	Mn	Cr	Zn	Cd
Compost doses 10 t ha^{-1}	95.2	19.5	35.4	349.73	29.1	3.4	7.4	0.26	86.98	3.12	0.12	1.73	0.01
Compost doses 20 t ha^{-1}	190.4	39	70.8	699.46	58.2	6.8	14.8	0.53	173.97	6.24	0.24	3.45	0.03

[1] electrical conductivity, [2] total carbon, [3] total nitrogen, [4] total carbon and total nitrogen ratio.

2.3. Observations of Plant Development and Plant Harvests

During the study period, observations of plant development and growth were carried out (Figure 1), and biometric measurements were made on 25 plants or shoots, which included: length (cm) and thickness of shoots (mm), number of shoots on one plant (pcs) and harvested biomass DMY (t ha^{-1}). The length, thickness of shoots and number of shoots per plant were carried out at the time of harvesting the plants for biomass. The yield of green and dry matter (t ha^{-1}) was determined on all plots. Plants for biomass were harvested after the end of vegetation in the March of the following year of the study (i.e., on 24 March 2010, 1 March 2011, 8 March 2012, 6 March 2013, 11 March 2014, 5 March 2016, 9 March 2016, and 8 March 2017).

Figure 1. *Sida* in various stages of development.

The energy value of dry plant matter was determined in the laboratory of the Department of Environmental Management, West Pomeranian University of Technology in Szczecin, using the KL-10 calorimeter manufactured by the Cooperative PRECYZJA from Bydgoszcz following PN-81/G-04513 as well as the technical and operational documentation of the KL-10 automatic calorimeter.

2.4. Statistical Analysis

The obtained study results were statistically processed by applying classic analysis of variance using the ANAWAR 5.3 software (developed by Professor Franciszek Rudnicki) and correlation analysis. Software ANAWAR 5.3 is used to analyse variance with the regression of source data from agricultural experiments. It contains computational programs for orthogonal data from single and multiple single-, double- and three-factor experiments. It considers the experimental systems most often used in experimental agricultural research. The significance of result diversity was determined by the Tukey test at the level of $p = 0.05$. Using Software Statistica 12.5, multivariate analysis was performed using principal component analysis (PCA).

2.5. Characteristics of Climatic Conditions during the Study

According to research [11,20–22], the amount of precipitation and temperature distribution is one of the most critical factors modifying the development of plants and, thus, their DMY. Meteorological data were obtained from the Agricultural Experimental Station in Lipnik for the years 2009–2016 and 1980–2008 (Table 2), and they indicate differences in air temperatures and the amount of precipitation during the study.

The average air temperature and the amount of atmospheric precipitation throughout the growing season during the study years were higher than in the corresponding period over the years (Table 2).

Within the study years, the warmest years were 2009, 2014, 2013 and 2016, in which the average air temperature during plant growth was 14.7, 16.7, 16.4 and 15.5 °C, which exceeded the average value over many years. The warmest months in 2009 were July and August, in which the average temperature was 19.4 and 19.6 °C; in 2014, they were July, August and September; in 2013, they were June and July; and in 2016, they were May–September, in which the average temperatures were higher than the average for the same period over many years.

In the years of the study, atmospheric precipitation was characterised by significant differentiated distribution in individual months and years. The most considerable precipitation was observed in 2010—755.1 mm; the lowest was in 2016—473.7 mm. In 2010, the highest rainfall was recorded during the growing season (IV–X), which was 481.1 mm; 20.0 mm less rainfall during plant vegetation was recorded in 2014, 122.1 and 109.6 mm higher than the multiannual average for this period, respectively. The lowest rainfall during the growing season was recorded in 2015—300.2 mm (Table 2), while the month with the lowest rainfall was November in 2011—1.0 mm, and the highest was August in

2010—184.4 mm. In 2015, 2016 and 2013, the sums of rainfall during the growing season were 283.6, 304.7 and 343.8 mm, respectively, and they were significantly lower than in 2010 and lower by 75.4, 54.3, 19.7 and 15.2 mm than in the same period of many years.

Table 2. Average monthly air temperature and monthly total rainfall in the years 2009–2016.

Year	Month														
	I	II	III	IV	V	VI	VII	VIII	IX	X	XI	XII	Total for the Year	IV–X
Temperature [°C]															
Multi-year Average	−1.1	−0.3	2.8	7.4	12.7	16.0	17.6	17.2	13.3	8.8	3.8	0.4	9.6	15.0
2009	−3.1	1.5	−3.9	12.3	13.4	15.4	19.4	19.6	14.7	7.8	6.7	−0.2	12.1	14.7
2010	−5.5	−0.6	3.8	8.7	11.1	17.0	22.2	18.5	13.2	7.5	4.7	−4.7	8.0	14.0
2011	0.7	−0.9	3.9	11.9	14.3	18.2	17.7	18.3	14.9	9.5	4.1	3.9	9.7	15.0
2012	1.5	−2.3	6.3	8.8	15.5	16.2	18.6	18.1	14.5	8.7	5.1	−0.7	9.2	14.3
2013	−3.5	4.0	3.5	11.4	17.5	20.3	21.2	18.5	14.8	11.1	5.2	3.0	10.6	16.4
2014	−0.5	1.7	6.4	11.1	14.0	16.9	21.8	21.5	20.2	11.1	6.2	2.0	11.0	16.7
2015	1.1	0.0	3.9	9.7	11.9	14.5	17.6	20.1	13.0	7.3	5.6	5.3	9.2	13.4
2016	−1.2	3.2	4.2	8.9	16.9	19.2	19.4	18.2	17.1	8.8	3.9	2.7	10.1	15.5
Precipitation [mm]															
Multi-year Average	54.6	31.6	25.5	20.8	88.1	112.5	50.4	35.9	43.9	45.8	37.8	37.7	584.6	397.4
2009	19.4	49.4	53.4	16.6	70.3	60.7	61.9	58.0	45.4	82.7	46.9	32.7	597.4	395.6
2010	36.1	21.2	43.8	16.8	91.6	10.6	86.7	184.4	56.3	34.7	100.3	72.6	755.1	481.1
2011	31.0	33.4	23.9	12.5	27.9	44.8	148.5	57.7	52.2	37.9	1.0	70.8	541.6	381.5
2012	64.7	41.1	18.0	32.4	21.1	45.8	103.4	90.2	25.1	53.5	40.5	39.1	574.9	371.5
2013	54.6	31.6	25.5	20.8	88.1	112.5	50.4	35.9	43.9	45.8	37.8	37.7	584.6	397.4
2014	39.6	14.6	25.3	39.1	94.0	45.0	100.6	88.1	59.3	35.0	7.7	74.0	622.3	461.1
2015	71.0	4.9	39.3	18.3	42.6	51.3	68.0	17.8	66.1	35.8	53.2	35.3	503.9	300.2
2016	32.6	34.5	25.7	25.7	43.7	70.6	68.7	41.2	9.7	45.1	50.5	25.7	473.7	304.7

3. Results and Discussion

3.1. Plant Density

The analysis of results of plant density on the surface in full performance years was relatively stable and was at the level obtained in the sowing year; hence, the paper presents only their average density from the years of study (Table 3).

Table 3. Effects of seeding amount and fertilization with compost on the *Sida* plant density during the years of full use (2010–2016).

Seeding Amount (kg ha^{-1})	Compost Fertilization (t ha^{-1})			Average Plant Density (Plants m^{-2})
	0	10	20	
	Plant Density (Plants m^{-2})			
1	6.8	12.0	13.1	10.6
2	10.3	12.7	13.7	12.2
3	11.1	14.7	16.7	14.2
Average	9.4	13.1	14.5	12.4
LSD$_{0.05}$ for:		2009–2016		
Compost fertilization—I		1.46		
Seeding amount—II		0.63		

It should be noted that emergence uniformity on the surface occurred at their seeding in the amount of 3 kg of seeds per hectare, and the lowest density and the least uniformity occurred at the objects where 1 kg of seeds was sown. This density status continued until the last year of the study. On average, from the years of the study, the density at seeding 2 kg of seeds was higher by 15.1%, and at seeding 3 kg of seeds it was higher by 39.9% compared to the density at objects where 1 kg of seeds was sown. The obtained plant density was similar to that recommended by Borkowska and Styk [31]. Often, the reason for the low plant density when establishing plantations by sowing, according to Tworkowski et al. [32], Kurucz et al. [33] and Packa et al. [34], is the low germination capacity of seeds in the sowing year, which results, among others, from the occurrence of "hard seeds" in the seeding material, characterised by the presence of an impermeable seed coat for water and gases. The density of plants planted directly into the ground was stable in the first and subsequent years of the study. Bury et al. [11] and Tworkowski et al. [32] found results of falling out in a significant number of plants sown into the ground in comparison planted ones (germination capacity, preparation of seeds, rainfall in the period of emergence and early development). The plant population of *Sida* should be 20–60 thousand plants per 1 ha when grown for biomass on good soils [11,31], thus showing that higher seeding of *Sida* seeds allows for higher DMY and better quality [8].

In the case of poor soils, the fertilization of the *Sida* crop positively affects the density of the sown plant. Molas et al. [35] indicated that fertilization in the amount of 20 N, 20 P and 40 K in kg ha^{-1} increased the plant density by more than 5 plants per 1 m^2 compared to the combination without fertilization. Therefore, the use of alternative nutrient sources such as compost or sewage sludge can also have beneficial effects, which was confirmed by the obtained results. The assessment of the impact of applied pre-sowing doses of compost (0, 10 and 20 t ha^{-1}) on plant density showed that the use of compost contributed to an increase in the density of *Sida* plants by 39.4% and 54.3%, respectively, in relation to the number of plants not fertilized with compost.

3.2. Plant Development and Morphology

Depending on the applied fertilization level and seeding amount in the first year of the study (year of plant establishment), plants produced from 1.1 to 2.2 shoots. In the years of full use of the plants (2010–2016), the number of shoots produced by plants was many times higher (Table 4), which is typical of this species [11,20]. In the case of the research of Tworkowski et al. in 2014 [32], plants sown from seeds in the first year of cultivation with mineral fertilization had only one shoot. In the following year, the number of shoots increased to 7, while in the case of seedlings, the number of shoots even reached 15 [36]. Research shows that in the case of plants from seedlings, more significant amounts of plant shoots are recorded, especially in the first and second years of cultivation [11,32]. A significant increase in the number of shoots per plant under the influence of compost fertilization was found in the subsequent years of use. On objects fertilized with compost at 10 and 20 t ha^{-1}, the number of shoots per plant was 14.6% and 23.6% higher, respectively, compared to the object not fertilized with the compost. Similar effects were also obtained in the case of mineral fertilization [31].

Applied standards for *Sida* seeding 1, 2 and 3 kg ha^{-1} in the years of full use contributed to a significant increase in the number of shoots on one plant. Plants produced more shoots on objects where a higher seeding standard was used by 13.5% and 23.7%, respectively, compared to the number of shoots produced by plants on the object with the lowest seeding standard. The results are consistent with those obtained by Tworkowski et al. [32], in which the number of stems on the plant grown from root cuttings and seedlings was more remarkable than those from seed seeding.

Table 4. Effects of seeding amount and fertilization with compost on the number of *Sida* shoots during the years of full use (2010–2016).

Seeding Amount (kg ha^{-1})	Compost Fertilization (t ha^{-1})			Average
	0	10	20	
	Number of Shoots			
1	7.9	9.0	9.9	8.9
2	8.9	10.4	11.1	10.1
3	9.8	11.3	12.1	11.1
Average	8.9	10.2	11.0	10.0
LSD$_{0.05}$ for:	2009–2016			
Compost fertilization—I	1.46			
Seeding amount—II	0.63			

The productivity of *Sida* crops and the quality of the biomass obtained for combustion were tested by Chołuj et al. [37], Szyszlak-Bargłowicz et al. [38], Borkowska et al. [39] Bilandžija et al. [40], and Możdżer et al. [41], who claim that the productivity and quality depend on many factors. One of them is the morphological structure (e.g., height and diameter of shoots) of the plant affecting the productivity of the biomass obtained. PCA analysis (Figure 2) showed the relationship between *Sida* morphological parameters: plant height, number of shoots and thickness, which modify the energy yield.

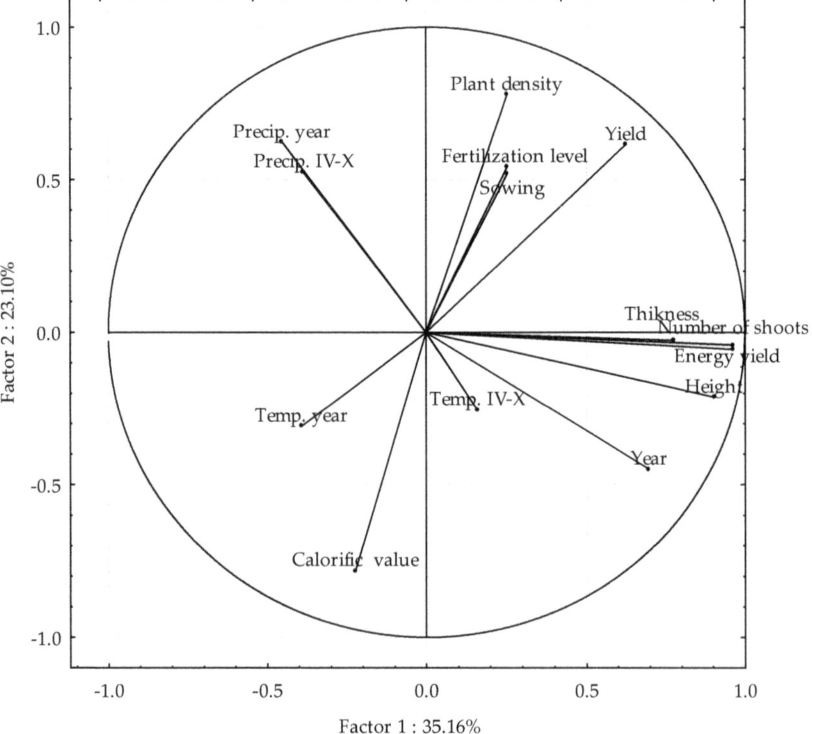

Figure 2. The principal component analysis (PCA) for *Sida* DMY, biometric parameters, energetic parameters and climatic conditions.

On objects not fertilized with compost, the average height of shoots was 249.3 cm, while the height of *Sida* plants on objects fertilized with compost in the amount of 10 and 20 t ha^{-1} (Table 5) was not significantly differentiated: only the trend of increasing the height of shoots was noted—by 1.5% and 2.8%. Similarly to the level of compost fertilization, the impact of the seed sowing amount on the height of *Sida* shoots was shaped, the differences between the objects were not statistically significant, and only a slight lengthening occurred with the increase in the seeding amount. Similarly, no significant effect of fertilization and seeding amount on the average thickness of shoots was found (Table 6). Veste et al. [28], in the studies on the effect of fertilization with compost, indicated an increase in the height of fertilized plants compared to the control, while in the first year of the experiments, the plants reached the height of 120 cm, and strong effects were also obtained in the fertilization with digestate [17]. The obtained average plant height results were similar to the results of Borkowska et al. [19] in the case of plants fertilized with nitrogen in a dose of 100 and 200 kg ha^{-1} and phosphorus in a dose of 39.28 and 52.38 kg ha^{-1} and the same level of K fertilization on light sandy loam. With similar fertilization but on better soil (composed of dust and clay), Borkowska et al. [17] obtained a higher average plant height, more than 290 cm in the fourth year of production, while the plants fertilized with sewage sludge in Croatia reached an average height of 310 cm [42], although weather conditions could also have a significant impact.

Table 5. Effects of seeding amount and fertilization with compost on the average height of the *Sida* shoots (cm) during the years of full use (2010–2016).

Seeding Amount (kg ha^{-1})	Compost Fertilization (t ha^{-1})			Average Height of Shoots (cm)
	0	10	20	
	Height of Shoots (cm)			
1	249.3	254.0	258.0	253.8
2	255.7	259.1	263.2	259.3
3	258.9	262.2	264.1	261.7
Average	254.6	258.4	261.7	258.3
LSD$_{0.05}$ for:		2009–2016		
Compost fertilization—I		i. d. *		
Seeding amount—II		i. d. *		

* insignificant difference.

Table 6. Effects of seeding amount and fertilization with compost on the average thickness of the *Sida* shoots (mm) during the years of full use (2010–2016).

Seeding Amount (kg ha^{-1})	Compost Fertilization (t ha^{-1})			Average Thickness of Shoots (mm)
	0	10	20	
	Thickness of Shoots (mm)			
1	16.2	16.2	16.0	16.1
2	16.0	15.9	15.9	16.0
3	15.9	16.0	15.8	15.9
Average	16.1	16.0	15.9	16.0
LSD$_{0.05}$ for:		2009–2016		
Compost fertilization—I		i. d. *		
Seeding amount—II		i. d. *		

* insignificant difference.

The assessment of the *Sida* shoots' thickness indicates that they were not significantly dependent on the factors studied (Table 6). Organic compost fertilizers applied to the ground (10 and 20 t ha^{-1}) and seeding amount influence a slight decrease in the thickness of produced shoots of the plant; on average, from the years of the research, the thickness of the shoots decreased by 0.2 and 0.3 mm, i.e., by 1%, 2% and 1.8% relating to the thickness of shoots produced on the object not fertilized with compost and with the lowest seeding amount. The number of shoots and their length and thickness influence the overall DMY [16,36].

3.3. Dry Matter Yield

The applied research factors (organic compost fertilization and seeding amount) affected the amount of DMY of *Sida* plants (Table 7). The applied levels of compost fertilization (10 and 20 t ha^{-1}) positively affected the obtained DMY. In all the years of the study, the increase in DMY was significant, and compared to the DMY from an object not fertilized with compost, it was was 24.9% and 50.7%, respectively, on average during all the years of observation (2009–2016).

Table 7. Effects of seeding amount and fertilization with compost on the DMY (t ha^{-1}) of *Sida* during the years of observation.

| Year | Compost Fertilization (t ha^{-1}) | Seeding Amount (kg ha^{-1}) | | | Average Dry Matter Yield (t ha^{-1}) | LSD$_{0.05}$ |
| | | 1 | 2 | 3 | | |
		Dry Matter Yield (t ha^{-1})				
2009	0	0.50	0.94	1.64	1.03	
	10	1.05	1.38	1.91	1.45	I—i.d. *
	20	1.24	1.69	2.01	1.65	II—0.34
	Average	0.93	1.34	1.85	1.37	
2010	0	5.88	8.63	12.69	9.07	
	10	8.00	11.63	16.38	12.00	I—0.89
	20	13.38	13.94	22.69	16.67	II—0.96
	Average	9.09	11.40	17.25	12.58	
2011	0	5.63	8.75	10.31	8.23	
	10	7.81	12.19	13.44	11.15	I—1.53
	20	9.06	13.75	16.25	13.02	II—1.86
	Average	7.50	11.56	13.33	10.80	
2012	0	3.75	5.00	6.25	5.00	
	10	6.25	8.00	10.00	8.08	I—1.79
	20	9.62	11.94	15.00	12.19	II—0.50
	Average	6.54	8.31	10.42	8.42	
2013	0	4.21	5.26	6.46	5.31	
	10	5.23	7.46	8.26	6.98	I—0.14
	20	5.68	8.24	9.24	7.72	II—0.18
	Average	5.04	6.99	7.99	6.67	
2014	0	6.12	8.46	10.28	8.29	
	10	7.21	9.24	11.28	9.24	I—0.18
	20	7.86	10.43	11.68	9.99	II—0.16
	Average	7.06	9.38	11.08	9.17	
2015	0	5.86	8.28	11.24	8.46	
	10	6.48	8.42	11.84	8.91	I—0.16
	20	7.26	9.36	12.24	9.62	II—0.15
	Average	6.53	8.69	11.77	9.00	
2016	0	5.26	7.68	9.23	7.39	
	10	6.12	8.24	10.41	8.26	I—0.16
	20	6.74	8.67	10.89	8.77	II—0.14
	Average	6.04	8.20	10.18	8.14	

Table 7. *Cont.*

Year	Compost Fertilization (t ha^{-1})	Seeding Amount (kg ha^{-1})			Average Dry Matter Yield (t ha^{-1})	LSD$_{0.05}$
		1	2	3		
		Dry Matter Yield (t ha^{-1})				
Average (2010–2016)	0	5.24	7.44	9.49	7.39	
	10	6.73	9.31	11.66	9.23	I—0.95
	20	8.51	10.90	14.00	11.14	II—0.33
	Average	6.83	9.22	11.72	9.25	

* insignificant difference.

When assessing the impact of seeding amount on the DMY, it should be stated that increasing the seeding amount significantly increased the DMY by 35.0% on average at the 2 kg ha^{-1} and 71.6% when seeding 3 kg ha^{-1} compared to the lowest seeding rate (Table 7). PCA analysis (Figure 2) showed that *Sida*'s DMY depends on the seeding amount and the subsequent plant density and fertilization. The influence of the amount of precipitation in individual years and growing seasons was also noticeable.

Analysis of the DMY of *Sida* plants in years of use shows that the lowest DMY was produced by plants in the sowing year (2009), which ranged from 0.5 to 2.0 t ha^{-1}. In the years of full use, the DMY was many times higher and ranged from 3.75 to 22.69 t ha^{-1}, depending on the factors studied and the year of the experiment. The highest DMY in the tested conditions was obtained in the first year of full use (2010), which on average for the examined factors was 12.58 t ha^{-1}, which was higher by 14.1% in the second, 33.1% in the third, 47.0% in the fourth, 27.1% in the fifth, 28.5% in the sixth and 35.3% in the seventh year of the full use.

The DMY of crops grown for energy purposes is influenced not only by agrotechnical factors (e.g., fertilization, soil tillage, cultivation) but also by weather in the subsequent years of plantation use [7,40–45]. The course of meteorological conditions affects the condition of plants; they can determine their performance and the quality of harvested DMY in subsequent years. The results of previous studies confirmed low DMYs in the first year and much higher DMYs in the second and subsequent years of use. Perennial species usually reach their full DMY in the third–fourth year after sowing or planting [40,41]. The annual DMY of *Sida* plant biomass ranges from 8.7 to 20.3 t ha^{-1} DM. In most studies, the annual capacity of *Sida* exceeded 10 t ha^{-1} after the first two years of growth, harvested from October [45] until February [32]. The diversity of cultivation technologies and variability of habitat conditions cause significant differences in the DMY of energy crops, including *Sida* [6,45–48]. Average crop DMYs under real EU production conditions are 10–12 t ha^{-1}, with fluctuations 6–15 t ha^{-1} [8,20,34,45,49–53]. In our research, the highest DMY of *Sida* was obtained in the second year after sowing, but in subsequent years, the obtained DMY coincided with those obtained by other authors [7,16,20,32,39]. Borkowska et al. [20] report that in the second year, under favourable habitat conditions, a significant DMY of *Sida* can be harvested, and in subsequent years, the DMYs reach maximum values, and the average DMYs from eight years of cultivation exceeded 11 t ha^{-1} DM. During the eight years of research—from the second to the ninth year of cultivation—the highest DMYs were obtained significantly in the eighth (2010), fifth (2006) and sixth (2007) years of use [20]. Results of the personal research in time-space were slightly different, but their DMY was comparable. The research confirms that compost from municipal waste with additional NPK fertilization can be successfully used to cultivate energy crops on poor and marginal soils. The increase in DMYs after the use of substitutes for mineral fertilization was also confirmed in other studies, although the effects were varied [17,23,27–29,41]. Veste et al. [28], in the combination of compost fertilization (which constitutes 20% of the substrate) and nitrogen fertilization (100 kg ha^{-1}), obtained an almost eight-fold increase in the DMY of *Sida* in the first year of cultivation, while in the case of using only compost (which constitutes 50% of the substrate), the increase in the DMY compared to control was almost two-fold. Interesting results were obtained by Barbosa et al. [27] and Nabel et al. [17], and Nabel et al. [18], who showed that fertilization with the digestate brings better results on poor soils (marginal) compared to the use of NPK fertilization with fertilizer doses of 160 kg N ha^{-1}. Similar relationships were also shown by earlier pot tests on light soils, where sewage sludge in the highest doses of 40 and 60 t ha^{-1} had the greatest positive effect on the mallow DMYs compared to mineral fertilization and compost from various sources [29]. Šurić et al. [42] also noted the positive effect of fertilization with sewage sludge and noted that these effects were significant when using sufficiently high doses of sludge (more than 10 t ha^{-1}).

3.4. Energy Yield and Calorific Value

The energy yield of *Sida* plants in light soil conditions was modified by the examined factors and years of research, but its height was closely correlated with the impact of the studied factors on the DMYs of plants (Table 7, Figure 2).

The applied compost fertilization increased the energy yield of the obtained biomass of *Sida* plants in all the years of the research. The average energy yield of plants fertilized with compost in the amount of 10 and 20 t·ha^{-1} was higher by 12.1% and 21.0% than the energy yield of plants not fertilized with compost (Table 8). The obtained energy yield results for plants under the influence of the applied seeding amount (1, 2 and 3 kg ha^{-1}) prove that it was significantly higher for plants from objects with higher seeding amount than for plants from objects with the lowest seeding amount. On average, from the research years, the increase in energy yield at objects with higher seeding amount increased by 14.9% and 22.6%, respectively. The cultivation and seeding amount method influences the energy yield and is directly related to the DMY [33]. According to Šiaudinis et al. [51] and Jankowski et al. [52], the energy yield of *Sida* plants is favourable for combustion compared to the biomass of other herbaceous plants, and according to Jablonowski et al. [8], its combustion properties are similar to those of wood biomass, while their values range from 105 to 236 GJ ha^{-1} [20,39,54–56]. The energy yield of *Sida* in our research was consistent with the results mentioned above which were obtained under conditions of different fertilization levels of plants with compost and different norms of *Sida* seeding. Energy yields determined for *Sida* by Šiaudinis et al. [51] and Jankowski et al. [52] are much lower than those reported by Jablonowski et al. [8]. The difference in energy yield that the authors reported was due to the amount of DMY obtained rather than its energy value. The most significant interest in *Sida* lies in its potential as a renewable energy source.

Table 8. Effects of seeding amount and fertilization with compost on the average energy yield (GJ ha^{-1}) of *Sida* during the years of full use (2010–2016).

Seeding Amount (kg ha^{-1})	Compost Fertilization (kg ha^{-1})			Average Energy Yield (GJ ha^{-1})
	0	10	20	
	Energy Yield (GJ ha^{-1})			
1	136.94	153.08	165.38	151.80
2	155.60	177.35	190.08	174.34
3	168.79	186.87	202.56	186.07
Average	153.78	172.43	186.01	170.74
LSD$_{0.05}$ for:	2009–2016			
Compost fertilization—I	6.03			
Seeding amount—II	11.30			

Therefore, research on this species was focused on its thermophysical and biochemical properties in the context of direct combustion and biogas production. The parameters determining its energy usefulness are higher heat values (HHVs) and lower heat values (LHVs).

In the literature, the HHVs for *Sida* were between 16.5 and 19.5 MJ kg^{-1} DM (average 18.4 MJ kg^{-1} DM), while the LHVs were between 14.0 and 17.2 MJ kg^{-1} (average 16.1 MJ kg^{-1} DM) [7,8,21,26,40,43].

The calorific value of the harvested biomass in the years of the study ranged from 16.8 to 17.4 MJ kg^{-1} DM, and no significant influence of the examined factors on its concentration in plants was found (Table 9). The obtained values of the calorific value of *Sida* were similar to the results indicated in the literature [8,39,40,43]. It is worth noting that the quality of the biomass improves when the harvest date is delayed, which affects the composition of the main elements (C, H, N, O, S), lignocellulose and moisture in plants [8,27,33] which determines the energy value, however, no significant impact of the method of establishing the cultivation and sowing density on the quality of the biomass was confirmed [32,33].

Table 9. Effects of seeding amount and fertilization with compost on the average calorific value (MJ kg^{-1} DM) of *Sida* during the years of full use (2010–2016).

Seeding Amount (kg ha^{-1})	Compost Fertilization (kg ha^{-1})			Average Calorific Value (MJ kg^{-1} DM)
	0	10	20	
	Calorific Value (MJ kg^{-1} DM)			
1	17.4	17.2	16.9	17.2
2	17.3	17.1	16.8	17.1
3	17.1	16.9	16.8	16.9
Average	17.3	17.1	16.8	17.1
LSD$_{0.05}$ for:	2009–2016			
Compost fertilization—I	i. d. *			
Seeding amount—II	i. d. *			

* insignificant difference.

4. Conclusions

The applied levels of compost fertilization (10 and 20 t ha^{-1}) increased the plant DMY in all the years of the research compared to plants not fertilized with compost, and their DMY was 24.9% and 50.7% higher, respectively, on average from the years of the study. Analysing the impact of the applied seeding amount (1, 2 and 3 kg ha^{-1}) on the DMY of *Sida* in light soil conditions, it should be stated that this factor significantly increased the crop DMY in all the years of the research and on average, it increased by 35.0% and 71.6% compared to the lowest seeding rate. Analysis of the DMY of *Sida* in the years of use shows that the lowest DMY was obtained by plants in the sowing year (2009), which is typical for *Sida*. In the years of full use, the DMY was many times higher and ranged from 3.75 to 22.69 t ha^{-1}, depending on the factors studied and the year of the research. The highest energy value of *Sida* was observed under the highest organic compost fertilization and highest seeding amount. It can therefore be concluded that this study confirms the suitability of establishment via sowing for cultivating *Sida* on light soils. Furthermore, the use of organic fertilization with urban green compost and nitrogen allows for a biomass yield at a level similar to mineral fertilization described in other studies [16,19,32,57].

Author Contributions: Conceptualization, T.K. and E.Ł.; methodology, T.K.; software, T.K.; validation, T.K., R.M. and G.J.; formal analysis, G.J.; investigation, T.K., Ł.Z., G.J. and R.M.; resources, T.K.; data curation, T.K.; writing—original draft preparation, T.K.; writing—review and editing, G.J.; visualisation, T.K.; supervision, T.K.; project administration, T.K.; funding acquisition, T.K. and Ł.Z. All authors have read and agreed to the published version of the manuscript.

Funding: This research received no external funding.

Institutional Review Board Statement: Not applicable.

Informed Consent Statement: Not applicable.

Data Availability Statement: West Pomeranian University of Technology in Szczecin.

Conflicts of Interest: The authors declare no conflict of interest.

References

1. National Research Council. *Biobased Industrial Products: Priorities for Research and Commercialization*; National Academy Press: Washington, DC, USA, 2000.
2. EC. Communication (2014) 0015 from the Commission to the European Parliament, The Council, The European Economic and Social Committee and the Committee of the Regions. A Policy Framework for Climate and Energy in the Period from 2020 to 2030, COM/2014/015 Final. Available online: https://www.eea.europa.eu/policy-documents/com-2014-15-final (accessed on 8 April 2022).
3. Directive 2018/2001/EU of the European Parliament and of the Council of 11th December 2018 on the Promotion of the Use of Energy from Renewable Sources (Text with EEA Relevance). Available online: https://eur-lex.europa.eu/legal-content/PL/TXT/?uri=celex%3A32009L0028 (accessed on 24 April 2021).
4. Keoleian, G.A.; Volk, T.A. Renewable Energy from Willow Biomass Crops: Life Cycle Energy, Environmental and Economic Performance. *Crit. Rev. Plant Sci.* **2007**, *24*, 385–406. [CrossRef]
5. Grzybek, A. Zasoby krajowe biopaliw i możliwości ich wykorzystania w aspekcie technicznym i organizacyjnym (National resources of biofuels and possibilities of their use in technical and organizational aspect). *Energetyka* **2006**, *9*, 8–11. (In Polish)
6. Dunnett, A.; Shah, N. Prospects for bioenergy. *J. Biobased Mater. Bioenergy* **2007**, *1*, 1–18. [CrossRef]
7. Borkowska, H.; Molas, R. Yield comparison of four lignocellulosic perennial energy crop species. *Biomass Bioenergy* **2013**, *51*, 145–153. [CrossRef]
8. Jablonowski, N.D.; Kollmann, T.; Meiller, M.; Dohrn, M.; Müller, M.; Nabel, M.; Zapp, P.; Schonhoff, A.; Schrey, S.D. Full assessment of Sida (*Sida hermaphrodita*) biomass as a solid fuel. *GCB Bioenergy* **2020**, *12*, 618–635. [CrossRef]
9. Marks-Bielska, R.; Bielski, S.; Pik, K.; Kurowska, K. The Importance of Renewable Energy Sources in Poland's Energy Mix. *Energies* **2020**, *13*, 4624. [CrossRef]
10. Pawłowski, R. Energia odnawialna w Polsce szansą na zwiększenie konkurencyjności sektora rolnego (The Renewable Energy in Poland as an Opportunity for Increasing the Competitiveness of the Agricultural Sector). *Roczniki Naukowe Stowarzyszenia Ekonomistów Rolnictwa i Agrobiznesu* **2008**, *4*, 325–326. (In Polish)
11. Bury, M.; Kitczak, T.; Możdżer, E.; Siwek, H.; Włodarczyk, M. *Uprawa ślazowca pensylwańskiego [Sida hermaphrodita (L.) Rusby] Wyniki produkcyjne, agrotechnika i wykorzystanie (Cultivation of Virginia Fanpetals [Sida hermaphrodita (L.) Rusby] Production Results, Agrotechnics and Use)*; Wydawnictwo Zachodniopomorskiego Uniwersytetu Technologicznego w Szczecinie: Szczecin, Poland, 2019; p. 111. (In Polish)
12. Bełdycka-Bórawska, A.; Bórawski, P.; Borychowski, M.; Wyszomierski, R.; Bórawski, M.B.; Rokicki, T.; Ochnio, L.; Jankowski, K.; Mickiewicz, B.; Dunn, J.W. Development of Solid Biomass Production in Poland, Especially Pellet, in the Context of the World's and the European Union's Climate and Energy Policies. *Energies* **2021**, *14*, 3587. [CrossRef]
13. Celińska, A. Charakterystyka różnych gatunków upraw energetycznych w aspekcie ich wykorzystania w energetyce zawodowej (Characteristics of various energy crops in aspect of use in power industry). *Polityka Energetyczna* **2009**, *12*, 29–35. (In Polish)
14. Igliński, B.; Piechota, G.; Buczkowski, R. Development of biomass in polish energy sector: An overview. *Clean Technol. Environ. Policy* **2015**, *17*, 317–329. [CrossRef]
15. Waliszewska, B.; Grzelak, M.; Gaweł, E.; Spek-Dźwigała, A.; Sieradzka, A.; Czekała, W. Chemical Characteristics of Selected Grass Species from Polish Meadows and Their Potential Utilization for Energy Generation Purposes. *Energies* **2021**, *14*, 1669. [CrossRef]
16. Nahm, M.; Morhart, C. Virginia mallow (*Sida hermaphrodita* (L.) Rusby) as perennial multipurpose crop: Biomass yields, energetic valorization, utilization potentials, and management perspectives. *GCB Bioenergy* **2018**, *10*, 393–404. [CrossRef]
17. Nabel, M.; Tenperton, V.M.; Poorter, H.; Lücke, A.; Jablonowski, N.D. Energizing marginal soils—The establishment of the energy crop *Sida hermaphrodita* as dependet on digestate fertilization, NPK, and legume intercropping. *Biomass Bioenergy* **2016**, *87*, 9–16. [CrossRef]
18. Nabel, M.; Schrey, S.D.; Poorter, H.; Koller, R.; Jablonowski, N.D. Effects of digestate fertilization on *Sida hermaphrodita*: Boosting biomass yields on marginal soils by increasing soil fertility. *Biomass Bioenergy* **2017**, *107*, 207–213. [CrossRef]
19. Borkowska, H.; Molas, R.; Kupczyk, A. Virginia Fanpetals (*Sida hermaphrodita* Rusby) cultivated on light soil; height of yield and biomass productivity. *Pol. J. Environ. Stud.* **2009**, *18*, 563–568.
20. Borkowska, H.; Molas, R.; Skiba, D. Plonowanie ślazowca pensylwańskiego w wieloletnim użytkowaniu (Virginia fanpetals yielding in multi-year use). *Acta Agrophysica* **2015**, *22*, 5–15. (In Polish)
21. Kurucz, E.; Antal, G.; Gábor, F.M.; Popp, J. Cost-effective mass propagation of Virginia fanpetals (*Sida hermaphrodita* (L.) Rusby) from seeds. *Environ. Eng. Manag. J.* **2014**, *13*, 2845–2852. [CrossRef]

22. Cumplido-Marin, L.; Graves, A.R.; Burgess, P.J.; Morhart, C.; Paris, P.; Jablonowski, N.D.; Facciotto, G.; Bury, M.; Martens, R.; Nahm, M. Two Novel Energy Crops: *Sida hermaphrodita* (L.) Rusby and *Silphium perfoliatum* L.—State of Knowledge. *Agronomy* **2020**, *10*, 928. [CrossRef]
23. Sienkiewicz, S.; Wierzbowska, J.; Kovacik, P.; Krzebietke, S.; Zarczynski, P. Digestate as a substitute of fertilizers in the cultivation of Virginia fanpetals. *Fresenius Environ. Bull.* **2018**, *27*, 3970–3976.
24. Krzywy-Gawrońska, E. The Effect of industrial wastes and municipal sewage sludge compost on the quality of Virginia fanpetals (Sida hermaphrodita Rusby) biomass Part 1. Macroelements content and their uptake dynamics. *Polish J. Chem. Technol.* **2012**, *14*, 9–15. [CrossRef]
25. Antonkiewicz, J.; Kołodziej, B.; Bielińska, E.J.; Gleń-Karolczyk, K. The use of macroelements from municipal sewage sludge by the multiflora Rose and the Virginia fanpetals. *J. Ecol. Eng.* **2018**, *9*, 1–13. [CrossRef]
26. Nabel, M.; Schrey, S.D.; Poorter, H.; Koller, R.; Nagel, K.A.; Temperton, V.M.; Dietrich, C.C.; Briese, C.; Jablonowski, N.D. coming late for dinner: Localized digestate depot fertilization for extensive cultivation of marginal soil with Sida Hermaphrodita. *Front. Plant. Sci.* **2018**, *9*, 1095. [CrossRef] [PubMed]
27. Barbosa, D.B.P.; Nabel, M.; Jablonowski, N.D. Biogas-digestate as nutrient source for biomass production of *Sida hermaphrodita*, *Zea Mays* L. and *Medicago Sativa* L. *Energy Procedia* **2014**, *59*, 120–126. [CrossRef]
28. Veste, M.; Halke, C.; Garbe, D.; Freese, D. Effect of nitrogen fertiliser and compost on photosynthesis and growth of Virginia fanpetals (Sida hermaphrodita Rusby) [Einfluss von Stickstoffdüngung und Kompost auf Photosynthese und Wachstum der Virginiamalve (Sida hermaphrodita Rusby)]. *J. Für Kult.* **2016**, *68*, 423–428. (In Germany) [CrossRef]
29. Ociepa-Kubicka, A.; Pachura, P. The use of sewage sludge and compost for fertilization of energy crops on the example Miscanthus and Virginia Mallow. *Annu. Set Environ. Ptection* **2013**, *15*, 2267–2278.
30. FAO. *International Union of Soil Sciences Working Group WRB World Reference Base for Soil Resources 2014*; Update 2015; FAO: Rome, Italy, 2014.
31. Borkowska, H.; Styk, B. *Ślazowiec pensylwański (Sida hermaphrodita Rusby). Uprawa i wykorzystanie (Virginia Fanpetals. Cultivation and Use)*; Wydawnictwo Akademii Rolniczej: Lublin, Poland, 2006; p. 69. (In Polish)
32. Tworkowski, J.; Szczukowski, S.; Stolarski, M.J.; Kwiatkowski, J.; Graban, Ł. Produkcyjność i właściwości biomasy ślazowca pensylwańskiego jako paliwa w zależności od materiału siewnego i obsady roślin (Productivity and properties of Virginia fanpetals biomass as fuel depending on the propagule and plant density). *Fragm. Agron.* **2014**, *31*, 115–125.
33. Kurucz, E.; Fári, M.G.; Antal, G.; Gabnai, Z.; Popp, J.; Bai, A. Opportunities for the production and economics of Virginia fanpetals (*Sida hermaphrodita*). *Renew. Sustain. Energy Rev.* **2018**, *90*, 824–834. [CrossRef]
34. Packa, D.; Kwiatkowski, J.; Graban, Ł.; Lajszner, W. Germination and dormancy of *Sida hermaphrodita* seeds. *Seed Sci. Technol.* **2014**, *42*, 1–15. [CrossRef]
35. Molas, R.; Borkowska, H.; Skiba, D. Development and yielding of Virginia fanpetals depending on some elments of agricultural practices. *Agron. Sci.* **2019**, *74*, 153–162. [CrossRef]
36. Šiaudinis, G.; Skuodienė, R.; Repšienė, R. The investigation of three potential energy crops: Common mugwort, Cup plant and Virginia mallow on western Lithuania's Albeluvisol. *Appl. Ecol. Environ. Res.* **2017**, *15*, 611–620. [CrossRef]
37. Chołuj, D.; Podlaski, S.; Wiśniewski, G.; Szmalec, J. Kompleksowa ocena 7 gatunków roślin wykorzystywanych na cele energetyczne. In *Uprawa roślin energetycznych a wykorzystanie rolniczej przestrzeni produkcyjnej w Polsce*; Harasim, A., Ed.; Instytut Uprawy Nawożenia i Gleboznawstwa Państwowy Instytut Badawczy: Puławy, Poland, 2008; Volume 11, pp. 81–89.
38. Szyszlak-Bargłowicz, J.; Zając, G.; Piekarski, W. Energy biomass characteristics of chosen plants. *Int. Agrophisics* **2012**, *26*, 175–179. [CrossRef]
39. Borkowska, H.; Molas, R.; Skiba, D.; Machaj, H. Plonowanie oraz wartość energetyczna ślazowca pensylwańskiego w zależności od poziomu nawożenia azotem (Yielding and energy value of Virginia fanpetals in relation to the level of nitrogen fertilization). *Acta Agrophysica* **2016**, *23*, 5–14. (In Polish)
40. Bilandžija, N.; Krička, T.; Matin, A.; Leto, J.; Grubor, M. Effect of Harvest Season on the Fuel Properties of *Sida hermaphrodita* (L.) Rusby Biomass as Solid Biofuel. *Energies* **2018**, *11*, 3398. [CrossRef]
41. Możdżer, E.; Siwek, H.; Włodarczyk, M.; Bury, M.; Kitczak, T. Influence of the Virginia Fanpetals Cultivation Method on Calorific Value, Content and Dynamics of Macronutrient Uptake. *J. Ecol. Eng.* **2020**, *21*, 120–128. [CrossRef]
42. Šurić, J.; Brandić, I.; Peter, A.; Bilandžija, N.; Leto, J.; Karažija, T.; Kutnjak, H.; Poljak, M.; Voća, N. Wastewater Sewage Sludge Management via Production of the Energy Crop Virginia Mallow. *Agronomy* **2022**, *12*, 1578. [CrossRef]
43. Borkowska, H.; Molas, R. Two extremely different crops, *Salix* and *Sida*, as sources of renewable bioenergy. *Biomass Bioenergy* **2012**, *36*, 234–240. [CrossRef]
44. Stolarski, M.; Wróblewska, H.; Szczukowski, S.; Tworkowski, J.; Cichy, W. Charakterystyka biomasy wierzby i ślazowca pensylwańskiego jako potencjalnego surowca przemysłowego (Characteristics of the biomass of Willow and Virginia fanpetals as a potential industrial raw material). *Fragm. Agron.* **2006**, *3*, 277–289. (In Polish)
45. Slepetys, J.; Kadziuliene, Z.; Sarunaite, L.; Tilvikiene, V.; Kryzeviciene, A. Biomass potential of plants grown for bioenergy production. In *Renewable Energy and Energy Efficiency. Proceedings of the International Scientific Conference*; Rivza, P., Ed.; Latvia University of Agriculture: Jeglava, Latvia, 2012; pp. 66–72.
46. Boehmel, C.; Lewandowski, I.; Claupein, W. Comparing annual and perennial energy cropping systems with different management intensities. *Agric. Syst.* **2008**, *96*, 224–236. [CrossRef]

47. Diltz, R.; Johnson, G. Sustainable land use for bioenergy in the 21st century. *Ind. Biotechnol.* **2011**, *7*, 436–447. [CrossRef]
48. Fazio, S.; Monti, A. Life cycle assessment of different bioenergy production systems including perennial and annual crops. *Biomass Bioenergy* **2011**, *35*, 4868–4878. [CrossRef]
49. Fischer, G.; Prieler, S.; Van Veldhuizen, H.; Lensink, S.M.; Londo, M.; De Wit, M. Biofuel production potentials in Europe: Sustainable use of cultivated land and pastures. Part 1: Land productivity potentials. *Biomass Bioenergy* **2010**, *34*, 159–172. [CrossRef]
50. Roszkowski, A. Energia z biomasy-efektywność, sprawność i przydatność energetyczna. Cz. 1 (Energy from biomass—Effectiveness, efficiency and energetic usability Part 1). *Probl. Inżynierii Rol.* **2013**, *1*, 97–124. (In Polish)
51. Šiaudinis, G.; Jasinskas, A.; Šarauskis, E.; Steponavicius, D.; Karčauskienė, D.; Liaudanskienė, I. The assessment of Virginia mallow (*Sida hermaphrodita* Rusby) and cup plant (*Silphium perfoliatum* L.) productivity, physico–mechanical properties and energy expenses. *Energy* **2015**, *93*, 606–612. [CrossRef]
52. Jankowski, K.J.; Dubis, B.; Budzyński, W.S.; Bórawski, P.; Bułkowska, K. Energy efficiency of crops grown for biogas production in a large-scale farm in Poland. *Energy* **2016**, *109*, 277–286. [CrossRef]
53. Stolarski, M.J.; Tworkowski, J.; Szczukowski, S.; Kwiatkowski, J.; Graban, L. Opłacalność i efektywność energetyczna produkcji biomasy ślazowca pensylwańskiego w zależności od stosowanego materiału siewnego (Cost-effectiveness and energy efficiency of the production of Pennsylvanian mallow biomass depending on the seed used). *Fragm. Agron.* **2014**, *31*, 96–106. (In Polish)
54. Stolarski, M.J.; Krzyżaniak, M.; Warmiński, K.; Olba-Zięty, E.; Penni, D.; Bordiean, A. Energy efficiency indices for lignocellulosic biomass production: Short rotation coppices versus grasses and other herbaceous crops. *Ind. Crops Prod.* **2019**, *135*, 10–20. [CrossRef]
55. Jankowski, K.J.; Dubis, B.; Sokólski, M.M.; Załuski, D.; Bórawski, P.; Szempliński, W. Biomass yield and energy balance of Virginia fanpetals in different production technologies in north-eastern Poland. *Energy* **2019**, *185*, 612–623. [CrossRef]
56. Franzaring, J.; Schmid, I.; Bäuerle, L.; Gensheimer, G.; Fangmeier, A. Investigations on plant functional traits, epidermal structures and the ecophysiology of the novel bioenergy species *Sida hermaphrodita* Rusby and *Silphium perfoliatum* L. *J. Appl. Bot. Food Qual.* **2014**, *87*, 36–45. [CrossRef]
57. Molas, R.; Borkowska, H.; Kupczyk, A.; Osiak, J. Virginia fanpetals (*Sida*) biomass can be used to produce high-quality bioenergy. *Agron. J.* **2018**, *110*, 24–29. [CrossRef]

 *agronomy*

Article

The True Costs and Benefits of Miscanthus Cultivation

Moritz Wagner [1,*], Bastian Winkler [2], Jan Lask [2], Jan Weik [2], Andreas Kiesel [2], Mirjam Koch [1], John Clifton-Brown [3] and Moritz von Cossel [2]

[1] Department of Applied Ecology, Hochschule Geisenheim University, 65366 Geisenheim, Germany
[2] Department of Biobased Resources in the Bioeconomy (340b), Institute of Crop Science, University of Hohenheim, 70599 Stuttgart, Germany
[3] iFZ Research Centre for Biosystems, Justus Liebig University, Heinrich-Buff-Ring 26, 35392 Gießen, Germany
* Correspondence: moritz.wagner@hs-gm.de

Abstract: Agroecosystems provide numerous ecosystem services (ESs) such as provisioning, regulating, habitat and cultural services. At the same time, the management of these agroecosystems can cause various negative impacts on the environment such as the generation of greenhouse gas emissions. However, the way humans manage agroecosystems often focuses only on the production of agricultural goods, which yield monetary benefits in the short term but do not include the positive and negative external effects on ESs. In order to enable a holistic assessment of the economic and environmental costs and benefits, the current study combines the production costs, the monetary value of the ESs provided and the monetization of the environmental impacts caused by the management of agroecosystems using the perennial crop miscanthus as an example. Depending on the scenario assessed, the cultivation of miscanthus leads to a net benefit of 140 to 3051 EUR ha^{-1} yr^{-1}. The monetary value of the ESs provided by the miscanthus cultivation thereby considerably outweighs the internal and external costs. The approach applied allows for a holistic assessment of the benefits and costs of agroecosystems and thus enables management decisions that are not only based on the biomass yield but include the various interactions with the environment.

Keywords: life-cycle assessment; ecosystem services; true cost accounting; monetization; bioeconomy; miscanthus

Citation: Wagner, M.; Winkler, B.; Lask, J.; Weik, J.; Kiesel, A.; Koch, M.; Clifton-Brown, J.; von Cossel, M. The True Costs and Benefits of Miscanthus Cultivation. *Agronomy* **2022**, *12*, 3071. https://doi.org/10.3390/agronomy12123071

Academic Editor: Agnes van den Pol-van Dasselaar

Received: 8 November 2022
Accepted: 2 December 2022
Published: 4 December 2022

Publisher's Note: MDPI stays neutral with regard to jurisdictional claims in published maps and institutional affiliations.

1. Introduction

Ecosystems provide numerous benefits for humans, such as food and air to breathe, without which survival would not be possible [1–3]. Agroecosystems play a crucial role in the preservation of the provision of these services, as in Germany for example, more than 50% of the land area is used for agriculture [4]. The incorporation of perennial crops such as miscanthus (*Miscanthus* ANDERSSON) or cup plant (*Silphium perfoliatum* L.) into the predominately annual monoculture cropping systems offers the chance to increase the provision of various ecosystem services (ESs), including water purification, pollination and biological plant protection in addition to the provision of biomass [5–8]. Von Cossel et al. (2020b) [9] showed for miscanthus, a multipurpose industrial perennial crop for providing biomass for bioenergy and biobased products [10,11], that the monetary value of the ESs provided can be more than three times the profit a farmer earns for just selling the biomass.

However, the way humans manage ecosystems is often focused only on the production of agricultural goods, including biomass, and less attention is paid to the complexity of the long-term factors underlying ESs, especially with regard to their mutual interactions (synergies, trade-offs, etc.) [12]. For example, agricultural activities often lead to a decline in biodiversity despite the potential importance of biodiversity for the resilience of the agroecosystem [13]. In addition, the existence of many indirect ecosystem factors, some of which are still unknown in their importance for provisioning ESs, like faunal species diversity, is just taken for granted [13]. The main reason is that in the management of conventional cropping systems, usually only ESs with monetary benefits in the short term are considered, such as biomass provision [14]. On the other side, despite levels being usually lower than annual cropping systems, miscanthus cultivation can cause various negative impacts on the environment, such as the generation of greenhouse gas emissions (GHG) through the combustion of fossil fuels or nutrient leaching through the application of mineral and organic fertilizers [15]. These emissions cause considerable external costs, which at the moment are usually disregarded or not adequately taken into account [16]. In summary, the cultivation of agroecosystem may lead to various positive and negative external effects, which are currently not included in the management process.

So far, true-cost-accounting approaches usually only focus on costs without taking benefits into account. Therefore, this paper aims to apply a new expanded approach by taking a more holistic look at both benefits and costs (the sum of which would result in "true costs and benefits") of cropping systems using miscanthus (*Miscanthus* ANDERSSON) cultivation as an example for a perennial, industrial crop. This assessment combines the production costs, the monetary value of the ESs provided as well as the monetization of the environmental impacts caused by the cultivation and the harvest of the biomass. The combination of these analyses allows for a holistic assessment of the true economic and environmental value provided by cropping systems in monetary terms.

2. Materials and Methods

2.1. Goal and Scope

The goal of the current study is the holistic assessment of the economic and environmental costs and benefits of miscanthus cultivation for society. The current study focuses, therefore, on crop cultivation, harvest, and transport of the biomass to the farm gate. The costs and benefits occurring in the further downstream value chain are not assessed. In addition, social costs and benefits are not included in this study. The economic costs of miscanthus cultivation are assessed based on the production costs. The environmental benefits are represented via the monetized ESs provided by miscanthus, including the revenue of the biomass sale. The environmental costs of miscanthus cultivation are assessed by conducting a life-cycle assessment (LCA) and monetizing the identified environmental impacts. This allows us to internalize these previously external costs. However, there are various monetization approaches available that differ substantially in their monetization factors, for example, due to differences in the selected cost approach or the area of reference [17,18]. Therefore, the influence of the selected monetization approach is critically analyzed and discussed.

2.2. Production Costs

The production costs of miscanthus cultivation in Germany are based on Winkler et al. (2020) [19] and comprise average machine, material, energy and labor costs as well as interest. Winkler et al. (2020) [19] calculated the production costs for two different cultivation systems (conventional and organic), two yield levels, field sizes and farm-field distances, as well four utilization pathways differing in harvest regimes and methods. In the present study, a conservative approach was selected, setting the field-farm distance at 10 km and the field size at 1 ha with an annual average dry matter (DM) yield of 15 Mg ha^{-1}. The miscanthus harvest was considered annually via direct cutting and chipping on the field by a forage harvester in March, which is the standard harvest procedure for miscanthus in Germany, as combustion is still the most common form of use [20,21].

In the production costs calculated by Winkler et al. (2020) [19], land costs were not included. According to the German Federal Office for Agriculture and Food, the annual lease prices for agricultural land per hectare in the year 2020 amounted to 375 EUR (BLE 2021). For a holistic assessment, the costs of land have to be included in the production costs.

2.3. Monetization of Ecosystem Services

The ESs assessed, as well as their monetary value, are based on von Cossel et al. (2020b) [9] (Figure 1). The revenues generated by the biomass sale are based on Winkler et al. (2020) [19], assuming biomass prices between 65 to 95 EUR per Mg chopped miscanthus material and a biomass yield of 15 Mg DM ha^{-1} yr^{-1}. In von Cossel et al. (2020b) [9], the calculation of the environmental benefits provided by the sequestration of CO_2 in the soil is based on a CO_2 emission certificate price of 26.83 EUR per Mg CO_2. In order to be consistent with the monetization factors used for the LCA results, the approach of avoidance cost was applied, as presented in Trinomics (2020) [22]. Therefore, an environmental benefit of 102.50 EUR per Mg CO_2 sequestered in the soil is applied in the current study. The CO_2 emissions, which can be substituted by the miscanthus-based products [23,24], are not included because the current study only focuses on the cultivation of miscanthus and not on the entire miscanthus-based value chain. In addition, the monetary values of both *N_2 fixation* and *nutrient recycling* are excluded from the current study because these two ESs lead to a reduction in the amount of mineral fertilizer required [25,26] which is already included in the production costs. Furthermore, the ES *waste treatment—reduced nutrient leaching* is based on a reduction in nitrate leaching when comparing the cultivation of miscanthus and maize [9]. As the current study focuses on the assessment of the costs and benefits of one cultivation system and does not apply a comparison, this ES is not included.

Figure 1. Overview of the main ecosystem services considered in this study in accordance with von Cossel et al. (2020b) [9]. Provisioning ecosystem services (in red): (1) raw material, (2) genetic resources, (3) fresh water/groundwater, (4) ornamental resources; Regulating ecosystem services (in orange): (5) air quality regulation, (6) climate regulation, (7) improvement of soil fertility, (8) erosion prevention, (9) moderation of extreme events; habitat ecosystem services (in green): (10) pollination and biocontrol; cultural ecosystem services (in blue): (11) aesthetic information and (12) recreation and tourism.

For several ESs, the monetary value in EUR h^{-1} yr^{-1} is given by von Cossel et al. (2020b) [9] as a range with minimum and maximum values. In the current study, the mean of these values is used to display the average benefit of the ESs provided. The influence of this assumption is analyzed in a scenario analysis, applying the minimum and maximum values.

2.4. Assessment and Monetization of the Environmental Impacts of Miscanthus Cultivation

In order to assess the environmental performance of the miscanthus cultivation a cradle-to-farm-gate LCA was conducted following the structure of the ISO standards 14040 and 14044 [27,28]. In the current study, the 16 impact categories and assessment methods were applied, which are included in the Product Environmental Footprint (PEF) methodology of the European Commission [29]. The selected functional unit (FU) is 1 ha under miscanthus cultivation with an average yield of 15 Mg DM ha^{-1} yr^{-1}. An area-based FU is chosen so that a consistent comparison is possible between production costs, benefits provided, in form of ESs, which are given on a hectare basis and the results of the LCA.

The data used for modeling the foreground system is based on the miscanthus cultivation system described in Winkler et al. (2020) [19], which also provides the basis for the calculation of the production costs (see Section 2.2). Summaries of the agricultural operations conducted during the cultivation period and the main in- and outputs are presented in Tables 1 and 2. Inputs that are only applied in the establishment or the harvest phase, such as pesticides or fertilizer, are converted to the entire cultivation period of 20 years. Nitrous oxide (N_2O), nitrate (NO_3^-) and phosphorus emissions due to the use of mineral fertilizers are modeled according to the recommendations of Pant and Zampori (2019) [29]. N_2O emissions from harvest residues were modeled according to IPCC (2019) [30]. The proportion of harvest residues in the form of leaves and stubbles is taken from Lask et al. (2021) [31]. Heavy metal emissions to agricultural soils caused by the application of fertilizers and pesticides are estimated based on Freiermuth (2006) [32]. It is assumed that 90% of the pesticides are released into agricultural soils, 9% into air and 1% to water [29]. Background data on emissions associated with the production of the input substrates such as fertilizers or pesticides are based on the ecoinvent database 3.8 using the cut-off system model [33]. In the current study, market datasets are used in order to include average transport impacts [33]. The software openLCA 1.10.3 is applied for the modeling and the calculation of the impacts using the integrated PEF method EF 3.0 (adapted).

Table 1. Agricultural operations during a 20-year miscanthus cultivation period (adapted from Winkler et al. (2020) [19]).

Agricultural Operation	Frequency per Cultivation Period
Plowing	2
Rotary harrowing	1
Planting	1
Mulching—first year	1
Herbicide spraying	2
Fertilizing	18
Harvesting	19

Table 2. Main inputs and outputs of miscanthus cultivation per year (adapted from Winkler et al. (2020) [19]).

Input/Output	Amount	Unit
N	47	kg ha^{-1} yr^{-1}
P	5	kg ha^{-1} yr^{-1}
K	82	kg ha^{-1} yr^{-1}
Herbicides	0.34	kg ha^{-1} yr^{-1}
Biomass dry matter yield	15	Mg ha^{-1} yr^{-1}

The monetization factors for the respective impact category are based on the central values stated in the report prepared by Trinomics (2020) [22] (see Table 3), as this is the only study that suggests a set of monetization factors which are explicitly meant to be used in combination with the PEF method applied in the current study [34]. In order to test the influence of the monetization factors, a sensitivity analysis was conducted, applying, in addition to central values, a low and high monetization value, as shown by Trinomics (2020) [22]. For terrestrial eutrophication, no satisfactory monetization approach is available at present, which could be applied to the PEF method at this early development stage [22,34]. Therefore, the external costs of this impact category are not included in the current study.

Table 3. Monetization factor for the impact categories assessed in EUR_{2018} per unit impact based on Trinomics (2020) [22].

Environmental Impact Category	Unit	Monetization Factor [EUR_{2018} per Unit Impact]
Acidification	mol H^+ eq.	0.344
Climate change	kg CO_2 eq.	0.1025
Ecotoxicity, freshwater	CTUe	0.0000382
Eutrophication, freshwater	kg P eq.	1.92
Eutrophication, marine	kg N eq.	3.21
Eutrophication, terrestrial	mol N eq.	-
Human toxicity, cancer	CTUh	902,616
Human toxicity, non-cancer	CTUh	163,447
Ionizing radiation	kBq U-235 eq.	0.0012
Land use	Pt	0.000175
Ozone depletion	kg CFC11 eq.	31.4
Particulate matter	disease inc.	784,126
Photochemical ozone formation	kg NMVOC eq.	1.19
Resource use, fossils	MJ	0.0013
Resource use, minerals and metals	kg Sb eq.	1.64
Water use	m^3 water eq.	0.00499

3. Results

The following sections describe the results of the analyses of (i) the monetary values provided by miscanthus cultivation and (ii) the environmental and economic costs of miscanthus cultivation.

3.1. Monetary Values of the ESs Provided by Miscanthus Cultivation

The estimated average monetary values of the ESs provided annually by cultivating miscanthus on 1 ha sum up to 3118 EUR (see Table 4). In the current study, the average monetary values for the ESs provided by miscanthus cultivation were applied, as explained in Section 2.3. In case the minimum monetary values of the ESs were used, that excluded location-specific ESs (e.g., flood plain management, erosion prevention, provision of drinking water through sediment passage), the ESs combined were worth 1985 EUR ha^{-1} yr^{-1}. Assuming the maximum monetary values of the ESs provided (including location-specific ESs) they would account for 4250 EUR ha^{-1} yr^{-1} [9].

Table 4. Single monetary values and total value of the ESs in EUR ha^{-1} yr^{-1} provided by miscanthus cultivation adapted from Winkler et al. (2020) [19] and von Cossel et al. (2020b) [9]. For those ESs in which ranges of variation are given in von Cossel et al. (2020b) [9], the arithmetic means were calculated. Some ESs (e.g., nutrient cycling, N$_2$ fixation) shown in von Cossel et al. (2020b) [9] were excluded in this study, as explained in Section 2.3.

ES Category	ES	Value (EUR ha^{-1} yr^{-1})
Provisioning services	Raw material	1200
	Genetic resources	18
	Fresh water/groundwater	56
	Ornamental resources	17
Regulating services	Air quality regulation	64
	Climate regulation	828
	Improvement of soil fertility	23
	Erosion prevention	22
	Moderation of extreme events	386
Habitat services	Pollination and biocontrol	50
Cultural services	Aesthetic information	429
	Recreation and tourism	27
Total	-	**3118**

3.2. Environmental and Economic Costs of Miscanthus Production

In the scenario described, the annual production costs amount to 1010 EUR ha^{-1}, including the lease price for agricultural land. Besides the land costs, the establishment of the miscanthus plantation in the first year of cultivation is one of the main cost drivers [19].

In Table 5, the LCA results of the miscanthus cultivation per ha in the analyzed impact categories are displayed.

Table 5. Environmental impact of miscanthus cultivation per environmental impact category and ha and year.

Environmental Impact Category	Impact Result	Unit
Acidification	22.98	mol H$^+$ eq.
Climate change	1248.01	kg CO$_2$ eq.
Ecotoxicity, freshwater	3.11×10^4	CTUe
Eutrophication, freshwater	0.42	kg P eq.
Eutrophication, marine	16.24	kg N eq.
Eutrophication, terrestrial	94.00	mol N eq.
Human toxicity, cancer	1.51×10^{-6}	CTUh
Human toxicity, non-cancer	3.49×10^{-5}	CTUh
Ionizing radiation	42.79	kBq U-235 eq.
Land use	5.10×10^5	Pt
Ozone depletion	8.82×10^{-5}	kg CFC11 eq.
Particulate matter	0.00015	disease inc.
Photochemical ozone formation	4.98	kg NMVOC eq.
Resource use, fossils	9949.30	MJ
Resource use, minerals and metals	0.02	kg Sb eq.
Water use	394.86	m^3 water eq.

In Table 6, the monetized environmental impacts are shown applying the low, central and high monetization factors. The costs of the environmental impacts are, to a great extent, caused by the impact categories climate change (fossil fuel combustion and fertilizer production, as well as fertilizer-induced emissions), land used for agricultural production, and particulate matter formation (fertilizer-induced ammonia emissions).

Table 6. Monetized environmental impact of miscanthus cultivation applying the monetization factors by Trinomics (2020) [22].

Environmental Impact Categories	Monetized Environmental Impacts (EUR$_{2018}$)		
	Low	Central	High
Acidification	4.04	7.90	37.16
Climate change	76.75	127.92	241.61
Ecotoxicity, freshwater	7.44×10^{-20}	1.19	5.85
Eutrophication, freshwater	0.11	0.81	0.92
Eutrophication, marine	52.13	52.13	52.13
Eutrophication, terrestrial	0	0	0
Human toxicity, cancer	0.26	1.36	4.21
Human toxicity, non-cancer	1.05	5.70	26.36
Ionizing radiation	0.03	0.05	1.97
Land use	44.39	89.29	178.07
Ozone depletion	2.01×10^{-3}	2.77×10^{-3}	0.01
Particulate matter	99.30	117.62	180.69
Photochemical ozone formation	4.34	5.93	9.47
Resource use, fossils	0	12.93	67.66
Resource use, minerals and metals	0	0.03	0.11
Water use	1.65	1.97	93.15
Total (EUR ha^{-1} yr^{-1})	**284.06**	**424.84**	**899.36**

Figure 2 shows the "Standard scenario" in which average ESs are provided by the miscanthus cultivation and central monetization factors are applied. The ES provided are divided into provisioning services, which are mainly dominated by revenues generated through the sale of the biomass, regulating and habitat services, and cultural services. In the standard scenario, the monetarized benefits of miscanthus cultivation considerably outweigh the economic and environmental costs resulting in a true benefit of 1762 EUR ha^{-1} yr^{-1}.

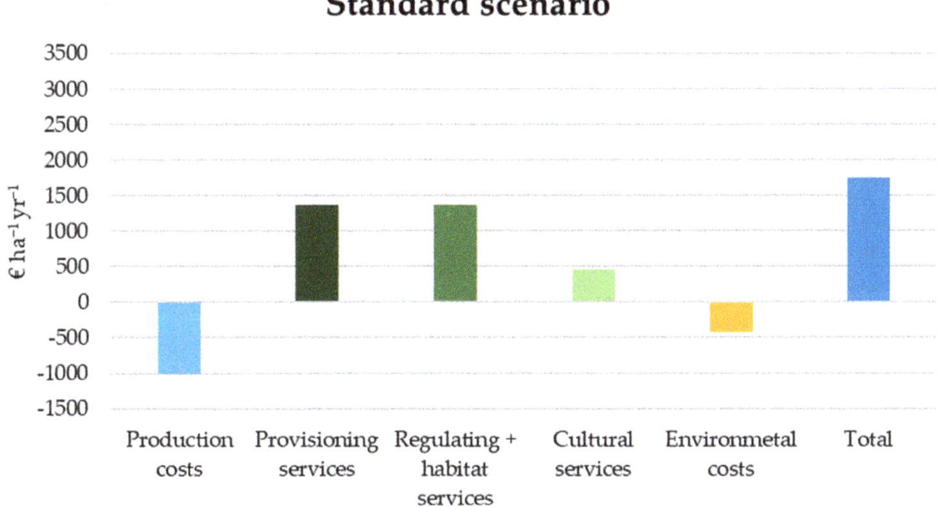

Figure 2. True costs and benefits in the standard scenario, assuming average ESs provision and central monetization factors for environmental impacts.

Figure 3 shows the "Best-case scenario", in which maximum ESs are provided by the miscanthus cultivation (including location-specific ESs) and low monetization factors are applied. The substantially higher location, specific ESs and the low environmental costs lead to a total benefit of miscanthus cultivation of 3051 EUR ha^{-1} yr^{-1} (see Figure 3).

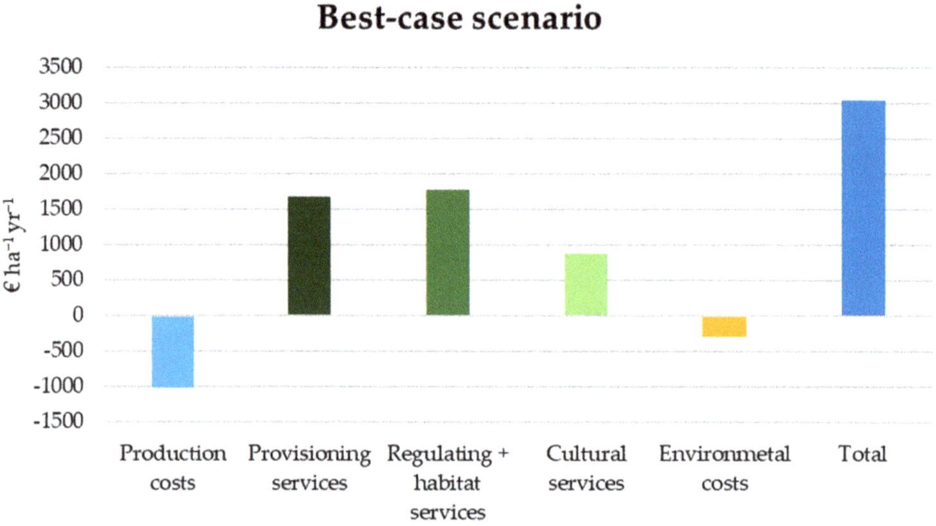

Figure 3. True costs and benefits in the best-case scenario assuming maximum ESs provision and low monetization factors for environmental impacts.

Figure 4 shows the "Worst-case scenario", in which minimum ESs are provided by the miscanthus cultivation (excluding location-specific ES) and high monetization factors are applied. This still leads, in total, to a benefit of 140 EUR ha^{-1} yr^{-1} (see Figure 4).

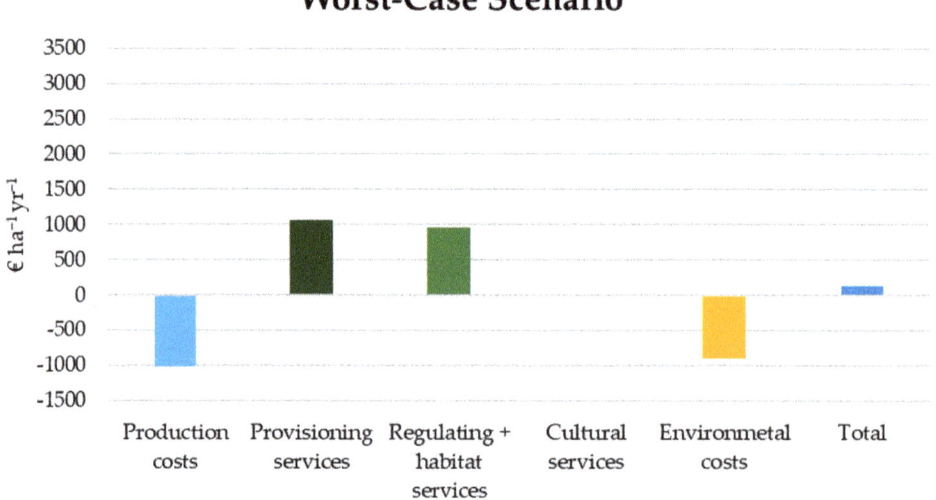

Figure 4. True costs and benefits in the worst-case scenario assuming minimum ESs provision and high monetization factors for environmental impacts.

4. Discussion

4.1. Discussion of the Data Used and the Methodologies Applied

In the following, the influence of the data used and the methodologies applied on the results are critically discussed. In particular, the biomass selling price is an important factor in the assessment of the economic costs and benefits of miscanthus cultivation. In the current study, a conservative biomass selling price of 80 EUR Mg^{-1} DM was applied. If the miscanthus is cultivated for other utilization pathways, the selling price could be substantially higher (105–600 EUR Mg^{-1} DM) [19]. As a result, the monetary value provided by the cultivation of 1 ha miscanthus would increase significantly. Besides the selected utilization pathway, also the annual fluctuations in biomass demand have a considerable influence on the biomass selling price. For a realistic evaluation of the costs and benefits of different cropping systems, it has, therefore, to be emphasized that it is crucial to use average data and to assess and discuss the uncertainty included in the results.

Besides the availability of reliable data, one significant barrier to the implementation of the approach described is the variety of methods available for assessing the economic costs, the environmental benefits and impacts, as well as their monetization, which hinders the comparability of the results. For example, no standardized life-cycle costing (LCC) framework is available for the agricultural and food sector [35]. According to Degieter et al. (2022) [35], it is crucial to include all cost categories (e.g., inputs, labor) and to report all methodological choices made during the preparation of the study to provide comparable and comprehensible results.

In addition to the method used to estimate the costs, the selected life-cycle impact assessment (LCIA) method and the chosen monetization approach significantly influence the results [17,18]. In order to evaluate this impact, the monetized environmental impacts of the miscanthus cultivation were also assessed in the current study applying the LCIA method Recipe 1.13 Midpoint (Hierarchist) and the monetization approach described in the Environmental Prices Handbook [36,37]. Using these two approaches, the monetized environmental impacts of 1 ha miscanthus cultivation amount to 1266 EUR ha^{-1} yr^{-1} (see Table S1 in the supplementary material) and are therefore slightly higher than the values applied in the worst-case scenario (899 EUR ha^{-1} yr^{-1}). Applying the costs assessed in the sensitivity analysis in the standard scenario, would still yield a significant benefit. Only the worst-case scenario would lead to net costs due to the cultivation of miscanthus.

The values from the study by von Cossel et al. (2020b) [9], which were used for the assessment of the monetary values of the ESs, are subject to a large degree of uncertainty. One reason is the large temporal and spatial variability of the assessed Ess, for instance, due to variations in the biomass yield. These might be caused by climate change-induced variations in growing conditions such as drought periods [38,39], changes in precipitation distributions [40], or frost damage due to lack of a snow cover [41]. In addition, temporal and spatial variations may also occur in the synergies and trade-offs between individual ESs. The age of the miscanthus stand has, for example, a significant influence on erosion control [42] due to better ground cover and biomass yield. After the establishment phase, both increase significantly. However, there are also trade-offs between different ESs, such as biomass yield and pollination. Gaps in the miscanthus stand could lead to an increased appearance of wild plants, which on the one hand, provide the fauna with an additional food spectrum [43,44]. However, on the other hand, the gaps could permanently worsen the biomass yield performance of miscanthus in the following years [19,44].

A more accurate and reliable assessment of the overall monetary value of the ESs provided by different agroecosystems, therefore, requires region- and year-specific calculation. Nevertheless, despite the uncertainty included in the calculation of the ESs monetary values, the overall values considered here in the best-, standard- and worst-case scenarios provide a reliable basis to serve as a benchmark for future studies, as also critically discussed by von Cossel et al. (2020b) [9].

In a combined assessment of environmental impacts and ESs, as was undertaken in the context of the present study, it is crucial to ensure that double counting of environmental impacts is avoided. One example of this is the ES *carbon sequestration* and the LCA impact category of *climate change*. Various LCA studies include the carbon sequestered in the soil in their LCIA results [45]. In case the carbon sequestered in the soil is already accounted for in the LCIA results in the impact category *climate change*, it cannot be accounted for in the monetization of the ESs. The same could be the case for the ES *improvement of the soil quality*, for which LCIA methods also already exist or are in development [46]. Alejandre et al. (2019) [47] analyzed which ESs can be evaluated via existing LCIA methods and which have to be evaluated by other ES assessment methods. Their study could be used to identify possible areas of overlap between ESs and LCA results and thus reduce the risk of double counting.

In addition, it has to be emphasized when discussing the different assessment approaches that the evaluation of the economic and environmental costs of miscanthus production includes the whole previous value chain (e.g., production of the input substrates). The evaluation of the environmental benefits, however, only focuses on the field level and excludes the upstream processes since there is not enough information available for a holistic ESs assessment across the entire whole value chain. Another methodological consideration is the inclusion of social costs and benefits in the future to complement the economic and environmental towards a holistic true cost accounting approach for the assessment of agricultural production systems.

4.2. Discussion of the Results and Applicability in Practice

The results of the current study demonstrate a clear total benefit of the miscanthus cultivation independently of the scenarios assessed. However, as also discussed in the section above, these results are associated with a high degree of uncertainty. Increased harmonization and standardization regarding the monetization of the LCA results and especially the assessment of ESs is needed to reduce this uncertainty and to ensure comparability within studies, but also with other true-cost accounting studies in the agricultural sector [48]. In order to use these results in the decision process or for the development of subsidies, comparable assessments for other cropping systems are needed. Only by comparing the local costs and benefits of different cropping systems well-founded decisions about the advantageousness of the respective systems can be made. Thereby it is crucial to apply a holistic view of the cropping systems under study, especially when evaluating the costs and benefits of annual crops, and to include crop interactions in the assessment, such as positive effects between different crops in a crop rotation [49].

In the current study, a cradle-to-farm gate approach was applied. However, depending on the goal of the study, other system boundaries may be more suitable because the selection of the system boundaries can have a significant influence on the results. Von Cossel et al. (2020b) [9], for example, showed in their publication that the substitution of the fossil alternatives by miscanthus-based isobutanol could lead to CO_2 savings of 19.1 Mg CO_2-eq. ha^{-1} a^{-1}. This would correspond to an additional benefit of 1958 EUR ha^{-1} yr^{-1}.

The results presented here demonstrate that ESs provided by miscanthus cultivation have a significant value besides the sole provision of biomass. However, these ESs are currently not included in the management process. One possibility to holistically include the ESs provided by the cultivation systems into the farmer's management process is to encourage sustainable agricultural practices via subsidies [50] or direct payments for ecosystem services (PES) [51]. Miscanthus, for example, has been included on the positive list for cultivation in ecological focus areas by the European Commission in 2018 [52]. For farmers, this means that they can receive an extra subsidy for cultivating miscanthus. The careful development of subsidies could thereby lead to crop selection and rotation planning, which is not only based on the sale price of the biomass but also on much-needed and wanted ESs. Knowing the true costs and benefits of different cropping systems could be a valuable basis for making decisions about the development of such subsidies.

5. Conclusions

The approach applied in the current study allows for a holistic assessment of the benefits and costs of agroecosystems through the inclusion of the monetary value of various ESs, the production costs as well as the monetized environmental impacts. For miscanthus, it could be shown that the monetary value of the ESs provided by its cultivation considerably outweigh the internal and external costs. This approach thereby enables management decisions, which are not only based on the biomass yield but include the various interactions with the environment. In addition, the results of such an approach provide valuable insights for the development of environmental incentives and the determination of the amount of payment farmers receive for environmental-friendly farming practices.

However, there is still considerable uncertainty associated with the results. Standardized ES assessment and monetization methods are required in order to enable sound comparison between different cultivation systems in terms of economic and environmental sustainability. Furthermore, the approach has to be applied using local data because ESs provided by agroecosystems can vary greatly locally, for example, in regard to erosion control or flood prevention.

Supplementary Materials: The following supporting information can be downloaded at: https://www.mdpi.com/article/10.3390/agronomy12123071/s1, Table S1: Monetized environmental impacts of miscanthus cultivation per hectare and year applying the ReCiPe (H) 1.13 methodology.

Author Contributions: Conceptualization, all authors; methodology, B.W., J.L., M.v.C. and M.W.; software, M.W.; investigation, J.L., M.v.C. and M.W.; writing—original draft preparation, B.W., J.L., M.v.C. and M.W.; writing—review and editing, all authors; visualization, M.v.C. and M.W. All authors have read and agreed to the published version of the manuscript.

Funding: This research received no external funding.

Data Availability Statement: Data available on request made to the corresponding author M.W.

Conflicts of Interest: The authors declare no conflict of interest.

References

1. Carpenter, S.R.; Mooney, H.A.; Agard, J.; Capistrano, D.; Defries, R.S.; Díaz, S.; Dietz, T.; Duraiappah, A.K.; Oteng-Yeboah, A.; Pereira, H.M.; et al. Science for managing ecosystem services: Beyond the Millennium Ecosystem Assessment. *Proc. Natl. Acad. Sci. USA* **2009**, *106*, 1305–1312. [CrossRef] [PubMed]
2. Costanza, R.; d'Arge, R.; de Groot, R.; Farber, S.; Grasso, M.; Hannon, B.; Limburg, K.; Naeem, S.; O'Neill, R.V.; Paruelo, J.; et al. The value of the world's ecosystem services and natural capital. *Nature* **1997**, *387*, 253–260. [CrossRef]
3. Millennium Ecosystem Assessment. *Ecosystems and Human Well-Being: Synthesis*; Island Press: Washington, DC, USA, 2005.
4. Statistisches Bundesamt. Land-und Forstwirtschaft, Fischerei: Bodenfläche nach Art der tatsächlichen Nutzung. Fachserie 3 Reihe 5.1. 2022, pp. 191–192. Available online: https://www.destatis.de/DE/Themen/Branchen-Unternehmen/Landwirtschaft-Forstwirtschaft-Fischerei/Flaechennutzung/Publikationen/Downloads-Flaechennutzung/bodenflaechennutzung-20305102 17005.html (accessed on 21 October 2022).
5. von Cossel, M.; Amarysti, C.; Wilhelm, H.; Priya, N.; Winkler, B.; Hoerner, L. The replacement of maize (*Zea mays* L.) by cup plant (*Silphium perfoliatum* L.) as biogas substrate and its implications for the energy and material flows of a large biogas plant. *Biofuels Bioprod. Biorefining* **2020**, *14*, 152–179. [CrossRef]
6. Emmerling, C.; Pude, R. Introducing Miscanthus to the greening measures of the EU Common Agricultural Policy. *GCB Bioenergy* **2017**, *9*, 274–279. [CrossRef]
7. Asbjornsen, H.; Hernandez-Santana, V.; Liebman, M.; Bayala, J.; Chen, J.; Helmers, M.; Ong, C.K.; Schulte, L.A. Targeting perennial vegetation in agricultural landscapes for enhancing ecosystem services. *Renew. Agric. Food Syst.* **2014**, *29*, 101–125. [CrossRef]
8. Weißhuhn, P.; Reckling, M.; Stachow, U.; Wiggering, H. Supporting Agricultural Ecosystem Services through the Integration of Perennial Polycultures into Crop Rotations. *Sustainability* **2017**, *9*, 2267. [CrossRef]
9. von Cossel, M.; Winkler, B.; Mangold, A.; Lask, J.; Wagner, M.; Lewandowski, I.; Elbersen, B.; Eupen, M.; Mantel, S.; Kiesel, A. Bridging the Gap Between Biofuels and Biodiversity Through Monetizing Environmental Services of Miscanthus Cultivation. *Earth's Future* **2020**, *8*, e2020EF001478. [CrossRef]
10. Anderson, E.; Arundale, R.; Maughan, M.; Oladeinde, A.; Wycislo, A.; Voigt, T. Growth and agronomy of Miscanthus x giganteus for biomass production. *Biofuels* **2011**, *2*, 71–87. [CrossRef]

11. Lewandowski, I.; Clifton-Brown, J.; Trindade, L.M.; van der Linden, G.C.; Schwarz, K.-U.; Müller-Sämann, K.; Anisimov, A.; Chen, C.-L.; Dolstra, O.; Donnison, I.S.; et al. Progress on Optimizing Miscanthus Biomass Production for the European Bioeconomy: Results of the EU FP7 Project OPTIMISC. *Front. Plant Sci.* **2016**, *7*, 1620. [CrossRef]

12. Bethwell, C.; Burkhard, B.; Daedlow, K.; Sattler, C.; Reckling, M.; Zander, P. Towards an enhanced indication of provisioning ecosystem services in agro-ecosystems. *Environ. Monit. Assess.* **2021**, *193*, 269. [CrossRef]

13. Gascon, C.; Brooks, T.M.; Contreras-MacBeath, T.; Heard, N.; Konstant, W.; Lamoreux, J.; Launay, F.; Maunder, M.; Mittermeier, R.A.; Molur, S.; et al. The importance and benefits of species. *Curr. Biol.* **2015**, *25*, R431–R438. [CrossRef] [PubMed]

14. de Groot, R.; Moolenaar, S.; de Vente, J.; de Leijster, V.; Ramos, M.E.; Robles, A.B.; Schoonhoven, Y.; Verweij, P. Framework for integrated Ecosystem Services assessment of the costs and benefits of large scale landscape restoration illustrated with a case study in Mediterranean Spain. *Ecosyst. Serv.* **2022**, *53*, 101383. [CrossRef]

15. Wagner, M.; Lewandowski, I. Relevance of environmental impact categories for perennial biomass production. *GCB Bioenergy* **2017**, *9*, 215–228. [CrossRef]

16. Pieper, M.; Michalke, A.; Gaugler, T. Calculation of external climate costs for food highlights inadequate pricing of animal products. *Nat. Commun.* **2020**, *11*, 6117. [CrossRef] [PubMed]

17. Arendt, R.; Bachmann, T.M.; Motoshita, M.; Bach, V.; Finkbeiner, M. Comparison of Different Monetization Methods in LCA: A Review. *Sustainability* **2020**, *12*, 10493. [CrossRef]

18. Pizzol, M.; Weidema, B.; Brandão, M.; Osset, P. Monetary valuation in Life Cycle Assessment: A review. *J. Clean. Prod.* **2015**, *86*, 170–179. [CrossRef]

19. Winkler, B.; Mangold, A.; von Cossel, M.; Clifton-Brown, J.; Pogrzeba, M.; Lewandowski, I.; Iqbal, Y.; Kiesel, A. Implementing miscanthus into farming systems: A review of agronomic practices, capital and labour demand. *Renew. Sustain. Energy Rev.* **2020**, *132*, 110053. [CrossRef]

20. Iqbal, Y.; Lewandowski, I. Inter-annual variation in biomass combustion quality traits over five years in fifteen Miscanthus genotypes in south Germany. *Fuel Process. Technol.* **2014**, *121*, 47–55. [CrossRef]

21. Lewandowski, I.; Clifton-Brown, J.; Kiesel, A.; Hastings, A.; Iqbal, Y. Miscanthus. In *Perennial Grasses for Bioenergy and Bioproducts*; Academic Press: London, UK, 2018.

22. Trinomics. *External Costs: Energy Costs, Taxes and the Impact of Government Interventions on Investments: Final Report*; Publications Office of the European Union: Luxembourg, 2020; ISBN 978-92-76-23128-8.

23. Kiesel, A.; Wagner, M.; Lewandowski, I. Environmental Performance of Miscanthus, Switchgrass and Maize: Can C4 Perennials Increase the Sustainability of Biogas Production? *Sustainability* **2017**, *9*, 5. [CrossRef]

24. Wagner, M.; Kiesel, A.; Hastings, A.; Iqbal, Y.; Lewandowski, I. Novel Miscanthus Germplasm-Based Value Chains: A Life Cycle Assessment. *Front. Plant Sci.* **2017**, *8*, 990. [CrossRef]

25. Liu, Y.; Ludewig, U. Nitrogen-dependent bacterial community shifts in root, rhizome and rhizosphere of nutrient-efficient Miscanthus x giganteus from long-term field trials. *GCB Bioenergy* **2019**, *11*, 1334–1347. [CrossRef]

26. Ruf, T.; Schmidt, A.; Delfosse, P.; Emmerling, C. Harvest date of Miscanthus x giganteus affects nutrient cycling, biomass development and soil quality. *Biomass Bioenergy* **2017**, *100*, 62–73. [CrossRef]

27. *ISO 14040:2006*; Environmental Management—Life Cycle Assessment—Principles and Framework. ISO: International Organization for Standardization: Geneva, Switzerland, 2006.

28. *ISO 14044:2006*; Environmental Management—Life Cycle Assessment—Requirements and Guidelines. ISO: International Organization for Standardization: Geneva, Switzerland, 2006.

29. Pant, R.; Zampori, L. *Suggestions for Updating the Organisation Environmental Footprint (OEF) Method*; Publications Office of the European Union: Luxembourg, 2019; ISBN 978-92-76-00654-1.

30. IPCC. *2019 Refinement to the 2006 IPCC Guidelines for National Greenhouse Gas Inventorie*; IPCC: Geneva, Switzerland, 2019; Available online: https://www.ipcc.ch/report/2019-refinement-to-the-2006-ipcc-guidelines-for-national-greenhouse-gas-inventories/ (accessed on 21 October 2022).

31. Lask, J.; Kam, J.; Weik, J.; Kiesel, A.; Wagner, M.; Lewandowski, I. A parsimonious model for calculating the greenhouse gas emissions of miscanthus cultivation using current commercial practice in the United Kingdom. *GCB Bioenergy* **2021**, *13*, 1087–1098. [CrossRef]

32. Freiermuth, R. Modell zur Berechnung der Schwermetallflüsse in der Landwirtschaftlichen Ökobilanz: SALCA-Schwermetall. 2006. Available online: https://www.agroscope.admin.ch/dam/agroscope/en/dokumente/themen/umwelt-ressourcen/produktionssysteme/salca-schwermetall.pdf.download.pdf/SALCA-Schwermetall.pdf (accessed on 21 October 2022).

33. Wernet, G.; Bauer, C.; Steubing, B.; Reinhard, J.; Moreno-Ruiz, E.; Weidema, B. The ecoinvent database version 3 (part I): Overview and methodology. *Int. J. Life Cycle Assess.* **2016**, *21*, 1218–1230. [CrossRef]

34. Amadei, A.M.; de Laurentiis, V.; Sala, S. A review of monetary valuation in life cycle assessment: State of the art and future needs. *J. Clean. Prod.* **2021**, *329*, 129668. [CrossRef]

35. Degieter, M.; Gellynck, X.; Goyal, S.; Ott, D.; de Steur, H. Life cycle cost analysis of agri-food products: A systematic review. *Sci. Total Environ.* **2022**, *850*, 158012. [CrossRef]

36. CE Delft. Environmental Prices Handbook: EU28 version. 2018. Available online: https://cedelft.eu/publications/environmental-prices-handbook-eu28-version/ (accessed on 21 October 2022).

37. Huijbregts, M.A.J.; Steinmann, Z.J.N.; Elshout, P.M.F.; Stam, G.; Verones, F.; Vieira, M.; Zijp, M.; Hollander, A.; van Zelm, R. ReCiPe2016: A harmonised life cycle impact assessment method at midpoint and endpoint level. *Int. J. Life Cycle Assess.* **2017**, *22*, 138–147. [CrossRef]

38. Hastings, A.; Clifton-Brown, J.; Wattenbach, M.; Mitchell, C.P.; Stampfl, P.; Smith, P. Future energy potential of Miscanthus in Europe. *GCB Bioenergy* **2009**, *1*, 180–196. [CrossRef]

39. Teuling, A.J. A hot future for European droughts. *Nat. Clim. Change* **2018**, *8*, 364–365. [CrossRef]

40. Von Cossel, M.; Wagner, M.; Lask, J.; Magenau, E.; Bauerle, A.; Von Cossel, V.; Warrach-Sagi, K.; Elbersen, B.; Staritsky, I.; Van Eupen, M.; et al. Prospects of Bioenergy Cropping Systems for A More Social-Ecologically Sound Bioeconomy. *Agronomy* **2019**, *9*, 605. [CrossRef]

41. Farrell, A.D.; Clifton-Brown, J.C.; Lewandowski, I.; Jones, M.B. Genotypic variation in cold tolerance influences the yield of Miscanthus. *Ann. Appl. Biol.* **2006**, *149*, 337–345. [CrossRef]

42. Mazur, A.; Kowalczyk-Juśko, A. The Assessment of the Usefulness of Miscanthus x giganteus to Water and Soil Protection against Erosive Degradation. *Resources* **2021**, *10*, 66. [CrossRef]

43. Kohli, L.; Daniel, O.; Schönholzer, F.; Hahn, D.; Zeyer, J. Miscanthus sinensis and wild flowers as food resources of *Lumbricus terrestris* L. *Appl. Soil Ecol.* **1999**, *11*, 189–197. [CrossRef]

44. Zimmermann, J.; Styles, D.; Hastings, A.; Dauber, J.; Jones, M.B. Assessing the impact of within crop heterogeneity ('patchiness') in young Miscanthus × giganteus fields on economic feasibility and soil carbon sequestration. *GCB Bioenergy* **2014**, *6*, 566–576. [CrossRef]

45. Nayak, A.K.; Rahman, M.M.; Naidu, R.; Dhal, B.; Swain, C.K.; Nayak, A.D.; Tripathi, R.; Shahid, M.; Islam, M.R.; Pathak, H. Current and emerging methodologies for estimating carbon sequestration in agricultural soils: A review. *Sci. Total Environ.* **2019**, *665*, 890–912. [CrossRef]

46. de Laurentiis, V.; Secchi, M.; Bos, U.; Horn, R.; Laurent, A.; Sala, S. Soil quality index: Exploring options for a comprehensive assessment of land use impacts in LCA. *J. Clean. Prod.* **2019**, *215*, 63–74. [CrossRef]

47. Alejandre, E.M.; van Bodegom, P.M.; Guinée, J.B. Towards an optimal coverage of ecosystem services in LCA. *J. Clean. Prod.* **2019**, *231*, 714–722. [CrossRef]

48. de Adelhart Toorop, R.; Yates, J.; Watkins, M.; Bernard, J.; de Groot Ruiz, A. Methodologies for true cost accounting in the food sector. *Nat. Food* **2021**, *2*, 655–663. [CrossRef]

49. Goglio, P.; Brankatschk, G.; Knudsen, M.T.; Williams, A.G.; Nemecek, T. Addressing crop interactions within cropping systems in LCA. *Int. J. Life Cycle Assess.* **2018**, *23*, 1735–1743. [CrossRef]

50. Piñeiro, V.; Arias, J.; Dürr, J.; Elverdin, P.; Ibáñez, A.M.; Kinengyere, A.; Opazo, C.M.; Owoo, N.; Page, J.R.; Prager, S.D.; et al. A scoping review on incentives for adoption of sustainable agricultural practices and their outcomes. *Nat. Sustain* **2020**, *3*, 809–820. [CrossRef]

51. Grima, N.; Singh, S.J.; Smetschka, B.; Ringhofer, L. Payment for Ecosystem Services (PES) in Latin America: Analysing the performance of 40 case studies. *Ecosyst. Serv.* **2016**, *17*, 24–32. [CrossRef]

52. European Commission. *Commission Delegated Regulation—Amending Delegated Regulation (EU) No 639/2014 as regards certain provisions on the greening practices established by Regulation (EU) No 1307/2013 of the European Parliament and of the Council*; Publications Office of the European Union: Luxembourg, 2018.

 *agronomy*

Article

Wood-Ash Fertiliser and Distance from Drainage Ditch Affect the Succession and Biodiversity of Vascular Plant Species in Tree Plantings on Marginal Organic Soil

Austra Zuševica [1,*], Santa Celma [1], Santa Neimane [2], Moritz von Cossel [3] and Dagnija Lazdina [1]

[1] Latvian State Forest Research Institute SILAVA, Riga St. 111, LV-2169 Salaspils, Latvia; santa.celma@silava.lv (S.C.); dagnija.lazdina@silava.lv (D.L.)
[2] Faculty of Biological and Environmental Sciences, University of Helsinki, 00014 Helsinki, Finland; santa.neimane@helsinki.fi
[3] Biobased Resources in the Bioeconomy (340b), Institute of Crop Science, University of Hohenheim, 70599 Stuttgart, Germany; mvcossel@gmx.de
* Correspondence: austra.zusevica@silava.lv; Tel.: +371-28145267

Abstract: Cutaway peatland is a marginal land, which without further management is an unfavourable environment for plant growth due to low bearing capacity, high acidity and unbalanced nutrient composition of the soil. After wood-ash application, the soil becomes enriched with P and K, creating better conditions for tree growth. In addition to being economically viable, tree plantations ensure long-term carbon storage and promote habitat restoration. In a three-year term, we studied how distance from a drainage ditch and three different doses of wood-ash—5, 10, and 15 tons per hectare—affect the diversity of vascular plants in a tree plantation on a cutaway peatland. Plant species richness, vegetation cover and composition were positively affected by the distance from the drainage ditch and application with fertiliser, but in most cases, fertiliser dose had no significant effect. Both cover and species diversity were not affected by the planted tree species. In a tree plantation, herbaceous plants provide soil fertility by decay and recycling, and reduce mineral leaching in the long term. Since vascular plants play an important role in both the development of habitats and tree growth, it is important to know how multiple factors influence the development of vegetation in tree plantations.

Keywords: cutaway peatlands; ecosystem services; peat; plant growth forms; reforestation; restoration; vegetation

Citation: Zuševica, A.; Celma, S.; Neimane, S.; von Cossel, M.; Lazdina, D. Wood-Ash Fertiliser and Distance from Drainage Ditch Affect the Succession and Biodiversity of Vascular Plant Species in Tree Plantings on Marginal Organic Soil. *Agronomy* **2022**, *12*, 421. https://doi.org/10.3390/agronomy12020421

Academic Editor: Nikos Tzortzakis

Received: 20 December 2021
Accepted: 5 February 2022
Published: 8 February 2022

Publisher's Note: MDPI stays neutral with regard to jurisdictional claims in published maps and institutional affiliations.

1. Introduction

It is estimated that peatlands occupy 2.84 percent [1] of the land area globally. These areas provide long-term carbon storage because the existing environmental conditions prevent plant material from decaying, causing accumulation in the ecosystem of a large amount of vegetation debris relative to the proportion of primary production [2]. During peat extraction, biomass accumulated as peat is removed from storage and used for either horticulture or energy production purposes [3]. In recent years, peat has been extracted in Latvia only for horticulture purposes [4]. After peat extraction, the previous mire ecosystem is completely changed and, without further management, the potential for habitat recovery is very low [5]. Without vegetation in cutaway peatland, no further carbon accumulation occurs, while oxidation reactions occur in the peat [6]. In Latvia there are 18,000 ha of abandoned milled peatlands, and the licenses for extraction of peat in peatland requires that the area must be restored or reclaimed after cessation of extraction [7]. Marginal land management also helps to mitigate greenhouse gases and later it is possible to gain financial income from land with low agricultural value [8]. Suitable uses of cutaway peatland include afforestation, rewetting, and use for the cultivation of crops or fodder plants [9]. In areas where ecological restoration is not possible because the water table cannot be raised, tree stands after soil improvement can be grown [10–12]. This process can be carried out in two ways: either by establishing a financially valuable tree plantation that has already shown a significant increase in growth and survival after wood-ash application, or by letting the area to restore naturally after soil improvement [13,14].

Although peat has large nitrogen reserves, which are beneficial for biomass production, vegetation naturally develops in cutaway peatland very slowly due to the following factors: fluctuating water level, lack of a viable seed bank in the ecosystem and unfavourable soil chemical properties, such as low pH values, and low levels of phosphorus and potassium, which adversely affect the fertility of the soil [6,15]. Drainage ditch systems in peatlands lower the water table, but the water table is not even in all drained areas, depending on the distance to the drainage ditch [16,17]. In mechanically managed peat fields, vegetation is usually sparse, but the natural occurrence of vascular species can be supported by the application of phosphorus fertilisers or wood-ash fertilisers that increase phosphorus and potassium uptake by plants [13,18]. The application of wood-ash fertiliser in cutaway peatland firstly increases biological activity, thereby increasing CO_2 emissions, but with a successful recovery of vegetation, a significant amount of carbon is accumulated in plants, thereby compensating for the emissions [19,20].

Many studies have shown that herbaceous plants are a key part of conservation of species diversity and maintaining the forest ecosystem [21–23]. In forest stands, understorey vegetation, together with tree litter, is the most important source of nutrients, which is particularly important during early ecosystem succession when there is a lack of nutrients in the soil and risk of soil erosion [24,25]. Herbaceous plants also play a key role in preventing nutrient leaching in tree stands after fertilisation [26]. In addition, perennial herbaceous plants have the highest mineral storage capacity in the spring period, which coincides with the time when mineral leaching from the soil is the highest [27]. It is important to add that plant composition mainly depends on soil properties, but forest soil properties are determined by the dominant tree species [28]. Understorey plant species' richness and biomass productivity are also influenced by overstorey tree species [29]. In comparison to deciduous tree stands, the litter in the coniferous forests has lower pH and the top layer of soil has limited plant-available nutrient content, but the pH of deeper layers of the soil does not differ significantly between stands [30].

In the case of secondary succession, vegetation plays an important role, both at the very beginning of succession, to store nutrients and reduce their leaching, and in the further stages, as one of the main sources of nutrients. In the case of cutaway peatland, it is important to clarify how combinations of factors such as the dose of wood-ash fertiliser and the distance from a drainage ditch, change the composition of natural plant communities, as this information is needed to determine the best management in terms of financial income and afforestation quality. This study addresses the following research questions: (1) Does application of wood-ash fertiliser in cutaway peatland affect the abundance and richness of naturally colonising vascular plant species and does increasing the amount of the dose of fertiliser increase plant diversity? (2) Does the distance from a drainage ditch affect the number and composition of species? (3) Does the planted tree species affect the composition of ground vegetation within the same wood-ash fertiliser group? Based on previous studies, it was assumed that increasing the amount of wood-ash fertiliser applied per hectare would increase the richness of vascular plants. As the drainage ditch is associated with higher soil moisture, which is one of the limiting factors in cutaway peatland, we hypothesised that the highest plant species number and abundance will be closer to the ditch. It is known that, in forests, the chemical composition of tree litter affects the chemical properties of soil, and thus the composition of understorey plant communities should be related to tree species.

2. Materials and Methods

Study Site and Design

The study site is located in central Latvia (N 56°43′41.35″ E 23°34′39.61″) in a cutaway peatland where active peat extraction is still ongoing in other parts of the area. Peat extraction was for horticulture. The residual peat layer consisted of acidic raised bog, fen, and transitional mire peat with variable depth of at least 50 cm. The upper part of the peat was acidic, moderately decomposed, raised bog peat [14].

The research field was established within the "Sustainable and responsible management and re-use of degraded peatlands in Latvia" (LIFE14 CCM/LV/001103) project [31]. At the beginning of the vegetation season in 2016, the study site was prepared for tree planting by sequentially performing milling, cleaning of drainage ditches, and wood-ash application [31]. Milling was performed to remove all vegetation, which mainly consisted of sparse *Phragmites australis*. The forest stand adjacent to one side of the study site and large remnants of wood exposed after the removal of the upper peat layer suggest that the study site had been a forest ecosystem at some time. The applied wood-ash was unprocessed and had a small particle size. To prevent the effect of wind on the spread of ash, water was added before application. The wood-ash consisted of K 24.7, Mg 18.2, Ca 120.4 and P 6.6 g·kg^{-1} [14]. The study site was fertilised in sectors with three different doses (5, 10, and 15 tons per hectare), and one control sector left without fertilisation (control). After the application, soil pH value changed from 3.5 in the control group to 5.9 in sectors with 15 tons per hectare [14]. All sectors were established in three replicates with size of 236 × 20 m. Drainage ditches separated each sector along the two longest sides. Peat extraction for horticulture in this site was carried out by vacuum harvesting, which requires an extensive network of shallow drainage ditches. After cleaning, the ditch dimensions were 50 cm wide and 100 cm deep; Figure 1. The total study site area was 8 ha. Each sector was divided into five parts with size of 900 m^2, where in four randomly selected parts, four economically significant tree species (*Pinus sylvestris*, *Alnus glutinosa*, Poplar clone *Vesten* (Biopoplar s.r.l., Cavallermaggiore, Italy) (*Populus* v. *Vesten*), and *Betula pendula*) were planted. In one part, no trees were planted, but natural reforestation from 2016 was observed. In each part, 95 trees were planted (equal to 1055 trees per hectare). In each part, three 2.5 × 3.5 m sampling plots were established—0.5 to 4 m, 4 to 7.5 m and 7.5 to 11 m from the drainage ditch. The total number of sampling plots was 180; Figure 2.

(a)   (b)

Figure 1. Drainage ditch in the study site: (**a**) drainage ditch before cleaning; (**b**) drainage ditch after cleaning.

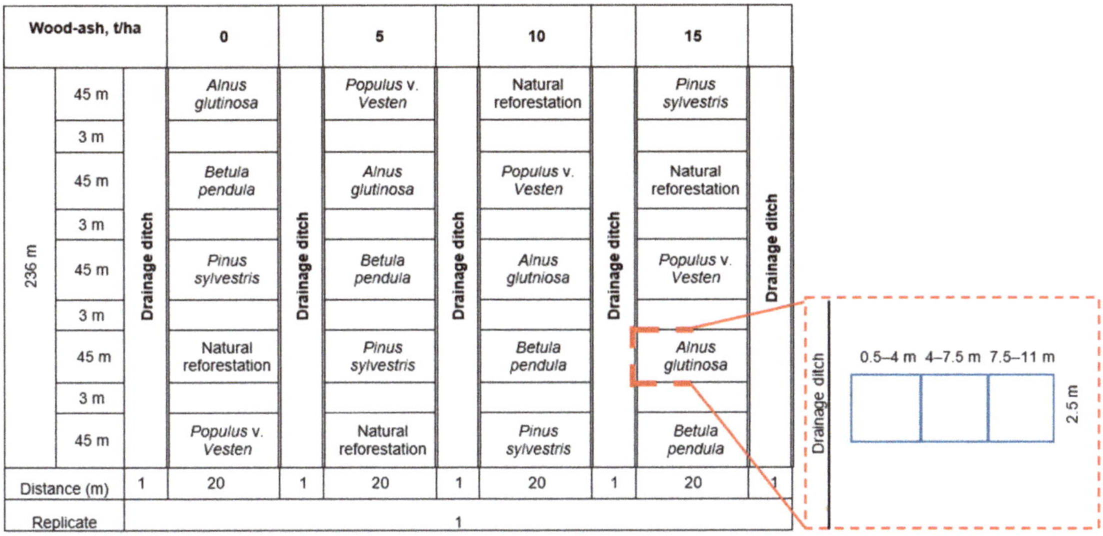

Wood-ash, t/ha		0		5		10		15	
	45 m	Alnus glutinosa		Populus v. Vesten		Natural reforestation		Pinus sylvestris	
	3 m								
	45 m	Betula pendula		Alnus glutinosa		Populus v. Vesten		Natural reforestation	
	3 m								
	45 m	Pinus sylvestris		Betula pendula		Alnus glutniosa		Populus v. Vesten	
	3 m								
	45 m	Natural reforestation		Pinus sylvestris		Betula pendula		Alnus glutinosa	
	3 m								
	45 m	Populus v. Vesten		Natural reforestation		Pinus sylvestris		Betula pendula	
Distance (m)	1	20	1	20	1	20	1	20	1
Replicate					1				

(Left margin label: 236 m. Vertical labels between columns: Drainage ditch. Inset detail plot: 0.5–4 m 4–7.5 m 7.5–11 m, 2.5 m)

Figure 2. Plan of one repetition in research area.

The vegetation was surveyed three times over a period of three years, identifying all species in the sampling plots once annually in the middle of the growing season. No vegetation management was implemented before and after the vegetation survey. The understorey species canopy cover (%) was recorded to the nearest 5% once during the vegetation season in July. We used Ellenberg's indicator values [32] for the Czech Republic to determine plant functional traits such as light, temperature, moisture, reaction, nutrients, and salinity [33]. Experimental studies have found that Ellenberg's indicator values for nutrients (N), moisture (M), and soil reaction (R) are well correlated with real field data, including for areas outside the Central Europe region [34], and therefore could be applied in this study. The LEDA Traitbase: A database of life-history traits of Northwest European flora was used to determine plant functional traits [35] for each species. Cover-weighted average Ellenberg's indicator values F_m were calculated by summing each species i indicator value F_i and weight W_i, calculated for species based on its percentage cover C_i : $W_i = f(C_i)$, for all species n in each sampling plot:

$$F_m = \sum_{i=1,n} F_i W_i / \sum_{i=1,n} W_i$$

The species were divided into groups based on plant life form [36]: hemicryptophyte, therophyte, phanerophyte, geophyte; and Universal Adaptive Strategy Theory (UAST) [37]: competitor, stress-tolerant, and ruderal. The mean percentage cover for plant life forms and USAT classes was calculated for each sampling plot.

The computer package R Statistics 4.0.5 for Windows was used for two-way analysis of variance (ANOVA), Principal Component Analysis (PCA) and Tukey HSD analyses [38,39]. The independent two-sample Student's t-test was used to determine significant differences in plant functional traits between years. The effects of fertiliser dose, distance from the drainage ditch, and planted tree species on species richness were tested using ANOVA. Before preforming ANOVA, the Shapiro–Wilk test showed that the data did not differ significantly from a normal distribution. The Tukey HSD test was used to determine significant differences in variables between treatments. Each variable of plant functional trait data (plant growth form, ecological strategies) was analysed individually using ANOVA and the Tukey HSD test and visualised with PCA.

PCA was used to determine plant functional parameters that were significantly affected by fertiliser or ditch effect. All sampling plots, including the control group, were included in PCA analyses. Before analyses, the Kaiser–Meyer–Olkin (KMO) test for sampling adequacy was used to determine if the data fit the assumptions of PCA. The KMO test output was 0.6. The analyses were performed using the rda function from the vegan package [40]. The decostand function with the Hellinger method was used to standardise the data, as the data were not linear. The scaling method was used to observe differences between plots. The cumulative value of the eigenvalues for the first two axes was 63 percent. The ordiellipse function was used to visualise groups of plots by drawing polygons from standard error of the (weighted) average of scores.

3. Results

Over the three-year period, 84 herbaceous and woody plant species were observed in the study area (see Table A1). The overall trend shows that in most sectors, number of species (richness) continued to increase over the three-year period; Figure 3. In all three years of the study, both the distance from the ditch (2nd year $p = 0.001$; 3rd year $p = 0.001$; 4th year $p = 0.001$) and the fertiliser dose (2nd year $p = 0.001$; 3rd year $p = 0.001$; 4th year $p = 0.001$) had significant effects, but the planted tree species did not significantly affect species richness. The only exception was in the second year after the application of wood-ash, a higher number of species occurred in plots where *Alnus glutinosa* was planted compared to plots with *Betula pendula* ($p = 0.02$). During all three years, species richness was higher in the fertilised plots compared with the control group (2nd year $p = 0.001$, 3rd year $p = 0.001$, 4th $p = 0.001$), while it did not differ between fertiliser doses. In the third and fourth year of the study, species richness was significantly higher in the plots 0.5–4 m from the ditch than that at 4–7.5 and 7.5–11 m (0.5–4 and 7.5–11 m from the ditch 2nd $p = 0.001$, 3rd $p = 0.001$, 4th $p = 0.001$; 0.5–4 and 4–7.5 m from the ditch 2nd $p = 0.001$, 3rd $p = 0.003$, 4th $p = 0.001$), but no significant difference was observed between the sampling plots 4–7.5 and 7.5–11 m from the ditch.

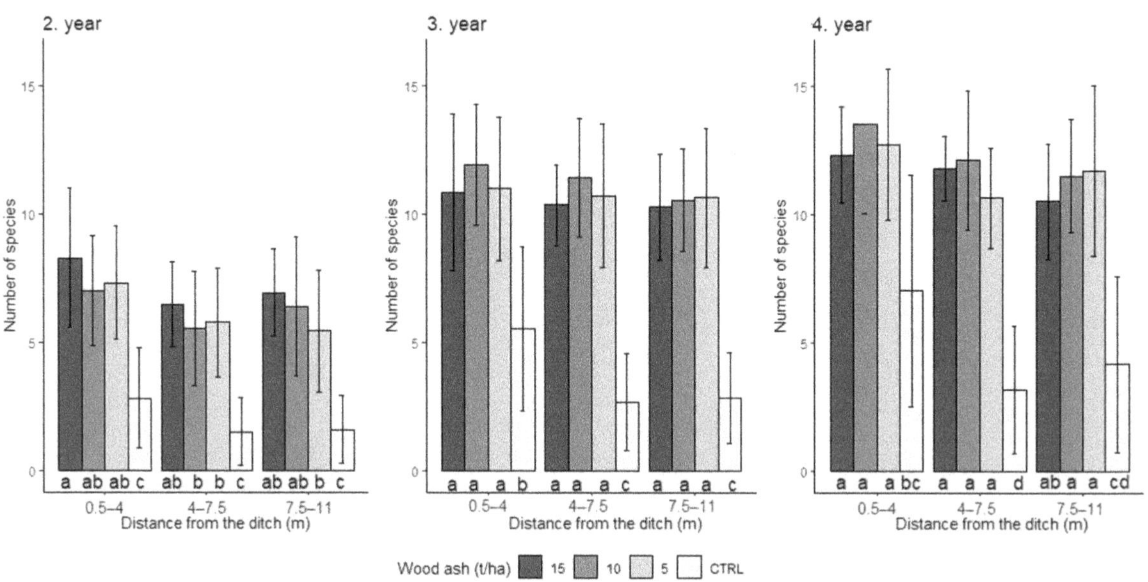

Figure 3. Species richness depending on the distance from the drainage ditch and wood-ash fertiliser dose during second to fourth vegetation season following the application of fertiliser. Different letters (a, b, c, d) indicate significant differences between groups, $p < 0.05$ according to the Tukey HSD test. Error bars represent standard deviation.

The total vegetation cover continued to increase over the three-year period and, similar to the number of species, developed more quickly in fertilised groups, regardless of wood-ash fertiliser dose Figure 4. In the fourth year after fertilisation, vegetation cover was significantly higher in plots 0.5–4 m from the ditch compared to 4–7.5 and 7.5–11 m from the ditch ($p = 0.001$)—Figure 4c.

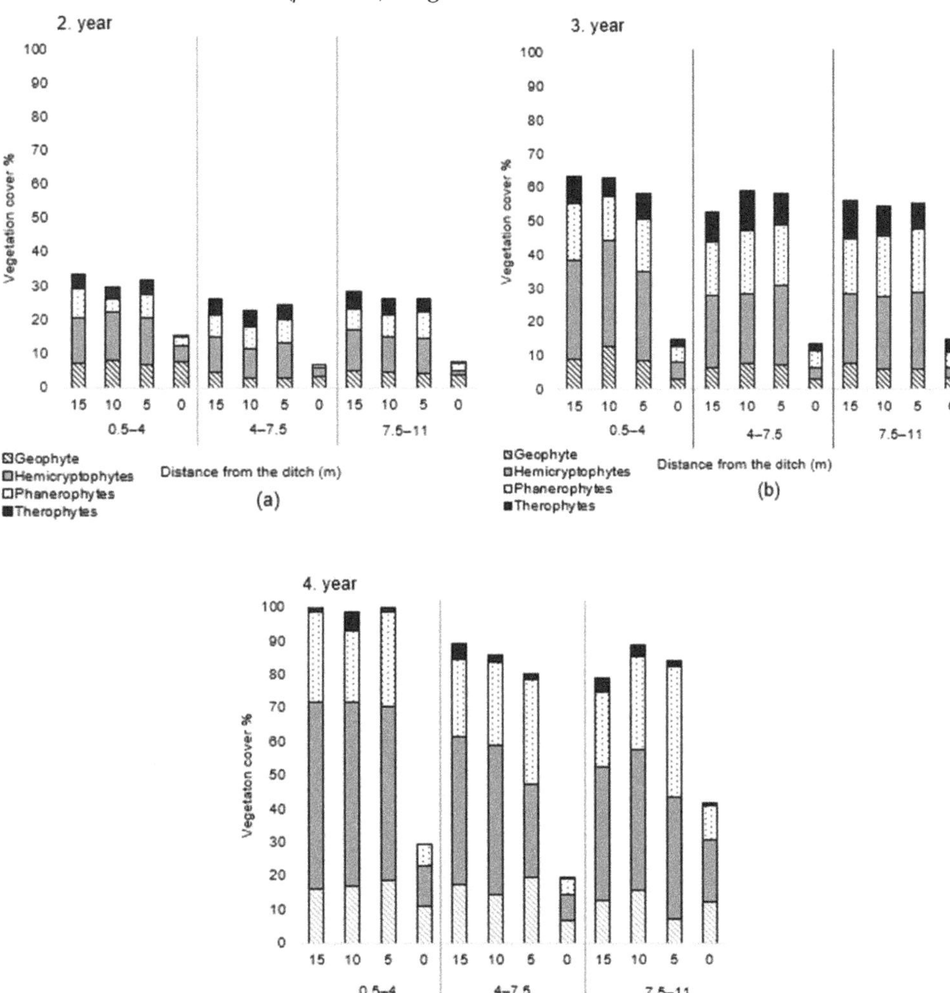

Figure 4. Plant growth form proportion from total vegetation cover depending on different doses of wood-ash fertiliser and distance from the drainage ditch: (**a**) second year after fertilisation; (**b**) third year after fertilisation; (**c**) fourth year after fertilisation.

Moreover, the structure of vegetation according to plant functional trait classes changed over the three-year ecological succession and was affected both by fertilisation and distance from the drainage ditch but was not affected by tree species; Figures A1–A11. During the three-year period, the cover of ruderal species in the plant community decreased, and the cover of competitor species and species with no specific plant adaptive strategy (CSR) increased; Tables 1 and 2. Higher cover of plants with the CSR strategy in the second year was found 0.5–4 m from the ditch, but during the three-year study, these strategies' plant cover significantly increased in plots 4–7.5 and 7.5–11 m from the ditch; Table 2. There was a tendency for geophyte cover to be lower under the highest fertiliser dose, independent of location; Figure A1. Although no significant changes occurred, the mean hemicryptophyte cover was higher in the plots 0.5–4 m from the ditch.

Table 1. Relative cover (proportion of total cover) of plants by strategy (UAST) and growth form and community weighted mean values for Ellenberg's indicator values in plots depending on the dose of wood-ash fertiliser during the observation period. Underlined text indicates significant differences in 3rd and 4th year of research with previous season and bold text indicates significant differences between 2nd and 4th year of research (Student's *t*-test ($p < 0.05$)).

	Plant Functional Traits	Wood-Ash Dose (t/ha)											
		0	5	10	15	0	5	10	15	0	5	10	15
		2. Year				3. Year				4. Year			
UAST (%)	Competitors	50.4	53.8	47.1	44.9	27.8	34.6	35.9	26.7	51.7	53.0	45.5	46.4
	Ruderals	5.5	9.2	8.2	12.3	7	6.4	3.5	12.4	0	1	1.7	1.4
	Competitors/Ruderals Ruderals	5.3	7.6	5.3	9.1	7.3	6.8	7.3	6.4	10.2	9.7	8.8	11.1
	Competitors/Stress tolerant Stress tolerant	6.2	11.6	11.7	10.4	13.5	17.4	16.9	17.9	14.6	13.2	16.6	13.5
	Competitors/Stress tolerant/Ruderals	17.9	15.7	19.9	22.6	39.6	31.3	32.0	32.0	22.6	20.6	27.3	27.5
Growth form (%)	Geophyte	23.7	26.7	24.6	13.2	15.5	16.2	18.6	12.5	19.2	23.8	22.9	17.5
	Therophyte	8.3	13.7	9.7	18.1	10.0	9.2	4.6	16.4	0.6	0.4	0.8	0.6
	Hemicryptophyte	30.5	35.7	40.9	40.3	33.7	42.6	42.2	38.6	43.3	46.1	51.3	47.0
	Phanerophyte	22.7	22.1	17.5	28.2	39.7	30.5	32.1	31.3	36.9	27.3	25.0	34.9
Ellenberg's value	Moisture	6.6	6.9	6.5	6.2	6.5	6.5	6.2	6.6	6.9	6.8	6.9	7.0
	Nitrogen	5.0	5.3	5.1	5.3	5.4	5.4	5.3	5.4	5.9	5.5	5.6	5.8

Table 2. Relative cover (proportion of total cover) of plants by strategy (UAST) and growth form and community weighted mean values for Ellenberg's indicator values in plots depending on the distance from the drainage ditch during the observation period. Underlined text indicates significant differences in 3rd and 4th year of study with the previous season and bold text indicates significant differences between 2nd and 4th Year (Student's t-test ($p < 0.05$)).

	Plant Functional Traits	Distance from Drainage Ditch (m)								
		0.5–4	4–7.5	7.5–11	0.5–4	4–7.5	7.5–11	0.5–4	4–7.5	7.5–11
		2. Year			3. Year			4. Year		
UAST (%)	Competitors	52.6	45.3	48.8	37.6	28.4	27.9	49.6	49	48.8
	Ruderals	9.1	10.2	8	3.1	9.7	9.2	1.9	1.1	0.8
	Competitors/Ruderals	4.6	8.5	7.8	8.5	6	6.5	8.5	10.8	10.7
	Competitors/Stress tolerant	11.6	10.3	9	20	15.1	14.7	15.6	13.7	14.1
	Competitors/Stress tolerant/Ruderals	19.4	20.2	17.9	28.1	36.1	36.8	25.5	25.4	23.8
Growth form (%)	Geophyte	27.9	16.4	21.3	17.7	14.74	14.6	22.7	21.8	18
	Therophyte	11	15	12.5	4.3	13.2	12.8	0.5	0.5	0.7
	Hemicryptophyte	40.6	37.6	34.1	48.2	34.8	43.9	51	44.7	45.1
	Phanerophyte	19.2	25.2	23.6	27.6	35.6	36.9	25.8	32	34.4
Ellenberg's value	Moisture	7	6.1	6.5	6.2	6.6	6.6	6.8	7	7
	Nitrogen	5.4	5	5.2	5.3	5.5	5.4	5.5	5.9	5.8

Between the second and fourth research years, community weighted mean Ellenberg's moisture values increased in all study areas, but significantly only in highest fertiliser doses. In the second year, the moisture value was higher in plots 0.5–4 m from the ditch, but in the fourth year of research, it increased significantly in plots 4–7.5 and 7.5–11 from the ditch and was similar in all locations. The community weighted mean nitrogen indicator value was lower in the control group in the second year, but in the fourth year of the research it increased significantly in both control and fertilised groups, except for 5 t/ha. Nitrogen value was higher in plots 0.5–4 m from the ditch in the second year, but significantly increased in plots 4–7.5 and 7.5–11 m from the ditch, and did not differ between locations in the fourth year of research. Overall, the relative cover of therophytes and plants with ruderal strategy decreased over time, whereas phanerophyte and hemicryptophyte cover and community weighed mean Ellenberg's indicator moisture and nitrogen value increased.

In the control group, the changes in plant functional properties over time were less evident than in the fertilised plots. In general, vegetation in the control plots clearly differed from all fertilised plots; Figures 5a and A12. Although here were some differences in plant functional traits between years and less frequently between fertiliser doses, the overall trend in the development of vegetation was similar in all fertilised plots. Greater relative cover of geophytes, plant species with high Ellenberg's moisture indicator value, and relative cover of plants with a competitor strategy were found in the control plots. Distance from the drainage ditch was not as important in the formation of vegetation structure as the effect of fertiliser; Figure 5b.

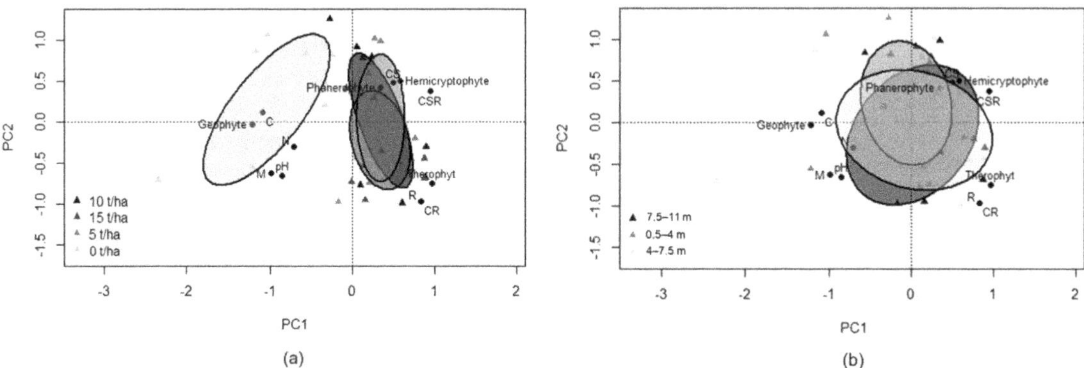

Figure 5. Principal Component Analysis of plant functional trait values that differ significantly: (**a**) between fertilised and non-fertilised plots; (**b**) depending on distance from the drainage ditch. Plant functional traits: Geophyte; Hemicryptophyte; Phanerophytes; Therophyte; C—competitors; S—stress tolerant; R—ruderal. Ellenberg's indicator values: N—Nitrogen, M—Moisture. Cumulative % variance for PC1 and PC2 explain 90.4%.

4. Discussion

It was found that both the wood-ash treatment and distance from the drainage ditch had a significant effect on plant species richness. In all years, there were significant differences in species richness between non-treated sampling plots and treated sampling plots, which shows that wood-ash fertiliser had a significant positive effect on recolonising species richness, even when used at a low dose. During the three-year period, species richness was higher in plots 0.5–4 m from the ditch, which is most likely due the higher soil moisture closer to the ditch, as water accessibility is one of the limiting factors for natural revegetation in cutaway peatland [6]. In meliorated peat soils in wet conditions, the soil moisture increases with distance to the ditch, but decreases in drier conditions [41]. As this area had been afforested, the hydrology of study area represented drier conditions. An abundant network of shallow drainage ditches, which were made during peat vacuum harvesting, may have a beneficial effect on revegetation in a tree stand because the water table depth is not as variable as in sites with a sparse ditch network. The effect of the dose of wood-ash fertiliser was greater than the effect of the distance from the ditch, as the effect of the latter was not very noticeable when fertilisation was applied. Natural revegetation can mitigate the negative effect of the melioration system, since plant root systems can decrease hydraulic conductivity and increase soil moisture [42,43]. The number of observed species also increased during the three-year period, which indicates that the vegetation structure is still changing from one ecosystem, a cutaway peatland, to a planted forest. To obtain a full picture of the revegetation under the influence of fertilisation and the distance from the ditch, it is necessary to observe vegetation over an even longer period, but with a wider interval of time between the surveys, since the changes will not be so noticeable within one growing season, but rather between several seasons.

Similar to the results for species richness, there were also major successional changes of plant functional traits in the treated plots, which were less evident in non-treated plots. Again, this confirms that wood-ash treatment accelerates the succession process in the cutaway peatland [18]. During all study times, phanerophyte cover was higher in the control plots, compared to the treated plots. In fertilised plots with higher dose (10 and 15 t/ha), plants with the CSR strategy, which are usually late successional species, had higher cover during all three years. In the control group and 5 t/ha, vegetation cover was sparser; therefore, in these plots, plants are exposed to a number of factors that may adversely affect revegetation—the erosion of the soil top layer, increased risk of the soil moisture level fluctuation, especially the top layers, and minerals leaching [6,26,44]. Wood-ash acts as a liming agent in acidic soils, therefore enhancing nutrient availability to plants. Wood-ash fertiliser may contribute understorey vegetation growth, which is more demanding for nutrients than phanerophytes and, after the first years, competes less with planted trees than naturally regrowing phanerophytes; Figure 6. In the control group, where vegetation was also sparse during the fourth season, geophytes such as *Phragmites australis*, *Tussilago farfara*, and *Taraxacum officinale* were more abundant than under higher fertiliser doses; Figure A1. These species indicate that, after four years, unfertilised cutaway peatland does not have optimal growth conditions because geophytes form underground organs for storage of water or nutrient reserves, thereby maintaining the availability of these stores under adverse environmental conditions [45]. This tendency was mostly significant in plots 0.5–4 m from the ditch because a large number of geophytes found in the study area grow not only under poor nutrient growing conditions, but also on sites where water is not a limiting factor, as in plots closer to a ditch [46,47].

(a)   (b)   (c)

Figure 6. Vegetation cover four years after treatment with wood-ash fertiliser in different doses: (**a**) control group; (**b**) 10 tons per hectare; (**c**) 15 tons per hectare.

The community weighted mean Ellenberg's indicator values showed significant differences between treated groups and the control group in moisture and nitrogen values, which both have a major role in natural vegetation formation in cutaway peatlands [6]. A higher Ellenberg's moisture value was observed in the non-fertilised plots in the second and third study year, but in the last study year it was higher in plots fertilised with 10 and 15 t/ha. In the first years in control plots, there tended to be a few ruderal species with a high Ellenberg's indicator value, such as *Phragmites australis* and *Tussilago farfara*, explaining the high moisture value in these plots. In fertilised plots with higher species richness, these values may be more reliable. As discussed before, vegetation cover development can raise soil moisture by decreasing water infiltration and soil conductivity. In fertilised plots, during the three-year period, species richness, total vegetation cover, and Ellenberg's moisture value increased. In the second year, the nitrogen indicator value was higher in fertilised plots, but in the fourth year it significantly increased both in fertilised and control groups. The distance from the drainage ditch has a greater influence on Ellenberg's nitrogen value than fertiliser. An experimental study showed that the Ellenberg's nitrogen value had a fairly poor correlation with soil N content, but represented overall productivity [48]. In the study area, vegetation cover and species richness were higher in plots 0.5–4 m from the ditch; Figure 4. It is possible that the Ellenberg's nitrogen values did not represent the amount of nitrogen in the soil, but the total productivity, which was higher in plots 0.5–4 m from the ditch.

Wood-ash fertiliser and distance from the drainage ditch have a more significant impact on the species richness and vegetation cover, but ecological successions have a stronger impact on species composition. In all the treated groups, during the first vegetation survey season, the dominated functional traits were typical of the early stages of primary successions—a large number of therophytes (annual plant species), in addition to plants with ruderal and stress-tolerant strategies. Such vegetation structure is usually combined with barren vegetation, which explains the lower CSR plant cover. During the four-year succession, in many parts, a large number of the therophytes were replaced by hemicryptophytes. As the vegetation structure stabilised, more plants that have no special strategy (CSR), thus indicating a later stage of habitat succession, were more common. During the time between the 2nd and 4th years after application of wood-ash, in fertilised plots the Ellenberg's moisture value increased due to the increase in the vegetation cover.

5. Conclusions

This study showed that both wood-ash fertiliser and distance from the drainage ditch, in addition to time after treatment application, have an impact on plant species richness, cover, and vegetation composition. Wood-ash fertiliser positively affects vascular plant species richness, but during the first growing years, there are no significant differences between treatment doses. Species richness is higher closer to the drainage ditch, most likely due to a higher water level, which is one of the limiting factors in a cutaway peatland. Consequently, it can be concluded that appropriate management of cutaway peatland provides the area with the nutrients it needs, thereby allowing colonisation of vegetation. In the first growing seasons, planted tree species have no significant effect on vegetation composition, richness, and distribution. Planted and naturally regrown trees on marginal land contribute to the overall development of the habitat by increasing the biodiversity due to interaction with other species. This restoration practice also increases the financial value of otherwise agronomically low-value land. Vascular plant diversity and cover is essential as it reduces nutrient leaching, thus improving growth conditions for planted trees. Along with the herbaceous species, tree species such as *Betula pendula*, *Betula pubescens*, and *Salix* spp. also naturally regrow in the area therefore the area has the potential to become a silvopastoral agroforestry system or naturally reforested site. Further research is needed to determine how different doses of fertiliser affect the development of vegetation over time and when planted tree species begin to affect the plant communities.

Author Contributions: Conceptualisation, D.L., S.N., A.Z.; methodology, D.L., S.N. and A.Z.; software, A.Z.; validation, D.L.; formal analysis, A.Z.; investigation, S.C., S.N., A.Z.; data curation, A.Z.; writing—original draft preparation, A.Z.; writing—review and editing, S.C., S.N., D.L., M.v.C.; visualisation, A.Z.; supervision, D.L., M.v.C.; project coordination, D.L.; funding acquisition, D.L. All authors have read and agreed to the published version of the manuscript.

Funding: This research received funding from the European Union's Horizon 2020 research and innovation program under grant agreement No 727698, the project "Sustainable and responsible management and re-use of degraded peatlands in Latvia" (LIFE14 CCM/LV/001103) and the European Regional Development Fund project Elaboration of innovative White Willow - perennial grass agroforestry systems on marginal mineral soils improved by wood ash and less demanded peat fractions amendments (1.1.1.1/19/A/112). The article processing charge was funded by the European Regional Development Fund project Elaboration of innovative White Willow - perennial grass agroforestry systems on marginal mineral soils improved by wood ash and less demanded peat fractions amendments (1.1.1.1/19/A/112).

Data Availability Statement: Data generated from this study is available upon request to the corresponding author.

Acknowledgments: We thank "Laflora" Ltd. for the chance to study the possibilities of improving the properties of peat soil with wood-ash.

Conflicts of Interest: The authors declare no conflict of interest.

Appendix A

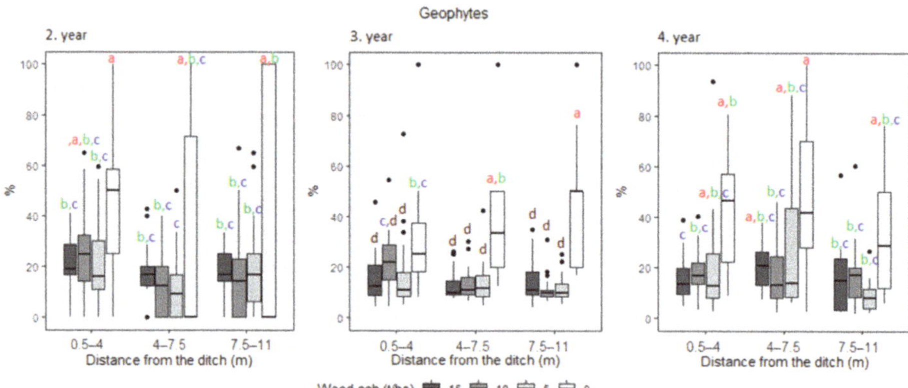

Figure A1. Relative cover of plants with geophyte growth form of the total vegetation cover, depending on the distance from the drainage ditch and the dose of wood-ash fertiliser. Different letters (a, b, c, d) indicate significant differences between groups, $p < 0.05$ according to the Tukey HSD test.

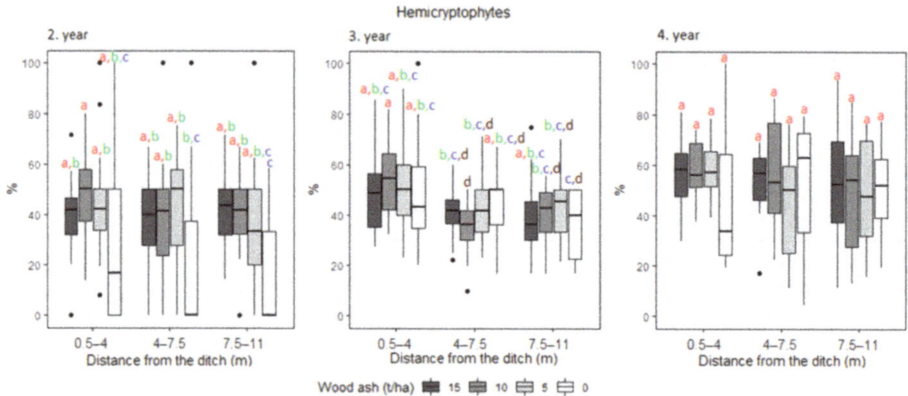

Figure A2. Relative cover of plants with hemicryptophyte growth form from the total vegetation cover, depending on the distance from the drainage ditch and the dose of wood-ash fertiliser. Different letters (a, b, c, d) indicate significant differences between groups, $p < 0.05$ according to the Tukey HSD test.

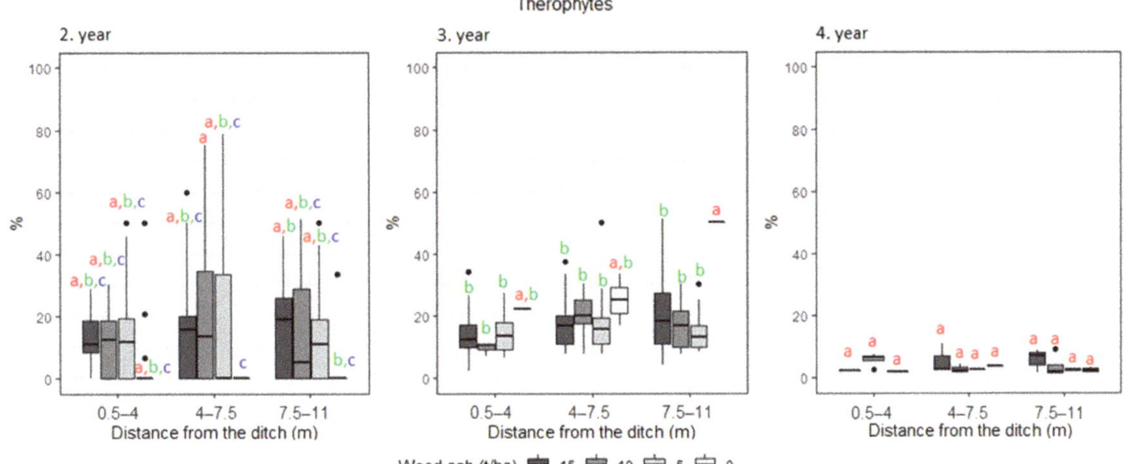

Figure A3. Relative cover of plants with therophyte growth form from the total vegetation cover, depending on the distance from the drainage ditch and the dose of wood-ash fertiliser. Different letters (a, b, c) indicate significant differences between groups, $p < 0.05$ according to the Tukey HSD test.

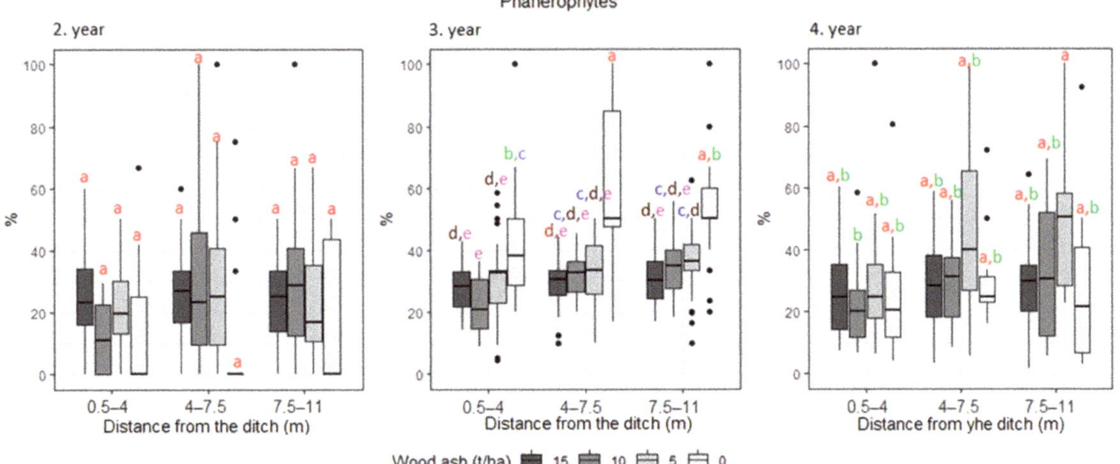

Figure A4. Percentage of plants with phanerophyte growth form from the total vegetation cover, depending on the distance from the drainage ditch and the dose of wood-ash fertiliser. Different letters (a, b, c, d, e) indicate significant differences between groups, $p < 0.05$ according to the Tukey HSD test.

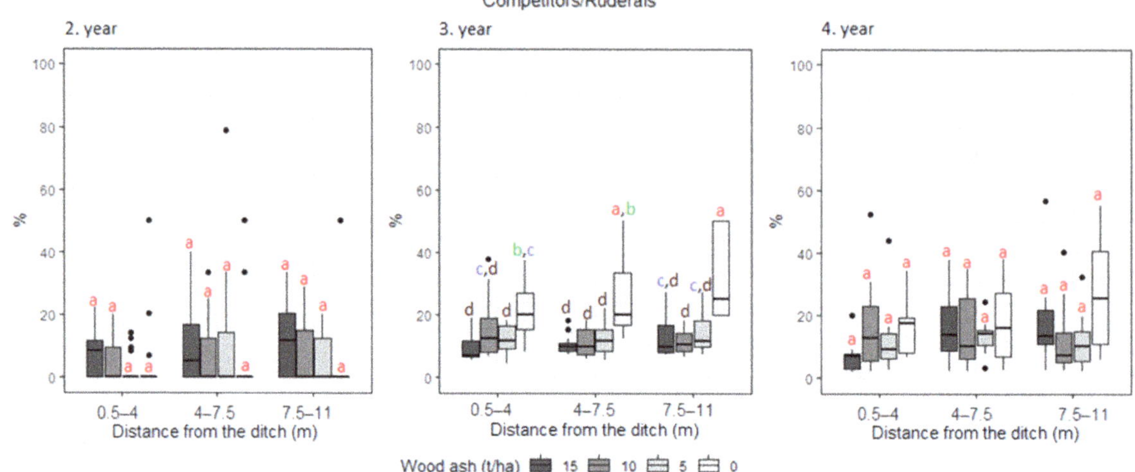

Figure A5. Relative cover of plants with competitor/ruderal universal adaptive strategy from the total vegetation cover, depending on the distance from the drainage ditch and the dose of wood-ash fertiliser. Different letters (a, b, c, d) indicate significant differences between groups, $p < 0.05$ according to the Tukey HSD test.

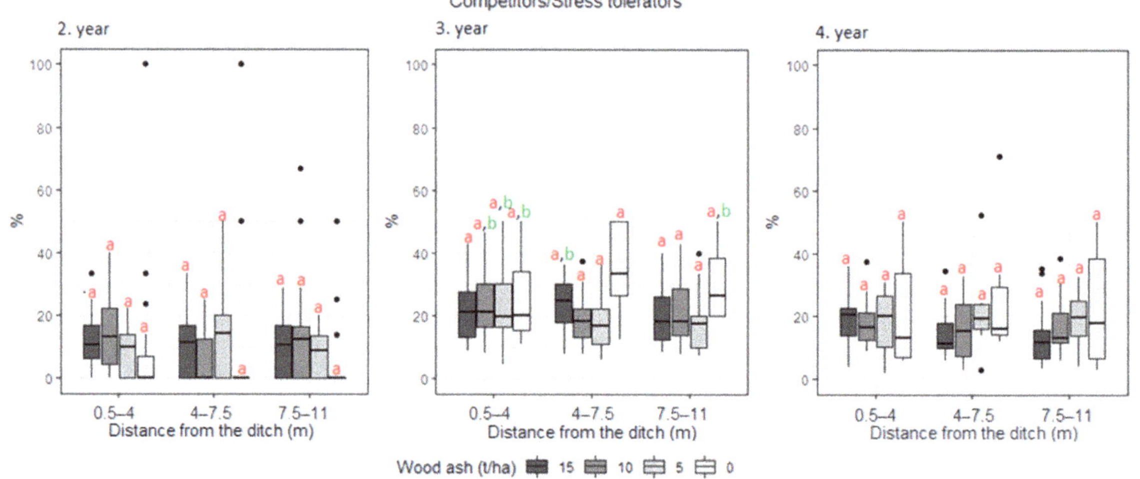

Figure A6. Relative cover of plants with competitor/stress tolerant universal adaptive strategy from the total vegetation cover, depending on the distance from the drainage ditch and the dose of wood-ash fertiliser. Different letters (a, b) indicate significant differences between groups, $p < 0.05$ according to the Tukey HSD test.

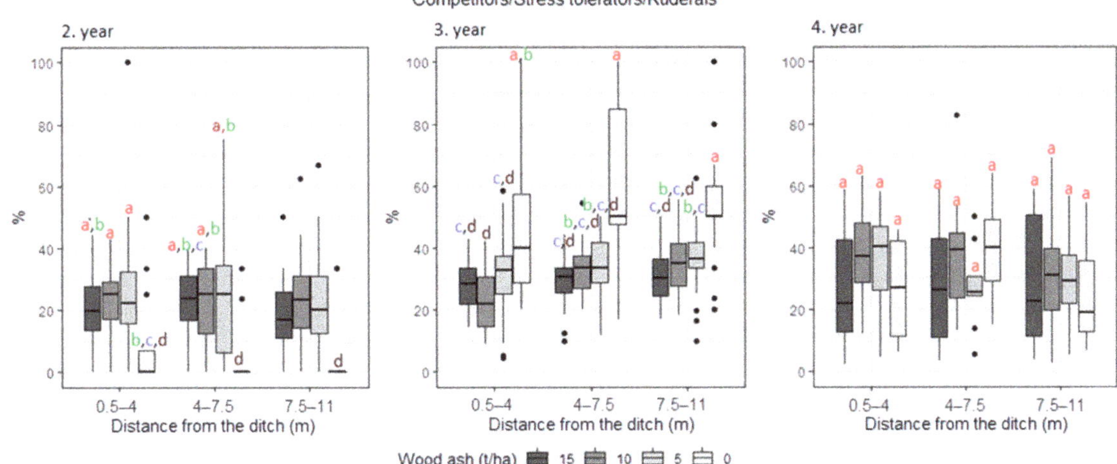

Figure A7. Relative cover of plants with competitor/stress tolerant/ruderal (no-specific strategy) universal adaptive strategy from the total vegetation cover, depending on the distance from the drainage ditch and the dose of wood-ash fertiliser. Different letters (a, b, c, d) indicate significant differences between groups, $p < 0.05$ according to the Tukey HSD test.

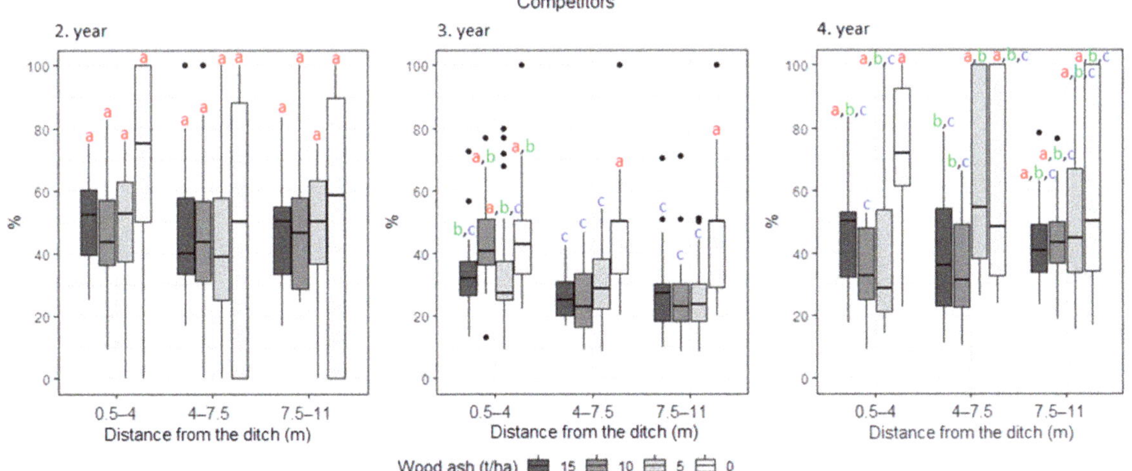

Figure A8. Relative of plants with competitor universal adaptive strategy from the total vegetation cover, depending on the distance from the drainage ditch and the dose of wood-ash fertiliser. Different letters (a, b, c) indicate significant differences between groups, $p < 0.05$ according to the Tukey HSD test.

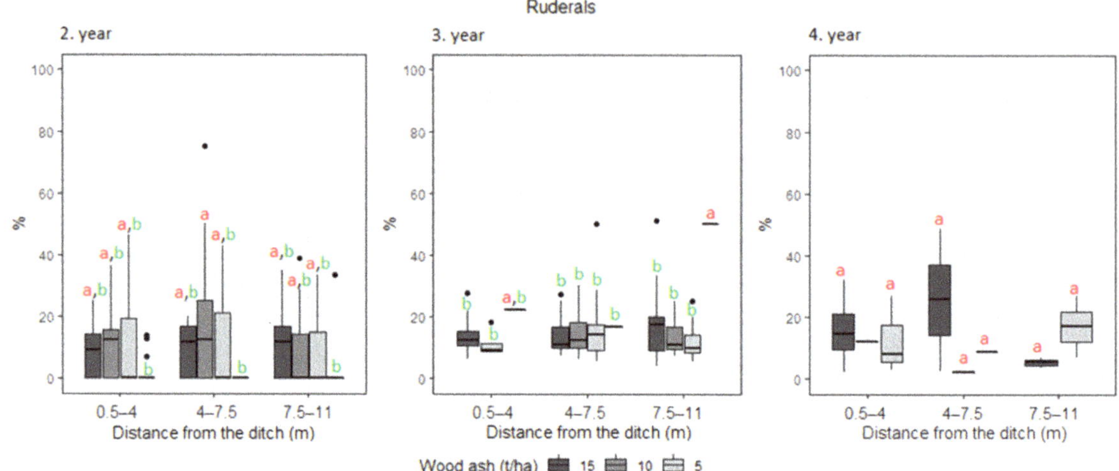

Figure A9. Relative of plants with ruderal universal adaptive strategy from the total vegetation cover, depending on the distance from the drainage ditch and the dose of wood-ash fertiliser. Different letters (a, b) indicate significant differences between groups, $p < 0.05$ according to the Tukey HSD test.

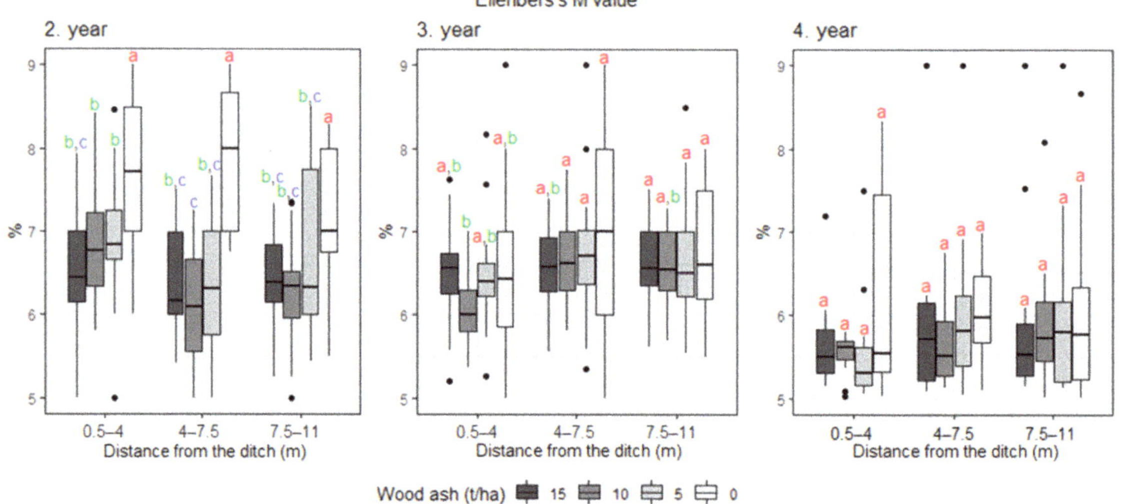

Figure A10. Community weighted mean Ellenberg's indicator Moisture value depending on the distance from the drainage ditch and the dose of wood-ash fertiliser. Different letters (a, b, c) indicate significant differences between groups, $p < 0.05$ according to the Tukey HSD test.

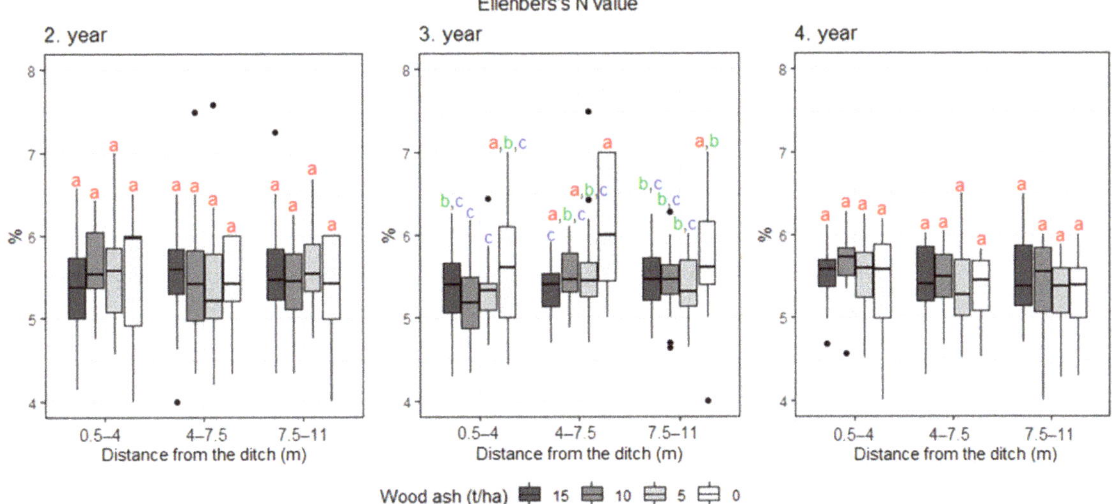

Figure A11. Community weighted mean Ellenberg's indicator Nitrogen value depending on the distance from the drainage ditch and the dose of wood-ash fertiliser. Different letters (a, b, c) indicate significant differences between groups, $p < 0.05$ according to the Tukey HSD test.

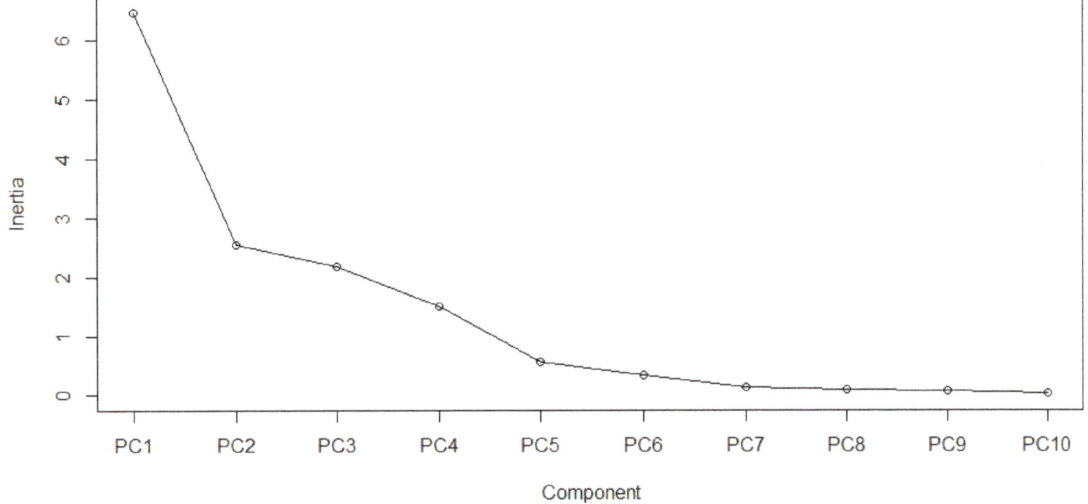

Figure A12. Scree plot from PCA analyses.

Table A1. Observed species in study area during the three-year study.

No.	Species	2019	2018	2017	No.	Species	2019	2018	2017
1.	*Agrostis capillaris*	x	x	x	43.	*Luzula pilosa*		x	
2.	*Arabidopsis thaliana*		x	x	44.	*Lycopus europaeus*	x	x	x
3.	*Arctium lappa*		x	x	45.	*Matricaria perforata*		x	x

Table A1. *Cont.*

No.	Species	2019	2018	2017	No.	Species	2019	2018	2017
4.	*Arctium tomentosum*		x		46.	*Mycelis muralis*			x
5.	*Barbara stricta*	x			47.	*Persicaria maculosa*	x		
6.	*Betula pendula*	x	x	x	48.	*Petasites hybridus*	x		
7.	*Betula pubescens*	x	x	x	49.	*Phragmites australis*	x	x	x
8.	*Bidens tripartita*	x	x	x	50.	*Picea abies*	x	x	x
9.	*Brassica campestris*		x	x	51.	*Picris hieracioides*	x	x	x
10.	*Calamagrostis canescens*	x	x	x	52.	*Pinus sylvestris*	x	x	x
11.	*Calluna vulgaris*		x	x	53.	*Plantago lanceolata*		x	
12.	*Carex cespitosa*	x	x	x	54.	*Plantago major*		x	x
13.	*Carex hirta*	x			55.	*Poa palustris*	x	x	x
14.	*carex pseudocyperus*	x			56.	*Polygonum* sp.			x
15.	*Carex vesicaria*	x			57.	*Polytrihum* sp.		x	x
16.	*Cerastium holosteoides*	x		x	58.	*Populus tremula*	x	x	x
17.	*Chamerion angustifolium*	x	x	x	59.	*Rubus idaeus*	x	x	x
18.	*Chenopodium album*	x		x	60.	*Rumex acetosa*	x	x	
19.	*Cirsium arvense*	x	x	x	61.	*Rumex acetosella*	x	x	x
20.	*Cirsium aucale*	x			62.	*Rumex longifolius*	x		
21.	*Cirsium oleraceum*	x		x	63.	*sagina nodosa*		x	
22.	*Cirsium palustre*			x	64.	*Salix alba*	x		
23.	*Crepis biennis*		x	x	65.	*Salix caprea*	x	x	x
24.	*Echinochloa crusgalli*		x	x	66.	*Salix myrsinifolia*	x		
25.	*Eirophorum polystachion*	x			67.	*Salix rosmarinifolia*	x		
26.	*Epilobium parviflorum*	x	x	x	68.	*Salix* sp.		x	x
27.	*Equisetum arvense*			x	69.	*Salix starkeana*	x		
28.	*Equisetum fluviatile*	x			70.	*Salix triandra*	x		
29.	*Equisetum sylvaticum*		x		71.	*Salix viminalis*	x		
30.	*Erigon canadesis*	x	x	x	72.	*Scirpus sylvaticus*			x
31.	*Eriophorum vaginatum*	x	x	x	73.	*Senecio sylvaticus*	x	x	
32.	*Eupatorium cannabinum*	x	x	x	74.	*Silene vulgaris*			x
33.	*Festuca rubra*	x			75.	*Solidago cannadensis*	x	x	x
34.	*Fragaria vesca*	x	x	x	76.	*Sonchus asper*		x	x
35.	*Frangula alnus*	x	x		77.	*Stellaria media*		x	x
36.	*Gnaphalium uliginosum*		x		78.	*Taraxacum officinale*	x	x	x
37.	*Hieracium pilosella*	x	x	x	79.	*Trifolium repens*			x
38.	*Juncus articulatus*	x	x	x	80.	*Tripleurospermum inodorum*	x		
39.	*Juncus effusus*	x	x	x	81.	*Tussilafgo farfara*	x	x	x
40.	*Juncus tenuis*	x	x	x	82.	*Typha latifolia*	x	x	x
41.	*Lamium album*		x	x	83.	*Utrica dioica*		x	
42.	*Linaria vulgaris*	x	x		84.	*Valeriana officinalis*		x	

References

1. Xu, J.; Morris, P.J.; Liu, J.; Holden, J. PEATMAP: Refining Estimates of Global Peatland Distriubution Based on a Meta-Analysis. *Catena* **2018**, *160*, 134–140. [CrossRef]
2. Clymo, R.S. The Limits to Peat Bog Growth. Philosophical Transactions of the Royal Society of London. *Ser. B Biol. Sci.* **1984**, *303*, 605–654. [CrossRef]
3. Yli-Petays, M.; Laine, J.; Vasander, H.; Tuittila, E.-S. Carbon Gas Exchange of a Re-Vegetated Cut-Away Peatland Five Decades after Abandonment. *For. Ecol. Manag.* **2007**, *12*, 177–190.
4. Lazdiņš, A.; Lupikis, A. Life Restore Project Contribution to the Greenhouse Gas Emission Accounts in Latvia. In *Sustainable and Responsible After-Use of Peat Extraction Areas*; Baltijas krasti: Riga, Latvia, 2019; pp. 21–52.
5. Huotari, N.; Tillman-Sutela, E.; Kauppi, A.; Kubin, E. Fertilization Ensures Rapid Formation of Ground Vegetation on Cut-Away Peatlands. *Can. J. For. Res.* **2006**, *37*, 874–883. [CrossRef]
6. Tuittila, E.-S.; Komulainen, V.-M.; Vasander, H.; Laine, J. Restored Cut-Away Peatland as a Sink for Atmospheric CO_2. *Oecologia* **1999**, *120*, 563–574. [CrossRef] [PubMed]
7. Pētersons, J.; Lazdiņš, A.; Kasakovskis, A. Life Restore Database on Areas Affected by Peat Extraction. In *Sustainable and Responsible After-Use of Peat Extraction Areas*; Baltijas krasti: Riga, Latvia, 2019; pp. 122–128.
8. Mehmood, M.A.; Ibrahim, M.; Rashid, U.; Nawaz, M.; Shafaqat, A.; Hussain, A.; Gull, M. Biomass Production for Bioenergy Using Marginal Lands. *Sustain. Prod. Consum.* **2017**, *9*, 3–21. [CrossRef]
9. Gong, J.; Shurpali, N.J.; Kellomäki, S.; Wang, K.; Zhang, C.; Salem, M.M.A. High Sensitivity of Peat Moisture Content to Seasonal Climate in Acutaway Peatland Cultivated with a Perennial Crop (*Phalarisarundinaceae*, L.): A Modeling Study. *Agric. For. Meteorol.* **2012**, *180*, 225–235. [CrossRef]
10. Ots, K.; Orru, M.; Tilk, M.; Kuura, L.; Aguraijuja, K. Afforestation of Cutaway Peatlands: Effect of Wood Ash on Biomass Formation and Carbon Balance. *For. Stud.* **2017**, *67*, 17–36. [CrossRef]
11. Padur, K.; Ilomets, M.; Põder, T. Identification of the Criteria for Decision Making of Cut-Away Peatland Reuse. *Environ. Manag.* **2017**, *59*, 505–521. [CrossRef]
12. Neimane, S.; Celma, S.; Butlers, A.; Lazdiņa, D. Conversion of an Industrial Cutaway Peatland to a Betulacea Family Tree Species Plantation. *Agron. Res.* **2019**, *17*, 741–753. [CrossRef]
13. Kikamägi, K.; Ots, K.; Kuznetsova, T.; Pototski, A. The Growth and Nutrients Status of Conifers on Ash-Treated Cutaway Peatland. *Trees Struct. Funct.* **2013**, *28*, 53–64. [CrossRef]
14. Neimane, S.; Celma, S.; Zusevica, A.; Lazdina, D.; Ievinsh, G. The Effect of Wood Ash Application on Growth, Leaf Morphological and Physiological Traits of Trees Planted in a Cutaway Peatland. *Mires Peat* **2021**, *27*, 12. [CrossRef]
15. Hytönen, J.; Aro, L. Biomass and Nutrition of Naturally Regenerated and Coppiced Birch on Cutaway Peatland During 37 Years. *Silva Fenn.* **2012**, *46*, 377–394. [CrossRef]
16. Wilson, L.; Wilson, J.M.; Holden, J.; Johnstone, I.; Armstrong, A.; Morris, M. Recovery of Water Tables in Welsh Blanket Bog after Drain Blocking: Discharge Rates, Time Scales and the Influence of Local Conditions. *J. Hydrol.* **2010**, *391*, 377–386. [CrossRef]
17. Gatis, N.; Luscombe, D.J.; Grand-Clement, E.; Hartley, I.P.; Anderson, K.; Smith, D.; Brazier, R.E. The Effect of Drainage Ditches on Vegetation Diversity and CO_2 Fluxes in a *Molinia caerulea*- Dominated Peatland. *Ecohydrology* **2016**, *9*, 407–420. [CrossRef]
18. Renou-Wilson, F.; Farrell, E.P. Reclaiming Peatlands for Forestry. In *Restoration of Boreal and Temperate Forests*; CRC Press: New York, NY, USA, 2004; pp. 541–557.
19. Strack, M.; Keith, A.M.; Xu, B. Growing Season Carbon Dioxide and Methane Exchange at a Restored Peatland on the Western Boreal Plain. *Ecol. Eng.* **2014**, *64*, 231–239. [CrossRef]
20. Silvan, N.; Hytönen, J. Impact of Ash-Fertilization and Soil Preparation on Soil Respiration and Vegetation Colonization on Cutaway Peatlands. *Am. J. Clim. Change* **2016**, *5*, 178–192. [CrossRef]
21. Kelemen, K.; Mihók, B.; Galhidy, L.; Standovár, T. Dynamic Response of Herbaceous Vegetation to Gap Opening in a Central European Beech Stand. *Silva Fenn.* **2012**, *46*, 53–65. [CrossRef]
22. Thrippleton, T.; Bugmann, H.; Kramer-Priewasser, K.; Snell, R.S. Herbaceous Understorey: An Overlooked Player in Forest Landscape Dynamics? *Ecosystems* **2016**, *19*, 1240–1254. [CrossRef]
23. Gilliam, F.S. Excess Nitrogen in Temperate Forest Ecosystems Decreases Herbaceous Layer Diversity and Shifts Control from Soil to Canopy Structure. *Forests* **2019**, *10*, 66. [CrossRef]
24. Augusto, L.; Dupouey, J.-L.; Ranger, J. Effects of Tree Species on Understory Vegetation and Environmental Conditions in Temperate Forests. *Ann. For. Sci.* **2003**, *60*, 823–831. [CrossRef]
25. Kou, M.; García-Fayos, P.; Hu, S.; Jiao, J. The Effect of *Robinia pseudoacacia* Afforestation on Soil and Vegetation Properties in the Loess Plateau (China): A Chronosequence Approach. *For. Ecol. Manag.* **2016**, *375*, 146–158. [CrossRef]
26. Huotari, N.; Tillman-Sutela, E.; Kubin, E. Ground Vegetation Has a Major Role in Element Dynamics in an Ash-Fertilized Cut-Away Peatland. *For. Ecol. Manag.* **2011**, *261*, 2081–2088. [CrossRef]
27. Mabry, C.M.; Golay, M.G.; Thompson, J.R. Seasonal Storage of Nutrients by Perennial Herbaceous Species in Undisturbed and Disturbed Deciduous Forests. *Appl. Veg. Sci.* **2009**, *11*, 37–44. [CrossRef]
28. Weigel, R.; Gilles, J.; Klisz, M.; Manthey, M.; Kreyling, J. Forest Understory Vegetation Is More Related to Soil than to Climate towards the Cold Distribution Margin of European Beech. *J. Veg. Sci.* **2019**, *30*, 746–755. [CrossRef]

29. Zhang, Y.; Chen, H.Y.H.; Tayor, A.R. Positive Species Diversity and Above-ground Biomass Relationships Are Ubiquitous across Forest Strata despite Interference from Overstory Trees. *Funct. Ecol.* **2017**, *31*, 419–426. [CrossRef]
30. Burgess-Conforti, J.R.; Moore, P.A., Jr.; Owens, P.R.; Miller, D.M.; Ashworth, A.J.; Hays, P.D.; Evans-White, M.A.; Anderson, K.R. Are Soils beneath Coniferous Tree Stands More Acidic than Soils beneath Deciduous Tree Stands? *Environ. Sci. Pollut. Res.* **2019**, *26*, 14920–14929. [CrossRef]
31. Lazdiņa, D.; Neimane, S.; Celma, S. Afforestation Demo Site. In *Sustainable and Responsible After-Use of Peat Extraction Areas*; Baltijas krasti: Riga, Latvia, 2019; pp. 208–220.
32. Ellenberg, H.; Weber, H.E.; Düll, R.; Wirth, V.; Werner, W.; Paulißen, D. *Zeigerwerte von Pflanzen in Mitteleuropa*, 2nd ed.; Verlag Erich GoltzeKG: Göttingen, Germany, 1991.
33. Chytrý, M.; Tichý, L.; Dřevojan, P.; Sádlo, J.; Zelený, D. Ellenberg-Type Indicator Values for the Czech Flora. *Preslia* **2018**, *90*, 83–103. [CrossRef]
34. Diekmann, M. Use and Improvement of Ellenberg's Indicator Values in Decidous Forests of Boreal-Nemoral Zone in Sweeden. *Ecography* **1995**, *18*, 178–189. [CrossRef]
35. Kleyer, M.; Bekker, R.M.; Knevel, I.C.; Bakker, J.P.; Sonnenschein, M.; Poschlod, P.; van Groenendael, J.; Klimeš, L.; Klimešová, J.; Klotz, S.; et al. The LEDA Traitbase: A Database of Life History Traits of Northwest European Flora. *J. Ecol.* **2008**, *96*, 1266–1274. [CrossRef]
36. Niklas, K. Life Forms. In *Encyclopedia of Ecology*; Elsevier: Amsterdam, The Netherlands, 2008; pp. 2160–2167.
37. Saugier, B. Plant Strategies and Vegetation Processes. *Plant Sci.* **2001**, *161*, 813. [CrossRef]
38. RStudio Team. *RStudio: Integrated Development for R*; RStudio: Boston, MA, USA, 2019.
39. R Core Team. *R: A Language and Environment*; R Foundation for Statistical Computing: Vienna, Austria, 2020.
40. Oksanen, J.; Blanchet, F.G.; Friendly, M.; Kindt, R.; Legendre, P.; McGlinn, D.; Minchin, P.R.; O'Hara, R.B.; Simpson, G.L.; Solymos, P.; et al. Vegan Community Ecology Package Version 2.5-7. November 2020. Available online: https://cran.r-project.org (accessed on 10 December 2021).
41. Bellamy, P.E.; Stephen, L.; Maclean, L.S.; Grant, M.C. Response of Blanket Bog Vegetation to Drain-Blocking. *Appl. Veg. Sci.* **2012**, *15*, 129–135. [CrossRef]
42. Price, J.; Heathwaite, A.L.; Baird, A.J. Hydrological Processes in Abandoned and Restored Peatlands: An Overview of Management Approaches. *Wetl. Ecol. Manag.* **2003**, *11*, 65–83. [CrossRef]
43. Leung, A.K.; Garg, A.; Coo, J.L.; Ng, C.W.W.; Hau, B.C.H. Effects of the Roots of *Cynodon dactylon* and *Schefflera heptaphylla* on Water Infiltration Rate and Soil Hydraulic Conductivity. *Hydrol. Process.* **2015**, *29*, 3342–3354. [CrossRef]
44. Campbell, D.R.; Lavoie, C.; Rochefort, L. Wind Erosion and Surface Stability in Abandoned Milled. *Can. J. Soil Sci.* **2002**, *82*, 89–95. [CrossRef]
45. Dafni, A.; Cohen, D.; Noy-Mier, I. Life-Cycle Variation in Geophytes. *Ann. Mo. Bot. Gard.* **1981**, *68*, 652–660. [CrossRef]
46. Marks, M.; Lapin, B.; Randall, J. *Phragmites australis* (*P. communis*): Threats, Management and Monitoring. *Nat. Areas J.* **1994**, *6*, 285–294. [CrossRef]
47. Myerscough, P.J.; Whitehead, F. Comparative Biology of *Tussilago farfara* L., *Chamaenerion angustifolium* (L.) Scop., *Epilobium montanum* L., and *Epilobium adenocaulon* Hausskn. II. Growth and Ecology. *New Phytol.* **1967**, *66*, 785–823. [CrossRef]
48. Schaffers, A.P.; Sýkora, K.V. Reliability of Ellenberg Indicator Values for Moisture, Nitrogen and Soil Reaction: A Comparison with Field Measurements. *J. Veg. Sci.* **2000**, *11*, 225–244. [CrossRef]

MDPI

Article

Role of Rhizosphere Soil Microbes in Adapting Ramie (*Boehmeria nivea* L.) Plants to Poor Soil Conditions through N-Fixing and P-Solubilization

Shenglan Wu [1,2], Hongdong Jie [1] and Yucheng Jie [1,*]

1   College of Agronomy, Hunan Agricultural University, Changsha 410128, China; wushenglan@hunau.edu.cn (S.W.); Jhd20210218@stu.hunau.edu.cn (H.J.)
2   Orient Science & Technology College of Hunan Agricultural University, Changsha 410128, China
*   Correspondence: fbfcjyc@hunau.edu.cn or ibfcjyc@vip.sina.com

Abstract: The N-fixing and P-solubilization functions of soil microbes play a vital role in plant adaptation to nutrient-deficiency conditions. However, their exact roles toward the adaptation of ramie to poor soil conditions are still not clear. To fill this research gap, the N-fixing and P-solubilization efficiencies of soils derived from the rhizosphere of several ramie genotypes with different levels of poor soil tolerance were compared. Correlations between the N-fixing, P-solubilization efficiency, and the poor soil tolerable index were analyzed to quantify their contributions towards the adaptation of ramie plants to poor soil conditions. To explore how the microorganisms affected the potential of N-fixing/P-solubilization, the activities of the nutrients related the soil enzymes were also tested and compared. The results of this study confirm the existence of N-fixing and P-solubilization bacteria in the ramie rhizosphere of the soil. The number of N-fixing bacteria varied from 3010.00 to 46,150.00 c.f.u. per gram dry soil for the ramie treatment, while it was only 110.00 c.f.u. per gram dry soil for treatment without ramie cultivation. The average P-solubilization efficiency of ramie treatment was almost five times higher than that of the control soil (0.65 vs. 0.13 mg mL^{-1}). The significant correlations between the poor soil tolerance index and the N-fixing bacteria number ($r = 0.829$)/nitrogenase activity ($r = 0.899$) suggest the significantly positive role of N-fixing function in the adaptation of ramie plants to poor soil. This is also true for P-solubilization, as indicated by the significant positively correlation coefficients between the ramie poor soil tolerance index and P-solubilization efficiency (0.919)/acid phosphatase activity (0.846). These characteristics would accelerate the application of "holobiont" breeding for improving ramie nutrient use efficiency.

Keywords: fibrous crop; rhizosphere soil; soil enzyme; tolerance strategy; infertile soil

Citation: Wu, S.; Jie, H.; Jie, Y. Role of Rhizosphere Soil Microbes in Adapting Ramie (*Boehmeria nivea* L.) Plants to Poor Soil Conditions through N-Fixing and P-Solubilization. *Agronomy* **2021**, *11*, 2096. https://doi.org/10.3390/agronomy11112096

Academic Editors: Moritz Von Cossel, Joaquín Castro-Montoya and Yasir Iqbal

Received: 30 September 2021
Accepted: 17 October 2021
Published: 20 October 2021

Publisher's Note: MDPI stays neutral with regard to jurisdictional claims in published maps and institutional affiliations.

1. Introduction

Ramie (*Boehmeria nivea* L.) is a perennial fiber yielding crop, which generally requires high inputs to achieve potential yield and maintain a good fiber quality [1]. The nutrient requirements for ramie production are generally two to four times higher than that of the normal field crops. For example, the recommended demands of N, P_2O_5, and K_2O for the production of 100 kg ramie fiber are 8.00 kg, 2.00 kg, and 9.00 kg [2], respectively, while they are only 2.56 kg, 0.77 kg, and 2.53 kg for the production of 100 kg corn grains [3]. Actually, during the farmer's production process, higher amounts of fertilizer are usually applied as most Chinese farmers believe the higher the fertilizer application, the higher the crop yield. Over-fertilization exists in almost all the agricultural production fields in China, which results in the crops' nutrient uptake efficiencies being generally lower than 30% or even 20% [4]. This over-fertilization has caused a series of problems such as surface and groundwater pollution, as well as increasing greenhouse gas emissions. To address these problems, the Chinese government has implemented a strategy of "agricultural transformation and upgrading". One important task of the strategy is to establish a sustainable agricultural production system with special emphasis on low input cropping systems.

When implementing such a system in ramie production, the key aspect is to minimize the use of production inputs, especially use of synthetic fertilizers [5]. One way to reduce fertilizer input is through optimization of the cultivation techniques [2]. However, it is still advised to breed new ramie varieties that are characterized by high NUE (nutrient utilization efficiency). Traditionally, plant breeding was conducted by altering the plant's own genomic information. Recently, a new perspective of "holobiont" breeding has emerged and been accepted [6], which claims "plant breeding goes microbial" [7]. The holobiont considerers the plant and its associated microbiome as an evolutionary unit, which together transmit the genetic information to next generation [8]. Therefore, future breeding should set the selection targets not only in terms of plant materials, but also for the associated microbes.

Soil microorganisms play a vital role in nutrient recycling [9]. During our previous research, the helpful roles of soil microbes in the adaptation of the ramie to poor soil conditions were confirmed [10]. This is generally achieved through enrichment of the beneficial bacteria and through a reduction of harmful fungi simultaneously. The current results indicate that ramie's NUE can be improved through holobiont breeding. However, the premise for achieving this potential lies in understanding how the microorganisms improve the ramie's NUE.

Therefore, the current study is designed with an overall aim to explore the role of soil microbes for improving the NUE of the ramie plant. The improvement in NUE by microorganisms is mainly evaluated through their potential for increasing the amount of nutrients available for plant uptake and subsequent utilization [9]. For example, in legume species, the high NUE is attributed to the N-fixing characteristic, which is also true for some non-legume species such as miscanthus [11,12]. Phosphate solubilizing microorganisms (PSM) can convert immobilized inorganic P to soluble organic P (i.e., P-solubilization), which increases the bioavailable P for plant uptake and utilization [13]. Our previous work already detected the N-fixing bacteria of *Bradyrhizobium* from the rhizosphere soil of ramie using sequencing technology [10]. However, the exact N-fixing potential is not clear. Therefore, the first objective (Objective I) of this study is to quantitatively assess the N-fixing potential of the ramie rhizosphere soil and to evaluate the contribution of the N-fixing characteristic in the poor soil tolerance of the ramie plant. Next to nitrogen, phosphorus is the second most important element for plant growth. Thus, the second objective (Objective II) of this study is to explore the role of P-solubilization, which is similar to N-fixing, and its potential contribution towards adaptation of the ramie plant to poor soil conditions. Another key aspect is the role of soil enzymes, which act as a bridge between the microorganism and N-fixing/P-solubilization [14]. To explore how the microorganisms affect the potential of N-fixing/P-solubilization (Objective III), the activities of the nutrient-related soil enzymes are also monitored and compared as a part of this study. Furthermore, based on the outcomes of the former research work [10], it is revealed that the harmful fungal communities (e.g., *Cladosporium*) can be enriched in the ramie rhizosphere soil, which can kill the beneficial bacteria and limit the N-fixing potential. Therefore, the fourth objective (Objective IV) is to verify the inhibitory effect of fungal communities on N-fixing bacterial communities in the rhizosphere.

2. Materials and Methods

The N-fixing and P-solubilization potential of soils derived from the rhizosphere of several ramie genotypes have different tolerance levels to poor soil conditions. The tolerance level is expressed by the overall plant field performance under poor soil conditions, as a higher field performance will have a stronger poor soil tolerance ability. The overall field performance was normalized (detail calculation process shown in [10]) and then expressed to a field performance index (NFPI). The N-fixing potential was quantified by the number of nitrogen-fixing bacteria in the soil and the related nitrogenase activity. The P-solubilization quantity was measured as the available phosphorus (AP) content in the TCP (tricalcium phosphate) liquid medium after microbial cultivation. Other than nitrogenase, the activity of acid phosphatase (S-ACP), urease (S-UE), and sucrase (S-SC) were also tested and compared in this study. Their contributions for helping the ramie plant adapt to poor soil conditions were evaluated through the correlation analysis between the soil enzyme activity and the corresponding NFPI value. Quantification of the anti-N-fixing-bacteria activity by fungal communities was conducted using the inhibition-zone assay.

2.1. Materials and Sampling Strategies

All of the above-described tests were conducted using the fresh soil samples collected in December 2020. Soil samples of four ramie genotypes and one blank control CK (i.e., without ramie cultivation) were compared. The four ramie genotypes, namely Xiangzhu XB (XZ-XB), Zhongzhu 1 (ZZ-1), Xiangzhu X2 (XZ-X2), and Xiangzhu 3 (XZ-3), were established in 2010 at the experimental station of Huarong (29°32′46″ N, 112°39′57″ E, 73 m a.s.l.). The CK was set as the weed free bare ground, which was adjacent to the ramie cultivation block. Without ramie cultivation, the CK block was dominated by the species of *Chrysanthemum indicum* L. and *Humulus scandens* (Lour.) Merr. The field soil is a poor sandy red soil that was found to have a total nitrogen content of 0.69 g/kg, available phosphorus content of 9.62 mg/kg, and exchangeable potassium content of 56.53 mg/kg. According to the field evaluation results [10], the poor soil tolerance ability of the four tested genotypes were shown as XZ-XB (*NFPI* = 0.953) > ZZ-1(*NFPI* = 0.701) > XZ-X2(*NFPI* = 0.452) > XZ-3 (*NFPI* = 0.000). More detailed information about the experiment design and the field performance of the genotypes can be found in our previously published work [10]. The rhizosphere soil samples of the ramie plants were collected in the field and transported to the laboratory in an ice box. The rhizosphere soil samples, defined as the soil remaining attached to the roots after shaking plants vigorously [15], were collected according to the "air shaking method". The rhizosphere soil samples of five plants within one block were randomly selected in an "S" pattern and then mixed to one composite rhizosphere sample. Five soil cores (0–20-cm depth) within the CK block were collected according to the "S" pattern and were mixed to be the composite CK sample. All the samples were sieved through 50-mesh sieves and stored at 4 °C until the start of the test (less than 30 days here).

2.2. Quantification of the Number of Soil N-Fixing Bacteria

The number of the N-fixing bacteria was quantified using the spread plate method, with a detailed procedure comprised of following steps:

(a) Preparation of the soil suspension by adding 10.00 g of the fresh soil sample to 90 mL sterile distilled water and then shaking at 30 °C at 150 rpm for 30 min. (b) Preparation of the diluted soil suspension by 10-, 10^2-, 10^3-, 10^4-, and 10^5-fold dilutions of the suspension, which was prepared by adding 1 mL previous-level-fold suspension to 9 mL of sterile distilled water. (c) Determination of the N-fixing bacteria number of 10^3-, 10^4-, and 10^5-fold diluted suspensions were plated onto Petri plates containing Ashby nitrogen-free solid medium. This was followed by 3 d incubation at 30 °C and the number of N-fixing bacteria in each plate was counted. Afterwards, the counted number was converted to the comparable number, expressed as log c.f.u. per gram of dry soil.

The above steps of each soil sample were repeated four times. The composition of the Ashby nitrogen-free solid medium included 10.00 g mannitol, 0.20 g KH_2PO_4, 0.20 g $MgSO_4$, 0.20 g NaCl, 0.30 g K_2SO_4, 5.0 g $CaCO_3$, 1000.00 mL distilled water, and 18.00 g agar. At first, these components were mixed and adjusted to a pH of 7.0, followed by sterilization at 121 °C for 30 min.

2.3. Quantitative Assay to Determine the P Solubilization Efficiency of Rhizosphere Microbes

In this assay, each 250 mL Erlenmeyer flasks containing 90.00 mL TCP liquid medium was inoculated with 10.00 mL soil suspension (see *Step A* in Section 2.2) and then shook at 30 °C at 150 rpm for 72 h. To eliminate the background effect, a control treatment was carried out simultaneously without the soil suspension and adding only the corresponding sterilized (at 121 °C for 30 min) suspension. At the end of incubation time, 5 mL cultures were sampled and centrifuged at 12,000 rpm for 5 min. Afterwards, the supernatants were used to determine the AP contents by the molybdenum blue method. The net P-solubilization quantity of each sample was the difference of AP content between the normal treatment and control treatment. Each soil sample had four replications. The composition of the TCP liquid medium included 0.30 g NaCl, 0.30 g MgSO$_4$·7H$_2$O, 0.50 g (NH$_4$)$_2$SO$_4$, 0.30 g KCl, 0.03 g FeSO$_4$·7H$_2$O, 0.03 g MnSO$_4$·4H$_2$O, 5.00 g Ca$_3$(PO$_4$)$_2$, 10.00 g glucose, and 1000.00 mL distilled water. In addition, prior to microbial cultivation, the pH of the aforementioned medium was adjusted to 7.0 and sterilized at 121 °C for 30 min.

2.4. Quantitative Estimation of the Soil Enzyme Activity

The nitrogenase activity was measured using an acetylene reduction assay (ARA) and was expressed by the conversion efficiency of acetylene (C$_2$H$_2$) to ethylene (C$_2$H$_4$). The detailed description of ARA method is as follows: (1) adding 0.50 mL soil suspension to each serum bottle (100 mL) containing 50.00 mL nitrogen-free Ashby liquid medium; (2) sealing the bottle by cotton plugs and incubation at 30 °C and 150 rpm for 72 h; (3) replacing cotton plugs with air tight serum stopper; (4) removing 5.00 mL atmospheric air from the tube and injecting same volume of acetylene; and (5) afterwards, continue with incubation for another 72 h and draw 1 mL gas sample from the tube to measure the C$_2$H$_4$ concentration using gas chromatography. Based on this measured concentration, the amount of C$_2$H$_4$ produced per gram of dry soil per 24 h is the nitrogenase activity, which was calculated according to the method described in the study of Haskett [16].

The activities of S-ACP, S-UE, and S-SC were determined using the Solarbio detection kits (Solarbio Technology Co., Ltd., Beijing, China) of BC0140, BC0120, and BC0240, respectively, as per the manufacturer's instructions. The S-UE activity was expressed as the amount of NH$_3$-N, the S-ACP activity as the amount of phenol, and the S-SC activity as the amount of reducing sugar produced per gram of dry soil after 24 h at 37 °C.

2.5. Antagonistic Test to Quantify the Effect of Fungal Communities on N-Fixing Bacteria

Quantification of the anti-N-fixing-bacteria activity by fungal communities was conducted using an inhibition-zone assay. In this assay, the dot culture of the fungal colony on the N-fixing bacteria grown in Petri plates was carried out. The inhibition efficiency was calculated as halo zone diameter/colony diameter (HD/CD). The procedure comprised the following: (a) enrichment of the N-fixing bacteria by adding 5.00 mL soil suspension to 95.00 mL sterilized liquid N-fixing bacteria enrichment medium (15.00 g glucose, 0.80 g KH$_2$PO$_4$, 0.20 g MgSO$_4$, 0.20 g NaCl, 1.00 g CaCO$_3$, 1.00 mL Na$_2$MoO$_4$ (mass fraction of 1%), 1.00 mL H$_3$BO$_3$ (mass fraction of 1%), 1.00 mL MnSO$_4$ (mass fraction of 1%), 1.00 mL FeSO$_4$·7H$_2$O (mass fraction of 1%) and 1000.00 mL distilled water) and then incubated at 30 °C and 150 rpm for 3 days, then taking 5.00 mL of the bacterial culture and enriching again by following the same procedure. (b) Preparation of the N-fixing bacteria to be cultured in petri plates by plating 1.00 mL dule-enriched N-fixing bacterial culture to the surface of petri plates containing a solid LB (Luria-Bertani) medium followed by incubation at 30 °C for 3 days. (c) Preparation of the fungal colony by plating 1 mL ten-fold diluted (10^{-1}) soil suspension on the surface of Petri plates containing a solid PAD medium, followed by incubation at 28 °C until the whole surface was covered by mycelium. Afterwards, a fungal colony was cut using a 5 mm cork borer. (d) Co-incubation of the fungal colony and N-fixing bacteria in Petri plates by spotting three fungal colonies and one control colony (prepared according to the same method for fungal colony with replacing 1 mL soil suspension to 1 mL distilled water) on the surface of N-fixing bacteria grown at four equidistant points near the Petri center in four directions, followed by co-incubation of the paired plates at 30 °C for 3 days. (e) At the end of incubation, the diameter of the halo and the colony were measured.

All of the above steps for each sample were carried out in four replications.

2.6. Statistical Analysis

Data are presented as mean $\pm$ SD. Differences in terms of the tested parameters as described above among the different ramie genotypes were analyzed using one-way ANOVA (analysis of variance) in SAS 9.4 software (SAS Institute, Cary, NC, USA). The mean values of these parameters were compared using Duncan's multiple range tests at both a $p < 0.05$ and $p < 0.01$ level. If $0.01 < p < 0.05$, the significance was marked by $p < 0.05$, otherwise by $p < 0.01$. A correlation analysis was performed between the activities of the four tested soil enzymes and the poor soil tolerance ability of ramie plants (expressed by the *NFPI*).

3. Results

3.1. Comparison of the N-Fixing and P Solubilization Efficiencies between Different Ramie Germplasms

The results shown in Figure 1 indicate that the N-fixing and P-solubilization efficiency of the ramie rhizosphere soil was significantly stronger ($p < 0.05$) than that in the blank control CK soil without ramie. The number of N-fixing bacteria in the ramie rhizosphere soil, which was pooled of 30,822.00 c.f.u. per gram dry soil for all the four genotypes, was 280-times higher than that of the CK (110.00 c.f.u. per gram dry soil) without ramie cultivation (Figure 1a). This was also true for the P-solubilization efficiency, as the AP content (Figure 1b) in the incubation culture of the ramie rhizosphere (four genotypes pooled) was almost five-times higher than that of the CK soil (0.65 vs. 0.13 mg mL^{-1}). To compare the tested genotypes, there is a general trend that genotypes with a better adaptability to poor soil conditions are characterized by a high N-fixing and P-solubilization efficiency. For example, XZ-XB, as the most tolerable genotype, had the second highest N-fixing bacteria number (46,150.00 c.f.u. per gram dry soil), which was just slightly lower than that (53,060.00 c.f.u. per gram dry soil) of XZ-X2 ($p > 0.01$), but significantly ($p < 0.01$) higher than that of ZZ-1 and XZ-3. In particular, XZ-XB showed 15.3-times more N-fixing bacteria (46,150.00 vs. 3010.00 c.f.u.) compared with the poorest tolerable genotype (XZ-3). The correlation results show that the N-fixing bacteria number was significantly positively correlated with the poor soil tolerance index NFPI ($r = 0.829, p < 0.01$). In terms of differences in the P-solubilization efficiency, the genotypes from a high to low P solubilization ability were XZ-XB (AP content of 0.85 mg mL^{-1}), ZZ-1 (0.85 mg mL^{-1}), XZ-3 (0.47 mg mL^{-1}), and XZ-X2 (0.41 mg mL^{-1}). Although the order of XZ-3 and XZ-X2 for P solubilization ability was against that of NFPI, the difference in the AP content between XZ-3 and XZ-X2 did not reach a significant level ($p > 0.01$). The positive correlation ($r = 0.919$) between the P-solubilization efficiency and NFPI was also significant ($p < 0.01$).

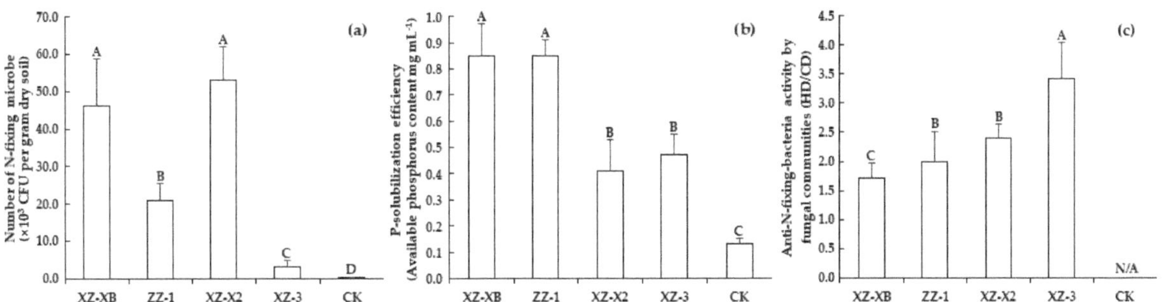

Figure 1. Variations of the rhizosphere soil content nitrogen-fixing microbe numbers (**a**), P-solubilization efficiency of the soil microbes (**b**) and anti-N-fixing-bacteria activity by fungal communities (**c**) between different ramie germplasms. Different capital letters within each measured trait indicate the least significant differences at a $p < 0.01$ level. N/A means not available. XZ-XB, Xiangzhu XB; ZZ-1, Zhongzhu 1; XZ-X2, Xiangzhu X2; XZ-3, Xiangzhu 3. CK is the control without ramie cultivation. HD/CD represents the halo zone diameter/colony diameter that measured in the antagonistic test.

3.2. Effects of Fungal Communities on N-Fixing Bacteria

The results of this study confirm that the ramie rhizosphere soil derived fungal communities have an inhibitory effect on N-fixing bacteria. The halo zone was observed for all of the ramie treatments except for the CK treatment (Figure 1c). This inhibitory effect weakened the ramie poor soil tolerance ability, as indicated by the significantly negative relationship ($r = -0.995$, $p < 0.01$) between HD/CD and NFPI. More precisely, the highest HD/CD ratio of 3.42 was observed in the rhizosphere soil of XZ-3 (the worst performing genotypes under poor soil condition), followed by XZ-X2 (2.39), ZZ-1 (2.00), and XZ-XB (1.72). This trend was opposite to the order of the poor soil tolerance ability. Additionally, the halo zone boundary of the XZ-XB treatment was not fixed and some sporadic N-fixing bacteria colonies were observed within the halo zone. This also indicates that the fungal communities from XZ-XB rhizosphere soil had a relatively weak inhibitory effect on the growth of N-fixing bacteria. In contrast, XZ-3 treatment had a large halo zone and obvious boundary, suggesting a strong inhibitory effect. However, the poor ramie soil tolerance was not fully attributed to the inhibitory effect of the fungal communities on the N-fixing bacteria. For example, the genotype of XZ-X2 had the second-highest HD/CD, but also the highest number for the N-fixing bacteria. Furthermore, the number of N-fixing bacteria in XZ-X2 was 2.5-times higher than that of ZZ-1, despite no significant differences in terms of the HD/CD ratio between these two genotypes.

3.3. Comparison of the Soil Enzyme Activity between Different Ramie Germplasms

The results presented in Figure 2 compare the differences in terms of the activities of nitrogenase, S-ACP, S-UE, and S-SC among the soils cultivated with different ramie genotypes and the control without ramie cultivation. The activities of the tested enzymes in the ramie cultivation treatments were all higher than that of the CK treatment. In particular, the difference in terms of nitrogenase activity was the most significant, as the average activity in ramie cultivation treatment (287.60 μ mol (C_2H_4) g^{-1} d^{-1}) was 15-times higher than that of the CK (19.13 μ mol (C_2H_4) g^{-1} d^{-1}), whereas the S-SC activity (23.39 mg (reducing sugar) g^{-1} d^{-1}) was 13.4-times higher than the CK treatment (1.75 mg (reducing sugar) g^{-1} d^{-1}). The S-ACP activity only had a difference of 1.86-times (64.19 vs. 34.41 μ mol (phenol) g^{-1} d^{-1}). For the activity of S-UE, the ramie treatment (four genotypes pooled) was only 34.3% higher (0.27 vs. 0.18 μg (NH_3-N) g^{-1} d^{-1}) than that of CK.

To compare the ramie genotypes, significant ($p < 0.01$) differences were observed in terms of nitrogenase, S-ACP, and S-SC activities, but not in case of the S-UE activity ($p > 0.05$). The nitrogenase activity showed a completely consistent trend with the poor soil tolerance ability of the ramie plant. The nitrogenase activity (Figure 2a) in the XZ-XB treatment (353.30 μ mol (C_2H_4) g^{-1} d^{-1}) was significantly ($p < 0.05$) higher than that of XZ-X2 treatment (259.33 μ mol (C_2H_4) g^{-1} d^{-1}) and the XZ-3 treatment (194.10 μ mol (C_2H_4) g^{-1} d^{-1}), whereas the difference was not significant for the ZZ-1 treatment (343.47 μ mol (C_2H_4) g^{-1} d^{-1}). The differences in S-ACP activity (Figure 2b) among the genotypes were generally consistent with the ramie's poor soil tolerance ability. The best performing XZ-XB treatment had the highest S-ACP activity of 75.45 μ mol (phenol) g^{-1} d^{-1}, whereas the lowest activity of 53.28 μ mol (phenol) g^{-1} d^{-1} was recorded in the worst performing XZ-3. In the better performing ZZ-1 treatment, the S-ACP activity was lower than the XZ-X2 treatment (63.72 vs. 64.30 μ mol (phenol) g^{-1} d^{-1}), but these differences were not significant. The S-UE activities (Figure 2c) of the four genotypes ranged from 0.25 to 0.29 μg (NH_3-N) g^{-1} d^{-1}, which were not significantly different. For the S-SC activity (Figure 2d), the XZ-X2 treatment had the highest activity of 43.02 mg (reducing sugar) g^{-1} d^{-1}, which was 36.50% higher ($p < 0.01$) than that of the second-highest treatment of ZZ-1 (27.34 mg (reducing sugar) g^{-1} d^{-1}). However, the best-performing XZ-XB had the lowest S-SC activity of only 8.51 mg (reducing sugar) g^{-1} d^{-1}.

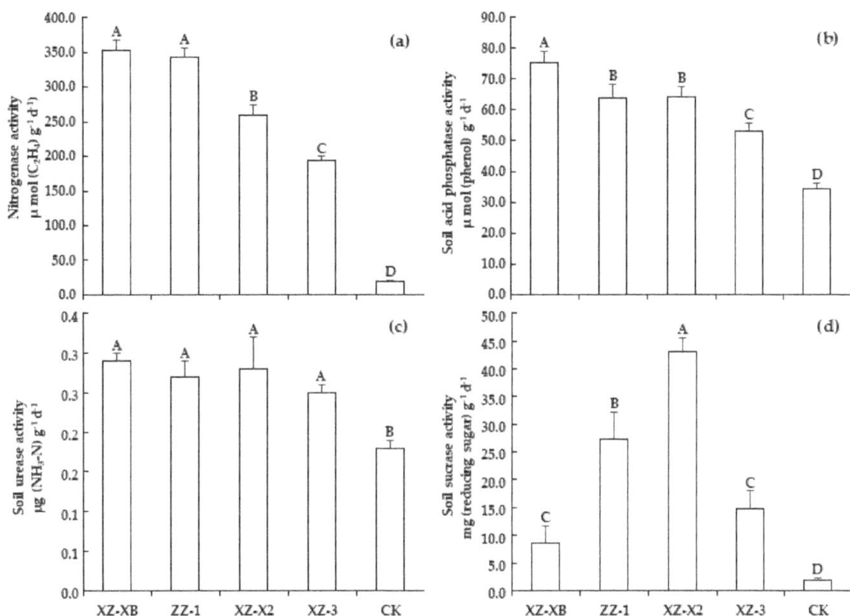

Figure 2. Comparison of the soil enzyme activity of nitrogenase (**a**), acid phosphatase (**b**), urease (**c**), and sucrose (**d**) between different ramie germplasms. Different capital letters within each measured trait indicate least significant differences at $p < 0.01$ level. XZ-XB, Xiangzhu XB; ZZ-1, Zhongzhu 1; XZ-X2, Xiangzhu X2; XZ-3, Xiangzhu 3. CK is the control without ramie cultivation.

The contribution of the soil enzyme to improve the ability of ramie to tolerate poor soil conditions was expressed by the relationship between the soil enzyme activity and the poor soil tolerance index NFPI. The results (Table 1) show that the adaptability of ramie to poor soil was significantly and positively correlated with the nitrogenase activity, S-ACP activity, and S-UE activity, but not with the S-SC activity ($r = 0.256$, $p = 0.421$). According to the Pearson correlation coefficient, the adaptability of the ramie plant to poor soil was mostly correlated with the nitrogenase activity, as indicated by the highest correlation coefficient of 0.899 ($p < 0.001$), followed by the S-ACP activity ($r = 0.846$, $p = 0.005$) and then the S-UE activity ($r = 0.698$, $p = 0.012$). Besides, a significantly negative correlation ($r = -0.995$, $p < 0.001$) was observed between the nitrogenase activity and HD/CD, indicating an inhibitory effect of the fungal communities on the N-fixing bacterial activity.

Table 1. Correlations between the soil enzyme activity (nitrogenase, acid phosphatase, urease, and sucrase) and the ability of ramie to tolerate poor soil, expressed by the normalized field performance index (NFPI).

	Nitrogenase Activity	Acid Phosphatase Activity	Urease Activity	Sucrase Activity
r	0.899	0.846	0.698	0.256
p	<0.001	0.005	0.012	0.421

Note: Detail information of NFPI is shown in the published study of [10]; r: Pearson's coefficients; p: p-values of the Pearson's coefficients.

4. Discussion

Historically, research in terms of N-fixing microbes was only conducted on legume species due to the specific characteristics of the rhizobia. However, recently, an increasing number of studies have also shown the existence of N-fixing bacteria on non-legume species, such as corn [17], miscanthus [12], and sugarcane [18]. N-fixing bacteria can reduce N_2 in the air to a form (mainly NH_4^+) that can be absorbed and utilized by plants, especially under the N-deficiency condition [19]. This explains the positive correlation between the poor soil tolerance ability of the ramie plant and the N-fixing efficiency of the rhizosphere microbes. The N-fixing efficiency is co-contributed by the N-fixing bacteria number and nitrogenase activity. The different N-fixing bacteria number between genotypes can be explained by the plant that controls the N-fixing bacteria number by controlling the type and amount of root exudates [20,21]. Besides, the results of present study indicate the harmful fungal communities is also a factor in determining the N-fixing bacteria number. Additionally, the N-fixing efficiency is more related with the nitrogenase activity. Wang [22] found bacterial strains with nearly 10-times difference of nitrogenase activity in potato rhizosphere soil. Although the N-fixing potential is confirmed by ramie, there is still a large gap compared with legume species. For example, the highest N-fixing potential in this study is only 10–20% of that by soybean [23]. In the future, more studies are required to close the gap. Moreover, this paper only generally describes the existence of N-fixing bacteria in the ramie rhizosphere soil and its positive contribution in help ramie plant adapting to the poor soil condition. To apply this characteristic in future holobiont breeding, it is necessary to identify the specific species of the N-fixing bacteria and then create a stable genetic holobiont group of ramie plant-nitrogen fixing bacteria. For nitrogen cycling, urease also plays an important role in affecting the hydrolysis process of urea [24]. However, there is no significant difference in the urease activity among the tested genotypes, indicating that ramie plants did not adapt to a poor soil environment by affecting the utilization of urea.

This study also found that the P-solubilization by soil microorganisms makes a positive contribution to the adaptation of the ramie to poor soil. The soil microorganisms can secrete organic acids, protons, polysaccharides, and other substances. These substances could accelerate the conversion of insoluble P (e.g., rock P) to soluble form, which prevents the phosphorus availability and absorption by plants [25,26]. This explains the positive role of soil microorganisms in promoting the growth of ramie under poor soil conditions. The P-solubilization efficiency of the microorganism is mainly controlled by the generated phosphatase activity [27]. In this paper, the P-solubilization efficiency of the microorganism is consistent with the corresponding acid phosphatase activity, which also proves the applicability of this view in ramie plants. This study also finds that although there are significant differences in sucrase activity among different ramie genotypes, it is not significantly correlated with the adaptability of the ramie to poor soil. As an important material for catalyzing the decomposition of organic matter, the sucrase activity is closely related to the soil fertility [28]. However, the soil fertility is mainly increased in terms of organic matter, but not in the mineral elements such as N, P, and K.

The results of this study confirm the N-fixing and P-solubilization potential of the ramie rhizosphere soil microbes, especially from the poor soil tolerable genotype. In addition, the negative effects of the rhizosphere soil fungal community on the N-fixing bacterial are confirmed. These results are summarized based on the artificial experiments that were conducted in the sterilized conditions with only one or a few microbial strains. However, in reality, soil microbes grown in an unsterilized condition encounter a more diverse microbial community. The diverse condition suggests a more serious interaction potential between different microbes [29]. This could weaken or also strengthen the microbe's effect tested in the artificial conditions. For this reason, a more realistic test of the N-fixing and P-solubilization potential of the ramie rhizosphere soil microbes is required. This can be conducted by isolating the N-fixing, P-solubilization microbes firstly, then adding the isolation inoculum to unsterilized soil and evaluating their potential on the growth and NUE improvement of the ramie plant. In future isolation research, the effect of the cultivation medium should be taken into consideration to get high selectivity and reliability. In this study, the Ashby medium was used in the N-fixing cultivation as it is one of the most common and suitable media for diazotrophs co-cultivation [30–32]. Actually, diazotroph is not only one kind of prokaryote, but includes several different kinds such as *Rhizobium*, *Ensifer*, *Azospirillum* [33,34]. Each kind of diazotroph has its own most suitable media as, in general, Ashby for *Azotobacter* [35] and yeast extract mannitol agar (YMA) for *Rhizobium* [31]. The exact species of the ramie rhizosphere contented N-fixing bacteria are still not confirmed. To isolate the N-fixing bacteria more effectively in the future study, different cultivation mediums, e.g., *Beijerinckia* medium and *Derxia* medium, should be compared.

5. Conclusions

This study confirms the existence of N-fixing and P-solubilization in the rhizosphere soil of ramie plants. These characteristics of rhizosphere soil microbes help ramie plants adapt to poor soil conditions. The N-fixing efficiency is co-contributed by the N-fixing bacteria number and strong nitrogenase activity. One reason for the low N-fixing efficiency of intolerable genotypes is that the fungal communities in the corresponding rhizosphere soil strongly reduce the nitrogenase activity, also in terms of N-fixing bacteria number.

Author Contributions: S.W.: Methodology, Investigation, Formal analysis, Writing—Original Draft Preparation; H.J.: Resources, Formal analysis, Visualization; Y.J.: Conceptualization, Supervision, Project Administration. All authors have read and agreed to the published version of the manuscript.

Funding: This study was financially supported by the National Natural Science Foundation of China (32071940), China's National Key R&D Program (2019YFD1002205-3 & 2017FY100604-02), Foundation for the Construction of Innovative Hunan (2020NK2028) and Research Project Founded by Hunan Education Department (20C0957).

Institutional Review Board Statement: Not applicable.

Informed Consent Statement: Not applicable.

Data Availability Statement: Not applicable.

Acknowledgments: We appreciate Yu Dai, Qiao Liu and Jing Xiao for their help in the process of experiment implementation and chemical analysis. We would also like to thank Shuai Xue for his assistance in experiment design and Yasir Iqbal for the language editing of this manuscript.

Conflicts of Interest: The authors declare no conflict of interest.

References

1. Tatar, Ö.; Ilker, E. Impact of different nitrogen and potassium application on yield and fiber quality of ramie (Boehmeria nivea). *Int. J. Agric. Biol.* **2010**, *14*, 369–372.
2. Liu, L.J.; Chen, H.Q. Effect of planting density and fertilizer application on fiber yield of ramie (*Boehmeria nivea*). *J. Integr. Agric.* **2012**, *11*, 1199–1206. [CrossRef]
3. Wang, P.; Wang, Y.M. Effects of different fertilizations on maize nutrient uptake in Hemi-dry-land. *Chin. J. Soil Sci.* **2009**, *40*, 1135–1138.
4. Miao, Y.; Stewart, B.A.; Zhang, F. Long-term experiments for sustainable nutrient management in China. A review. *Agron. Sustain. Dev.* **2011**, *31*, 397–414. [CrossRef]
5. Jiao, X.Q.; Mongol, N. The transformation of agriculture in China: Looking back and looking forward. *J. Integr. Agric.* **2018**, *17*, 755–764. [CrossRef]
6. Gopal, M.; Gupta, A. Microbiome selection could spur next-generation plant breeding strategies. *Front. Microbiol.* **2016**, *7*, 1971. [CrossRef] [PubMed]
7. Wei, Z.; Jousset, A. Plant breeding goes microbial. *Trends Plant Sci.* **2012**, *22*, 555–558. [CrossRef] [PubMed]
8. Kroll, S.; Agler, M.T. Genomic dissection of host-microbe and microbe-microbe interactions for advanced plant breeding. *Curr. Opin. Plant Biol.* **2017**, *36*, 71–78. [CrossRef] [PubMed]
9. Baligar, V.C.; Fageria, N.K. Nutrient use efficiency in plants. *Commun. Soil Sci. Plan.* **2001**, *32*, 921–950. [CrossRef]
10. Wu, S.; Xue, S. Seasonal nutrient cycling and enrichment of nutrient-related soil microbes aid in the adaptation of ramie (*Boehmeria nivea* L.) to nutrient-deficient conditions. *Front. Plant Sci.* **2021**, *12*, 644904. [CrossRef]
11. Xu, Y.; Zheng, C. Quantitative assessment of the potential for soil improvement by planting Miscanthus on saline-alkaline soil and the underlying microbial mechanism. *GCB Bioenergy* **2021**, *3*, 1191–1205. [CrossRef]
12. Liu, Y.; Ludewig, U. Nitrogen-dependent bacterial community shifts in root, rhizome and rhizosphere of nutrient-efficient *Miscanthus x giganteus* from long-term field trials. *GCB Bioenergy* **2019**, *11*, 1334–1347. [CrossRef]
13. Sharma, S.B.; Sayyed, R.Z. Phosphate solubilizing microbes: Sustainable approach for managing phosphorus deficiency in agricultural soils. *SpringerPlus* **2013**, *2*, 587. [CrossRef]

14. Nannipieri, P.; Trasar-Cepeda, C. Soil enzyme activity: A brief history and biochemistry as a basis for appropriate interpretations and meta-analysis. *Biol. Fertil. Soils* **2018**, *54*, 11–19. [CrossRef]
15. Chaudhary, D.R.; Gautam, R.K. Nutrients, microbial community structure and functional gene abundance of rhizosphere and bulk soils of halophytes. *Appl. Soil Ecol.* **2015**, *91*, 16–26. [CrossRef]
16. Haskett, T.L.; Knights, H.E. A simple in situ assay to assess plant-associative bacterial nitrogenase activity. *Front. Microbiol.* **2021**, *12*, 690439. [CrossRef] [PubMed]
17. Wang, C.; Bai, Y. Maize aerial roots fix atmospheric N_2 by interacting with nitrogen fixing bacteria. *Sci. Sin. Vitae* **2019**, *49*, 89–90.
18. de Lima, D.R.M.; dos Santos, I.B. Genetic diversity of N-fixing and plant growth-promoting bacterial community in different sugarcane genotypes, association habitat and phenological phase of the crop. *Arch. Microbiol.* **2021**, *203*, 1089–1105. [CrossRef]
19. Rocha, K.F.; Kuramae, E.E. Microbial N-cycling gene abundance is affected by cover crop specie and development stage in an integrated cropping system. *Arch. Microbiol.* **2020**, *202*, 2005–2012. [CrossRef] [PubMed]
20. Dennis, P.G.; Miller, A.J. Are root exudates more important than other sources of rhizodeposits in structuring rhizosphere bacterial communities? *FEMS Microbiol. Ecol.* **2010**, *72*, 313–327. [CrossRef] [PubMed]
21. Li, Y.; Pan, F. Response of symbiotic and asymbiotic nitrogen-fixing microorganisms to nitrogen fertilizer application. *J. Soils Sediment.* **2019**, *19*, 1948–1958. [CrossRef]
22. Wang, W.; Yang, S. Isolation and characterization of nitrogen fixing and phosphate solubilizing bacteria from potato rhizosphere of reclaimed cropland. *Nat. Sci. Ed.* **2019**, *47*, 127–133.
23. Wang, J.; Xu, Y. Effects of rhizobium, soybean variety, soil type on nitrogenase activity. *J. Northeast Agric. Univ.* **2008**, *39*, 36–39.
24. Fu, Q.; Abadie, M. Effects of urease and nitrification inhibitors on soil N, nitrifier abundance and activity in a sandy loam soil. *Biol. Fertil. Soils* **2020**, *56*, 185–194. [CrossRef] [PubMed]
25. Sharma, A.; Saha, T.N. Efficient microorganism compost benefits plant growth and improves soil health in Calendula and Marigold. *Hortic. Plant J.* **2017**, *3*, 67–72. [CrossRef]
26. Adetunji, A.T.; Lewu, F.B. The biological activities of β-glucosidase, phosphatase and urease as soil quality indicators: A review. *J. Plant Nutr. Soil Sci.* **2017**, *17*, 794–807. [CrossRef]
27. Widdig, M.; Schleuss, P.M. Nitrogen and phosphorus additions alter the abundance of phosphorus-solubilizing bacteria and phosphatase activity in grassland soils. *Front. Environ. Sci.* **2019**, *7*, 185. [CrossRef]
28. Guo, Q.Q.; Xiao, M.R. Polyester microfiber and natural organic matter impact microbial communities, carbon-degraded enzymes, and carbon accumulation in a clayey soil. *J. Hazard. Mater.* **2021**, *405*, 124701. [CrossRef] [PubMed]
29. Ordoñez, Y.M.; Fernandez, B.R. Bacteria with phosphate solubilizing capacity alter mycorrhizal fungal growth both inside and outside the root and in the presence of native microbial communities. *PLoS ONE* **2016**, *11*, e0154438. [CrossRef]
30. Stella, M.; Suhaimi, M. Selection of suitable growth medium for free-living diazotrophs isolated from compost. *J. Trop. Agric. Food Sci.* **2010**, *38*, 211–219.
31. Ramírez, M.D.A.; España, M. Genetic diversity and characterization of symbiotic bacteria isolated from endemic Phaseolus cultivars located in contrasting agroecosystems in Venezuela. *Microbes Environ.* **2021**, *36*, ME20157. [CrossRef]
32. Wickramasinghe, W.R.K.D.W.K.V.; Girija, D. Multi-phasic nitrogen fixing plant growth promoting rhizobacteria as biofertilizer for rice cultivation. *Res. J. Agric. Sci.* **2021**, *12*, 399–404.
33. Soumare, A.; Diedhiou, A.G. Exploiting biological nitrogen fixation: A route towards a sustainable agriculture. *Plants* **2020**, *9*, 1011. [CrossRef] [PubMed]
34. Gopalakrishnan, S.; Srinivas, V. Nitrogen fixation, plant growth and yield enhancements by diazotrophic growth-promoting bacteria in two cultivars of chickpea (Cicero arietinum L.). *Biocatal. Agric. Biotechnol.* **2017**, *11*, 116–123. [CrossRef]
35. Jiménez, D.J.; Montaña, J.S. Characterization of free nitrogen fixing bacteria of the genus Azotobacter in organic vegetable-grown Colombian soils. *Braz. J. Microbiol.* **2011**, *42*, 846–858. [CrossRef]

Article

Phytoremediation of Cadmium-, Lead-, and Nickel-Polluted Soils by Industrial Hemp

Giorgio Testa *, Sebastiano Andrea Corinzia, Salvatore Luciano Cosentino and Barbara Rachele Ciaramella

Department of Agriculture, Food and Environment (Di3A), University of Catania, Via Valdisavoia 5, 95123 Catania, Italy; andrea.corinzia@unict.it (S.A.C.); cosentin@unict.it (S.L.C.); barbara.ciaramella@phd.unict.it (B.R.C.)
* Correspondence: gtesta@unict.it

Abstract: The restoration of polluted soils is crucial for ecosystem recovery services. Evidently, phytoremediation is a biological and sustainable technique that includes the use of plants to remediate heavy-metal-contaminated land; the plants should be tolerant to the contamination and capable of uptake or immobilization of the heavy metals in the soil. Moreover, defining an economically efficient approach to the remediation of a contaminated area, with the possibility of further utilization of phytoremediation biomass, renders energy crops a great option for this technique. Energy crops, in fact, are known for their ability to grow with low agricultural input, and later, the biomass product can be used to produce biofuels, bioenergy, and bioproducts in a sustainable and renewable way, creating economic potential, especially when these crops are cultivated in marginal lands. The aim of this work is to test two monoecious industrial hemp varieties in different levels of Cd, Pb, and Ni in soil. Both varieties were tolerant to levels of Cd and Pb contamination that were higher than the limit for commercial and industrial use, while Ni showed a significant effect at all the tested concentrations. The variety Futura 75 performed better than Kc Dora in terms of productivity and tolerance.

Keywords: heavy metal; contaminated soil; *Cannabis sativa* L.; phytoextraction

Citation: Testa, G.; Corinzia, S.A.; Cosentino, S.L.; Ciaramella, B.R. Phytoremediation of Cadmium-, Lead-, and Nickel-Polluted Soils by Industrial Hemp. *Agronomy* **2023**, *13*, 995. https://doi.org/10.3390/agronomy13040995

Academic Editor: Marjana Regvar

Received: 7 March 2023
Revised: 21 March 2023
Accepted: 27 March 2023
Published: 28 March 2023

1. Introduction

The challenges that agriculture will face in the near future will be determined by the growth of the world's population, which is directly linked to the increase in the use of natural resources, the finite availability of agricultural land, and climate change, which is leading to higher temperatures and greater variability in precipitation, with an increase in extreme weather events [1]. Therefore, lands that are suitable for food production can be hindered by soil contamination [2]. In this context, the adoption of sustainable farming systems to restore ecosystems while sequestering atmospheric carbon will be necessary to overcome these challenges [3].

Human activities are the primary source of soil contamination with heavy metals. For example, the residues from mining, pesticides, and herbicides that are used in agricultural activities; residues from the petroleum industry or its derivates; residues from battery production; and the inappropriate discard of electronic components are some of the human actions that result in soil contamination with heavy metals [4]. Among all the contaminants that can compromise the quality of the soil, heavy metals can be hazardous for human health and the ecosystem in general, despite some of the heavy metals being used by humans and animals as micronutrients due to the process of bioaccumulation in the food chain and the impossibility of degradation [5]. The most common heavy-metal pollution that has originated from agriculture concerns Zn, As, Cd, Pb, Cu, Se, and U. Soil contamination of Cd, Pb, As, and Hg originates from mining and smelting activities, while Cd, Hg, Cr, As, Cu, Co, Ni, and Zn contamination originates from waste disposal. Another source of heavy metal contamination is atmospheric deposition in proximity to urban areas, which is relevant for As, Pb, Cu, Cd, Cr, Zn, and Hg. Overall, the most common cases of soil pollution from heavy metals concern As, Pb, Cr, Hg, Cu, Cd, and U [6]

Furthermore, excess absorption of heavy metals by humans and animals can cause serious health problems, for example, by damaging the nervous system or generate tumors [7]. Metal toxicity is due to the ability of these metals to alter biological mechanisms at the cellular and the molecular level. For example, Cr, Be, As, V, Cu, Ni, are genotoxic, i.e., they cause DNA mutation. Pb and Cd increase the incidence of tumors and cancers indirectly, decreasing the efficiency of the immune system in repairing chemical damage affecting the DNA [8,9]. A European commission report estimated a total number of 2.5 million potentially contaminated sites in Europe, and it is expected that 340,000 of these sites are contaminated and likely to require remediation, showing the significance of this problem [10]. The most frequent contaminants are heavy metals, affecting 35% of European soils [11,12].

Soil decontamination can be attained by following different paths: using chemical, physical, or biological techniques or a mix of them [13]. Phytoremediation is a biological technique for decontaminating the soil that uses plants to extract or stabilize the contaminants [14]. The plants are selected based on several criteria, such as tolerance to heavy metals, high biomass yield, deep and extensive root system, and awareness of using low agronomic inputs [15].

Many energy crops meet these requirements, and the biomass that they obtain on contaminated soils can be used as a feedstock for energy production (heat, biofuels, biogas) or in the bioproducts field (textile, paper, mats, bioplastics) with low environmental and health risks [16].

Currently, the use of land to cultivate crops for bioenergy has become an important policy objective, set out in RED II (Renewable Energy Directive, 2018 EU) [17]. Several industrial crops have been evaluated, such as giant reed [18], switchgrass [19], castor [20], safflower [15], camelina [21], flax [22], and kenaf [23]. Among all these crops, hemp appears to show a phytoremediation potential, with the possibility of reusing the biomass in several methods of conversion [24].

Various studies on industrial hemp (*Cannabis sativa* L.) have demonstrated the ability of this plant to accumulate toxic trace metals such as lead, cadmium, magnesium, copper, chromium, and cobalt and, therefore, reclaim contaminated soil while offering different end uses for its biomass. Mihoc et al. (2012), Canu et al. (2022), De Vos et al. (2023), and Shi et al. (2012) observed that hemp can offer a sustainable and economic solution for soil decontamination [25–28].

Historically, hemp has been grown for its long bast fibers and seeds, although, in modern times, it can also be grown for energy production [29]. Its high cellulose content renders hemp an attractive annual crop for second-generation bioethanol production [30].

Hemp can be grown under various agroecological conditions, varying in temperature, photoperiod, and soil water availability, by choosing planting date and variety according to the local condition [31,32]. In addition, hemp varieties can be classified according to several attributes such as geographic origin, end use (fiber or seed), ripening time, and reproductive system (dioecious or monoecious) [31].

As reported by the European Environmental Agency in the industrial pollution profiles of countries, the most abundant heavy metals from industrial waste in Italy, considering the period from 2007 to 2016, were cadmium, lead, and nickel [33,34]. For this reason, this research aims to evaluate the adaptability of two monoecious industrial hemp varieties: Futura 75, a French late-ripening cultivar which is one of the most cultivated varieties of industrial hemp in South Europe due to its excellent acclimatization to high temperatures; and KC Dora, a Hungarian variety that can achieve high biomass and seed yield in a broad spectrum of climatic conditions, including those of the Mediterranean area.

Both varieties were tested under three different levels of cadmium, lead, and nickel soil contamination in order to assess their phytoremediation potential and the effects of the pollutants on the yield of hemp in the southern Mediterranean area.

2. Materials and Methods

A two-year experiment (2020/2021) was carried out at the Department of Agriculture, Food and Environment—University of Catania (Sicily, Italy). In a block-randomized experimental design, the following factors were studied in pots with three replications in order to evaluate the tolerance of two varieties of *Cannabis sativa* L. (Futura 75 and KC Dora) in soils contaminated with three heavy metals (Ni, Cd, Pb) that were applied in the soil as nitrate (Cd (NO_3), Pb (NO_3), Ni (NO_3)). The amount of the single contaminant in the soil was decided according to the Italian law limit, which was referred to in sites for commercial and industrial use, as reported in D.Lgs 3 April 2006 n.152 (2006) [35] (Table 1).

Table 1. Heavy metal concentration at the legal limit for commercial and industrial sites and the levels of contamination applied to the experimental pots.

Contaminant	Cd	Pb	Ni
Legal limit (mg kg^{-1})	15	1000	500
Concentration I (mg kg^{-1})	60	1000	500
Concentration II (mg kg^{-1})	90	1500	1000
Concentration III (mg kg^{-1})	120	2000	1500
Concentration IV (mg kg^{-1})	150		

The non-contaminated soil was investigated as a control group.

The soil (Andisol, USDA) that was used was taken from the area of Mount Etna and was sampled at a depth of 30 cm.

At the start of the experiments, the soil was analyzed by collecting 1 kg of soil that had been dried in an oven (Herather, Thermo Fisher Scientific Inc., Waltham, MA, USA), at a temperature ranging from 25 to 30 °C and then sieved through a 2 mm mesh.

The sample size was measured, and electrical conductivity was measured in 1:1 soil/distilled water suspensions after 1 h by using conductivity electrodes (Hydros 21, Meter Group Inc., Pullman, WA, USA).

For the measurement of pH (H_2O), a pH meter P.H. 7 Vio (XS Instruments, Carpi, Italy) was used. Soil organic matter was determined via the Walkley–Black procedure [36].

Quantification of the total metal content (Cd, Ni and Pb) of the soil was performed by using the aqua regia digestion samples according to ISO 11466 (ISO, 1998) [37], and after filtration, the heavy metals in the soil were detected by flame atomic absorption spectrometry (AAnalyst 200 AA spectrometer, PerkinElmer, Waltham, MA, USA).

Furthermore, heavy metal bioavailability in the soil was determined according to ISO 17402 [38], by using an EDTA (Merck KGaA, Darmstadt, Germany) concentration of 0.05 M, pH 7.5 (close to soil pH) to a volumetric ratio of 1:20 in 1 g of soil, which was agitated for 24 h. Atomic absorption spectrometry was performed on the filtrate solution to quantitatively determine the available heavy metals.

Seeds were germinated in petri dishes, and each germinated seed was planted in peat pots and was transferred two weeks later to a contaminated pot (three plants per pot). Throughout the growing cycle, the seedlings were kept in well water. A nearby weather station recorded the main meteorological parameters. Over the 2 growing seasons (April–September), the range of the minimum temperatures was 6.7 °C to 19.8 °C and 5.3 °C to 21.1 °C in the 1st and 2nd years, respectively, while the range of maximum temperatures was 14.9 °C to 31.9 °C and 17.4 °C to 35.7 °C in the 1st and 2nd years, respectively.

The plants in each of the pots were harvested and fractionated into stems, leaves, and seeds. The biomass was then weighed and dried in an oven at 65 °C to a constant weight.

Roots were also collected, washed with ultrapure water to remove soil particles, freshly weighed, and oven-dried at 65 °C to obtain a stable weight.

After each sample was ground with a mill on a 1 mm sieve (IKA M20), 1 g of biomass was combusted in a muffle furnace at 550 °C for 5 h. Digestion of the biomass samples for heavy metals was performed with 10 mL of 1:1 nitric acid solution (65% nitric acid, Merck KGaA, Darmstadt, Germany).

Atomic absorption spectrometry (AAnalyst 200 AA Spectrometer, PerkinElmer, Waltham, MA, USA) was used to quantify the total heavy metals in the extract [19].

Data Analysis

The tolerance index (*TI*), bioconcentration factor (*BCF*), accumulation index (*mAI*), and translocation factor (*TF*) were calculated in order to evaluate the tolerance of the two industrial hemp varieties [19].

The tolerance index (*TI*) was calculated to assess the tolerance of the plants at the increasing levels of contaminants in the soil [14,19,39]. The *TI* was obtained by dividing the dry aboveground biomass of contaminated plants (g pot^{-1}) by the dry aboveground biomass of control plants (g pot^{-1}).

$$TI = \frac{dry\ aboveground\ biomass\ weight\ of\ contaminated\ plants,\ \mathrm{g\ pot^{-1}}}{dry\ aboveground\ biomass\ weight\ of\ control\ plants,\ \mathrm{g\ pot^{-1}}} \tag{1}$$

The modified accumulation index (*mAI*) was calculated to assess the ability of the plant to absorb the heavy metal from the soil [14,19]. It was obtained via the ratio between the metal accumulation in the contaminated plant (mg kg^{-1}) and the heavy metal accumulation in the control plants (mg kg^{-1}).

$$mAI = \frac{metal\ accumulation\ in\ the\ contaminated\ plants,\ \mathrm{mg\ kg^{-1}}}{metal\ accumulation\ in\ the\ control\ plants,\ \mathrm{mg\ kg^{-1}}} \tag{2}$$

The ability of the plant to uptake and accumulate the metal in the biomass was determined via the modified bioconcentration factor (*mBCF*). Soil bioavailable metal content, as determined by EDTA extraction, represents the level of heavy metal potentially extracted by the plant. Thus, this factor may represent the ability of the metal to be translocated in plants [14,19,40]. This was determined as the relationship between the heavy metal in the plant fraction (mg kg^{-1}) and the bioavailable metal in the soil (mg kg^{-1}).

$$mBCF = \frac{metal\ concentration\ in\ the\ plant\ fraction,\ \mathrm{mg\ kg^{-1}}}{bioavailable\ metal\ concentration\ in\ the\ soil,\ \mathrm{mg\ kg^{-1}}} \tag{3}$$

The translocation factor (*TF*) is expressed as the relationship between the concentration of metal in the aboveground fraction of the plant (mg kg^{-1}) and the concentration of metal in the root fraction of the plant (mg kg^{-1}) [19,41]. It was established as the concentration of metals in the aboveground plant fraction (mg kg^{-1}) divided by the concentration in the belowground plant fraction (mg kg^{-1}).

$$TF = \frac{metal\ concentration\ in\ the\ aboveground\ plant\ fraction,\ \mathrm{mg\ kg^{-1}}}{metal\ concentration\ in\ the\ belowground\ plant\ fraction,\ \mathrm{mg\ kg^{-1}}} \tag{4}$$

Potentially suitable for phytoextraction are plants with *mBCF* and *TF* indices greater than one (>1) [42].

Data were statistically analyzed by using R-4.2.3 software (R Core Team, 2013). The pollutants and their levels were treated as the main factors, and Tuckey's HSD test was used to isolate the means. The normality of the residual distribution was tested by using the Shapiro test. Differences in productivity and heavy metal concentrations between years were tested by using ANOVA.

Person's correlation matrix, based on the yields of the biomass fractions and the heavy metal concentrations in the fractions of the plants, was applied to interpret and visualize the multivariate data [15,43].

3. Results

3.1. Soil Characterization

The soil was characterized as sandy soil (Andisol, USDA), with neutral pH, low nitrogen, and high iron content (Table 2).

Table 2. Physical and chemical characteristics of the soil.

Physical Characteristics	
Clay (%)	3.0
Silt (%)	4.1
Sand (%)	92.9
Texture	Sandy
Conductivity (μS/cm)	34.2
Chemical Characteristics	
pH	7.4
Organic matter (%)	0.86
Fe (mg kg^{-1})	23.6
P (mg kg^{-1})	7
Mn (mg kg^{-1})	0.1
Cu (mg kg^{-1})	21.8

Soil bioavailable Cd, Ni, and Pb concentrations at the sowing time showed no differences between the pots that were used for Futura 75 and KC Dora (Table 3).

Table 3. Total and available heavy metal (mg kg^{-1}) in the soil.

			Total	Available
			H.M. in soil (mg kg^{-1})	H.M. in soil (mg kg^{-1})
Cd		Control	1.7 ± 0.1	1.1 ± 0.1
		60	59.0 ± 2.3	36.0 ± 2.1
		90	88.2 ± 1.4	55.3 ± 2.0
		120	119.4 ± 1.4	80.2 ± 2.3
		150	150.5 ± 2.6	112.9 ± 1.5
Pb		Control	39.6 ± 0.0	19.3 ± 0.0
		1000	1075.5 ± 46.9	570.9 ± 7.1
		1500	1546.6 ± 11.9	1116.2 ± 57.9
		2000	1808.1 ± 32.3	1465.9 ± 53.7
Ni		Control	40.3 ± 5.7	8.7 ± 1.9
		500	508.2 ± 43.1	331.0 ± 14.4
		1000	1047.3 ± 44.5	753.6 ± 29.5
		1500	1491.5 ± 18.7	1153.9 ± 16.1

In Cd-contaminated soil, the bioavailability ranged from 60.2% at the lowest level of Cd-contamination (Cd$_{60}$) to 75.0% at the highest level of contamination. The bioavailability of Ni in soil underwent a considerable increase from a low to a high level of contamination, ranging from 21.7% to 77.4%. In Pb-contaminated soil, the bioavailability ranged from 48.7% to 81.1%.

3.2. Morphological Measurement

The two studied hemp varieties differed in morphology but showed similar behavior in response to the heavy metal contamination (Table 4). All the plants of both Futura 75 and KC Dora varieties that were sown in uncontaminated soil survived until harvesting, while the plant survival rate decreased at high levels of contamination, particularly at Cd$_{150}$ and Ni$_{1500}$, with the rate of survival approaching 50%. In uncontaminated soil, Futura

75 grew taller than KC Dora. Cd contamination did not reduce plant height and basal diameter, except for the highest concentration (Cd_{150}) in Futura 75 and at concentrations higher than 120 mg/kg in KC Dora. Ni-contamination induced the largest plant height and basal diameter reduction in both varieties. Both varieties were little affected by the lowest level of Pb contamination (Pb_{1000}), but a significant reduction in plant height and basal diameter was observed at the two higher concentrations (Pb_{1500}, Pb_{2000}).

Table 4. Plant survival per pot, height of the plant, and basal diameter. Multiple comparisons between means were performed within the different morphological measurements. Different letters indicate significant differences between the means (according to HSD at $p \leq 0.05$).

Variety	Cont	Conc.	Plant Survival (%)	Average Height (cm)	Avarage Diameter (mm)
	Control		100 a	81.9 ± 9.6 a	4.8 ± 1.0 a
	Cd	60	100 a	88.3 ± 7.3 a	4.6 ± 0.2 a
	Cd	90	93 ab	75.4 ± 7.7 a	4.6 ± 0.3 a
	Cd	120	73 ab	80.1 ± 10.3 a	4.5 ± 0.8 a
	Cd	150	57 b	72.3 ± 7.3 a	4.3 ± 0.4 a
Futura 75	Ni	500	93 ab	66.3 ± 4.1 a	3.6 ± 0.3 a
	Ni	1000	87 ab	63.0 ± 2.3 a	3.7 ± 0.3 a
	Ni	1500	53 b	64.9 ± 18.6 a	3.7 ± 0.8 a
	Pb	1000	87 ab	78.3 ± 16.4 a	4.8 ± 1.5 a
	Pb	1500	80 ab	60.4 ± 3.4 a	3.7 ± 0.6 a
	Pb	2000	73 ab	63.2 ± 9.5 a	3.5 ± 0.6 a
	Control		100 a	77.9 ± 10.2 a	4.2 ± 1.9 a
	Cd	60	93 a	76.7 ± 5.1 a	4.9 ± 0.5 a
	Cd	90	93 a	77.3 ± 14.9 a	4.5 ± 1.1 a
	Cd	120	73 ab	65.3 ± 2.3 a	4.1 ± 0.8 a
	Cd	150	6 ab	55.3 ± 16.2 a	3.6 ± 1.3 a
KC Dora	Ni	500	87 a	71.0 ± 8.4 a	4.4 ± 0.9 a
	Ni	1000	87 a	62.1 ± 12.7 a	4.0 ± 0.5 a
	Ni	1500	47 b	47.4 ± 11.4 a	3.0 ± 0.6 a
	Pb	1000	80 ab	76.5 ± 8.9 a	4.8 ± 0.9 a
	Pb	1500	87 a	75.4 ± 11.4 a	4.9 ± 0.2 a
	Pb	2000	73 ab	69.9 ± 4.6 a	3.8 ± 2.5 a

3.3. Plant Biomass Production

Biomass production can be observed in Figure 1. The two hemp varieties did not differ in biomass productivity on uncontaminated soil. However, in heavy-metal-contaminated soil, Futura 75 showed greater tolerance than KC Dora, in particular at Cd_{150}, Ni_{1500}, and Pb_{2000}, for which the biomass yield reduction in comparison with the uncontaminated control was 32%, 38%, and 38%, respectively, for Futura 75 and 47%, 71%, and 44%, respectively, for KC Dora. Both industrial hemp varieties recorded the greatest reduction in biomass yield in Ni-contaminated soil.

Regarding the biomass production, a significant difference was observed in both varieties for the dry weight of stems and leaves, whereas a not significant difference was observed in the dry weight of the roots and seeds (Table 5).

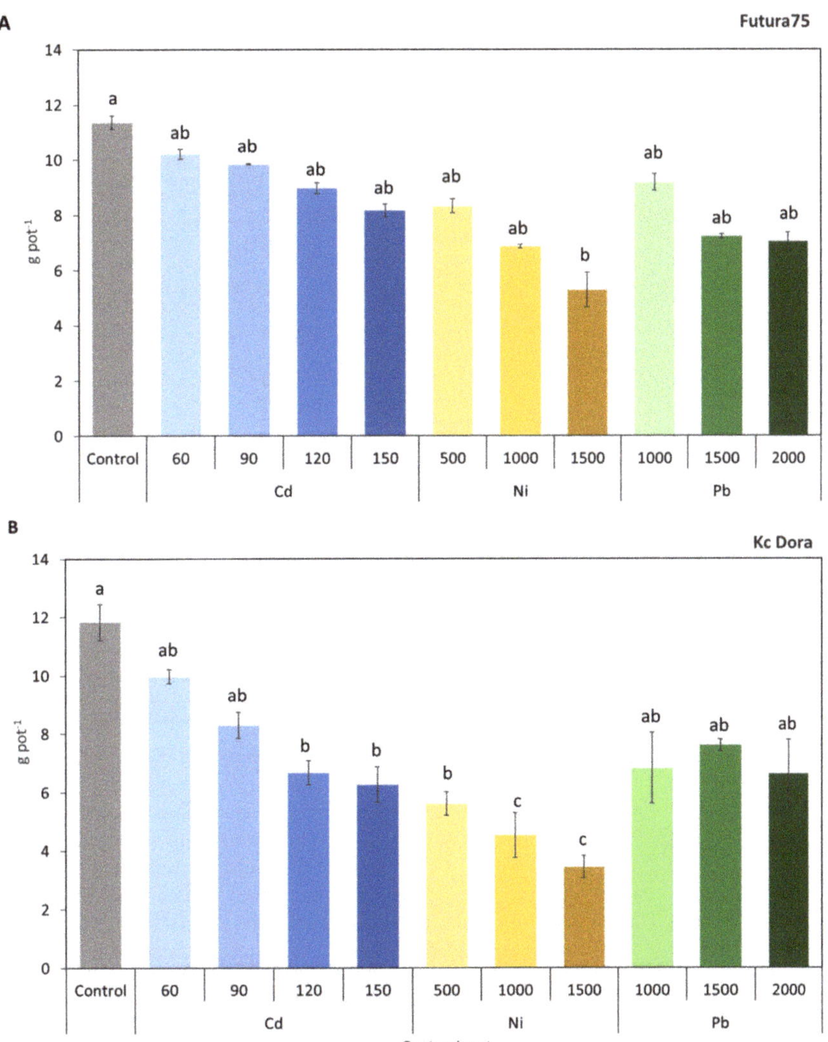

Figure 1. Aboveground biomass of Futura 75 (**A**) and KC Dora (**B**). Different letters indicate significant differences between the means (according to HSD at $p \leq 0.05$).

The biomass of the stems was significantly reduced at the concentration of Ni_{1500} and Pb_{2000} for Futura 75 and Ni_{1000} and Ni_{1500} for KC Dora. The production of leaves in the two varieties was affected by the concentration of the heavy metals: a significant reduction was observed in Cd_{120} for Futura 75 and in Ni_{1500} for KC Dora. Seed yield ranged between 0.4 and 1.5 g pot^{-1} for Futura 75, while in KC Dora, seed yield ranged between 0.3 and 1.2 g pot^{-1}. In both varieties, the highest productivity of seeds was recorded in the untreated pots.

Table 5. Average of the weight of the different compounds of the biomass (roots, stems, leaves, and seeds) in relation to different contaminants and concentrations. The multiple comparisons were performed within the fractions of the plants. Different letters indicate significant differences between the means (according to HSD at $p \leq 0.05$).

Variety	Cont.	Conc.	Average Roots Biomass (g)	Average Stems Biomass (g)	Average Leaves Biomass (g)	Average Seeds Biomass (g)
Futura 75	Control		1.6 a	6.4 ab	3.5 a	1.5 a
	Cd	60	3.2 a	6.9 a	2.7 ab	0.6 a
		90	1.4 a	5.2 ab	3.8 ab	0.9 a
		120	1.2 a	5.3 ab	3.3 b	0.4 a
		150	1.3 a	4.3 ab	3.1 ab	0.7 a
	Ni	500	1.1 a	4.0 ab	3.8 ab	0.6 a
		1000	1.2 a	4.1 ab	1.9 ab	0.9 a
		1500	0.9 a	3.0 b	1.6 ab	0.7 a
	Pb	1000	1.2 a	5.7 ab	2.7 ab	0.7 a
		1500	1.6 a	3.9 ab	2.5 ab	0.8 a
		2000	1.0 a	3.4 b	2.8 ab	0.8 a
KC Dora	Control		2.0 a	6.6 a	4.0 a	1.2 a
	Cd	60	1.6 a	5.5 ab	3.3 ab	1.3 a
		90	1.6 a	5.2 ab	2.3 ab	0.8 a
		120	0.9 a	4.2 ab	1.7 ab	0.8 a
		150	1.1 a	3.9 ab	2.1 ab	0.3 a
	Ni	500	0.9 a	2.8 ab	2.2 ab	0.6 a
		1000	0.7 a	2.5 b	1.5 ab	0.6 a
		1500	0.6 a	2.0 b	1.2 b	0.3 a
	Pb	1000	1.8 a	4.2 ab	2.0 ab	0.7 a
		1500	1.6 a	4.3 ab	2.3 ab	1.0 a
		2000	1.6 a	3.8 ab	2.0 ab	0.9 a

3.4. The Concentration of Heavy Metals in the Different Parts of the Plants

At low levels of cadmium contamination, the highest Cd concentration among plant organs in Futura 75 was observed in the leaves. At high levels of contamination, above Cd_{120}, the plants decreased the translocation of the heavy metal from the roots toward the aboveground organs, leading to a higher concentration of cadmium in the roots. KC Dora showed a larger translocation tendency for cadmium than Futura 75, which led to similar concentrations in roots and leaves at all levels of soil contamination. Cadmium concentration in the aboveground organs did not increase linearly with the concentration in the soil, suggesting the existence of a limitation factor for the translocation. Cadmium concentration in the seeds was lower than 3 μg g^{-1} at any level of soil contamination.

Futura 75 showed a higher nickel uptake and translocation than KC Dora: nickel concentration in the plant tissues was higher in Futura 75 than in KC Dora in roots, leaves, stems, and seeds. A significant difference was observed in all the concentrations. Regarding the aboveground biomass, the highest concentration was observed in the leaves of Futura 75, with a concentration of 26%, 57%, and 87% for Ni_{500}, Ni_{1000}, and Ni_{1500}, respectively. In comparison, the concentration of Ni in the leaves increased in KC Dora, with a percentage of 16%, 30%, and 31% in Ni_{500}, Ni_{1000}, and Ni_{1500}.

Lead translocation potential from the roots to the aboveground organs was low for both Futura 75 and KC Dora. Both varieties showed higher lead concentration in the roots, reaching over 100 μg g^{-1} at Pb_{2000}. Lead concentration was lower in the aboveground organs, staying below 40 μg g^{-1} in the stem and the leaves and below 20 μg g^{-1} in the seeds at the highest level of lead soil contamination for both varieties. The concentration of the contaminants can be observed for cadmium in Figure 2, for nickel in Figure 3, and for lead in Figure 4.

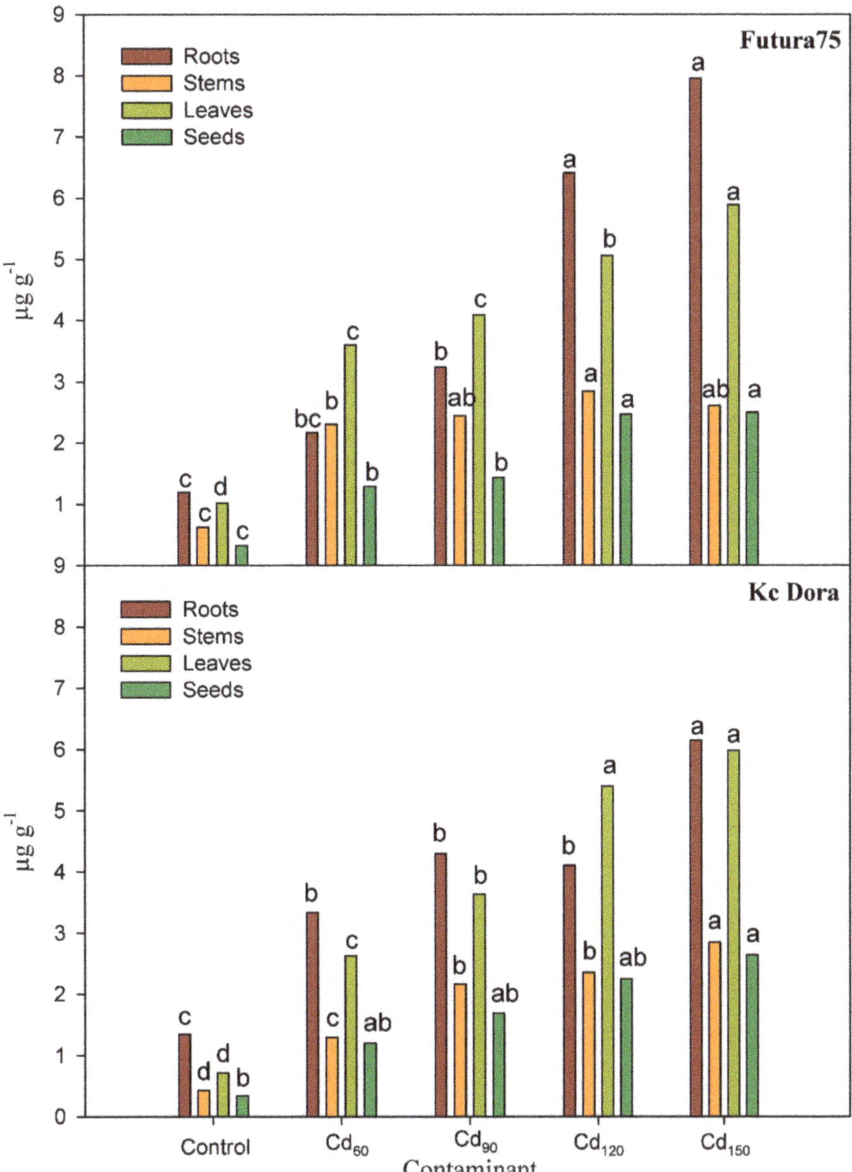

Figure 2. Concentration of cadmium ($\mu g\ g^{-1}$) in roots, stems, leaves, and seeds in Futura 75 and KC Dora. The comparisons were performed within the fractions of the plants. Different letters indicate significant differences between the means (according to HSD at $p \leq 0.05$).

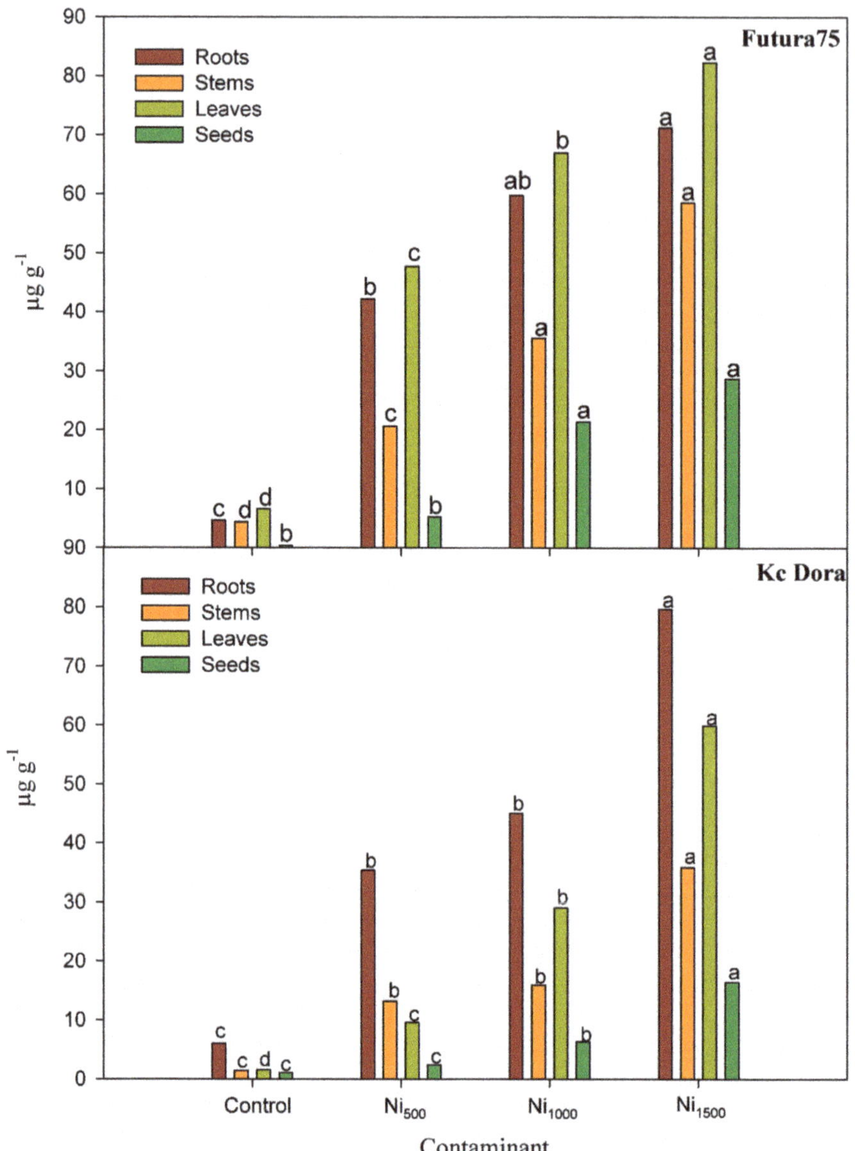

Figure 3. Concentration of nickel ($\mu g\,g^{-1}$) in roots, stems, leaves, and seeds in Futura 75 and KC Dora. The comparisons were performed within the plant fractions. Different letters indicate significant differences between the means (according to HSD at $p \leq 0.05$).

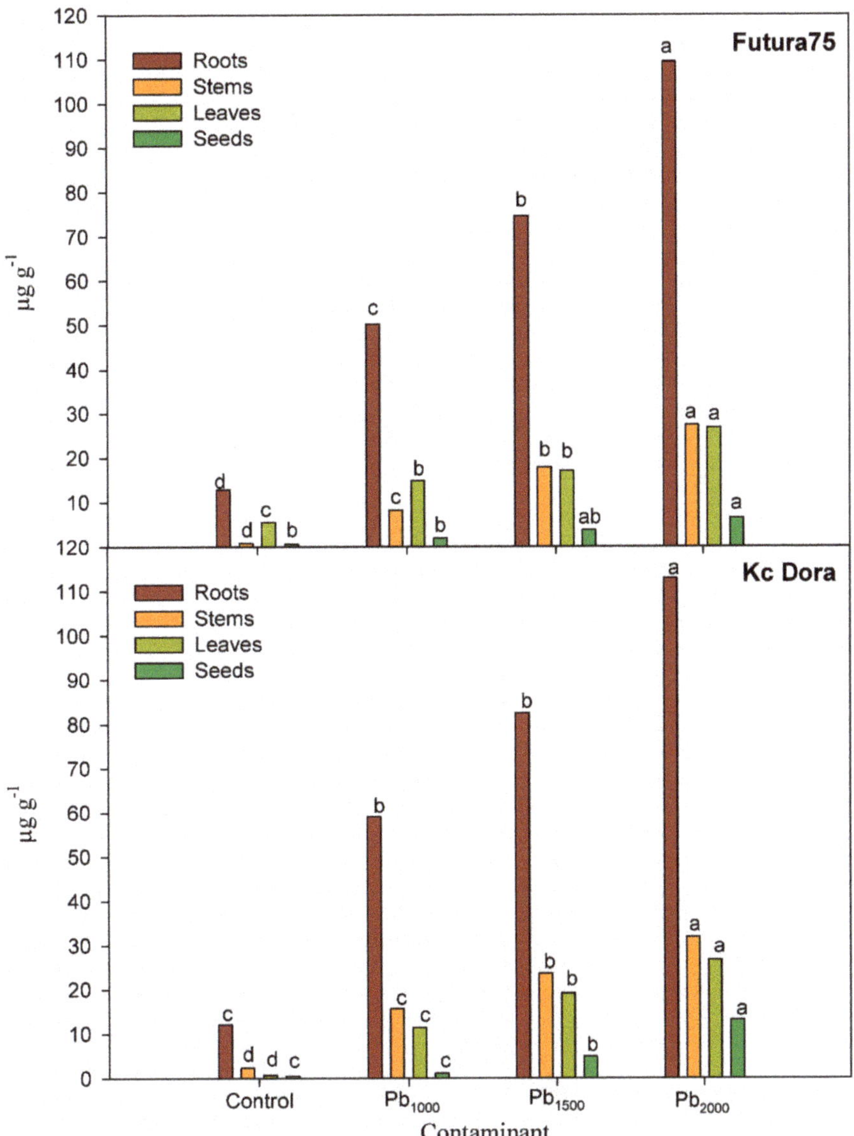

Figure 4. Concentration of lead (μg g^{-1}) in roots, stems, leaves, and seeds in Futura 75 and KC Dora. The comparisons were performed within the fractions of the plant. Different letters indicate significant differences between the means (according to HSD at $p \leq 0.05$).

3.5. Evaluating the Tolerance and the Potential Phytoextraction by Phytoremediation Index and Factors

The several indices and factors can be calculated to evaluate the adaptability to soil contamination (*TI*) and the phytoextraction potential (*mAI*, aboveground and belowground *mBCF* and *TF*) (Table 6). The TI shows the adaptability of the two industrial hemp varieties for growing in soils that were contaminated with progressive levels of cadmium, nickel, and lead.

Table 6. Phytoremediation indices and factors of phytoremediation extraction of Futura 75 and Kc Dora.

Varieties	H.M.—Conc		*TI*	*mAI*	*mBCF* Aboveground	*TF*	*mBCF* Belowground
Futura 75	Cd	60	0.90	3.66	0.28	3.32	0.09
		90	0.87	4.06	0.20	2.47	0.08
		120	0.79	5.26	0.16	1.62	0.10
		150	0.72	5.58	0.13	1.38	0.09
	Ni	500	0.73	6.55	0.34	1.76	0.20
		1000	0.60	10.90	0.26	2.07	0.13
		1500	0.46	14.92	0.16	2.38	0.07
	Pb	1000	0.81	3.68	0.05	0.50	0.10
		1500	0.64	5.71	0.03	0.52	0.07
		2000	0.62	8.94	0.05	0.55	0.08
KC Dora	Cd	60	0.84	3.45	0.14	1.54	0.09
		90	0.63	5.03	0.14	1.74	0.08
		120	0.57	6.72	0.15	2.44	0.06
		150	0.53	8.40	0.15	2.04	0.07
	Ni	500	0.62	6.02	0.12	0.71	0.17
		1000	0.51	12.27	0.11	1.14	0.10
		1500	0.35	26.79	0.15	1.41	0.11
	Pb	1000	0.75	8.12	0.04	0.49	0.09
		1500	0.64	13.39	0.04	0.58	0.07
		2000	0.56	20.19	0.06	0.60	0.09

The tolerance index decreased for the increasing level of soil contamination for both hemp varieties and all the heavy metals that were tested. The lowest TI score was observed at Ni_{1500} (0.46 and 0.35 for Futura 75 and KC Dora, respectively). Futura 75 showed higher TI than KC Dora for all the heavy metals at all the levels of contamination.

The mAI, which assesses the amount of the heavy metal uptake, increased for the increasing level of soil contamination for Futura 75 and KC Dora, indicating that the plants can phytoextract a higher amount of heavy metals from soil with high heavy metal concentrations. The highest mAI score was observed in KC Dora at Ni_{1500} e Pb_{2000}. KC Dora showed higher values of mAI than Futura 75. The comparison of aboveground and belowground mBCF gives insight into the heavy metal partitioning between plant organs. Both factors tend to decrease at high contamination levels. Under cadmium and nickel contamination, Futura 75 showed a higher aboveground mBCF than KC Dora, suggesting a better suitability for the uptake and removal of the heavy metal from the soil.

Under lead and nickel contamination, both Futura 75 and KC Dora had increasing TF scores for increasing soil concentrations. Under cadmium contamination, only KC Dora had increasing TF scores for the increasing soil Cd concentration, while the TF of Futura 75 decreased.

3.6. Correlation of the Main Factor between the Two Varieties of Industrial Hemp

A multivariate analysis was carried out to assess the effect of metal contaminants at different concentrations on variables for cadmium, nickel, and lead (Figures 5–7).

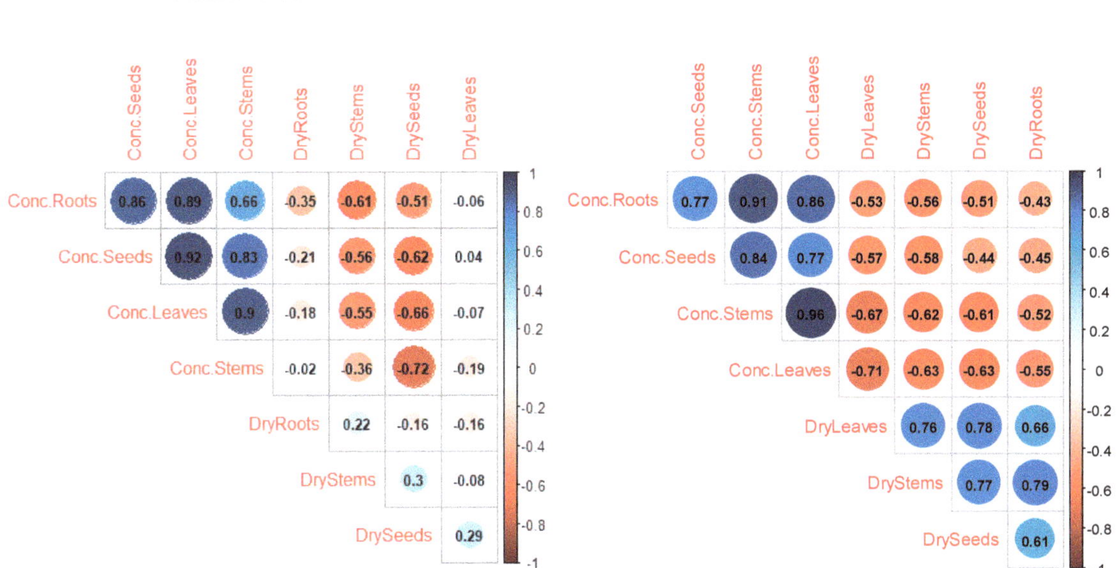

Figure 5. Correlation matrix of Cd of Futura75 and KC Dora, using as the variable the biomass yield of roots, stems, leaves, and seeds, and the concentration of the heavy metal measured in each part of the plant. The numbers within the circles represent the Pearson correlation coefficient.

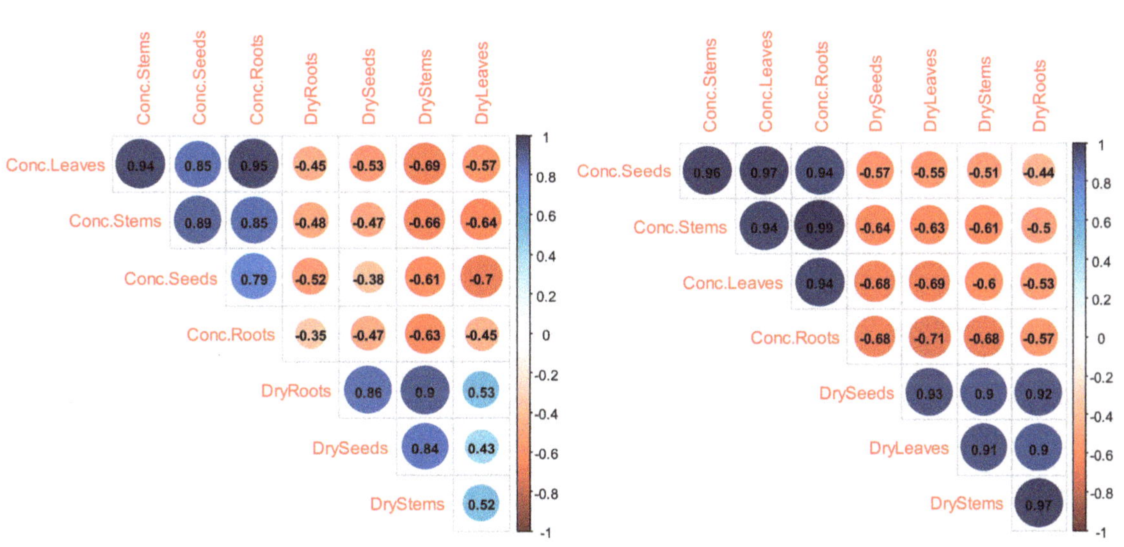

Figure 6. Correlation matrix of Ni of Futura 75 and KC Dora, using as the variable the biomass yield of roots, stems, leaves, and seeds, and the concentration of the heavy metal measured in each part of the plant. The numbers within the circles represent the Pearson correlation coefficient.

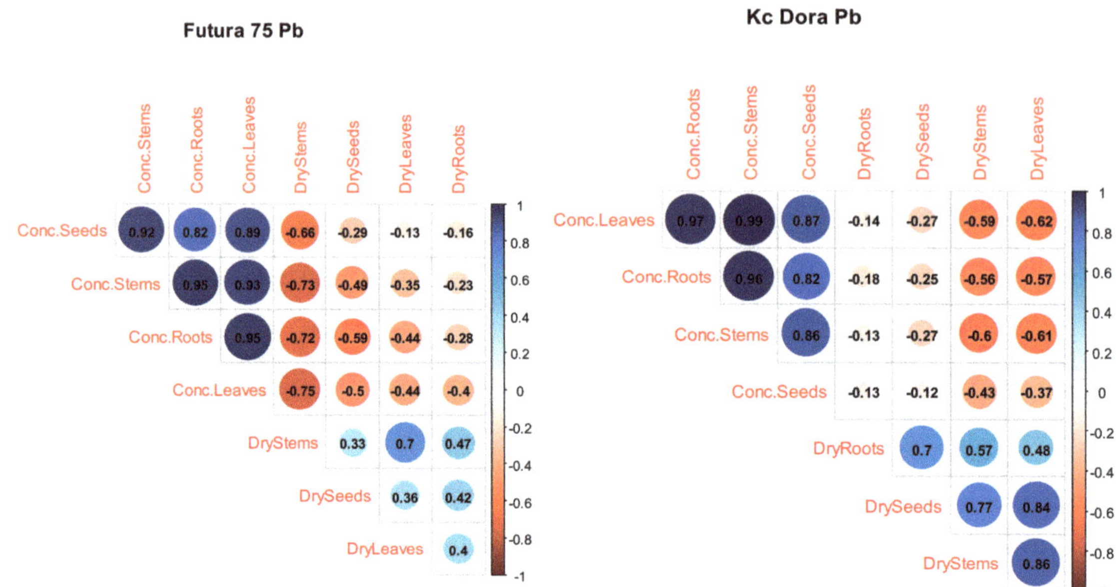

Figure 7. Correlation matrix of Pb of Futura 75 and KC Dora, using as the variable the biomass yield of roots, stems, leaves, and seeds, and the concentration of the heavy metal measured in each part of the plant. The numbers within the circles represent the Pearson correlation coefficient.

Specifically, the components of biomass yield (stems, leaves, and roots) were correlated with each other and negatively correlated with the contaminant concentration in the plant fractions (stems, leaves, seeds, and roots yield).

In fact, in cadmium-contaminated soil, in Futura 75, only the biomass of the stems and the seeds was strongly negatively affected by the concentration of cadmium in the different parts of the plant. In contrast, in KC Dora, all the biomass of the plants was strongly negatively correlated with the concentration in the various parts of the plants. Similar behavior was obtained in nickel-contaminated soil in the correlation matrix of Futura 75 and KC Dora. However, in lead-contaminated soil, the biomass of stems and leaves in KC Dora was strongly negatively correlated with the concentration of Pb in the different parts of the plant. In Futura 75, the stem biomass was strongly negatively correlated with the concentration of the heavy metal in the plant.

4. Discussion

Industrial hemp can be grown in most of the world for its high environmental adaptability [31]. The selection of the best-suited genotype for a specific environment, climatic condition, and agronomic management is crucial for crop success [32,44]. Various studies carried out on *C. sativa* have shown its potential as an accumulator for different toxic traces of metals such as lead, cadmium, magnesium, copper, chromium, and cobalt, which pose a great risk to the ecological system [24,30], making it possible to reclaim contaminated soil while it yields fiber and/or seeds [29].

All over the world, for the problem of soil contamination, hemp can provide a solution that is both economical and sustainable [22,25].

In this study, the productivity of stems in both varieties of hemp was affected by the increasing level of heavy metal, while no significant difference was observed in the seed production. However, low levels of contamination were not detrimental to the overall aboveground biomass; morphologic parameters were not affected by the heavy metal in the soil.

A similar result was observed by De Vos et al. (2023) [27], Pietrini et al. (2019) [45], and Guidi Nissim et al. (2018) [46], who reported no differences in stem height and stem diameter between the control and plants that were cultivated in a low level of soil contamination.

The present study found that Futura 75, a late ripening variety, was more tolerant than KC Dora, an early ripening variety, to high concentrations of cadmium, lead, and nickel [26,31,47].

Cadmium is considered to be one of the most phytotoxic heavy metals [27]. Linger et al. (2002) [36] showed that the photosynthetic pathway in hemp was affected by cadmium indirectly, with the uptake of water and ions by the plant, and directly in the chloroplast apparatus after entering the leaf cells. Cd concentrations up to 72 mg kg^{-1} (soil) had no negative effect on the germination of hemp. Shi et al. (2012) [28] compared 18 hemp accessions cultivated on cadmium-contaminated soils for biodiesel production. It was found that below 25 mg of cadmium per kg of dry soil, most varieties of hemp could grow quite well. Under this condition, the tolerance factor observed in hemp was high (68.6–92.3%), and the ability to store cadmium in the aerial fraction of biomass was suitable for phytoextraction, indicating that the production of this crop can be an alternative to valorize and remediate cadmium-contaminated soils.

Hemp productivity was less affected by lead contamination when compared with the highest concentration of cadmium or nickel. The translocation of lead from roots to the aerial biomass was low; therefore, the highest concentration was observed in the roots. A similar result was observed by Ahmad et al. (2016) [48] and Angelova et al. (2004) [49], who reported Pb concentrations in hemp plants in the following order: roots > stems > leaves > seeds; and by Pietrini et al. (2019) [45], who reported that hemp tends to accumulate lead mainly in the roots, with minimal translocation to the aboveground biomass, which explains the relatively low BCF for Pb that was observed in the present study.

Nickel soil contamination induced the highest reduction in biomass production among the heavy metals that were tested. Ferrarini et al. (2021) [50] reported that hemp had a reduced yield in soil that was contaminated by nickel (>500 mg kg^{-1}). Zhao et al. (2022) [22] reported a reduction in germination and biomass production even at low nickel concentrations (110 and 220 mg kg^{-1}), and both higher concentration in plant organs and higher translocation factor (TF) than the value observed for lead.

For cadmium and nickel, with the exception of Ni$_{500}$, the translocation factor was higher than 1, indicating the high suitability of hemp for the phytoextraction processes, thanks to the accumulation of the heavy metals in the aerial part of the plant.

Although the soil analysis indicated that the bio-availability of cadmium was low, the actual availability of cadmium can increase over time due to the low tendency of this metal to form complexes, while the bio-availability of lead and nickel have a higher complex rate, which reduces the bio-availability.

However, the high tolerance of hemp toward certain heavy metals in the soil renders this plant a suitable alternative for contaminated soil valorization and remediation [27].

5. Conclusions

This research highlighted the different phytoextraction capabilities among the two industrial hemp varieties and demonstrated the capability of industrial hemp to translocate metals from the soils to the aerial parts of the plants, suggesting a good potential for the phytoextraction process. Hemp showed the ability to complete its life cycle until seed ripening in heavily contaminated soils.

The two varieties were tolerant to levels of Cd and Pb contamination above the limit for commercial and industrial use, while Ni showed a significant effect at all the concentrations tested. Futura 75 performed better than Kc Dora in terms of productivity and tolerance.

The low heavy metal concentration in hemp seeds enables the utilization of this plant as a source of oil for bioenergy conversion purposes, avoiding the concerns about contaminant dispersion. The remaining biomass such as stems and leaves can be further valorized through conversion into bioenergy, raising the interest of industrial hemp. Future investigation on the bioconversion processes and on the economic viability of the entire supply chain would be useful to assess the suitability of the entire phytoremediation process.

Author Contributions: Conceptualization, G.T. and B.R.C.; methodology, G.T., B.R.C. and S.L.C.; formal analysis, B.R.C. and S.A.C.; investigation, B.R.C., G.T. and S.A.C.; data curation, B.R.C., G.T. and S.A.C.; writing—original draft preparation, B.R.C., G.T. and S.A.C.; writing—review and editing, B.R.C., G.T. and S.A.C.; supervision, S.L.C. and G.T.; project administration, S.L.C. and G.T.; funding acquisition, S.L.C. and G.T. All authors have read and agreed to the published version of the manuscript.

Funding: This research received no external funding.

Data Availability Statement: Not applicable.

Conflicts of Interest: The authors declare no conflict of interest.

References

1. Shen, X.; Dai, M.; Yang, J.; Sun, L.; Tan, X.; Peng, C.; Ali, I.; Naz, I. A Critical Review on the Phytoremediation of Heavy Metals from Environment: Performance and Challenges. *Chemosphere* **2022**, *291*, 132979. [CrossRef] [PubMed]
2. Scordia, D.; Cosentino, S.L. Perennial Energy Grasses: Resilient Crops in a Changing European Agriculture. *Agriculture* **2019**, *9*, 169. [CrossRef]
3. Tóth, G.; Hermann, T.; Da Silva, M.R.; Montanarella, L. Heavy Metals in Agricultural Soils of the European Union with Implications for Food Safety. *Environ. Int.* **2016**, *88*, 299–309. [CrossRef] [PubMed]
4. Jaskulak, M.; Grobelak, A.; Vandenbulcke, F. Modelling Assisted Phytoremediation of Soils Contaminated with Heavy Metals –Main Opportunities, Limitations, Decision Making and Future Prospects. *Chemosphere* **2020**, *249*, 126196. [CrossRef] [PubMed]
5. Citterio, S.; Santagostino, A.; Fumagalli, P.; Prato, N.; Ranalli, P.; Sgorbati, S. Heavy Metal Tolerance and Accumulation of Cd, Cr and Ni by *Cannabis sativa* L. *Plant Soil* **2003**, *256*, 243–252. [CrossRef]
6. Khalid, S.; Shahid, M.; Khan, N.; Murtaza, B.; Bibi, I.; Dumat, C. A Comparison of Technologies for Remediation of Heavy Metal Contaminated Soils. *J. Geochemical. Explor.* **2017**, *182*, 247–268. [CrossRef]
7. Manno, E.; Varrica, D.; Dongarrà, G. Metal Distribution in Road Dust Samples Collected in an Urban Area Close to a Petrochemical Plant at Gela, Sicily. *Atmos. Environ.* **2006**, *40*, 5929–5941. [CrossRef]
8. Beyersmann, D.; Hartwig, A. Carcinogenic Metal Compounds: Recent Insight into Molecular and Cellular Mechanisms. *Arch. Toxicol.* **2008**, *82*, 493–512. [CrossRef]
9. Cristaldi, A.; Oliveri, G.; Hea, E.; Zuccarello, P. Environmental Technology & Innovation Phytoremediation of Contaminated Soils by Heavy Metals and PAHs. A Brief Review. *Environ. Technol. Innov.* **2017**, *8*, 309–326. [CrossRef]
10. EEA (European Environment Agency). *Progress in Management of Contaminated Sites*; European Environment Agency: Copenhagen, Denmark, 2014; ISBN 9789279348464.
11. Fagnano, M.; Visconti, D.; Fiorentino, N. Agronomic Approaches for Characterization, Remediation, and Monitoring of Contaminated Sites. *Agronomy* **2020**, *10*, 1335. [CrossRef]
12. Sur, I.M.; Micle, V.; Polyak, E.T.; Gabor, T. Assessment of Soil Quality Status and the Ecological Risk in the Baia Mare, Romania Area. *Sustainability* **2022**, *14*, 3739. [CrossRef]
13. Shah, V.; Daverey, A. Phytoremediation: A Multidisciplinary Approach to Clean up Heavy Metal Contaminated Soil. *Environ. Technol. Innov.* **2020**, *18*, 100774. [CrossRef]
14. Barbosa, B.; Boléo, S.; Sidella, S.; Costa, J.; Duarte, M.P.; Mendes, B.; Cosentino, S.L.; Fernando, A.L. Phytoremediation of Heavy Metal-Contaminated Soils Using the Perennial Energy Crops *Miscanthus* spp. and *Arundo donax* L. *Bioenergy Res.* **2015**, *8*, 1500–1511. [CrossRef]
15. Ciaramella, B.R.; Corinzia, S.A.; Cosentino, S.L.; Testa, G. Phytoremediation of Heavy Metal Contaminated Soils Using Safflower. *Agronomy* **2022**, *12*, 2302. [CrossRef]
16. Fernando, A.L.; Duarte, M.P.; Vatsanidou, A.; Alexopoulou, E. Environmental Aspects of Fiber Crops Cultivation and Use. *Ind. Crops Prod.* **2015**, *68*, 105–115. [CrossRef]
17. EU Directive (EU) 2018/2001 of the European Parliament and of the Council on the Promotion of the Use of Energy from Renewable Sources. *Off. J. Eur. Union* **2018**, *2018*, 82–209.
18. Papazoglou, E.G.; Arundo Donax, L. Stress Tolerance under Irrigation with Heavy Metal Aqueous Solutions. *Desalination* **2007**, *211*, 304–313. [CrossRef]
19. Gomes, L.; Costa, J.; Moreira, J.; Cumbane, B.; Abias, M.; Santos, F.; Zanetti, F.; Monti, A.; Fernando, A.L. Switchgrass and Giant Reed Energy Potential When Cultivated in Heavy Metals Contaminated Soils. *Energies* **2022**, *15*, 5538. [CrossRef]
20. Bauddh, K.; Singh, R.P. Growth, Tolerance Efficiency and Phytoremediation Potential of *Ricinus communis* (L.) and *Brassica juncea* (L.) in Salinity and Drought Affected Cadmium Contaminated Soil. *Ecotoxicol. Environ. Saf.* **2012**, *85*, 13–22. [CrossRef]
21. Dimitriu, D. Restoration of heavy metals polluted soils case study—Camelina. *AgroLife Sci. J.* **2014**, *3*, 29–38.

22. Zhao, X.; Guo, Y.; Papazoglou, E.G. Screening Flax, Kenaf and Hemp Varieties for Phytoremediation of Trace Element-Contaminated Soils. *Ind. Crops Prod.* **2022**, *185*, 115121. [CrossRef]
23. Chen, P.; Chen, T.; Li, Z.; Jia, R.; Luo, D.; Tang, M.; Lu, H.; Hu, Y.; Yue, J.; Huang, Z. Transcriptome Analysis Revealed Key Genes and Pathways Related to Cadmium-Stress Tolerance in Kenaf (*Hibiscus cannabinus* L.). *Ind. Crops Prod.* **2020**, *158*, 112970. [CrossRef]
24. Golia, E.E.; Bethanis, J.; Ntinopoulos, N.; Kaffe, G.G.; Komnou, A.A.; Vasilou, C. Investigating the Potential of Heavy Metal Accumulation from Hemp. The Use of Industrial Hemp (*Cannabis sativa* L.) for Phytoremediation of Heavily and Moderated Polluted Soils. *Sustain. Chem. Pharm.* **2023**, *31*, 100961. [CrossRef]
25. Mihoc, M.; Pop, G.; Alexa, E.; Radulov, I. Nutritive Quality of Romanian Hemp Varieties (*Cannabis sativa* L.) with Special Focus on Oil and Metal Contents of Seeds. *Chem. Cent. J.* **2012**, *6*, 122. [CrossRef]
26. Canu, M.; Mulè, P.; Spanu, E.; Fanni, S.; Marrone, A.; Carboni, G. Hemp Cultivation in Soils Polluted by Cd, Pb and Zn in the Mediterranean Area: Sites Characterization and Phytoremediation in Real Scale Settlement. *Appl. Sci.* **2022**, *12*, 3548. [CrossRef]
27. De Vos, B.; De Souza, M.F.; Michels, E.; Meers, E. Industrial Hemp (*Cannabis sativa* L.) Field Cultivation in a Phytoattenuation Strategy and Valorization Potential of the Fibers for Textile Production. *Environ. Sci. Pollut. Res.* **2023**. *Online ahead of print.* [CrossRef]
28. Shi, G.; Liu, C.; Cui, M.; Ma, Y.; Cai, Q. Cadmium Tolerance and Bioaccumulation of 18 Hemp Accessions. *Appl. Biochem. Biotechnol.* **2012**, *168*, 163–173. [CrossRef]
29. Tang, K.; Struik, P.C.; Yin, X.; Calzolari, D.; Musio, S.; Thouminot, C.; Bjelková, M.; Stramkale, V.; Magagnini, G.; Amaducci, S. A Comprehensive Study of Planting Density and Nitrogen Fertilization Effect on Dual-Purpose Hemp (*Cannabis sativa* L.) Cultivation. *Ind. Crops Prod.* **2017**, *107*, 427–438. [CrossRef]
30. Rheay, H.T.; Omondi, E.C.; Brewer, C.E. Potential of Hemp (*Cannabis sativa* L.) for Paired Phytoremediation and Bioenergy Production. *GCB Bioenergy* **2021**, *13*, 525–536. [CrossRef]
31. Salentijn, E.M.J.; Zhang, Q.; Amaducci, S.; Yang, M.; Trindade, L.M. New Developments in Fiber Hemp (*Cannabis sativa* L.) Breeding. *Ind. Crops Prod.* **2015**, *68*, 32–41. [CrossRef]
32. Wang, X.; Li, Q.X.; Heidel, M.; Wu, Z.; Yoshimoto, A.; Leong, G.; Pan, D.; Ako, H. Comparative Evaluation of Industrial Hemp Varieties: Field Experiments and Phytoremediation in Hawaii. *Ind. Crops Prod.* **2021**, *170*, 113683. [CrossRef]
33. Fiorentino, N.; Mori, M.; Cenvinzo, V.; Duri, L.G.; Gioia, L.; Visconti, D.; Fagnano, M. And Degraded Land on Er Al. *Ital. J. Agron.* **2018**, *13*, 34–44.
34. Fagnano, M. The Ecoremed Protocol for an Integrated Agronomic Approach to Characterization and Remediation of Contaminated Soils. Definition of a Site as Contaminated: Problems Related to Agricultural Soils. *Ital. J. Agron.* **2018**, *13*, 1–68.
35. della Repubblica, I.P. DECRETO LEGISLATIVO 3 Aprile 2006, n. 152 Norme in Materia Ambientale. *Gazz. Uff.* **2006**, *1*, 172.
36. Nelson, D.W.; Sommers, L.E. Total Carbon, Organic Carbon, and Organic Matter. In *Methods of Soil Analysis*; John Wiley & Sons, Ltd.: Hoboken, NJ, USA, 1983; pp. 539–579; ISBN 9780891189770.
37. *ISO 11047*; Soil Quality—Determination of Cadmium, Chromium, Cobalt, Copper, Lead, Manganese, Nickel and Zinc—Flame and Electrothermal Atomic Absorption Spectrometric Methods. Vernier: Geneva, Switzerland, 1998.
38. *61010-1©Iec2001*; ISO 707: 2008 International Standard International Standard. Vernier: Geneva, Switzerland, 2003.
39. Yadav, S.K.; Juwarkar, A.A.; Kumar, G.P.; Thawale, P.R.; Singh, S.K.; Chakrabarti, T. Bioaccumulation and Phyto-Translocation of Arsenic, Chromium and Zinc by *Jatropha curcas* L.: Impact of Dairy Sludge and Biofertilizer. *Bioresour. Technol.* **2009**, *100*, 4616–4622. [CrossRef]
40. Mattina, M.J.I.; Lannucci-Berger, W.; Musante, C.; White, J.C. Concurrent Plant Uptake of Heavy Metals and Persistent Organic Pollutants from Soil. *Environ. Pollut.* **2003**, *124*, 375–378. [CrossRef]
41. Malik, R.N.; Husain, S.Z.; Nazir, I. Heavy Metal Contamination and Accumulation in Soil and Wild Plant Species from Industrial Area of Islamabad, Pakistan. *Pakistan J. Bot.* **2010**, *42*, 291–301.
42. Salt, D.E.; Blaylock, M.; Kumar, N.P.B.A.; Dushenkov, V.; Ensley, B.D.; Chet, I.; Raskin, I. Phytoremediation: A Novel Strategy for the Removal of Toxic Metals from the Environment Using Plants. *Bio/Technology* **1995**, *13*, 468–474. [CrossRef]
43. Pidlisnyuk, V.; Erickson, L.; Stefanovska, T.; Popelka, J.; Hettiarachchi, G.; Davis, L.; Trögl, J. Potential Phytomanagement of Military Polluted Sites and Biomass Production Using Biofuel Crop Miscanthus x Giganteus. *Environ. Pollut.* **2019**, *249*, 330–337. [CrossRef]
44. Cosentino, S.L.; Testa, G.; Scordia, D.; Copani, V. Sowing Time and Prediction of Flowering of Different Hemp (*Cannabis sativa* L.) Genotypes in Southern Europe. *Ind. Crops Prod.* **2012**, *37*, 20–33. [CrossRef]
45. Pietrini, F.; Passatore, L.; Patti, V.; Francocci, F.; Giovannozzi, A.; Zacchini, M. Morpho-Physiological and Metal Accumulation Responses of Hemp Plants (*Cannabis sativa* L.) Grown on Soil from an Agro-Industrial Contaminated Area. *Water* **2019**, *11*, 808. [CrossRef]
46. Guidi Nissim, W.; Palm, E.; Mancuso, S.; Azzarello, E. Trace Element Phytoextraction from Contaminated Soil: A Case Study under Mediterranean Climate. *Environ. Sci. Pollut. Res.* **2018**, *25*, 9114–9131. [CrossRef]
47. Cosentino, S.L.; Riggi, E.; Testa, G.; Scordia, D.; Copani, V. Evaluation of European Developed Fibre Hemp Genotypes (*Cannabis sativa* L.) in Semi-Arid Mediterranean Environment. *Ind. Crops Prod.* **2013**, *50*, 312–324. [CrossRef]
48. Afzal, O.; Hassan, F.; Ahmed, M.; Shabbir, G.; Ahmed, S. Determination of Stable Safflower Genotypes in Variable Environments by Parametric and Non-Parametric Methods. *J. Agric. Food Res.* **2021**, *6*, 100233. [CrossRef]

49. Angelova, V.; Ivanova, R.; Delibaltova, V.; Ivanov, K. Bio-Accumulation and Distribution of Heavy Metals in Fibre Crops (Flax, Cotton and Hemp). *Ind. Crops Prod.* **2004**, *19*, 197–205. [CrossRef]
50. Ferrarini, A.; Fracasso, A.; Spini, G.; Fornasier, F.; Taskin, E.; Fontanella, M.C.; Beone, G.M.; Amaducci, S.; Puglisi, E. Bioaugmented Phytoremediation of Metal-Contaminated Soils and Sediments by Hemp and Giant Reed. *Front. Microbiol.* **2021**, *12*, 645893. [CrossRef]

Review

A Critical Review on Lignocellulosic Biomass Yield Modeling and the Bioenergy Potential from Marginal Land

Jan Haberzettl [†], Pia Hilgert [†] and Moritz von Cossel [*]

Biobased Resources in the Bioeconomy (340b), Institute of Crop Science, University of Hohenheim, Fruwirthstr. 23, 70599 Stuttgart, Germany; jan.haberzettl@uni-hohenheim.de (J.H.); pia.hilgert@uni-hohenheim.de (P.H.)
* Correspondence: mvcossel@gmx.de
† Authors contributed equally to this work.

Abstract: Lignocellulosic biomass from marginal land is needed for a social–ecologically sustainable bioeconomy transition. However, how much biomass can be expected? This study addresses this question by reviewing the limitations of current biomass yield modeling for lignocellulosic crops on marginal land and deriving recommendations to overcome these limitations. It was found that on the input side of biomass yield models, geographically limited research and the lack of universally understood definitions impose challenges on data collection. The unrecognized complexity of marginal land, the use of generic crop growth models together with data from small-scale field trials and limited resolution further reduce the comparability of modeling results. On the output side of yield models, the resistance of modeled yields to future variations is highly limited by the missing incorporation of the risk of land use changes and climatic change. Moreover, several limitations come with the translation of modeled yields into bioenergy yields: the non-specification of conversion factors, a lack of conversion capacities, feedstock yield–quality tradeoffs, as well as slow progress in breeding and the difficulty of sustainability criteria integration into models. Intensified political support and enhancement of research on a broad range of issues might increase the consistency of future yield modeling.

Keywords: bioeconomy; black locust; eucalyptus; giant reed; miscanthus; reed canary grass; Siberian elm; switchgrass; poplar; willow

Citation: Haberzettl, J.; Hilgert, P.; von Cossel, M. A Critical Review on Lignocellulosic Biomass Yield Modeling and the Bioenergy Potential from Marginal Land. *Agronomy* **2021**, *11*, 2397. https://doi.org/10.3390/agronomy11122397

Academic Editor: Susanne Barth

Received: 23 October 2021
Accepted: 22 November 2021
Published: 25 November 2021

Publisher's Note: MDPI stays neutral with regard to jurisdictional claims in published maps and institutional affiliations.

1. Introduction

Global agriculture in the 21st century is facing a multitude of challenges, especially the drastically growing demand for food, fodder and industrial biomass for an increasing population [1]. This population growth is expected to increase the need for food and animal feed from today's 2.1 to 3 billion t in 2050. Compared to 2007, a rise in the global food production by 70% until 2050 is necessary [2]. This growing demand for biomass, the intensifying impacts of climate change and incremental water scarcity require more sustainable agricultural production systems. It is estimated that the expansion of agricultural production until 2050—of which 80% is expected to take place in developing countries—demands higher yields, intensified cropping and land expansions of around 70 million ha globally. As the future availability of arable land is expected to decline in developed countries, the necessary land expansion will need to take place mostly in developing countries, especially in sub-Saharan Africa and Latin America [3]. At the same time, the transition of the global fossil-based economy towards a bio-based economy, a so-called 'bioeconomy', demands industrial biomass in large quantities for conversion into bioenergy and bio-based materials [4]. Projections suggest a total biomass demand of 6.7 to 13.4 billion t p.a. in 2050, amounting to a 198 to 396% increase compared to 2011 (3.4 billion t p.a.) [5].

A highly promising strategy to avoid the accretive competition of food and non-food crops for land is the cultivation of lignocellulosic biomass on marginal agricultural land (hereinafter referred to as "marginal land"). Agricultural land that is not suitable for the cultivation of food crops is used for growing selected industrial crops [6]. This becomes even more promising when considering that in sub-Saharan Africa and Latin America—the hotspots of future land expansions—a high share of the land potential is subject to physical–chemical limitations or lacking input factors and thus is not suitable for high-demanding food or fodder crops [3]. Two second-generation bioenergy feedstocks are of special interest for cultivation on marginal land: non-edible oilseed crops (e.g., *Millettia pinnata* L.; *Jatropha* L.) and lignocellulosic biomass crops (e.g., *Miscanthus* (*Miscanthus* Andersson)) [7]. In addition to lignocellulosic crops' adaptability to marginal land, further important benefits are their high energy yield and density per unit biomass and volume [8]. Furthermore, the low cultivation costs and reduced environmental impacts of perennial lignocellulosic crops are outstanding [9]. The high energy density is also the reason why lignocellulosic crops are recommended to not be primarily used for the production of platform chemicals for bio-based materials [9]. The cultivation of annual and perennial lignocellulosic crops on marginal land might allow an expanded production of bioenergy without endangering food security [10]. The International Energy Agency evaluates the cultivation of lignocellulosic feedstock and their transformation into biofuels as one of the best options for reducing greenhouse gas emissions. Lignocellulosic biomass can be cultivated on soils of different quality, while providing a remarkably high biomass output. This is one of the reasons for the increasing political support for the conversion of lignocellulosic feedstock to bioenergy: for instance, in the United Kingdom, double Renewable Transport Fuel Certificates are granted for lignocellulosic biofuels [11].

The growing political and economic interest in lignocellulosic feedstock cultivation on marginal land necessitates precise biomass and bioenergy yield estimations and forecasts. Reliable and science-based biomass and bioenergy estimates and projections are essential for societal, political and economic decision-making, as well as for the development, adaption and refinement of climate mitigation scenarios and strategies [12]. This review holistically assesses the current practice of lignocellulosic crop yield modeling within the scope of marginal land (Figure 1). Thereby, the focus is on the cultivation of lignocellulosic crops on marginal land for a future bioeconomy, specifically the conversion into bioenergy.

Firstly, the question will be addressed whether globally and trans-nationally consistent definitions and data available for the cultivation of lignocellulosic crops on marginal land exist.

Secondly, this review aims to reveal the predominant shortcomings limiting the informative power of biomass yield modeling on marginal land and the translation of lignocellulosic biomass yields into bioenergy potentials.

This review does not focus on socio-economic restrictions (e.g., food sovereignty), which could limit the use of the land, the biomass obtained and the energy yield. Nevertheless, reference is made to key techno-economic aspects when applicable, providing a holistic view on the underlying issues. From an economic perspective, this review only considers low- and medium-input cultivation practices on marginal land [13] and thus excludes high-input cropping systems.

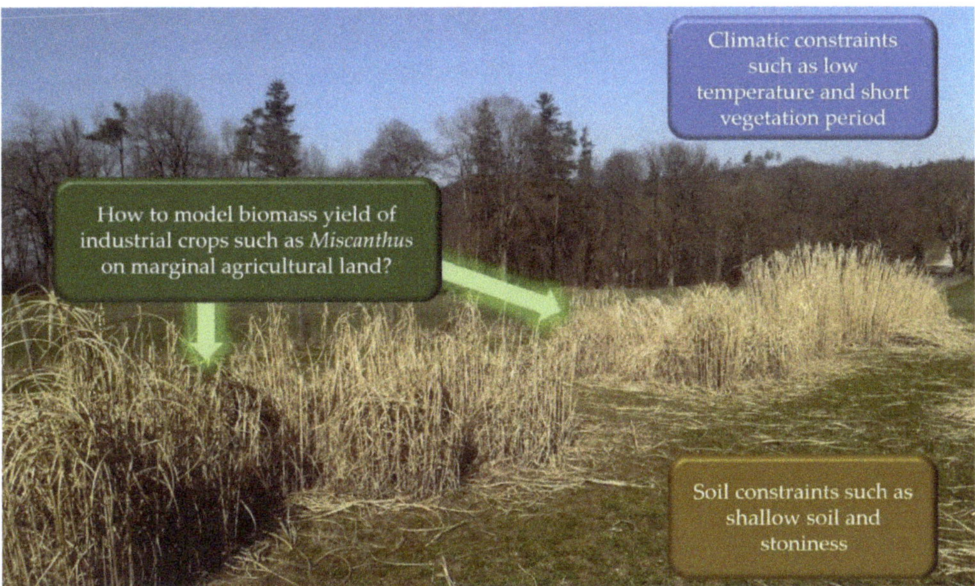

Figure 1. Schematic overview of the subject of this study using the example of the cultivation of different *Miscanthus* Andersson genotypes whose performance has been tested on a shallow stony soil in southwestern Germany since 2014. Clear differences in aboveground biomass growth can be seen here (photograph was taken in winter 2019–2020). These result not only from climatic and soil physical site influences but also from the physiological and morphological characteristics of the genotypes.

Regarding the geographical boundaries, urban marginal lands are excluded from the review. Within the scope of this research, a literature review was performed. Literature on lignocellulosic crops, industrial biomass feedstock, marginal land and crop-specific modeling studies on lignocellulosic feedstock were researched via Scopus, Google Scholar, Wiley and Web of Science. To derive shortcomings and limitations from a broad number of conducted yield modeling studies (Table 1), the authors not only researched general lignocellulosic yield modeling approaches, but also crop-specific modeling such as yield modeling of *Miscanthus* [14] (Figure 1) and *Populus* L. on marginal land [15].

In the following sections, first a contextualization of lignocellulosic crops and bioenergy will be given, including an overview on applied definitions, as well as statistical data for the global cultivation and usage. Secondly, the main characteristics of marginal land are summarized, and the ways these terms are defined in different disciplines on marginal land are analyzed. In the third part, shortcomings and limitations within the current yield modeling practice will be assessed, revealing major unknown factors today's methodology does not take into account.

2. Contextualization of Lignocellulosic Crops, Bioenergy and Marginal Land

2.1. Definitions of Lignocellulosic Crops

Lignocellulosic crops are characterized by a content of about 80% lignocellulose (the sum of celluloses, hemicelluloses, pectins and lignins), and thus could be defined as a subgroup of industrial crops, as they are not suitable for human or animal consumption and are thus cultivated exclusively for industrial use [16,17]. The European Technology and Innovation Platform for Bioenergy defines lignocellulosic crops as species containing varying contents of lignin, different chain lengths and varying degrees of polymerization [18]. The three constituents are cellulose, hemicellulose and lignin, together building the so-called microfibril [19]. Lignocellulosic crops are the most abundant renewable feedstock on this planet, and compared to crude oil, biofuels from lignocelluloses show enormous cost advantage potentials [20] Lignocellulosic crops can either be used as feedstock for renewable energy production or the lignocellulosic fibers can be used for the production of various (bio-based) materials and platform chemicals (e.g., organic acids, furfurals, sugar alcohols or 5-Hydroxymethylfurfural) [10,20,21]. Regarding the biochemical production of fuels, ethanol and butanol are the most interesting pathways [22]. For both energetic and chemical application, cellulose and hemicellulose need to be converted chemically or by (hemi-)cellulolytic enzymes to produce sugars (predominantly glucose and xylose). These sugars are then fermented by yeasts or bacteria to either ethanol/butanol or other platform chemicals [23]. A second possible bioenergy pathway—producing electricity and heat as well as liquid and gaseous fuels—is thermochemical conversion, especially via combustion, gasification, pyrolysis, liquefaction, carbonization and co-firing [24]. In the production of bioenergy from lignocellulose, lignin is the main by-product. Among other options, the production of phenols (e.g., vanillin or ferulic acid) and carbon additives or the direct usage of lignin as an additive (e.g., in the paper industry) is possible [23].

Nearly all lignocellulosic crops are perennial crops, so a re-cultivation or sowing is not necessary after cutting and harvesting for a certain number of growing cycles [17]. Lignocellulosic crops include perennial herbaceous grasses, also called grass-like crops (e.g., *Miscanthus*, *Panicum virgatum* L. (switchgrass), *Phalaris arundinacea* L. (reed canary grass), *Arundo donax* L. (giant reed), *Lolium perenne* L. (perennial rye grass)), as well as fast-growing tree species, so-called short rotation coppice or woody biomass (e.g., *Salix* L. (willow), *Eucalyptus* L'Hér (eucalyptus), *Paulownia* (Siebold & Zucc.), *Populus* L. (poplar)) [17,25]. The latter are usually cultivated in relatively short rotation cycles, and more precisely, either in coppicing systems (the stump is left for regrowth) or replanted after each harvest [10]. The European Technology and Innovation Platform for Bioenergy also includes wood from forestry in their definition of lignocellulosic crops, but does not consider it as a sustainable feedstock for bioenergy [18]. From a biochemical point of view, lignocellulosic crops can be distinguished by lignin and nutrient content; higher in lignin but lower in nutrients is woody biomass, mostly low in lignin and showing a higher nutrient content is associated with herbaceous biomass (mostly grasses) [26].

2.2. Statistics and Forecasts

Neither the Food and Agriculture Organization of the United Nations (FAO) nor other governmental or non-governmental organizations record coherent data for the cultivation and/or usage of lignocellulosic biomass on a global scale. The FAO collects data for the cultivation and harvest of different industrial and non-industrial crops, but usually does not consider their final usage (e.g., food, feed or energy) [27]. The FAO lists crops in different categories (e.g., primary crops, fiber crops), and data are collected at farm level, not considering any processing steps [28]. Having a more detailed view on fiber crops, the FAO includes different lignocellulosic crops, for instance, China grass, Indian hemp, agave or Mauritius flax [29]. As fiber-rich crops are unsuited for human or animal consumption, their industrial usage can be assumed, but there is no final evidence included in the FAO metadata [17]. Moreover, only estimations of global land use for industrial lignocellulosic feedstock cultivation are available, of which a large share is outdated and/or inconsistent [30]. Piotrowski et al. estimated a global biomass production (agriculture and forestry) of 12.1 billion tons in 2011, of which 16% were used for heat and power generation, 10% for bio-based materials and 1% for biofuels [5]. The latter is in accordance with the International Energy Agency (IEA), which estimated that in 2010, less than 1% of global agricultural land was used for cultivating feedstock for biofuels [31]. Nevertheless, these numbers might have changed since then due to rapidly increasing biofuel conversion capacities. Estimations of the land used for bioenergy feedstock cultivation resulted in a 2.7% share of global land in 2008. It is obvious that this percentage is based on outdated data and furthermore varies throughout the literature [5,30]. Data availability for feedstock production considering the type of energy carrier is similarly of poor quality (with an exception for biogas) [17,32].

When it comes to statistical data for the national or regional cultivation of industrial and/or lignocellulosic crops, the data are also insufficient. A positive exception is the European Union (EU); Eurostat not only tracks and documents the total national area dedicated to industrial crops, but also the production of industrial crops on the EU country-specific NUTS 2 scale (basic regions). According to Eurostat, the used agricultural area in the EU consists of 61% arable land, which contains a 7% share of land dedicated to industrial crops [33]. Eurostat provides data for industrial crops in 13 specific crop groups (e.g., rape and turnip rap, other oilseed crops, fiber flax). Some of these crops, parts of them or their by-products can be characterized as lignocellulosic crops, for instance, cotton, hemp, fiber flax, oil palm frond and empty fruit bunch as well as tobacco stalks and other residues [34]. In 2008, approximately 5.5 million hectares within the EU-27 were dedicated to the cultivation of bioenergy feedstock, having a 1% share in the total used agricultural area, equaling 50,000 to 60,000 ha of total landmass [17]. The largest share of the used agricultural area was dedicated to oil crops (82%; for biodiesel), followed by sugar and starch crops (11%; for bioethanol). The largest areas of industrial lignocellulosic cropping are in the United Kingdom (mainly willow and *Miscanthus*), Sweden (willow and reed canary grass), Finland (reed canary grass), Germany (*Miscanthus* and willow), Spain and Italy (*Miscanthus* and poplar) [17]. The literature shows that there are almost no further statistics of industrial lignocellulosic crops available for European countries [17]. One of the countries with sufficient data availability is the United Kingdom, publishing yearly data on crops grown as bioenergy feedstock, recording, for instance, the cultivation of *Miscanthus*, short rotation coppice and straw crops. In 2018, 1.9% of the United Kingdom's arable land was used for the cultivation of bioenergy feedstock (94,000 ha), of which 29% was dedicated to biofuel feedstock. A total of 7000 ha were dedicated to *Miscanthus* (0.1% of arable land) and 3000 ha (<0.1%) were used for the cultivation of short rotation coppice [11].

When it comes to the supply of industrial biomass for the future bioeconomy, the largest feedstock volume will be demanded by the global bioenergy sector, playing an essential role in a future low-carbon economy. For the year 2020, a usage of 84 million tons of energy crops for the production of bioenergy was estimated for Europe [35]. This equals more than a doubling in comparison to 2012 (40 million tons). Looking at the status quo regarding lignocellulosic feedstock, in 2019 only 10 million out of 5505 million liters of bioethanol produced in the EU came from lignocelluloses. In 2019, there were only two European refineries with a total capacity of 60 million liters of cellulosic ethanol production [36]. Regarding estimations of the future demand for industrial (including lignocellulosic) energy crops, a wide-ranging spectrum of biomass-based energy potential estimates is available in the literature. Starting from 2017's global energy use of 1500 EJ, the IEA estimates a possible add-on of 100 to 300 EJ from bio-based resources in 2060. A total of 60 to 100 EJ could be potentially derived from agricultural land, which does not conflict with food safety, leads to only low land use change emissions and complies with a range of sustainability criteria. The IEA considers an additional primary biomass supply of 145 EJ in 2060 as necessary to achieve the Paris climate mitigation goals. This agricultural biomass supply is evaluated as challenging, but achievable [37]. The Intergovernmental Panel on Climate Change assumes a bioenergy contribution of 120 to 160 EJ in their 2050 scenarios [38]. Studies consistently show that globally large areas of marginal and/or degraded land are available and are not suited for food or feed production, but can be used to feed the growing demand for (lignocellulosic) bioenergy feedstock [39].

2.3. Definitions of Marginal Land

The characteristic aspects of marginal land depend highly on the disciplinary terminology, which differs between the environmental biological–ecological, economic, political (legal) and social perspectives. The adjective 'marginal' commonly refers to something 'situated at a margin or border', which is therefore 'not of central importance' and 'close to the lower limit of qualification, acceptability, or function: barely exceeding the minimum requirements' [40]. A clear definition of 'marginal' is context-dependent and subjective to the overall aim of declaration [41]. Marginal land in the context of cultivating industrial crops for bioenergy or biochemical production without threatening food production is often associated with unused, under-utilized, idle, spare, abandoned, degraded, fallow or set-aside land [42]. These descriptions highlight the low quality of marginal land due to challenging climate conditions and soil characteristics, which limit its productivity and therein its suitability for food crop cultivation. This generic definition is relatively consistent across studies with different scopes and disciplines, while the working definitions of marginal land across different studies vary depending on the geographic location in focus, the background of the authors and the aim of the study [43].

2.3.1. Food vs. Fuel Definition

A very broad definition of marginal land is 'land that has bio-physical and/or socio-economic constraints for food production' [44]. Land with easily improvable soil conditions by measures such as irrigation, fertilization or drainage is usually excluded in marginal land definitions to avoid competition with food on that land [41,44]. Marginal land is therefore closely associated with low input and reduced management as economic efficiency is the main determinant of the suitability of cultivation on marginal land [45]. This concept of low input and reduced management has to be distinguished from low-input framing systems, which aim to close input and output cycles to maximize the use of resources produced on the farm site, while minimizing off-farm inputs such as purchased fertilizers and pesticides [46]. Compared to the first definition of von Cossel et al. [44], which highlights the lower productivity of food crops on marginal land, Shortall classifies marginal land as a place 'where food production cannot take place because the land is not productive enough'. Both of these definitions are normative, stating that the land is not suitable for efficient food crop cultivation but assuming that it is technically possible and economically feasible to produce industrial crops there [42]. These definitions, furthermore, indirectly assume that farmers would be willing to dedicate marginal land to industrial crops instead of cultivating these on prime land to avoid indirect land use change (iLUC), a frequently cited negative impact of energy crop cultivation [42,47].

Elbersen et al. go even further and define marginal land as land that has limited agricultural productivity due to negative human interventions that have made this land highly sensitive to degradation [41]. Some authors also include fallow land, which is arable but not cultivated during one cropping season, in their definition of marginal land as it is 'assumed to be kept out of food production in the future with regard to its lower-than-average quality' [48].

2.3.2. Environmental and Biological Definition

There are two main fields of environmental definitions on marginal land: one deals with the soil and climatic conditions (including natural or man-made conditions) of the land, while the other focuses on issues beyond the given conditions and assesses the ecological importance of marginal land. The environmental conditions that make a soil marginal are well defined and were extensively researched with respect to the growth of industrial crops in recent decades [9,41,44]. Most prominent soil and climate conditions leading to marginal land are—amongst others—drought/dryness (<200 mm per growth season), low temperatures (<5 °C), excessive soil moisture/waterlogging, soil texture, shallow rooting horizons (<35–80 cm), soil quality (chemical conditions: salinity > 4 dS m^{-1}, sodicity, acidity pH < 4), soil contamination (natural or human-made toxicity by pollutants, e.g., heavy metals or calcium) and steep slopes (>15–30°) [9,41,44,49]. Especially combinations of these limiting soil conditions, so called negative synergies, impose challenges for the cultivation of industrial crops on this type of land. There are few combinations of the individual factors that cancel each other out, resulting in positive synergies with improved conditions, and the interaction of some characteristics is still unclear [49]. The negative synergies of soil characteristics often favor soil erosion by wind and rain, for example, in Mediterranean countries where high temperatures, limited annual rainfall, steep slopes and low vegetative soil cover prevail [12]. This imposes challenging conditions on the farmers and requires adequate knowledge regarding the right time for tillage and fertilization to minimize soil erosion. Apart from the environmental conditions, biological–ecological conditions are important for assessing the marginality of land as natural vegetation and existing ecosystems can provide habitats for agricultural fauna and contribute to biodiversity [50].

2.3.3. Socio-Economic Definition

A broad economic definition characterizes marginal land by its often poor infrastructure, which leads to limited market access of the goods that could be produced on that land [41] and thereby affects mostly rural areas in regions with difficult accessibility. In more concrete economic terms, marginal land can be utilized 'at the margin of economic viability' [51], meaning that the profit obtained from these lands is close to zero. This definition is prescriptive compared to the environmental ones, as it suggests that under the given set of conditions this land should be used for industrial crop cultivation rather than for food crop cultivation to increase its economic viability [42]. The economic perspective on marginal land is not directly based on the fertility or the conditions of the soil but rather on the relation of inputs and outputs to and from the land. From this perspective, the degree of marginality can only be assessed based on the comparison of different crop production systems on this land as they have varying break-even points due to different inputs and outputs [52]. This understanding of marginality implies that food crops that could be grown on that land might not be cultivated there when a better, more economically beneficial alternative is present, leading to land use change [47].

2.3.4. Political and Legal Definition

As national tendencies and strategies for the mitigation of fossil fuel emissions in the form of energy consumption differ, the boundaries of marginal land change from country to country. For the national assessment of greenhouse gas emissions from land use, land use change and forestry, national subcategories of the six main Intergovernmental Panel on Climate Change categories are determined and evaluated [53]. Regulations regarding the distance between agricultural production and urban areas as well as fresh water sources further limit the availability of marginal land but are often not accounted for, causing an overestimation of marginal land [54]. These different definitions make a comparison between countries difficult [9], and even within Europe a common definition has not been established so far [55]. However, recent studies on marginal land in the EU are trying to establish common ground [41,44,48,49]. In 2019, the EU established criteria for identifying high indirect land use change (iLUC) feedstocks to decrease their use for bioenergy [56]. High iLUC feedstocks describe feedstocks where 'significant expansion of the production area into land with high-carbon stock is observed' [56], which are therefore a threat to the environment. The new directive only allows for an increase in the share of the high iLUC feedstocks' contribution towards the national renewable energy targets for biofuels and biomass from food and feed crops of 1%, based on 2020's national contribution of food and feed crops to the final energy consumption in transportation by road and rail (ibid.). By 2030, feedstocks with a high risk of iLUC are not allowed to be included in the national calculation of renewable energy in the transportation sector anymore. Nevertheless, the member states are still allowed to import and use high iLUC risk feedstocks [56]. To avoid iLUC, land where no feed and food crops can be grown could be used for the cultivation of industrial crops for bioenergy [57].

2.3.5. Social Definition

The social dimension of marginal areas is very diverse and, like the ecologic-biological dimension, often neglected [43]. Marginal land can serve many purposes other than crop cultivation, such as biodiversity conservation (providing a balanced environment to live in), subsistence agriculture (which generates income and can reduce poverty), educational purposes, the provision of firewood and food in some parts of the world (allowing for a human livelihood) as well the provision of ecosystem services and maintenance of cultural heritage [42,43,58]. Due to lacking measurement methods and the intangibility of many of these social values, their assessment is usually only performed on a very reduced regional scale, if at all [58].

3. Limitations of Current Yield Modeling Approaches on Marginal Land

Several studies assessed in this paper conduct or combine previous yield modeling results to evaluate regional, national or global biomass and/or bioenergy potentials (Table 1). A recent example is the study of Pancaldi and Trindade, concluding that between 28 and 85% (for the 6.7 billion t scenario [5]) and between 14 and 42% (13.4 billion t scenario [5]) of the global biomass demand in 2050 could be met by lignocellulosic crops [59]. The base of this calculation was the yield modeling of Nijsen et al., calculating a global average yearly yield of lignocellulosic crops on marginal land of 7.9 tons per hectare as well as an overall bioenergy potential between 150 and 190 EJ per year [60]. In this review, the input data of biomass yield models for marginal land, their results (output), as well as their general methodology and the interpretation of results were critically analyzed.

Table 1. Overview of biomass yield modeling studies included in this study.

Crops.	Geographical Boundary	Date	Marginal Land Definition	Sources of Input for Biomass Yield Modeling	References
Switchgrass, Giant reed, *Miscanthus*	Mediterranean Basin (Greece and Italy)	1993/7/8 and 2004–2014	Water scarcity, salt and nutrient stress	Soil assessment: not specified Climate assessment: between 2005 and 2014 based on calculation of indices for rainfall distribution according to Monti and Venturi 2007 Yield: empirical data, compared by means of an ANOVA analysis Yield model: No projection of yields (only ex post analysis)	[61]
Low-input high-diversity mixtures of native perennials, *Miscanthus*, Switch-grass	Africa, China, Europe, India, South America and United States	Not specified	Abandoned land as well as mixed crop and vegetation land	Soil assessment: Harmonized World Soil Database (FAO/IIAS 2009) Topography assessment: Global Terrain Slope (Global Agro-Ecological Zones (GAEZ) 2008) Climate assessment: humidity and temperature (Natural Resources Conservation Service NRCS 2001 and New et al. 2000) Yield: based on empirical knowledge and expert opinions Yield model: Fuzzy logic modeling (FLM) to estimate productivity and net energy gain from marginal land	[62]
Poplar, Black locust	Germany	Not specified (part of EU SEEMLA project 2016–2018)	Soil physical and chemical parameters, which give a Soil Quality Rating score < 40	Soil assessment: Muencheberg Soil Quality Rating system based on data from European Soil Database and a geographic information system (GIS) toolset Topography assessment: National Aeronautics and Space Administration (NASA) Shuttle Radar Topography Mission (SRTM), European Environment Agency (EEA) Climate assessment: WorldClim–Global Climate Data, Institute for Veterinary Public Health Yield: empirical data of surrounding fields Yield model: Soil Quality Rating and GIS tool to calculate marginal land availability in Europe	[55]
Black locust, Pine	Greece				
Miscanthus, Poplar, Willow	Ukraine				

Table 1. *Cont.*

Crops.	Geographical Boundary	Date	Marginal Land Definition	Sources of Input for Biomass Yield Modeling	References
Switchgrass, *Miscanthus*, Poplar	Southern Appalachian Mountain region (United States)	2008–2012	Land that is currently not used for food production	Soil assessment: based on United States Department of Agriculture (USDA) Soil Survey Topography assessment: based on literature Climate assessment: *Current*: averages and standard deviations for 30 years (1981–2001) based on DayMet dataset *Future*: from global circulation model, adjusted for the IPCC's medium- and high-emissions scenarios (representative concentration pathway, RCP 4.5 and RCP 8.5) Yield: literature data (of growth period of 4 years 2008–2012), Yield model: process-based crop growth model Agricultural Land Management Alternative with Numerical Assessment Criteria (ALMANAC), comparison between current and future yields by means of an ANOVA analysis	[63]
Switchgrass	Great Plains (United States)	2010–2012	Land with a crop indemnity lower than USD 2,157,068	Soil assessment: based on literature, available soil water capacity from NRCS Topography assessment: United States Geological Survey's (USGS) National Elevation Dataset, USGS compound topographic index Climate assessment: based on literature, USGS irrigation map, USDA Natural Resources, Conservation Service (NRCS), Soil Survey Geographic Database (SSURGO) Yield: derived from satellite-derived growing season Normalized Difference Vegetation Index (NDVI) for 2010–2012, USGS crop mask (from USDA National Agricultural Statistics Service Cropland Data Layer) Economic suitability: USDA county-level crop indemnity map	[64]

Table 1. *Cont.*

Crops.	Geographical Boundary	Date	Marginal Land Definition	Sources of Input for Biomass Yield Modeling	References
Miscanthus	Hesse (Germany)	Not specified	Produces low yields, similar to set-aside farm land	Soil assessment: Integrated Administration and Control System for Payments in the Context of the EU Common Agricultural Policy (EC 2007), German Soil Rating Survey ("Bodenschätzung"), Topography assessment: Hessian Agency for the Environment and Geology (Hessisches Landesamt für Umwelt und Geologie), Digital Elevation Model Climate assessment: Hessisches Landesamt für Umwelt und Geologie, National Weather Service of Germany. Yield: estimates at field level, literature and expert knowledge Yield model: yield function of the economic GIS-based model ProLand (Prognosis of Land use)	[65]
Miscanthus	Denmark, Germany, Greece, Ireland, Italy, Netherlands, Portugal, Sweden, Turkey, United Kingdom, US	Not specified	Not specified (just labeled 'different agro-ecological environments')	Soil assessment: texture class, soil depth and water holding capacity from literature of experiments and FAO's digital soil map of the word Topography assessment: slope class from literature of experiments and FAO's digital soil map of the word Climate assessment: from global gridded databases, average monthly weather information (1961–1990) on radiation, temperature, vapor pressure, wind speed, precipitation and rainfall days Yield: Field experiments from literature (used annual biomass yields after 3rd year of growth) Yield model: LINPAC (modified LINTUL model for Perennial and Annual Crops), including sensitivity analysis	[25]
Willow	Canada, Finland, Germany, Sweden, United Kingdom, US				
Reed canary grass	Czech Republic, Finland, Lithuania, Sweden, US				
Eucalyptus	Australia, Brazil, China, Congo, India, New Zealand, South Africa, US				

Table 1. *Cont.*

Crops.	Geographical Boundary	Date	Marginal Land Definition	Sources of Input for Biomass Yield Modeling	References
Russian olive, Euphrates poplar, Siberian elm	Aral Sea Basin (Uzbekistan)	2003–2005	Increased salinity, low ground water availability and reduced irrigation water availability	Soil assessment: based on literature and empirical data Topography assessment: based on literature Climate assessment: based on literature Yield: empirical data, compared by means of an ANOVA analysis Yield model: No projection of yields (only ex post analysis)	[66]
Miscanthus	Loess Plateau (China)	2010	Reduced soil cover by severely eroded soil, landscape degradation and nutrient depletion in the soil	Soil assessment: based on literature and empirical data Topography assessment: based on literature Climate assessment: based on data from Xifeng Meteorological Station (next to the experimental site), from the Data Sharing Infrastructure of Earth System Science Yield: empirical data Yield model: based on the radiation model by Monteith 1977, modified for *Miscanthus* field trials by Beale and Long [67]; Clifton-Brown et al. [68]	[69]
Miscanthus	Italy and Greece	Not specified	Not specified	Soil assessment: literature, expert knowledge Topography assessment: literature, expert knowledge Climate assessment: literature, expert knowledge Yield: literature, expert knowledge Yield model: No projection of yields (only ex post analysis) Economic suitability: Analysis of strengths, weaknesses, opportunities and threats Social suitability: input output analysis	[70]

Table 1. *Cont.*

Crops.	Geographical Boundary	Date	Marginal Land Definition	Sources of Input for Biomass Yield Modeling	References
Switchgrass, *Miscanthus*	US	1989–2008	Abandoned land, land with mixed vegetation and marginal productivity (based on Cai et al. [62])	Soil assessment: based on data from Food and Agriculture Organization (FAO)/Civil Service Reform Committee digitalization of the FAO/United Nations Educational, Scientific and Cultural Organization (UNESCO) soil map of the world Topography assessment: based on data from NASA Shuttle Radar Topography Mission Climate assessment: based on data from European Centre for Medium-Range Weather Forecasts, National Oceanic and Atmospheric Administration Mauna Loa CO_2 record Yield: calculated based on model (based on N and C dynamics) Yield model: AgTEM including ecophysiological, biogeochemical and management-related processes into the Terrestrial Ecosystem Model framework	[71]
Miscanthus, Switchgrass, Giant Reed, Reed canary Grass, Cardoon, Willow, Poplar, Eucalyptus	Europe	Not specified	Low quality land, where only non-competitive yields for rotational food and feed crops can be achieved	Soil assessment: based on data from literature (European Soil classification) Topography assessment: based on data from literature (European Soil classification) Climate assessment: based on data from literature Yield: calculated based on the Aqua Crop model for low, medium and high management practices Yield model: Aqua Crop model from FAO Economic suitability: ABC cost model (for minimum cost prices of feedstock production)	[45]

Table 1. *Cont.*

Crops.	Geographical Boundary	Date	Marginal Land Definition	Sources of Input for Biomass Yield Modeling	References
Miscanthus, Switchgrass, Jatropha	China	Miscanthus 2009–2010, for switchgrass and jatropha not specified	Land that is not a forest, an environmental reserve, a residential area and that is not currently used as cropland or pastoral land	Soil assessment: HWSD (2000–2016), Data Center for Resources and Environmental Sciences, Chinese Academy of Sciences (RESDC, 2015), Institute of Soil Science, Chinese Academy of Sciences (ISSCAS) Topography assessment: NASA Shuttle Radar Topography Mission (SRTM) Climate assessment: Climatic Research Unit Time Series (CRU TS; 2000–2016), China Meteorological Administration (CMA), General circulation model (GCM) Yield: *Miscanthus'* yields (as input for MiscanFor) for 2009–2010 from expert knowledge Calculated based on the respective yield model Yield model: MiscanFor (for *Miscanthus*, Hastings et al. [72]), GIS-based Environmental Policy Integrated Climate Model (GEPIC, for switchgrass; Liu et al. [73]), GAEZ Model (for *Miscanthus*, switchgrass and jatropha; International Institute for Applied Systems Analysis (IIASA)/FAO, 2012)	[74]

3.1. Overview on Yield Modeling Practice and Approaches for Marginal Land

Most studies assessed in this review focus on a select few regions/countries, mainly China, Germany, Greece, Italy, Sweden and the United States, where the cultivation potential of the most common perennial crops including *Miscanthus*, switchgrass, poplar and willow is assessed. The studies frequently use crop yield data from 2000 to 2015 as input parameters, while for the climatic data usually more historical information from 1960 to 1990 is used as this time horizon is commonly provided within the different databases. Several databases are used in yield models to provide all information needed in order to adequately calculate the biomass growth. This includes, for instance, topological data, including soil classification according to different limiting factors, as well as data on crop demands and their development along the growth cycle [65]. This information for each parameter is then transferred onto the respective map of the region(s) under study. Thereby, several maps with different information are generated. Through laying the different maps on top of each other and analyzing overlapping areas, the suitable marginal area for lignocellulosic crop cultivation is determined [45]. Some of the studies additionally assessed estimated future yields, for instance, under different climatic conditions [63]. Another common approach is the comparison of the average biomass yield for a specific region (calculated by a generic model) with empirical regional data to highlight important input parameters that have to be adjusted in the generic models to enable a suitable assessment [25]. For this purpose, these studies make use of generic or crop-specific biomass yield models that go beyond the creation and stacking of maps. In cases where the suitability of cultivating different crops on the same marginal area is compared, authors often have to refer to different generic biomass yield models as only few crop-specific models are currently available (cf. Zhang et al. using MiscanFor, GAEZ and GEPIC [74]).

If different crops in different regions are to be compared, often primary data that have been collected over short or medium time spans are inserted into equations for biomass growth calculation. The limited long-term data availability makes modeling future situations with the current biomass yield models challenging as the increased variance between the different locations and crops cannot be represented sufficiently [61]. The same holds true for models that take multiple ecological, economic and social constraints into consideration, especially if the assessment takes place at field or farm level [65,75].

Depending on the focus of the study, additional disciplinary perspectives might be included: Ramirez-Almeyda et al. [45] extended their study to account for the economic costs associated with cultivating switchgrass and *Miscanthus* in the Mediterranean. Gu and Wylie [64] also assessed the possibility to grow switchgrass on marginal land in the Great Plains (United States) from an economic perspective [64]. Based on the value of the indemnity that has to be paid when cultivating common crops on that land, a threshold was derived that serves as an indicator for the suitability of marginal land for switchgrass cultivation. Harvolk et al. [65], assessed the technical potential to cultivate *Miscanthus* on an area in a small municipality in Hesse (Germany) under consideration of different ecological situations.

Most of the studies reviewed use different sources from the literature, varying databases and sometimes expert opinions when sufficient data are lacking. Due to their different foci, assumptions and boundaries, the results obtained differ widely and provide an adequate snapshot of the current heterogeneity within the field of biomass yield modeling.

3.2. Overview on Limitations and Shortcomings

A summary of the identified challenges and limitations on the input and output side of biomass yield modeling is given in Figure 2. All these factors decrease the comparability of the results from different studies, limiting their interpretation and contextualization [76]. A typical yield modeling sequence starts with the definition of scope and the basic terms and continues with the selection of the model and the input parameters. The resilience of the modeling results can be controlled by means of a comparison with future variations. In addition, several yield modeling studies translate the biomass yields into bioenergy potentials. The final predictions of the yield modeling have to be constantly revised by making use of new models or by updating existing ones. In both cases, the modeling cycle is repeated.

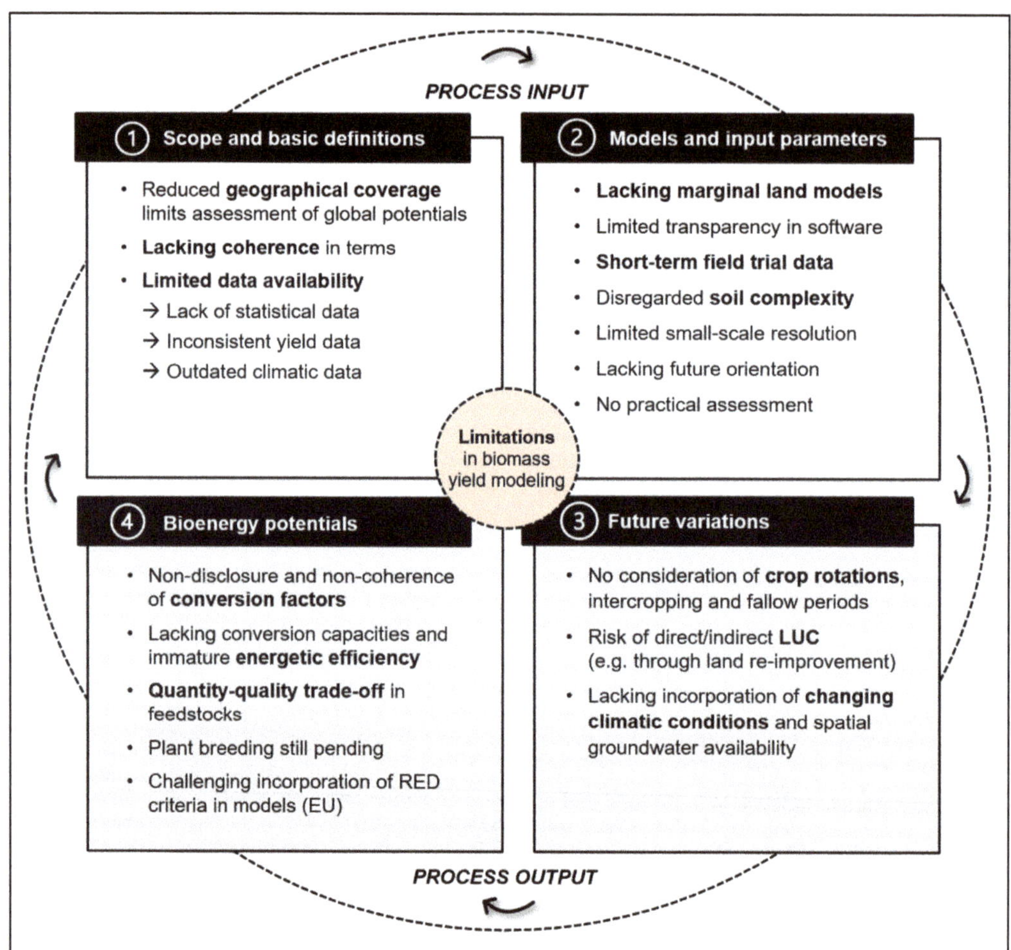

Figure 2. Overview on the most relevant limitations on input and output side of the modeling cycle (RED = Renewable Energy Directive, LUC = land use change).

3.3. Challenges on the Input Side of Biomass Yield Modeling

On the input side, limitations of the current yield modeling practice (including the models applied) can be grouped into two categories: (i) the scope and basic definitions underlying the biomass yield assessment, and (ii) the models and input parameters applied in the modeling (Figure 2). The second category is subdivided into shortcomings related to general aspects of the model while a second category focuses specifically on the challenges of the input parameters used by the models.

3.3.1. Scope and Basic Definitions

The scope of the current research undertaken in biomass yield modeling is geographically narrow and focuses mostly on member states of the Organization for Economic Co-operation and Development, especially North America, Europe and China [58,77].

Models and approaches developed there are presently state-of-the-art and applied in studies conducted all over the world. Due to the varying environmental conditions within and between different countries, models calibrated with data from the Northern Hemisphere are only limitedly suitable for other regions such as Africa or Latin America [25]. Many countries are therefore still unconsidered in biomass yield modeling research. The lack of geographical coverage limits the holistic assessment of the global potential of biomass yield from lignocellulosic crops and might lead to forgone potentials.

To ensure comparable, realistic and detailed results from biomass yield leading to suitable decisions on the spatial and temporal distribution of biomass cultivation, the underlying basic definitions of the input parameters have to be coherent [42]. Currently, there are no standardized globally accepted definitions of the most important inputs to the modeling process including marginal land and industrial and lignocellulosic crops. These definitions are highly contextual and study-dependent, and even though most of them share a common core, variances in the wording and the conscious inclusion or exclusion of certain aspects lead to very different perceptions and interpretations [43]. For example, while a common guideline to assess marginal land in Europe exists [49], the amount of land classified as marginal varies widely between studies. This implies that the validity and reliability of studies calculating the global marginal land availability are reduced because they do not account for differences in local definitions. Instead, they often apply the same definition to all locations, which can result in an overestimation or underestimation of the actual potential. To avoid this problem, some studies assess marginal land from the perspective of the current use of the land and the potential to change this land use to favor lignocellulosic crop cultivation. This approach does not consider the reasons for a certain current land use and will lead to inadequate results, which can only be corrected if biophysical data are additionally considered [55]. In the case of crops, the lack of coherent definitions leads to vague distinctions of the different types as few classification criteria exist [78]. In some statistical assessments, for example, corn is classified as an industrial crop for animal feed but it can also be destined for human consumption (with or without refinement), which would then convert it into a food crop. The same holds true for other multipurpose crops such as soybean, sorghum and cotton. Without a precise, internationally agreed definition (or a fixed set of definitions) of industrial, lignocellulosic and food crops, the underlying competition for land between these crops cannot be evaluated.

The lack of definitions further imposes challenges on the data collection of marginal land availability and biomass yield achievability. On a global scale, accurate statistical data are often missing [79], and even between the different European countries the quality and quantity of datasets vary widely (for instance on soil conditions) [55]. This leads to situations where the input parameters date back to different years, increasing the uncertainty and inaccuracy of the final result, especially for climate data where values from the 1960 to 1990s are often used. Comparisons of studies thereby become more challenging, and especially for promising lignocellulosic crops such as *Miscanthus*, very few long-term yields are available at national or international level. This might be due to the recent interest in *Miscanthus* as a bioenergy crop and the complex interactions between a number of factors such as planting method, species and site conditions [45]. Not only for cultivation on marginal land, but also for growth on arable land, no data are (publicly) available for *Miscanthus* from Eurostat, FAO or USDA. Assessing if yields on marginal land are significantly lower than on arable land is nearly impossible then and might lead to less accurate definitions of marginal land, their estimated future availability and the amount of biomass yields that can be achieved from that land. Overall, the contextualization of biomass yield modeling results becomes more difficult if the availability of representative, high-quality data, which is based on the same definitions at regional, national and global scale, is limited [61].

3.3.2. Models and Input Parameters

Modeling approaches also need to shift their focus towards a more detailed and holistic evaluation of biomass growth, as currently used biomass yield models are mostly generic. Generic models such as the frequently used GAEZ or GEPIC model are based on average growth data for a variety of crops [74]. Zhang et al. in their study on *Miscanthus* growth on marginal land in China calculated an average annual dry matter yield of 0.318 Mg ha^{-1} with the GAEZ model, while the MiscanFor model predicted annual dry matter yields of 14.6 Mg ha^{-1}. MiscanFor is a software for biomass yield modeling (developed in Europe) that is specifically calibrated for the *Miscanthus* genotype *Miscanthus* × *giganteus* [74]. The differences in the yield might not only be caused by the different types of models, but it is reasonable to assume that the more detailed MiscanFor estimations lead to more precise results [61]. Nevertheless, no specially tailored models for biomass growth on marginal land exist currently [55]. The results obtained from the applied models are thus based on information from arable land. Some marginal land parameters—such as specific soil and climatic conditions—can be inserted into models, yet they cannot be represented in their whole complexity.

Besides having arable land as a basis for biomass yield modeling, most crop growth models are compiled software packages that do not provide open access to the source code used [80]. The model users can therefore not trace and reproduce the calculations performed by the model. This reduced transparency increases model uncertainty as the influence of the model's internal structure on the simulation output cannot be fully assessed and the chosen input parameters might only be limitedly related to the calculated output [81]. Semenov and Porter [82] additionally highlight the limited range of data on which the calculations were based and the increased amount of assumptions and hypotheses underlying the biomass yield models. These input uncertainties together with the model uncertainty increase the output uncertainty and the reliability of the results [83]. Holzworth et al. [84] mention the lack of focus on universal software platforms as a reason for the reduced efficiency in agricultural modeling. This also leads to an increased gap between the software industry and agricultural production researchers [84]. This lacking connection between the two most important actors within the biomass yield model sphere promotes inadequate representations of the growth process. For example, the MiscanFor model calculates biomass yields of *Miscanthus* based on yields of mature rhizomes, which are usually only achieved after three to five years [79]. In the first years after the establishment, *Miscanthus* yields are much lower, which should be adequately represented in the models to avoid an overestimation of the results [61]. However, the relatively stable yields after the establishing period allow predictions of biomass yields from perennial crops if adequate data for all input parameters are available [25]. The sequence in which the different input parameters and their thresholds accounting for the marginality of the land are inserted into the model is also important. Different data compiling methods and pre-selections alter the amount and characteristics of marginal land estimated [74]. The points mentioned above provide reasons for the very limited comparability of the results obtained from biomass yield models and the huge variations in marginal land, yield and energy potential estimations.

The concrete input parameters used in biomass yield models on marginal land are often grounded on short-term field trial data [85]. These data are not representative, as only certain management practices and few soil characteristics are included, which insufficiently account for the complexity of marginal land [45]. In addition, the whole growth cycle cannot be assessed by means of three-year trials and the use of average yield data further reduces the accuracy of the model outputs [79]. This puts the importance of local environmental and climatic conditions into focus [74]. One of the most important parameters is the soil, as it provides the basis for biomass growth. The interactions within the soil and between the soil and the other environmental parameters are highly complex and site-specific. The combination of individual marginal soil characteristics can create positive, negative or unclear synergies [49]. Unclear synergies were reported for the combinations of excess soil moisture and stoniness, organic soil texture and steep slope, stoniness and heavy clay texture, sandy texture and heavy clay as well as heavy clay and steep slope [49]. As the interactions between soil, crop and climate are very complex, it is difficult to assess the performance of a crop on marginal land, especially when negative synergies are present. These mutual influences increase the complexity of soil systems even further, making a detailed assessment of the soil necessary for an adequate representation in the model. Many models neglect this complexity by determining a land's marginality based on only one threshold or a few averaged values. For instance, on marginal land where low crop management and no/very limited irrigation are prevailing, the various soil conditions influencing the water holding capacity (e.g., slope, soil texture and soil cover) need to be assessed in conjunction with each other to assess if sufficiently high yields can be achieved [76]. *Miscanthus*, as one of the most promising future bioenergy crops, is often given as an example for cultivation on marginal land due to its reduced environmental and management demands. Nevertheless, *Miscanthus* has a low resistance to drought and abiotic stress [86]. This reduces its suitability to be grown on marginal land, for example, on steep slopes in the warm Mediterranean regions, because abiotic stresses are often present in combinations, for example, drought coming along with heat, or in successions as in the case of waterlogging followed by drought [87]. As not all crops are equally suitable for the same soil and environmental conditions, the assessment of only one type of crop on marginal land might not be sufficient to identify the best land–environment–crop fit [74]. As important as a detailed environmental assessment is, it is often practically difficult to model certain physical constraints such as damages from strong winds. Apart from modeling the damage, a probability for the damaging event as well as a recovery period with lower growth would additionally have to be modeled, which further increases complexity. Depending on the specific location, for example, coastal areas, it is useful and necessary to take these aspects into consideration [88]. Overall, the degree of detail of each study depends on its objective as well as time and financial constraints.

In addition to input data, the modeling process needs to be conducted at high resolution [74]. The resolution of the results depends on the lowest resolution of the input data and studies highlight the assessment at a 1 km^2 scale as most adequate [55,65]. For example, Richter et al. [89] calculated an average dry matter yield of 9.6 t ha^{-1} for *Miscanthus* cultivated in the United Kingdom, while Aylott et al. [90] modeled a combined short rotation coppice yield (willow and poplar) of 9.7 DM t ha^{-1} for the same area. Bauen et al. [88], however, using a 1 km^2 grid resolution, predicted an average yield of 11.9 DM Mg ha^{-1} for the United Kingdom when selecting the highest yielding crop (*Miscanthus*, poplar or willow) for each grid. Schorling et al. [91] in their study on *Miscanthus* yields on marginal land in Germany also obtained more accurate results compared to a Europe-wide study assessing the same. Their detailed geographical, climatic and geological data input allowed for the avoidance of an overlap in the types of areas (e.g., rural villages and marginal land) and accounted in detail for areas where the soil composition changed [91].

The time dependency of the yields calculated by the models is an additional source of uncertainty [55]. Changes in the climatic data have a severe influence on the yields, but few studies predict future yields based on future data, instead using past data, though valid predictions exist, e.g., from Intergovernmental Panel on Climate Change for climatic data [74]. Most studies therefore claim to make an ex ante assessment but make use of ex post data, which provide only limited suitability and reduce the study's validity. This insufficient long-term orientation of the input data is caused by a combination of all the aforementioned factors.

With regard to the extensiveness and inclusion of different perspectives, biomass yield models have to be improved. Many studies model the technical potential of biomass cultivation on marginal land but lack a practical orientation. Yields are thereby calculated based on input environmental data (especially area) that do not take the social, cultural and ecological value of the land into consideration [77]. The estimated yields from these models are therefore theoretically and technically available, but not practically feasible as they might come along with great socio-economic changes and distortions. Therefore, calculating the technical potential provides a rough overview of the spatial distribution of biomass cultivation potentials and highlights areas that might be worth further investigation [74]. To avoid an overestimation of biomass yields and to adequately assess the long-term feasibility of perennial crop cultivation on marginal land, taking socio-economic and political criteria into consideration is necessary [61].

Marginal lands provide ecosystem services that are positively or negatively influenced by the crop cultivated and the applied management system [92]. These changes drastically affect the local population, their lifestyle and livelihoods and should therefore be assessed as part of a realistic biomass yield model. In addition, the land that is considered marginal is frequently privately owned, so including this land in the calculation is only suitable if the owner is willing to cultivate the selected crop on their land [45]. Apart from land availability, crop cultivation also requires labor. Beyond the recurring crop management activities such as fertilization and harvesting, labor is especially needed for the initial field preparation and the planting of the *Miscanthus* rhizomes as this is usually performed by hand [93]. Recently, many people moved from rural areas, where marginal land is located, to cities, leading to a decrease in the rural labor force, which in the long run can impose challenges on the crop cultivation on marginal land [3].

Biomass growth is an investment, and planting costs, in particular for crops such as *Miscanthus* that cannot be sown, are high and likely to increase if perennial crops are cultivated on marginal soils [45]. These planting costs are sunk costs, which cannot be recovered by the farmer and might deter marginal landowners from investing in *Miscanthus* cultivation, especially because the yield obtained is rather lower in the first years, which increases risk. Considering the economic perspective of biomass growth on marginal land is also important because the stable long-term yields (starting from the fourth or fifth year of cultivation) reduce the farmers' flexibility to grow other crops when the market conditions for the perennial crops negatively change [45].

The aspects considered above are closely related to the political and institutional environment. Policy support for the industrial use of biomass is needed to increase the cultural and social acceptability of biomass as an energy feedstock [74]. To convince the local society that crop cultivation on marginal land is useful, more long-term data over the whole growth cycle of the crop must be generated and made available. This is difficult when the funding for most studies is limited to two to three years [55]. Furthermore, current research is mostly conducted on field trial areas with a limited amount of different environments available, which further decreases the validity and thereby society's trust in the obtained results [74]. Another challenge is the limited transnational cooperation, which reduces the exchange of experience and the potential to develop a standardized framework that is applied consistently [55]. In addition to data generation, the implementation and promotion of the studies' recommendations throughout the different sectors is an important task where political guidance can still be improved [57].

3.4. Challenges on the Output Side of Yield Modeling

The following chapter will shed light on the limitations and shortcomings related to the output side of yield modeling (Figure 2). Therefore, the suitability of selected yield modeling to incorporate future variations and issues related to the translation of yield modeling results into bioenergy potentials and the incorporation of Renewable Energy Directive-related sustainability criteria into models will be critically assessed.

3.4.1. Challenges Due to Future Variations

Yield modeling studies are often conducted for only a certain crop or species on a defined area of land, ignoring future variations (Figure 2). The modeling results cannot be considered as a sufficient statistical base for biomass yield predictions, as also marginal land, crop rotation and partially also intercropping can play an important role [12,13]. The failure to incorporate the necessity of crop rotations and the potentials of intercropping is a common problem of yield modeling studies. *Miscanthus*, for instance, cannot be grown alone over decades with consistently high yields [74]. That a long-term mono-cropping of lignocellulosic crops should be avoided is part of the Renewable Energy Directive and thus a criterion for the sustainable production of biofuel feedstocks. To support agro-biodiversity within the EU, a mix of at least three perennial crops (covering both herbaceous crops and short rotation coppice) per region is required according to the guideline EU 2018/2001 [94]. With this in mind, the extensive modeling of single crops or single crop categories for whole regions or even whole agricultural zones within the EU is unfavorable [25,65,70]. Furthermore, it needs to be considered that farmers might also leave marginal land fallow within their lignocellulosic crop rotation systems [57]. Marginal land might not be agriculturally used for one cropping season to recover, or because of several socio-economic decisions (e.g., lacking market opportunities). The critical issue is that this practice would influence the overall yield, which would then differ from the modeled yield (if no crop rotations, intercropping or fallow periods are assigned). Especially in the case of biomass and bioenergy potential predictions for several decades, the lack of modeling crop rotations might subvert the mid- and long-term validity. This becomes even more alarming when considering that crop rotations were neither mentioned nor incorporated in any of the studies assessed within the scope of this research, even though there are some promising tools available such as CropSyst (http://modeling.bsyse.wsu.edu/CS_Suite/index.html (accessed on 22 November 2021)).

Depending on the definition of marginal land—especially if fallow land is declared as marginal land—and depending on the intensity of crowding out effects, iLUC and direct land use changes (dLUC) can occur. Firstly, these might undermine the ecological benefits of lignocellulosic crops cultivated on marginal land. Secondly, they might subvert the results of biomass yield modeling. Crowding out effects might occur when marginal land was originally defined as economically marginal, but a higher demand for food crops revokes economic marginality. Furthermore, there is the possibility to use marginal land as food crop land after a certain period of energy crop cultivation, during which the soil quality has improved sufficiently. Zhang et al. suggest a period of 15 years as a sufficient timespan for soil amelioration [74]. It is reported in the literature that the cultivation of perennials can reduce soil erosion, capture nutrients, stabilize the soil through rooting, maintain a more firm structure, provide a habitat for wildlife and boost biodiversity, and moreover store carbon in the soil (soil-carbon sequestration) [92]. These beneficial effects might lead to a sufficient remediation of marginal land and allow future food production. This would reduce the mid- and long-term biomass potential on the affected marginal land. These possible shifts are not at all incorporated in the assessed biomass yield modeling studies, even though the risk of drastic changes in the future land availability is high. Two conceivable scenarios might come with the remediation of marginal land: in the first case, improved soils continue to be used for lignocellulosic feedstock cultivation. Assuming an increasing demand for food and feed, food production might be displaced to other areas, resulting in iLUC. In the second case, the cultivation of lignocellulosic crops might be relocated to new land, causing dLUC. This presupposes that the originally environmentally marginal land has sufficiently improved and is again used for food cultivation. If in both scenarios the sum of LUC equals the sum of re-improved land, the land available for biomass cultivation might remain unchanged. Nevertheless, significant greenhouse gas emissions might occur due to land use changes [95]. If the land dedicated to lignocellulosic crops shifts to new (marginal or non-marginal) land with divergent environmental conditions, the originally calculated biomass yields might be incorrect.

The most severe limitation might be the lack of sensitivity of biomass yield models to future environmental variations, especially regarding highly complex climatic changes. For common food crops, significantly decreasing yields are already reported as an effect of extreme weather, temperature and precipitation events [96,97]. Nevertheless, varying climatic conditions affect crops in different ways, as they can be either positive, negative or outbalance each other. Especially for lignocellulosic crops, it is not known yet how quickly they can adjust to shifts in climatic conditions [74]. The main problem with modeling systems is the uncertainty inherent in the construction of the future meteorological scenario used as input for the models [82]. This becomes even more problematic when taking into account that precise climatic and ecological data at regional and/or local level would be necessary to properly map environmental changes [88]. For instance, short rotation coppice demands sufficient groundwater levels. Future shifts in groundwater must be investigated and integrated via parameters into models to adequately assess yield potentials of trees on marginal land [74,98]. Once again, the spatial distribution of crop cultivation plays an important role to assess shifts in groundwater [88]. However, not only the environmental conditions but also the crops cultivated might change in the next few decades. Graves et al. observed a tendency of shifting from herbaceous-grassy-like biomass to woody-tree-like biomass due to global climatic change. One of the possible reasons for this systematic change in the agricultural pattern of farmers is the increasing productivity of C3 plants with rising temperatures, compared to C4 plants, where the productivity-increasing effect is limited [63]. This tendency still needs to be statistically proven, but nonetheless adds to the general uncertainty inherent in the methodology of mid- and long-term yield predictions.

In conclusion, yield modeling for lignocellulosic crops on marginal land can lead to drastic over- or underestimations of the exploitable biomass potential. Furthermore, the depiction of actual and historical climatic developments in yield models is already challenging, and the incorporation of future changes is an additional challenge.

3.4.2. Limitations of Bioenergy Potential Assessments

Studies show that the global energy demand in 2030 could be fully covered by the conversion of biomass grown on non-arable land [99,100]. This and other bioenergy projections rely on quantitative data of biomass potentials, often gathered through yield modeling. In fact, one of the main applications of the results of yield modeling studies on marginal lands is the calculation and assessment of bioenergy potentials.

The Compilation of Reliable Quantitative Data

The focus on bioenergy potentials of lignocellulosic crops grown on marginal land can be explained by the increasing need for reliable, quantitative data on the potential of biomass and bioenergy within the scope of climate change mitigation. In the Special Report on 1.5 Degrees [101] and the Special Report on Climate Change and Land [95], an even broader view on bioenergy was taken. Amongst other issues, socio-economic and environmental limitations of biomass cultivation were assessed in detail by the Intergovernmental Panel on Climate Change, especially in the Special Report on Climate Change and Land. This is an important indication of the increasing relevance of qualitative risk assessments and quantitative bioenergy potential analyses, providing a basis for climate mitigation strategies. Often, a bioenergy yield estimation is performed as part of a yield modeling study (Figure 2). For instance, Scagline-Mellor et al. model yields for *Miscanthus* and switchgrass on marginal land for the eastern part of the US [102]. The biomass yields per hectare are used to further model bioethanol yields per hectare. Regarding marginal land within the area of Boston, outgoing from a crop yield of 42,130 tons of poplar (per growing season), a bioenergy yield of 830 TJ (higher heating value, HVV) is calculated in a study of Saha and Eckelmann [103]. Mehmood et al. list 15 crops (e.g., *Miscanthus*, switchgrass, reed canary grass and agave) and summarize studies conducted and yields modeled, as well as their geographical boundaries [76]. Furthermore, the listed biomass yields are connected to bioenergy potentials, citing relevant biomass and bioenergy yield modeling studies. However, it was not always clear how the conversion from biomass yields to bioenergy potentials was performed [76]. Panoutsou and Chiaramonti evaluated the positive impacts of cultivation and conversion of *Miscanthus* into bioenergy (by means of a combined heat and power plant and fast pyrolysis) on the social and economic situation (employment and income) of people in the southern part of Italy and Greece [70]. Despite the growing data and research base, there are several limitations in the modeling of bioenergy potentials.

Smith and Porter concur that there has been a significant improvement in the quantification of mitigation potentials of bioenergy in the Intergovernmental Panel on Climate Change reports [104], but uncertainties are still tremendous. Different variations in data—for instance, assumptions on land availability, yield improvements over time, efficiency increases in conversion technology and optimization of infrastructure—come with an enormous level of uncertainty. A good example for this uncertainty is the study of Hoogwijk et al., assessing the geographical and technical potential of bioenergy crops on abandoned and low-productive land for the time span between 2050 and 2100. Their analysis is based on the IPCC's Special Report on Emission Scenarios, computing bioenergy potentials between 130 to 410 EJ/year for 2050 and 240 to 850 EJ/year for 2100 [105]. The authors conclude that—based on the geographical and technical potential—the potential of bioenergy from low-productive and abandoned land could hypothetically be several times as high as the energy supply through crude oil at the beginning of the 2000s. Moreover, the study takes different socio-economic and environmental developments into account. For instance, the global agricultural area between 1970 and 2100 (in Gha) is simulated, incorporating various scenarios for population, GDP, land management factors, diet and trade. The different scenarios result in a hypothetical global agricultural area in 2100 between 1.5 Gha and 6.5 Gha. This discrepancy again dramatically influences the exploitable biomass and bioenergy yield, and thereby the climate mitigation potential from bioenergy. These uncertainties ask for an overview of the most relevant limitations of yield modeling and its connection to the simulation of bioenergy potentials.

The Conversion from Biomass to Bioenergy Potential

If a transformation of modeled yields into energy potentials is performed within or attached to biomass yield modeling on marginal land, a conversion factor needs to be applied to convert the biomass yield (in t per ha) into bioenergy potentials (in GJ, MJ or EJ per ha). The biomass yield therefore needs to be multiplied by the energy yield exploitable within the dedicated bioenergy pathway [106]. The wide range of possible bioenergy pathways includes the co-combustion of biomass with coal in electricity generation plants, the conversion of biomass into cellulosic ethanol and electricity (as a by-product), the conversion of biomass into gasoline, diesel synfuels and electricity via integrated gasification and Fischer–Tropsch hydrocarbon synthesis (IGCC-FT). Different conversion pathways of low-input high-diversity (LIHD) mixtures of native grassland perennials come with highly differing energy yields, ranging from 18.1 GJ/ha/y for electricity to 17.8 GJ/ha/y for cellulosic ethanol and electricity to 28.4 GJ/ha/y for gasoline via IGCC-FT [106]. Unfortunately, in several studies analyzed, either the conversion factor (biomass–bioenergy) is not given nor explained, or the pursued energy carrier is missing. In the global assessment of bioenergy potentials on marginal land by Nijsen et al., neither the conversion factor is given, nor the energy carrier is declared [60]. In the study of Harvolk et al., the conversion of biomass to thermal energy is mentioned, but neither the assumed conversion technology nor the conversion factor was applied [65]. Quin et al. converted the biomass yield of *Miscanthus* and switchgrass into liters of ethanol following a two-sided approach [71]: following current and potential biomass-to-biofuel conversion efficiencies as well as the parameters of Lynd et al. [107]. Unfortunately, Lynd et al. simply derived an average bioenergy yield of biomass energy crops of 105.4 gallons ethanol/dry ton (approx. 399.0 L), without further specification or classification into crops or crop categories [107]. Quin et al. applied this average ratio on *Miscanthus* and switchgrass [71], while for instance Scagline-Mellor et al. calculated significantly different, and among themselves also slightly diverging, ethanol yields (i.e., 453 L/dry ton of *Miscanthus* and 450 L/dry ton switchgrass) [102]. Furthermore, the potential values adduced by Quin et al. can only be obtained if appropriate technologies are available [107], of which several are not state-of-the-art today, and their future realization is still unclear. The uncertainty along with the methodologies applied by Lynd et al. and Quin et al. are significant. Zhan et al. transformed within their study a biomass potential into a technical energy potential, defined by the authors as the available energy content potential provided by the biomass production per grid cell. The technical potential was calculated by multiplying the yield by the crop-specific higher heating value (HVV) [74]. The declaration of the methodological approach by Zhang et al. is sufficient. Nevertheless, the calculation of the technical potential excludes energy losses through conversion [108]. As Zhang et al. furthermore analyze the hypothetical, proportional contribution of the calculated energy yield to China's energy demand, the involvement of a technical potential in the calculation without a conversion step highly skews the results. All given issues regarding the non-declaration or inconsistency of the conversion factor complicate the comparison and interpretation of bioenergy potentials.

The Conversion Capacities for Lignocellulosic Crops

Another important limitation undermining the meaningfulness of modeled bioenergy yields is the lack of capacities for the conversion of lignocellulosic crops. Ethanol from lignocellulosic crops (CH_3CH_2OH) is chemically identical to first-generation ethanol, but is produced through cellulose hydrolysis, which is a more complex process, requiring highly sophisticated production plants. Taking the EU as an example, in 2019 only 10 out of 5505 million liters of bioethanol production came from cellulosic feedstock. There were only two European refineries with a total capacity of 60 million liters of cellulosic ethanol production per year [36]. The first commercialization ventures for the production of lignocellulosic-based ethanol can be observed in Europe, especially two commercial cellulosic ethanol plants: one in Romania (run by Clariant International Ltd.) and another one in Slovakia (operated by Enviral). Furthermore, a pre-commercial demonstration plant in Germany is run by Clariant International Ltd. [109,110]. On a global scale, the largest cellulosic ethanol plants are located in the United States (DuPont, Iowa, 83,000 t/y; POET-DSM Advanced Biofuels, Iowa, 75,000 t/y; Abengoa Bioenergy Biomass of Kansa, 75,000 t/y), in Brazil (GranBio, Alagoas, 65,000 t/y) and in China (Longlive Bio-technology Co. Ltd., Shandong, 60,000 t/y) [111]. Unfortunately, only the plants in Brazil and China are still in operation. The DuPont facility was sold in 2019 and is currently converted to a natural gas plant [112], the facility of POET-DSM Advanced Biofuels is currently idle [113] and the operator of the plant in Kansas is bankrupt [114]. Furthermore, a large amount of demonstration and flagship plants testing different conversion pathways of lignocellulosic crops to ethanol or butanol can be found in various countries, for instance in Italy, Denmark, Spain and Finland [22]. As estimated in the 2020s market report of the Global Industry Analysts Inc., the global market for cellulosic ethanol summed up to USD 631.7 million in 2020 and is expected to grow to USD 6.6 billion at a rate of 39.8% between 2020 and 2027. The growth of the cellulosic ethanol market (2020–2027) is forecasted highest for China (46.5% share in the Compound Annual Growth Rate (CAGR)), followed by Canada (37.4%), Japan (33.2%) and Germany (35.6%) [115]. Nevertheless, the idle and bankrupt large-scale plants in the United States show the vulnerability of this capital- and research-intensive business field. Furthermore, the cellulosic bioethanol sector often consists of fragmented markets and is characterized by geographically differing market regulations [77]. The spectacular CAGR of the global cellulosic ethanol sector should not intend to divert attention from the fact that there are no mature markets for cellulosic ethanol yet. The systematic lack of transportation infrastructure and logistics and the currently limited access of smallholder farmers to the bioenergy feedstock market [63] are other socio-economic issues to be considered when mapping global bioenergy potentials. Without governmental support and/or subsidies, sufficient conversion capacities and an appropriate spatial distribution (with short distances to the cultivation sites), the exploitable biomass and bioenergy potentials calculated in various studies will remain only theoretical potentials. In conclusion, their informative and predictive value is highly limited.

The Energetic Efficiency of Lignocellulosic Ethanol

Furthermore, there is still a significantly immature energetic efficiency of lignocellulosic ethanol. For the conversion, either acid or enzyme hydrolysis can be applied, and both have been intensively researched since the 1970s in the United States and Europe. Besides investments into conversion capacities, the enzymatic pathways are currently optimized through strain development and novel strain discovery, as well as innovative feedstock pretreatment (e.g., ionic liquid pretreatment) [111]. Suitable technologies for the biochemical conversion of lignocellulosic feedstock to biofuels are still under development and not economically competitive to fossil fuels yet [116]. The technological and commercial maturity of the conversion technology is currently evaluated as poor [88]. The energetic efficiency of biofuels can be assessed through the energy return on invest ratio (EROI), calculating the ratio between the energy delivered by a fuel and the energy invested in the production and delivery of this energy [117]. The input energy either includes only non-renewable energy or non-renewable and renewable energy (e.g., electricity and steam produced from lignin) [118]. Hall et al. conclude from 74 assessed EROIs for bioethanol, of which 33 values were below a 5:1 ratio, that—besides optimal values obtained in the tropics—most ethanol EROI values are at or below 3:1. This ratio is defined by Hall et al. as the minimal value for societal usefulness. Thus, most bioethanol pathways assessed are not socially desirable regarding their energetic efficiency [117]. The Natural Resources Defense Council and Climate Solutions in cooperation with Hammerschlag et al. assessed three studies for cellulosic ethanol between 1994 and 2005 [119,120]. While only considering non-renewable energy input, the resulting EROIs of 4.40 (corn stover), 4.55 (poplar) and 6.61 (various) show that bioethanol from cellulosic crops came with significantly higher EROIs than bioethanol from corn (ranges from 0.84 to 1.65). Nevertheless, other assessments conclude drastically lower EROIs for cellulosic ethanol: for instance, 0.2 for ethanol from switchgrass [121] and 0.64 for ethanol from wood [122]. Latter numbers are based on a techno-pessimistic approach, which assumes fossil fuel inputs are used to produce distillation steam instead of energy from lignin combustion. This approach is highly criticized in the literature [117,123]. From the calculations of Barel et al., it can be concluded that even if renewable process energy is considered, the EROI of cellulosic ethanol from switchgrass is lower than that of gasoline [118]. Based on these calculations, it can hypothetically be concluded that lignin-poor lignocellulosic crops might come with lower EROIs than lignin-rich feedstocks, but there is no evidence in the literature. Murphy et al. list several insecurities and limitations of the EROI methodology, pointing out that the quantitative results highly depend on a range of parameters (e.g., boundary and co-products) [124]. Nevertheless, the numbers mentioned show that efficiency in the production of cellulosic ethanol is still immature and highly depends on the crops cultivated, the agricultural management practices and the geographical location [125]. The importance of the crop choice is in line with Baral et al., assessing drastic differences in the EROI of yellow poplar and switchgrass [118]. From the highly diverging and partially outdated literature available, no general and definite statement on energetic advantages of cellulosic ethanol to other biofuels and energy carriers can be made.

The Tradeoff between Yield and Feedstock Quality

Another shortcoming links the production of lignocellulosic biomass with the conversion to bioenergy, especially biofuels. There is a tradeoff between high yields and a high feedstock quality. It is elaborated that high agricultural yields might result in a feedstock with a high water, ash or salt content [4]. Ash and certain inorganic elements in particular lead to difficulties in processing. Energy-intensive drying might be necessary for wet feedstock, and corrosion, slugging or plugging of the conversion reactor might occur in the presence of minerals [126]. Not only the biomass quality in general but also its quantitative and qualitative constancy over time play an important role for conversion in biorefineries. A constant and qualitatively-reliable feedstock supply is essential for conversion plants [45]. Furthermore, the final application of the lignocellulosic biomass determines the cell wall ideotype. The extraction of target molecules is still cost-intensive, as the loosening and fractionation of cell wall components require intensive pre-treatments [127,128]. Thus, one of the major quality parameters of lignocellulosic feedstock is the easy destructibility of cell walls [127,129]. Moreover, the relative content of desired molecules within the cell walls in consistency with the dedicated end-use of the feedstock is vitally important [130]. For combustion, a high lignin, high cellulose and high hemicellulose content is preferable, as those factors increase the calorific value [126]. Feedstock with a low lignin, high cellulose and high hemicellulose, low cross-linking of cellulose–hemicellulose, low crystallinity index, low cellulose–lignin branching and reduced polymerization is known for good suitability for ethanol production [129,131,132]. Thus, high yields alone are not a sufficient parameter for assessing biomass potentials. If the end-use-dependent feedstock quality, quantity and constancy over seasons do not meet the requirements of the conversion pathways, the bioenergy potentials of lignocellulosic biomass cannot be realized.

The Knowledge on the Crops' Genetic Potentials

Most perennial grasses, which are used as bioenergy feedstock, are undomesticated crops, collected from wild environments and tested in field trials. Hence, some are still in the first stage of breeding programs [133], or even novel orphan crops without any previous genetic improvement. This applies especially with regard to biomass-related characteristics [134–137]. Consequently, the biomass yields and qualities of many perennial grasses are highly variable and in many cases drastically lower than the crops' genetic potentials [138]. For the achievement of the climate mitigation objectives of the Paris agreement, a yearly 145 EJ biomass increment until 2060 will be necessary. The International Energy Agency calls the cultivation of high yield energy crops a key element for the time span 2017 to 2025 to reach the energy potentials [37]. For an intensive cultivation of, e.g., *Miscanthus*, further optimization of the plant is necessary [139,140]. Differing environmental conditions require an accurate selection of suitable species. For instance, the Renewable Energy Directive includes criteria for assessing the availability of land for bioenergy-dedicated biomass. Following this sustainability approach, an avoidance of negative impacts on water resources needs to be included in the crop choice [94]. Thus, for instance, in Mediterranean areas, a variety of highly drought-tolerant (lignocellulosic) crops is required [45], which must be developed through breeding programs. A whole range of new crops tailored to marginal environments needs to be bred and tested [59]. Without a general intensification in breeding and an eco-physical adoption of lignocellulosic crops to specific environmental conditions, high energy potentials calculated and projected by the International Energy Agency, Intergovernmental Panel on Climate Change and others might not be realistic.

The Demand for Higher Sustainability

Within the recast of the Renewable Energy Directive, the European Commission updated goals and regulations for the production of renewable energy [141]. The production of feedstocks and biofuels has to comply with several sustainability criteria to be eligible for financial support by public authorities and to be credited to the national renewable energy targets [142]. The Renewable Energy Directive contains several criteria that can be applied to the assessment of land availability for dedicated biomass crops, and in particular, to the selection of marginal sites for the cultivation of lignocellulosic bioenergy feedstock. The end-user roadmap of Dees et al. [94] summarizes rules that come with the criteria defined in the Renewable Energy Directive. Amongst others, land selection for bioenergy feedstock within the EU has to meet the following requirements: only using lands that have been registered as agricultural lands since 1990, the exclusion of permeable grasslands, the sole usage of surplus, marginal and polluted lands to avoid LUC, no usage of fallow land if the fallow land share (of total arable land) declines to <10%, the avoidance of monocultures and the consideration of a maximum slope limit to perennial plantations. The CAPRI model is the only model available that incorporates the diverse regional circumstances regarding land use changes between 2020 and 2030 within the EU 28 [57]. Nevertheless, even the CAPRI baseline needs to be further adapted to include all Renewable Energy Directive criteria, as, for instance, the rule on fallow land is not considered [94]. It can be assumed that not all yield modeling studies scoping the EU incorporate the Renewable Energy Directive criteria within the assessment of marginal land availability and the cultivation of lignocellulosic crops. This especially applies to yield modeling conducted before the resolution of the directive EU 2018/2001 in 2018. Unfortunately, the compliance of recent studies with the Renewable Energy Directive criteria is not assessed within the scope of this research. If a non-compliance of yield models with the EU directive 2018/2001 should be the case, biomass and bioenergy potentials calculated in those scenarios might be unusable. This aspect might also affect yield modeling outside of the EU. Coming with the updated Renewable Energy Directive, biofuel feedstocks with a high iLUC risk—also originating from non-EU countries—are gradually phased out [143]. It is furthermore assumed that there are yield modelings performed for non-EU areas without incorporating an iLUC risk assessment in coherence with the Renewable Energy Directive regulations.

4. Recommendations and Milestones for Reliable Future Predictions

The first recommendations for biomass yield models/modeling improvement were derived based on the assessment of shortcomings and limitations, which lead to input and model uncertainty on the one side and output uncertainty regarding the biomass and bioenergy potential on the other side. The order of the recommendations equals the order of the limitations assessed in part one of this study. A graphical overview of all derived recommendations and milestones is shown in Figure 3.

4.1. General Recommendation Regarding the Scope and Definitions

To provide sufficient and reliable data to develop models and calculate biomass yields, a globally standardized set of definitions on marginal land, industrial and ligno-cellulosic crops is necessary. These definitions should provide sufficient guidance and a regulated framework that is applicable to countries globally and limits possibilities for national/regional adjustments and interpretation. Based on these definitions, data collection must take place at regional, national and international levels to achieve reliable and adequate results for all regions and to ultimately be able to meet the global biomass demand for non-food purposes. A strong focus on developing countries is thereby required as these countries are often abundant in marginal land [58].

4.2. Recommendations for Models and Input Parameters

The obtained data on lignocellulosic biomass growth and their collection process need to be documented in a detailed way and made available, for example, on statistical databases of the FAO, to provide reference for future studies and assessments. In biomass yield modeling, the future focus should be on the development of crop-specific models such as MiscanFor, which are calibrated based on the genotype-specific demand of the crop [45]. The development of models specifically for marginal land, in which the different environmental (mainly soil) parameters can be represented in detail, is also necessary. To establish these types of models, an integrated, interdisciplinary approach is necessary that brings modeling and plant experts together to adequately represent the different growth stages and underlying calculations in the models [65]. In addition, it is important to increase software and model transparency, making the assumptions and calculations in the background accessible for the model users [81].

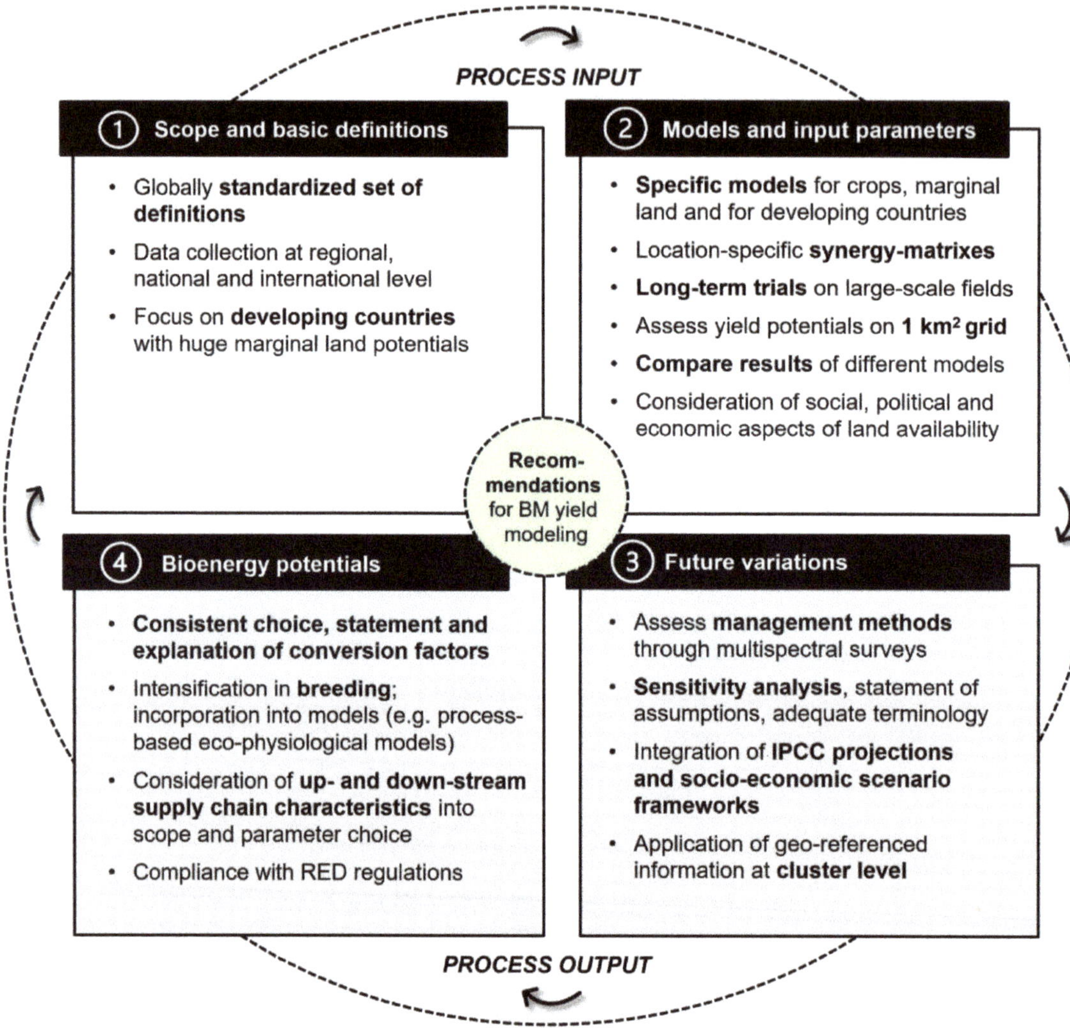

Figure 3. Overview on the most relevant recommendations and milestones for future biomass (BM) yield modeling.

To account for the complexity and diversity of marginal land, a detailed soil and environmental assessment is recommended, especially for European studies [49]. Furthermore, it has to be determined if the same synergies are applicable for other regions, and whether more regional synergy matrixes have to be developed and applied. Until then, the use of a selection of adequate, locally representative soil and environmental parameters is suggested [61]. These should be combined with yield results from long-term trials on commercial-sized fields under similar climatic and management conditions to represent the crop yield development along the growth cycle [74] and further expand the database on crop yield expectations. A comparison of yields from different crops [88] and different crop varieties on the assessed marginal land provides more in-depth insights into the realistic yield range as the varieties' yields can divert significantly from the crop average [45]. If the best crop among several theoretically feasible options should be selected, similar assumptions and the avoidance of average yield data are key aspects [61,79]. Yield calculations best take place on a small scale, using grids with a 1 km² size to identify the best land–environment–crop fit that provides the highest yield per 1 km² and thereby increases the overall modeled yields. If no crop-specific model is available, a comparison of the results from different models, e.g., GAEZ and GEPIC, helps narrow down the spatial and timely range of the yield potential [74]. Furthermore, a comparison of the modeling results with data obtained from several field trials, for example, from different locations across the country for which the biomass yield potential is modeled, could help benchmark the model results and provide a suitable option for comparison [88]. Clearly stating the assumptions made and presenting the limitations of the chosen analytical method further contribute to more robust and reproducible results [41]. Based on this, a transparent quantification of the uncertainty of the individual input parameters as well as of the overall results increases the comparability of the results and the reliability of the study [61]. A sensitivity analysis identifying the most influential parameters on the yield outcome is useful to estimate the impact of wrong assumptions on the model results [25].

To provide a holistic assessment of the practical biomass yield potential on marginal land, it is crucial to include the social, political and economic characteristics related to the use of that land. A local assessment of the non-environmental conditions of the marginal land is important as the social, economic and political conditions vary widely between regions, and local decision-making is necessary to turn theoretical into practical yield potential [65]. Here, political guidance and support for feedstock cultivation on marginal land is most important as cultural and social acceptability of the usage of biomass as energy feedstock can only be achieved by transparently informing people on the advantages and disadvantages and by listening to their fears [74]. Simultaneously, local, regional, national and supranational governments (such as the EU) have to actively promote the cultivation of bioenergy crops on marginal land, for instance, by making use of coherent energy, environmental and agricultural policies [77]. Long-term, future-oriented strategies and adjustments in the Common Agricultural Policy (CAP) to ensure compliance with environmental and social requirements to avoid, for example, indirect and direct land use change are the main political instruments that could be used [55]. However, not only in the primary sector but also in the feedstock-to-bioenergy conversion sector incentives such as subsidies are necessary to extend the value chains to an industrial scale, connecting rural areas where the marginal land is located with the conversion plant operators and finally with the consumers [45]. Beyond the social and political assessment, taking the economic perspective into consideration is important, especially as future research and development is likely to reduce the establishment costs and increase the yields of perennial crops, which enhances the economic attractiveness of biomass production on marginal land considerably [88]. This potential can only be used if the accompanying infrastructure is adequately established and value chains to exploit the regional biomass potential are developed and fortified [45].

The variety of input parameters and underlying assumptions increases the complexity of biomass yield modeling on marginal land. This complexity must be reduced to a manageable and understandable amount while leaving the basic interactions and their impact dimensions unchanged. Therefore, a careful, concise and transparent documentation and reasoning for the selected input parameters is decisive to provide high-quality, realistic and comparable results that can be interpreted in a meaningful way.

Agricultural management strategies for low-input systems need to be in line with sustainable cultivation practices to yield environmentally sustainable produced biomass with a high quality, for example, by incorporating agricultural practices such as crop rotations into agricultural systems on marginal land [44]. Thus, it might be beneficial for the soil and the nutrient balance to grow crops in a certain sequence or even intercrop them. These practices and their impacts on overall biomass yields must be represented in yield modeling. Therefore, regionally specific data from the respective farmers might be necessary [91]. Multispectral surveys need to be conducted to precisely assess time series of crop cultivation and rotation [91]. Intensive input on a regional and local level is required to correctly depict farmers' actual and future agricultural operation and management strategies.

4.3. Recommendations for Consideration of Future Variations

Uncertainties and fluctuations in biomass and bioenergy yield modeling cannot be fully eliminated. Models always contain uncertainty and are a simplification of reality. This is the reason why assumptions made must be clearly stated, limitations of the methodology and input data must be highlighted and the inherent uncertainty in modeling results must be analyzed in future studies [41]. A sensitivity analysis of the yield modeling results and a transparent description of assumptions are crucial. Moreover, sensitivity regarding the use of terms is necessary. The results of yield modeling should only be published as 'forecasts' or 'predictions' if the projection is the most likely one. This needs to be analyzed through a deterministic model and a comparison of a sufficiently large number of scenarios [144].

Furthermore, there is a need for biophysical models, which can cover various ecological parameters and their dynamic changes over time. A good example is the assessment of global bioenergy potentials of *Miscanthus*, conducted by Shepherd et al. [79]. The MiscanFor model was therefore not only extended to incorporate the RCP 2.6 climate scenario of the Intergovernmental Panel on Climate Change, but also the SSP2 socio-economic scenario of the Intergovernmental Panel on Climate Change, gaining anticipating weight to at least some extent. There is an urgent need to integrate environmental scenarios into yield models via a direct embedding or through baseline extensions, especially for climate and groundwater predictions. To comply with the current state of research and to apply internally consistent data, the Intergovernmental Panel on Climate Change projections and scenario frameworks should be starting point for the incorporation of 'future' into models. This accounts for socio-economic scenarios, land use and land cover change scenarios, environmental scenarios (e.g., carbon dioxide, water resources, acidifying compounds), climatic scenarios, sea-level rise scenarios as well as their interactions [144]. Global data need to be combined with national, regional, or in the best case, even local datasets on a 1 km^2 scale. An assessment of, for instance, NUTS 2 resolution levels in the EU does not provide sufficiently detailed information on the regional and local environmental conditions. If, for instance, the CARPI model is used, the option of integrating geo-referenced information at cluster level (1 km^2 grid cell) must be chosen to increase specificity [145]. For yield modeling on marginal land in the EU, a whole range of models can and should be used and interconnected for regulation-consistent scenarios: non-carbon dioxide emissions and pollutants (GAINS model), land use change and forestry (GLOBIUM/G4M model), agriculture (GAPRI), energy, including transport and processes (especially PRIMES biomass supply model, PRIMES energy system model) and overall framework assumptions (Prometheus model, GEM-E3 model) [146].

4.4. Milestones for Improving Bioenergy Potential Assessments

It is highly indispensable that applied conversion factors, projected bioenergy pathways and the specific conversion technologies need to be stated and explained in future bioenergy-potential projecting yield modelings. The conversion factors and technologies incorporated in the modeling of bioenergy scenarios must be time- and location-specific. Thus, conversion factors need to represent the technological state of the art (for status quo assessments) or realistic future technological advancements (for mid- and long-term projections), respectively. The choice of bioenergy carriers (e.g., fuels, heat/power) needs to be consistent with national and socio-economic patterns, markets and regulations. Furthermore, conversion factors must account for conversion and delivery losses to depict realistic end-use values.

A rapid and targeted expansion of conversion capacities, suitable infrastructure and supply chains need to come along with the progressing projection and incorporation of biomass and bioenergy potentials into governmental and non-governmental climate mitigation strategies. The conversion plants need to be adequately spatially distributed, considering land use patterns and regional biomass potentials, and have to come with sufficient infrastructural connections to local markets and up- as well as downstream supply chains. At the same time, innovative and efficiency-boosting pretreatments and bio- as well as thermochemical conversion technologies need to be developed, tested in demonstration plants and upscaled to commercial level. A promising approach is, for instance, the sunliquid® technology developed by Clariant International Ltd. Thereby, process-specific enzymes and simultaneous C5 and C6 fermentation of (ligno-)cellulosics (so far only wheat and other cereal straw) are applied to boost the commercial performance of cellulosic ethanol production [109]. An expansion of governmental subsidies for research and development, as well as financial support for the upscaling of innovative concepts, has to be fostered globally. Furthermore, holistic Life Cycle Sustainability Assessments [147], energetic efficiency analysis and techno-economic assessments of (ligno-)cellulosic ethanol life cycle chains [148] need to evaluate the economic, ecological and social impacts and performance of biofuels from lignocellulosic feedstock. The bare suitability of marginal land for bioenergy feedstock production does not imply that cultivation and production are automatically sustainable, but on the contrary, the whole value chain needs to be assessed [55]. In conclusion, only a scientific, qualitative and quantitative sustainability assessment can lead to objective and holistic judgements of the performance of lignocellulosic biofuels or different conversion pathways, respectively.

The opportunities and potentials coming with intensive breeding of perennial grasses are evaluated as enormous [149]. The extensive knowledge on plant biology and genetics, as well as the large toolbox for analytical and genomic approaches, can result in innovative, high-yielding and low-demanding perennial crops, tailored to diverse agroecological systems. Higher yields, a higher resource use efficiency (especially nutrients and water) and a better exploitability of soil come with higher efforts in the breeding of C4 grasses [129,149]. It is obvious that intensive research, breeding and field trials need to be fostered within the next few years. The breeding should not only focus on single crops, but a whole range of species, as agroecological environments differ and the demands of biorefineries are broad. It needs to be highlighted that due to the evolutionary relation of C4 grasses, advances in the breeding of one crop can boost the development of other crops [129]. Focusing on *Miscanthus* again, this recommendation is fully applicable. Twenty species of *Miscanthus* are known, and the genus holds significant potential for adaptations to the environmental conditions or the assigned conversion pathway. The market potential of lignocellulosic ethanol from *Miscanthus* and other grasses can be scaled up through current and future breeding efforts, especially the development of new hybrids, which come with higher biomass yields and streamlined degradability [139,140]. A promising but highly complex approach [137] is the engineering of a C4 photosynthesis in C3 crops to further increase biomass yields of highly productive C3 species (e.g., giant reed or tall wheatgrass) [150]. Consequently, innovative, molecular breeding for lignocellulosic crops must be taken into account to exploit genetic resources, for instance, via next generation sequencing, high-throughput genotyping, molecular breeding, marker-assisted selection and genomic selection [138]. A more challenging but highly promising approach is the development and improvement of methods to analyze cell wall compositions and nanoscale structures [151]. The exploration of molecules in cell walls, their chemically specific imaging tags and their development over the life cycle of plants (from cell wall formation, over maturation, transformation, dehydration and processing into feedstocks) can support the predictive modeling of feedstock qualities and quantities. New findings are necessary to develop and improve advanced feedstocks and optimize their processing pathways. Ultimately, these research and breeding methods must be both accurate and relatively inexpensive, allowing the handling of large amounts of samples in breeding programs [151]. The United States Department of Energy published a plant-physiological, genetic and biotechnological roadmap for lignocellulosic crops in 2006, including technical milestones to increase the market potential of cellulosic ethanol within 15 years—for instance, the optimization of cell wall composition to increase the content of fermentable sugar and the discovery of genetic regulatory factors that determine the synthesis and deposition of lignin [151]. In future research, the achievement of those milestones as well as limitations and future objectives need to be assessed for a wide range of lignocellulosic crops in well-coordinated field experimentations worldwide, as also noted by Reinhardt et al. [152]. If current and future breeding efforts are successful, the achieved and updated yield levels promptly need to be incorporated into yield modeling, so realistic biomass potentials for future decades can be computed. Process-based eco-physiological models could be combined with genomics to make progress in plant phenotyping [153]. Connections between controlled-conditions phenotyping and crop performance in the field can be made to reduce model uncertainty. Furthermore, yield modeling for newly domesticated varieties, genotypes and their performance on marginal land urgently need to be assessed, providing meaningful predictions of potentials.

The breeding approaches assessed will highly support the optimization of the lignocellulose composition of the feedstock, which is necessary to allow an efficient conversion into biofuels. Researchers performing yield modeling should also see a mission in assessing the potentials of different bioenergy value chains. The final use and conversion technology of the lignocellulosic feedstock highly influences the crop choice. Thus, the incorporation of socio-economic factors and the technological state-of-the-art for biomass conversion need to be taken into account in yield modeling. Crops in yield models shall not exclusively be chosen regarding the best yield performance, but also under consideration of the upstream supply chain's requirements.

The overall goal of yield modeling on marginal land can be defined as the assessment of biomass yield potentials under consideration of environmental, crop-specific and (partially) socio-economic and other constraints. Thus, yield modeling needs to take environmental regulations into account, depicting societal judgements on current and future agricultural practices [12]. A good example is the reflected Renewable Energy Directive (EU 2018/2001), which not only affects yield modeling practices inside, but also outside of the EU. Future yield modeling needs to be performed in consistency with all relevant sustainability criteria and rules that are applicable to the scope and boundary of the modeling [57]. Even the baseline in the CAPARI model was in compliance with EU policies on bioenergy targets (based on the PRIMES energy model), and a further upgrading was necessary to incorporate all relevant Renewable Energy Directive criteria.

5. Conclusions

Marginal land is defined primarily by limiting soil and climatic conditions that a variety of lignocellulosic industrial crops can tolerate. Applied in several studies, yield modeling aims to precisely predict future yields. The modeled yields are further transformed into bioenergy potentials, representing tangible energy contributions of lignocellulosic crops. However, a massive lack of globally or trans-nationally coherent definitions of lignocellulosic crops inevitably leads to inconsistency in statistical data on current and future yields. Several other key limitations also reduce the informative power of yield modeling studies.

This study shows that there are no sufficient data available to precisely model lignocellulosic biomass on marginal land. Even though there are several modeling approaches, an increased number of parameters and various data sources available for the modeling of lignocellulosic feedstocks, their suitability for the modeling of biomass cultivation on marginal land is still limited. Several limitations and shortcomings were assessed to point out a multitude of data and methodology limitations on both the input and output side of yield modeling. These issues derived from a review of several yield modeling studies prove that, currently, yields of lignocellulosic biomass on marginal land are not modeled precisely. The relevance of lignocellulosic crops for the growing global bioeconomy was substantiated with political incentives, forecasts of organizations (e.g., the International Energy Agency) and statistical data, buttressing the promising potentials of lignocellulosic crops as a bioenergy feedstock.

The need for modeling the biomass potentials on marginal land was confirmed by analyzing the significant need of biomass and bioenergy within the next few decades, and the partially insufficient data basis for the calculation of those biomass potentials. The cultivation of perennial lignocellulosic crops on marginal land comes with promising advantageous benefits. The increase in biodiversity, ecosystem services such as pollination [154,155] and pest suppression through perennials [156] is not only reflected on the areas they take place but also on neighboring agricultural land, for instance, used for food crop cultivation. A potential yield increase by up to 25% on annual croplands is reported [157]. This, together with the production of lignocellulosic feedstock for bioenergy, can significantly improve food security, boost rural development, create employment opportunities and deliver sustainable energy sources, improving living standards of rural communities [158].

Ultimately, this review has demonstrated the complexity of biomass yield modeling on the one side and the potentials of the methodology on the other side. Several crucial shortcomings limiting the use of biomass yield model approaches for lignocellulosic crops on marginal land were derived. The rapidly increasing demand for food and non-food crops asks for transparent, multi-dimensional and highly adaptable models, producing clear and meaningful scenarios of exploitable biomass and bioenergy potentials. Quantitative data on the supply and demand of biomass and bioenergy are an essential base for international negotiations of climate mitigation agreements, strategic sustainability goals and their practical implementation into (trans-)national policies. Nevertheless, the biomass potentials resulting from yield modeling on marginal land shall not be interpreted as easily exploitable bioenergy potentials, but rather as first drafts of future bioenergy supply chains.

Author Contributions: Conceptualization: J.H., P.H. and M.v.C.; Methodology: J.H., P.H. and M.v.C.; Validation: J.H., P.H. and M.v.C.; Investigation: J.H. and P.H.; Writing—original draft: J.H., P.H. and M.v.C.; Writing—review and editing: M.v.C.; Resources: J.H. and P.H.; Visualization: J.H. and P.H.; Funding acquisition: M.v.C.; Supervision: M.v.C. All authors have read and agreed to the published version of the manuscript.

Funding: This research received funding from the Federal Ministry of Education and Research (01PL16003) and the University of Hohenheim.

Data Availability Statement: Datasets for this research are included in the references.

Acknowledgments: The authors are very grateful to Iris Lewandowski for making this work possible. Special thanks go to Eva Lewin for improving the language quality of the manuscript.

Conflicts of Interest: The authors declare that they have no known competing financial interests or personal relationships that could have appeared to influence the work reported in this paper. The funders had no role in the design of the study; in the collection, analyses, or interpretation of data; in the writing of the manuscript, or in the decision to publish the results.

References

1. Department of Economic and Social Affairs, Population Division, United Nations. World Population Prospects 2019 Highlights. 2019. Available online: https://population.un.org/wpp/Publications/Files/WPP2019_Highlights.pdf (accessed on 19 October 2021).
2. Tomlinson, I. Doubling Food Production to Feed the 9 Billion: A Critical Perspective on a Key Discourse of Food Security in the UK. *J. Rural Stud.* **2011**, *29*, 81–90. [CrossRef]
3. High Level Expert Forum, FAO Global Agriculture towards 2050. 2009. Available online: http://www.fao.org/fileadmin/templates/wsfs/docs/Issues_papers/HLEF2050_Global_Agriculture.pdf (accessed on 19 October 2021).
4. Lewandowski, I.; Lippe, M.; Castro Montoya, J.; Dicköfer, U. Agricultural Production. In *Bioeconomy: Shaping the Transition to a Sustainable, Biobased Economy*; Lewandowski, I., Ed.; Springer: Berlin/Heidelberg, Germany; New York, NY, USA, 2018; pp. 99–123. ISBN 978-3-319-68151-1.
5. Piotrowski, S.; Carus, M.; Essel, R. Global Bioeconomy in the Conflict between Biomass Supply and Demand. *Ind. Biotechnol.* **2015**, *11*, 308–315. [CrossRef]
6. Abideen, Z.; Ansari, R.; Khan, M.A. Halophytes: Potential Source of Ligno-Cellulosic Biomass for Ethanol Production. *Biomass Bioenergy* **2011**, *35*, 1818–1822. [CrossRef]
7. Hadar, Y. Sources for Lignocellulosic Raw Materials for the Production of Ethanol. In *Lignocellulose Conversion: Enzymatic and Microbial Tools for Bioethanol Production*; Faraco, V., Ed.; Springer: Berlin/Heidelberg, Germany, 2013; pp. 21–38. ISBN 978-3-642-37861-4.

8. Chiaramonti, D.; Giovannini, A.; Janssen, R.; Mergner, R. *Lignocellulosic Ethanol Process and Demonstration*; WIP Renewable Energies: München, Germany, 2013. Available online: https://www.wip-munich.de/biolyfe-handbook/1_7_1_BIOLYFE_Handbook_Part-I.pdf (accessed on 19 October 2021).

9. Milbrandt, A.; Overend, R.P. *Assessment of Biomass Resources from Marginal Lands in APEC Economies*; National Renewable Energy Lab. (NREL): Golden, CO, USA, 2009. Available online: https://www.nrel.gov/docs/fy10osti/46209.pdf (accessed on 19 October 2021).

10. IEA. Bioenergy Mobilising Sustainable Bioenergy Supply Chains: Opportunities for Agriculture. *Summary and Conclusions from the IEA Bioenergy ExCo77 Workshop*. 2016. Available online: https://www.ieabioenergy.com/blog/publications/ws20-mobilising-sustainable-bioenergy-supply-chains-opportunities-for-agriculture/ (accessed on 19 October 2021).

11. Department for Environment, Food and Rural Affairs. *Crops Grown For Bioenergy in the UK: 2018*; Department for Environment Food & Rural Affairs: London, UK, 2019. Available online: https://assets.publishing.service.gov.uk/government/uploads/system/uploads/attachment_data/file/856695/nonfood-statsnotice2018-08jan20.pdf (accessed on 22 November 2021).

12. von Cossel, M.; Wagner, M.; Lask, J.; Magenau, E.; Bauerle, A.; Von Cossel, V.; Warrach-Sagi, K.; Elbersen, B.; Staritsky, I.; Van Eupen, M.; et al. Prospects of Bioenergy Cropping Systems for a More Social-Ecologically Sound Bioeconomy. *Agronomy* **2019**, *9*, 605. [CrossRef]

13. Von Cossel, M.; Lewandowski, I.; Elbersen, B.; Staritsky, I.; Van Eupen, M.; Iqbal, Y.; Mantel, S.; Scordia, D.; Testa, G.; Cosentino, S.L.; et al. Marginal Agricultural Land Low-Input Systems for Biomass Production. *Energies* **2019**, *12*, 3123. [CrossRef]

14. Xue, S.; Lewandowski, I.; Wang, X.; Yi, Z. Assessment of the Production Potentials of Miscanthus on Marginal Land in China. *Renew. Sustain. Energy Rev.* **2016**, *54*, 932–943. [CrossRef]

15. Schweier, J.; Becker, G. Economics of Poplar Short Rotation Coppice Plantations on Marginal Land in Germany. *Biomass Bioenergy* **2013**, *59*, 494–502. [CrossRef]

16. Cheah, W.Y.; Sankaran, R.; Show, P.L.; Ibrahim, T.N.B.; Chew, K.W.; Culaba, A.; Chang, J.-S. Pretreatment Methods for Lignocellulosic Biofuels Production: Current Advances, Challenges and Future Prospects. *Biofuel Res. J.* **2020**, *7*, 1115–1127. [CrossRef]

17. Janssen, R.; Khawaja, C. Sustainable Supply of Non-Food Biomass for a Resource Efficient Bioeconomy. A Review Paper on the State-of-the-Art. 2014. Available online: http://www.s2biom.eu/images/Publications/S2biom_review_state-of_the_art_Final.pdf (accessed on 19 October 2021).

18. ETIP. Bioenergy Lignocellulosic Crops for Production of Advanced Biofuels. Available online: https://www.etipbioenergy.eu/value-chains/feedstocks/agriculture/lignocellulosic-crops (accessed on 19 October 2021).

19. Hao, L.C.; Sapuan, S.M.; Hassan, M.R.; Sheltami, R.M. 2—Natural fiber reinforced vinyl polymer composites. In *Natural Fibre Reinforced Vinyl Ester and Vinyl Polymer Composites*; Sapuan, S.M., Ismail, H., Zainudin, E.S., Eds.; Woodhead Publishing Series in Composites Science and Engineering; Woodhead Publishing: Sawston, UK, 2018; pp. 27–70. ISBN 978-0-08-102160-6.

20. Ge, X.; Chang, C.; Zhang, L.; Cui, S.; Luo, X.; Hu, S.; Qin, Y.; Li, Y. Chapter Five—Conversion of Lignocellulosic Biomass Into Platform Chemicals for Biobased Polyurethane Application. In *Advances in Bioenergy*; Li, Y., Ge, X., Eds.; Elsevier: Amsterdam, The Netherlands, 2018; Volume 3, pp. 161–213.

21. Shankaran, D.R. Chapter 14—Cellulose Nanocrystals for Health Care Applications. In *Applications of Nanomaterials*; Mohan Bhagyaraj, S., Oluwafemi, O.S., Kalarikkal, N., Thomas, S., Eds.; Micro and Nano Technologies; Woodhead Publishing: Sawston, UK, 2018; pp. 415–459. ISBN 978-0-08-101971-9.

22. European Biofuels Technology Platform, Bioenergy Value Chain 5: Sugar to Alcohols. 2016. Available online: https://etipbioenergy.eu/images/EIBI-5-sugar-to-alcohols.pdf (accessed on 19 October 2021).

23. Ding, H.; Xu, F. Chapter 20—Production of Phytochemicals, Dyes and Pigments as Coproducts in Bioenergy Processes. In *Bioenergy* Research: Advances and Applications; Gupta, V.K., Tuohy, M.G., Kubicek, C.P., Saddler, J., Xu, F., Eds.; Elsevier: Amsterdam, The Netherlands, 2014; pp. 353–365. ISBN 978-0-444-59561-4.

24. Patel, M.; Zhang, X.; Kumar, A. Techno-Economic and Life Cycle Assessment on Lignocellulosic Biomass Thermochemical Conversion Technologies: A Review. *Renew. Sustain. Energy Rev.* **2016**, *53*, 1486–1499. [CrossRef]

25. Jing, Q.; Conijn, S.J.G.; Jongschaap, R.E.E.; Bindraban, P.S. Modeling the Productivity of Energy Crops in Different Agro-Ecological Environments. *Biomass Bioenergy* **2012**, *46*, 618–633. [CrossRef]

26. Elbersen, H.W.; Lammens, T.M.; Alankangas, E.A.; Annevelink, E.; Harmsen, P.F.H.; Elbersen, B.S. Lignocellulosic Biomass Quality: Matching Characteristics With Biomass Conversion Requirements. In *Modeling and Optimization of Biomass Supply Chains: Top down and Bottom up Assessment for Agricultural, Forest and Waste Feedstock*; Academic Press: Cambridge, MA, USA, 2017; pp. 55–78. [CrossRef]

27. FAO FAOSTAT. Available online: http://www.fao.org/faostat/en/#home (accessed on 19 October 2021).

28. FAO. *Food Balance Sheets. A Handbook*; Food and Agriculture Organization of the United Nations: Rome, Italy, 2011. Available online: http://www.fao.org/3/x9892e/x9892e00.htm (accessed on 19 October 2021).

29. FAO Definition and Classification of Commodities. Fibres of Vegetal and Animal Origin. Available online: http://www.fao.org/waicent/faoinfo/economic/faodef/fdef09e.htm (accessed on 19 October 2021).

30. Raschka, A.; Carus, M. *Stoffliche Nutzung von Biomasse Basisdaten für Deutschland, Europa und die Welt*; Nova-Institut GmbH: Hürth, Germany, 2012. Available online: https://www.iwbio.de/fileadmin/Publikationen/IWBio-Publikationen/Stoffliche_Nutzung_von_Biomasse_nova.pdf (accessed on 19 October 2021).

31. Berndes, G. Bioenergy, Land Use Change and Climate Change Mitigation. 2010. Available online: https://www.ieabioenergy. com/wp-content/uploads/2013/10/Bioenergy-Land-Use-Change-and-Climate-Change-Mitigation-Background-Technical-Report.pdf (accessed on 19 October 2021).

32. Edrisi, S.A.; Abhilash, P.C. Exploring Marginal and Degraded Lands for Biomass and Bioenergy Production: An Indian Scenario. *Renew. Sustain. Energy Rev.* **2016**, *54*, 1537–1551. [CrossRef]

33. Eurostat, Archive: Main Annual Crop Statistics. Available online: https://ec.europa.eu/eurostat/statistics-explained/index. php?title=Main_annual_crop_statistics&oldid=258003 (accessed on 19 October 2021).

34. Eurostat Eurostat—Data Explorer—Industrial Crops by NUTS 2 Regions. Available online: https://appsso.eurostat.ec.europa. eu/nui/submitViewTableAction.do (accessed on 19 October 2021).

35. Scarlat, N.; Dallemand, J.-F.; Monforti-Ferrario, F.; Nita, V. The Role of Biomass and Bioenergy in a Future Bioeconomy: Policies and Facts. *Environ. Dev.* **2015**, *15*, 3–34. [CrossRef]

36. Phillips, S.; Flach, B.; Lieberz, S.; Bolla, S. Biofuels Annual EU-28. 2019. Available online: https://apps.fas.usda.gov/newgainapi/ api/report/downloadreportbyfilename?filename=Biofuels%20Annual_The%20Hague_EU-28_7-15-2019.pdf (accessed on 19 October 2021).

37. International Energy Agency (IEA). Technology Roadmap: Delivering Sustainable Bioenergy 2017. Available online: https: //www.ieabioenergy.com/publications/technology-roadmap-delivering-sustainable-bioenergy/ (accessed on 19 October 2021).

38. Intergovernmental Panel on Climate Change (IPCC). Special Report on Renewable Energy and Climate Mitigation. 2011. Available online: https://www.ipcc.ch/2011/06/28/special-report-on-renewable-energy-sources-and-climate-change-mitigation-srren/ (accessed on 19 October 2021).

39. Fritsche, U.; Eppler, U.; Fehrenbach, H.; Giergrich, J. Linkages between the Sustainable Development Goals (SDGs) and the GBEP Sustainability Indicators for Bioenergy (GSI)—Technical Paper for the GBEP Task Force on Sustainability. 2018. Available online: http://www.globalbioenergy.org/fileadmin/user_upload/gbep/docs/Indicators/IINAS_IFEU__2018__Linkages_SDGs_ and_GSIs_-_final.pdf (accessed on 19 October 2021).

40. Merriam-Webster Dictionary, Definition of MARGINAL. Available online: https://www.merriam-webster.com/dictionary/ marginal (accessed on 19 October 2021).

41. Elbersen, B.; Van Eupen, E.; Mantel, S.; Verzandvoort, S.; Boogaard, H.; Mucher, S.; Cicarreli, T.; Elbersen, W.; Bai, Z.; Iqbal, Y.; et al. *Methodological Approaches to Identify and Map Marginal Land Suitable for Industrial Crops in Europe*; WUR: Wageningen, The Netherlands, 2018.

42. Shortall, O.K. "Marginal Land" for Energy Crops: Exploring Definitions and Embedded Assumptions. *Energy Policy* **2013**, *62*, 19–27. [CrossRef]

43. Lewis, S.; Kelly, M. Mapping the Potential for Biofuel Production on Marginal Lands: Differences in Definitions, Data and Models across Scales. *IJGI* **2014**, *3*, 430–459. [CrossRef]

44. von Cossel, M.; Iqbal, Y.; Scordia, D.; Cosentino, S.L.; Elbersen, B.; Staritsky, I.; Van Eupen, M.; Mantel, S.; Prysiazhniuk, O.; Maliarenko, O.; et al. D4.1 Low-Input Agricultural Practices for Industrial Crops on Marginal Land. *Zenodo* **2019**. [CrossRef]

45. Ramirez-Almeyda, J.; Elbersen, B.; Monti, A.; Staritsky, I.; Panoutsou, C.; Alexopoulou, E.; Schrijver, R.; Elbersen, W. Assessing the Potentials for Nonfood Crops. In *Modeling and Optimization of Biomass Supply Chains: Top down and Bottom up Assessment for Agricultural, Forest and Waste Feedstock*; Panoutsou, C., Ed.; Elsevier: Cambridge, CA, USA, 2017; pp. 219–251. ISBN 978-0-12-812303-4.

46. Parr, J.F.; Stewart, B.A.; Hornick, S.B.; Singh, R.P. Improving the Sustainability of Dryland Farming Systems: A Global Perspective. In *Advances in Soil Science*; Singh, R.P., Parr, J.F., Stewart, B.A., Eds.; Advances in Soil Science; Springer: New York, NY, USA, 1990; Volume 13, pp. 1–8. ISBN 978-1-4613-8984-2.

47. Searchinger, T.; Heimlich, R.; Houghton, R.A.; Dong, F.; Elobeid, A.; Fabiosa, J.; Tokgoz, S.; Hayes, D.; Yu, T.-H. Use of U.S. Croplands for Biofuels Increases Greenhouse Gases through Emissions from Land-Use Change. *Science* **2008**, *319*, 1238–1240. [CrossRef]

48. Krasuska, E.; Cadórniga, C.; Tenorio, J.L.; Testa, G.; Scordia, D. Potential Land Availability for Energy Crops Production in Europe. *Biofuels Bioprod. Bioref.* **2010**, *4*, 658–673. [CrossRef]

49. Confalonieri, R.; Jones, B.; Van Diepen, K.; Van Orshoven, J. *Scientific Contribution on Combining Biophysical Criteria Underpinning the Delineation of Agricultural Areas Affected by Specific Constraints: Methodology and Factsheets for Plausible Criteria Combinations*; Terres, J.M., Hagyo, A., Wania, A., Eds.; European Comission-Joint Research Centre, Institute for Environment and Sustainability: Luxembourg, 2014; ISBN 978-92-79-44340-4. Available online: http://publications.jrc.ec.europa.eu/repository/bitstream/JRC926 86/lbna26940enn.pdf (accessed on 19 October 2021).

50. Gopalakrishnan, G.; Negri, M.C.; Wang, M.; Wu, M.; Snyder, S.W.; LaFreniere, L. Biofuels, Land, and Water: A Systems Approach to Sustainability. *Environ. Sci. Technol.* **2009**, *43*, 6094–6100. [CrossRef]

51. Strijker, D. Marginal Lands in Europe—Causes of Decline. *Basic Appl. Ecol.* **2005**, *6*, 99–106. [CrossRef]

52. Peterson, G.M.; Galbraith, J.K. The Concept of Marginal Land. *J. Farm. Econ.* **1932**, *14*, 295. [CrossRef]

53. Plassmann, K. Direct and Indirect Land Use Change. In *Biokerosene*; Kaltschmitt, M., Neuling, U., Eds.; Springer: Berlin/Heidelberg, Germany, 2018; pp. 375–402. ISBN 978-3-662-53063-4.

54. Young, A. Is There Really Spare Land? A Critique of Estimates of Available Cultivable Land in Developing Countries. *Environ. Dev. Sustain.* **1999**, *1*, 3–18. [CrossRef]

55. Gerwin, W.; Repmann, F.; Galatsidas, S.; Vlachaki, D.; Gounaris, N.; Baumgarten, W.; Volkmann, C.; Keramitzis, D.; Kiourtsis, F.; Freese, D. Assessment and Quantification of Marginal Lands for Biomass Production in Europe Using Soil-Quality Indicators. *SOIL* **2018**, *4*, 267–290. [CrossRef]

56. European Commission. Report From The Commission to the European Parliament, The Council, The European Economic and Social Committee and the Committee of the Regions on the Status of Production Expansion of Relevant Food and Feed Crops Worldwide. 2019. Available online: https://ec.europa.eu/energy/sites/ener/files/documents/report.pdf (accessed on 22 November 2021).

57. Ramirez-Almeyda, J. Lignocellulosic Crops in Europe: Integrating Crop Yield Potentials with Land Potentials. Ph.D. Thesis, University of Bologna, Bologna, Italy, 2017. Available online: http://amsdottorato.unibo.it/7854/ (accessed on 19 October 2021).

58. Eisentraut, A. SuStainable Production of Second-Generation Biofuels-Potential and Perspectives in Major Economies and Developing Countries. 2010. Available online: https://www.oecd-ilibrary.org/energy/sustainable-production-of-second-generation-biofuels_5kmh3njpt6r0-en (accessed on 19 October 2021).

59. Pancaldi, F.; Trindade, L.M. Marginal Lands to Grow Novel Bio-Based Crops: A Plant Breeding Perspective. *Front. Plant. Sci.* **2020**, *11*, 227. [CrossRef]

60. Nijsen, M.; Smeets, E.; Stehfest, E.; Vuuren, D.P. An Evaluation of the Global Potential of Bioenergy Production on Degraded Lands. *Glob. Chang. Biol. Bioenergy* **2012**, *4*, 130–147. [CrossRef]

61. Alexopoulou, E.; Zanetti, F.; Scordia, D.; Zegada-Lizarazu, W.; Christou, M.; Testa, G.; Cosentino, S.L.; Monti, A. Long-Term Yields of Switchgrass, Giant Reed, and Miscanthus in the Mediterranean Basin. *Bioenerg. Res.* **2015**, *8*, 1492–1499. [CrossRef]

62. Cai, X.; Zhang, X.; Wang, D. Land Availability for Biofuel Production. *Environ. Sci. Technol.* **2011**, *45*, 334–339. [CrossRef] [PubMed]

63. Graves, R.A.; Pearson, S.M.; Turner, M.G. Landscape Patterns of Bioenergy in a Changing Climate: Implications for Crop Allocation and Land-Use Competition. *Ecol. Appl.* **2016**, *26*, 515–529. [CrossRef]

64. Gu, Y.; Wylie, B.K. Mapping Marginal Croplands Suitable for Cellulosic Feedstock Crops in the Great Plains, United States. *GCB Bioenergy* **2017**, *9*, 836–844. [CrossRef]

65. Harvolk, S.; Kornatz, P.; Otte, A.; Simmering, D. Using Existing Landscape Data to Assess the Ecological Potential of Miscanthus Cultivation in a Marginal Landscape. *GCB Bioenergy* **2014**, *6*, 227–241. [CrossRef]

66. Khamzina, A.; Lamers, J.P.A.; Vlek, P.L.G. Tree Establishment under Deficit Irrigation on Degraded Agricultural Land in the Lower Amu Darya River Region, Aral Sea Basin. *For. Ecol. Manag.* **2008**, *255*, 168–178. [CrossRef]

67. Beale, C.V.; Long, S.P. Can Perennial C4 Grasses Attain High Efficiencies of Radiant Energy Conversion in Cool Climates? *Plant. Cell Environ.* **1995**, *18*, 641–650. [CrossRef]

68. Clifton-Brown, J.C.; Lewandowski, I.; Andersson, B.; Basch, G.; Christian, D.G.; Kjeldsen, J.B.; Jørgensen, U.; Mortensen, J.V.; Riche, A.B.; Schwarz, K.-U. Performance of 15 Miscanthus Genotypes at Five Sites in Europe. *Agron. J.* **2001**, *93*, 1013–1019. [CrossRef]

69. Liu, W.; Yan, J.; Li, J.; Sang, T. Yield Potential of Miscanthus Energy Crops in the Loess Plateau of China. *Glob. Chang. Biol. Bioenergy* **2012**, *4*, 545–554. [CrossRef]

70. Panoutsou, C.; Chiaramonti, D. Socio-Economic Opportunities from Miscanthus Cultivation in Marginal Land for Bioenergy. *Energies* **2020**, *13*, 2741. [CrossRef]

71. Qin, Z.; Zhuang, Q.; Cai, X. Bioenergy Crop Productivity and Potential Climate Change Mitigation from Marginal Lands in the United States: An Ecosystem Modeling Perspective. *GCB Bioenergy* **2015**, *7*, 1211–1221. [CrossRef]

72. Hastings, A.; Clifton-Browon, J.; Wattenbach, M.; Mitchell, C.P.; Smith, P. The Development of MISCANFOR, a New Miscanthus Crop Growth Model: Towards More Robust Yield Predictions under Different Climatic and Soil Conditions. *GCB Bioenergy* **2009**, *1*, 154–170. [CrossRef]

73. Liu, J.; Williams, J.R.; Zehnder, A.J.B.; Yang, H. GEPIC—Modelling Wheat Yield and Crop Water Productivity with High Resolution on a Global Scale. *Agric. Syst.* **2007**, *94*, 478–493. [CrossRef]

74. Zhang, B.; Hastings, A.; Clifton-Brown, J.C.; Jiang, D.; Faaij, A.P.C. Modeled Spatial Assessment of Biomass Productivity and Technical Potential of *Miscanthus × Giganteus*, *Panicum Virgatum* L., and *Jatropha* on Marginal Land in China. *GCB Bioenergy* **2020**, *12*, 328–345. [CrossRef]

75. Cobuloglu, H.I.; Büyüktahtakın, İ.E. A Mixed-Integer Optimization Model for the Economic and Environmental Analysis of Biomass Production. *Biomass Bioenergy* **2014**, *67*, 8–23. [CrossRef]

76. Mehmood, M.A.; Ibrahim, M.; Rashid, U.; Nawaz, M.; Ali, S.; Hussain, A.; Gull, M. Biomass Production for Bioenergy Using Marginal Lands. *Sustain. Prod. Consum.* **2017**, *9*, 3–21. [CrossRef]

77. Pulighe, G.; Bonati, G.; Colangeli, M.; Morese, M.M.; Traverso, L.; Lupia, F.; Khawaja, C.; Janssen, R.; Fava, F. Ongoing and Emerging Issues for Sustainable Bioenergy Production on Marginal Lands in the Mediterranean Regions. *Renew. Sustain. Energy Rev.* **2019**, *103*, 58–70. [CrossRef]

78. Wiggins, S.; Henley, G.; Keats, S. Competitive or Complementary? Industrial Crops and Food Security in Sub-Saharan Africa. *ODI Rep.* **2015**. Available online: https://odi.org/en/publications/competitive-or-complementary-industrial-crops-and-food-security-in-sub-saharan-africa/ (accessed on 19 October 2021).

79. Shepherd, A.; Littleton, E.; Clifton-Brown, J.; Martin, M.; Hastings, A. Projections of Global and UK Bioenergy Potential from Miscanthus × Giganteus—Feedstock Yield, Carbon Cycling and Electricity Generation in the 21st Century. *GCB Bioenergy* **2020**, *12*, 287–305. [CrossRef]

80. Peng, R.D. Reproducible Research in Computational Science. *Science* **2011**, *334*, 1226–1227. [CrossRef] [PubMed]

81. Foster, T.; Brozović, N.; Butler, A.P.; Neale, C.M.U.; Raes, D.; Steduto, P.; Fereres, E.; Hsiao, T.C. AquaCrop-OS: An Open Source Version of FAO's Crop Water Productivity Model. *Agric. Water Manag.* **2017**, *181*, 18–22. [CrossRef]

82. Semenov, M.A.; Porter, J.R. Climatic Variability and the Modelling of Crop Yields. *Agric. For. Meteorol.* **1995**, *73*, 265–283. [CrossRef]

83. Jiang, R.; Wang, T.; Shao, J.; Guo, S.; Zhu, W.; Yu, Y.; Chen, S.; Hatano, R. Modeling the Biomass of Energy Crops: Descriptions, Strengths and Prospective. *J. Integr. Agric.* **2017**, *16*, 1197–1210. [CrossRef]

84. Holzworth, D.P.; Snow, V.; Janssen, S.; Athanasiadis, I.N.; Donatelli, M.; Hoogenboom, G.; White, J.W.; Thorburn, P. Agricultural Production Systems Modelling and Software: Current Status and Future Prospects. *Environ. Model. Softw.* **2015**, *72*, 276–286. [CrossRef]

85. Lewandowski, I.; Clifton-Brown, J.; Kiesel, A.; Hastings, A.; Iqbal, Y. Miscanthus. In *Perennial Grasses for Bioenergy and Bioproducts*; Alexopoulou, E., Ed.; Elsevier: Amsterdam, The Netherlands, 2018; pp. 35–59. ISBN 978-0-12-812900-5.

86. Clifton-Brown, J.; Lewandowski, I. Water Use Efficiency and Biomass Partitioning of Three Different Miscanthus Genotypes with Limited and Unlimited Water Supply. *Ann. Bot.* **2000**, *86*, 191–200. [CrossRef]

87. Mickelbart, M.V.; Hasegawa, P.M.; Bailey-Serres, J. Genetic Mechanisms of Abiotic Stress Tolerance That Translate to Crop Yield Stability. *Nat. Rev. Genet.* **2015**, *16*, 237–251. [CrossRef]

88. Bauen, A.W.; Dunnett, A.J.; Richter, G.M.; Dailey, A.G.; Aylott, M.; Casella, E.; Taylor, G. Modelling Supply and Demand of Bioenergy from Short Rotation Coppice and Miscanthus in the UK. *Bioresour. Technol.* **2010**, *101*, 8132–8143. [CrossRef]

89. Richter, G.M.; Riche, A.B.; Dailey, A.G.; Gezan, S.A.; Powlson, D.S. Is UK Biofuel Supply from Miscanthus Water-Limited? *Soil Use Manag.* **2008**, *24*, 235–245. [CrossRef]

90. Aylott, M.J.; Casella, E.; Tubby, I.; Street, N.R.; Smith, P.; Taylor, G. Yield and Spatial Supply of Bioenergy Poplar and Willow Short-Rotation Coppice in the UK. *New Phytol.* **2008**, *178*, 358–370. [CrossRef] [PubMed]

91. Schorling, M.; Enders, C.; Voigt, C.A. Assessing the Cultivation Potential of the Energy Crop *Miscanthus × Giganteus* for Germany. *GCB Bioenergy* **2015**, *7*, 763–773. [CrossRef]

92. Blanco-Canqui, H. Growing Dedicated Energy Crops on Marginal Lands and Ecosystem Services. *Soil Sci. Soc. Am. J.* **2016**, *80*, 845–858. [CrossRef]

93. McIsaac, G.F.; David, M.B.; Mitchell, C.A. Miscanthus and Switchgrass Production in Central Illinois: Impacts on Hydrology and Inorganic Nitrogen Leaching. *J. Environ. Qual.* **2010**, *39*, 1790–1799. [CrossRef] [PubMed]

94. Dees, M.; Elbersen, B.; Fitzgerald, J.; Vis, M.; Anttila, P.; Forsell, N.; Ramirez-Almeyda, J.; García Galindo, D.; Glavonjic, B.; Staritsky, I.; et al. *A Spatial Data Base on Sustainable Biomass Cost-Supply of Lignocellulosic Biomass in Europe—Methods & Data Sources*; S2BIOM Project Report 1.6; University of Freiburg: Breisgau, Germany, 2017; 176p. [CrossRef]

95. Intergovernmental Panel on Climate Change (IPCC). Special Report on Climate Change and Land. 2019. Available online: https://www.ipcc.ch/site/assets/uploads/2019/11/SRCCL-Full-Report-Compiled-191128.pdf (accessed on 19 October 2021).

96. Haberl, H.; Erb, K.-H.; Krausmann, F.; Bondeau, A.; Lauk, C.; Müller, C.; Plutzar, C.; Steinberger, J.K. Global Bioenergy Potentials from Agricultural Land in 2050: Sensitivity to Climate Change, Diets and Yields. *Biomass Bioenergy* **2011**, *35*, 4753–4769. [CrossRef]

97. Powell, J.P.; Reinhard, S. Measuring the Effects of Extreme Weather Events on Yields. *Weather Clim. Extrem.* **2016**, *12*, 69–79. [CrossRef]

98. Wicke, B. *Bioenergy Production on Degraded and Marginal Land: Assessing Its Potentials, Economic Performance, and Environmental Impacts for Different Settings and Geographical Scales*; Utrecht University: Utrecht, The Netherlands, 2011. Available online: /paper/Bioenergy-production-on-degraded-and-marginal-land-Wicke/b6b235cafcdb8139039fa28ff3c0be17369df4b5 (accessed on 19 October 2021).

99. Metzger, J.O.; Hüttermann, A. Sustainable Global Energy Supply Based on Lignocellulosic Biomass from Afforestation of Degraded Areas. *Naturwissenschaften* **2009**, *96*, 279–288. [CrossRef]

100. Skevas, T.; Swinton, S.M.; Hayden, N.J. What Type of Landowner Would Supply Marginal Land for Energy Crops? *Biomass Bioenergy* **2014**, *67*, 252–259. [CrossRef]

101. Intergovernmental Panel on Climate Change. In *Global Warming of 1.5 °C*; World Meteorological Organization: Geneva, Switzerland, 2019. Available online: https://www.ipcc.ch/sr15/ (accessed on 19 October 2021).

102. Scagline-Mellor, S.; Griggs, T.; Skousen, J.; Wolfrum, E.; Holásková, I. Switchgrass and Giant Miscanthus Biomass and Theoretical Ethanol Production from Reclaimed Mine Lands. *Bioenerg. Res.* **2018**, *11*, 562–573. [CrossRef]

103. Saha, M.; Eckelman, M.J. Geospatial Assessment of Potential Bioenergy Crop Production on Urban Marginal Land. *Appl. Energy* **2015**, *159*, 540–547. [CrossRef]

104. Smith, P.; Porter, J.R. Bioenergy in the IPCC Assessments. *GCB Bioenergy* **2018**, *10*, 428–431. [CrossRef]

105. Hoogwijk, M.; Faaij, A.; Eickhout, B.; Devries, B.; Turkenburg, W. Potential of Biomass Energy out to 2100, for Four IPCC SRES Land-Use Scenarios. *Biomass Bioenergy* **2005**, *29*, 225–257. [CrossRef]

106. Tilman, D.; Hill, J.; Lehman, C. Carbon-Negative Biofuels from Low-Input High-Diversity Grassland Biomass. *Science* **2006**, *314*, 1598–1600. [CrossRef]

107. Lynd, L.R.; Laser, M.S.; Bransby, D.; Dale, B.E.; Davison, B.; Hamilton, R.; Himmel, M.; Keller, M.; McMillan, J.D.; Sheehan, J.; et al. How Biotech Can Transform Biofuels. *Nat. Biotechnol.* **2008**, *26*, 169–172. [CrossRef]

108. BASIS – Biomass Availability and Sustainability Information System. Report on Conversion Efficiency of Biomass. 2015. Available online: http://www.basisbioenergy.eu/fileadmin/BASIS/D3.5_Report_on_conversion_efficiency_of_biomass.pdf (accessed on 19 October 2021).

109. Clariant International Ltd. Clariant and Enviral Announce First License Agreement on Sunliquid® Cellulosic Ethanol Technology. Available online: https://www.clariant.com/en/Corporate/News/2017/09/Clariant-and-Enviral-announce-first-license-agreement-on-sunliquid-cellulosic-ethanol-technology (accessed on 19 October 2021).

110. Clariant International Ltd. Clariant to Build Flagship Sunliquid® Cellulosic Ethanol Plant in Romania. Available online: https://www.clariant.com/en/Corporate/News/2017/10/Clariant-to-build-flagship-sunliquid-cellulosic-ethanol-plant-in-Romania (accessed on 19 October 2021).

111. ETIP. Bioenergy Cellulosic Ethanol. Available online: https://www.etipbioenergy.eu/value-chains/products-end-use/products/cellulosic-ethanol#ce1 (accessed on 19 October 2021).

112. Witsch, K. Biosprit: So Kann der Durchbruch in Europa Gelingen. Available online: https://www.handelsblatt.com/unternehmen/energie/erneuerbare-energien-umstrittener-biosprit-wie-der-durchbruch-auch-in-europa-gelingen-kann/25427964.html (accessed on 19 October 2021).

113. Mayer, A. Amid Biofuels Uncertainties, Iowa Cellulosic Ethanol Plant Halts Production. Available online: https://www.iowapublicradio.org/agriculture/2019-11-19/amid-biofuels-uncertainties-iowa-cellulosic-ethanol-plant-halts-production (accessed on 19 October 2021).

114. Grzincic, B.; Abengoa Affiliates Can't Recoup $70 Million in Loans to Bankrupt Biomass Plant—10th Circuit. Reuters 2020. Available online: https://www.reuters.com/article/abengoa-affiliates-cant-recoup-70-millio-idUSL1N2CO0BZ (accessed on 22 November 2021).

115. Global Industry Analysts, Inc. Cellulosic Ethanol—Global Market Trajectory & Analytics. Available online: https://www.researchandmarkets.com/reports/5029784/cellulosic-ethanol-global-market-trajectory-and (accessed on 19 October 2021).

116. Chundawat, S.P.S.; Beckham, G.T.; Himmel, M.E.; Dale, B.E. Deconstruction of Lignocellulosic Biomass to Fuels and Chemicals. *Annu. Rev. Chem. Biomol. Eng.* **2011**, *2*, 121–145. [CrossRef]

117. Hall, C.A.S.; Lambert, J.G.; Balogh, S.B. EROI of Different Fuels and the Implications for Society. *Energy Policy* **2014**, *64*, 141–152. [CrossRef]

118. Baral, A.; Bakshi, B.R.; Smith, R.L. Assessing Resource Intensity and Renewability of Cellulosic Ethanol Technologies Using Eco-LCA. *Environ. Sci. Technol.* **2012**, *46*, 2436–2444. [CrossRef] [PubMed]

119. NRDC. Ethanol: Energy Well Spent—A Survey of Studies Published Since 1990. 2006. Available online: https://www.nrdc.org/sites/default/files/ethanol.pdf (accessed on 19 October 2021).

120. Hammerschlag, R. Ethanol's Energy Return on Investment: A Survey of the Literature 1990−Present. *Environ. Sci. Technol.* **2006**, *40*, 1744–1750. [CrossRef] [PubMed]

121. Patzek, T. A Probabilistic Analysis of the Switchgrass Ethanol Cycle. *Sustainability* **2010**, *2*, 3158–3194. [CrossRef]

122. Pimentel, D.; Patzek, T.W. Ethanol Production Using Corn, Switchgrass, and Wood; Biodiesel Production Using Soybean and Sunflower. *Nat. Resour. Res.* **2005**, *14*, 65–76. [CrossRef]

123. Cleveland, C.J.; Hall, C.A.S.; Herendeen, R.A. Energy Returns on Ethanol Production. *Science* **2006**, *312*, 1746–1748. [CrossRef] [PubMed]

124. Murphy, D.J.; Hall, C.A.S.; Dale, M.; Cleveland, C. Order from Chaos: A Preliminary Protocol for Determining the EROI of Fuels. *Sustainability* **2011**, *3*, 1888–1907. [CrossRef]

125. Robertson, P.G.; Landis, D.A.; Khanna, M. The Sustainability of Cellulosic Biofuels 2012. Available online: https://lter.kbs.msu.edu/wp-content/uploads/2012/06/sustainability-of-cellulosic-biofuels.pdf (accessed on 19 October 2021).

126. van der Weijde, T.; Kiesel, A.; Iqbal, Y.; Muylle, H.; Dolstra, O.; Visser, R.G.F.; Lewandowski, I.; Trindade, L.M. Evaluation of *Miscanthus Sinensis* Biomass Quality as Feedstock for Conversion into Different Bioenergy Products. *GCB Bioenergy* **2017**, *9*, 176–190. [CrossRef]

127. Isikgor, F.H.; Becer, C.R. Lignocellulosic Biomass: A Sustainable Platform for the Production of Bio-Based Chemicals and Polymers. *Polym. Chem.* **2015**, *6*, 4497–4559. [CrossRef]

128. Kralova, K.; Masarovičová, E. Plants for the Future. *Ecol. Chem. Engineering. S Chem. I Inżynieria Ekologiczna. S* **2006**, *13*, 1179–1207.

129. van der Weijde, T.; Alvim Kamei, C.L.; Torres, A.F.; Vermerris, W.; Dolstra, O.; Visser, R.G.F.; Trindade, L.M. The Potential of C4 Grasses for Cellulosic Biofuel Production. *Front. Plant Sci.* **2013**, *4*, 107. [CrossRef] [PubMed]

130. Trindade, L.M.; Dolstra, O.; van Loo, E.N.; Visser, R.G.F. *Plant Breeding and Its Role in a Biobased Economy*; Routledge: London, UK, 2010; ISBN 978-1-84407-770-0.

131. da Costa, R.M.F.; Pattathil, S.; Avci, U.; Winters, A.; Hahn, M.G.; Bosch, M. Desirable Plant Cell Wall Traits for Higher-Quality Miscanthus Lignocellulosic Biomass. *Biotechnol. Biofuels* **2019**, *12*, 85. [CrossRef] [PubMed]

132. Torres, A.F.; Visser, R.G.F.; Trindade, L.M. Bioethanol from Maize Cell Walls: Genes, Molecular Tools, and Breeding Prospects. *GCB Bioenergy* **2015**, *7*, 591–607. [CrossRef]

133. Zegada-Lizarazu, W.; Parrish, D.; Berti, M.; Monti, A. Dedicated Crops for Advanced Biofuels: Consistent and Diverging Agronomic Points of View between the USA and the EU-27. *Biofuels Bioprod. Bioref.* **2013**, *7*, 715–731. [CrossRef]

134. Dauber, J.; Brown, C.; Fernando, A.L.; Finnan, J.; Krasuska, E.; Ponitka, J.; Styles, D.; Thrän, D.; Van Groenigen, K.J.; Weih, M.; et al. Bioenergy from "Surplus" Land: Environmental and Socio-Economic Implications. *BioRisk* **2012**, *7*, 5–50. [CrossRef]

135. Jones, M.B.; Finnan, J.; Hodkinson, T.R. Morphological and Physiological Traits for Higher Biomass Production in Perennial Rhizomatous Grasses Grown on Marginal Land. *GCB Bioenergy* **2015**, *7*, 375–385. [CrossRef]

136. Zegada-Lizarazu, W.; Elbersen, H.W.; Cosentino, S.L.; Zatta, A.; Alexopoulou, E.; Monti, A. Agronomic Aspects of Future Energy Crops in Europe. *Biofuels Bioprod. Bioref.* **2010**, *4*, 674–691. [CrossRef]

137. Zhu, X.-G.; Chang, T.-G.; Song, Q.-F.; Finnan, J.; Barth, S.; Mårtensson, L.-M.; Jones, M.B. A Systems Approach Guiding Future Biomass Crop Development on Marginal Land. In Proceedings of the Perennial Biomass Crops for a Resource-Constrained World; Barth, S., Murphy-Bokern, D., Kalinina, O., Taylor, G., Jones, M., Eds.; Springer International Publishing: Cham, Switzerland, 2016; pp. 209–224.

138. Allwright, M.R.; Taylor, G. Molecular Breeding for Improved Second Generation Bioenergy Crops. *Trends Plant Sci.* **2016**, *21*, 43–54. [CrossRef]

139. van der Weijde, T.; Torres, A.F.; Dolstra, O.; Dechesne, A.; Visser, R.G.F.; Trindade, L.M. Impact of Different Lignin Fractions on Saccharification Efficiency in Diverse Species of the Bioenergy Crop Miscanthus. *Bioenerg. Res.* **2016**, *9*, 146–156. [CrossRef]

140. Purdy, S.J.; Maddison, A.L.; Cunniff, J.; Donnison, I.; Clifton-Brown, J. Non-Structural Carbohydrate Profiles and Ratios between Soluble Sugars and Starch Serve as Indicators of Productivity for a Bioenergy Grass. *AoB Plants* **2015**, *7*, plv032. [CrossRef] [PubMed]

141. European Comission, Directive (EU). 2018/2001 of the European Parliament and of the Council of 11 December 2018 on the Promotion of the Use of Energy from Renewable Sources (Text with EEA Relevance). *OJL* **2018**, *328*, 82–209. Available online: https://eur-lex.europa.eu/legal-content/EN/TXT/?uri=uriserv:OJ.L_.2018.328.01.0082.01.ENG (accessed on 19 October 2021).

142. European Commission. Sustainable and Optimal Use of Biomass for Energy in the EU beyond 2020—Annexes of the Final Report. 2017. Available online: https://ec.europa.eu/energy/studies/sustainable-and-optimal-use-biomass-energy-eu-beyond-2020_en (accessed on 19 October 2021).

143. European Commission (EC). Sustainability Criteria for Biofuels Specified. Available online: https://ec.europa.eu/commission/presscorner/detail/en/MEMO_19_1656 (accessed on 19 October 2021).

144. Carter, T.R.; La Rovere, E.L. Developing and Applying Scenarios. 2018. Available online: https://www.ipcc.ch/site/assets/uploads/2018/03/wg2TARchap3.pdf (accessed on 19 October 2021).

145. European Commission. Methodology Underlying the CAPRI Model. Available online: https://ec.europa.eu/clima/sites/default/files/strategies/analysis/models/docs/capri_model_methodology_en.pdf (accessed on 19 October 2021).

146. European Commission. Modelling Tools for EU Analysis. Available online: https://ec.europa.eu/clima/policies/strategies/analysis/models_en (accessed on 19 October 2021).

147. Lask, J.; Wagner, M.; Trindade, L.M.; Lewandowski, I. Life Cycle Assessment of Ethanol Production from Miscanthus: A Comparison of Production Pathways at Two European Sites. *GCB Bioenergy* **2018**, *11*, 269–288. [CrossRef]

148. Li, Q.; Hu, G. Techno-Economic Analysis of Biofuel Production Considering Logistic Configurations. *Bioresour. Technol.* **2016**, *206*, 195–203. [CrossRef]

149. Crews, T.; Cattani, D. Strategies, Advances, and Challenges in Breeding Perennial Grain Crops. *Sustainability* **2018**, *10*, 2192. [CrossRef]

150. Schuler, M.L.; Mantegazza, O.; Weber, A.P.M. Engineering C_4 Photosynthesis into C_3 Chassis in the Synthetic Biology Age. *Plant J.* **2016**, *87*, 51–65. [CrossRef]

151. Houghton, J.; Weatherwax, S.; Ferrell, J. *Breaking the Biological Barriers to Cellulosic Ethanol: A Joint Research Agenda*; EERE Publication and Product Library: Washington, DC, USA, 2006. Available online: https://www.osti.gov/biblio/1218382 (accessed on 22 November 2021).

152. Reinhardt, J.; Hilgert, P.; Von Cossel, M. A Review of Industrial Crop Yield Performances on Unfavorable Soil Types. *Agronomy* **2021**, *11*, 2382. [CrossRef]

153. White, J.W. Combining Ecophysiological Models and Genomics to Decipher the GEM-to-P Problem. *NJAS Wagening. J. Life Sci.* **2009**, *57*, 53–58. [CrossRef]

154. Bennett, A.B.; Isaacs, R. Landscape Composition Influences Pollinators and Pollination Services in Perennial Biofuel Plantings. *Agric. Ecosyst. Environ.* **2014**, *193*, 1–8. [CrossRef]

155. Carlsson, G.; Mårtensson, L.-M.; Prade, T.; Svensson, S.-E.; Jensen, E.S. Perennial Species Mixtures for Multifunctional Production of Biomass on Marginal Land. *GCB Bioenergy* **2017**, *9*, 191–201. [CrossRef]

156. Werling, B.P.; Meehan, T.D.; Robertson, B.A.; Gratton, C.; Landis, D.A. Biocontrol Potential Varies with Changes in Biofuel–Crop Plant Communities and Landscape Perenniality. *GCB Bioenergy* **2011**, *3*, 347–359. [CrossRef]

157. Liere, H.; Kim, T.N.; Werling, B.P.; Meehan, T.D.; Landis, D.A.; Gratton, C. Trophic Cascades in Agricultural Landscapes: Indirect Effects of Landscape Composition on Crop Yield. *Ecol. Appl.* **2015**, *25*, 652–661. [CrossRef]

158. Valentine, J.; Clifton-Brown, J.; Hastings, A.; Robson, P.; Allison, G.; Smith, P. Food vs. Fuel: The Use of Land for Lignocellulosic 'next Generation' Energy Crops That Minimize Competition with Primary Food Production. *GCB Bioenergy* **2012**, *4*, 1–19. [CrossRef]

Review

A Review of Industrial Crop Yield Performances on Unfavorable Soil Types

Jana Reinhardt [†], Pia Hilgert [†] and Moritz Von Cossel *

Biobased Resources in the Bioeconomy (340b), Institute of Crop Science, University of Hohenheim, Fruwirthstr. 23, 70599 Stuttgart, Germany; reinhardtja21@gmail.com (J.R.); pia.hilgert@uni-hohenheim.de (P.H.)
* Correspondence: mvcossel@gmx.de
† Authors contributed equally to this work.

Abstract: Industrial crop cultivation on marginal agricultural land limits indirect land-use change effects that pose a threat to food security. This review compiles results from 91 published crop-specific field trial datasets spanning 12 relevant industrial crops and discusses their suitability for cultivation on unfavorable soil types (USTs). It was shown that the perennial species *Miscanthus* (*Miscanthus* Andersson) and reed canary grass (*Phalaris arundinacea* L.) performed well on USTs with both high clay and/or high sand contents. Information on stoniness (particles sizes > 2 mm), where mentioned, was limited. It was found to have only a small impact on biological yield potential, though it was not possible to assess the impact on mechanization as would be used at a commercial scale. For soils with extreme clay or sand contents, half of the crops showed moderate suitability. The large yield variations within and between crops revealed large knowledge gaps in the combined effects of crop type and agronomy on USTs. Therefore, more field trials are needed on diverse USTs in different climates with better equipment and more consistent measurements to improve the accuracy of potential yield predictions spatially and temporally. Additionally, larger trials are needed to optimize cultivation and harvesting.

Keywords: camelina; cardoon; crambe; cup plant; giant reed; hemp; sorghum; switchgrass; poplar; willow

Citation: Reinhardt, J.; Hilgert, P.; Von Cossel, M. A Review of Industrial Crop Yield Performances on Unfavorable Soil Types. *Agronomy* **2021**, *11*, 2382. https://doi.org/10.3390/agronomy11122382

Academic Editor: Carlos Cantero-Martínez

Received: 8 November 2021
Accepted: 20 November 2021
Published: 24 November 2021

Publisher's Note: MDPI stays neutral with regard to jurisdictional claims in published maps and institutional affiliations.

1. Introduction

A large proportion of marginal farmlands in Europe are characterized by unfavorable soil types (USTs). However, there are no precise estimates of UST land areas in the literature, as USTs and stoniness are often mentioned in connection with other limiting site factors, such as low rooting depth (LRD) [1,2]. However, LRD and USTs account for nearly half of marginal farmland [1]. This is of great agricultural relevance because the soil type directly influences basic agronomic factors, such as water storage capacity and the plant nutrient availability of the soil. Moreover, the soil type affects the infiltration, runoff, and movement of water in the soil [3]. The grain size fractions of the soil are divided into coarse soil (with equivalent diameters of > 2 mm) and fine soil (equivalent diameter of $\leq$ 2 mm) [3,4].

Sandy and loamy-sandy soils have a low water storage capacity [5] and, therefore, reduced soil fertility [3]. For these reasons, there is a loss of yield on sandy soils for food crops. For example, at a site near Jyndevad (Denmark), where the sand content is 88%, maize achieved a 4-year average biomass yield of 10.9 Mg DM ha^{-1} [6]. This is considerably lower than yields that can be achieved on non-sandy soils of 20 Mg ha^{-1} and more [6].

The tiny particles in clay soils both bind water tightly and hinder root penetration, making it harder for plants to establish rapidly [3]. In this study, sites are classified as marginal if they contain soils with the soil type clay (UST-CL) or sand (UST-S) according to the definition of [7] (if soils consist of either $\geq$ 40% of the grain size fraction clay (silty clay, loamy clay, clay) or $\geq$ 70% of the grain size fraction sand (loamy sand)).

High proportions of coarse fragments or stones > 2 mm in the soil can also have negative effects on plant growth. A share of more than 15% coarse soil reduces the water holding capacity of soil by at least 40% and aggravates seasonal droughts in most European climate zones. In addition, a high proportion of coarse fragments, such as stones, can severely restrict the space available for root development, which in turn reduces the absorption of water and nutrients. Moreover, coarse fragments larger than 10 cm damage tillage equipment, and boulders could be a barrier to tillage in general [3]. For this reason, a site is classified as a marginally unfavorable soil type and stoniness-stones (UST-St) if its soil has a share of $\geq$ 15% of coarse soil.

The relevance of agricultural marginality constraints can be expressed by the area of agricultural land affected. The assumed total area of available European marginal cropland that is characterized solely by UST is < 1 Mha [1], but the actual area is likely much larger, as already noted above. However, this would correspond to tremendous potential for the production of non-food biomass [8], which is urgently required to successfully and promptly manage the transition to a bioeconomy [9,10]. Therefore, among many socio-economic challenges [11–15], the success of cultivating industrial crops under UST conditions is critical, with biomass yield being one of the most important components [16, 17]. Furthermore, there is still much uncertainty about the link between USTs and both biomass yield and quality of industrial crops [1,18]. Hence, Gerwin et al. [1] call for this to be investigated more thoroughly in the future, which is also currently being done in projects such as MAGIC [2]. Therefore, this study focuses on the following main research question: How do industrial crops perform in terms of biomass yield on European marginal farmland characterized by UST?

2. Material and Methods

In this study, the biomass yield of industrial crops on farmland that is not marginal (e.g., favorable climate and fertile soils) [19,20] was used as a reference for the industrial crops' yield performance on marginal farmland. This was intended to enable a first relevant insight into the future potential of biomass production on marginal farmland. The selection and evaluation of industrial crops were performed as explained in the following sections (Figure 1).

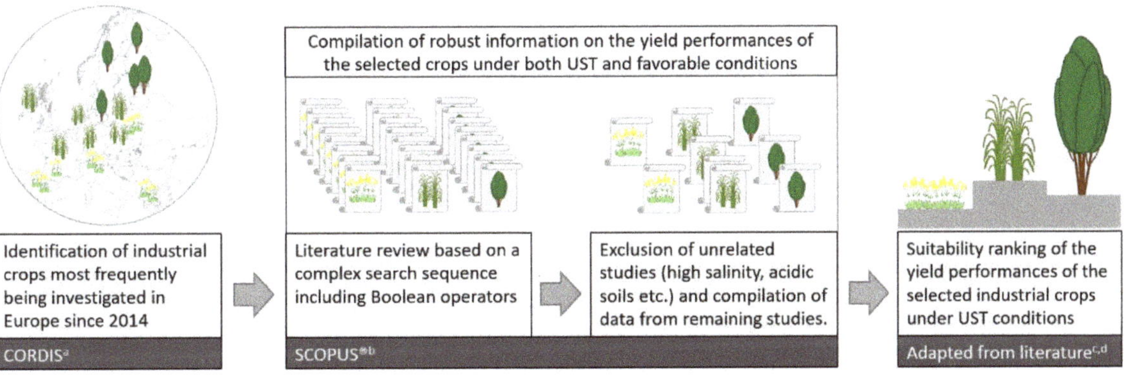

Figure 1. Schematic overview of the methodological approach of this study ([a] [21], [b] [22], [c] [8]), [d] [2]).

2.1. Identification of Most Relevant Industrial Crops in Europe

For identifying the most relevant industrial crops in Europe, only those industrial crops were selected that were involved at least four times in one of the EU projects that started or ended in the period 1 January 2014 to 6 December 2019. Using the Community Research and Development Information Service (CORDIS) [21], 24 EU projects were found and considered for this purpose. In this way, twelve industrial crops were found relevant (Table A1), which is mostly in line with the crop selections in other studies [2,23].

2.2. Literature Search on the Performance of Industrial Crops under UST Conditions

An extensive literature search was conducted in the Scopus® database (Elsevier B.V., Amsterdam, The Netherlands) to obtain robust information on the yield performance of the selected industrial crops under both UST and favorable conditions. For this purpose, the advanced search function was used in Scopus® using a complex search sequence including Boolean operators [23]. After documents were identified in which the plant name occurred in the title, abstract, or keywords. For each plant species, both the English trivial name and the botanical name were used in this process. Subsequently, articles containing terms such as "model", "gis", or "gene" were identified and excluded to reduce the search results to original articles with field trial results. In the remaining search results, articles with the marginality factor UST were searched for using common synonyms (adverse soil texture, sandy soil, stoniness, etc.). The studies found were then further selected to determine the location of the trial, any agronomic information, and harvest time and type (whole crop biomass or just seed). Soil properties and weather conditions had to be sufficiently specified to exclude, for example, sites with high salinity, heavy metal contamination, steep slope, or acidic soils. This is because if UST occurs in combination with drought, the effect of drought on yield performance might overshadow the effect of USTs. Therefore, the yield-UST combination has been placed in a broader perspective here and is discussed in more detail when interpreting anomalies in the yield performances of the crops. Non-European sites were excluded only if their climatic conditions were not comparable to any of the European agro-ecological zones.

2.3. Suitability Ranking

Based on the yield data for each industrial crop trial and further meta-data in the literature, a classification of the cultivation suitability according to a classification of Ramirez-Almeyda et al. [8] was conducted. If no sites were found that met the criteria described above, the classification was estimated from indirectly related literature. This information is then given in brackets. The classifications are denoted as follows: 4 = very good, a much higher yield than on favored sites; 3 = good, the yields of the sites were approximately equivalent to the average yields on favored sites; 2 = average, a lower yield was shown compared to the average yields on favored sites; 1 = low, much lower yields observed compared to yields on favorable sites. If the plant is classified as 0, it means that it is considered unsuitable, i.e., it cannot grow on sites with UST [2,8].

3. Results and Discussion

According to the methodology described above, 91 crop-specific field trial datasets were found that comply with the threshold values (Table A1). However, since only EU projects were searched for relevant industrial crops in this study (in order to narrow the focus of the study somewhat), it cannot be ruled out that other industrial crops not taken into account might also be relevant for European growing conditions, but have not yet been considered in EU projects. High variations in dry matter yield performances were revealed for most of the crops and each of the three major USTs, sandy, clayey, and stony soil (Figure 2).

Some sites and the respective tests on them, which also meet the criteria of this paper, could not be identified using the methodology described. The reason for this is that in order to classify a site into marginal, certain information about the site is required. However, this information is often missing in papers. The selected sites per plant with the respective yields and information on agricultural marginality constraints can be found in Table A1. In the following Sections 3.1–3.12, the yield performances of the selected industrial crops are presented and discussed as yield per year (biomass, stalk, oil, grain, and ethanol yields). Only the aboveground biomass of the crops is considered.

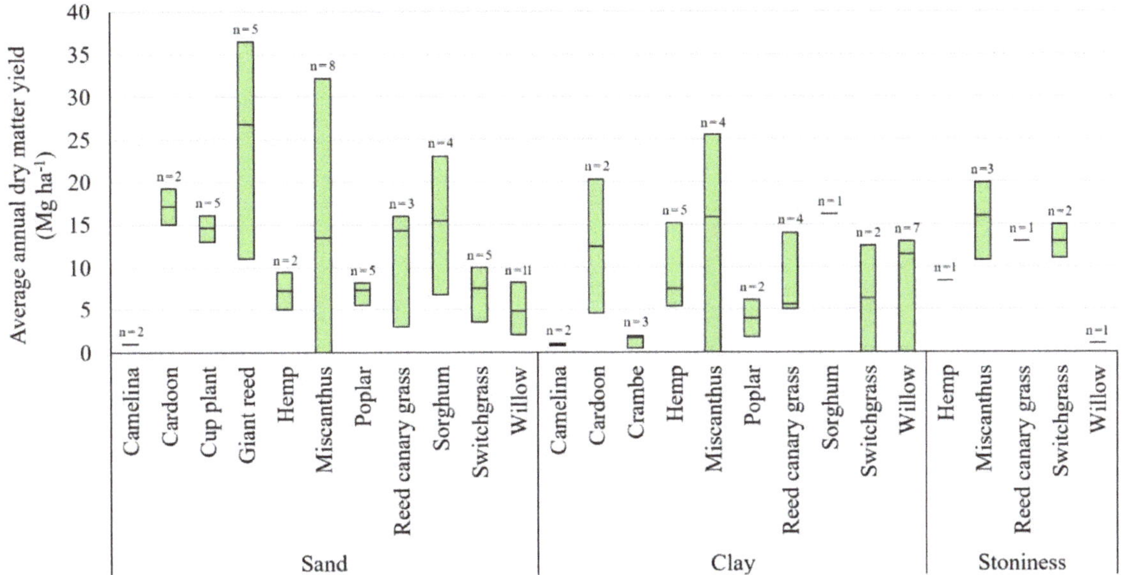

Figure 2. Overview of annual average dry matter yield performances (minimum and maximum values) of the selected industrial crops grown in field trials under unfavorable soil types sandy (Sand), clayey (Clay), and stony soil (Stoniness), finding 3–13 studies per crop (total 91 studies from 16 countries worldwide). The bars show the ranges between the minimum, median, and maximum values per culture, and "*n*" is the number of values used for the calculation. For *Camelina* and *Crambe*, grain yields are shown, for hemp, stem yields are shown, for all other crops, it's the whole aboveground dry matter.

3.1. Camelina (Camelina sativa L. Crantz)

On favorable sites, *Camelina* has an average grain yield of about 1.5 Mg ha^{-1}, usually ranging from 0.9 to 2.2 Mg ha^{-1} [24]. Under very favorable conditions, the grain yield can be up to 3 Mg ha^{-1} [25]. The average oil content of the grains is 42% [26], usually ranging from 35–45% [24]. The thousand-grain mass is between 0.8 and 1.8 g [27]. At a favorable location in Austria, an average oil yield of 807 kg ha^{-1}, a grain yield of 1.9 Mg ha^{-1}, and a thousand grain mass of 1.4 g was achieved [28]. The results of a trial in Poland on a fertile site showed a grain yield of 1.7 Mg ha^{-1} to 2.2 Mg ha^{-1}, oil content of 39.3–42.2%, and an oil yield of 690–930 kg ha^{-1} [29]. The amount of the yield is mainly influenced by the environment and, to a lesser extent, by the genotype [30].

Two locations, each meeting the criteria UST-clay (El centro and West Side) and UST-sand (Güterfelde and Wilmersdorf), were found (Tables 1 and A1).

Table 1. Grain yields per year of *Camelina* (*Camelina sativa* L. Crantz) under unfavorable soil types (USTs) conditions (S = sand, Cl = clay).

Marginality Factor	Grain Yield [Mg ha^{-1}]	Place	Country	Source
	Summer-annual *Camelina*			
UST-S	0.91	Güterfelde	Germany	[31]
UST-S	1.01	Wilmersdorf	Germany	[32]
	Winter-annual *Camelina*			
UST-Cl	0.73	El centro	USA	[33]
UST-Cl	1.02	West Side	USA	[33]

Summer *Camelina* was grown on the Güterfelde and Wilmersdorf sites and winter *Camelina* on the El centro and West Side sites. The grain yields on all four sites are lower

(Table 1) compared to the stated average grain yield of 1.5 Mg ha^{-1}; however, the yields of West Side, Güterfelde, and Wilmersdorf are in the yield range on favorable sites [24]. The oil content in Güterfelde was 40%, which corresponds to the figures in Panacea [26]. The oil yield of 364 kg ha^{-1} at this site is lower than the oil yield of 807 kg ha^{-1} found by Vollmann et al. [28]. In Güterfelde and Wilmersdorf, the soil type is loamy sand. The grain yield on the site Güterfelde comes from a trial with different fertilization variants. The stated yield was achieved without fertilization. It was found that the variant without fertilization leads to a nitrogen deficiency on this site, which is why the leaves become small, pale yellow-green, and ripen prematurely [31]. A higher fertilization (120 N kg ha^{-1}) level led to a higher grain yield of 1.8 Mg ha^{-1}. The oil content decreased significantly with the increase of the nitrogen fertilizer to 38% [31]. In the trial, years 2003 and 2004, Wilmersdorf was affected by a severe drought in spring, and in 2005, heavy rainfall occurred in May [32]. The yield was probably, therefore, lower compared to the average yield on favored sites [24]. According to the classification in Von Cossel et al. [2,34], *Camelina* is well suited for cultivation on sandy soils. This is also confirmed by the results found under sufficient nitrogen and water supply.

In El centro, the grain yield of 0.73 Mg ha^{-1} was lower than the grain yield on the West Side with 1.02 Mg ha^{-1}. This difference in grain yield can be explained by the fact that the clay content of 57.5% at El centro is higher than that of 45.4% at West Side.

Consequently, *Camelina* can be cultivated on a wide variety of soils, with the exception of heavy clay soils [27,29,35,36]. With 57.5% clay content, the El centro soil is still clay soil according to the FAO classification of soil types but close to the border with heavy clay soil, which may explain the lower yield. According to the classification in Von Cossel et al. [2,34], *Camelina* has a good cultivation tendency for clay soils and a medium cultivation tendency on heavy clay soils. However, in view of the above results, the suitability of *Camelina* on UST-CL is moderate to good.

3.2. Cardoon (Cynara cardunculus L.)

In Europe, cardoon achieves an above-ground biomass yield of 10–30 Mg ha^{-1} DM and a grain yield of 0.6–4.3 Mg ha^{-1} under different irrigation and fertilization options [37–41]. On favorable sites, cardoon achieves an average aboveground DM yield of 20 Mg ha^{-1} DM [42]. This is also shown by Fernández et al. [38], who indicate that cardoon produced a DM yield of 24.2 Mg ha^{-1}, averaged over four different favored sites in Europe. A field trial over three years in Catania, Italy, on a favored site, investigated the yield of different genotypes. A maximum yield of 36.2 Mg ha^{-1} aboveground DM biomass and 2.8 Mg ha^{-1} grain yield was achieved. On average, over all of the test years and genotypes, 20.6 Mg ha^{-1} above-ground DM biomass, 1.37 Mg ha^{-1} grain yield, and a thousand-grain mass of 29.3 g were determined [43]. In an experiment by Angelini et al. [44], cardoon was grown for eleven years on a favored site without irrigation. An average yield of 12.6 Mg ha^{-1} DM was achieved over all of the eleven growing years. The yield results under UST conditions are only from trials determined from the 1st to 3rd year of cultivation. In comparison, Angelini et al. [44] reported an average above-ground DM yield of cardoon of 19.4 Mg ha^{-1} in the 2nd to 4th year.

Three different locations were found for this agricultural marginality constraint (Tables 2 and A1).

Table 2. Grain and biomass yields per year of cardoon (*Cynara cardunculus* L.) under varying unfavorable soil types (USTs) (DR = drought, S = sand, CL = clay, SSL = steep slope, DM = dry matter).

Marginality Factor	DM Yield [Mg ha^{-1}]	Grain Yield [Mg ha^{-1}]	Place	Country	Source
DR; UST-S	15.1		Enna	Italy	[45]
UST-S	19.3		Enna	Italy	[45]
UST-CL	20.3	1.2	Catania	Italy	[46]
SSL, UST-CL	4.6		Daganzo	Spain	[47]

The Enna site was classified as marginal due to its sandy soil. The Catania site is marginal due to a clay content of 45% in the soil, and the Daganzo site is also marginal due to its clay soil. The Enna site achieved an average yield of 19.3 Mg ha^{-1} DM in the first three years of cultivation (first year 26.1, second year 22.5, and third year 9.2 Mg ha^{-1}) [38,43]. The Catania location also produced a yield comparable to a favorable location with 20.3 Mg ha^{-1} DM and 1.2 Mg ha^{-1} grain yield. Five different genotypes were grown on fields in Catania, with globe artichoke × wild cardoon in the second year with a DM yield of 27.6 Mg ha^{-1} and globe artichoke × cultivated cardoon in the third year with 27.7 Mg ha^{-1}. Only the Daganzo site had a much lower DM yield of 4.6 Mg ha^{-1}. This was probably due to the negative effects of the combination of steep slope (SSL) and UST-CL. UST-CL soils have a low water infiltration capacity. On sloping terrain, low infiltration leads to rainwater run-off, which in turn leads to reduced soil moisture and thus increases the risk of drought [5].

After a classification in Von Cossel et al. [2,34], cardoon has: medium cultivation suitability for sandy soils as well as for clay soils, very good cultivation suitability for silt soils, and is unsuitable for heavy clay soils. Țîței [48] reports that the cultivation of cardoon on clay and heavy clay soils is not recommended. In contrast, the yield on the Catania site indicates that cardoon is relatively well-suited for cultivation on clay soils. The yield of cardoon on the Enna site additionally contradicts the assessment in Von Cossel et al. [2,34] and shows a good suitability for cultivation on sandy soils.

3.3. Crambe (Crambe abyssinica Hochst. Ex R.E.Fr.)

Only two sites were found for the plant *Crambe*, and these were found for the agricultural marginality constraint UST-CL (Tables 3 and A1). The grain yield of *Crambe* can range from 0.2–3.2 Mg ha^{-1}, depending on soil, climate conditions, and crop management [49].

At favorable locations in Poland, *Crambe* achieved a grain yield of 1.2–3.2 Mg ha^{-1} [50]. In Legnaro in northern Italy, *Crambe* achieved a grain yield of 2.7 Mg ha^{-1}, an oil yield of 837 kg ha^{-1}, and a thousand-grain mass of 6.12 g [51]. In Europe, *Crambe* achieved an average grain yield of 2.4 Mg ha^{-1} and an average oil yield of 846 kg ha^{-1} [52]. The average oil content of the grains was 38% [53] and the thousand-grain mass was 6–10 g [52].

Table 3. Grain and oil yields per year of *Crambe* (*Crambe abyssinica* Hochst. Ex R.E.Fr.) on marginal agricultural lands with unfavorable soil types (UST) (CL = clay).

Marginality Factor	Grain Yield [Mg ha^{-1}]	Oil Yield [kg ha^{-1}]	Place	Country	Source
UST-CL	0.49	104	Cascavel-Paraná	Brazil	[54]
UST-CL	1.7		Cascavel-Paraná	Brazil	[55]

The site in Rosa et al. [54] was classified as marginal, as it is an Oxisol soil with a high clay content. In Viana et al. [55], the soil was described as Red Latosol with a high clay content. The clay content in Rosa et al. [54] was much lower than in Viana et al. [55] and on favored sites [52,56,57]. The oil yield was also significantly lower than the oil yield on favored sites [51,52]. The grain yield in Viana et al. [55] was within the grain yield range of favored locations in Poland [57] but was almost 1 Mg ha^{-1} lower compared to the average European grain yield of 2.4 Mg ha^{-1} [52]. In Zanetti et al. [51], a field trial in Pozzallo Italy with winter-annual *Crambe* was described. With a clay content of 38%, this site does not meet the described criteria of a marginal agricultural land, but has a high clay content. With a grain yield of 1.9 Mg ha^{-1}, it is comparable with favorable sites in Poland [50]. However, it is below the average grain yield of *Crambe* in Europe [52]. The thousand-grain weight of 5.8 g is below that of favored sites (even if only slightly), as is the oil yield of 500 kg ha^{-1} [51,52]. According to Zhu [49], *Crambe* needs fertile soils with a moderately coarse to fine texture to achieve high yields. According to Von Cossel et al. [2,34], *Crambe* is very suitable for clay and sandy soils. In contrast, the results described above show a rather mediocre suitability of *Crambe* for cultivation on UST-CL, most likely due to a combination of limitations in the field trials. Furthermore, *Crambe* can only grow poorly on stony soils [52], with the main limitation being low fertility, and higher susceptibility to drought events may further reduce the suitability of stony soils for *Crambe*.

3.4. Cup Plant (Silphium perfoliatum L.)

Under good growing conditions, cup plant DM yields from the 2nd year of cultivation onwards are 11–22 Mg ha^{-1} [58–60]. Usually, the DM yield is about 15 Mg ha^{-1} [59]. In a study by the Chamber of Agriculture of Lower Saxony, a DM yield of 15.8 Mg ha^{-1} was determined for cup plants in the 3rd year of harvest (average over three trial locations) [61].

Five locations were found for the agricultural marginality constraint UST-S (Tables 4 and A1).

Table 4. Annual biomass yields of cup plant (*Silphium perfoliatum* L.) grown under unfavorable soil type (UST) conditions (S = sand, DM = dry matter).

Marginality Constraint	DM Yield [Mg ha^{-1}]	Place	Country	Source
UST-S	14.6	Werlte	Germany	[62]
UST-S	13.0	Danube-Gravel plain	Germany	[60]
UST-S	13.0	Altitude Bavarian Forest	Germany	[60]
UST-S	14.6	Lipnik	Poland	[63]
UST-S	16.1	Brunswick	Germany	[64]

The determined yields were between 13 and 16.1 Mg DM ha^{-1} for the 2nd year of cultivation. This corresponds to the yields on favorable sites [59]. At the Lipnik site, a high yield of 19.8 Mg DM ha^{-1} was achieved in the 3rd year of cultivation. The review of Gansberger et al. [59] concludes that the soil type has no clear influence on the yield. The cup plant is tolerant to different soil types and has, therefore, a good cultivation suitability for sites with UST-S. However, maximum yields can only be achieved on nutrient-rich soils with good water availability [58–60].

3.5. Giant Reed (Arundo donax L.)

It was reported that 50% of all DM yield data of giant reed are in the range of 25–40 Mg ha^{-1} [65] (Figure 2). Under favorable growing conditions, giant reed shows very high yields of 33.8 Mg ha^{-1} [66] to 37.7 Mg ha^{-1} [67]. This is in line with Pilu et al. [68], who report DM yields of 36–55 Mg ha^{-1}. Four marginal agricultural lands with the factor UST-S were found for giant reed (Tables 5 and A1).

Table 5. Annual biomass yields of giant reed (*Arundo donax* L.) on marginal agricultural lands prone to unfavorable soil types (USTs) (DR = drought, S = sand, DM = dry matter).

Marginality Constraint	Year of Cultivation	DM Yield [Mg ha^{-1}]	Place	Country	Source
UST-S; DR	2.-3.	26.8	Enna	Italy	[45]
UST-S	2.–3.	36.5	Enna	Italy	[45]
UST-S	2.	29.1	Citra	USA	[69]
UST-S	2.–3.	17.0	San Piero a Grado	Italy	[70]
UST-S	1.–8.	11.0	Casale Monferrato	Italy	[71]

Compared to the other sites with this factor, the DM yield was lowest in Casale Monterrato with 11 Mg ha^{-1}. This site was the only one not fertilized. Furthermore, the sand content of this soil is very high at 90% [71]. This is probably why the yield was lower.

At the San Piero a Grade site, the sand content is 77.8%, and the conditions in summer 2011 were extremely dry. Giant reed was not irrigated in this trial, unlike the other sites with this factor. The high sand content of the soil reduces its ability to retain water, which has an effect on plant development, especially in dry summers when precipitation is low [70]. Probably for this reason, the DM yield of giant reed was lower with 17 Mg ha^{-1} compared to the Citra and Enna sites and compared to favorable sites [67,68]. At 29.1 Mg ha^{-1} DM and 26.8 Mg ha^{-1} DM, respectively, the yield at the Citra and Enna sites was only slightly below that of favored sites [66–68]. At the Enna site, the DM yield with sufficient irrigation was 36.5 Mg ha^{-1}, which is very similar to the yield of Angelini et al. [67] with 37.7 Mg ha^{-1} on a clay soil, i.e., a favored site. With higher fertilization and sufficient irrigation, a significantly higher DM yield of 42 Mg ha^{-1} can be achieved on the Enna site compared to the experimental variant with low fertilization and little irrigation [45].

Giant reed can grow in all soil types, from clay to sandy, and also on soils with a high stone content [72]. However, the best conditions are sandy and peaty soils with a sufficiently high groundwater level [73]. According to the division made by von Cossel et al. [2,34], giant reed is well suited for clay soils, medium for sand, and very well-suited to clay soils. Parenti et al. [74] classify the suitability for cultivation as follows: very good for a sandy or loamy sandy soil, medium for a soil with more than 50% clay content, and good for a stony soil. The results confirm that giant reed is well suited for cultivation on sandy soils, provided that sufficient precipitation falls or that irrigation is applied.

3.6. Hemp (Cannabis sativa L.)

On favorable sites, fiber hemp achieves an average stalk yield of 10 Mg ha^{-1} DM [75]. Struik et al. [76] investigated five different varieties of hemp grown for fiber on three favored sites in Europe (Italy, The United Kingdom, and The Netherlands) over three years with different fertilization and plant density variants. A maximum above-ground biomass yield of 22.5 Mg ha^{-1} DM and a stalk yield of 18.5 Mg ha^{-1} DM were achieved. The average aboveground DM biomass yield over all of the locations, years, varieties, and trial variants was 14 Mg ha^{-1} with a stalk DM yield of 11 Mg ha^{-1}. Furthermore, on a sandy loamy soil in Foulum, Denmark, an above-ground biomass yield of 13–15 Mg ha^{-1} DM was achieved [6]. For industrial processing of the fibers, the optimum stem length was 200–250 cm [75]. Six different locations were found for UST (Tables 6 and A1).

Table 6. Annual stalk and whole above-ground biomass yields of hemp (*Cannabis sativa* L.) on various types of unfavorable soil types (USTs) (S = sand, ST = stoniness, CL = clay, DM = dry matter).

Marginality Factor	DM Stem Yield [Mg ha^{-1}]	DM Yield [Mg ha^{-1}]	Plant Height [cm]	Place	Country	Source
UST-S		5		Jyndevad	Denmark	[6]
		Monoecious				
UST-CL	7.1		182	Jokioinen (Tu)	Finland	[77]
UST-CL	5.4		130	Jokioinen (Tt)	Finland	[78]
UST-ST	8.3			Michamps	Belgium	[79]
UST-S		9.4		Potsdam-Bornim	Germany	[80]
		Dioecious				
UST-CL	8.0		208	Jokioinen (Tu)	Finland	[77]
UST-CL	7.4		183	Jokioinen (Tt)	Finland	[78]
UST-CL	15.2	17.8	277	Randwijk	Netherlands	[81]

The Michamps site was classified as marginal, as it has very stony soil. Five different monoecious varieties were cultivated on it, which had an average yield of 8.3 Mg ha^{-1} DM and a grain yield of 0.26 Mg ha^{-1} over one year. The variety Epsilon 68 achieved the highest yield on this site with 10.5 Mg ha^{-1} DM [79]. The stem yield was somewhat lower compared to yields on favored sites, although for the variety Epsilon 68, only very little [76]. The suitability of growing hemp on stony soil is therefore good.

For the unfavorable soil type sand, the sites Potsdam-Bormin (loamy sand) and Jyndevad (sandy soil) were found. The soil of Jyndevad has a very high sand content (88%) and achieved an above-ground biomass yield of 5 Mg ha^{-1} DM. This is very bad in comparison to the favored site Foulum, which was also investigated by Manevski et al. [6] and achieved an above-ground biomass yield of 13–15 Mg ha^{-1} DM. Jyndevad was described in Manevski et al. [6] and classified as unsuitable for hemp cultivation, as only a low yield was achieved. The monoecious varieties Fedora 19 and Fedrina 74 were grown at the Potsdam-Bornim site. Under the fertilization variant 0 N kg ha^{-1}, an average above-ground biomass yield of 9.4 Mg ha^{-1} DM was achieved [80]. This is also lower than on a favored site [6,76]. With a fertilization of 150 N kg ha^{-1}, the yield increased to 11.8 Mg ha^{-1} DM. Overall, the yields on sites with UST-S show a medium suitability for cultivation, but hemp is only slightly suitable for cultivation on soils with very high sand content.

The Randwijk and Jokioninen sites were classified as marginal agricultural lands due to the heavy clay soil, and the Jokioninen site due to its silty clay soil. In addition to this agricultural marginality constraint, Jokioinen and Jokioinen met the criteria of location factor low temperature (LT) [2,34]. With 15.2 Mg ha^{-1} DM stalk yield and 17.8 Mg ha^{-1} DM above-ground biomass yield, the Randwijk site yield was approximately twice as high as Jokioinen and Jokioinen and also higher than the average yield on favorable sites [76]. Three different dioecious varieties were grown on the Randwijk site and various monoecious and dioecious varieties on the Jokioinen and Jokioinen sites. These two locations probably yielded lower yields compared to Randwijk, as the location factors LT and UST-CL have a negative synergy [5]. Hemp thrives best on medium-heavy soils, especially on silty loam, clay loam, and sandy loam [82]. Furthermore, Bócsa et al. [75] list the following soils as suitable for hemp cultivation: loamy sand, highly loamy sand, sandy loam, loam, and heavy loam. This is somewhat in line with Von Cossel et al. [2,34], especially in the early growth phases, hemp reacts sensitively to poor soil structures [76]. Even though Von Cossel et al. [2,34] classify heavy clay soils as having low suitability for cultivation, Randwijk's yield shows that soil with heavy clay can also achieve high yields. Hemp is therefore well suited for cultivation in UST-CL locations.

3.7. Miscanthus (Miscanthus Andersson)

Under favorable conditions, 50% of all *Miscanthus* (*Miscanthus* Andersson) DM yields are in the range of 13–28 Mg ha^{-1} DM (Figure 2) [65]. With irrigation, yields of over 30 Mg ha^{-1} DM can be achieved on sites in Southern Europe (average temperatures of 15.4 °C) [83]. In Central and Northern Europe (from Austria to Denmark), where global irradiation and average temperatures are lower (7.3–8.0 °C), yields without irrigation are typically 10–25 Mg ha^{-1} DM [83]. In a long-term test of Angelini et al. [67] without irrigation on a favored site near Pisa, the average yield over the 2nd–12th year of cultivation was 28.7 Mg ha^{-1} DM. As with Amaducci et al. [71], the yield of the first year of cultivation of this experiment was much lower than in the following years. From the first to second year, their yield increased; in the 3rd–8th year, the highest yields were achieved; and from the 9th–12th year, the yield decreased [67]. The complete establishment of a *Miscanthus* population took three to five years [83,84].

For *Miscanthus*, 14 locations were found where field trials were conducted under UST conditions (Tables 7 and A1). Of these, seven locations meet the criteria of a UST-S, three locations meet the criteria of a UST-ST, and five locations meet the criteria of a UST-CL.

Table 7. Annual biomass yields of Miscanthus (*Miscanthus* Andersson) on various unfavorable soil types (USTs) (DR = drought, S = sand, CL = clay, ST = stoniness, DM = dry matter, LRD = low rooting depth).

Marginal Factor	Year of Cultivation	DM Yield [Mg ha^{-1}]	Place	Country	Source
DR; UST-S	2.–3.	17.8	Enna	Italy	[45]
UST-S	2.–3.	24.4	Enna	Italy	[45]
UST-S	1.–8.	12.5	Casale Monferrato	Italy	[71]
UST-S	4.–19.	9.7	Foulum	Denmark	[85]
UST-S	2.–3.	0.0		Sweden	[86]
UST-S	2.–3.	32.1		Portugal	[86]
UST-S; LRD	4.	8.9	Durmersheim	Germany	[87]
UST-S	2.	14.4	Gutenzell	Germany	[87]
UST-ST;	1.	10.8	Near Leak Kill	USA	[88]
UST-ST; LRD	4.	16	Aberystwyth	Great Britain	[89]
UST-CL; UST-ST	1.–2.	19.9	Hohenheim "Unterer Lindenhof"	Germany	[90]
UST-CL	6.–8.	7.4	Gütersleben	Germany	[91]
UST-CL	3.	25.5		Germany	[86]
UST-CL	3.	24.3	Stuttgart, Ihingen court	Germany	[87]
UST-CL	2.–3	0.02	Rainton Bridge	Great Britain	[92]

The Hohenheim site fulfills the criteria of UST-CL and UST-ST; it has a high clay and stone content. Coarse fragments lead to high wear of tillage equipment, and clay also contributes to poorer tillage. On the other hand, the poor soil quality of clay can be positively influenced by a high stone content, since better aeration and better water infiltration is possible. In addition, the high stone content can accelerate the warming of the soil in spring, depending on the local conditions [5]. With a DM yield of 19.9 Mg ha^{-1}, the Hohenheim site is in the upper half of the yield range of *Miscanthus* on favored sites in northern and central Europe [83]. It should be noted that this yield was determined in the first two years of cultivation. Therefore, it can be assumed that in the following years, higher or lower yields can be achieved at this location. This suggests that *Miscanthus* is well suited for cultivation on sites that meet the two factors UST-CL and UST-ST. The Near Leck Kill site has a soil with 31.7% rock content. The Aberystwyth site also has a high stone content of 35% and has a depth of only 50 cm [89]. In addition, the Aberystwyth site meets the criteria of LRD (Reinhardt, Hilgert, and Von Cossel, 2021), while the Near Leck Kill site also has SSL. Despite this, *Miscanthus* achieved a yield of 10.8 Mg ha^{-1} in Near Leck Kill and 16 Mg ha^{-1} DM in Aberystwyth. However, this is a scientific yield from a small quadrat in a favorable year.

Two locations in Germany, examined by Clifton-Brown et al. [86] and Lewandowski et al. [87] (Table 7), achieved very similar yields of 25.5 and 24.3 Mg ha^{-1} DM (both in the 3rd year of cultivation). Both locations are characterized by soils with UST-CL. These are in accordance with the upper-end yield range for Central Europe mentioned by Lewandowski et al. [83]. At 7.4 Mg ha^{-1} DM, the yield at Gütersleben (also UST-CL) was significantly below that of the two locations mentioned above. This was probably due to the lower annual precipitation of 603 mm. At the German location of Clifton-Brown et al. [86] (Table 7), the annual PR was 687 mm and 779 mm at the Stuttgart site. At the Rainton Bridge site, an extremely low yield of 0.02 Mg ha^{-1} DM was achieved due to the presence of not only the UST-CL factor, but also low soil drainage (LSD) conditions. This was due to poor establishment caused by the UST at this site. With developments in agronomy (e.g., a stone burier), it may well be possible to establish crops on such a site.

The yields determined on the UST-S sites vary widely, ranging from 0 to 32.1 Mg ha^{-1} DM. The Swedish site in Clifton-Brown et al. [86] did not produce any yields, as *Miscanthus* did not survive the first winter after planting. The reasons for this were probably (i) the low soil temperature which fell below the plant's required minimum, and (ii) that the in vitro plants may not grow sufficient rhizomes for the plant to overwinter well. On the site in Portugal, which also meets the factor UST-S, high average yields of 32.1 Mg ha^{-1} DM were achieved in the second and third years of cultivation. This site was irrigated. In the 3rd year of cultivation, the yield was 36.4 Mg ha^{-1} DM, which is comparable to the data of Lewandowski et al. [83] for a site in Southern Europe with irrigation, and higher compared to an unirrigated site in Southern Europe [67]. Compared to the location in Portugal, the yield in Denmark was much lower, with 9.7 Mg ha^{-1} DM. With 657 mm annual rainfall and no irrigation, the water supply of *Miscanthus* in Denmark was lower compared to Portugal, and in addition, the annual average temperature was only 8 °C in Denmark, compared to 15.4 °C in Portugal. Furthermore, the sand content in Portugal was lower with 76.2% compared to Denmark with 84.4%. The Casale Monferrato site (in Italy) achieved a lower yield than Portugal with 12.5 Mg ha^{-1} DM due to rainfed conditions. This could also be due to a high sand content of 90% in Casale Monferrato. Another site with sandy soil is Gutenzell. On this site, *Miscanthus* achieved a yield of 14.4 Mg ha^{-1} DM, which is lower than in Portugal, but within the yield range of sites in Central Europe [83]. The Enna site, along with Portugal, achieved the highest yield of all UST-S sites found, with 24.4 Mg ha^{-1} DM, under an irrigation of 75% ET (Table A1).

Following Von Cossel et al. [2,34], *Miscanthus* showed low suitability for cultivation on sandy soils (depending on precipitation and irrigation conditions), good suitability for cultivation on clay soils, and medium suitability for cultivation on heavy clay soils. Ramirez-Almeyda et al. [8] indicate that *Miscanthus* is moderately suitable for cultivation on sandy and clay soils (provided they make it through the first few months after planting) and poorly suitable for cultivation on heavy clay soils. The most important requirement that *Miscanthus* places on the soil is its water holding capacity [93]. Although it is easier to establish on light soils, it produces higher yields on heavy soils over several years [83]. Even though Panacea [94] indicates that *Miscanthus* can only grow poorly on stony soils, the published plot trials show this is not always the case. These sites each meet the criteria for another marginal factor and still achieved mediocre yields. This suggests that *Miscanthus* is well suited for cultivation on sites with only UST-ST. In addition, the sites found indicate that this plant is very suitable for cultivation on UST-CL sites in Central Europe with a good water supply. For UST-S sites in Southern Europe, *Miscanthus* shows good suitability for cultivation, provided sufficient irrigation is provided. A very high sand content (>75%) in the soil results in lower and more variable annual yields than a site with a reduced sand fraction.

3.8. Poplar (Populus L.)

The yield results from the systematic literature search with the corresponding indication of the respective agricultural marginality constraints are listed in Table 8. In order to better assess the suitability of poplar cultivation on marginal agricultural lands, the yield figures determined are compared with those on favored sites. In a systematic literature search and meta-analysis of Laurent et al. [65] on poplar yields, it was found that 50% of all yield data are in the range of 7–10 Mg ha^{-1} DM (Figure 2). Berendonk et al. [61] indicate that a yield of 8–12 Mg ha^{-1} DM can be expected for short rotation plantations. On a favorable location in Mira in Northern Italy, a yield of 20 Mg ha^{-1} DM in the second rotation and 15 Mg ha^{-1} DM in the first rotation was achieved [95]. Seven sites with UST were found for poplar, five of which are UST-S and two of which are UST-CL (Tables 8 and A1).

Table 8. Annual biomass yields of poplar (*Populus* L.) under unfavorable soil type (USTs) conditions (DR = drought, S = sand, CL = clay, LSD = low soil drainage, DM = dry matter).

Marginal Factor	Age of Tree (Trunk)	DM Yield [Mg ha^{-1}]	Place	Country	Source
DR; UST-CL	2	1.8	Rutigliano	Italy	[96]
UST-CL	2	6.1	Bigarello	Italy	[95]
UST-S	2	7.2	Casale Monferrato	Italy	[71]
UST-S	3	7.3	Soria	Spain	[97]
UST-S	2	7.7	Potsdam-Bornim	Germany	[80]
UST-S	4	5.5	Olsztyn	Poland	[98,99]
LSD; UST-S	2	8.1	Lochristi	Belgium	[100,101]

The Lochchristi site also shows low soil drainage (LSD), while the Rutigliano site meets the criteria for DR. The Casale Monferrato site has a very high sand content of over 90%. Here, poplar achieved a yield of 7.2 Mg ha^{-1} DM. At the Soria site, a very similar yield of 7.3 Mg ha^{-1} DM was obtained over three rotations (9 Mg ha^{-1} DM in the first and second rotation and 4 Mg ha^{-1} DM in the third) [97]. At the Potsdam-Bornim site, a yield of 7.7 Mg ha^{-1} DM was achieved. The yield of the genotype Japan 105 was over 9 Mg ha^{-1} DM [80]. The Olsztyn site achieved the lowest yield of 5.5 Mg ha^{-1} DM of all UST-S sites investigated with poplar. It also has a very high sand content of 90% of the soil. On this site, however, a yield of 10.6 Mg ha^{-1} DM could be achieved with lignin and mineral fertilization [98,99]. The Lochchristi site achieved the highest yield of all UST-S sites at 8.1 Mg ha^{-1} DM. The yields of four of the five sites found are slightly below the yield in Berendonk et al. [61]. An exception is Lochchristi; the yield achieved on this corresponds to the yield on favored sites [61].

Eight different genotypes were grown on the Bigarello site. These yielded an average of 6.1 Mg ha^{-1} DM over two rotations, which was slightly below the yield of Berendonk et al. [61] for poplar. The highest yield in the second rotation was achieved by the genotype 83,148,041 (*P. x canadensis*) with 11.1 Mg ha^{-1}. The Rutigliano site achieved a yield of 1.8 Mg ha^{-1} DM, which is much lower than the yield stated in Berendonk et al. [61]. Optimum conditions for poplar are soils with a maximum clay content of 30% [95].

According to Von Cossel et al. [2,34], poplar has a low suitability for cultivation in sandy soils, very good to medium suitability for clay soils and, low suitability for heavy clay soils. Ramirez-Almeyda et al. [8] also classify the suitability for cultivation in this way. Only heavy clay soils are classified as unsuitable for poplar cultivation. For soils with UST-CL, this information can be confirmed with the yields of the identified sites. They show a medium suitability for cultivation. The yields of the locations found for UST-S do not match the classifications of Ramirez-Almeyda et al. [8] and Von Cossel et al. [34]. It can be assumed that poplar also has medium suitability for cultivation in soils with UST-S, provided it has a high water table or high precipitation.

3.9. Reed Canary Grass (Phalaris arundinacea L.)

Reed canary grass achieves biomass yields of 5–13 Mg ha^{-1} DM under favorable conditions [93,102,103]. For example, reed canary grass achieved an average yield of 8 Mg ha^{-1} DM (yields of 3.9–13.8 Mg ha^{-1} DM were measured) on four different favored sites in the Czech Republic from the 2nd–6th year of cultivation [104]. Eight locations were found that meet the criteria for UST (Tables 9 and A1).

Table 9. Annual biomass yields of reed canary grass (*Phalaris arundinacea* L.) under unfavorable soil types (USTs) conditions (S = sand, CL = clay, ST = stoniness, DM = dry matter, LRD = low rooting depth).

Marginal Factor	Year of Cultivation	DM Yield [Mg ha^{-1}]	Place	Country	Source
UST-S	3.–7.	3	Willsboro (lS)	USA	[105]
UST-CL	3.–7.	5	Willsboro (SCL)	USA	[105]
UST-S	2.–4.	14.3	Jyndevad	Denmark	[6]
UST-S	2.–4.	16	Foulum	Denmark	[6]
UST-CL	2.–3.	14	Carlow (Farrell's Field)	Ireland	[106]
UST-S; UST-ST; LRD	2.–3.	13	Carlow (Far Avenue Meadow)	Ireland	[106]
UST-CL	2.–3.	5.5	Rainton Bridge	Great Britain	[92]
UST-CL	1.–2.	5.6	Burlington	USA	[107]

Of these, three are UST-S, four are UST-CL, and one is UST-ST. The Carlow site (Far Avenue Meadow) also meets the criteria for the LRD factor. The Willsboro (SCL), Rainton Bridge, and Burlington sites also have low soil drainage. Willsboro (lS) also has the factor UST-S. Furthermore, Willsboro (lS) and Willsboro (SCL) fulfill the criteria of factor LT. The soil at the Jyndevad site has a sand content of 88% and a yield of 14.3 Mg ha^{-1}. The Foulum site has a slightly lower sand content of 78% and achieved a higher yield of 16 Mg ha^{-1}. Both yields are higher than the yield on favored sites [93]. At the Willsboro (lS) site, a low yield of 3 Mg ha^{-1} DM was achieved. With NPK fertilization, however, the yield could be increased to 9 Mg ha^{-1}. The Carlow site (Far Avenue Meadow) achieved a yield of 13 Mg ha^{-1} DM, which corresponds to the yield on favored sites [93].

The soil at the Carlow site (Farrels field) has a clay content of over 45% and yielded 14 Mg ha^{-1} DM [106]. This yield is higher than the yield on favored sites [17]. On Willsboro (SCL), a yield of 5 Mg ha^{-1} DM could be achieved without fertilization and 11 Mg ha^{-1} with N-fertilization [105]. Furthermore, in Burlington, a higher fertilization level with 253 kg N ha^{-1} resulted in a higher crop yield of 10.5 Mg ha^{-1} DM [107].

The assessments of the suitability for the cultivation of reed canary grass in Ramirez-Almeyda et al. [8] and in Von Cossel et al. [2,34] agree. They state that this crop's suitability for cultivation on sand is low, on loam very good, on clay moderate, and on heavy clay low. The results of this work partly contradict these statements. They rather indicate good suitability for cultivation on soils with high sand content and also on soils with high clay content. That the soil type influences the yield and that soil with less than 15% clay has higher yields than clay soil [108] could not be confirmed with the yield results of the locations found. A greenhouse trial in Freiberg showed that reed canary grass could also grow in pots with 100% sand and be fertilized with an NPK solution [109,110]. Reed canary grass can grow in a variety of soils [104] and shows good suitability for cultivation on sites with UST-S, UST-CL, and UST-ST, as long as enough water and nutrients are available.

3.10. Sorghum (Sorghum bicolor L. Moench)

Under favorable conditions, *Sorghum* reaches above-ground DM yields of 15–20 Mg DM ha^{-1} [111]. For example, on a favored site in Cadriano, Italy, fiber and sweet *Sorghum* yielded an average of 20.8 Mg DM ha^{-1} [112]. Three sites were found, which meet the criteria of a UST-S and one site, which meets the criteria of a UST-CL (Tables 10 and A1).

Table 10. Annual stalk and biomass yields of *Sorghum* (*Sorghum bicolor* L. Moench) on marginal sites characterized by unfavorable soil types (USTs) (LT = low temperature; DR = drought, S = sand, DM = dry matter, tE = theoretical ethanol).

Marginality Constraints	DM Yield [Mg ha^{-1}]	Stem DM Yield [Mg ha^{-1}]	tE Yield [L ha^{-1}]	Place	Country	Source
LT; DR; UST-S	6.7	4.5	2135	Wushen	China	[113]
DR; UST-S	13.9	11.8	2491	Ganquika	China	[114]
DR; UST-S	17.0		2150	Big Spring	USA	[115]
UST-S	23.0		3000	Big Spring	USA	[115]
UST-S	16.2		4212	El Centro	USA	[116]

Ganquika, like the Wushen site, also meets the criteria of a DR site. The Big Spring site achieved a yield of 23 Mg ha^{-1} DM in the trial version with irrigation. This yield is comparable to the yield on favored sites [111] and is slightly higher than the yield achieved in Cadriano, Italy (favored location) [112]. The theoretical ethanol yield at Big Spring was 3000 L ha^{-1}. This is comparable to a favorable location on a silty clayey loamy soil with 3209 L ha^{-1} [117]. Compared to the favored site in Weslaco, USA, with a clayey loamy soil, on which a theoretical ethanol yield of 2601 L ha^{-1} was achieved [118], the amount of ethanol produced at Big Spring is higher. The biomass yield at Ganquika was lower compared to Big Spring and also compared to favored sites [111]. The theoretical ethanol yield (2491 L ha^{-1}) was also lower in Ganquika compared to Big Spring. Presumably, the yield on Ganquika is lower because the location factors DR and UST-S show a negative synergy [5]. For this reason, and because the Wushen site also has the factor LT, the yield at this site is much lower than at the favored sites [111]. The ethanol content of Wushen, at 2135 L ha^{-1}, is also below that of Big Spring. At the El Centro site, *Sorghum* was grown in one year over three growing seasons, with a yield of 16.2 Mg ha^{-1} DM in the second growing season [116]. This yield is more or less equal to that obtained in favored locations [111,112].

As shown, *Sorghum* can grow in a wide variety of soils, including clay and sandy soils [119]. Clay soils are ideal for the cultivation of *Sorghum* [119–121]. According to the division elaborated by Von Cossel et al. [2,34], *Sorghum* is unsuitable for cultivation on sandy soils and is very suitable for cultivation on clay and heavy clay soils. The yield results at the sites with UST-S contradict this classification, whereas it is confirmed by the results at the sites with UST-CL. *Sorghum* has good suitability for cultivation on sites with UST-S and very good suitability for cultivation on sites with UST-CL.

3.11. Switchgrass (Panicum virgatum L.)

For switchgrass grown under favorable conditions, 50% of all DM yields are in the range of 7–12 Mg ha^{-1} (Figure 2) [65]. In Europe, switchgrass DM yields range from 5–23 Mg ha^{-1} [93]. According to Parrish et al. [122], switchgrass produces DM yields of about 15 Mg ha^{-1} on favored sites over a longer period of time. Panacea [123] reported switchgrass DM yields of 10–25 Mg ha^{-1} on favorable sites. Nine locations were found for the agricultural marginality constraint UST (Tables 11 and A1).

Table 11. Annual biomass yields of switchgrass (*Panicum virgatum* L.) on different types of marginal agricultural land characterized inter alia by unfavorable soil types (USTs) (DR = drought, S = sand, CL = clay, ST = stoniness, DM = dry matter).

Marginal Factor	Year of Cultivation	DM Yield [Mg ha^{-1}]	Ecotype	Place	Country	Source
UST-CL	3.–5.	0	Upland	Rainton Bridge	Great Britain	[92]
UST-CL	3.–7.	12.5	Upland	Willsboro (SCL)	USA	[105]
UST-S	3.–7.	10	Upland	Willsboro (IS)	USA	[105]
UST-S; DR	5., 6., 8., 9.	9.8	Upland	Ganqika	China	[114,124]
UST-S	1.–3.	4.5	Upland	Becker	USA	[125]
UST-S	1.–3.	3.5	Lowland	Becker	USA	[125]
UST-S	1.–8.	7.5	Lowland	Casale Monferrato	Italy	[71]
UST-ST	2.	15	Lowland	Mason	USA	[126]
UST-ST	1.	11		Near Leak Kill	USA	[88]

Of these, two fulfill the marginality thresholds of UST-CL, five of UST-S, and two of UST-ST. The Near Leak Kill site has a stone content of 31.7% in the soil and additionally an SSL. Switchgrass achieved a yield of 11 Mg ha^{-1} DM in the first year of cultivation. This corresponds to the yields on favored sites [123]. Switchgrass grown on sites with only UST-ST could achieve even higher yields in further growing years than in the first growing year on the Near Leck Kill site. On the Mason site, the lowland variety Nebraska 29 yielded 15 Mg ha^{-1} DM. This corresponds to the yield on favored sites [122]. The described results confirm the statement of Parenti et al. [74] that switchgrass has a good cultivation suitability for sites with UST-ST.

On the Willsboro (SCL) with UST-CL and Willsboro (IS) with UST-S sites, the Cave-in-Rock (Upland) variety was cultivated. The yield on Willsoboro (IS) was 10 Mg ha^{-1} DM, which is slightly below the yield of 12.5 Mg ha^{-1} DM on Willsboro (SCL). Higher yields were produced on both sites with NPK fertilization: Willsboro (IS) 12.5 Mg ha^{-1} DM and Willsoboro (SCL) 14.5 Mg ha^{-1} DM [105]. The Ganquika site has UST-S and DR. At this site, the Blackwell upland variety achieved a yield of 9.8 Mg ha^{-1} DM. This is slightly lower than the yields on favored sites [123]. At the Becker site, the Shawnee and Sunburst upland varieties and the Liberty lowland variety were cultivated. In the first three years of cultivation, Liberty achieved a mediocre yield of 3.5 Mg ha^{-1} DM without fertilization, while Shawnee and Sunburst yielded 4.5 Mg ha^{-1} DM. With a fertilization of 112 kg ha^{-1} N, the yield of Liberty at 7 Mg ha^{-1} and of Sunburst and Shawnee at 11 Mg ha^{-1} DM was significantly higher than without fertilization. The marginal agricultural land Becker also has low soil moisture. Lowland varieties like Liberty are sensitive to water stress, whereas upland varieties are better adapted to dry conditions [125]. The soil of the Casale Monferrato site has a 90% sand content. The Lowland variety Alamo produced a yield of 7.5 Mg ha^{-1} DM on this site. This yield is slightly lower than the yield on the other site [123].

According to Parenti et al. [74], switchgrass is very suitable for sandy soils and only slightly suitable for clay soils. Von Cossel et al. [34] and Ramirez-Almeyda et al. [8] indicate that switchgrass has medium cultivation suitability for sandy soils, very good suitability for clay soils, and medium suitability for clay soils. Switchgrass is declared unsuitable for cultivation on heavy clay soils Ramirez-Almeyda et al. [8]. In general, the ecotype Upland is suitable for sandy soils, and the Lowland ecotype is only moderately suitable for clay soils. Upland varieties are moderately suitable for cultivation on clay soils.

3.12. Willow (Salix L.)

For willow, 50% of all DM yield data are in the range of 8–13 Mg ha^{-1} [65]. According to [61], a DM yield of 8–12 Mg ha^{-1} is expected for short rotation plantations. Under optimal conditions, willow can achieve DM yields of 20–30 Mg ha^{-1} DM [42]. On a favored site in Poland, willow produced an average yield of 13.7 Mg ha^{-1} DM and a maximum yield of 16 Mg ha^{-1} DM [127].

For willow grown under UST conditions, 19 locations were found (Tables 12 and A1).

Table 12. Annual biomass yields of willow (*Salix* L.) on marginal sites prone to unfavorable soil types (USTs) (S = sand, CL = clay, ST = stoniness, DM = dry matter).

Marginality Factor	Age of Tree (Trunk)	DM Yield [Mg ha^{-1}]	Place	Country	Source
UST-S	3	6.7	St. Lawrence (sL)	Canada	[128]
UST-CL	3	3.1	Saskatoon	Canada	[129,130]
UST-CL	3	12.8	St. Lawrence (SCL)	Canada	[128]
UST-CL	3	13.0	Alma	Canada	[131]
UST-CL	2	2.5	La Morandiere	Canada	[131]
UST-CL	3	11.5	Beloeil	Canada	[131]
UST-CL	3	11.5	La Pocatiere	Canada	[131]
UST-S	2	4.8	Casale Monferrato	Italy	[71]
UST-S	2	6.0	Potsdam-Bornim	Germany	[80]
UST-S	4	5.1	Olsztyn	Poland	[98,99]
UST-S	3	2.9	Lezany	Poland	[127]
UST-S	3	4.0	Hojmark	Denmark	[132]
UST-S	2	4.0	Odum	Denmark	[132]
UST-S	3	8.2	Foulum	Denmark	[132]
UST-S	3	5.3	Jyndevad	Denmark	[132]
UST-S	2	2.0	Saint-Roch-de-l Achigan, Sandy field	Canada	[133]
UST-ST	2	1.0	Saint-Roch-de-l Achigan, Rocky field	Canada	[133]
UST-CL	2	0.0	Rainton Bridge	Great Britain	[92]
UST-S	3	4.2	Foersom	Denmark	[132]

Of these, eleven have a UST-S, seven a UST-CL, and one a UST-ST. The yields on sites with UST-S are in the range of 2 to 8.2 Mg DM ha^{-1}. With the exception of the Foulum site with a yield of 8.2 Mg ha^{-1} DM, all yields of the sites with UST-S are lower than those on favorable sites [61]. At the Saint Roch (Sandy) site, the unfertilized trial variant achieved 2 Mg DM ha^{-1}, while the other variant with a fertilization of 75 kg N ha^{-1} produced a significantly higher yield of 3 Mg DM ha^{-1} [133]. This site has the third-highest sand content of all sites found, at 89%. The Foersom site has the highest sand content (95%) and achieved a DM yield of 4.2 Mg ha^{-1}. The results of all sites with UST-S show that a high sand content leads to yield losses in comparison to favored sites. It can therefore be assumed that willow is suitable for medium cultivation at sites with UST-S.

The yields on sites with UST-CL range from 0 to 13 Mg DM ha^{-1}. The Rainton Bridge site has LSD in addition to UST-CL. The Tora (SW910007) and Torhild (SW930725) genotypes of *Salix schwerinii x Salix viminalis* cultivated on it grew very poorly. For this reason, a very low DM yield of 0.04 Mg ha^{-1} was achieved [92]. The Beloeil and La Pocatiere sites both have a UST-CL only and achieved a DM yield of 11.5 Mg ha^{-1} each, which is comparable to the yield in Berendonk et al. [61] and corresponds to the yield mentioned above. At the Alma site, willow achieved a yield of 13 Mg DM ha^{-1} despite UST-CL and additional LSD. These results show that willow is well suited for cultivation on sites with UST-CL.

The soil of the Saint Roch (Rocky) site contains 30% stones (1–5 cm in size). *Salix miyabeana* Seeman SX64 and SX61 were grown on it, and a DM yield of 1 Mg ha^{-1} (without fertilization) was achieved. With a fertilization of 75 kg N ha^{-1}, a significantly higher yield of 1.5 Mg DM ha^{-1} was achieved on this site [133]. Both yield values are far below those on favored sites [61]. On Saint Roch (Rocky), willow produced little biomass because the soil was too difficult for the roots of this plant to penetrate due to its high stone content [133]. Sites with a UST-ST are only slightly suitable for the cultivation of willow.

According to the division into the categories established by Von Cossel et al. [2,34], willow is suitable for growing on sandy soils, very well suited on clay soils, moderately suited on clay soils, and poorly suited on heavy clay soils. The categorization reported by Ramirez-Almeyda et al. [8] is in accordance with Von Cossel et al. [2,34], except for heavy clay soil, which is considered unsuitable for willow cultivation.

4. Suitability Ranking of Industrial Crops for Cultivation on USTs

Based on the results from the literature review, some contrasting objective cultivation suitability values were derived for the selected industrial crops on unfavorable soil types sandy, clayey, and stony soil (Table 13).

Table 13. Classification of cultivation suitability per crop and soil type: 4 = very good suitability for cultivation (dark green), 3 = good suitability for cultivation (light green), 2 = moderate suitability for cultivation (orange), 1 = poor suitability for cultivation (red). Suitability values in brackets were derived from other literature because no references were found that provide the exact combination of the crop and soil type (UST = unfavorable soil type, S = sand, CL = clay, ST = stones, Lo = lowland, Up = upland). Where no alternative literature was found either, the field was left blank as no conclusions about suitability could be made.

Crop		UST		
		S	CL	ST
Camelina (*Camelina sativa* L. Crantz)		3	2–3	
Cardoon (*Cynara cardunculus* L.)		3	3	
Crambe (*Crambe abyssinica* Hochst. Ex R.E.Fr.)		(3)	2	(1)
Cup plant (*Silphium perfoliatum* L.)		3		
Giant reed (*Arundo donax* L.)		3	2	(3)
Hemp (*Cannabis sativa* L.)		2	3	3
Miscanthus (*Miscanthus* Andersson)		3	4	3
Poplar (*Populus* L.)		2	2	
Reed canary grass (*Phalaris arundinacea* L.)		3	3	3
Sorghum (*Sorghum bicolor* L. Moench)		3	3	
Switchgrass (*Panicum virgatum* L.)	Lo	2	(2)	3
	Up	3	2	3
Willow (*Salix* L.)		2	3	1

Industrial crops whose suitability for cultivation on sites with a certain marginality is rated as good or very good may be recommended for cultivation. Good and very good suitability for cultivation means that yields comparable to favorable locations can be achieved on marginal sites. The expected yields are listed below.

The following plants are well suited for cultivation on sites with a marginality due to UST-S: *Camelina*, cardoon, giant reed, *Miscanthus*, upland switchgrass, reed canary grass, cup plant, and *Sorghum* showed well suitability for sites characterized by sandy soil (UST-S) (Table 13). The highest yields can be expected with the cultivation of giant reed, cardoon, and *Sorghum* under UST-S conditions. Other cultivation parameters, such as the temperature and the sunshine duration, are also important. Considering these, the best recommendation would be to grow cardoon and giant reed in southern Europe, *Sorghum* in southern and central Europe, reed canary grass in central and northern Europe, whereas *Miscanthus*, camelina, and switchgrass could grow on UST-S sites throughout Europe [134]. UST-S conditions, however, can place great demands on agricultural practices, depending on the location, as already mentioned to some extent in the previous sections. For example, great importance must be placed on site-specific fertilization and water supply, as sandy soils tend to be low in nutrients and have low field capacity. Depending on other site conditions, such as steep slopes or close proximity to groundwater protection areas, the requirement of fertilization or irrigation may preclude the cultivation of the industrial crops mentioned above, even if they are suitable in principle.

On marginal agricultural land characterized by UST-CL (clayey soil), the following industrial crops show good suitability for cultivation: hemp, cardoon, *Sorghum*, reed canary grass, and willow (Table 13). For the industrial crop *Miscanthus*, a very good suitability for cultivation was revealed. But again, other growth conditions than the soil type are, of course, important as well. Thus, high yields of *Miscanthus* on sites with UST-CL can only be expected under warm climatic conditions with a long growing season and sufficient water supply. Hemp is suitable for cultivation throughout Europe, but care must be taken to select only varieties approved in the EU [135] in order to meet not only the growing conditions but also the respective legal requirements. Cardoon is suitable for cultivation in southern Europe, *Sorghum* in southern and central Europe, reed canary grass and willow in central and northern Europe, and *Miscanthus* throughout Europe [134].

For UST-ST, based on the results of the literature review, the selection of suitable industrial crops is much lower compared to UST-S and UST-CL (Table 13). Only the industrial crops hemp, *Miscanthus*, switchgrass, and reed canary grass are considered suitable, with the lowest DM yield level expected for hemp (Figure 2).

5. Outlook

The sparse data on UST-ST (Table 13, Figure 2) is a good example of how much information is still lacking on the adaptive capabilities and resulting performance spectra of the known industrial crops on marginal agricultural land. In order to avoid data patchworks, such as those uncovered in this review, in the future, research institutions should be better networked and coordinated with each other. This applies to research networks that serve food security, for example, the European Consortium for Open Field Experimentation (ECOFE), and those that serve biomass security. Biomass security here refers to the long-term availability of biomass types other than those used for food and feed supply. Even though biomass security does not have the same priority as food security, a sustainable transition to a bioeconomy may only be possible if long-term planning research networks that deal with industrial crops are created. A carefully thought-out and well-coordinated network of field trial stations such as ECOFE [136] may be necessary to provide a clearer picture of where which industrial crops are performing well, especially considering the currently accelerating dynamics of climate change [137,138] and the growing conflicts of interest over the use of available land [139]. Therefore, it should be thoroughly investigated whether such networks should become part of a permanent state-funded infrastructure in order to run as successfully as possible. It is true that there are already a large number of long-term field trials with industrial crops in the EU [140,141], some of which could be passed on from one project to the next. However, this is not the rule, with the consequence that many field trials cannot be monitored continuously over several years or, in the worst case, have to be abandoned despite their great potential for use as long-term studies. In addition, the number of participating institutions that can conduct representative field trials is severely limited in such short-term projects (duration of up to five years). This is because, in addition to crop-related tasks, there are always a variety of other areas of work that must be included in the proposal (e.g., sustainability assessment, value chain development, dissemination, etc.) without which an acceptance of the proposals is not realistic. As already mentioned, it is important to cover as broadly as possible the marginality constraints to identify relevant combined effects and to develop possible agronomic solutions, including, for example, the site-specific selection of the most suitable crop or crop rotation. An important first approach could be to establish a good database of all trial sites (at EU and global level) with accurate information on the crops and genotypes used, the soil and climatic characteristics of the sites, the exact yields per year, and the exact management measures. This will allow better comparisons and informed decisions on where to grow which crops, or where there are synergies in field trials and possible networks that could be formed to start joint long-term planning.

The present study should thus be considered as only one part of the necessary basis to enable sustainable biomass production from industrial cropping systems on marginal agricultural land characterized by UST conditions in the future. The development of site-specific industrial cropping systems must always take into account the overall site conditions, including environmental conditions (especially other marginality constraints) and socio-economic requirements.

Author Contributions: Conceptualization: J.R. and M.V.C.; methodology: J.R. and M.V.C.; validation, J.R., P.H. and M.V.C.; investigation: J.R.; writing—original draft: J.R., P.H. and M.V.C.; writing—review and editing: M.V.C.; resources: J.R.; visualization: J.R. and M.V.C.; funding acquisition, M.V.C.; supervision: M.V.C. All authors have read and agreed to the published version of the manuscript.

Funding: This research received funding from the European Union's Horizon 2020 research and innovation program under grant agreement No 727698, the Federal Ministry of Education and Research (01PL16003), and the University of Hohenheim.

Institutional Review Board Statement: Not applicable.

Informed Consent Statement: Not applicable.

Data Availability Statement: Data sets for this research are included in the references.

Acknowledgments: The authors are very grateful to Iris Lewandowski for making this work possible. Special thanks to Eva Lewin for the linguistic improvement of this study.

Conflicts of Interest: The authors declare that they have no known competing financial interest or personal relationships that could have appeared to influence the work reported in this paper. The funders had no role in the design of the study; in the collection, analyses, or interpretation of data; in the writing of the manuscript, or in the decision to publish the results.

Appendix A

Table A1. Overview of growing season and meteorological conditions data from the field trial datasets used in this study.

Crop	Reference	Field Trial Start	Field Trial End	Sowing	Harvest	Length of Vegetation Period (Days)	Annual Average Temperature (°C)	Vegetation Period Average Temperature (°C)	Annual Precipitation (mm)	Precipitation within Vegetation Period (mm)	Vegetation Period	Annual Irrigation (mm)
Giant reed	[45]	2002	2007	10 April 2002	February/March		x	x	x	463.64	March–January	137
Giant reed	[45]	2002	2007	10 April 2002	February/March		x	x	x	463.64	March–January	443
Giant reed	[69]	2008	2010	November 2008	End of November		x	24	1228	741	April–November	417.5
Giant reed	[70]	2009	2011	7 May 2009	3 July 2011		x	x	626.4	315.3	April–November	0
Giant reed	[71]	2007	2014	April 2007	End of winter		13.6	18.2	818	519	x	70
Camelina	[31]	1993	1995	22 March 1993, 11 March 1994, 13 March 1995	30 July 1993, 19 July 1994, 27 July 1995		8.6	x	595	x	x	0
Camelina	[32]	2003	2005	Spring	x		x	x	x	x	x	0
Camelina	[33]	2013	2015	2 October 2013, 13 November 2014	12 March 2014, 14 April 2015		x	x	x	30	x	465
Camelina	[33]	2012	2015	28 November 2012, 28 October 2013, 28 October 2014	06 May 2013, 01 May 2014, 13 May 2015		x	x	x	105	x	133
Hemp	[6]	2013	2013	May	September	x	x	x	800.0	x	x	175
Hemp	[77]	2003	2004	23 and 17 May	October	x	x	11.7	x	454.6	May–October	0
Hemp	[78]	1995	1995	31 May	Sept	x	x	13.2	x	370.1	May–September	0
Hemp	[79]	2008	2008	23 April to 16 July	3 October to 12 November	x	x	x	x	675.3	April–November	0
Hemp	[80]	1994	1999	x	x	x	x	9.3	x	523.0	x	0
Hemp	[81]	1991	1992	20 April 1991, 24 April 1992	16 September 91, 14 September 92	x	x	x	x	x	x	0
Crambe	[54]	2011	x	August 2011	x	x	19	x	1600	x	x	0
Crambe	[55]	2012	x	April 2012	x	x	17.91	x	1800	445	April–August	0
Crambe	[51]	2006	2008	End November 2006 and 2007	May 2007 and 2008	x	x	16.4	x	397	November–May	0
Cardoon	[45]	2002	2007	30 May2002	August/September	x	x	x	x	541.45	August–July	53
Cardoon	[45]	2002	2007	30 May2002	August/September	x	x	x	x	541.45	August–July	155
Cardoon	[46]	1998	2002	25 September 1998	End of July/early August	x	x	x	x	x	x	50
Cardoon	[47]	2009	2012	December 2009	September	x	x	14.1	x	485.6	September 10–August 11	0
Miscanthus	[45]	2002	2007	10 April 2002	February/March	x	x	x	x	489.0	March–January	148

Table A1. *Cont.*

Crop	Reference	Field Trial Start	Field Trial End	Sowing	Harvest	Length of Vegetation Period (Days)	Annual Average Temperature (°C)	Vegetation Period Average Temperature (°C)	Annual Precipitation (mm)	Precipitation within Vegetation Period (mm)	Vegetation Period	Annual Irrigation (mm)
Miscanthus	[71]	2007	2014	April 2007	End of winter	x	13.6	18.2	818.0	519.0	x	70
Miscanthus	[85]	1993	2012	May 1993	x	x	8.0	x	657.0	x	x	0
Miscanthus	[86]	1997	1999	17 June	12 November, 24 November, 18 October	x	7.9	13.0	696.0	359.0	April–September	0
Miscanthus	[86]	1997	1999	21 March	27 October, 15 October, 6 October	x	15.4	19.6	665.0	156.0	April–September	396
Miscanthus	[87]	1991	1995	May 1991	February 1995	x	10.5	x	801.0	576.0	March–October	First two years
Miscanthus	[87]	1993	1995	May 1993	February 1995	x	8.3	x	988.0	687.0	March–October	0
Miscanthus	[88]	2012	2013	Spring 2012	Early November 2013	x	9.2	x	1060.0	x	x	x
Miscanthus	[89]	2012	2015	18 May 2012	February–April	x	9.7	12.6	1038.0	475.0	April–September 2012–2015, (Lewandowski et al. 2016: annual temp. and annual precipitation)	0
Miscanthus	[90]	2016	2017	x	Mid–September to mid-October 2016 and 2017	x	10.0	x	830.0	x	x	0
Miscanthus	[91]	1994	1997	x	x	x	9.5	x	603.0	x	x	0
Miscanthus	[86]	1997	1999	21 May	11 November, 23 November, 21 November	x	7.9	13.3	687.0	425.0	April–September	0
Miscanthus	[87]	1992	1995	May 1992	February 1995	x	9.7	x	779.0	599.0	March–October	In first Year
Miscanthus	[92]	2007	2010	Spring 2007, 20 cm	February–March 2009; February 2010	x	9.3	x	630.0	x	x	0
Switchgrass	[92]	2007	2010	Spring 2007, 20 cm	February–March 2009; February 2010	x	9.3	x	630	x	x	0
Switchgrass	[105]	2009	2012	2006	Early October	150 days freeze-free growing season	x	x	x	699	March–October	0

Table A1. *Cont.*

Crop	Reference	Field Trial Start	Field Trial End	Sowing	Harvest	Length of Vegetation Period (Days)	Annual Average Temperature (°C)	Vegetation Period Average Temperature (°C)	Annual Precipitation (mm)	Precipitation within Vegetation Period (mm)	Vegetation Period	Annual Irrigation (mm)
Switchgrass	[105]	2009	2012	2006	Early October	150 days freeze-free growing season	x	x	x	698	March–October	0
Switchgrass	[114]	2012	2016	2008	7 October	x	5.8	x	451.1	x	x	0
Switchgrass	[125]	2012	2015	May 2012	Post–frost	x	5.9	x	704	512	May–October	150 (in establishment year)
Switchgrass	[125]	2012	2015	May 2012	Post–frost	x	5.9	x	704	512	May–October	150 (in establishment year)
Switchgrass	[71]	2007	2014	April 2007	End of winter	x	13.6	18.2	818	519	x	70
Switchgrass	[126]	2008	2011	May 2009	September/October 2010	x	x	x	177	x	x	120
Switchgrass	[88]	2012	2013	Spring 2012	Early November 2013	x	9.2	x	1060	x	x	x
Reed canary grass	[105]	2009	2012	2006	Early July and early October	150 days freeze-free growing season	x	x	x	698	March–October	0
Reed canary grass	[105]	2009	2012	2006	Early July and early October	151 days freeze-free growing season	x	x	x	699	March–October	0
Reed canary grass	[6]	2012	2015	May 2012	2012 (August and October), 2013 (May, August, and October), 2014 (May, July, August, October), 2015 (May, October)	x	x	x	1000	x	x	20–50
Reed canary grass	[6]	2012	2015	May 2012	2012 (August and October), 2013 (May, August, and October), 2014 (May, July, August, October), 2015 (May, October)	x	x	x	800	x	x	150–200

Table A1. *Cont.*

Crop	Reference	Field Trial Start	Field Trial End	Sowing	Harvest	Length of Vegetation Period (Days)	Annual Average Temperature (°C)	Vegetation Period Average Temperature (°C)	Annual Precipitation (mm)	Precipitation within Vegetation Period (mm)	Vegetation Period	Annual Irrigation (mm)
Reed canary grass	[106]	2012	2014	24 May 2012	2013 (4–7 June, 29–31 July, 24–23 September), 2014 (27–29 June, 29–31 July, 18 September) (A cut took place at the end of the 2012 growth period)	x	x	12.2	x	370	March–September	0
Reed canary grass	[106]	2012	2014	24 May 2012	2013 (4–7 June, 29–31 July, 24–23 September), 2014 (27–29 June, 29–31 July, 18 September)	x	x	12.2	x	370	March–September	0
Reed canary grass	[92]	2007	2009	Spring 2007, 20 cm	November–December 08; Oct 09	x	9.3	x	630	x	x	0
Reed canary grass	[107]	1996	1997	x	3 cut (6–9 June, 22–23 July, 13 September/13 October	x	x	16	x	494	May–October	0
Poplar	[96]	2010	2012	March 2010, 22 cm long and 1.5–2 cm diameter	March 2012	x	x		530	x	x	In first Year 175
Poplar	[95]	2003	2007	Spring 2003	2005 and 2007	x	13.6	x	639.5	189	May–August	70
Poplar	[71]	2007	2014	April 2007	End of years 2, 4, 6	x	13.6	18.2	818	519	May–August	70
Poplar	[97]	2006	2014	March 2006, 25 cm long	End of years 2008, 2011 and 2014	x	10.4	x	494	x	x	In the dry months May to September
Poplar	[80]	1994	1999	x	Two-year rotation cycle (1995/1996), (1997/1998), (1999/2000)	x	9.3	x	523	x	x	0
Poplar	[98]	2010	2013	Early April 2010	December 2013	x	7.7	13.85	694	521	May–October	0
Poplar	[101]	2010	2014	7–10 April 2010	February 2012, 2014	x	10	14	730	430	x	0
Willow	[128]	1995	2000	Spring 1995	November 1997, November 2000	182 frost-free days	6.4	17	954	427	May–September	0

Table A1. *Cont.*

Crop	Reference	Field Trial Start	Field Trial End	Sowing	Harvest	Length of Vegetation Period (Days)	Annual Average Temperature (°C)	Vegetation Period Average Temperature (°C)	Annual Precipitation (mm)	Precipitation within Vegetation Period (mm)	Vegetation Period	Annual Irrigation (mm)
Willow	[129]	2007	2013	Spring 2007, 25 cm	Coppiced at the end of the first growing season, and first-rotation biomass was harvested 3 years later	112	2	x	375	x	x	0
Willow	[128]	1995	2000	Spring 1995	November 1997, November 2001	182 frost-free days	6.4	17	954	427	May–September	0
Willow	[131]	2012	2014	Spring 2011	2014 (cut back in fall 2011)	x	3.7	15.2	810	465	May–September	0
Willow	[131]	2012	2014	Spring 2012 (due to high mortality at the first planting in 2011 had to be planted again)	2014	x	1.5	13.7	714	423	May–September	0
Willow	[131]	2012	2014	Spring 2011	2014 (cut back in fall 2011)	x	7.3	17.9	952	468	May–September	0
Willow	[131]	2012	2014	Spring 2012	2014 (cut back in fall 2011)	x	5.4	15.7	728	383	May–September	0
Willow	[71]	2007	2014	April 2007	End of year 2, 4, 6	x	13.6	18.2	818	519	x	70
Willow	[80]	1994	1999	x	Two-year rotation cycle (1995/1996), (1997/1998), (1999/2000)	x	x	9.3	x	523	x	0
Willow	[98]	2010	2013	Early April 2010, at 25 cm length and 0.9–1.1 cm diameter	December 2013	x	7.7	13.85	694	521	May–October	0
Willow	[127]	2008	2012	20–21 April 2008	Early January 2012; end of 2008 vegetation period was harvested to increase the number of shoots	x	7.8	x	650	x	x	0
Willow	[132]	2010	2012	5 May 2010, 20 cm	13 November 2012	x	8	13.9	896	503	May–September	0
Willow	[132]	2010	2012	21 May 2010, 20 cm	19 November 2012 (cut-back after first growth year)	x	7.8	13.8	596	320	May–September	0
Willow	[132]	2010	2012	10–17 May 2010, 20 cm	11 February 2013	x	7.5	13.6	673	341	May–September	0
Willow	[132]	2010	2012	10 May 2010, 20 cm	5 February 2013	x	8.1	14.5	858	453	May–September	0

Table A1. *Cont.*

Crop	Reference	Field Trial Start	Field Trial End	Sowing	Harvest	Length of Vegetation Period (Days)	Annual Average Temperature (°C)	Vegetation Period Average Temperature (°C)	Annual Precipitation (mm)	Precipitation within Vegetation Period (mm)	Vegetation Period	Annual Irrigation (mm)
Willow	[133]	2010	2011	2010	Early December 2011	x	x	x	x	x	x	0
Willow	[133]	2010	2011	2010	Early December 2011	x	x	x	x	x	x	0
Willow	[92]	2007	2010	Spring 2007	January 2008, December 2009–February 2010	x	9.3	x	630	x	x	0
Willow	[132]	2010	2012	6 May 2010	13 November 2012	x	8	13.9	896	503	May–September	0
Cup plant	[62]	2016	2017	2016	October 2017	x	9	x	768	x	x	0
Cup plant	[60]	2014	2015	2014	Early September 2015	x	x	14.5	x	356	March–September	0
Cup plant	[60]	2014	2015	2014	Early September 2015	x	x	13.6	x	383	March–September	0
Cup plant	[63]	2016	2018	May 16	Early October 2017 and 2018	x	8.2	16	536	434	April–October	0
Cup plant	[64]	2012	2013	x	x	x	x	x	x	x	x	215
Sorghum	[113]	2013	2014	7 May 2013, 12 May 2014	September	Frost-free period 130–160	6.2	x	348.3	x	x	156
Sorghum	[114]	2012	2013	7 May	7 October	x	5.8	x	451.1	x	x	0
Sorghum	[115]	2009	2010	3 June 2009, 24 May 2010	21 August 2009, 19 August 2010	x	17.1	27	549	255	May–August	0
Sorghum	[115]	2009	2010	4 June 2009, 24 May 2010	22 August 2009, 19 August 2010	x	17.1	27	594	255	May–August	219
Sorghum	[116]	2012	2012	5 June	14 August	x	x	x	20	x	x	Yes

References

1. Gerwin, W.; Repmann, F.; Galatsidas, S.; Vlachaki, D.; Gounaris, N.; Baumgarten, W.; Volkmann, C.; Keramitzis, D.; Kiourtsis, F.; Freese, D. Assessment and Quantification of Marginal Lands for Biomass Production in Europe Using Soil-Quality Indicators. *Soil* **2018**, *4*, 267–290. [CrossRef]
2. Von Cossel, M.; Iqbal, Y.; Scordia, D.; Cosentino, S.L.; Elbersen, B.; Staritsky, I.; Van Eupen, M.; Mantel, S.; Prysiazhniuk, O.; Maliarenko, O.; et al. *Low-Input Agricultural Practices for Industrial Crops on Marginal Land*; University of Hohenheim: Stuttgart, Germany, 2018. [CrossRef]
3. Rossiter, D.; Schulte, R.; van Velthuizen, H.; Le-Bas, C.; Nachtergaele, F.; Jones, R.; van Orshoven, J. *Updated Common Bio-Physical Criteria to Define Natural Constraints for Agriculture in Europe*; Terres, J., Toth, T., Van Orshoven, J., Eds.; EUR, Scientific and Technical Research Series; Publications Office: Luxembourg, 2014; Volume 26638, ISBN 92-79-38190-3.
4. Scheffer, F.; Schachtschabel, P.; Blume, H.-P.; Brümmer, G.W.; Horn, R.; Kandeler, E.; Kögel-Knabner, I.; Kretzschmar, R.; Stahr, K.; Thiele-Bruhn, S.; et al. 16. Auflage. In *Lehrbuch der Bodenkunde*; Spektrum Akademischer Verlag: Heidelberg, Germany, 2010; ISBN 978-3-8274-1444-1.
5. Confalonieri, R.; Jones, B.; Van Diepen, K.; Van Orshoven, J. Scientific Contribution on Combining Biophysical Criteria Underpinning the Delineation of Agricultural Areas Affected by Specific Constraints. In *Methodology and Factsheets for Plausible Criteria Combinations*; Terres, J., Hagyo, A., Wania, A., Eds.; JRC Publications Office: Luxembourg, 2014. [CrossRef]
6. Manevski, K.; Lærke, P.E.; Jiao, X.; Santhome, S.; Jørgensen, U. Biomass Productivity and Radiation Utilisation of Innovative Cropping Systems for Biorefinery. *Agric. For. Meteorol.* **2017**, *233*, 250–264. [CrossRef]
7. Jahn, R.; Blume, H.P.; Asio, V.B.; Spaargaren, O.; Schad, P. *Guidelines for Soil Description*, 4th ed.; FAO: Rome, Italy, 2006; ISBN 978-92-5-105521-2.
8. Ramirez-Almeyda, J.; Elbersen, B.; Monti, A.; Staritsky, I.; Panoutsou, C.; Alexopoulou, E.; Schrijver, R.; Elbersen, W. Chapter 9-Assessing the Potentials for Nonfood Crops. In *Modeling and Optimization of Biomass Supply Chains*; Panoutsou, C., Ed.; Academic Press: Cambridge, MA, USA, 2017; pp. 219–251, ISBN 978-0-12-812303-4.
9. Dahmen, N.; Lewandowski, I.; Zibek, S.; Weidtmann, A. Integrated Lignocellulosic Value Chains in a Growing Bioeconomy: Status Quo and Perspectives. *GCB Bioenergy* **2019**, *11*, 107–117. [CrossRef]
10. Lange, L.; Connor, K.O.; Arason, S.; Bundgård-Jørgensen, U.; Canalis, A.; Carrez, D.; Gallagher, J.; Gøtke, N.; Huyghe, C.; Jarry, B.; et al. Developing a Sustainable and Circular Bio-Based Economy in EU: By Partnering Across Sectors, Upscaling and Using New Knowledge Faster, and For the Benefit of Climate, Environment & Biodiversity, and People & Business. *Front. Bioeng. Biotechnol.* **2021**, *8*, 1456. [CrossRef]
11. Panoutsou, C.; Singh, A. A Value Chain Approach to Improve Biomass Policy Formation. *GCB Bioenergy* **2020**, *12*, 464–475. [CrossRef]
12. Fernando, A.L.; Rettenmaier, N.; Soldatos, P.; Panoutsou, C. Sustainability of Perennial Crops Production for Bioenergy and Bioproducts. In *Perennial Grasses for Bioenergy and Bioproducts*; Alexopoulou, E., Ed.; Academic Press: Cambridge, MA, USA, 2018; pp. 245–283, ISBN 978-0-12-812900-5.
13. Ferreira, J.A.; Brancoli, P.; Agnihotri, S.; Bolton, K.; Taherzadeh, M.J. A Review of Integration Strategies of Lignocelluloses and Other Wastes in 1st Generation Bioethanol Processes. *Process. Biochem.* **2018**, *75*, 173–186. [CrossRef]
14. Ubando, A.T.; Felix, C.B.; Chen, W.-H. Biorefineries in Circular Bioeconomy: A Comprehensive Review. *Bioresour. Technol.* **2020**, *299*, 122585. [CrossRef]
15. Vogelpohl, T. Transnational Sustainability Certification for the Bioeconomy? Patterns and Discourse Coalitions of Resistance and Alternatives in Biomass Exporting Regions. *Energy Sustain. Soc.* **2021**, *11*, 3. [CrossRef]
16. Baum, G. Betriebswirtschaftliche Betrachtung der Wildpflanzennutzung für Biogasbetriebe 2019. Available online: https://baden-wuerttemberg.nabu.de/imperia/md/content/badenwuerttemberg/vortraege/baum_betriebswirtschaftl_wildpflanzen_f_r_biogas_ver_ffentlichung.pdf (accessed on 23 November 2021).
17. Winkler, B.; Mangold, A.; Von Cossel, M.; Clifton-Brown, J.; Pogrzeba, M.; Lewandowski, I.; Iqbal, Y.; Kiesel, A. Implementing Miscanthus into Farming Systems: A Review of Agronomic Practices, Capital and Labour Demand. *Renew. Sustain. Energy Rev.* **2020**, *132*, 110053. [CrossRef]
18. Haberzettl, J.; Hilgert, P.; Von Cossel, M. A Critical Review on Lignocellulosic Biomass Yield Modeling and the Bioenergy Potential from Marginal Land. *Agronomy* **2021**, *11*. (in press).
19. BMEL. *Agrarexporte Verstehen-Fakten Und Hintergründe*; BMEL: Berlin, Germany, 2018. Available online: https://www.bmel.de/SharedDocs/Downloads/DE/Broschueren/Agrarexporte-verstehen.pdf;jsessionid=A391E0A1B2016894D23176E8A8D57E2E.internet2852?__blob=publicationFile&v=5 (accessed on 23 November 2021).
20. UBA (Umweltbundesamt). Toward Ecofriendly Farming. Available online: https://www.umweltbundesamt.de/en/topics/soil-agriculture/toward-ecofriendly-farming (accessed on 20 October 2021).
21. European Commission CORDIS EU Research Results. Available online: https://cordis.europa.eu/projects/en (accessed on 23 November 2021).
22. Elsevier. How Can I Best Use the Advanced Search? Available online: https://service.elsevier.com/app/answers/detail/a_id/11365/c/10546/supporthub/scopus/ (accessed on 20 October 2021).

23. von Cossel, M.; Lewandowski, I.; Elbersen, B.; Staritsky, I.; Van Eupen, M.; Iqbal, Y.; Mantel, S.; Scordia, D.; Testa, G.; Cosentino, S.L.; et al. Marginal Agricultural Land Low-Input Systems for Biomass Production. *Energies* **2019**, *12*, 3123. [CrossRef]

24. Moser, B.R. Camelina (*Camelina sativa* L.) Oil as a Biofuels Feedstock: Golden Opportunity or False Hope? *Lipid Technol.* **2010**, *22*, 270–273. [CrossRef]

25. Panacea. *Scientific Papers and Training Materials-Panacea. Summary Factsheet Camelina Sativa*; Agricultural University of Athens: Athens, Greece, 2020; Available online: http://www.panacea-h2020.eu/wp-content/uploads/2019/10/SUMMARY-FACTSHEET_camelina.pdf (accessed on 20 October 2021).

26. Panacea. *Scientific Papers and Training Materials-Panacea. Summary Factsheet HEAR*; Agricultural University of Athens: Athens, Greece, 2020; Available online: http://www.panacea-h2020.eu/wp-content/uploads/2019/10/SUMMARY-FACTSHEET_HEAR.pdf (accessed on 20 October 2021).

27. Zubr, J. Oil-Seed Crop: Camelina Sativa. *Ind. Crop. Prod.* **1997**, *6*, 113–119. [CrossRef]

28. Vollmann, J.; Moritz, T.; Kargl, C.; Baumgartner, S.; Wagentristl, H. Agronomic Evaluation of Camelina Genotypes Selected for Seed Quality Characteristics. *Ind. Crop. Prod.* **2007**, *26*, 270–277. [CrossRef]

29. Krzyżaniak, M.; Stolarski, M.J.; Tworkowski, J.; Puttick, D.; Eynck, C.; Załuski, D.; Kwiatkowski, J. Yield and Seed Composition of 10 Spring Camelina Genotypes Cultivated in the Temperate Climate of Central Europe. *Ind. Crop. Prod.* **2019**, *138*, 111443. [CrossRef]

30. Zanetti, F.; Eynck, C.; Christou, M.; Krzyżaniak, M.; Righini, D.; Alexopoulou, E.; Stolarski, M.J.; Van Loo, E.N.; Puttick, D.; Monti, A. Agronomic Performance and Seed Quality Attributes of Camelina (*Camelina sativa* L. Crantz) in Multi-Environment Trials across Europe and Canada. *Ind. Crop. Prod.* **2017**, *107*, 602–608. [CrossRef]

31. Agegnehu, M.; Honermeier, B. Effects of Seeding Rates and Nitrogen Fertilization on Seed Yield, Seed Quality and Yield Components of False Flax (Camelina Sativa Crtz.). *Die Bodenkult.* **1997**, *15*, 1.

32. Paulsen, H.M. Organic mixed cropping systems with oilseeds 1. Yields of mixed cropping systems of legumes or spring wheat with false flax (*Camelina sativa* L. Crantz). *Landbauforsch. Völkenrode* **2007**, *57*, 107–117.

33. George, N.; Hollingsworth, J.; Yang, W.-R.; Kaffka, S. Canola and Camelina as New Crop Options for Cool-Season Production in California. *Crop. Sci.* **2017**, *57*, 693–712. [CrossRef]

34. Von Cossel, M.; Wagner, M.; Lask, J.; Magenau, E.; Bauerle, A.; Von Cossel, V.; Warrach-Sagi, K.; Elbersen, B.; Staritsky, I.; van Eupen, M.; et al. Prospects of Bioenergy Cropping Systems for A More Social-Ecologically Sound Bioeconomy. *Agronomy* **2019**, *9*, 605. [CrossRef]

35. Zubr, J. Qualitative Variation of Camelina Sativa Seed from Different Locations. *Ind. Crop. Prod.* **2003**, *17*, 161–169. [CrossRef]

36. Zanetti, F.; Alberghini, B.; Marjanović Jeromela, A.; Grahovac, N.; Rajković, D.; Kiprovski, B.; Monti, A. Camelina, an Ancient Oilseed Crop Actively Contributing to the Rural Renaissance in Europe. A Review. *Agron. Sustain. Dev.* **2021**, *41*, 2. [CrossRef]

37. Boari, F.; Schiattone, M.I.; Calabrese, N.; Montesano, F.F.; Cantore, V. Effect of Nitrogen Management on Wild and Domestic Genotypes of Cardoon for Agro-Energy Purpose. *Acta. Hortic.* **2016**, 15–22. [CrossRef]

38. Fernández, J.; Curt, M.D.; Aguado, P.L. Industrial Applications of *Cynara cardunculus* L. for Energy and Other Uses. *Ind. Crop. Prod.* **2006**, *24*, 222–229. [CrossRef]

39. Francaviglia, R.; Bruno, A.; Falcucci, M.; Farina, R.; Renzi, G.; Russo, D.E.; Sepe, L.; Neri, U. Yields and Quality of *Cynara cardunculus* L. Wild and Cultivated Cardoon Genotypes. A Case Study from a Marginal Land in Central Italy. *Eur. J. Agron.* **2016**, *72*, 10–19. [CrossRef]

40. Gominho, J.; Curt, M.D.; Lourenço, A.; Fernández, J.; Pereira, H. *Cynara cardunculus* L. as a Biomass and Multi-Purpose Crop: A Review of 30 Years of Research. *Biomass Bioenergy* **2018**, *109*, 257–275. [CrossRef]

41. Grammelis, P.; Malliopoulou, A.; Basinas, P.; Danalatos, N.G. Cultivation and Characterization of Cynara Cardunculus for Solid Biofuels Production in the Mediterranean Region. *Int. J. Mol. Sci.* **2008**, *9*, 1241–1258. [CrossRef]

42. Panoutsou, C.; Singh, A.; Christensen, T. *D7.1–Good Practices*; Zenodo: Genève, Switzerland, 2018. [CrossRef]

43. Foti, S.; Mauromicale, G.; Raccuia, S.A.; Fallico, B.; Fanella, F.; Maccarone, E. Possible Alternative Utilization of Cynara Spp.: I. Biomass, Grain Yield and Chemical Composition of Grain. *Ind. Crop. Prod.* **1999**, *10*, 219–228. [CrossRef]

44. Angelini, L.G.; Ceccarini, L.; Nassi o Di Nasso, N.; Bonari, E. Long-Term Evaluation of Biomass Production and Quality of Two Cardoon (*Cynara cardunculus* L.) Cultivars for Energy Use. *Biomass Bioenergy* **2009**, *33*, 810–816. [CrossRef]

45. Mantineo, M.; D'Agosta, G.M.; Copani, V.; Patanè, C.; Cosentino, S.L. Biomass Yield and Energy Balance of Three Perennial Crops for Energy Use in the Semi-Arid Mediterranean Environment. *Field Crop. Res.* **2009**, *114*, 204–213. [CrossRef]

46. Mauromicale, G.; Ierna, A. Biomass and Grain Yield in *Cynar cardunculus* L. Genotypes Grown in a Permanent Crop with Low Input. *Acta Hortic.* **2004**, 593–598. [CrossRef]

47. Curt, M.D.; Mosquera, F.; Sanz, M.; Sánchez, J.; Sánchez, G.; Esteban, B.; Fernández, J. Effect of Land Slope on Biomass Production of *Cynara cardunculus* L. In Proceedings of the 20th EU Biomass Conference and Exhibition, Milan, Italy, 18–22 June 2012; pp. 186–190.

48. Țîței, V. The Evaluation of the Biomass Quality of Cardoon, Cynara Cardunculus, and Prospects of Its Use in Moldova. *Sci. Papers. Ser. A Agron.* **2019**, *62*, 159–166.

49. Zhu, L.-H. Crambe (Crambe abyssinica). In *Industrial Oil Crops*; McKeon, T., Hayes, D., Hildebrand, D., Weselake, R., Eds.; Elsevier Inc.: Cambridge, MA, USA, 2016; pp. 195–205, ISBN 978-1-893997-98-1.

50. Krzyżaniak, M.; Stolarski, M.J. Life Cycle Assessment of Camelina and Crambe Production for Biorefinery and Energy Purposes. *J. Clean. Prod.* **2019**, *237*, 117755. [CrossRef]
51. Zanetti, F.; Scordia, D.; Vamerali, T.; Copani, V.; Dal Cortivo, C.; Mosca, G. Crambe Abyssinica a Non-Food Crop with Potential for the Mediterranean Climate: Insights on Productive Performances and Root Growth. *Ind. Crop. Prod.* **2016**, *90*, 152–160. [CrossRef]
52. Falasca, S.L.; Flores, N.; Lamas, M.C.; Carballo, S.M.; Anschau, A. Crambe Abyssinica: An Almost Unknown Crop with a Promissory Future to Produce Biodiesel in Argentina. *Int. J. Hydrog. Energy* **2010**, *35*, 5808–5812. [CrossRef]
53. Zoz, T.; Steiner, F.; Zoz, A.; Castagnara, D.D.; Witt, T.W.; Zanotto, M.D.; Auld, D.L. Effect of Row Spacing and Plant Density on Grain Yield and Yield Components of Crambe Abyssinica Hochst. *SCA* **2018**, *39*, 393. [CrossRef]
54. Rosa, H.A.; Secco, D.; Santos, R.F.; Ciotti de Marins, A.; Fornasari, C.H.; Veloso, G. Structuring potential of cover crops in a clayey oxisol and their effect on crambe grain yield and oil content. *RSA* **2018**, *19*, 160. [CrossRef]
55. Viana, O.H.; Santos, R.F.; de Oliveira, R.C.; Secco, D.; de Souza, S.N.M.; Tokura, L.K.; da Silva, T.R.B.; Gurgacz, F. Crambe ("Crambe Abyssinica" H.) Development and Productivity under Different Sowing Densities. *Aust. J. Crop. Sci.* **2015**, *9*, 690.
56. Krzyżaniak, M.; Stolarski, M.J.; Graban, Ł.; Lajszner, W.; Kuriata, T. Camelina and Crambe Oil Crops for Bioeconomy—Straw Utilisation for Energy. *Energies* **2020**, *13*, 1503. [CrossRef]
57. Stolarski, M.J.; Krzyżaniak, M.; Kwiatkowski, J.; Tworkowski, J.; Szczukowski, S. Energy and Economic Efficiency of Camelina and Crambe Biomass Production on a Large-Scale Farm in North-Eastern Poland. *Energy* **2018**, *150*, 770–780. [CrossRef]
58. Chrestensen, N.L. Anbauanleitung Für Die Durchwachsene Silphie (Silphium Perfoliatum). 2012. Available online: http://www.rekulta.org/fileadmin/downloads/Ziel3-Projektstatus/2013_07_10_Feldtag/Silphie_Anbauanleitung_2013.pdf (accessed on 20 October 2021).
59. Gansberger, M.; Montgomery, L.F.R.; Liebhard, P. Botanical Characteristics, Crop Management and Potential of *Silphium perfoliatum* L. as a Renewable Resource for Biogas Production: A Review. *Ind. Crop. Prod.* **2015**, *63*, 362–372. [CrossRef]
60. Hartmann, A.; Lunenberg, T. Yield Potential of Cup Plant under Bavarian Cultivation Conditions. *J. Fur Kult.* **2016**, *68*, 385–388. [CrossRef]
61. Berendonk, C.; Dahlhoff, A.; Dickeduisberg, M.; Dissemond, A.; Erhardt, N.; Gruber, W.; Hartmann, H.-B.; Holz, J.; Jacobs, G.; Kasten, P.; et al. *Nachwachsende Rohstoffe Vom Acker*; Landwirtschaftskammer Nordrhein-Westfalen: Münster, Germany, 2012.
62. Facciotto, G.; Bury, M.; Chiocchini, F.; Cumplido Marín, L.; Czyż, H.; Graves, A.; Kitczak, T.; Martens, R.; Morhart, C.; Paris, P. Performance of Sida Hermaphrodita and Silphium Perfoliatum in Europe: Preliminary Results. Proceedings of 26th European Biomass Conference and Exhibition Proceedings, Copenhagen, Denmark, 14 May 2018; pp. 350–352. [CrossRef]
63. Siwek, H.; Włodarczyk, M.; Możdżer, E.; Bury, M.; Kitczak, T. Chemical Composition and Biogas Formation Potential of Sida Hermaphrodita and Silphium Perfoliatum. *Appl. Sci.* **2019**, *9*, 4016. [CrossRef]
64. Schittenhelm, S.; Schoo, B.; Schroetter, S. Yield physiology of biogas crops: Comparison of cup plant, maize, and lucerne-grass. *J. Für Kult.* **2016**, *68*, 378–384.
65. Laurent, A.; Pelzer, E.; Loyce, C.; Makowski, D. Ranking Yields of Energy Crops: A Meta-Analysis Using Direct and Indirect Comparisons. *Renew. Sustain. Energy Rev.* **2015**, *46*, 41–50. [CrossRef]
66. Scordia, D.; Testa, G.; Cosentino, S.L. Perennial Grasses as Lignocellulosic Feedstock for Second-Generation Bioethanol Production in Mediterranean Environment. *Ital. J. Agron.* **2014**, *9*, 84. [CrossRef]
67. Angelini, L.G.; Ceccarini, L.; Di Nassi o Nasso, N.; Bonari, E. Comparison of *Arundo conax* L. and Miscanthus × Giganteus in a Long-Term Field Experiment in Central Italy: Analysis of Productive Characteristics and Energy Balance. *Biomass Bioenergy* **2009**, *33*, 635–643. [CrossRef]
68. Pilu, R.; Manca, A.; Landoni, M. Arundo Donax as an Energy Crop: Pros and Cons of the Utilization of This Perennial Plant. *Maydica* **2013**, *58*, 54–59.
69. Erickson, J.E.; Soikaew, A.; Sollenberger, L.E.; Bennett, J.M. Water Use and Water-Use Efficiency of Three Perennial Bioenergy Grass Crops in Florida. *Agriculture* **2012**, *2*, 325–338. [CrossRef]
70. Di Nassi o Nasso, N.; Roncucci, N.; Bonari, E. Seasonal Dynamics of Aboveground and Belowground Biomass and Nutrient Accumulation and Remobilization in Giant Reed (*Arundo donax* L.): A Three-Year Study on Marginal Land. *Bioenergy Res.* **2013**, *6*, 725–736. [CrossRef]
71. Amaducci, S.; Facciotto, G.; Bergante, S.; Perego, A.; Serra, P.; Ferrarini, A.; Chimento, C. Biomass Production and Energy Balance of Herbaceous and Woody Crops on Marginal Soils in the Po Valley. *GCB Bioenergy* **2017**, *9*, 31–45. [CrossRef]
72. Impagliazzo, A.; Mori, M.; Fiorentino, N.; Di Mola, I.; Ottaiano, L.; Gianni, D.; Nocerino, S.; Fagnano, M. Crop Growth Analysis and Yield of a Lignocellulosic Biomass Crop (Arundo Donax L.) in Three Marginal Areas of Campania Region. *Ital. J. Agron.* **2017**, *12*, 755. [CrossRef]
73. Hanzhenko, O. *Catalogue for Bioenergy Crops and Their Suitability in the Categories of MagLs*; European Union: Brussels, Belgium, 2016; Available online: http://seemla.eu/wp-content/uploads/2019/02/D2.2-Catalogue-for-bioenergy-crops-and-their-suitability-in-the-categories-of-MagLs.pdf (accessed on 20 October 2021).
74. Parenti, A.; Lambertini, C.; Monti, A. Areas with Natural Constraints to Agriculture: Possibilities and Limitations for The Cultivation of Switchgrass (*Panicum virgatum* L.) and Giant Reed (*Arundo donax* L.) in Europe. In *Land Allocation for Biomass Crops: Challenges and Opportunities with Changing Land Use*; Li, R., Monti, A., Eds.; Springer International Publishing: Cham, Switzerland, 2018; pp. 39–63, ISBN 978-3-319-74536-7.

75. Bócsa, I.; Karus, M.; Lohmeyer, D. Vollst. überarb. und erg. 2. Aufl. In *Der Hanfanbau*; Landwirtschaftsverlag: Münster-Hiltrup, Germany, 2000; ISBN 3-7843-3066-5.

76. Struik, P.C.; Amaducci, S.; Bullard, M.J.; Stutterheim, N.C.; Venturi, G.; Cromack, H.T.H. Agronomy of Fibre Hemp (*Cannabis Sativa* L.) in Europe. *Ind. Crop. Prod.* **2000**, *11*, 107–118. [CrossRef]

77. Pahkala, K.; Pahkala, E.; Syrjälä, H. Northern Limits to Fiber Hemp Production in Europe. *J. Ind. Hemp* **2008**, *13*, 104–116. [CrossRef]

78. Sankari, H.S.; Mela, T.J.N. Plant Development and Stem Yield of Non-Domestic Fibre Hemp (*Cannabis Sativa* L.) Cultivars in Long-Day Growth Conditions in Finland. *J. Agron. Crop. Sci.* **1998**, *181*, 153–159. [CrossRef]

79. Faux, A.-M.; Draye, X.; Lambert, R.; d'Andrimont, R.; Raulier, P.; Bertin, P. The Relationship of Stem and Seed Yields to Flowering Phenology and Sex Expression in Monoecious Hemp (*Cannabis Sativa* L.). *Eur. J. Agron.* **2013**, *47*, 11–22. [CrossRef]

80. Scholz, V. The Growth Productivity, and Environmental Impact of the Cultivation of Energy Crops on Sandy Soil in Germany. *Biomass Bioenergy* **2002**, *23*, 81–92. [CrossRef]

81. Van der Werf, H.M.G.; Wijlhuizen, M.; Schutter, J.A.A. Plant Density and Self-Thinning Affect Yield and Quality of Fibre Hemp (*Cannabis Sativa* L.). *Field Crop. Res.* **1995**, *40*, 153–164. [CrossRef]

82. Ranalli, P. Agronomical and Physiological Advances in Hemp Crops. In *Advances in Hemp Research*; CRC Press: Boca Raton, FL, USA, 1999; ISBN 978-0-429-07626-8.

83. Lewandowski, I.; CLIFTON-BROWN, J.C.; Scurlock, J.M.O.; Huisman, W. Miscanthus: European Experience with a Novel Energy Crop. *Biomass Bioenergy* **2000**, *19*, 209–227. [CrossRef]

84. Bufe, C.; Korevaar, H. *Evaluation of Additional Crops for Dutch List of Ecological Focus Area*; Wageningen Research Foundation (WR) Business Unit Agrosystems Research: Lelystad, The Netherlands, 2018; Volume 793.

85. Larsen, S.U.; Jørgensen, U.; Kjeldsen, J.B.; Lærke, P.E. Long-Term Miscanthus Yields Influenced by Location, Genotype, Row Distance, Fertilization and Harvest Season. *Bioenergy Res.* **2014**, *7*, 620–635. [CrossRef]

86. Clifton-Brown, J.C.; Lewandowski, I.; Andersson, B.; Basch, G.; Christian, D.G.; Kjeldsen, J.B.; Jørgensen, U.; Mortensen, J.V.; Riche, A.B.; Schwarz, K.-U. Performance of 15 Miscanthus Genotypes at Five Sites in Europe. *Agron. J.* **2001**, *93*, 1013–1019. [CrossRef]

87. Lewandowski, I.; Kicherer, A. Combustion Quality of Biomass: Practical Relevance and Experiments to Modify the Biomass Quality of Miscanthus x Giganteus. *Eur. J. Agron.* **1997**, *6*, 163–177. [CrossRef]

88. Saha, D.; Rau, B.M.; Kaye, J.P.; Montes, F.; Adler, P.R.; Kemanian, A.R. Landscape Control of Nitrous Oxide Emissions during the Transition from Conservation Reserve Program to Perennial Grasses for Bioenergy. *GCB Bioenergy* **2017**, *9*, 783–795. [CrossRef]

89. Kalinina, O.; Nunn, C.; Sanderson, R.; Hastings, A.F.S.; van der Weijde, T.; Özgüven, M.; Tarakanov, I.; Schüle, H.; Trindade, L.M.; Dolstra, O.; et al. Extending Miscanthus Cultivation with Novel Germplasm at Six Contrasting Sites. *Front. Plant Sci.* **2017**, *8*, 563. [CrossRef] [PubMed]

90. Mangold, A.; Lewandowski, I.; Möhring, J.; Clifton-Brown, J.; Krzyżak, J.; Mos, M.; Pogrzeba, M.; Kiesel, A. Harvest Date and Leaf:Stem Ratio Determine Methane Hectare Yield of Miscanthus Biomass. *GCB Bioenergy* **2019**, *11*, 21–33. [CrossRef]

91. Kahle, P.; Beuch, S.; Boelcke, B.; Leinweber, P.; Schulten, H.-R. Cropping of Miscanthus in Central Europe: Biomass Production and Influence on Nutrients and Soil Organic Matter. *Eur. J. Agron.* **2001**, *15*, 171–184. [CrossRef]

92. Lord, R.A. Reed Canarygrass (*Phalaris arundinacea*) Outperforms Miscanthus or Willow on Marginal Soils, Brownfield and Non-Agricultural Sites for Local, Sustainable Energy Crop Production. *Biomass Bioenergy* **2015**, *78*, 110–125. [CrossRef]

93. Lewandowski, I.; Scurlock, J.M.O.; Lindvall, E.; Christou, M. The Development and Current Status of Perennial Rhizomatous Grasses as Energy Crops in the US and Europe. *Biomass Bioenergy* **2003**, *25*, 335–361. [CrossRef]

94. Panacea. *Scientific Papers and Training Materials-Panacea. Summary Factsheet Miscanthus*; Agricultural University of Athens: Athens, Greece, 2020; Available online: http://www.panacea-h2020.eu/wp-content/uploads/2019/10/SUMMARY-FACTSHEET_miscanthusV2.pdf (accessed on 20 October 2021).

95. Paris, P.; Mareschi, L.; Sabatti, M.; Pisanelli, A.; Ecosse, A.; Nardin, F.; Scarascia-Mugnozza, G. Comparing Hybrid Populus Clones for SRF across Northern Italy after Two Biennial Rotations: Survival, Growth and Yield. *Biomass Bioenergy* **2011**, *35*, 1524–1532. [CrossRef]

96. Navarro, A.; Facciotto, G.; Campi, P.; Mastrorilli, M. Physiological Adaptations of Five Poplar Genotypes Grown under SRC in the Semi-Arid Mediterranean Environment. *Trees* **2014**, *28*, 983–994. [CrossRef]

97. Fernández, M.J.; Barro, R.; Pérez, J.; Losada, J.; Ciria, P. Influence of the Agricultural Management Practices on the Yield and Quality of Poplar Biomass (a 9-Year Study). *Biomass Bioenergy* **2016**, *93*, 87–96. [CrossRef]

98. Stolarski, M.J.; Krzyżaniak, M.; Szczukowski, S.; Tworkowski, J.; Załuski, D.; Bieniek, A.; Gołaszewski, J. Effect of Increased Soil Fertility on the Yield and Energy Value of Short-Rotation Woody Crops. *Bioenergy Res.* **2015**, *8*, 1136–1147. [CrossRef]

99. Stolarski, M.J.; Krzyżaniak, M.; Tworkowski, J.; Szczukowski, S.; Niksa, D. Analysis of the Energy Efficiency of Short Rotation Woody Crops Biomass as Affected by Different Methods of Soil Enrichment. *Energy* **2016**, *113*, 748–761. [CrossRef]

100. Navarro, A.; Portillo-Estrada, M.; Arriga, N.; Vanbeveren, S.P.P.; Ceulemans, R. Genotypic Variation in Transpiration of Coppiced Poplar during the Third Rotation of a Short-Rotation Bio-Energy Culture. *GCB Bioenergy* **2018**, *10*, 592–607. [CrossRef]

101. Verlinden, M.S.; Broeckx, L.S.; Ceulemans, R. First vs. Second Rotation of a Poplar Short Rotation Coppice: Above-Ground Biomass Productivity and Shoot Dynamics. *Biomass Bioenergy* **2015**, *73*, 174–185. [CrossRef]

102. Tilvikiene, V.; Kadziuliene, Z.; Dabkevicius, Z.; Venslauskas, K.; Navickas, K. Feasibility of Tall Fescue, Cocksfoot and Reed Canary Grass for Anaerobic Digestion: Analysis of Productivity and Energy Potential. *Ind. Crop. Prod.* **2016**, *84*, 87–96. [CrossRef]
103. Usťak, S.; Šinko, J.; Muňoz, J. Reed Canary Grass (*Phalaris arundinacea* L.) as a Promising Energy Crop. *J. Cent. Eur. Agric.* **2019**, *20*, 1143–1168. [CrossRef]
104. Strašil, Z. Evaluation of Reed Canary Grass (*Phalaris arundinacea* L.) Grown for Energy Use. *Res. Agr. Eng.* **2012**, *58*, 119–130. [CrossRef]
105. Cherney, J.H.; Ketterings, Q.M.; Davis, M.; Cherney, D.J.R.; Paddock, K.M. Management of Warm- and Cool-Season Grasses for Biomass on Marginal Lands: I. Yield and Soil Fertility Status. *Bioenergy Res.* **2017**, *10*, 959–968. [CrossRef]
106. Meehan, P.; Burke, B.; Doyle, D.; Barth, S.; Finnan, J. Exploring the Potential of Grass Feedstock from Marginal Land in Ireland: Does Marginal Mean Lower Yield? *Biomass Bioenergy* **2017**, *107*, 361–369. [CrossRef]
107. Carter, J.E.; Jokela, W.E.; Bosworth, S.C. Grass Forage Response to Broadcast or Surface-Banded Liquid Dairy Manure and Nitrogen Fertilizer. *Agron. J.* **2010**, *102*, 1123–1131. [CrossRef]
108. Jensen, E.F.; Casler, M.D.; Farrar, K.; Finnan, J.M.; Lord, R.; Palmborg, C.; Valentine, J.; Donnison, I.S. Reed Canary Grass: From Production to End Use. In *Perennial Grasses for Bioenergy and Bioproducts*; Alexopoulou, E., Ed.; Academic Press: Cambridge, MA, USA, 2018; ISBN 978-0-12-812900-5.
109. Reinhardt, J. Einfluss von organischen Säuren auf die Bioverfügbarkeit von Germanium im Pflanze-Boden System. Bachelor Thesis, Technical University of Freiberg, Germany, 2016.
110. Wiche, O.; Tischler, D.; Fauser, C.; Lodemann, J.; Heilmeier, H. Effects of Citric Acid and the Siderophore Desferrioxamine B (DFO-B) on the Mobility of Germanium and Rare Earth Elements in Soil and Uptake in *Phalaris arundinacea*. *Int. J. Phytoremediation* **2017**, *19*, 746–754. [CrossRef]
111. Panacea. *Scientific Papers and Training Materials-Panacea. Summary Factsheet Sorghum*; Agricultural University of Athens: Athens, Greece, 2020; Available online: http://www.panacea-h2020.eu/wp-content/uploads/2019/10/SUMMARY-FACTSHEET_sorgho.pdf (accessed on 20 October 2021).
112. Barbanti, L.; Grandi, S.; Vecchi, A.; Venturi, G. Sweet and Fibre Sorghum (*Sorghum Bicolor* (L.) Moench), Energy Crops in the Frame of Environmental Protection from Excessive Nitrogen Loads. *Eur. J. Agron.* **2006**, *25*, 30–39. [CrossRef]
113. Diallo, B.; Li, M.; Tang, C.; Ameen, A.; Zhang, W.; Xie, G.H. Biomass Yield, Chemical Composition and Theoretical Ethanol Yield for Different Genotypes of Energy Sorghum Cultivated on Marginal Land in China. *Ind. Crop. Prod.* **2019**, *137*, 221–230. [CrossRef]
114. Fu, H.M.; Meng, F.Y.; Molatudi, R.L.; Zhang, B.G. Sorghum and Switchgrass as Biofuel Feedstocks on Marginal Lands in Northern China. *Bioenergy Res.* **2016**, *9*, 633–642. [CrossRef]
115. Cotton, J.; Burow, G.; Acosta-Martinez, V.; Moore-Kucera, J. Biomass and Cellulosic Ethanol Production of Forage Sorghum Under Limited Water Conditions. *Bioenergy Res.* **2013**, *6*, 711–718. [CrossRef]
116. Oikawa, P.Y.; Jenerette, G.D.; Grantz, D.A. Offsetting High Water Demands with High Productivity: Sorghum as a Biofuel Crop in a High Irradiance Arid Ecosystem. *GCB Bioenergy* **2015**, *7*, 974–983. [CrossRef]
117. Smith, G.A.; Buxton, D.R. Temperate Zone Sweet Sorghumethanol Production Potential. *Bioresour. Technol.* **1993**, *43*, 71–75. [CrossRef]
118. Wiedenfeld, R.P. Nutrient Requirements and Use Efficiency by Sweet Sorghum. *Energy Agric.* **1984**, *3*, 49–59. [CrossRef]
119. Zegada-Lizarazu, W.; Monti, A. Are We Ready to Cultivate Sweet Sorghum as a Bioenergy Feedstock? A Review on Field Management Practices. *Biomass Bioenergy* **2012**, *40*, 1–12. [CrossRef]
120. Panoutsou, C.; Singh, A. *Training Materials for Agronomists and Students*; Imperial College: London, UK, 2019.
121. Zeller, F.J. Sorghum (*Sorghum Bicolor* L. Moench): Utilization, Genetics, Breeding. *Bodenkultur* **2000**, *51*, 71–85.
122. Parrish, D.J.; Fike, J.H. The Biology and Agronomy of Switchgrass for Biofuels. *Crit. Rev. Plant Sci.* **2005**, *24*, 423–459. [CrossRef]
123. Panacea. *Scientific Papers and Training Materials-Panacea. Summary Factsheet Switchgrass*; Agricultural University of Athens: Athens, Greece, 2020; Available online: http://www.panacea-h2020.eu/wp-content/uploads/2019/10/SUMMARY-FACTSHEET_switchgrass.pdf (accessed on 20 October 2021).
124. Ameen, A.; Tang, C.; Han, L.; Xie, G.H. Short-Term Response of Switchgrass to Nitrogen, Phosphorus, and Potassium on Semiarid Sandy Wasteland Managed for Biofuel Feedstock. *Bioenergy Res.* **2018**, *11*, 228–238. [CrossRef]
125. Sawyer, A.; Rosen, C.; Lamb, J.; Sheaffer, C. Nitrogen and Harvest Management Effects on Switchgrass and Mixed Perennial Biomass Production. *Agron. J.* **2018**, *110*, 1260–1273. [CrossRef]
126. Porensky, L.M.; Davison, J.; Leger, E.A.; Miller, W.W.; Goergen, E.M.; Espeland, E.K.; Carroll-Moore, E.M. Grasses for Biofuels: A Low Water-Use Alternative for Cold Desert Agriculture? *Biomass Bioenergy* **2014**, *66*, 133–142. [CrossRef]
127. Krzyżaniak, M.; Stolarski, M.J.; Szczukowski, S.; Tworkowski, J.; Bieniek, A.; Mleczek, M. Willow Biomass Obtained from Different Soils as a Feedstock for Energy. *Ind. Crop. Prod.* **2015**, *75*, 114–121. [CrossRef]
128. Labrecque, M.; Teodorescu, T.I. High Biomass Yield Achieved by Salix Clones in SRIC Following Two 3-Year Coppice Rotations on Abandoned Farmland in Southern Quebec, Canada. *Biomass Bioenergy* **2003**, *25*, 135–146. [CrossRef]
129. Amichev, B.Y.; Hangs, R.D.; Bélanger, N.; Volk, T.A.; Vujanovic, V.; Schoenau, J.J.; van Rees, K.C.J. First-Rotation Yields of 30 Short-Rotation Willow Cultivars in Central Saskatchewan, Canada. *Bioenergy Res.* **2015**, *8*, 292–306. [CrossRef]
130. Amichev, B.Y.; Volk, T.A.; Hangs, R.D.; Bélanger, N.; Vujanovic, V.; van Rees, K.C.J. Growth, Survival, and Yields of 30 Short-Rotation Willow Cultivars on the Canadian Prairies: 2nd Rotation Implications. *New For.* **2018**, *49*, 649–665. [CrossRef]

131. Lafleur, B.; Lalonde, O.; Labrecque, M. First-Rotation Performance of Five Short-Rotation Willow Cultivars on Different Soil Types and Along a Large Climate Gradient. *Bioenergy Res.* **2017**, *10*, 158–166. [CrossRef]
132. Larsen, S.U.; Jørgensen, U.; Lærke, P.E. Willow Yield Is Highly Dependent on Clone and Site. *Bioenergy Res.* **2014**, *7*, 1280–1292. [CrossRef]
133. Pray, T.; Guidi Nissim, W.; St-Arnaud, M.; Labrecque, M. Investigating the Effect of a Mixed Mycorrhizal Inoculum on the Productivity of Biomass Plantation Willows Grown on Marginal Farm Land. *Forests* **2018**, *9*, 185. [CrossRef]
134. Pari, L.; Scarfone, A. *Inventory on Available Harvesting Technology for Industrial Crops on Marginal Lands*; CREA: Rome, Italy, 2019. [CrossRef]
135. Scordia, D.; Cosentino, S.L. Perennial Energy Grasses: Resilient Crops in a Changing European Agriculture. *Agriculture* **2019**, *9*, 169. [CrossRef]
136. Stützel, H.; Brüggemann, N.; Inzé, D. The Future of Field Trials in Europe: Establishing a Network Beyond Boundaries. *Trends Plant Sci.* **2016**, *21*, 92–95. [CrossRef]
137. Boers, N. Observation-Based Early-Warning Signals for a Collapse of the Atlantic Meridional Overturning Circulation. *Nat. Clim. Chang.* **2021**, *11*, 680–688. [CrossRef]
138. Teuling, A.J. A Hot Future for European Droughts. *Nat. Clim. Chang.* **2018**, *8*, 364. [CrossRef]
139. Winkler, K.; Fuchs, R.; Rounsevell, M.; Herold, M. Global Land Use Changes Are Four Times Greater than Previously Estimated. *Nat. Commun.* **2021**, *12*, 2501. [CrossRef]
140. Reinhardt, J.; Hilgert, P.; von Cossel, M. Biomass Yield of Selected Herbaceous and Woody Industrial Crops across Marginal Agricultural Sites with Shallow Soil. *Agronomy* **2021**, *11*, 1296. [CrossRef]
141. Reinhardt, J.; Hilgert, P.; Von Cossel, M. Yield Performance of Dedicated Industrial Crops on Low-Temperature Characterized Marginal Agricultural Land in Europe—A Review. *Biofuels Bioprod. Biorefining.* (in press). [CrossRef]

Article

Spatiotemporal Changes in the Geographic Imbalances between Crop Production and Farmland-Water Resources in China

Dajing Li [1,2], Hongqi Zhang [1] and Erqi Xu [1,*]

[1] Key Laboratory of Land Surface Pattern and Simulation, Institute of Geographic Sciences and Natural Resources Research, Chinese Academy of Sciences, Beijing 100101, China; lidj.18b@igsnrr.ac.cn (D.L.); zhanghq@igsnrr.ac.cn (H.Z.)
[2] University of Chinese Academy of Sciences, Beijing 100049, China
* Correspondence: xueq@igsnrr.ac.cn

Abstract: Agricultural production is constrained by farmland and water resources, especially in China with limited per capita resources. Understanding of the geographic changes between national crop production and resource availability with the spatial shift of crop production has been limited in recent decades. To solve this issue, we quantified the changes in geographic relationships between crop production and farmland-water resources in China from 1990 to 2015 by a spatial imbalance measurement model. Results found a clear spatial concentration trend of crop production in China, which increased the pressure on the limited farmland and water resources in the main production areas. The geographic imbalances between the total production of crops and farmland resources ($\sum SMI\_PF$) alleviated slightly, whereas that of water resources ($\sum SMI\_PW$) increased by 9.12%. The rice production moved toward the north of the country with less water but abundant farmland resources, which led to a decrease of 1.34% in $\sum SMI\_PF$ and an increase of 14.20% in $\sum SMI\_PW$. The shift of wheat production to the south was conducive to alleviating the pressure on water resources, but the production concentration still increased the demand for farmland and water resources, resulting in an increase in $\sum SMI\_PF$ and $\sum SMI\_PW$ by 39.96% and 10.01%, respectively. Of the five crops, adjustments to the spatial distribution of corn production had the most significant effect on reducing pressure on farmland and water resources and $\sum SMI\_PF$ and $\sum SMI\_PW$ decreased by 11.23% and 1.43%, respectively. Our results provided a reference for adjustments in crop production distribution and for policy formulation to sustainably utilize farmland and water resources.

Keywords: geographic imbalance; crop production; resource pressure; water resources; China

Citation: Li, D.; Zhang, H.; Xu, E. Spatiotemporal Changes in the Geographic Imbalances between Crop Production and Farmland-Water Resources in China. *Agronomy* **2022**, *12*, 1111. https://doi.org/10.3390/agronomy12051111

Academic Editors: Moritz Von Cossel, Joaquín Castro-Montoya and Yasir Iqbal

Received: 26 March 2022
Accepted: 29 April 2022
Published: 3 May 2022

Publisher's Note: MDPI stays neutral with regard to jurisdictional claims in published maps and institutional affiliations.

1. Introduction

Crops are a product of the intensive utilization of farmland and water resources [1–3]. Nowadays, land degradation and water shortages have directly threatened food security and ecosystem health in many countries around the world [4–6]. For example, the water resources in some countries, such as Kuwait, Saudi Arabia, and the United Arab Emirates, have been insufficient to meet the needs of agricultural production [7]. In order to mitigate the threat of farmland-water resources to agricultural production, the United Nations (UN) put forward the 2030 Agenda for Sustainable Development in 2015. Specifically, goal 2.4 proposes to improve land and soil quality, goal 6 proposes to ensure availability and sustainable management of water, and goal 12 proposes to ensure sustainable production patterns [8]. In China, cereal production accounts for 21% of total world production, but China controls only 9% of the world's farmland and 6% of the world's fresh water [9–11]. The contradiction between crop production and farmland-water resources is thus very obvious [12,13]. In addition, the spatial distribution of farmland and water resources in China is unbalanced [14]. Northern China accounts for 60% of the nation's farmland but only 19% of its water resources [15,16]. The geographic imbalances between farmland and water resources thus greatly limit China's agricultural development [17]. In recent years, crop production has increased and regional imbalances have become increasingly obvious [18–20]. This has gradually intensified the pressures on farmland and water resources in the main crop-producing areas, which now directly threatens China's food security [21]. Therefore, it is critical to clarify to what extent crop production matches available farmland and water resources, as well as their spatiotemporal variations, in order to formulate food production policies and the sustainable utilization of limited resources.

Between 1960 and 2018, global cereal production increased by a factor of 2.38 due to the increased use of "green revolution" technologies and investment in farmland, water, and other natural resources [9,22]. However, most of the countries with a high demand for crop production are also countries with a shortage of farmland and water resources [23], which leads to the overexploitation of groundwater. In order to meet the demand for agricultural irrigation, some countries had to resort to alternative sources such as desalinated seawater [24]. The demand for increased crop production comes mainly from lower- and middle-income countries, whereas the per capita farmland (usually of low quality) available for crop production is less than half of that in high-income countries [23]. The geographic imbalances between food production and farmland-water resources thus pose a great challenge to global food security and the sustainable development of economies and society in general [25]. In China, with the rapid development and expansion of urbanization and industrialization, the area available for farming is continually decreasing [26,27], whereas crop production is increasing each year [19], which increases pressure on farmland, thereby endangering the stable production of crops [28]. As a consequence, the spatiotemporal changes in crop production further change the degree of geographic imbalances between crop production and available farmland [29]. China's main crop production area is located in the north of the country where water resources are scarce and cannot support crop production in some areas for extended periods of time [30,31] even requiring cross-regional water diversion to meet increasing demand [32]. For example, groundwater depletion in some areas of the Huang-Huai-Hai Plain in northern China has reached 20–100 mm/a [33]. In addition, with an increasing population and further urban expansion in the years ahead, the demand for farmland and water resources will continue to increase, and crop production will face even greater challenges. If China maintains its current level of food consumption, the increase in population will lead to increases of 6.5% and 7.1% in the demand for farmland and irrigation water by 2032, respectively [34].

Previous studies of the relationship between crop production and farmland-water resources have focused more on the efficiency of the utilization of those resources [35–37], the impact of the changes in those resources on food production [38,39], and on resource demand by crop production through the concepts of virtual water and virtual land [40–42]. In contrast, these studies cannot quantify the degree of geographic imbalance between crop production evolution and farmland-water resources. Li et al. [29] and Chai et al. [43] analyzed the geographic imbalances of grain production and farmland resources but that of water resources has not been clarified. The spatial heterogeneity of water resources is obvious in China and ignoring the matching degree between crop production and water resources will cause environmental problems. In addition, most previous studies have examined crop production as a whole [44], whereas studies of how different types of crops are matched with available farmland and water resources are still rare. Different crops have different growing habits [45–47] and their spatial distributions in the main production areas are, therefore, also quite different leading to different geographic imbalances with farmland and water resources. These differences are often ignored in the overall study of total crop production, which means that the results of those studies cannot provide a sound basis for effective decision making in relation to crop production. Therefore, it would be of great theoretical and practical significance to identify the geographic imbalances between the production of different types of crops and farmland-water resources in order to reveal the possible prospects for crop production and promote the sustainable development of agriculture.

As a large and populous country, China's crop production is very important to the world's food security and economic development. With this in mind, this study selected the Chinese mainland as the study area and carried out the following analyses: (1) the spatiotemporal changes in five kinds of crop production at the county scale between 1990 and 2015 were identified; and (2) the geographic imbalances between those five kinds of crop production and farmland-water resources were analyzed. Our study aimed to provide a reference for adjustments in food production and distribution and for the formulation of policies for protecting farmland and water resources.

2. Materials and Methods

2.1. Study Area

The study area chosen was the Chinese mainland. Rice, wheat, corn, soybean, and tubers are the main crops. Most regions of China experience rainfall in the hot season, which provides good conditions for growth. The annual total hours of sunshine increase from southeast to northwest across the country and both the frost-free period and the average temperature of ≥ 10 °C increase from north to south. The spatial distribution of precipitation is therefore very different in different regions of the country. Specifically, southern China, where the climate is humid and water is abundant, produces mainly rice with a high demand for water; northern China, where the rainy season is short and concentrated mainly in summer, produces corn, wheat, soybeans, etc. Referring to a previous study [48], combined with multisource information such as geographical location, natural resource endowment, agricultural production conditions, ecological environment safety, etc., the study area was divided into ten subregions, as shown in Figure 1.

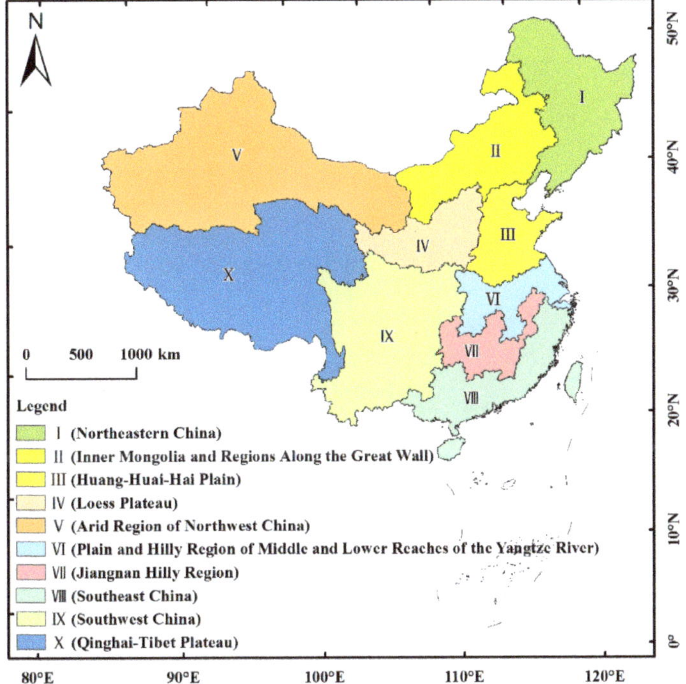

Figure 1. Location of the study area (Chinese mainland).

2.2. Data Collection and Processing

For this study, statistical data at the county scale for the period 1990–2015 were used. Data for the yields of rice, wheat, corn, soybean, tubers, and for the areas of available farmland were obtained from the Ministry of Agriculture and Rural Affairs of the People's Republic of China. Data for water resources were taken from the China Statistical Yearbook and the Water Resources Bulletin for each of China's provinces and for parts of cities. The amount of water resources at the county scale was calculated from the county-scale population and the per capita water resources of the city where it is located. The amount of water resources here refers to the total amount of surface and underground water produced by local precipitation, representing the stock of available water resources, which is used to reflect the abundance of regional water resources. To eliminate the effects of administrative division adjustments on the results, based on the county administrative divisions of China in 2015, the data for 1990 were revised. In addition, we merged some municipal districts, to give, finally, 2367 county-level units.

2.3. Research Methods

2.3.1. Concentration Ratio of Crop Production (CRCP)

CRCP is the ratio of the crop production of each county to the total crop production of the whole country. It is given by Equation (1).

$$CRCP_j = \frac{P_{ij}}{\sum_{i=1}^{n} P_{ij}} \tag{1}$$

where $CRCP_j$ is the CRCP of crop j and P_{ij} is the yield of crop j in the ith county.

2.3.2. Spatial Gravity Center Model

The spatial gravity center model originated from the field of physics and has been widely used in studies of crop production [16]. Since the CRCP of different crops experienced significant changes between 1990 and 2015, the gravity center model was used to analyze the direction and distance of the gravity center deviation of those crops. The method of calculation is as follows.

$$x_j = \sum_{i=1}^{n} \left(x_i \times P_{ij} \right) / P_{ij} \tag{2}$$

$$y_j = \sum_{i=1}^{n} \left(y_i \times P_{ij} \right) / P_{ij} \tag{3}$$

where (x_j, y_j) are the barycentric coordinates of crop j and (x_i, y_i) are the coordinates of the geological center of county i.

$$d_j = \sqrt{(x_{t2} - x_{t1})^2 + (y_{t2} - y_{t1})^2} \tag{4}$$

where d_j is the distance of the gravity center deviation of crop j, and (x_{t1}, y_{t1}) and (x_{t2}, y_{t2}) are the barycenter coordinates in 1990 and 2015, respectively.

2.3.3. Geographic Imbalances Analysis

Referring to previous studies [25], the spatial mismatch index (SMI) was used to analyze the geographic imbalances between farmland and water resources, between crop production and farmland, and between crop production and water resources. The various SMIs can be obtained from the following formulas.

$$SMI\_WF_i = \left(\frac{F_i}{\sum_{i=1}^{n} F_i} - \frac{W_i}{\sum_{i=1}^{n} W_i} \right) \times 100 \tag{5}$$

where SMI_WF_i is the geographic imbalance between the farmland and water resources of county i; F_i is the area of farmland in county i; and W_i is the total volume of water resources of county i. The larger the absolute value of SMI_WF_i, the higher the degree of geographic imbalance.

$$SMI\_PF_{ij} = \left(\frac{P_{ij}}{\sum_{i=1}^{n} P_{ij}} - \frac{F_{ij}}{\sum_{i=1}^{n} F_{ij}} \right) \times 100 \qquad (6)$$

$$SMI\_PW_{ij} = \left(\frac{P_{ij}}{\sum_{i=1}^{n} P_{ij}} - \frac{W_{ij}}{\sum_{i=1}^{n} W_{ij}} \right) \times 100 \qquad (7)$$

$$\sum SMI\_PF_j = \sum_{i=1}^{n} \left| SMI\_PF_{ij} \right| \qquad (8)$$

$$\sum SMI\_PW_j = \sum_{i=1}^{n} \left| SMI\_PW_{ij} \right| \qquad (9)$$

where SMI_PF_{ij} is the geographic imbalance between crop j's production and the area of farmland in county i; SMI_PW_{ij} is the geographic imbalance between crop j's production and the volume of water resources in county i; P_{ij} has the same meaning as in Equation (1); F_i and W_i have the same meanings as in Equation (5); and $\sum SMI\_PF_j$ ($\sum SMI\_PW_j$) is the degree of total geographic imbalance between crop production and farmland area (water resources) of crop j in China or a subregion. The larger the values of $\sum SMI\_PF_j$ ($\left| SMI\_PF_{ij} \right|$) and $\sum SMI\_PW_j$ ($\left| SMI\_PW_{ij} \right|$), the higher the degree of geographic imbalance. In addition, Jenks Natural Breaks Classification was used to judge whether crop production and farmland (water resources) were balanced.

3. Results

3.1. Spatial Distribution of and Spatiotemporal Changes in Crop Production

3.1.1. Spatial Distribution of Crop Production

In 2015, the CRCP of the total of the five crops showed obvious spatial heterogeneities between the different counties (Figure 2a). The counties with high CRCPs were concentrated in zones I, III, VI, and IX. These areas have flat terrain, fertile soil, and good growing conditions, accounting for 72.89% of the total yields of the five crops and were the areas with the highest crop production in China. The areas of low crop production are concentrated in zone X, central to and east of zone V, and northwest of zone IX, and some counties did not produce any crops.

The ratios of corn, rice, wheat, tubers, and soybean yields to the total yield of the five crops were 38.22%, 32.59%, 21.41%, 5.86%, and 1.92%, respectively. The main production areas of different crops showed obvious differences (Figure 2b–f). The main corn production areas were located in zones I and III of northern China, accounting for 35.92% and 25.84% of total corn production, respectively. This is consistent with the growth requirements of corn, which include short periods of sunshine and resistance to moderate, but not high, temperatures. The areas of high rice production were concentrated in zones VI, VII, and VIII of southeastern China, which accounted for 65.91% of the total. Rice production was also relatively high in zone I and the northern part of zone IX. China's wheat production was relatively concentrated: 63.35% of the total production was located in zone III, and another 15.69% was located in the northern part of zone VI. A proportion of 40.33% of soybean was distributed in zone I, and counties with high CRCPs were scattered in eastern China. In addition, the areas of high tuber production were located in zone IX, which accounted for 41.77% of total tuber production in China.

3.1.2. Spatiotemporal Changes in Crop Production in the Period 1990–2015

From the total of the five crops (Figure 3a), the counties where CRCP increased by more than 0.02 percentage points (pp) were located mainly in zones I and III. In those 25 years, the CRCPs in these two zones increased by 6.14 pp and 2.73 pp, respectively, which was significantly higher than the national average. These two regions have been greatly favored by the government's crop policy, coupled with improvements to farmland providing low or medium levels of production, which has led to a general increase in

CRCP. The counties where CRCP decreased by more than 0.02 pp were located mainly in zones VIII, VI, and northeast of zone IX, with reduced production levels of 5.19 pp, 3.20 pp, and 3.10 pp, respectively, due mainly to urban expansion and adjustments to the industrial structure in southern China. In addition, the changes in CRCP in areas of low crop production were relatively stable, including in zones IV and X.

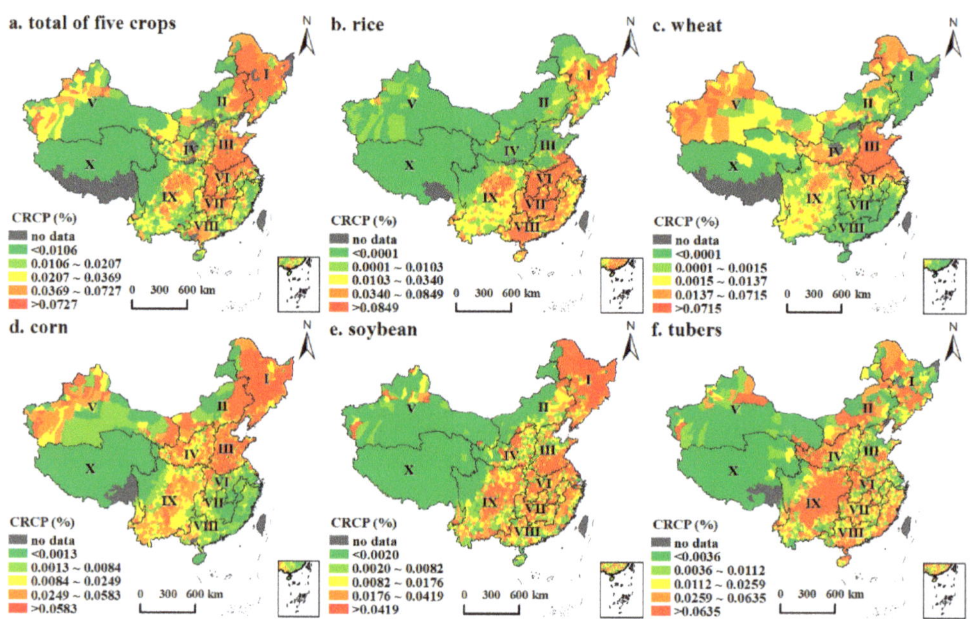

Figure 2. Spatial distribution of crop production in 2015 (CRCP represents the concentration ratio of crop production).

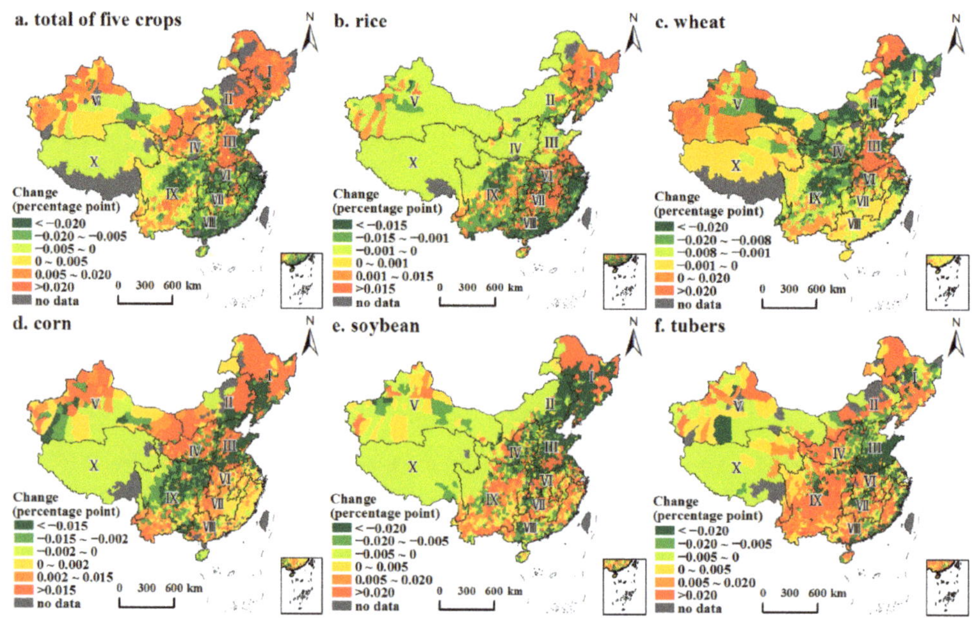

Figure 3. Spatial distribution of crop production changes in the period 1990–2015.

In terms of the different crops (Figure 3b–f), the dominant rice-producing areas were still located in southern China, but the CRCP in zone VIII decreased by 6.84 pp in the period 1990–2015. At the same time, the CRCP of rice increased by 7.15 pp in zone I, where the soil was fertile and irrigation water was pure. In contrast, the regional trend of wheat production was more obvious. In terms of the wheat, the counties with more than a 0.02 pp increase in CRCP were located mainly in zones III and the north of zone VI, and the CRCP of wheat in these two regions increased by 13.44 pp and 2.25 pp, respectively. Zones I and III in northern China have been areas of high corn production, but the CRCP in the south of zone III has been greatly reduced, resulting in a 7.01 pp decrease in zone III. On the other hand, the proportion of corn grown in northern China has increased. Counties with a CRCP increase of more than 0.015 pp were distributed mainly in the center and north of zone I, the southwest and northeast of zone II, the north and east of zone V, and the west of zone IV. With respect to soybean, counties with a CRCP decrease of more than 0.02 pp were located mainly in the center and north of zone III and the south of zone I. The CRCP of soybean decreased by 7.20 pp in zone III, which was due mainly to adjustments to the structure of the crop in exchange for a rapid increase in wheat production. The quantity of soybean grown in the north of zones I and IX increased rapidly, and the CRCP in zone IX increased by 6.24 pp. With respect to tubers, the counties with a significant increase or decrease in CRCP were located mainly in zone IX (increased by 14.88 pp) and zone III (decreased by 25.84 pp), respectively.

3.2. Trajectory of the Gravity Center of Crop Production and the Geographic Imbalances between Farmland and Water Resources

In 2015, the northern region of China was rich in farmland and short of water resources, whereas the southern region experienced the opposite (Figure 4). The spatial matching of farmland and water resources was unbalanced, which greatly limited crop production. Between 1990 and 2015, the gravity center of the total production of the five crops moved north (Figure 5), with a moving distance of 219.47 km in the direction of abundant farmland, fewer water resources, and increasing $|SMI\_WF|$. This result shows that the focus of crop production in China has gradually shifted northward from the southern rice-based growing area, with its relatively abundant water resources and developed economy, to the northern wheat- and corn-based growing area, with its relatively poor water resources and underdeveloped economy.

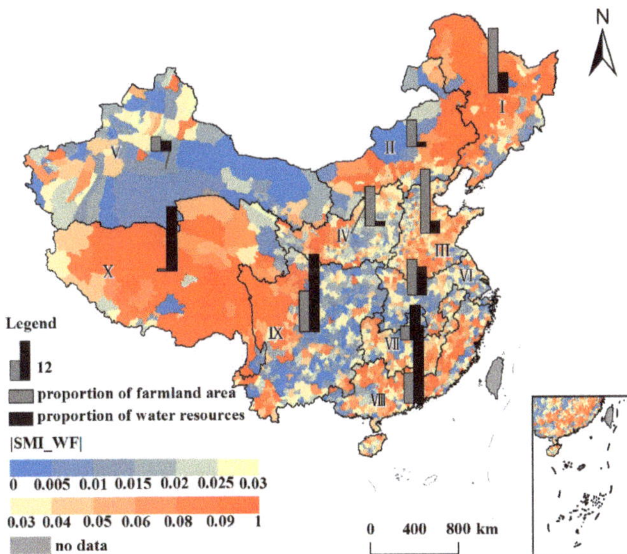

Figure 4. Geographic imbalances between farmland and water resources in 2015.

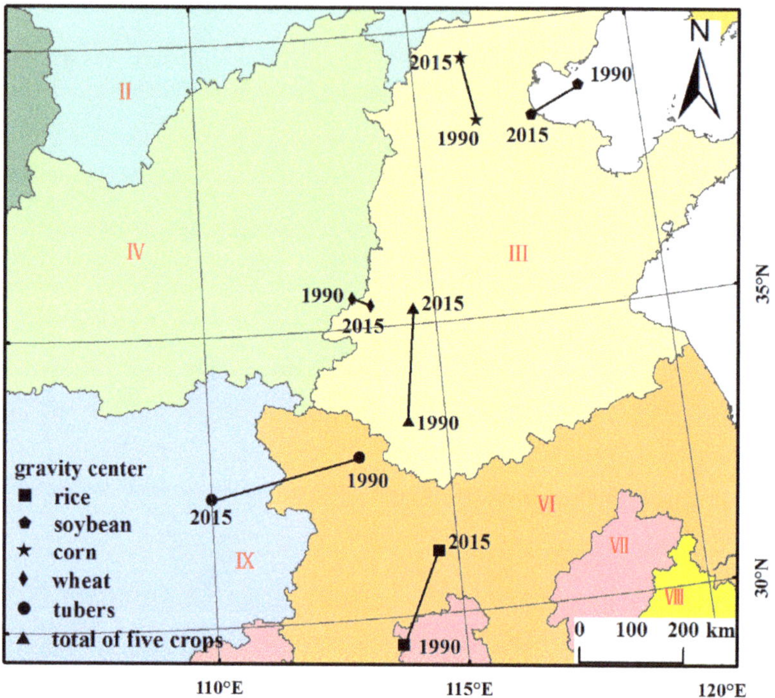

Figure 5. Trajectory of the gravity center of crop production in the period 1990–2015.

With respect to the different crops, the distance that the gravity center moved was ranked as follows: tubers > rice > corn > soybean > wheat. The gravity centers of tuber and soybean production moved toward the southwest by 295.14 and 106.78 km, respectively, moving in the direction of abundant water resources, less farmland, and reducing |SMI_WF|. These changes were attributed mainly to the significant increase in the CRCPs of tubers and soybean in zone IX. The gravity center of rice and corn production moved toward the north, in the direction of fewer water resources, abundant farmland, and increasing |SMI_WF|. The moving distance of the rice and corn gravity centers were 197.74 and 124.77 km, respectively. Compared with other crops, the moving distance of the gravity center of wheat was the smallest, by only 37.60 km to the southeast, which was due mainly to the significant increase in the CRCP of wheat in the north of zone VI. On the whole, wheat production gradually moved in the direction of abundant water resources, less farmland, and decreasing |SMI_WF|.

3.3. Geographic Imbalances between Crop Production and Farmland-Water Resources

3.3.1. Geographic Imbalances between Crop Production and Available Farmland

Based on the Jenks Natural Breaks Classification, the geographic imbalances between crop production and available farmland are shown in Figure 6. In the period 1990–2015, $\sum SMI\_PF$ decreased slightly from 46.94 in 1990 to 45.26 in 2015. The gravity center of the total production of the five crops moved toward the north with abundant farmland, which ensured that $\sum SMI\_PF$ decreased slightly against the background of increasing crop production.

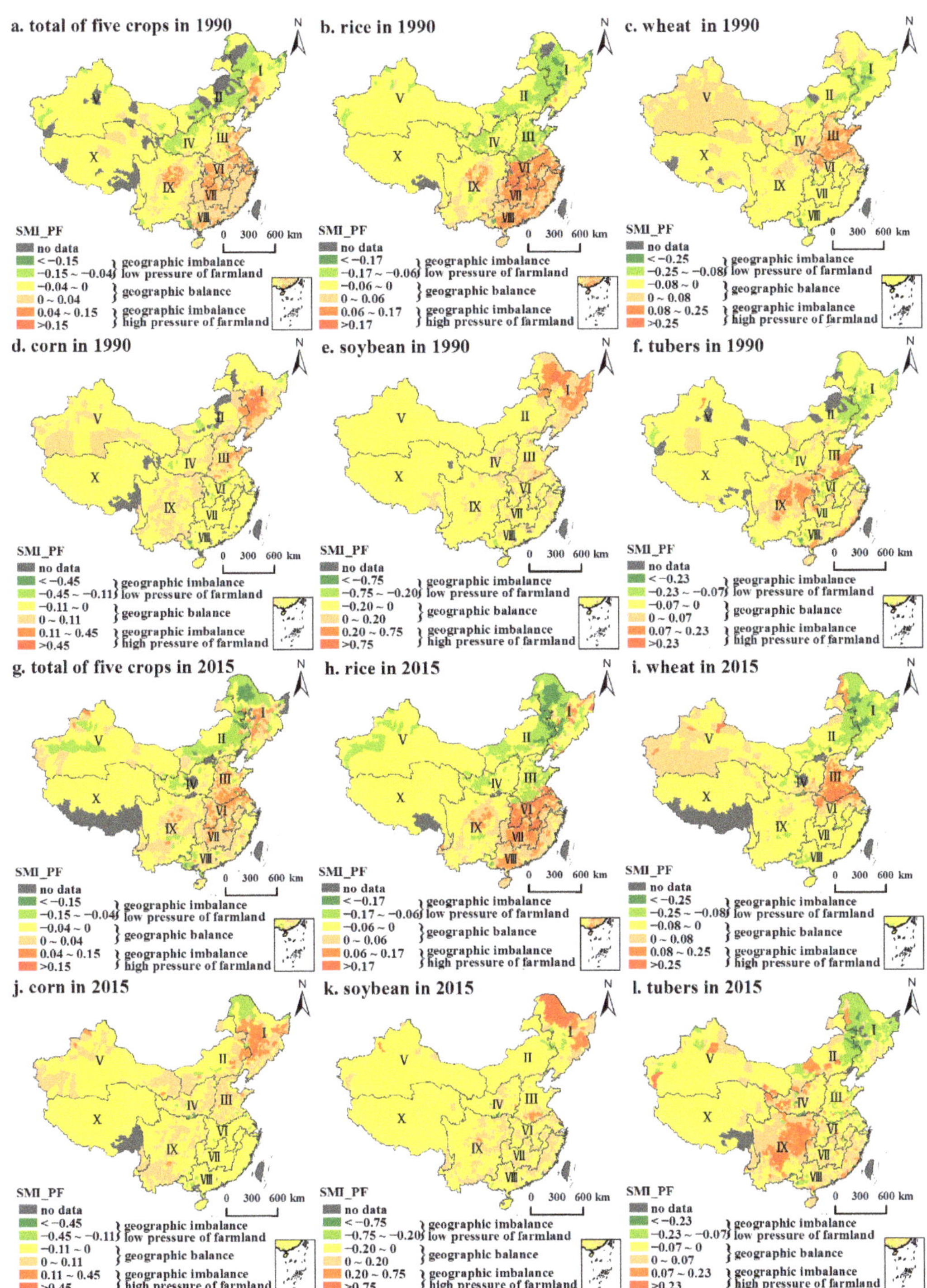

Figure 6. Geographic imbalances between crop production and farmland resources in the period 1990–2015.

From the perspective of the different crops, in the period 1990–2015 the $\sum SMI\_PF$ from high to low was: rice > tubers > wheat > corn > soybean and wheat > rice > tubers > soybean > corn, respectively. In the period 1990–2015, the $\sum SMI\_PF$ of wheat, soybean, and tubers increased. The $\sum SMI\_PF$ of wheat increased by 39.96%, which was due mainly to the large increase in wheat production during that period and its concentration in zone III, resulting in a significantly increased pressure on farmland in zone III. The $\sum SMI\_PF$ of soybean increased by 27.16%. Although the gravity center of soybean production moved toward the southwest, its main area of production was still in zone I. The CRCP of soybean in the southern zone I increased significantly, which increased the $\sum SMI\_PF$ of soybean significantly. The $\sum SMI\_PF$ of tubers increased by 12.22%, which was due mainly to the large increase in the CRCP of tubers in zone IX, which had a relative shortage of farmland. The geographic imbalance in corn production was thereby improved and the $\sum SMI\_PF$ decreased by 11.23%, which was due mainly to the shift of the gravity center of production to the north, which had rich available farmland. Except for the center and west of zone I, the degree of balance of corn production with available farmland was relatively good in almost all regions. The $\sum SMI\_PF$ of rice in 2015 was basically the same as that in 1990 and only decreased by 1.34%. The counties with a decrease in farmland pressure on rice production were distributed in zone VIII and the counties with an increase in farmland pressure were distributed in the center of zone I.

3.3.2. Geographic Imbalances between Crop Production and Water Resources

Between 1990 and 2015 (Figure 7), $\sum SMI\_PW$ changed faster than $\sum SMI\_PF$. The value of $\sum SMI\_PW$ increased from 100.47 in 1990 to 109.63 in 2015, an increase of 9.12%. The shift of the gravity center of the total production of the five crops to the north increased the pressure on water resources and the values of $\sum SMI\_PW$ in zones I and III increased by 38.29% and 13.50%, respectively.

For the different crops, the rankings of $\sum SMI\_PW$ in 1990 and 2015 were basically the same, namely, corn > wheat > soybean > tubers > rice and wheat > corn > soybean > tubers > rice, respectively. Although the shift of the wheat gravity center to the south was beneficial for reducing the $\sum SMI\_PW$, the $\sum SMI\_PW$ finally increased by 10.01% due to the large increase in wheat production in China and the increasing concentration of wheat growing in zone III with relatively few water resources. With respect to the rice, the $\sum SMI\_PW$ increased by 14.20%. This change was due mainly to the large increase in the CRCP of rice in the center of zone I because of water shortages, which increased the pressure on water resources there. At the same time, the northward shift of the rice gravity center increased the $\sum SMI\_PW$. The $\sum SMI\_PW$ of tubers and soybean decreased by 9.54% and 4.71%, respectively, which was related to the shift of the gravity center of these two crops to the water-rich south. In addition, the $\sum SMI\_PW$ of corn changed little, only reducing by 1.43%. The main reason for this was that the adjustment to the spatial distribution of corn occurred mainly in northern China, and there was no large cross-regional adjustment between the north and the south.

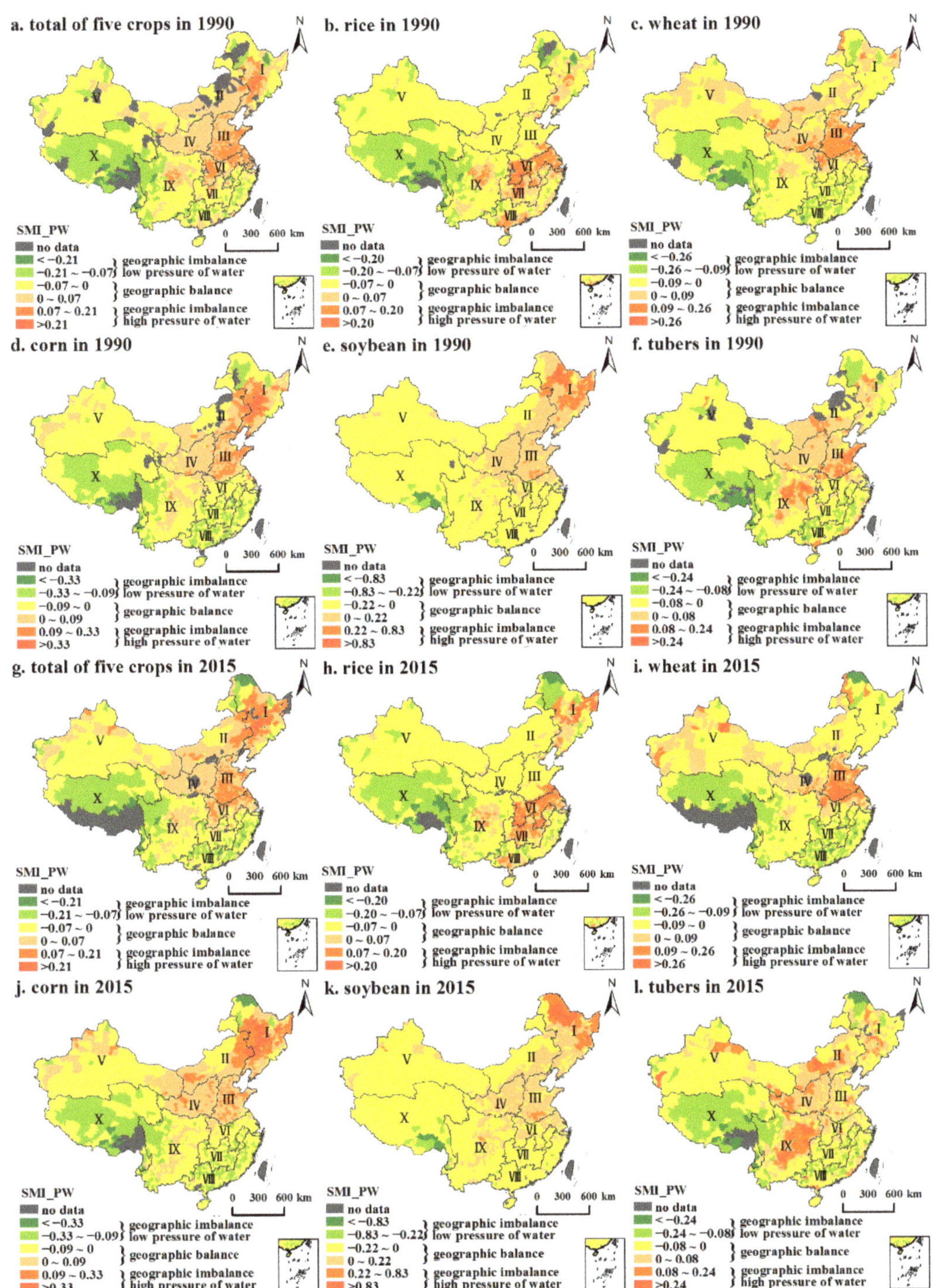

Figure 7. Geographic imbalances between crop production and water resources in the period 1990–2015.

4. Discussion

4.1. Changes in Crop Production and Its Relationship to Farmland and Water Resources

4.1.1. Total Production of the Five Crops

Between 1990 and 2015, the spatial distribution of crop production in China changed considerably. Our results showed that the gravity center of the total production of the five crops moved northward from 1990 to 2015, which was consistent with previous research results [19,49]. We further counted the crop production at the provincial scale from 2015 to 2020 and found that the CRCP of Heilongjiang, Inner Mongolia, Shandong, Henan, Liaoning, and Hebei provinces in northern China increased by 1.11 pp, 0.75 pp, 0.55 pp, 0.43 pp, 0.30 pp, and 0.26 pp, respectively, indicating that the northward shift of the gravity center of crop production has not changed from 2015 to 2020. The spatial centralization of crop production increased the consumption of farmland and water resources in the main producing areas. The spatial pattern of crop production changed from "crop transportation from south to north" to "crop transportation from north to south" [50], which further changed the geographic relationships between crop production and farmland-water resources. In terms of farmland resources, the geographic imbalances have been alleviated by the crop production barycenter's shift to northern China, which is the difference between our results and Li et al. [29]. The rich farmland resources in northern China make it possible for crop transfer and it is also conducive to the efficient use of farmland resources. Moreover, as a result of support from specific policies in northern China, such as the "market-oriented acquisition + subsidy" policy for corn, crop production in northern China increased. In terms of water resources, the overall pattern of water resources has not changed in China. The increase in precipitation in the north and northwest alleviated the demands of agricultural production for water resources to a certain extent. However, the large-scale expansion of farmland in the north increased the demand for water resources. In our paper, the geographic imbalances between the total production of the five crops and water resources aggravated, which has often been ignored in previous studies and should be paid attention to [43].

The southern plain is the best area for matching the degree of farmland with water resources in China. However, with the rapid economic development in the south, a large area of farmland became construction land [51,52], which weakened the status of crop production and did not give full play to the advantages of a region possessing rich water resources. In fact, the shift of the gravity center of crop production to the farmland-rich north further aggravated the contradiction between the supply and demand for water resources in northern China. In 2015, 44.93% of China's total production of five crops was produced in zones I and III but only 9.57% of the nation's water resources were to be found there, which led to serious groundwater overexploitation [53]. For example, Cao et al. [54] showed that the rate of depletion of recoverable groundwater in zone III averaged 4 km^3/a (30 mm/a) between 1970 and 2008. In the long run, if there is a drought, food security in northern China will decline and the transportation of crops from north to south will become unsustainable, which will directly threaten China's food security [55].

4.1.2. Different Types of Crops

At the same time, the spatiotemporal changes in total crop production are also accompanied by changes in the distribution of crops in different regions [45]. Our study shows that, of the five crops, adjustments to the spatial distribution of corn production had the most significant effect on reducing the pressure on farmland and water resources. Between 1990 and 2015, the $\sum SMI\_PF$ and $\sum SMI\_PW$ of corn decreased by 11.23% and 1.43%, respectively, which was due mainly to the increase in the CRCP of corn in zones I, II, IV, and V. It should be noted that the ecological environment of these areas was fragile: once farmland and water resources were damaged, it was difficult to restore them, thus implying that we should pay particular attention to the appropriate development of these valuable and vulnerable resources. The northward shift of the gravity center of rice production was due mainly to the increase in the CRCP in zone I, which has been confirmed by previous studies [56]. The water demand for rice is about 2–3 times that of wheat or corn [57]. This characteristic of rice not only increased the pressure on water resources in the north, but it also did not take advantage of the more plentiful water resources in the south, which increased the $\sum SMI\_PW$ by 9.86%.

For wheat, restricted by water resources, the CRCP increased significantly in the north of zone VI, which helped reduce the pressure on water resources in zone III. However, between 1990 and 2015, China's wheat production increased from 94.78 million tons to 135.74 million tons, and the CRCP of wheat in zone III, even with its water shortages, increased from 49.97% to 63.41%, resulting in an increase in both $\sum SMI\_PF$ and $\sum SMI\_PW$. The pressure of wheat production in China on farmland and water resources was still great during this period. The gravity centers of soybean and tuber production moved toward the southwest mainly because the CRCPs of soybean and tubers increased by 6.24 pp and 14.88 pp, respectively, in zone IX, whereas they decreased by 7.20 pp and 25.84 pp, respectively, in zone III. It should be pointed out that the degree of balance between farmland and water resources in zone IX was better than that in zone III (Figure 4). Therefore, the changes in the spatial distributions of soybean and potato were beneficial to the production of these two crops.

We further calculated the changes in the geographical imbalances on the provincial scale in the periods 1990–2015 and 1990–2020 and compared them with the results of the county scale from 1990 to 2015. It was found that the change trends of the three results were similar. With respect to the $\sum SMI\_PF$ (Figure 8a–c), the change trends of the three results were completely consistent. With respect to the $\sum SMI\_PW$ (Figure 8d–f), comparing the differences between the two periods on the provincial scale, it was found that the change trends of rice, wheat, soybean, and total yield were the same, except for corn and tubers. However, the changes in corn and tubers were small, which can be regarded as relatively stable. A comparison of the differences between the county and provincial scales from 1990 to 2015, showed that the change trends in rice, wheat, soybean, tubers, and total yield were the same, except for corn. From 1990 to 2015, the $\sum SMI\_PW$ of corn increased by 4.90 on the provincial scale and decreased by 2.16 on the county scale. The changes in the two spatial scales were small and can be regarded as relatively stable. Given all of this, our results on the county scale from 1990 to 2015 reflect the changes in the geographic imbalances between crop production and farmland-water resources in China over a long time.

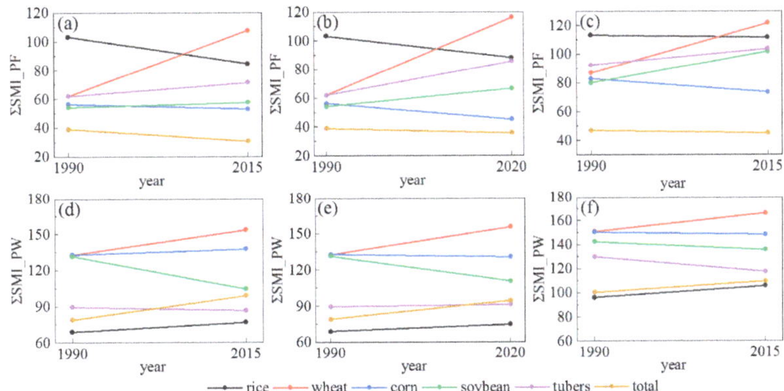

Figure 8. Changes in geographical imbalance. (**a**) Changes of $\sum SMI\_PF$ on a provincial scale from 1990 to 2015; (**b**) Changes of $\sum SMI\_PF$ on a provincial scale from 1990 to 2020; (**c**) Changes of $\sum SMI\_PF$ on a county scale from 1990 to 2015; (**d**) Changes of $\sum SMI\_PW$ on a provincial scale from 1990 to 2015; (**e**) Changes of $\sum SMI\_PW$ on a provincial scale from 1990 to 2020; (**f**) Changes of $\sum SMI\_PW$ on a county scale from 1990 to 2015. ($\sum SMI\_PF$: Geographic imbalances between crop production and farmland resources; $\sum SMI\_PW$: Geographic imbalances between crop production and water resources).

4.2. Geographic Imbalances between Crop Production and Farmland-Water Resources in Different Subregions: Policy Suggestions

The changes in $|SMI\_PF|$ and $|SMI\_PW|$ for the different crops show a clear spatial heterogeneity (Figure 9). With respect to crop production, we divided ten subregions into three categories and put forward the below policy suggestions for the sustainable utilization of farmland and water resources. Zones I and III are rich in farmland resources and are the main crop-producing areas in China. Zones VI, VII, VIII, and IX are rich in water resources and are the main areas for growing rice and tubers. The ecological environments of zones II, IV, V, and X are fragile, and crop production is low.

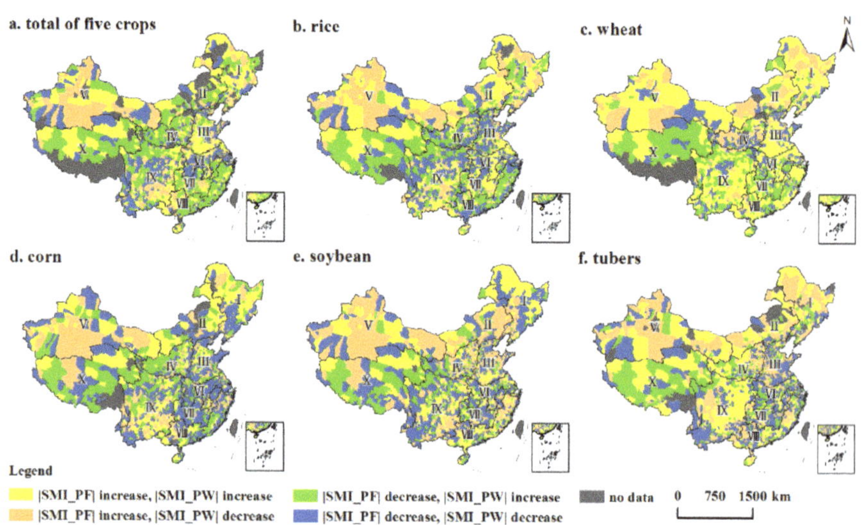

Figure 9. Changes in the degree of geographic imbalance between crop production and farmland and water resources in the period 1990–2015.

Zones I and III are the main crop-production areas in China and play an important role in ensuring China's food security. Between 1990 and 2015, the CRCP of the total production of the five crops in zones I and III increased by 6.14 pp and 2.73 pp, respectively, but this growth was at the cost of a significant expansion in the use of farmland, water, and other natural resources for agricultural purposes. In zone I, which is the main production area for soybean and corn, the $|SMI\_PF|$ and $|SMI\_PW|$ of soybean and corn increased in the north and decreased in the south. During the same period, the reserve farmland in zone I has gradually reduced, and the method of increasing crop production by increasing the area of farmland has become unsustainable [58,59]. It has become inevitable to improve farmland by low or medium levels of production and thus increase crop production per unit area. At the same time, the structure of crop production in zone I changed significantly and the CRCP of rice increased, which increased the pressure on water resources. In 2015, the proportion of agricultural water consumption in Heilongjiang reached 87.95%. The concentration of crop production further threatened the security of water resources. Therefore, the planting scale of rice with a large water consumption per unit area should be carefully controlled and irrigation control measures should be adopted in zone I. With respect to zone III, as the main area of wheat production, both $|SMI\_PF|$ and $|SMI\_PW|$ increased during the period 1990–2015. Particularly, the proportion of agricultural water consumption in Hebei and Shandong provinces located in zone III was 72.28% and 67.34%, respectively. The shortage of water resources and the high proportion of agricultural water consumption has led to the overexploitation of groundwater. In zone III, we should strictly protect high-yielding farmland, comprehensively promote high-efficiency water-saving technology, and appropriately reduce the area of wheat planting in those areas with serious groundwater overexploitation. Given all of this, in order to alleviate the pressure on water resources in zones I and III, the northward shift of the gravity center of crop production can be appropriately reversed in terms of regional agricultural function positioning. By ensuring that zone I and zone III are the main crop-production areas, the exploration of agricultural production potential in areas with abundant water resources should be strengthened.

Zones VI, VII, and VIII are the main rice-producing areas in China. These zones have abundant precipitation and high temperatures, which are conducive to the large-scale planting of rice. Between 1990 and 2015, the CRCP of rice decreased in zone VIII, which did not give full play to the advantages of rich water resources, resulting in the inefficient use of water resources. Zones VI, VII, and VIII have a high level of economic development; thus, we should strictly control the unbridled expansion of construction land and appropriately increase the area set aside for rice planting. Due to the long rainy season and the threat of floods in these zones, we should strengthen water conservation and improve flood resistance capacity. In zone IX, water resources and climatic conditions are relatively good, which makes it suitable for growing crops. In 2015, the CRCPs of rice, corn, soybean, and tubers were 14.93%, 10.16%, 13.49%, and 41.77%, respectively. However, the crop yield per unit area and the multiple cropping index of this area were low, which is inconsistent with the superior natural conditions. In contrast, the CRCP, $\sum SMI\_PF$, and $\sum SMI\_PW$ of tubers increased significantly in zone IX. Due to the lack of flat terrain, the possibilities for expanding farmland are limited. Therefore, the principal means of increasing crop production in zone IX is to carry out intensive management and improve the crop yield per unit area and the multiple cropping index. In addition, as zone IX is mostly a hilly plateau [60], improving water conservation measures is also a key to increased crop production.

Zones II, IV, V, and X are located in the northwest of China and are much affected by sandstorms. The limited development potential of farmland and water resources in these zones is not conducive to the growth of crops. In 2015, the CRCPs of the total production of the five crops in zones II, IV, V, and X were 4.36%, 5.21%, 3.48%, and 0.30%, respectively, which were low-yielding areas of crops. These four zones possess fragile ecological environments and their ability to self-regulate is weak. Overexploitation is leading to the unsustainable utilization of farmland and water resources and will thus increase the risk of deterioration of the ecological environment. Therefore, it is very important to ensure sustainable crop production that does not lead to environmental degradation. Specifically, the proportion of agricultural water consumption in Xinjiang province located in Zone V was as high as 94.66%. Zone V has a dry climate, large evaporation, and a high proportion of agricultural water consumption, which makes it suitable for planting crops with low water consumption such as tubers. For zone IV, which experiences serious soil erosion [61], the areas with sloping farmland need to be strictly controlled. The CRCP of corn in zone II increased by 3.29 pp, and, as a consequence, the contradiction between the demand for and the supply of limited water resources became much more apparent. The authorities responsible for Zone II need to pay more attention to developing a sustainable system of water-saving agriculture. With respect to zone X, the yield of five crops and farmland accounts for 0.30% and 0.55% of the national totals, respectively, and here, self-sufficient crop production needs to be developed.

4.3. Limitations and Prospects

Although our paper has identified the geographic imbalances between crop production and farmland-water resources, there are still the following limitations. First, data for the yields of rice, wheat, corn, soybean, and tubers were obtained at the county scale. There is a large amount of data within the county unit from the national scale and the update of the statistical yearbook of county scale data is lagging. Thus, the time horizon of our paper is somewhat outdated. However, our paper focuses on the geographic imbalances between crop production and farmland-water resources for an extended period, i.e., from 1990 onward and the results could meet research needs. Second, the spatiotemporal changes in the geographic imbalances have been identified but the driving factors have not been discussed. Finally, considering the availability of data, we have selected the total volume of water resources to reflect the abundance of regional water resources. In fact, irrigation, climate, crop types, and other factors will also affect water resources usage for crop growth, which is not analyzed in detail in our paper. Even so, our paper is still a good contribution to the study of farmland-water resources and agricultural production capacity in China.

5. Conclusions

In China, crop production plays a central role in ensuring the welfare of 1.4 billion people. The distribution of farmland-water resources is uneven in China and current understanding does not allow for a clear assessment of the geographic imbalances between the production of different types of crops and natural resources. In our paper, the spatial distribution of crop production and the geographic imbalances between crop production and farmland-water resources were examined using the spatial gravity center model and spatial mismatch index. From 1990 to 2015, the gravity center of the total production of crops moved to the north with abundant farmland resources. This change slightly alleviated the geographical imbalances between the total production of crops and farmland resources but exacerbated that of water resources. The geographical imbalances of different crops showed different change trends. In $\sum SMI\_PF$, wheat, soybean, and tuber increased by 39.96%, 27.16%, and 12.22%, respectively, and corn and rice decreased by 10.31% and 1.34%, respectively. In $\sum SMI\_PW$, wheat and rice increased by 10.01% and 9.86%, respectively, and tuber, soybean, and corn decreased by 9.54%, 4.71%, and 1.43%, respectively. Northeastern China and Huang-Huai-Hai Plain were the main crop production areas in China and the growth in production was at the cost of a significant expansion in the use of farmland, water, and other natural resources for agricultural purposes. The concentration of crop production increased the $\sum SMI\_PW$ of the two regions and water resources became the key factor in crop production. Our results reflect a good representation of the agricultural problems in China and provide operable recommendations for the efficient use of farmland-water resources and maintaining food security. In future studies, we intend to analyze the driving factors of the geographic imbalances in order to provide a scientific reference point for the sustainable development of agriculture in China.

Author Contributions: Conceptualization, methodology, D.L. and E.X.; validation, formal analysis, D.L.; resources, H.Z.; writing—original draft preparation, review, and editing, D.L.; supervision, project administration, funding acquisition, H.Z. and E.X. All authors have read and agreed to the published version of the manuscript.

Funding: This research was jointly supported by the Second Tibetan Plateau Scientific Expedition and Research Program (STEP) (2019QZKK0603), National Natural Science Foundations of China (41601095), and Youth Innovation Promotion Association CAS (2021052).

Institutional Review Board Statement: Not applicable.

Informed Consent Statement: Not applicable.

Data Availability Statement: The data presented in this study are available on request from the corresponding author.

Acknowledgments: We are thankful to the Second Tibetan Plateau Scientific Expedition and Research Program, the National Natural Science Foundations of China, and the Youth Innovation Promotion Association CAS for their financial support of the present study.

Conflicts of Interest: The authors declare no conflict of interest.

References

1. Dong, F.; Yan, Q.; Wu, L.; Yang, F.; Wang, J.; Zhang, J.; Yan, S. Soil water storage and maize (*Zea mays* L.) yield under straw return and tillage practices. *Crop Sci.* **2022**, *62*, 382–396. [CrossRef]
2. Porkka, M.; Gerten, D.; Schaphoff, S.; Siebert, S.; Kummu, M. Causes and trends of water scarcity in food production. *Environ. Res. Lett.* **2016**, *11*, 015001. [CrossRef]
3. Wang, Y.; Cao, Y.; Feng, G.; Li, X.; Zhu, L.; Liu, S.; Coulter, J.A.; Gao, Q. Integrated Soil-Crop System Management with Organic Fertilizer Achieves Sustainable High Maize Yield and Nitrogen Use Efficiency in Northeast China Based on an 11-Year Field Study. *Agronomy* **2020**, *10*, 1078. [CrossRef]
4. He, G.; Wang, Z.; Ma, X.; He, H.; Cao, H.; Wang, S.; Dai, J.; Luo, L.; Huang, M.; Malhi, S. Wheat Yield Affected by Soil Temperature and Water under Mulching in Dryland. *Agron. J.* **2017**, *109*, 2998–3006. [CrossRef]
5. Mgolozeli, S.; Nciizah, A.D.; Wakindiki, I.I.C.; Mudau, F.N. Innovative Pro-Smallholder Farmers' Permanent Mulch for Better Soil Quality and Food Security Under Conservation Agriculture. *Agronomy* **2020**, *10*, 605. [CrossRef]
6. Smith, P.; Calvin, K.; Nkem, J.; Campbell, D.; Cherubini, F.; Grassi, G.; Korotkov, V.; Anh Le, H.; Lwasa, S.; McElwee, P.; et al. Which practices co-deliver food security, climate change mitigation and adaptation, and combat land degradation and desertification? *Glob. Chang. Biol.* **2020**, *26*, 1532–1575. [CrossRef]
7. Kummu, M.; de Moel, H.; Porkka, M.; Siebert, S.; Varis, O.; Ward, P.J. Lost food, wasted resources: Global food supply chain losses and their impacts on freshwater, cropland, and fertiliser use. *Sci. Total Environ.* **2012**, *438*, 477–489. [CrossRef]
8. UN. Transforming our World: The 2030 Agenda for Sustainable Development. 2015. Available online: https://sustainabledevelopment.un.org/post2015/transformingourworld/publication (accessed on 25 April 2022).
9. FAO. FAO Database: Agriculture Production. Food and Agriculture Organization of the United Nations, Rome. 2020. Available online: www.fao.org/faostat/en/#data/QC (accessed on 26 April 2020).
10. He, J.; Liu, Y.; Yu, Y.; Tang, W.; Xiang, W.; Liu, D. A counterfactual scenario simulation approach for assessing the impact of farmland preservation policies on urban sprawl and food security in a major grain-producing area of China. *Appl. Geogr.* **2013**, *37*, 127–138. [CrossRef]
11. Xu, Y.; Li, J.; Wan, J. Agriculture and crop science in China: Innovation and sustainability. *Crop J.* **2017**, *5*, 95–99. [CrossRef]
12. Guo, L.; Liu, M.; Tao, Y.; Zhang, Y.; Li, G.; Lin, S.; Dittert, K. Innovative water-saving ground cover rice production system increases yield with slight reduction in grain quality. *Agric. Syst.* **2020**, *180*, 102795. [CrossRef]
13. Shi, W.; Tao, F.; Liu, J. Changes in quantity and quality of cropland and the implications for grain production in the Huang-Huai-Hai Plain of China. *Food Secur.* **2013**, *5*, 69–82. [CrossRef]
14. Piao, S.; Ciais, P.; Huang, Y.; Shen, Z.; Peng, S.; Li, J.; Zhou, L.; Liu, H.; Ma, Y.; Ding, Y.; et al. The impacts of climate change on water resources and agriculture in China. *Nature* **2010**, *467*, 43–51. [CrossRef] [PubMed]
15. Kang, S.; Hao, X.; Du, T.; Tong, L.; Su, X.; Lu, H.; Li, X.; Huo, Z.; Li, S.; Ding, R. Improving agricultural water productivity to ensure food security in China under changing environment: From research to practice. *Agric. Water Manag.* **2017**, *179*, 5–17. [CrossRef]
16. Sun, S.; Wu, P.; Wang, Y.; Zhao, X. The virtual water content of major grain crops and virtual water flows between regions in China. *J. Sci. Food Agric.* **2013**, *93*, 1427–1437. [CrossRef]
17. Dalin, C.; Qiu, H.; Hanasaki, N.; Mauzerall, D.L.; Rodriguez-Iturbe, I. Balancing water resource conservation and food security in China. *Proc. Natl. Acad. Sci. USA* **2015**, *112*, 4588–4593. [CrossRef]
18. Deng, A.; Chen, C.; Feng, J.; Chen, J.; Zhang, W. Cropping system innovation for coping with climatic warming in China. *Crop. J.* **2017**, *5*, 136–150. [CrossRef]
19. Wang, J.; Zhang, Z.; Liu, Y. Spatial shifts in grain production increases in China and implications for food security. *Land Use Policy* **2018**, *74*, 204–213. [CrossRef]
20. Wei, X.; Zhang, Z.; Wang, P.; Tao, F. Recent patterns of production for the main cereal grains: Implications for food security in China. *Reg. Environ. Change* **2017**, *17*, 105–116. [CrossRef]
21. Yan, H.; Liu, F.; Qin, Y.; Niu, Z.; Doughty, R.; Xiao, X. Tracking the spatio-temporal change of cropping intensity in China during 2000–2015. *Environ. Res. Lett.* **2019**, *14*, 035008. [CrossRef]
22. FAO. The Future of Food and Agriculture—Trends and Challenges. Rome. 2017. Available online: https://www.fao.org/3/i6583e/i6583e.pdf (accessed on 2 May 2022).
23. FAO. The State of the World's Land and Water Resources for Food and Agriculture (SOLAW)—Managing Systems at Risk. Food and Agriculture Organization of the United Nations, Rome and Earthscan, London. 2011. Available online: https://www.fao.org/3/i1688e/i1688e00.htm (accessed on 2 May 2022).
24. Aznar-Sanchez, J.A.; Belmonte-Urena, L.J.; Velasco-Munoz, J.F.; Valera, D.L. Aquifer Sustainability and the Use of Desalinated Seawater for Greenhouse Irrigation in the Campo de Níjar, Southeast Spain. *Int. J. Environ. Res. Public Health* **2019**, *16*, 898. [CrossRef]
25. Fader, M.; Gerten, D.; Krause, M.; Lucht, W.; Cramer, W. Spatial decoupling of agricultural production and consumption: Quantifying dependences of countries on food imports due to domestic land and water constraints. *Environ. Res. Lett.* **2013**, *8*, 014046. [CrossRef]
26. Deng, X.; Huang, J.; Rozelle, S.; Zhang, J.; Li, Z. Impact of urbanization on cultivated land changes in China. *Land Use Policy* **2015**, *45*, 1–7. [CrossRef]

27. Jiang, Y.; Yin, X.; Wang, X.; Zhang, L.; Lu, Z.; Lei, Y.; Chu, Q.; Chen, F. Impacts of global warming on the cropping systems of China under technical improvements from 1961 to 2016. *Agron. J.* **2021**, *113*, 187–199. [CrossRef]
28. Long, H.; Ge, D.; Zhang, Y.; Tu, S.; Qu, Y.; Ma, L. Changing man-land interrelations in China's farming area under urbanization and its implications for food security. *J. Environ. Manag.* **2018**, *209*, 440–451. [CrossRef]
29. Li, T.; Long, H.; Zhang, Y.; Tu, S.; Ge, D.; Li, Y.; Hu, B. Analysis of the spatial mismatch of grain production and farmland resources in China based on the potential crop rotation system. *Land Use Policy* **2017**, *60*, 26–36. [CrossRef]
30. Chen, H.; Wang, J.; Huang, J. Policy support, social capital, and farmers' adaptation to drought in China. *Glob. Environ. Chang.* **2014**, *24*, 193–202. [CrossRef]
31. Wang, J.; Li, Y.; Huang, J.; Yan, T.; Sun, T. Growing water scarcity, food security and government responses in China. *Glob. Food Secur. Agric.* **2017**, *14*, 9–17. [CrossRef]
32. Wang, S.; Hu, Y.; Yuan, R.; Feng, W.; Pan, Y.; Yang, Y. Ensuring water security, food security, and clean water in the North China Plain-Conflicting strategies. *Curr. Opin. Environ. Sust.* **2019**, *40*, 63–71. [CrossRef]
33. Wada, Y.; van Beek, L.P.H.; van Kempen, C.M.; Reckman, J.W.T.M.; Vasak, S.; Bierkens, M.F.P. Global depletion of groundwater resources. *Geophys. Res. Lett.* **2010**, *37*, L20402. [CrossRef]
34. He, G.; Zhao, Y.; Wang, L.; Jiang, S.; Zhu, Y. China's food security challenge: Effects of food habit changes on requirements for arable land and water. *J. Clean. Prod.* **2019**, *229*, 739–750. [CrossRef]
35. Cao, X.; Wang, Y.; Wu, P.; Zhao, X.; Wang, J. An evaluation of the water utilization and grain production of irrigated and rain-fed croplands in China. *Sci. Total Environ.* **2015**, *529*, 10–20. [CrossRef] [PubMed]
36. Kan, Z.; Liu, Q.; He, C.; Jing, Z.; Virk, A.L.; Qi, J.; Zhao, X.; Zhang, H. Responses of grain yield and water use efficiency of winter wheat to tillage in the North China Plain. *Field Crops Res.* **2020**, *249*, 107760. [CrossRef]
37. Li, J.; Zhang, Z.; Liu, Y.; Yao, C.; Song, W.; Xu, X.; Zhang, M.; Zhou, X.; Gao, Y.; Wang, Z.; et al. Effects of micro-sprinkling with different irrigation amount on grain yield and water use efficiency of winter wheat in the North China Plain. *Agric. Water Manag.* **2019**, *224*, 105736. [CrossRef]
38. Chen, H.; Yu, C.; Li, C.; Xin, Q.; Huang, X.; Zhang, J.; Yue, Y.; Huang, G.; Li, X.; Wang, W. Modeling the impacts of water and fertilizer management on the ecosystem service of rice rotated cropping systems in China. *Agric. Ecosyst. Environ.* **2016**, *219*, 49–57. [CrossRef]
39. Liu, G.; Zhang, L.; Zhang, Q.; Musyimi, Z. The response of grain production to changes in quantity and quality of cropland in Yangtze River Delta, China. *J. Sci. Food Agric.* **2015**, *95*, 480–489. [CrossRef]
40. Meier, T.; Christen, O.; Semler, E.; Jahreis, G.; Voget-Kleschin, L.; Schrode, A.; Artmann, M. Balancing virtual land imports by a shift in the diet. Using a land balance approach to assess the sustainability of food consumption. Germany as an example. *Appetite* **2014**, *74*, 20–34. [CrossRef]
41. Sun, S.; Wang, Y.; Engel, B.A.; Wu, P. Effects of virtual water flow on regional water resources stress: A case study of grain in China. *Sci. Total Environ.* **2016**, *550*, 871–879. [CrossRef]
42. Wu, S.; Ben, P.; Chen, D.; Chen, J.; Tong, G.; Yuan, Y.; Xu, B. Virtual land, water, and carbon flow in the inter-province trade of staple crops in China. *Resour. Conserv. Recycl.* **2018**, *136*, 179–186. [CrossRef]
43. Chai, J.; Wang, Z.; Yang, J.; Zhang, L. Analysis for spatial-temporal changes of grain production and farmland resource: Evidence from Hubei Province, central China. *J. Clean. Prod.* **2019**, *207*, 474–482. [CrossRef]
44. Li, Y.; Li, X.; Tan, M.; Wang, X.; Xin, L. The impact of cultivated land spatial shift on food crop production in China, 1990–2010. *Land Degrad. Dev.* **2018**, *29*, 1652–1659. [CrossRef]
45. Chu, Y.; Shen, Y.; Yuan, Z. Water footprint of crop production for different crop structures in the Hebei southern plain, North China. *Hydrol. Earth Syst. Sci.* **2017**, *21*, 3061–3069. [CrossRef]
46. Ha, N.; Feike, T.; Angenendt, E.; Xiao, H.; Bahrs, E. Impact of farm management diversity on the environmental and economic performance of the wheat-maize cropping system in the North China Plain. *Int. J. Agric. Sustain.* **2015**, *13*, 350–366. [CrossRef]
47. Zhuang, J.; Xu, S.; Li, G.; Zhang, Y.; Wu, J.; Liu, J. The Influence of Meteorological Factors on Wheat and Rice Yields in China. *Crop Sci.* **2018**, *58*, 837–852. [CrossRef]
48. Xu, E. Zoning of Agricultural Resource and Environment in China. *Strateg. Study Chin. Acad. Eng.* **2018**, *20*, 57–62. [CrossRef]
49. Zhong, H.; Liu, Z.; Wang, J. Understanding impacts of cropland pattern dynamics on grain production in China: A integrated analysis by fusing statistical data and satellite-observed data. *J. Environ. Manag.* **2022**, *313*, 114988. [CrossRef] [PubMed]
50. Pan, J.; Chen, Y.; Zhang, Y.; Chen, M.; Fennell, S.; Luan, B.; Wang, F.; Meng, D.; Liu, Y.; Jiao, L.; et al. Spatial-temporal dynamics of grain yield and the potential driving factors at the county level in China. *J. Clean. Prod.* **2020**, *255*, 120312. [CrossRef]
51. Li, J.; Li, Z. Physical limitations and challenges to Grain Security in China. *Food Secur.* **2014**, *6*, 159–167. [CrossRef]
52. Lu, W.; Chen, N.; Qian, W. Modeling the effects of urbanization on grain production and consumption in China. *J. Integr. Agric.* **2017**, *16*, 1393–1405. [CrossRef]
53. Jiao, L. Water Shortages Loom as Northern China's Aquifers Are Sucked Dry. *Science* **2010**, *328*, 1462–1463. [CrossRef]
54. Cao, G.; Zheng, C.; Scanlon, B.R.; Liu, J.; Li, W. Use of flow modeling to assess sustainability of groundwater resources in the North China Plain. *Water Resour. Res.* **2013**, *49*, 159–175. [CrossRef]
55. Zhang, Q.; Yu, H.; Sun, P.; Singh, V.P.; Shi, P. Multisource data based agricultural drought monitoring and agricultural loss in China. *Glob. Planet. Chang.* **2019**, *172*, 298–306. [CrossRef]

56. Xin, F.; Xiao, X.; Dong, J.; Zhang, G.; Zhang, Y.; Wu, X.; Li, X.; Zou, Z.; Ma, J.; Du, G.; et al. Large increases of paddy rice area, gross primary production, and grain production in Northeast China during 2000–2017. *Sci. Total Environ.* **2020**, *711*, 135183. [CrossRef] [PubMed]
57. Bouman, B.A.M.; Humphreys, E.; Tuong, T.P.; Barker, R. Rice and water. *Adv. Agron.* **2007**, *92*, 187–237. [CrossRef]
58. Jin, X.; Zhang, Z.; Wu, X.; Xiang, X.; Sun, W.; Bai, Q.; Zhou, Y. Co-ordination of land exploitation, exploitable farmland reserves and national planning in China. *Land Use Policy* **2016**, *57*, 682–693. [CrossRef]
59. Li, W.; Wang, D.; Liu, S.; Zhu, Y.; Yan, Z. Reclamation of Cultivated Land Reserves in Northeast China: Indigenous Ecological Insecurity Underlying National Food Security. *Int. J. Environ. Res. Public Health* **2020**, *17*, 1211. [CrossRef] [PubMed]
60. Xu, H.; Zhang, J.; Wei, Y.; Dai, J.; Wang, Y. Bedrock erosion due to hoeing as tillage technique in a hilly agricultural landscape, southwest China. *Earth Surf. Proc. Land* **2020**, *45*, 1418–1429. [CrossRef]
61. Liang, Y.; Jiao, J.; Tang, B.; Cao, B.; Li, H. Response of runoff and soil erosion to erosive rainstorm events and vegetation restoration on abandoned slope farmland in the Loess Plateau region, China. *J. Hydrol.* **2020**, *584*, 124694. [CrossRef]

Review

Mineral-Ecological Cropping Systems—A New Approach to Improve Ecosystem Services by Farming without Chemical Synthetic Plant Protection

Beate Zimmermann [1,*], Ingrid Claß-Mahler [1,*], Moritz von Cossel [2,*], Iris Lewandowski [2], Jan Weik [2], Achim Spiller [3], Sina Nitzko [3], Christian Lippert [4], Tatjana Krimly [4], Isabell Pergner [4], Christian Zörb [5], Monika A. Wimmer [5], Markus Dier [5], Frank M. Schurr [6], Jörn Pagel [6], Adriana Riemenschneider [6], Hella Kehlenbeck [7], Til Feike [7], Bettina Klocke [7], Robin Lieb [7], Stefan Kühne [7], Sandra Krengel-Horney [7], Julia Gitzel [7], Abbas El-Hasan [8], Stefan Thomas [8], Martin Rieker [8], Karl Schmid [9], Thilo Streck [10], Joachim Ingwersen [10], Uwe Ludewig [11], Günter Neumann [11], Niels Maywald [11], Torsten Müller [12], Klára Bradáčová [12], Markus Göbel [12], Ellen Kandeler [13], Sven Marhan [13], Romina Schuster [13], Hans-W. Griepentrog [14], David Reiser [14], Alexander Stana [14], Simone Graeff-Hönninger [15], Sebastian Munz [15], Dina Otto [15], Roland Gerhards [16], Marcus Saile [16], Wilfried Hermann [17], Jürgen Schwarz [7], Markus Frank [18], Michael Kruse [19], Hans-Peter Piepho [20], Peter Rosenkranz [21], Klaus Wallner [21], Sabine Zikeli [22], Georg Petschenka [23], Nicole Schönleber [1], Ralf T. Vögele [8] and Enno Bahrs [1,*]

Citation: Zimmermann, B.; Claß-Mahler, I.; von Cossel, M.; Lewandowski, I.; Weik, J.; Spiller, A.; Nitzko, S.; Lippert, C.; Krimly, T.; Pergner, I.; et al. Mineral-Ecological Cropping Systems—A New Approach to Improve Ecosystem Services by Farming without Chemical Synthetic Plant Protection. *Agronomy* **2021**, *11*, 1710. https://doi.org/10.3390/agronomy11091710

Academic Editor: Jie Zhao

Received: 6 August 2021
Accepted: 24 August 2021
Published: 27 August 2021

Publisher's Note: MDPI stays neutral with regard to jurisdictional claims in published maps and institutional affiliations.

[1] Farm Management (410b), University of Hohenheim, Schwerzstr. 44, 70599 Stuttgart, Germany; nicole.schoenleber@uni-hohenheim.de

[2] Biobased Resources in the Bioeconomy (340b), Institute of Crop Science, University of Hohenheim, Fruwirthstr. 23, 70599 Stuttgart, Germany; iris_lewandowski@uni-hohenheim.de (I.L.); jan.weik@uni-hohenheim.de (J.W.)

[3] Marketing for Food and Agricultural Products, Department of Agricultural Economics and Rural Development, University of Göttingen, Platz der Göttinger Sieben 5, 37073 Göttingen, Germany; a.spiller@agr.uni-goettingen.de (A.S.); snitzko@uni-goettingen.de (S.N.)

[4] Production Theory and Resource Economics (410a), University of Hohenheim, Schwerzstr. 44, 70599 Stuttgart, Germany; christian.lippert@uni-hohenheim.de (C.L.); t.krimly@uni-hohenheim.de (T.K.); isabell.pergner@uni-hohenheim.de (I.P.)

[5] Plant Product Quality (340e), University of Hohenheim, Emil Wolff Str. 25, 70599 Stuttgart, Germany; christian.zoerb@uni-hohenheim.de (C.Z.); m.wimmer@uni-hohenheim.de (M.A.W.); markus.dier@uni-hohenheim.de (M.D.)

[6] Landscape and Plant Ecology (320a), University of Hohenheim, Ottilie-Zeller-Weg 2, 70599 Stuttgart, Germany; frank.schurr@uni-hohenheim.de (F.M.S.); joern.pagel@uni-hohenheim.de (J.P.); a.riemenschneider@uni-hohenheim.de (A.R.)

[7] Julius Kühn Institute (JKI)—Federal Research Centre for Cultivated Plants, Institute for Strategies and Technology Assessment, Stahnsdorfer Damm 81, 14532 Kleinmachnow, Germany; hella.kehlenbeck@julius-kuehn.de (H.K.); til.feike@julius-kuehn.de (T.F.); bettina.klocke@julius-kuehn.de (B.K.); Robin.Lieb@julius-kuehn.de (R.L.); stefan.kuehne@julius-kuehn.de (S.K.); sandra.krengel@julius-kuehn.de (S.K.-H.); Julia.Gitzel@julius-kuehn.de (J.G.); juergen.schwarz@julius-kuehn.de (J.S.)

[8] Phytopathology (360a), Institute of Phytomedicine, University of Hohenheim, Otto-Sander-Str. 5, 70599 Stuttgart, Germany; aelhasan@uni-hohenheim.de (A.E.-H.); stefan.thomas@uni-hohenheim.de (S.T.); m.rieker@uni-hohenheim.de (M.R.); ralf.voegele@uni-hohenheim.de (R.T.V.)

[9] Crop Biodiversity and Breeding Informatics (350b), University of Hohenheim, Fruwirthstr. 21, 70599 Stuttgart, Germany; karl.schmid@uni-hohenheim.de

[10] Biogeophysics (310d), University of Hohenheim, Emil Wolff Str. 27, 70599 Stuttgart, Germany; thilo.streck@uni-hohenheim.de (T.S.); jingwer@uni-hohenheim.de (J.I.)

[11] Nutritional Crop Physiology (340h), Institute of Crop Science, University of Hohenheim, Fruwirthstr. 20, 70599 Stuttgart, Germany; u.ludewig@uni-hohenheim.de (U.L.); gd.neumann@uni-hohenheim.de (G.N.); Niels.Maywald@uni-hohenheim.de (N.M.)

[12] Fertilization and Soil Matter Dynamics (340i), University of Hohenheim, Fruwirthstr. 20, 70599 Stuttgart, Germany; torsten.mueller@uni-hohenheim.de (T.M.); klara.bradacova@uni-hohenheim.de (K.B.); markus.goebel@uni-hohenheim.de (M.G.)

[13] Soil Biology (310b), Institute of Soil Science and Land Evaluation, University of Hohenheim, Emil-Wolff-Str. 27, 70599 Stuttgart, Germany; ellen.kandeler@uni-hohenheim.de (E.K.); sven.marhan@uni-hohenheim.de (S.M.); romina.schuster@uni-hohenheim.de (R.S.)

14   Technology in Crop Production (440d), University of Hohenheim, Garbenstr. 9, 70599 Stuttgart, Germany; hw.griepentrog@uni-hohenheim.de (H.-W.G.); david.reiser@uni-hohenheim.de (D.R.); a.stana@uni-hohenheim.de (A.S.)

15   Cropping Systems and Modelling, Institute of Crop Science, University of Hohenheim, Fruwirthstr. 23, 70599 Stuttgart, Germany; simone.graeff@uni-hohenheim.de (S.G.-H.); s.munz@uni-hohenheim.de (S.M.); dina.otto@uni-hohenheim.de (D.O.)

16   Weed Science (360b), University of Hohenheim, Otto-Sander-Str. 5, 70599 Stuttgart, Germany; roland.gerhards@uni-hohenheim.de (R.G.); marcus.saile@uni-hohenheim.de (M.S.)

17   Agricultural Experiment Station, University of Hohenheim (400), Schwerzstraße 21, 70599 Stuttgart, Germany; wilfried.hermann@uni-hohenheim.de

18   Institute of Applied Agriculture (IAAF), Nuertingen Geislingen University, Marktstr. 16, 72622 Nürtingen, Germany; markus.frank@hfwu.de

19   Seed Science and Technology (350d), University of Hohenheim, Fruwirthstr. 21, 70599 Stuttgart, Germany; michael.kruse@uni-hohenheim.de

20   Biostatistics (340c), University of Hohenheim, Fruwirthstr. 23, 70599 Stuttgart, Germany; piepho@uni-hohenheim.de

21   Apicultural State Institute, University of Hohenheim, August-von-Hartmann-Str. 13, 70599 Stuttgart, Germany; peter.rosenkranz@uni-hohenheim.de (P.R.); klaus.wallner@uni-hohenheim.de (K.W.)

22   Center for Organic Farming (309), University of Hohenheim, Fruwirthstr. 14-16, 70599 Stuttgart, Germany; sabine.zikeli@uni-hohenheim.de

23   Applied Entomology (360c), University of Hohenheim, Otto-Sander-Str. 5, 70593 Stuttgart, Germany; georg.petschenka@uni-hohenheim.de

*   Correspondence: beate.zimmermann@uni-hohenheim.de (B.Z.); ingrid.classmahler@uni-hohenheim.de (I.C.-M.); moritz.cossel@uni-hohenheim.de (M.v.C.); bahrs@uni-hohenheim.de (E.B.)

Abstract: The search for approaches to a holistic sustainable agriculture requires the development of new cropping systems that provide additional ecosystem services beyond biomass supply for food, feed, material, and energy use. The reduction of chemical synthetic plant protection products is a key instrument to protect vulnerable natural resources such as groundwater and biodiversity. Together with an optimal use of mineral fertilizer, agroecological practices, and precision agriculture technologies, a complete elimination of chemical synthetic plant protection in mineral-ecological cropping systems (MECSs) may not only improve the environmental performance of agroecosystems, but also ensure their yield performance. Therefore, the development of MECSs aims to improve the overall ecosystem services of agricultural landscapes by (i) improving the provision of regulating ecosystem services compared to conventional cropping systems and (ii) improving the supply of provisioning ecosystem services compared to organic cropping systems. In the present review, all relevant research levels and aspects of this new farming concept are outlined and discussed based on a comprehensive literature review and the ongoing research project "Agriculture 4.0 without Chemical-Synthetic Plant Protection".

Keywords: food security; pesticide-free agriculture; biological control; nutrient efficiency; resistance breeding; equidistant seeding; precision farming; life cycle assessment; sustainable intensification; agroecological intensification

---

1. Introduction

Global population growth and rising yield risks pose an increasing challenge to global food security [1–3]. At the same time, natural livelihoods are threatened by accelerated climate change, rising biodiversity loss, and increasing disruption of nutrient cycles [4]. In recent decades, agricultural cropland intensification and expansion have also led to a significant environmental degradation in many regions of the world [5–7]. The global productive agricultural area is decreasing due to numerous reasons such as urbanization, water scarcity, and soil degradation [8,9]. Furthermore, increasing competition between different land uses can be observed; for instance, between the production of biomass (food, feed, fiber, and fuel) and the provisioning of other ecosystem services [10–12]. The question therefore arises as to what possible solutions there are for securing the world's food supply while at the same time reducing environmental damage. In addition to reducing food waste and changing dietary habits [13–15], agriculture is challenged to develop long-term sustainable, site-appropriate cropping systems that are able to meet local and global requirements in terms of environmental protection and food security.

The aim of this review is to describe the development of a new farming concept for moderate climates that may significantly improve the environmental performance of agroecosystems while safeguarding yields and product quality. The main characteristic of this new farming concept is the complete refrainment from the use of chemical synthetic plant protection products (CSPs). The exclusion of CSPs is expected to add value to food products from this new farming concept, which may be appreciated by a growing consumer demand for ecologically sustainable products. In an appropriate price segment, this may create a bridge between established conventional and organic products, making it easier for producers and consumers to opt for more sustainable production and consumption. In view of the above, first, different agricultural farming concepts are characterized below and the need for further development is outlined. Second, the idea of new cropping systems following the new farming concept mentioned above is presented. Such new cropping systems are currently being developed and tested in field trials at several locations in Germany as part of the joint research project "Agriculture 4.0 without Chemical-Synthetic Plant Protection" [16]. These new cropping systems are referred to below as "mineral-ecological cropping systems" (MECSs). The cultivation measures characterizing these cropping systems focus on improving the overall ecosystem services. At the level of cultivation measures, the potential provision of ecosystem services [17–19] by the new MECSs is analyzed and compared to alternative cultivation measures applied in organic and conventional cropping systems. Expectations for these new cropping systems are discussed from economic, ecological, and social perspectives, based on literature and expert knowledge of the research consortium. This contribution is intended to stimulate further research on MECSs under varying natural and economic conditions.

2. Characterization of Farming Concepts

Despite the recent emergence of land-independent food production systems, such as sky farming and urban farming, future global food security will continue to rely predominantly on land-based farming systems [1]. During the past century, various forms of land cultivation have emerged that differ in numerous ways, but coexist. Basically, all farming systems can be classified as either conventional or organic. Whereas conventional farming rather focuses on maximizing yields with the help of more or less industrialized processes, organic farming is oriented toward the use of natural regulatory processes (Figure 1). Within the two basic concepts of conventional and organic agriculture, there is a broad spectrum of conventional and organic farming systems that rely to varying degrees on industrialized or natural process control. Furthermore, they often integrate different sub-concepts with a partial or holistic scope.

The main characteristics of industrialized process control are a high degree of technology and specialization, in addition to a high input of energy and external means of production (Table 1). Production processes are comprehensively controlled, for instance with the aid of synthetic fertilizers, synthetic plant protection products or genetic engineering. The most highly industrialized form entails the land-independent production of plants or animals in closed facilities with automated control of light, water, and climatic conditions. Examples of highly industrialized farming systems are maize or soybean monocultures in North and South America, palm oil plantations in Asia, and vertically highly integrated forms of animal husbandry, especially in the poultry sector, in addition to indoor growing of fruit and vegetables. Natural process control is characterized by highly diversified crop rotations and site-adapted, resistant varieties. Natural cycles are largely closed by means of on-farm nutrient production and by largely avoiding external means of production.

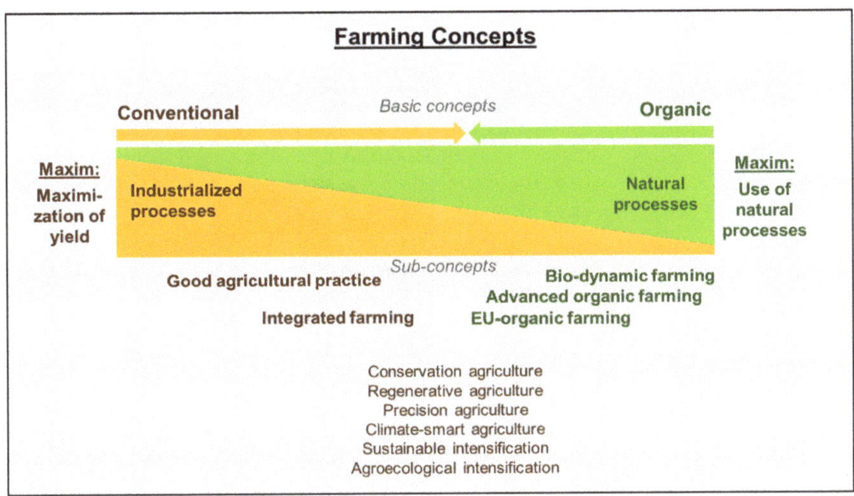

Figure 1. Schematic representation of farming concepts.

Table 1. Main characteristics of industrialized and natural process control in farming systems (modified in line with [20–23].

Industrialized Process Control	Natural Process Control
- monotone crop rotations	- diverse crop rotations
- low degree of heterogeneity in agricultural landscape	- high degree of heterogeneity in agricultural landscape
- high-yield varieties	- site-adapted, resistant varieties
- high external inputs	- low external inputs
- open cycles	- towards closed cycles
- high capital intensity	- low capital intensity
- low labor intensity	- high labor intensity
- high degree of mechanization	- low degree of mechanization
- high degree of specialization	- low degree of specialization
- high vertical integration	- covers a large part of the value chain
- comprehensive control of agricultural production processes	- higher use of natural regulatory processes for agricultural production

In their extreme forms, these two concepts of process control are only realized in a small proportion of farming systems. With regard to conventional and organic cropping systems, their characteristics are manifold and an increasing blending of the two approaches to process control can be observed. Specialized, highly technical arable farms are increasingly integrating measures to promote environmental sustainability, such as landscape elements, precision farming, and eco-schemes [24,25]. In organic farms, elements of industrialized process control, such as an increasing degree of mechanization and specialization, in addition to trends toward global processing and marketing structures, can be observed [26,27]. Examples include organic strawberry and grape monocultures with increasing input dependence in California [23,26], and rather industrially produced and marketed organic products, which are perceived as "organic-light" [28,29]. However, holistic sustainable farming systems not only fulfill defined minimum standards, e.g., by omitting chemical pesticides and synthetic fertilizers ("substitution approach"). They integrate ecological, social and cultural sustainability aspects [27,30,31], e.g., by promoting heterogeneous agricultural land [22,23] and regional value chains [29]. A truly sustainable agriculture is also demanded by "Organic 3.0", a vision of the global organic movement [27].

Globally, conventional farming systems account for the largest share of agricultural production and land use. Organic agriculture amounts to only 1.5% globally, but 7.7% in the EU and 9.1% in Germany with a growing tendency [32]. Within conventional farming, progress towards more sustainable farming systems can be observed. Globally, an estimated 29% of all farms practice sustainable intensification methods, such as integrated farming or conservation agriculture, on 9% of farmland [24]. Within the EU, conventional farming is based on the standard of "Good Agricultural Practice", which is defined by legally regulated minimum requirements. These relate to the use of plant protection products [33] or health management in animal husbandry. A further reduction or minimization of ecological and health risks is the goal of "Integrated Farming" [34]. "Integrated Farming" uses both chemical and organic inputs for nutrient supply and plant protection, but use is based on economic thresholds for damage. It takes advantage of the natural strengths of plants, such as resistance to drought or disease. An important pillar of integrated farming is "Integrated Pest Management" [33], which has been mandatory in the EU since January 2014. The characteristics of organic agriculture [30] range from the more pragmatically oriented EU regulations for organic agriculture [35] to anthroposophically oriented biodynamic agriculture [36]. Overall, there are several organic farming associations that defined advanced sub-concepts of organic farming including higher standards compared to EU regulations [37,38].

Within these sub-concepts of conventional and organic farming systems, there are multiple approaches to increase ecological or economic sustainability. These approaches are either process oriented or result oriented. They focus on individual cultivation measures, on the use of specific technologies, or on particular ecological or economic goals. Additionally, they may take into account partial aspects, the entire cropping system, or even the entire food sector. All these approaches to the optimization of farming systems focus on the preservation of natural resources and the promotion of ecological or economic sustainability. "Conservation Agriculture" aims to maintain and enhance soil fertility through reduced tillage, year-round greening, and diversification of varieties and crop rotations [39]. The main goals of "Regenerative Agriculture" are to build up humus, improve soil health, increase biodiversity, and promote plant–soil interactions. Key farming practices include eliminating or minimizing tillage, permanent greening, and organic fertilization [40]. "Precision Agriculture" seeks to minimize agricultural inputs by applying plant- and site-specific crop management using modern agricultural technologies, including digitization [41].

More holistic approaches to achieve global food and environmental goals underlie the concepts of "Sustainable Intensification" and "Agroecological Intensification". The overall objective of "Sustainable Intensification" is to achieve a yield increase without taking up additional land or harming the environment. It is relatively open and does not privilege any particular vision or method of agricultural production [42–44]. It focuses on increasing resource efficiency, including the use of technology [45]. "Sustainable Intensification" is guided by the concept of "land sparing" to preserve natural landscapes. "Agroecological Intensification" is more explicitly defined and focuses on understanding, strengthening and using biological and ecological processes by applying multiple agroecological practices [44,46]. Agroecological approaches connect scientific ecological disciplines and farm management [26,43]. This is because healthy ecosystems provide a range of services that help to maintain yield stability, pest and pathogen control, nutrient cycling, and resilience. Biodiversity plays a key role in this [42]. "Agroecological Intensification" is based on the concept of "land sharing" [45]. The implementation of agroecological approaches necessitates a fundamental redesign of farming systems considering both participatory approaches and adaptation to local conditions [26,31,42]. Climate-Smart Agriculture is an integrated management approach that addresses the interlinked challenges of food security and accelerated climate change [47,48]. Climate-Smart Agriculture is based on the concept of "Sustainable Intensification" [49]. The Climate-Smart Agriculture approach pursues three objectives: sustainably increasing productivity and incomes, adapting to climate change, and reducing greenhouse gas emissions [39]. The measures used to achieve these goals are highly variable [47]. "Climate-Smart Agriculture is not a set of practices that can be universally applied, but rather an approach that involves different elements embedded in local contexts" [39]. Overall, there is an ongoing contrasting debate about whether high technology-based or ecology-based practices are the most appropriate agricultural production practices to achieve the goal of higher yet sustainable food production [46].

Examples of private initiatives show how such approaches to improve sustainability of cropping systems can be implemented at a local level. In southern Germany, marketing communities such as "KraichgauKorn" [50] and "BlütenKORN" [51] are associations of farmers, mills, and bakeries that commit to specific cultivation measures to provide particular ecosystem services. They define guidelines for the entire value chain and add value to the ecosystem services they provide by placing appropriately labeled products on the market. A further example of a farmer's association producing and marketing more sustainable food products is "IP-Suisse" [52].

Studies on the ecosystem services of different farming systems are often based on comparisons between conventional and organic farming. Thus, numerous studies confirm that organic farming provides higher-regulating ecosystem services than conventional farming [14,53–58]. This is especially true for area-based considerations, which are normally taken as the reference for ecosystem services. In terms of output-related environmental efficiency, conventional agriculture performs better in most studies because of higher yields [57,59,60]. Numerous meta-studies show that yields are lower in organic farming than in conventional farming, due to nutrient deficiencies, damage from diseases, pests, or weeds. The average yield gaps range from 19 to 25% for all crops studied globally [57,58,61–63]. There are major differences between individual sites, crops, and specific cultivation methods [57,63]. For example, the yield gap of up to 40% for wheat and barley is above average, whereas for maize it is below average at around 15% [63]. Furthermore, it can be seen that the yield gap widens in some cases with increasing yields in conventional cultivation [62]. In Germany, the yield gap is up to 45% [14] or 50% [64]. When comparing the results of the Federal Ministry of Food and Agriculture (BMEL) farm network, yield differences of up to 50% are discernible depending on crop and year [64]. According to Treu et al. [65], organic farming in Germany requires 45% more land than conventional farming, even assuming reduced meat consumption and thus a lower land requirement. Overall, the reported yield gap tends to underestimate the actual yield difference in most studies by making comparisons at the crop level. Therefore, De Ponti et al. [62] call for an accurate productivity analysis of organic and conventional practices at a higher system level. This is intended to help adequately account for specific nutrient availability when organic farms (i) have additional rotations with nutrient-accumulating crops or (ii) use farm manure across farms.

Overall, however, it is the individual cultivation measures that produce specific ecosystem services, rather than the conventional or organic orientation of a cropping system. These include, for example, crop rotation and the type of fertilization, crop protection, or soil cultivation [55,57,59]. Thus, agricultural cropping systems are mainly characterized by the composition of their underlying cultivation practices. The optimization of cropping systems in terms of related ecosystem services therefore seems most feasible at the level of cultivation measures (Table 2). In addition, site factors such as landscape structure, which are important determinants of biodiversity regardless of cropping practices, determine the level of ecosystem services provided by agricultural landscapes [54,56,66,67].

3. Implications for the Further Development of Agricultural Cropping Systems

Different cropping systems provide different ecosystem services (provisioning, regulating, habitat, and cultural services [17,18]). Due to multiple trade-offs, individual ecosystem services of a cropping system cannot be maximized simultaneously. An increase in yield often leads to a decrease in regulating services and vice versa [68]. Therefore, the merits of different cropping systems cannot be assessed in general terms, but must always be considered in the context of local or global requirements for individual ecosystem services [39,69]. For example, certain mandatory local environmental requirements, such as groundwater or biodiversity protection, may justify very extensive cropping systems that provide lower yields, but a high level of regulating ecosystem services. Conversely, global food security and environmental goals may justify more intensive cropping systems that deliver higher yields and, in some cases, higher environmental efficiency, especially when land-use change effects are taken into account. Overall, the major challenge of agroecosystem management is to promote multiple ecosystem services in a manner that enhances their global provisioning by reducing trade-offs and increasing synergies [68,70,71]. In this context, numerous studies have concluded that there is no single optimal cropping system and that existing cropping systems must evolve [39,57,59]. If organic farming is to secure the world's food supply, yields must be increased significantly without causing additional harm to the environment [14,15,57,59]. The main challenges here are nutrient deficiencies, diseases, and pest and weed infestation. In contrast, for conventional farming systems, a reduction of chemical pesticide and fertilizer input, and their emissions, constitutes the greatest challenge when it comes to reducing damage to the environment [57]. To complement the intensification of organic agriculture and the greening of conventional agriculture [14,15,66], there are recommendations to remove the sharp boundaries between organic and conventional farming by developing "hybrid" farming systems that combine different technologies and farming practices from organic and conventional agriculture [63]. These hybrid farming systems already exist in many forms (Figure 1).

In general, the question arises regarding how the global food supply can be secured in the future [11]. In principle, an expansion of agricultural land, yield increases, a more efficient use of food and a change in human diets can contribute to improving the world's food supply. According to Niggli and Riedel [15], even an expansion of organic farming is feasible if food waste is reduced and animal-based foods in the human diet are partially replaced by plant-based foods that require less land. However, as long as there is no significant global change in dietary structure and no increased efficiency in food utilization, organic farming will only be able to make a limited contribution to global food security due to its lower productivity [63,72]. Conventional farming causes severe damage to ecosystems in some cases. The numerous hybrid farming concepts often demonstrate only minor ecological advantages. Furthermore, apart from a few local initiatives, they rarely succeed in placing a clearly distinguishable product with ecological valorization options on the market.

At the political level, the aim is also to develop environmentally friendly agricultural farming systems. As part of the EU's Green Deal, the EU Commission has formulated goals for the future direction of agriculture in the EU in its Farm to Fork strategy. The main pillars are a reduction in the use of synthetic chemical pesticides and of nutrient losses by at least 50% by the year 2030 [73]. In this context, the European research alliance "Towards a chemical pesticide free agriculture" was formed in 2020 [74]. Its aim is to create a roadmap for the development of European agriculture towards agriculture without any chemical pesticides [74]. Within the framework of the EU, in addition to at the national and regional level, various regulations and support programs have been established to promote environmentally friendly agriculture. According to the EU Framework Directive on the sustainable use of CSPs [75], all Member States have implemented National Action Plans (NAP) to reduce risks and impacts of pesticide use on human health and the environment. As part of the French NAP "Ecophyto" [76], a network of farmers has been established to test and evaluate possibilities for the reduction of chemical pesticide use [77]. The concern of reducing chemical pesticide use is also reflected in the EU framework for the Common Agricultural Policy (CAP) period 2023–2026. Here, the German strategy plan for the upcoming EU CAP period foresees the promotion of the abandonment of chemical pesticide use as a possible measure within the eco-schemes [78]. In addition, the "Insect Protection Action Program" restricts the application of CSPs at the national level in Germany [79]. At the local level, e.g., in the state of Baden Württemberg, the reduction of chemical pesticide use and the conversion to organic farming is promoted by the "Biodiversity Enhancement Act" [80].

4. Development of Mineral-Ecological Cropping Systems

The idea of mineral-ecological farming is to establish a new farming concept apart from conventional and organic farming that is appropriate to meet both future environmental and global food requirements (Figure 2). In MECSs, in accordance with [35] and in conjunction with [81], the use of CSPs must be completely avoided. At the same time, all yield-relevant cultivation measures are to be optimized to safeguard yields. In the design of this new cropping system, new and existing technologies are combined with agroecological practices [26] to promote natural regulatory processes, and to also optimize mineral fertilization and non-chemical curative crop protection. This aims at improving the overall ecosystem services of agricultural landscapes based on (i) improved provision of regulating ecosystem services compared to conventional cropping systems and (ii) improved supply of provisioning ecosystem services compared to organic cropping systems.

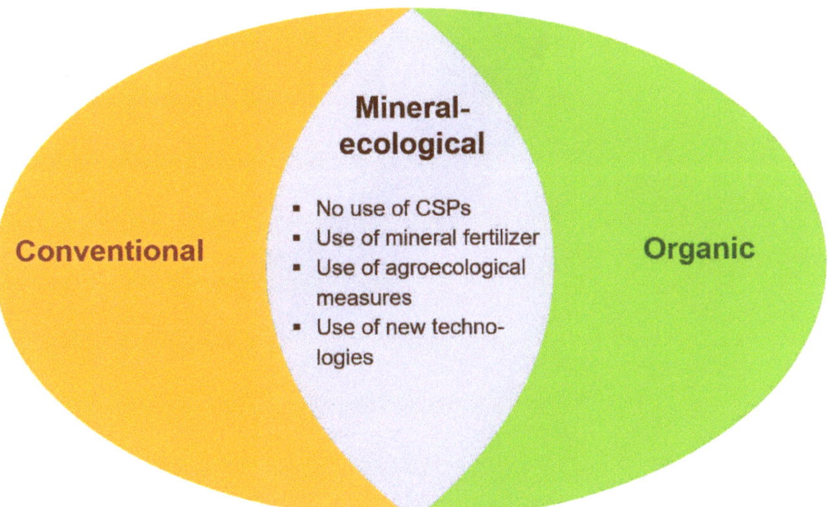

Figure 2. Complementing conventional and organic farming by mineral-ecological farming (CSPs: Chemical synthetic plant protection products).

The design, implementation, and evaluation of MECSs need to take into account various aspects at different levels (Figure 3). Multi-year system field trials are needed to capture crop rotation and long-term effects of cropping systems. Only a holistic approach will allow an adequate comparison of MECSs with conventional and organic cropping systems. This includes studies at the farm, regional, processor, and consumer levels with respect to success criteria and possible adaptations. Finally, MECSs and their contribution to improved ecosystem services in comparison to conventional cropping systems needs to be evaluated. In the following, various key aspects of MECSs (Figure 3) based on both scientific literature and the approach of the project "Agriculture 4.0 without Chemical-Synthetic Plant Protection" are outlined and discussed.

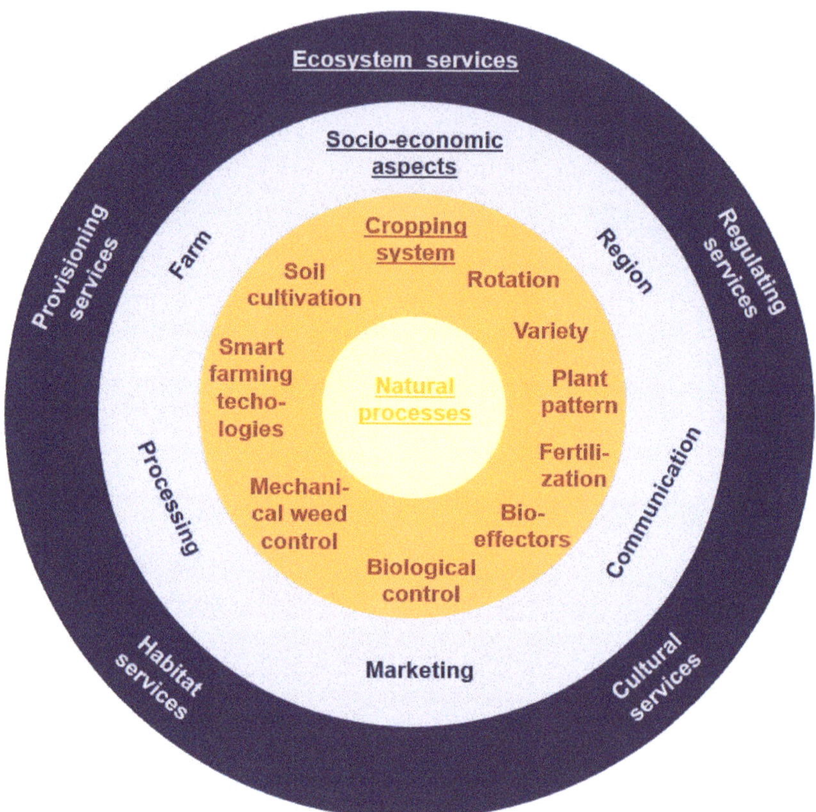

Figure 3. Schematic representation of essential levels and aspects to be considered in the development, implementation, and evaluation of mineral-ecological cropping systems.

4.1. Exclusion of Chemical Synthetic Plant Protection Products in Mineral-Ecological Cropping Systems

Refraining from CSPs is a key tool in MECSs for improving numerous ecosystem services (Table 2). In particular, positive effects on biodiversity, pollination, biological control, soil fertility, and climate regulation can be expected. Furthermore, increasing biodiversity may have positive effects on the cultural services of agricultural landscapes. However, yield losses due to disease and weed or pest infestation can affect regulating services, such as reduced nutrient efficiency [82], and thus lead to nutrient leaching, i.e., negative effects on nutrient cycles and water quality. Nonetheless, a reduction of the active ingredient contamination of water bodies can be expected by refraining from the use of CSPs [83]. In addition to reduced pesticide exposure, increased pathogen exposure is expected, leading to quality degradation and health hazards. A problem related to fungal diseases is the production of mycotoxins and, in particular, those produced by *F. graminearum* pose a risk for humans and livestock because they can cause poisoning and fertility and growth disorders [84–86]. Overall, the abandonment of chemical pesticides is expected to have predominantly positive effects in terms of regulating habitat, and cultural services, but at the same time, provisioning services are expected to be significantly impaired in terms of yield levels and yield stability [31,87,88].

Scientific evidence on yield effects associated with the absence of CSPs in MECSs is very limited. Although the yield differences between organic and conventional cropping systems have been investigated in numerous studies, it is not known to what extent the lower yields in organic farming are due to the absence of CSPs or to other factors. There is only fragmentary knowledge on the consequences of not using CSPs. This knowledge either builds on data from individual sites [87,89,90], only considers individual ecosystem services, primarily yield [91,92], or merely removes CSPs from the system without making further adjustments to the cropping system [88,93]. A detailed analysis of the factors influencing yield formation is a central pillar in the development of MECSs. In this regard, it is crucial to assess and develop optimal combinations of cultivation measures in order to keep yield and quality losses as low as possible while dispensing with CSPs [94].

4.2. Mineral-Ecological Cropping Systems from a Production Technology Perspective

When CSPs are excluded, the yield performance of agricultural cropping systems can be improved through a variety of agroecological and technical measures [26]. In this context, the development of MECSs focuses on optimizing all yield-relevant cultivation measures to manage the absence of CSPs. This includes both direct and indirect yield-impacting measures, and mixed forms designed to provide a wide range of ecosystem services (Table 2). The ecosystem services of individual cultivation measures, and of bundles of measures, are very complex and there are numerous interactions and trade-offs that cannot be fully explored in this review. Therefore, Table 2 only lists essential ecosystem services that are expected in connection with MECSs cultivation measures and that are described in the literature. A large number of these cultivation measures benefit from the use of precision agriculture technologies. A wide range of existing and new technologies can be applied, investigated, and further developed according to the specific requirements of the MECSs (Table 3). Indirect yield-impacting measures encompass all cultivation measures that promote natural regulatory processes, such as diverse crop rotation, the use of site-adapted, resistant varieties, and an optimal spatial distribution of plants in the field, for instance, in the form of equidistant seeding. All of these agroecological measures are intended to improve numerous regulating and habitat services, thereby helping to minimize yield fluctuations and losses. In organic cropping systems, for example, yields can be enhanced by optimizing cultivation methods, such as cultivation of mixed crops and diversification of crop rotations [61]. Biodiversity-enhancing measures, such as the promotion of diverse agroecosystems and structurally rich agricultural landscapes, lead to an improvement in many regulating services (pest regulation, pollination, and nutrient cycling), and thus to positive complementary or synergistic effects on crop yields [95–97]. Furthermore, improving soil fertility promotes plant growth and yield formation (provisioning services), and regulates diseases and pests (regulating services) [66]. In addition to agroecological cultivation measures, optimized mineral fertilization with macro- and micronutrients is a key measure to directly promote yield performance in MECSs. As nitrogen is applied as placed, stabilized ammonium, it is expected to generate positive effects on numerous regulating services compared to conventional fertilization (Table 2). Micronutrients and bioeffectors can positively influence yield performance both directly and indirectly through their plant-strengthening action. Furthermore, mechanical and biological methods of curative plant protection with different effects on regulating services are well-known cultivation measures for weed, pest, and pathogen control that directly influence yield (Table 2).

Table 2. Expected ecosystem services [a] of the main cultivation measures applied in MECSs [b] (+: positive effect expected; −: negative effect expected; o: no clear effect expected).

Objective	Promotion of Natural Regulatory Processes				Direct Yield Promotion and Promotion of Natural Regulatory Processes			Direct Yield Promotion	
Point of Application	Plant Protection	Plant			Fertilization and Plant Strengthening			Curative Plant Protection	
Category of Action	Pesticide Choice	Rotation Choice	Variety Choice	Plant Pattern	Fertilizer Choice		Effector Choice	Weed Control	Pest and Desease Control
MECSs Cultivation Measure	No Use of CSPs	Diverse Rotation	Resistant Varieties	Equidistant Seeding [c]	Miner. Fertilizer (Macro and Micro Nutrients) [c]	Placed, ammonium [c]	Bioeffectors [c]	Harrow, How [c]	Biocontrol Agents [c]
Provisioning services									
Yield	−[31,87,88,91,92,98–102]	+[97,103,104]	+[82,88,99,100,105–107]	+[108–111]	+[112]	+[99,113]	+[114–116]	+[117]	+[118,119]
Product quality	+[103,120–124] −[84–86,124]	+	+[82,106,107]	+[125,126]	+[127]	+	+[114–116]	+	+[128,129]
Regulating services									
Climate change mitigation	+[87] −[130]	+	+[82]		+[131] −[113]	o[113] +[131]	+[131]		+[132] −[133]
Water regulation and quality	+[83,134,135] −[130]	+[97,104]	+[82,105]	+[109]	+[121] −[136]	+[131]		−[137]	+[138]
Erosions prevention		+		+[108,110]				−	+[139]
Soil fertility	+[123,140–142]	+[23,97,104]		+[108,110]	+[112,136] −		+	−[137]	+[143–146]
Nutrient cycles (efficiency)	−[82]	+[97,104]	+[82]	+[108–111]	+[131]	+[113,131]	+[131]	−[137]	+[147–150]
Pollination	+[151–153]	+[22,23]							+[154]
Weed supression	−[92,155]	+		+[108,110,111]				+[117,156,157]	+[158,159]
Biological control	+[140,160–162]	+[23,98,163]	+[82,105]	+[164–166]	−[159]			+	+[144,159,167,168]
Habitat services									
Life cycles and genetic diversity	+[22,123,140–142,169,170]	+[22,23,160,163]		+[125,126]	−[23,163]				
Cultural services									
Recreation and tourism	+	+[171]							

[a] According to [17,18]. [b] Assessment of ecosystem services of individual cultivation measures applied in mineral-ecological cropping systems compared to alternative cultivation measures common in organic or conventional cropping systems: No use of CSPs—use of CSPs; diverse crop rotation—less diverse crop rotation; predominantly resistant varieties—predominantly high-yielding varieties; equidistant seeding—normal seeding; use of mineral fertilizer—no use of mineral fertilizer; placed ammonium fertilization—classical nitrogen fertilization; use of bioeffectors—no use of bioeffectors; use of mechanical or biological plant protection measures—no use of mechanical or biological plant protection measures. The color differentiation indicates whether a cultivation measure is more likely to be assigned to the organic (green), the conventional (yellow), or none (black) of the two farming concepts exclusively or clearly. The expected effects of the individual cultivation measures on selected ecosystem services are based on expert assessments and the literature (+: positive effect expected; −: negative effect expected; o: no clear effect expected; no symbol: no clear effect expected or no indication). [c] Automation through the use of precision agriculture technologies enables the optimal implementation of these cultivation measures.

Table 3. Precision agriculture technologies suitable for mineral-ecological cropping systems (MECSs).

Area of Application	Seeding	Fertilization	Plant Protection		
			Weed Control	Desease Monitoring	Desease Control
Precision Agriculture Technology	A sowing unit is mounted on a mobile robot platform [172]. The robot uses different sensors to navigate autonomously and to adapt the sowing parameters to the actual soil conditions [172] and crop requirements.	The modified sowing unit can be used for insoil fertilization. The robot uses image-based row recognition to precisely place the fertilizer between the crop rows [173].	A real-time camera-based automatic guidance system is used to steer hoeing blades in the center between the crop row [174,175].	Unmanned aerial vehicles (UAVs) are combined with hyperspectral imaging sensors to analyse the plants spectral signature for pathogen detection [176–178]. Machine learning and AI are used to analyse sensor data, allowing automatic detection and detection of subtle changes in early stages of pathogen development [179].	UAVs are used for the application of biological control agents (BCAs). UAV-application can easily schedule the release of BCAs exactly when and where they are needed according to their modes of action for better control of plant pathogens [180].
Research focus to optimize cultivation measures in MECSs	• Investigation and modification of technology to improve quality of incorporation of seeds into soils.	• Investigation of effectiveness and efficiency of in soil fertilization during vegetation period in grains. • Site-specific and individual plant-adapted fertilization by variable rate technology.	• Accuracy of steering and efficacy of weed control by camera-guided inter-row hoeing. • Crop response and crop yield in cereals, soybean, and maize with different row spacings, growth stages, driving speeds and hoeing elements.	• Mapping of problematic areas in the field for precision farming. • Refinement of forecast methods. • Objective assessment of disease severity and plant development in MECSs.	• Effective BCAs with al-ready elucidated modes of action will be evaluated under field conditions. • Investigation of different application schemes to establish BCAs in the field according to monitored forecast results.
Expected benefits	• Higher field emergence with less seeds. • More even seeding depth and soil coverage of seeds. • More even spatial crop plant distribution according to site and variety-based requirements. • Better aeration of crop plant stands and equal access of plants to resources. • Quality over quantity through autonomous work.	• Better access of crop plants to micro- and macronutrients, and bioeffectors. • Optimize fertilizer use and minimize losses. • Improving nutrient use efficiency.	• Higher weed control efficacy. • Less crop damage due to higher selectivity. • Higher labor efficiency due to higher driving speeds.	• Early and precise detection of disease symptoms and abiotic stress factors. • Assessment of disease severity and identification of disease species through specific changes within the plants spectral signature. • Drone-based measurements permit high throughput coupled with variable resolution. • Automated and objective data assessment through data analysis methods.	• Drone-based biocontrol offers a new tool for effective and sustainable control of plant diseases. • Minimizing yield losses quantitatively and qualitatively through combating pathogen infestation and minimize the risk of food safety hazards (e.g., mycotoxins).

4.2.1. Cultivation Measures to Promote Natural Regulatory Processes

Cultivation measures that promote natural regulatory processes are one focus of MECSs (Table 2). These include among other measures diversified crop rotations, the use of site-adapted resistant varieties, and optimized plant distribution in the field.

Given the numerous options for process control, tight crop rotations based on the most competitive crops are possible in conventional cropping systems. Without CSPs, diversification of crop rotations is one means of reducing weed, disease, and pest pressure. Diverse crop rotations promote pest regulation, soil fertility, and biodiversity [23] (Table 2). MECSs also prioritize the most competitive crops, but combine them with less competitive ones, including catch crops that provide diverse ecological services in return. Compared to organic farming, MECSs have the great advantage of not relying on legumes for nitrogen supply. To implement preventive plant protection, to promote soil fertility, and to optimize nutrient cycles and natural regulatory processes, MECSs are dependent on the diverse, resilient crop rotation of cereals and leaf crops, winter and spring crops, and main and catch crops. Conventional crop rotations of corn and winter cereals must be widened by integrating summer cereals, protein crops, and catch crops. Although legumes are less important in MECSs than in organic farming, they may be worth being integrated in terms of delivering multiple ecosystem services, such as climate change mitigation and improvement of soil fertility, nutrient cycles, water quality, biocontrol, and habitat and cultural services [181]. It is necessary to consider not only individual crops, but also entire crop rotations, because the crop rotation effects of different pre-crop–post-crop combinations must be quantified and evaluated [182]. This must be undertaken based on long-term field trials, which identify the effects of different crop rotation combinations on yield and further ecosystem services [183].

In the absence of CSPs, the breeding of resistant varieties is of particular importance to increase resilience in MECSs and positively influence yields. For a long time, resistance breeding has played a central and successful role in the development of varieties for conventional cropping systems. Therefore, not only yield, quality, and marketing opportunities, but also resistances against multiple pathogens are key factors in the legal protection of varieties and their selection for cultivation [184]. Winter wheat varieties released in Germany are good examples, because the observed yield increase over past decades also resulted from, among other factors, strongly improved pathogen resistance [88,100]. Cultivation systems without CSPs require the perpetual development of varieties with durable resistances against multiple pathogens [185]. Multi-resistant varieties are mainly developed by marker-assisted introgression of different resistance genes (pyramiding), which will be complemented in the near future by targeted genome editing of novel resistance alleles [186–188]. A second component are improved systems for the selection of resistant varieties by combining digital tools for pathogen monitoring and real-time resistance breeding (e.g., by genome editing). Such systems aim to slow the co-evolution of pathogens with their crop hosts by increasing the diversity of resistance genes in cultivated varieties at a geographic scale. A proof of concept is the rice and rice blast pathosystem in Asia [189,190]. It demonstrates the potential and future contribution of resistance breeding to refrain from CSPs while maintaining high yield levels.

In addition to crop rotation and cultivar resistance to pathogens and pests, stand conditions of crops are of crucial importance to reduce the risk of infection and weed pressure in arable farming. Accordingly, the optimization of the spatial distribution of cultivated plants in MECSs plays a central role. Depending on the sowing pattern, variety characteristics, and crop management, different crop development and crop architecture (temporal and spatial development of the crops) will emerge. This influences not only light and nutrient conditions of the crop, but also the microclimate (water availability, temperature, and humidity dynamics) and thus weed, pest, and disease pressure, in addition to yield formation. Plants are often not optimally distributed spatially in the field [110,191]. As a result, plants within a crop stand have different degrees of access to required resources. Therefore, their individual development will vary accordingly. Plants evenly distributed in the field can make better use of their potential and resources, and are also more solidly anchored in the soil. This has advantages for mechanical weed control measures, such as harrowing. Uniform plant distribution is achieved with so-called equidistant seeding. This is expected to achieve stronger weed suppression [108], and provide a better microclimate with a lower risk of infection by fungal pathogens [166,192]. At the same time, it is expected to optimize potential yield and nutrient efficiency, thus leading to optimal use of mineral fertilizer. In addition to positive effects on soil erosion and soil water balance, equidistant plant distribution is expected to lead to earlier crop closure [110]. In response to spatial plant distribution, plant species or even cultivars develop different phenotypes (plasticity) [193]. Both plasticity and morphology are mainly influenced by light quality, specifically the red:far-red ratio (RFR), especially during early growth stages [194,195]. In equidistant seeding, the change in photosynthetically active radiation (PAR) and RFR is expected to increase the number of branches in soybean and the number of cobs per plant in maize. The altered plant architecture and more even distribution of plants leads to increased competitiveness for light with weeds [125,126]. The complex interactions (light, competition, plant architecture, and physiology) in the crop, in addition to changes in crop management (variety and seeding pattern), can be depicted in functional-structural plant models (FSPM) [196].

To assess the impact of crop architecture on pathogen establishment, microclimate measurements can be combined with numerical simulations using 3D CFD (Computational Fluid Dynamics). The change in microclimate by equidistant seeding may mitigate the risk of infection. How much the crop warms up or how fast it dries after rainfall largely depends on aerodynamic properties such as displacement height and roughness length. These, in turn, are determined by factors such as seeding density, seeding pattern, leaf area, leaf width, leaf inclination, and the variability of growth height [165]. The aerodynamics in and above the crop canopy are seen to be a crucial control to reduce disease pressure from fungal diseases. The key factor is the turbulent mixing in the upper canopy layers. Flow simulations using a virtual wind tunnel can help in investigation of the fine structure of the turbulent exchange in the crop canopy and the adjacent parts of the atmospheric surface layer. This data can be used in the NoahMP Grecos plant growth and land surface model [164] to simulate crop temperature and humidity dynamics, and estimate the risk of infestation for altered spatial plant distribution.

4.2.2. Cultivation Measures for Direct Yield Increase, Plant Strengthening, and Plant Protection

Cultivation measures that indirectly support yield formation by promoting natural regulatory processes need to be complemented by direct measures for yield enhancement, plant strengthening, and plant protection, such as optimized mineral fertilization in combination with bioeffectors and micronutrients, mechanical weed control, and biological pest and disease control.

To achieve similar yield levels in MECSs compared to those of conventional cropping systems, optimal nutrient application is essential. In addition, the possibilities of nutrient combinations must be optimized because they can act prophylactically against fungal, bacterial, and animal pests through infestation-suppressing and resistance-increasing effects. Moreover, essential and beneficial plant nutrients, such as silicon (Si), zinc (Zn), and manganese (Mn), with proven protective effects against abiotic and biotic stress factors [197–199], in addition to plant growth-promoting bioeffectors (microorganisms and natural compounds such as algal extracts), can be applied by means of inoculation and mineral fertilization. In this context, continued development and selective use of ammonium depots (e.g., with the Cultan technique or the use of fertilizers stabilized by means of nitrification inhibitors using in-soil fertilization), in addition to targeted application of calcium cyanamide, bioeffectors, and micronutrients with adapted soil and foliar applications, is relevant. In addition to providing adequate nutrient supply and protective functions, these measures are used to ensure optimal nutrient balances. In contrast to organic fertilizers, individual mineral nutrients can be applied in a targeted, plant-available form as needed, and some forms of nitrogen also have a certain pathogen suppression potential [200,201]. Protective effects of silicon [202,203], micronutrients [106,204] or inoculation with beneficial microorganisms (bioeffectors) [107] have proved their benefits for various agricultural crops, but the extent to which these can replace conventional CSPs is not clear.

In non-chemical weed control strategies such as in MECSs (Tables 2 and 3), multiple weed suppression strategies are required to secure crop yields. These include crop rotations with spring and winter seeding crops, including cover crops to reduce the density of problematic weed species such as blackgrass (*Alopecurus myosuroides*) [205]. Inversion tillage with a plow significantly reduces weed infestation compared to reduced tillage practices [206]. A false seedbed is also a suitable method to reduce weeds [207]. Curative weed control can be undertaken by harrowing, hoeing, and other physical methods that can be performed between and within rows. Hoes can uproot and cover larger weeds very effectively [156,208]. Manual steering of hoes between rows can be made more precise by automatic steering systems using GNSS (Global Navigation Satellite System) techniques and optical sensors. When automatically steered hoes are used, control success against weeds within rows is higher than with conventional hoes because automatically steered hoes can go faster (10 km/h) and closer ($\pm$2 cm). In addition, there is less damage to the crop.

Biological control agents (BCAs) can be an alternative to CSPs to control plant pests and diseases [167]. An indirect mode of action of BCAs is the induction of plant defense reactions. This will put plants in a so-called priming state. Priming describes a state in which the plant is prepared more quickly and more resiliently to deal with possible pathogen infection [209]. The detection and treatment of pathogen infections at an early stage is crucial for effective pathogen control. This requires innovative technologies for sensor-based pathogen monitoring and applying BCAs (Table 3), in addition to appropriate formulations and methods of application that attain successful establishment of BCAs in the field.

In addition to curative plant protection measures, without CSPs, the discussion about optimal soil management takes on new importance in the context of prophylactic plant protection. More intensive tillage (plough) can make a substantial contribution to yield stabilization by reducing the pressure of diseases, weeds, and pests, especially in MECSs. At the same time, however, negative effects on soil erosion in silt-dominated soil textures, and on soil organic matter content and nutrient cycles, can occur (Table 2). In this context, the effects of different soil management measures in conventional farming, organic farming, and MECSs must be investigated with respect to crop yield, product quality (e.g., *F. graminearum* infestations), and ecological indicators.

4.2.3. Use of Precision Agriculture Technologies

Many of the cultivation measures of MECSs will benefit from the use of precision farming technologies (Table 3). Innovative technologies such as autonomous vehicles, drone-based monitoring and application methods, and automated hoeing technology allow the optimization of seeding and fertilization, in addition to the early detection and treatment of plant pathogens, pests, and weeds, making MECSs effective and efficient.

Automated, camera-controlled methods of hoeing technology offer advantages, especially for specific sowing patterns such as equidistant seeding. More precise crop row detection can be undertaken using image analysis technology. Hoeing blades can be adjusted hydraulically/electronically and, in agricultural crops, weeds can be effectively suppressed with such hoeing technology. Camera-based methods offer the possibility of precisely detecting plant rows and also individual plants, and of using the information to steer machines and equipment [174,175]. With a camera-assisted inter-row hoe with automatic side shifting, the effectiveness of weed control between and within rows in soybean and maize can be increased to 85%, compared to 70% for machine hoeing with manual guidance [210]. Even at a row spacing of 12.5–15 cm, certain camera-guided hoes can be used in cereals [174].

Many specific requirements of MECSs in terms of spatial distribution for seed placement, fertilizer application, and mechanical weed control can be met with the help of GNSS-controlled sensors, actuators, and autonomous vehicles that facilitate precise georeferenced seed and fertilizer distribution, and hoe steering. In recent years, it has been shown that soil fertilization with fertilizer depots in the soil, applied at specific rates, depths, and distances from the plant, can increase nutrient efficiency [211]. This is even more important given the increasingly dry periods. Therefore, fertilizers applied on the surface are increasingly at a disadvantage because precipitation is needed to make the nutrients available to the plants. Automated, highly uniform distribution of the plants and selective fertilizer applications into the soil are effective in increasing crop production, resource efficiency, and weed suppression. Existing autonomous platforms with the appropriate sensors and actuators can deliver a high level of automation.

For successful control of plant pathogens in MECSs using BCAs, an optimized application in terms of time and space is necessary. One potential technology for early detection of plant pathogens is the use of drone-based sensors that generate georeferenced image data. By combining hyperspectral cameras with modern data analysis methods, and comparing pathogen detection via sensor technology and molecular and conventional methods, it is possible to generate procedures for an early detection of plant pathogens and identify their location within the plant canopy for BCA application in the field (Table 3). The capabilities of sensor-based pathogen detection and quantification under controlled conditions have previously been demonstrated in several studies [212–214]. Multiple detection methods are currently being developed to establish a monitoring system for the detection of plant pathogens, which are expected to occur more frequently when CSPs are abandoned. Molecular methods enable the detection of pathogens within the plant, but also on crop residues or in the soil. They enable a holistic assessment of pathogen pressure in MECSs.

4.2.4. Impact on Natural Regulatory Processes

Soil organisms play an important role in the maintenance of different soil functions, i.e., subsequent supply of plant-available nutrients [215], detoxification/mineralization of organic pollutants [142], and stabilization of the soil structure [216]. Symbiotic interactions between soil microorganisms and various crops (e.g., mycorrhizae) protect crops from pathogenic fungi and enhance stress tolerance relating to drought [217]. However, soil microorganisms are influenced to a marked degree by crop management [218,219]. Pesticides usually evoke at least a short-term negative response in soil microorganisms and many soil animals, e.g., earthworms [220]. Accordingly, earthworms and beneficial species are expected to profit from the absence of CSPs. However, it is unclear to which extent this positive effect is relativized by the eventual need for more intensive soil tillage. Equidistant plant spacing is expected to lead to a homogeneous distribution not only of crop roots, but also of resources for soil organisms, and thus to improved efficiency of microbial transformation processes compared to conventional plant spacing.

Reducing the spread of plant pathogens plays a critical role in pathogen management. In conventional cropping systems, monogenic resistances in crop varieties and single mode of action pesticides exert strong selection pressure on pathogen populations, usually prompting the rapid development of resistances [221]. Sustainable management of pathogen populations includes the deceleration of pathogen evolution by (i) diversifying cropping practices (e.g., more complex crop rotations and small-scale cultivation); (ii) the use of multiple resistances on a polygenic basis in breeding; (iii) the cultivation of mixed varieties or mixed cultivation of different crops; and (iv) the development of new pest control methods such as BCAs (Table 2). This can result in the less frequent occurrence and slower spread of new resistance mutations in pathogen populations. In MECSs, the short-term application of CSPs in acute situations is not possible. Therefore, cultivation methods and plant breeding are of particular importance in MECSs. Intensive pathogen monitoring offers the possibility of predicting the epidemiology of pathogens. DNA sequencing, digital technologies, and machine learning techniques enable high temporal and spatial resolution monitoring of pathogen populations. This provides an important basis for designing cropping systems and setting breeding goals [222,223]. In particular, it needs to be investigated whether MECSs have a sustainable and positive effect on the spread of resistant and aggressive fungal pathogens, or if the exclusion of CSPs increases pathogen diversity, resulting in negative effects such as a faster evolutionary adaptation of pathogen populations.

The contribution of MECSs to increase biodiversity and strengthen natural pest control takes on a central role in the research and configuration of this new cropping system. Bengtsson et al. [54] and Tuck et al. [56] observed an increase of, on average, 30% in species diversity in organic cropping systems compared to conventional cropping systems. In this context, the influence of farm management varies by type and scale, and loses significance as landscape diversity increases. Burel et al. [67] state that a minimum amount of semi-natural land is required for biodiversity to be impacted at all by the type of management because many species cannot become established without sufficient habitats or seed stocks. When there are high proportions of semi-natural elements of about 20%, biodiversity is highly independent of the type of management [160]. In contrast, the management form (varied crop rotation, reduced input of nutrients and chemical pesticides, minimum soil tillage, etc.) can significantly enhance biodiversity in regions with a proportion of semi-natural areas of between 5 and 20%, which applies to a large proportion of arable land. This suggests the need for a cross-scale and cross-process approach to representing landscape system processes. Individual parts of this system can be studied very well and precisely with analytical and empirical methods. However, due to the many scale transitions and the complexity of the system, observations of the entire system are almost only possible with models [224].

Overall, biodiversity is of prominent importance in the promotion of numerous ecosystem services with complementary and synergistic effects [95,96]. This is particularly the case for yield formation [22], pollination services [225], and natural pest regulation [226]. Against this backdrop, spatial heterogeneity is a key factor for biodiversity [22,227]. MECSs are expected to have positive effects on biodiversity across spatial scales, and to be reinforced by an optimal design of landscape structure, thereby enhancing natural pest regulation and crop yields from local to regional scales. The extent to which stable predator–prey relationships support natural pest regulation in arable farming must be investigated at a small scale. Predatory flies are particularly suited as a new indicator of functional biodiversity and for analyzing the effects of different cropping systems on predator–prey relationships. They have a small range of activity compared to other antagonists, and their populations are extremely susceptible to any disturbances. As a study model, predatory flies of the genus *Platypalpus* may be suitable, because they are important natural antagonists of crop-damaging flies and midges [228–230]. Because the larvae of predatory flies develop in the soil [229], active soil life is conducive to the abundance and diversity of species of these natural antagonists. Furthermore, the direct and indirect application of CSPs (e.g., limitation of prey or habitat changes) affects the abundance and effectiveness of these beneficial insects. Moreover, additional effects on predator prey relationships generated by crop management, such as crop rotation, spatial distribution of crop plants, and fertilization, need to be evaluated. At the landscape scale, the occurrence and spatial distribution of pests and antagonists depend on (i) field size, (ii) landscape structure, and (iii) temporal land-use dynamics. These effects on dynamic interaction networks need to be considered, to allow major benefits of natural pest control in large-scale MECSs to be predicted. In this context, synergistic interactions between MECSs and other biodiversity-enhancing agri-environmental schemes, in particular the establishment of perennial flower strips, need to be quantified.

4.3. Mineral-Ecological Cropping Systems from the Perspective of Yield and Product Quality

Refraining from the use of CSPs can severely affect crop yield and product quality, depending on disease and pest infestation or weed pressure. In MECSs, efforts must be made to keep these effects as low as possible through optimal design of the cropping system. The potential crop yield losses due to pests and pathogens in wheat production systems are estimated to be as high as 70% [41,99,101]. However, due to conventional crop protection, the actual losses in agricultural practice amount to less than 20% [98]. In MECSs, refraining from the use of CSPs is expected to reduce yield levels and, above all, to increase spatial and temporal yield variability. In particular, pests and pathogens occurring epidemically and featuring a high yield loss potential are relevant in this context [88]. Yield losses lead not only to economic problems, but also to lower N-uptake, and thus reduced nutrient use efficiency and increased leaching risk. Therefore, the extent to which yields in MECSs can be stabilized by adjustments to the cropping system (equidistant seeding, crop variety selection, fertilizer management, etc.) needs to be investigated. The use of simulation models may serve to supplement the limited data on crop yield losses caused by pests and pathogens, to scale them up, and to evaluate scenarios on future or changed climatic, production, and cropping system conditions [231]. Process-based crop models, such as DSSAT-CROPSIM, simulate the soil–plant–atmosphere system, including abiotic stress due to water and nutrient deficiency. For a robust simulation of biotic stress, a sufficient empirical experimental database is essential for the calibration and validation of the crop model, and for the parameterization of specific pests, such as mites and nematodes, and their yield effects [232]. The validated model can be used to simulate the effects of not using CSPs in virtual experiments on yield, N-uptake, nitrogen use efficiency, and N-leaching risk. Inter alia, the impact of cultivar choice can be simulated in the model by considering differences in phenology and resistance to specific pests.

In cropping systems without the use of CSPs, it can be expected that plants will be exposed to increased stress due to intensified weed and pathogen infestation if no other adaptation measures are implemented. In both cases, imbalances in the supply of nutrients, photosynthates, and water for plant metabolism can occur. These, in turn, affect the quality of the harvested products [124]. This mainly concerns products whose quality depends on the composition of primary metabolites (proteins, organic acids, sugars), because their distribution is significantly affected by source/sink ratios in the plant, and rapidly change under stress [233]. In cereals, for example, the composition and temporal development of the storage proteins, which are essential for the baking quality of flour, are markedly affected by the nutrient and water supply [127]. Although the external (sensoric) quality is demanded by consumers, and the internal quality by the processing industry, there are health aspects that give grounds for concern, e.g., increased fungal infestation (mycotoxins) of the harvested material. Compared to conventional cropping systems, individual quality parameters (especially protein composition) of harvest products are expected to change in MECSs as a result of the increased stress level. Furthermore, it is expected that, in the case of cereals, the temporal course of storage protein incorporation between the flowering and grain-filling phases will change, and thus also affect the "final quality" of the grains. The successful establishment and acceptance of mineral-ecological cultivation systems is only possible if these systems deliver products of sufficient quality in the long term, which also meet the technical requirements for product processing. It is expected that optimized fertilization will lead to a stabilization of plant metabolism, and thus to improved resilience against biotic and abiotic stressors. Equidistant plant spacing should result in improved nutrient and water appropriation capacity, and, in particular, stable product quality under drought stress. Whether the increased stress in MECSs can be countered by resistant crop varieties, optimized fertilization, and equidistant seeding, resulting in sufficient product qualities, has to be investigated.

4.4. Mineral-Ecological Cropping Systems from a Socio-Economic Perspective

Economic efficiency and acceptance by farmers and consumers are major success factors for the establishment of MECSs. In terms of economics, MECSs will differ from both conventional and organic cropping systems [94,234]: Organic cropping systems are characterized by higher unit costs compared to conventional ones and, depending on the organic price premium, by higher reliance on government support. Against this background, analyses of unit costs and achievable market prices for products from MECSs including risk-related aspects and the preferences of stakeholders along the value-chain need careful consideration.

4.4.1. Economic Aspects at Farm Level

In MECSs, it is expected that not only the contribution margins, but also the contribution margin variances change as a result of higher yield fluctuations compared to conventional cropping systems. Hence, it is likely that rational farmers will try to adjust their production program in such a manner as to maximize the total contribution margin while maintaining a certain overall risk. This will depend on the risk tolerance of the individual actor. Linear programming models are common activity-analytical operating models in which the optimal production program of farms can be determined in each case by maximizing the total contribution margin subject to constraints [235]. These models are common activity-analytical operating models, but they do not take the total risk into account. To identify optimal risk-efficient cropping systems, the expected-value-variance criterion can be applied within the framework of quadratic risk programming [236,237]. For this purpose, farm models must be developed that allow the selection of cropping practices while maintaining acceptable economic risks (measured as the estimated total variance of the optimized operating profit). Stochastic risk analysis can be used to generate realistic variances.

4.4.2. Economic Aspects at Regional Level

It is not only in conventional cropping systems that the large-scale use of identical plant varieties and pesticide active ingredients is leading to increasing resistance of pathogens. In organic farming and MECSs, too, resistance to pathogens can be expected to diminish with large-scale use. If large-scale cultivation of identical varieties increases, the likelihood of progressively volatile crop yields due to pest and pathogen impacts will rise [238]. Already, the rapid development of pest and pathogen resistance to pesticide-active ingredients and the rapid loss of varietal resistance are playing an increasingly important role in arable agriculture, nationally and globally. Thus, resistance management is expected to gain further importance in the future [239]. It is expected that active ingredient efficiencies and varietal resistance will take on even more of a common property character, and that resistance and yield management will be improved by means of coordinated collective action by farmers based on individually negotiated solutions. In line with the "new institutional economics" [240], a concept for targeted collective action with respect to the preservation of crop variety resistance and stability of crop yields is being developed. In this concept, farmers will optimize the preservation of crop variety efficacy against pests and pathogens, and crop yield stability, by spatially and temporally coordinating suitable cropping and cultivation measures, in addition to suitable crop variety selection. This approach is based on the theory of self-organized and self-managed forms of collective action [241]. Voluntary agreements by farmers concerning a spatially defined unit are key factors. Despite the fact that information about ecosystem services other than yield and quality of the crops is still mostly unclear, they aim to achieve a Pareto optimal result as an incentive to conclude a negotiation. This is because the long-term surplus profit due to reduced costs and increased crop yields achieved through collective negotiation solutions can be high. Such cooperative approaches to achieving positive operational and environmental effects are also pursued by the EU in the context of recent EU CAP reform proposals [242].

4.4.3. Perspective of the Agricultural Sector

A new MECS brings with it many uncertainties for farmers. Numerous factors will decide whether this new cropping system will be adopted and implemented by agricultural operations. It is assumed that a new agricultural production regime without any CSPs can only be implemented on a wide scale if the guiding attitudes, social norms, and restrictions are recognized by practitioners in agricultural operations and specifically considered in corresponding strategies, for instance, related to advisory services or financial support schemes. In the agricultural sector, innovations have been debated almost entirely as the responsibility of the individual farms. However, the goal of dispensing with chemical pesticides is to a large extent being pushed on agriculture from within society. This raises the question of whether and how alliances between stakeholders in society and agricultural operations can promote the roll-out of agriculture based on a MECS as a concerted approach for a (particular) region. It is of decisive importance whether it will be possible to develop strategies that can be implemented on a partnership basis by various social actors (e.g., when agricultural operations receive appropriate support from local consumers). Key factors influencing acceptance and implementation of a MECS require deeper insights into the patterns of impact and leverages of promoting acceptance and implementation.

4.4.4. Perspective of Society and the Food Chain

The debate between the agricultural sector and civil society organizations regarding the future of agricultural systems (e.g., animal welfare) is highly polarized. MECSs will only be successful if they are not only accepted by farmers, but also trusted by key stakeholders and the food chain, and succeed in generating a willingness to pay among consumers. The existence of substantial barriers is highlighted by the example of "integrated cultivation", which (except for some minor exceptions in Switzerland) has failed to become established as a market segment among consumers [243]. Therefore, both consumer willingness to pay and acceptance by food producers and food retailers in the food chain must be determined in order to analyze the market barriers to the introduction of mineral-organic products. With respect to an introduction of barriers from a society perspective, the attitudes towards and trust in the new MECS among central stakeholders must be considered. According to the current state of research, consumer knowledge about CSPs is low and primarily based on mass media reporting. According to a study by the German Federal Institute for Risk Assessment (BfR), a skeptical attitude dominates: 67% of citizens consider CSPs to be harmful to humans even under normal conditions of use; three quarters consider CSPs to be unnecessary for food production. Furthermore, the majority of respondents suspect that there are regulatory deficits in application monitoring and pesticide residue controls [244,245]. Several studies show that consumers' pesticide-related concerns are associated with a greater preference for organic food [246–248]. The omitted use of CSPs or the absence of pesticide residues is a significant characteristic of organic food from a consumer's point of view [249–251]. Hence, positive effects to cater for the needs of many consumers for pesticide-free food have been derived particularly for organic agriculture and the sales of organic food [252]. To date, in Germany no studies have been conducted on consumer willingness to pay for or buy foods from agricultural systems that are specifically characterized by the absence of CSPs. Previous studies in various countries show that the highest proportion of consumers surveyed has a majority willingness (MW) to pay up to 10% more for pesticide-free foods than for conventional products. This finding is evident in studies from Canada (67.1% of consumer MW of 1–10%, [253]), the United States (66.1% of consumer MW of 5–10%, Ott, 1990; 30% of consumer MW of up to 10%, [254]), and Italy (34% of consumer MW of 6–10%, [255]). Although some of these studies are over twenty years old, they indicate that consumers show a positive willingness to pay for products from MECSs. Numerous studies on organic food show that expectations of organic products go beyond the foregoing of CSPs. This means that products from MECSs could possibly occupy a mid-market position. However, it is assumed that foods from MECSs can be clearly classified, valued, and accordingly positioned on the market by consumers as a consequence of the consistent exclusion of CSPs. Against this backdrop, the assumption would also seem to be that consumers view the absence of chemical pesticides as an important indicator of the naturalness of foods and that naturalness is of major or growing relevance as a criterion when shopping for food [256]. It remains to be seen whether consumers are less willing to pay for products from MECSs than for organic food. This attitude would, nonetheless, suffice to secure a successful positioning in the mid-market segment given the significantly lower additional costs of the cropping system. It is postulated for the food chain that the food retail trade will prefer products without any CSPs for risk and reputation reasons. Studies on the classification and estimation of social discourse are particularly complex, because they are driven not only by interests, but also by strategic positions, tactical calculations, and opportunities [257,258].

4.5. Mineral-Ecological Cropping Systems from the Perspective of Ecosystem Services

As shown in Table 2, individual agricultural cultivation measures can have a variety of positive or negative effects on the different ecosystem services of agricultural landscapes. Due to the ecotoxicity potential of CSPs, their avoidance is expected to have a positive impact on biodiversity, species diversity, and abundance, both in agricultural landscapes and in agricultural soils [123,140,141,259]. In addition, positive ecological effects can be expected from the measures specifically investigated in MECSs, particularly from the fertilization strategies. The application of ammonium depots and nitrification inhibitors can reduce the amount of nitrogen fertilizer used, thereby decreasing both the risk of nitrate leaching, and greenhouse gas emissions from the field and from the upstream fertilizer production [131]. Promoting biodiversity and natural ecosystem processes also enhances many aspects of cultural services (see Table 2) [171,260–262]. However, refraining from CSPs can also lead to a deterioration in crop supply services and environmental disadvantages if yield losses cannot be compensated by crop management measures as described above in Section 4.2 A significant decrease in crop yield because of the system conversion would have a negative impact on the environmental efficiency of crop production, because the environmental burdens would be attributed to lower output [130,261–265]. As a result, the total environmental burdens, for instance in the categories of eutrophication, acidification, and climate change, would be expected to increase [130]. Lower crop yields would also have further indirect effects, such as additional demand for agricultural land [266,267]. Globally, the expansion of agricultural land, along with other areas used by man (settlement areas, etc.) is one of the strongest drivers of biodiversity loss [268]. When switching from conventional cropping systems to MECSs, a variety of trade-offs within ecological services, and between production, quality, and ecological targets, are to be expected. Given the complex nature of causal relationships, the effects of individual cultivation measures on ecosystem services cannot be assessed separately, but must be evaluated as a package of measures of a cropping system [182,269]. Life cycle assessment (LCA) is an effective tool for the comparative assessment of cropping systems [270,271]. Firstly, it can be used to allocate the various emissions generated along the value chain of MECSs, for instance in the production of operating resources and land use, to specific environmental impacts. In this manner, the diverse environmental impacts of mineral-ecological, conventional, and organic cropping systems can be quantified and compared. Secondly, "hot spots" within the cropping systems can be identified, which provide indications of particularly effective approaches for optimization of the supply of ecosystem services [266]. Therefore, ecological analyses can help identify combinations of agroecological practices and modern production techniques that maintain the productivity while reducing emissions. This can help avoid negative environmental impacts, ideally without any loss of crop yield or product quality, and then develop strategies to balance trade-offs between economic and ecological goals in MECSs [272].

However, the quantification of some ecosystem services is beyond the traditional scope of LCAs, particularly those of soil quality and biodiversity [273,274]. A comprehensive evaluation of MECSs in comparison to conventional and organic cropping systems must consider the latest developments in methodology. For the measurement of soil quality, the LANCA method provides appropriate characterization factors for five soil-specific impact categories [275]. "Countryside species-area relationships" can be used to measure biodiversity in LCAs [276]. To demonstrate the advantages and disadvantages of modern agricultural technologies and agricultural measures on biodiversity, it is necessary to extend the scope of the assessment and further investigate the relationship between biodiversity and ecosystem services [277]. Therefore, the assessment methods of ecosystem services must be further developed for a comprehensive evaluation of MECSs through the collaboration of life cycle assessors, and particularly that of economists, ecologists, and soil scientists.

5. Conclusions

Through continuous further development and optimization of cropping systems, agriculture must continue to secure future global food supplies while, at the same time, preserving natural livelihoods. In addition to conventional and organic systems, advanced cropping systems are needed to improve the ecosystem services of agricultural landscapes. Depending on local and global requirements, different cropping systems may be beneficial. The individual ecosystem services (provisioning, regulating, habitat, and cultural services) must be balanced in the local and global context. The development of mineral-ecological cropping systems follows a new farming concept that aims at minimizing trade-offs between different ecosystem services and promoting synergies. This applies not only to the agricultural area under consideration, but also to interactions with areas and structures outside the agricultural landscape, especially with regard to pollutant inputs and land use changes, as well as to natural regulation processes. Future analyses of these new cropping systems should focus on investigating the extent to which it is possible to improve the ecosystem services of agricultural landscapes by establishing mineral-ecological cropping systems with optimized mineral fertilization, yet without the use of chemical synthetic plant protection products.

Author Contributions: Conceptualization, B.Z., I.C.-M., E.B., I.L., M.v.C.; methodology, B.Z., I.C.-M., M.v.C., E.B., I.L.; data curation, B.Z., I.C.-M., M.v.C., I.L., J.W., A.S. (Achim Spiller), S.N., C.L., T.K., I.P., C.Z., M.A.W., M.D., F.M.S., J.P., A.R., H.K., T.F., B.K., R.L., S.K., S.K.-H., J.G., A.E.-H., S.T., M.R., K.S., T.S., J.I., U.L., G.N., N.M., T.M., K.B., M.G., E.K., S.M. (Sven Marhan), R.S., H.-W.G., D.R., A.S. (Alexander Stana), S.G.-H., S.M. (Sebastian Munz), D.O., R.G., M.S., W.H., J.S., M.F., M.K., H.-P.P., P.R., K.W., S.Z., G.P., N.S., R.T.V., E.B.; writing—original draft preparation, B.Z., I.C.-M., E.B., I.L., M.v.C.; writing—review and editing, B.Z., I.C.-M., M.v.C., I.L., J.W., A.S. (Achim Spiller), S.N., C.L., T.K., I.P., C.Z., M.A.W., M.D., F.M.S., J.P., A.R., H.K., T.F., B.K., R.L., S.K., S.K.-H., J.G., A.E.-H., S.T., M.R., K.S., T.S., J.I., U.L., G.N., N.M., T.M., K.B., M.G., E.K., S.M. (Sven Marhan), R.S., H.-W.G., D.R., A.S. (Alexander Stana), S.G.-H., S.M. (Sebastian Munz), D.O., R.G., M.S., W.H., J.S., M.F., M.K., H.-P.P., P.R., K.W., S.Z., G.P., N.S., R.T.V., E.B.; visualization, B.Z., I.C.-M., M.v.C., E.B., I.L.; supervision, E.B., R.T.V., I.L., A.S. (Achim Spiller), C.L., C.Z., F.M.S., H.K., T.F., S.K., K.S., T.S., U.L., G.N., T.M., E.K., H.-W.G., S.G.-H., R.G., W.H., J.S.; project administration, E.B., R.T.V., N.S., I.C.-M., B.Z.; funding acquisition, E.B., H.K., A.S. (Achim Spiller). All authors have read and agreed to the published version of the manuscript.

Funding: This research was funded by Bundesministerium für Bildung und Forschung (BMBF), grant number 031B0731A. The APC was funded by Bundesministerium für Bildung und Forschung (BMBF), grant number 031B0731A.

Data Availability Statement: Data sharing is not applicable to this article.

Conflicts of Interest: The authors declare no conflict of interest. The funders had no role in the design of the study; in the collection, analyses, or interpretation of data; in the writing of the manuscript, or in the decision to publish the results.

References

1. Godfray, H.C.J.; Beddington, J.R.; Crute, I.R.; Haddad, L.; Lawrence, D.; Muir, J.F.; Pretty, J.; Robinson, S.; Thomas, S.M.; Toulmin, C. Food Security: The Challenge of Feeding 9 Billion People. *Science* **2010**, *327*, 812–818. [CrossRef] [PubMed]
2. FAO. *Transforming Food and Agriculture to Achieve the SDGs–20 Interconnected Actions to Guide Decision-Makers*; Food and Agriculture Organization of the United Nations: Rome, Italy, 2018; ISBN 978-92-5-130626-0.
3. IPBES. *Summary for Policymakers of the Global Assessment Report on Biodiversity and Ecosystem Services of the Intergovernmental Science-Policy Platform on Biodiversity and Ecosystem Services*; Díaz, S., Settele, J., Brondízio, E., Ngo, H., Guèze, M., Agard, J., Arneth, A., Balvanera, P., Brauman, K., Butchart, S., Eds.; IPBES Secretariat: Bonn, Germany, 2019; ISBN 978-3-947851-13-3.
4. Rockström, J.; Steffen, W.; Noone, K.; Persson, Å.; Chapin, F.S., III; Lambin, E.F.; Lenton, T.M.; Scheffer, M.; Folke, C.; Shellnhuber, H.J.; et al. A Safe Operating Space for Humanity. *Nature* **2009**, *461*, 472–475. [CrossRef] [PubMed]
5. Zabel, F.; Delzeit, R.; Schneider, J.M.; Seppelt, R.; Mauser, W.; Václavík, T. Global Impacts of Future Cropland Expansion and Intensification on Agricultural Markets and Biodiversity. *Nat. Commun.* **2019**, *10*, 2844. [CrossRef] [PubMed]
6. Kopittke, P.M.; Menzies, N.W.; Wang, P.; McKenna, B.A.; Lombi, E. Soil and the Intensification of Agriculture for Global Food Security. *Environ. Int.* **2019**, *132*, 105078. [CrossRef] [PubMed]
7. Benton, T.G.; Bieg, C.; Harwatt, H.; Pudasaini, R.; Wellesley, L. *Food System Impacts on Biodiversity Loss Three Levers for Food System Transformation in Support of Nature*; Chatham House: London, UK, 2021.
8. Barthel, S.; Isendahl, C.; Vis, B.N.; Drescher, A.; Evans, D.L.; van Timmeren, A. Global Urbanization and Food Production in Direct Competition for Land: Leverage Places to Mitigate Impacts on SDG2 and on the Earth System. *Anthr. Rev.* **2019**, *6*, 71–97. [CrossRef]
9. Gardi, C.; Panagos, P.; Liedekerke, M.V.; Bosco, C.; Brogniez, D.D. Land Take and Food Security: Assessment of Land Take on the Agricultural Production in Europe. *J. Environ. Plan. Manag.* **2015**, *58*, 898–912. [CrossRef]
10. Lambin, E.F.; Meyfroidt, P. Global Land Use Change, Economic Globalization, and the Looming Land Scarcity. *Proc. Natl. Acad. Sci. USA* **2011**, *108*, 3465–3472. [CrossRef]
11. Tilman, D.; Balzer, C.; Hill, J.; Befort, B.L. Global Food Demand and the Sustainable Intensification of Agriculture. *Proc. Natl. Acad. Sci. USA* **2011**, *108*, 20260–20264. [CrossRef]
12. Lambin, E.F.; Meyfroidt, P. *Trends in Global Land-Use Competition*; The MIT Press: Cambridge, MA, USA, 2014; ISBN 978-0-262-32212-6.
13. Foley, J.A.; Ramankutty, N.; Brauman, K.A.; Cassidy, E.S.; Gerber, J.S.; Johnston, M.; Mueller, N.D.; O'Connell, C.; Ray, D.K.; West, P.C.; et al. Solutions for a Cultivated Planet. *Nature* **2011**, *478*, 337–342. [CrossRef]
14. Haller, L.; Moakes, S.; Niggli, U.; Riedel, J.; Stolze, M.; Thompson, M. *Entwicklungsperspektiven Der Ökologischen Landwirtschaft in Deutschland*; Umweltbundesamt: Dessau-Roßlau, Germany, 2020.
15. Niggli, U.; Riedel, J. Entwicklungsperspektiven Für Die Ökologische Landwirtschaft in Deutschland–Vorstellung Der Szenarien. In Proceedings of the UBA-Workshop zum Sachverständigengutachten, Berlin, Germany, 22–23 September 2020.
16. NOcsPS. LaNdwirtschaft 4.0 Ohne Chemisch-Synthetischen PflanzenSchutz. 2021. Available online: https://nocsps.uni-hohenheim.de (accessed on 6 March 2021).
17. De Groot, R.; Brander, L.; Van Der Ploeg, S.; Costanza, R.; Bernard, F.; Braat, L.; Christie, M.; Crossman, N.; Ghermandi, A.; Hein, L. Global Estimates of the Value of Ecosystems and Their Services in Monetary Units. *Ecosyst. Serv.* **2012**, *1*, 50–61. [CrossRef]
18. De Groot, R.; Brander, L.; Solomonides, S. *Update of Global Ecosystem Service Valuation Database (ESVD)*; Wageningen University & Research: Wageningen, The Netherlands, 2020.
19. Costanza, R.; d'Arge, R.; De Groot, R.; Farber, S.; Grasso, M.; Hannon, B.; Limburg, K.; Naeem, S.; O'neill, R.V.; Paruelo, J. The Value of the World's Ecosystem Services and Natural Capital. *Nature* **1997**, *387*, 253. [CrossRef]
20. Vogt, G. Geschichte des Ökologischen Landbaus Im Deutschsprachigen Raum. *Ökologie Landbau* **2001**, *118*, 47–49.
21. Lauk, C. Sozial-Ökologische Charakteristika von Agrarsystemen. Ein Globaler Überblick Und Vergleich. In *Social Ecology Working Paper*; Institute of Social Ecology: Vienna, Austria, 2005.
22. Batáry, P.; Gallé, R.; Riesch, F.; Fischer, C.; Dormann, C.F.; Mußhoff, O.; Császár, P.; Fusaro, S.; Gayer, C.; Happe, A.-K.; et al. The Former Iron Curtain Still Drives Biodiversity–Profit Trade-Offs in German Agriculture. *Nat. Ecol. Evol.* **2017**, *1*, 1279–1284. [CrossRef]
23. Kremen, C.; Merenlender, A.M. Landscapes That Work for Biodiversity and People. *Science* **2018**, *362*. [CrossRef]
24. Pretty, J.; Benton, T.G.; Bharucha, Z.P.; Dicks, L.V.; Flora, C.B.; Godfray, H.C.J.; Goulson, D.; Hartley, S.; Lampkin, N.; Morris, C.; et al. Global Assessment of Agricultural System Redesign for Sustainable Intensification. *Nat. Sustain.* **2018**, *1*, 441–446. [CrossRef]
25. Renwick, A.; Schellhorn, N. A perspective on land sparing versus land sharing. In *Learning from Agri-Environment Schemes in Australia*; Ansell, D., Gibson, F., Salt, D., Eds.; Australian National University Press: Acton, ACT, Australia, 2016; pp. 117–125. ISBN 978-1-76046-015-0.
26. Altieri, M.A.; Nicholls, C.I.; Montalba, R. Technological Approaches to Sustainable Agriculture at a Crossroads: An Agroecological Perspective. *Sustainability* **2017**, *9*, 349. [CrossRef]
27. Arbenz, M.; Gould, D.; Stopes, C. ORGANIC 3.0—The Vision of the Global Organic Movement and the Need for Scientific Support. *Org. Agr.* **2017**, *7*, 199–207. [CrossRef]
28. Adams, D.C.; Salois, M.J. Local versus Organic: A Turn in Consumer Preferences and Willingness-to-Pay. *Renew. Agric. Food Syst.* **2010**, *25*, 331–341. [CrossRef]

29. Kratochvil, R. Biologischer Landbau Und Nachhaltige Entwicklung: Kongruenzen, Differenzen Und Herausforderungen. *Bio-Landbau Österreich Im Int. Kontext* **2005**, *2*, 85–104.

30. IFOAM Definition of Organic Agriculture. 2021. Available online: http://www.ifoam.bio/en/organic-landmarks/definition-organic-agriculture (accessed on 6 February 2021).

31. Wezel, A.; Herren, B.G.; Kerr, R.B.; Barrios, E.; Gonçalves, A.L.R.; Sinclair, F. Agroecological Principles and Elements and Their Implications for Transitioning to Sustainable Food Systems. A Review. *Agron. Sustain. Dev.* **2020**, *40*, 40. [CrossRef]

32. Willer, H.; Schlatter, B.; Trávníček, J.; Kemper, L.; Lernoud, J. *The World of Organic Agriculture. Statistics and Emerging Trends 2020*, 1st ed.; FiBL: Frick, Switzerland; IFOAM: Bonn, Germany, 2020.

33. FAO; WHO. *The International Code of Conduct on Pesticide Management*; FAO: Rome, Italy; WHO: Geneva, Switzerland, 2014.

34. EISA. *European Integrated Farming Framework. A European Definition and Characterisation of Integrated Farming (IF) as Guideline for Sustainable Development of Agriculture*; European Initiative for Sustainable Development in Agriculture: Erkelenz, Germany, 2012; Available online: http://www.sustainable-agriculture.org/wp-content/uploads/2012/08/EISA_Framework_english_new_wheel_170212.pdf (accessed on 25 August 2021).

35. EU Verordnung. *(EG) Nr. 834/2007 Des Rates Vom 28. Juni 2007 Über Die Ökologische/Biologische Produktion Und Die Kennzeichnung von Ökologischen/Biologischen Erzeugnissen Und Zur Aufhebung Der Verordnung (EWG) Nr. 2092/91*; European Commission: Brussels: Bruxelles, Belgium, 2007.

36. Demeter Richtlinien. *Erzeugung Und Verarbeitung. Richtlinien Für Die Zertifizierung "Demeter" Und "Biodynamisch"*; Demeter: Darmstadt, Germany, 2021.

37. IFOAM IFOAM. Organics International. 2021. Available online: https://www.ifoam.bio/ (accessed on 21 June 2021).

38. IFOAM IFOAM. Organics Europe. 2021. Available online: https://www.organicseurope.bio/ (accessed on 21 June 2021).

39. FAO. *Climate-Smart Agriculture*; FAO: Rome, Italy, 2021.

40. Newton, P.; Civita, N.; Frankel-Goldwater, L.; Bartel, K.; Johns, C. What Is Regenerative Agriculture? A Review of Scholar and Practitioner Definitions Based on Processes and Outcomes. *Front. Sustain. Food Syst.* **2020**, *4*, 194. [CrossRef]

41. EPRS. *Präzisionslandwirtschaft Und DieZukunft Der Landwirtschaft in Europa*; European Parliament: Brussels, Belgium, 2016.

42. Pretty, J.; Bharucha, P. Current Approaches to Sustainable Food and Agriculture. In *Sustainable Food and Agriculture*; Campanhola, C., Pandey, S., Eds.; Academic Press: Cambridge, MA, USA, 2019; pp. 169–170. ISBN 978-0-12-812134-4.

43. Pretty, J.; Bharucha, Z.P. Sustainable Intensification in Agricultural Systems. *Ann. Bot.* **2014**, *114*, 1571–1596. [CrossRef]

44. Wezel, A.; Soboksa, G.; McClelland, S.; Delespesse, F.; Boissau, A. The Blurred Boundaries of Ecological, Sustainable, and Agroecological Intensification: A Review. *Agron. Sustain. Dev.* **2015**, *35*, 1283–1295. [CrossRef]

45. Mockshell, J.; Kamanda, J. Beyond the Agroecological and Sustainable Agricultural Intensification Debate: Is Blended Sustainability the Way Forward? *Int. J. Agric. Sustain.* **2018**, *16*, 127–149. [CrossRef]

46. Wezel, A.; Casagrande, M.; Celette, F.; Vian, J.-F.; Ferrer, A.; Peigné, J. Agroecological Practices for Sustainable Agriculture. A Review. *Agron. Sustain. Dev.* **2014**, *34*, 1–20. [CrossRef]

47. Worldbank Climate-Smart Agriculture. Available online: https://www.worldbank.org/en/topic/climate-smart-agriculture (accessed on 27 January 2021).

48. CSA. How Is Climate-Smart Agriculture Different from Other Sustainable Agriculture Approaches? 2021. Available online: https://csa.guide/csa/how-is-it-different (accessed on 6 February 2021).

49. Campbell, B.M.; Thornton, P.; Zougmoré, R.; van Asten, P.; Lipper, L. Sustainable Intensification: What Is Its Role in Climate Smart Agriculture? *Curr. Opin. Environ. Sustain.* **2014**, *8*, 39–43. [CrossRef]

50. KraichgauKorn Marktgemeinschaft KraichgauKorn. 2021. Available online: https://kkhomepage.kraichgaukorn.de/ (accessed on 6 February 2021).

51. Blütenkorn Für Mensch und Natur Die Zukunft nachhaltiger Landwirtschaft. 2021. Available online: https://bluetenkorn.de/ (accessed on 2 June 2021).

52. IP-Suisse IP-SUISSE Bauern Denken an Morgen. 2021. Available online: https://www.ipsuisse.ch/ (accessed on 6 March 2021).

53. Sanders, J.; Heß, J. *Leistungen des Ökologischen Landbaus Für Umwelt Und Gesellschaft*; Thünen Report; Johann Heinrich von Thünen-Institut: Braunschweig, Germany, 2019.

54. Bengtsson, J.; Ahnström, J.; Weibull, A.-C. The Effects of Organic Agriculture on Biodiversity and Abundance: A Meta-Analysis. *J. Appl. Ecol.* **2005**, *42*, 261–269. [CrossRef]

55. Hole, D.G.; Perkins, A.J.; Wilson, J.D.; Alexander, I.H.; Grice, P.V.; Evans, A.D. Does Organic Farming Benefit Biodiversity? *Biol. Conserv.* **2005**, *122*, 113–130. [CrossRef]

56. Tuck, S.L.; Winqvist, C.; Mota, F.; Ahnström, J.; Turnbull, L.A.; Bengtsson, J. Land-Use Intensity and the Effects of Organic Farming on Biodiversity: A Hierarchical Meta-Analysis. *J. Appl. Ecol.* **2014**, *51*, 746–755. [CrossRef] [PubMed]

57. Tuomisto, H.L.; Hodge, I.D.; Riordan, P.; Macdonald, D.W. Does Organic Farming Reduce Environmental Impacts?–A Meta-Analysis of European Research. *J. Environ. Manage.* **2012**, *112*, 309–320. [CrossRef]

58. Mondelaers, K.; Aertsens, J.; Van Huylenbroeck, G. A Meta-analysis of the Differences in Environmental Impacts between Organic and Conventional Farming. *Br. Food J.* **2009**, *111*, 1098–1119. [CrossRef]

59. Seufert, V.; Ramankutty, N. Many Shades of Gray—The Context-Dependent Performance of Organic Agriculture. *Sci. Adv.* **2017**, *3*, e1602638. [CrossRef]

60. Riedel, J.; Niggli, U. Ökologische Und Soziale Vorzüglichkeit versus Globale Ökoeffizienz. In Proceedings of the UBA-Workshop zum Sachverständigengutachten, Berlin, Germany, 27 July 2020.
61. Ponisio, L.C.; M'Gonigle, L.K.; Mace, K.C.; Palomino, J.; de Valpine, P.; Kremen, C. Diversification Practices Reduce Organic to Conventional Yield Gap. *Proc. R. Soc. B Biol. Sci.* **2015**, *282*, 20141396. [CrossRef]
62. De Ponti, T.; Rijk, B.; van Ittersum, M.K. The Crop Yield Gap between Organic and Conventional Agriculture. *Agric. Syst.* **2012**, *108*, 1–9. [CrossRef]
63. Seufert, V.; Ramankutty, N.; Foley, J.A. Comparing the Yields of Organic and Conventional Agriculture. *Nature* **2012**, *485*, 229–232. [CrossRef]
64. *BMEL Buchführung Der Testbetriebe*; Bundesministerium für Ernährung, Landwirtschaft und Verbraucherschutz: Berlin, Germany, 2019.
65. Treu, H.; Nordborg, M.; Cederberg, C.; Heuer, T.; Claupein, E.; Hoffmann, H.; Berndes, G. Carbon Footprints and Land Use of Conventional and Organic Diets in Germany. *J. Clean. Prod.* **2017**, *161*, 127–142. [CrossRef]
66. Bommarco, R.; Kleijn, D.; Potts, S.G. Ecological Intensification: Harnessing Ecosystem Services for Food Security. *Trends Ecol. Evol.* **2013**, *28*, 230–238. [CrossRef] [PubMed]
67. Burel, F.; Aviron, S.; Baudry, J.; Le Féon, V.; Vasseur, C. The Structure and Dynamics of Agricultural Landscapes as Drivers of Biodiversity. In *Landscape Ecology for Sustainable Environment and Culture*; Fu, B., Jones, K.B., Eds.; Springer: Dordrecht, The Netherlands, 2013; pp. 285–308. ISBN 978-94-007-6530-6.
68. Raudsepp-Hearne, C.; Peterson, G.D.; Bennett, E.M. Ecosystem Service Bundles for Analyzing Tradeoffs in Diverse Landscapes. *Proc. Natl. Acad. Sci. USA* **2010**, *107*, 5242–5247. [CrossRef]
69. Wolters, V.; Isselstein, J.; Stützel, H.; Ordon, F.; von Haaren, C.; Schlecht, E.; Wesseler, J.; Birner, R.; von Lützow, M.; Brüggemann, N. Nachhaltige Ressourceneffiziente Erhöhung Der Flächenproduktivität: Zukunftsoptionen Der Deutschen Agrarökosystemforschung Grundsatzpapier Der Dfg Senatskommission Für Agrarökosystemforschung. *J. Für Kult.* **2014**, *7*, 225–236.
70. Reid, W.V.; Mooney, H.A.; Cropper, A.; Capistrano, D.; Carpenter, S.R.; Chopra, K.; Dasgupta, P.; Dietz, T.; Duraiappah, A.K.; Hassan, R.; et al. *Ecosystems and Human Well-Being: General Synthesis*; Sarukhán, J., Whyte, A., Eds.; Island Press: Washington, DC, USA, 2005.
71. Rodríguez, J.; Beard, J.; Bennett, E.; Cumming, G.; Cork, S.; Agard, J.; Dobson, A.; Peterson, G. Trade-Offs across Space, Time, and Ecosystem Services. *Ecol. Soc.* **2006**, *11*. [CrossRef]
72. Meemken, E.-M.; Qaim, M. Organic Agriculture, Food Security, and the Environment. *Annu. Rev. Resour. Econ.* **2018**, *10*, 39–63. [CrossRef]
73. European Commission. *Factsheet: From Farm to Fork: Our Food, Our Health, Our Planet, Our Future*; European Commission: Brussels, Belgium, 2020.
74. ERA European Research Alliance. *Towards a Chemical Pesticide Free Agriculture*; European Research Alliance: Paris, France, 2021.
75. European Commission Directive. 2009/128/EC of the European Parliament and of the Council of 21 October 2009 Establishing a Framework for Community Action to Achieve the Sustainable Use of Pesticides. *Off. J. Eur. Union* **2009**, *52*, 71–86.
76. Lamichhane, J.R.; Messéan, A.; Ricci, P. Chapter Two—Research and innovation priorities as defined by the Ecophyto plan to address current crop protection transformation challenges in France. In *Advances in Agronomy*; Sparks, D.L., Ed.; Academic Press: Cambridge, MA, USA, 2019; Volume 154, pp. 81–152.
77. Lapierre, M.; Sauquet, A.; Julie, S. Providing Technical Assistance to Peer Networks to Reduce Pesticide Use in Europe: Evidence from the French Ecophyto Plan. 2019. hal-02190979v1. Available online: https://hal.archives-ouvertes.fr/hal-02190979v1 (accessed on 25 August 2021).
78. Bundesrat Entwurf Eines Gesetzes Zur Durchführung Der Im Rahmen Der Gemeinsamen Agrarpolitik Finanzierten Direktzahlungen (GAP-Direktzahlungen-Gesetz–GAPDZG). 2021. Available online: https://www.bundesrat.de/SharedDocs/beratungsvorgaenge/2021/0301-0400/0301-21.html (accessed on 25 August 2021).
79. Delbrück, K.; Nürnberg, M. *Aktionsprogramm Insektenschutz Gemeinsam Wirksam Gegen Das Insektensterben*; Bundesministerium für Umwelt, Naturschutz und nukleare Sicherheit (BMU): Berlin, Germany, 2019.
80. MLR Gesetzesnovelle Zur Stärkung Der Biodiversität. 2020. Available online: https://mlr.baden-wuerttemberg.de/de/unsere-themen/biodiversitaet-und-landnutzung/biodiversitaetsgesetz/ (accessed on 31 May 2021).
81. European Commission. Regulation (EC) No 1107/2009 of the European Parliament and of the Council of 21 October 2009 Concerning the Placing of Plant Protection Products on the Market and Repealing Council Directives 79/117/EEC and 91/414/EEC. *Off. J. Eur. Union* **2009**, *52*, 1–50.
82. Voss-Fels, K.P.; Stahl, A.; Wittkop, B.; Lichthardt, C.; Nagler, S.; Rose, T.; Chen, T.-W.; Zetzsche, H.; Seddig, S.; Majid Baig, M.; et al. Breeding Improves Wheat Productivity under Contrasting Agrochemical Input Levels. *Nat. Plants* **2019**, *5*, 706–714. [CrossRef]
83. Strassemeyer, J.; Daehmlow, D.; Dominic, A.R.; Lorenz, S.; Golla, B. SYNOPS-WEB, an Online Tool for Environmental Risk Assessment to Evaluate Pesticide Strategies on Field Level. *Crop. Prot.* **2017**, *97*, 28–44. [CrossRef]
84. Tola, M.; Kebede, B. Occurrence, Importance and Control of Mycotoxins: A Review. *Cogent Food Agric.* **2016**, *2*, 1191103. [CrossRef]
85. Zain, M.E. Impact of Mycotoxins on Humans and Animals. *J. Saudi Chem. Soc.* **2011**, *15*, 129–144. [CrossRef]
86. Bryden, W.L. Mycotoxin Contamination of the Feed Supply Chain: Implications for Animal Productivity and Feed Security. *Anim. Feed Sci. Technol.* **2012**, *173*, 134–158. [CrossRef]

87. Feike, T.; zu Eisenbach, L.R.F.; Lieb, R.; Gabriel, D.; Sabboura, D.; Shawon, A.R.; Wetzel, M.; Klocke, B.; Krengel-Horney, S.; Schwarz, J. Einfluss von Pflanzenschutzstrategie und Bodenbearbeitung auf den CO_2-Fußabdruck von Weizen. *JFK* **2020**, *72*, 311–326. [CrossRef]

88. Laidig, F.; Feike, T.; Hadasch, S.; Rentel, D.; Klocke, B.; Miedaner, T.; Piepho, H.P. Breeding Progress of Disease Resistance and Impact of Disease Severity under Natural Infections in Winter Wheat Variety Trials. *Appl. Genet.* **2021**, *134*, 1281–1302. [CrossRef]

89. Deike, S.; Pallutt, B.; Christen, O. Investigations on the Energy Efficiency of Organic and Integrated Farming with Specific Emphasis on Pesticide Use Intensity. *Eur. J. Agron.* **2008**, *28*, 461–470. [CrossRef]

90. Herwig, N.; Felgentreu, D.; Hommel, B. Auswirkungen von natürlichen Standortbedingungen und ackerbaulichen Maßnahmen auf Bodenorganismen–Erhebungen in den Langzeitversuchen des Julius Kühn-Instituts in Dahnsdorf (Hoher Fläming, Land Brandenburg). *JFK* **2020**, *72*, 327–337. [CrossRef]

91. Klocke, B.; Wagner, C.; Schwarz, J. Erkenntnisse und Perspektiven eines 23-jährigen Dauerfeldversuches zum integrierten Pflanzenschutz gegen pilzliche Schaderreger im Winterweizen. *JFK* **2020**, *72*, 265–278. [CrossRef]

92. Schwarz, J.; Klocke, B.; Wagner, C.; Krengel, S. Untersuchungen zum notwendigen Maß bei der Anwendung von Pflanzen-schutzmitteln in Winterweizen in den Jahren 2004 bis 2016. *Gesunde Pflanz Pflanzenschutz Verbrauch. Umweltschutz* **2018**, *70*, 119–127. [CrossRef]

93. Jahn, M.; Wagner, C.; Sellmann, J. Ertragsverluste durch wichtige Pilzkrankheiten in Winterweizen im Zeitraum 2003 bis 2008–Versuchsergebnisse aus 12 deutschen Bundesländern. *J. Für Kult.* **2012**, *64*, 273–285.

94. Lechenet, M.; Deytieux, V.; Antichi, D.; Aubertot, J.-N.; Bàrberi, P.; Bertrand, M.; Cellier, V.; Charles, R.; Colnenne-David, C.; Dachbrodt-Saaydeh, S.; et al. Diversity of Methodologies to Experiment Integrated Pest Management in Arable Cropping Systems: Analysis and Reflections Based on a European Network. *Eur. J. Agron.* **2017**, *83*, 86–99. [CrossRef]

95. Garibaldi, L.A.; Andersson, G.K.S.; Requier, F.; Fijen, T.P.M.; Hipólito, J.; Kleijn, D.; Pérez-Méndez, N.; Rollin, O. Complementarity and Synergisms among Ecosystem Services Supporting Crop Yield. *Glob. Food Secur.* **2018**, *17*, 38–47. [CrossRef]

96. Dainese, M.; Martin, E.A.; Aizen, M.A.; Albrecht, M.; Bartomeus, I.; Bommarco, R.; Carvalheiro, L.G.; Chaplin-Kramer, R.; Gagic, V.; Garibaldi, L.A.; et al. A Global Synthesis Reveals Biodiversity-Mediated Benefits for Crop Production. *Sci. Adv.* **2019**, *5*, eaax0121. [CrossRef]

97. Tamburini, G.; Bommarco, R.; Wanger, T.C.; Kremen, C.; van der Heijden, M.G.A.; Liebman, M.; Hallin, S. Agricultural Diversification Promotes Multiple Ecosystem Services without Compromising Yield. *Sci. Adv.* **2020**, *6*, eaba1715. [CrossRef] [PubMed]

98. Oerke, E.-C. Crop Losses to Pests. *J. Agric. Sci.* **2006**, *144*, 31–43. [CrossRef]

99. Singh, R.P.; Singh, P.K.; Rutkoski, J.; Hodson, D.P.; He, X.; Jørgensen, L.N.; Hovmøller, M.S.; Huerta-Espino, J. Disease Impact on Wheat Yield Potential and Prospects of Genetic Control. *Annu. Rev. Phytopathol.* **2016**, *54*, 303–322. [CrossRef]

100. Zetzsche, H.; Friedt, W.; Ordon, F. Breeding Progress for Pathogen Resistance Is a Second Major Driver for Yield Increase in German Winter Wheat at Contrasting N Levels. *Sci. Rep.* **2020**, *10*, 20374. [CrossRef] [PubMed]

101. Savary, S.; Willocquet, L.; Pethybridge, S.J.; Esker, P.; McRoberts, N.; Nelson, A. The Global Burden of Pathogens and Pests on Major Food Crops. *Nat. Ecol. Evol.* **2019**, *3*, 430–439. [CrossRef]

102. Hossard, L.; Philibert, A.; Bertrand, M.; Colnenne-David, C.; Debaeke, P.; Munier-Jolain, N.; Jeuffroy, M.H.; Richard, G.; Makowski, D. Effects of Halving Pesticide Use on Wheat Production. *Sci. Rep.* **2014**, *4*, 4405. [CrossRef] [PubMed]

103. Christen, O. Ertrag, Ertragsstruktur und Ertragsstabilität von Weizen, Gerste und Raps in unterschiedlichen Fruchtfolgen. *Pflanzenbauwissenschaften* **2001**, *5*, 33–39.

104. Albizua, A.; Williams, A.; Hedlund, K.; Pascual, U. Crop Rotations Including Ley and Manure Can Promote Ecosystem Services in Conventional Farming Systems. *Appl. Soil Ecol.* **2015**, *95*, 54–61. [CrossRef]

105. Kehlenbeck, H.; Rajmis, S. Was Bleibt Unterm Strich. *DLG Mitt.* **2018**, *18*, 56–57.

106. Cabot, C.; Martos, S.; Llugany, M.; Gallego, B.; Tolrà, R.; Poschenrieder, C. A Role for Zinc in Plant Defense against Pathogens and Herbivores. *Front. Plant. Sci.* **2019**, *10*, 1171. [CrossRef]

107. Panpatte, D.G.; Jhala, Y.K.; Shelat, H.N.; Vyas, R.V. Pseudomonas fluorescens: A Promising Biocontrol Agent and PGPR for Sustainable Agriculture. In *Microbial Inoculants in Sustainable Agricultural Productivity: Vol. 1: Research Perspectives*; Singh, D.P., Singh, H.B., Prabha, R., Eds.; Springer: New Delhi, India, 2016; pp. 257–270. ISBN 978-81-322-2647-5.

108. Weiner, J.; Griepentrog, H.-W.; Kristensen, L. Suppression of Weeds by Spring Wheat Triticumaestivum Increases with Crop Density and Spatial Uniformity. *J. Appl. Ecol.* **2001**, *38*, 784–790. [CrossRef]

109. Tao, Z.; Ma, S.; Chang, X.; Wang, D.; Wang, Y.; Yang, Y.; Zhao, G.; Yang, J. Effects of Tridimensional Uniform Sowing on Water Consumption, Nitrogen Use, and Yield in Winter Wheat. *Crop. J.* **2019**, *7*, 480–493. [CrossRef]

110. Olsen, J.M.; Griepentrog, H.-W.; Nielsen, J.; Weiner, J. How Important Are Crop Spatial Pattern and Density for Weed Suppression by Spring Wheat? *Weed Sci.* **2012**, *60*, 501–509. [CrossRef]

111. Kottmann, L.; Hegewald, H.; Feike, T.; Lehnert, H.; Keilwagen, J.; Hörsten, D.; Greef, J.; Wegener, J.-K. Standraumoptimierung im Getreideanbau durch Gleichstandsaat. *J. Für Kult.* **2019**, *71*, 90–94. [CrossRef]

112. Geisseler, D.; Scow, K.M. Long-Term Effects of Mineral Fertilizers on Soil Microorganisms–A Review. *Soil Biol. Biochem.* **2014**, *75*, 54–63. [CrossRef]

113. Deppe, M.; Well, R.; Kücke, M.; Fuß, R.; Giesemann, A.; Flessa, H. Impact of CULTAN Fertilization with Ammonium Sulfate on Field Emissions of Nitrous Oxide. *Agric. Ecosyst. Environ.* **2016**, *219*, 138–151. [CrossRef]

114. Herrmann, M.N.; Wang, Y.; Hartung, J.; Hartmann, T.; Zhang, W.; Nkebiwe, P.M.; Chen, X.; Müller, T.; Yang, H. Effects of Bioeffectors on Crop Growth and Quality Indicators: A Global Network Meta-Analysis. *Agron. Sustain. Dev.* **2021**. (under preparation).

115. Lekfeldt, J.D.S.; Nkebiwe, P.M.; Symanczik, S.; Mäder, M.; Bar-Tal, A.; Biró, B.; Bradáčová, K.; Brecht, J.; Caniullan, P.C.; Choudhary, K.K.; et al. To BE, or Not to BE–A Meta-Analysis on the Effectiveness of Bioeffectors (BEs) on Maize, Wheat and Tomato Performance from Greenhouse to Field Scales across Europe and Israel. **2021**. (under preparation).

116. Bradáčová, K.; Florea, A.S.; Bar-Tal, A.; Minz, D.; Yermiyahu, U.; Shawahna, R.; Kraut-Cohen, J.; Zolti, A.; Erel, R.; Dietel, K.; et al. Microbial Consortia versus Single-Strain Inoculants: An Advantage in PGPM-Assisted Tomato Production? *Agronomy* **2019**, *9*, 105. [CrossRef]

117. Weber, J.F.; Kunz, C.; Gerhards, R. Chemical and Mechanical Weed Control in Soybean (Glycine Max). *Jul. Kühn-Arch.* **2016**, *452*, 171.

118. Pascale, A.; Vinale, F.; Manganiello, G.; Nigro, M.; Lanzuise, S.; Ruocco, M.; Marra, R.; Lombardi, N.; Woo, S.L.; Lorito, M. Trichoderma and Its Secondary Metabolites Improve Yield and Quality of Grapes. *Crop. Prot.* **2017**, *92*, 176–181. [CrossRef]

119. Giotis, C.; Markellou, E.; Theodoropoulou, A.; Kostoulas, G.; Wilcockson, S.; Leifert, C. The Effects of Different Biological Control Agents (BCAs) and Plant Defence Elicitors on Cucumber Powdery Mildew (Podosphaera Xanthii). *Org. Agr.* **2012**, *2*, 89–101. [CrossRef]

120. Barański, M.; Srednicka-Tober, D.; Volakakis, N.; Seal, C.; Sanderson, R.; Stewart, G.B.; Benbrook, C.; Biavati, B.; Markellou, E.; Giotis, C.; et al. Higher Antioxidant and Lower Cadmium Concentrations and Lower Incidence of Pesticide Residues in Organically Grown Crops: A Systematic Literature Review and Meta-Analyses. *Br. J. Nutr.* **2014**, *112*, 794–811. [CrossRef]

121. Gomiero, T. Food Quality Assessment in Organic vs. Conventional Agricultural Produce: Findings and Issues. *Appl. Soil Ecol.* **2018**, *123*, 714–728. [CrossRef]

122. Langenkämper, G.; Zörb, C.; Seifert, M.; Mäder, P.; Fretzdorff, B.; Betsche, T. Nutritional Quality of Organic and Conventional Wheat. *J. Appl. Bot. Food Qual.* **2006**, *80*, 150–154.

123. Meena, R.S.; Kumar, S.; Datta, R.; Lal, R.; Vijayakumar, V.; Brtnicky, M.; Sharma, M.P.; Yadav, G.S.; Jhariya, M.K.; Jangir, C.K.; et al. Impact of Agrochemicals on Soil Microbiota and Management: A Review. *Land* **2020**, *9*, 34. [CrossRef]

124. Zörb, C.; Steinfurth, D.; Gödde, V.; Niehaus, K.; Mühling, K.H.; Zörb, C.; Steinfurth, D.; Gödde, V.; Niehaus, K.; Mühling, K.H. Metabolite Profiling of Wheat Flag Leaf and Grains during Grain Filling Phase as Affected by Sulfur Fertilisation. *Funct. Plant. Biol.* **2011**, *39*, 156–166. [CrossRef]

125. Datta, A.; Ullah, H.; Tursun, N.; Pornprom, T.; Knezevic, S.Z.; Chauhan, B.S. Managing Weeds Using Crop Competition in Soybean (*lycine max* (L.) Merr.). *Crop. Prot.* **2017**, *95*, 60–68. [CrossRef]

126. Mhlanga, B.; Chauhan, B.S.; Thierfelder, C. Weed Management in Maize Using Crop Competition: A Review. *Crop. Prot.* **2016**, *88*, 28–36. [CrossRef]

127. Zörb, C.; Becker, E.; Merkt, N.; Kafka, S.; Schmidt, S.; Schmidhalter, U. Shift of Grain Protein Composition in Bread Wheat under Summer Drought Events. *J. Plant. Nutr. Soil Sci.* **2017**, *180*, 49–55. [CrossRef]

128. Leneveu-Jenvrin, C.; Charles, F.; Barba, F.J.; Remize, F. Role of Biological Control Agents and Physical Treatments in Maintaining the Quality of Fresh and Minimally-Processed Fruit and Vegetables. *Crit. Rev. Food Sci. Nutr.* **2020**, *60*, 2837–2855. [CrossRef]

129. Siroli, L.; Patrignani, F.; Serrazanetti, D.I.; Gardini, F.; Lanciotti, R. Innovative Strategies Based on the Use of Bio-Control Agents to Improve the Safety, Shelf-Life and Quality of Minimally Processed Fruits and Vegetables. *Trends Food Sci. Technol.* **2015**, *46*, 302–310. [CrossRef]

130. Lask, J.; Martínez Guajardo, A.; Weik, J.; Von Cossel, M.; Lewandowski, I.; Wagner, M. Comparative Environmental and Economic Life Cycle Assessment of Biogas Production from Perennial Wild Plant Mixtures and Maize (*Zea mays* L.) in Southwest Germany. *GCB Bioenergy* **2020**. (under review). [CrossRef]

131. Hillier, J.; Brentrup, F.; Wattenbach, M.; Walter, C.; Garcia-Suarez, T.; Mila-i-Canals, L.; Smith, P. Which Cropland Greenhouse Gas Mitigation Options Give the Greatest Benefits in Different World Regions? Climate and Soil-Specific Predictions from Integrated Empirical Models. *Glob. Chang. Biol.* **2012**, *18*, 1880–1894. [CrossRef]

132. Das, S.; Ho, A.; Kim, P.J. Editorial: Role of Microbes in Climate Smart Agriculture. *Front. Microbiol.* **2019**, *10*, 2756. [CrossRef]

133. Lu, X.; Siemann, E.; He, M.; Wei, H.; Shao, X.; Ding, J. Climate Warming Increases Biological Control Agent Impact on a Non-Target Species. *Ecol. Lett.* **2015**, *18*, 48–56. [CrossRef]

134. Syafrudin, M.; Kristanti, R.A.; Yuniarto, A.; Hadibarata, T.; Rhee, J.; Al-onazi, W.A.; Algarni, T.S.; Almarri, A.H.; Al-Mohaimeed, A.M. Pesticides in Drinking Water—A Review. *Int. J. Environ. Res. Public Health* **2021**, *18*, 468. [CrossRef]

135. De Souza, R.M.; Seibert, D.; Quesada, H.B.; de Jesus Bassetti, F.; Fagundes-Klen, M.R.; Bergamasco, R. Occurrence, Impacts and General Aspects of Pesticides in Surface Water: A Review. *Process. Saf. Environ. Prot.* **2020**, *135*, 22–37. [CrossRef]

136. Singh, B.; Craswell, E. Fertilizers and Nitrate Pollution of Surface and Ground Water: An Increasingly Pervasive Global Problem. *SN Appl. Sci.* **2021**, *3*, 518. [CrossRef]

137. Colnenne-David, C.; Grandeau, G.; Jeuffroy, M.-H.; Dore, T. Ambitious Environmental and Economic Goals for the Future of Agriculture Are Unequally Achieved by Innovative Cropping Systems. *Field Crop. Res.* **2017**, *210*, 114–128. [CrossRef]

138. Ndlela, L.L.; Oberholster, P.J.; Van Wyk, J.H.; Cheng, P.H. Bacteria as Biological Control Agents of Freshwater Cyanobacteria: Is It Feasible beyond the Laboratory? *Appl. Microbiol. Biotechnol.* **2018**, *102*, 9911–9923. [CrossRef] [PubMed]

139. Kerdraon, L.; Laval, V.; Suffert, F. Microbiomes and Pathogen Survival in Crop Residues, an Ecotone Between Plant and Soil. *Phytobiomes J.* **2019**, *3*, 246–255. [CrossRef]

140. Geiger, F.; Bengtsson, J.; Berendse, F.; Weisser, W.W.; Emmerson, M.; Morales, M.B.; Ceryngier, P.; Liira, J.; Tscharntke, T.; Winqvist, C.; et al. Persistent Negative Effects of Pesticides on Biodiversity and Biological Control Potential on European Farmland. *Basic Appl. Ecol.* **2010**, *11*, 97–105. [CrossRef]

141. Maeder, P.; Fliessbach, A.; Dubois, D.; Gunst, L.; Fried, P.; Niggli, U. Soil Fertility and Biodiversity in Organic Farming. *Science* **2002**, *296*, 1694–1697. [CrossRef] [PubMed]

142. Rodriguez-Campos, J.; Dendooven, L.; Alvarez-Bernal, D.; Contreras-Ramos, S.M. Potential of Earthworms to Accelerate Removal of Organic Contaminants from Soil: A Review. *Appl. Soil Ecol.* **2014**, *79*, 10–25. [CrossRef]

143. Rashid, M.I.; Mujawar, L.H.; Shahzad, T.; Almeelbi, T.; Ismail, I.M.I.; Oves, M. Bacteria and Fungi Can Contribute to Nutrients Bioavailability and Aggregate Formation in Degraded Soils. *Microbiol. Res.* **2016**, *183*, 26–41. [CrossRef]

144. Contreras-Cornejo, H.A.; Macías-Rodríguez, L.; del-Val, E.; Larsen, J. Ecological Functions of *Trichoderma* Spp. and Their Secondary Metabolites in the Rhizosphere: Interactions with Plants. *FEMS Microbiol. Ecol.* **2016**, *92*, fiw036. [CrossRef] [PubMed]

145. Halifu, S.; Deng, X.; Song, X.; Song, R. Effects of Two Trichoderma Strains on Plant Growth, Rhizosphere Soil Nutrients, and Fungal Community of Pinus Sylvestris Var. Mongolica Annual Seedlings. *Forests* **2019**, *10*, 758. [CrossRef]

146. Maji, D.; Singh, M.; Wasnik, K.; Chanotiya, C.S.; Kalra, A. The Role of a Novel Fungal Strain Trichoderma Atroviride RVF3 in Improving Humic Acid Content in Mature Compost and Vermicompost via Ligninolytic and Celluloxylanolytic Activities. *J. Appl. Microbiol.* **2015**, *119*, 1584–1596. [CrossRef] [PubMed]

147. Mitra, D.; Anđelković, S.; Panneerselvam, P.; Senapati, A.; Vasić, T.; Ganeshamurthy, A.N.; Chauhan, M.; Uniyal, N.; Mahakur, B.; Radha, T.K. Phosphate-Solubilizing Microbes and Biocontrol Agent for Plant Nutrition and Protection: Current Perspective. *Commun. Soil Sci. Plant. Anal.* **2020**, *51*, 645–657. [CrossRef]

148. Matsumoto, L.S.; dos Santos, I.M.O.; Barazetti, A.R.; Simões, G.C.; Farias, T.N.; Andrade, G. Effects of Biological Control Agents on Arbuscular Mycorrhiza Fungi Rhizophagus Clarus in Soybean Rhizosphere. *Agron. Sci. Biotechnol.* **2017**, *3*, 29. [CrossRef]

149. Ahmed, E.; Holmström, S.J.M. Siderophores in Environmental Research: Roles and Applications. *Microb. Biotechnol.* **2014**, *7*, 196–208. [CrossRef]

150. Sivasakthi, S.; Usharani, G.; Saranraj, P. Biocontrol Potentiality of Plant Growth Promoting Bacteria (PGPR)-Pseudomonas Fluorescens and Bacillus Subtilis: A Review. *AJAR* **2014**, *9*, 1265–1277. [CrossRef]

151. Catarino, R.; Bretagnolle, V.; Perrot, T.; Vialloux, F.; Gaba, S. Bee Pollination Outperforms Pesticides for Oilseed Crop Production and Profitability. *Proc. R. Soc. B Biol. Sci.* **2019**, *286*, 20191550. [CrossRef]

152. Potts, S.G.; Biesmeijer, J.C.; Kremen, C.; Neumann, P.; Schweiger, O.; Kunin, W.E. Global Pollinator Declines: Trends, Impacts and Drivers. *Trends Ecol. Evol.* **2010**, *25*, 345–353. [CrossRef] [PubMed]

153. Marja, R.; Kleijn, D.; Tscharntke, T.; Klein, A.-M.; Frank, T.; Batáry, P. Effectiveness of Agri-Environmental Management on Pollinators Is Moderated More by Ecological Contrast than by Landscape Structure or Land-Use Intensity. *Ecol. Lett.* **2019**, *22*, 1493–1500. [CrossRef]

154. Peñuelas, J.; Farré-Armengol, G.; Llusia, J.; Gargallo-Garriga, A.; Rico, L.; Sardans, J.; Terradas, J.; Filella, I. Removal of Floral Microbiota Reduces Floral Terpene Emissions. *Sci. Rep.* **2014**, *4*, 6727. [CrossRef]

155. Pallutt, B. 30 Jahre Dauerfeldversuche zum Pflanzenschutz. *JFK* **2010**, *62*, 230–237. [CrossRef]

156. Dierauer, H.U.; Stöppler-Zimmer, H. *Unkrautregulierung Ohne Chemie*; Verlag Ulmer: Stuttgart, Germany, 2014; ISBN 3-8001-4096-9.

157. Zhou, M.; Liu, C.; Wang, J.; Meng, Q.; Yuan, Y.; Ma, X.; Liu, X.; Zhu, Y.; Ding, G.; Zhang, J.; et al. Soil Aggregates Stability and Storage of Soil Organic Carbon Respond to Cropping Systems on Black Soils of Northeast China. *Sci. Rep.* **2020**, *10*, 265. [CrossRef]

158. Harding, D.P.; Raizada, M.N. Controlling Weeds with Fungi, Bacteria and Viruses: A Review. *Front. Plant. Sci.* **2015**, *6*, 659. [CrossRef]

159. Raymaekers, K.; Ponet, L.; Holtappels, D.; Berckmans, B.; Cammue, B.P.A. Screening for Novel Biocontrol Agents Applicable in Plant Disease Management–A Review. *Biol. Control.* **2020**, *144*, 104240. [CrossRef]

160. Tscharntke, T.; Klein, A.M.; Kruess, A.; Steffan-Dewenter, I.; Thies, C. Landscape Perspectives on Agricultural Intensification and Biodiversity—Ecosystem Service Management. *Ecol. Lett.* **2005**, *8*, 857–874. [CrossRef]

161. Krauss, J.; Gallenberger, I.; Steffan-Dewenter, I. Decreased Functional Diversity and Biological Pest Control in Conventional Compared to Organic Crop Fields. *PLoS ONE* **2011**, *6*, e19502. [CrossRef]

162. El-Wakeil, N.; Gaafar, N.; Sallam, A.; Volkmar, C. *Side Effects of Insecticides on Natural Enemies and Possibility of Their Integration in Plant Protection Strategies*; IntechOpen: London, UK, 2013; ISBN 978-953-51-0958-7.

163. Garratt, M.P.D.; Wright, D.J.; Leather, S.R. The Effects of Farming System and Fertilisers on Pests and Natural Enemies: A Synthesis of Current Research. *Agric. Ecosyst. Environ.* **2011**, *141*, 261–270. [CrossRef]

164. Ingwersen, J.; Högy, P.; Wizemann, H.D.; Warrach-Sagi, K.; Streck, T. Coupling the Land Surface Model Noah-MP with the Generic Crop Growth Model Gecros: Model Description, Calibration and Validation. *Agric. For. Meteorol.* **2018**, *262*, 322–339. [CrossRef]

165. Goudriaan, J. Simulation of Micrometeorology of Crops, Some Methods and Their Problems, and a Few Results. *Agric. For. Meteorol.* **1989**, *47*, 239–258. [CrossRef]

166. Vidal, T.; Boixel, A.-L.; Durand, B.; de Vallavieille-Pope, C.; Huber, L.; Saint-Jean, S. Reduction of Fungal Disease Spread in Cultivar Mixtures: Impact of Canopy Architecture on Rain-Splash Dispersal and on Crop Microclimate. *Agric. For. Meteorol.* **2017**, *246*, 154–161. [CrossRef]

167. El-Hasan, A.; Buchenauer, H. Actions of 6-Pentyl-Alpha-Pyrone in Controlling Seedling Blight Incited by Fusarium Moniliforme and Inducing Defence Responses in Maize. *J. Phytopathol.* **2009**, *157*, 697–707. [CrossRef]
168. Benjamin, E.O.; Wesseler, J.H.H. A Socioeconomic Analysis of Biocontrol in Integrated Pest Management: A Review of the Effects of Uncertainty, Irreversibility and Flexibility. *NJAS Wagening. J. Life Sci.* **2016**, *77*, 53–60. [CrossRef]
169. Goulson, D. Pesticides Linked to Bird Declines. *Nature* **2014**, *511*, 295–296. [CrossRef] [PubMed]
170. Niggli, U.; Riedel, J.; Brühl, C.; Liess, M.; Schulz, R.; Altenburger, R.; Märländer, B.; Bokelmann, W.; Heß, J.; Reineke, A.; et al. Pflanzenschutz und Biodiversität in Agrarökosystemen. *Ber. Über Landwirtsch. Z. Für Agrarpolit. Und Landwirtsch.* **2020**, *98*, 1–39. [CrossRef]
171. Huth, E.; Paltrinieri, S.; Thiele, J. Bioenergy and Its Effects on Landscape Aesthetics–A Survey Contrasting Conventional and Wild Crop Biomass Production. *Biomass Bioenergy* **2019**, *122*, 313–321. [CrossRef]
172. Reiser, D.; Martín-López, J.M.; Memic, E.; Vázquez-Arellano, M.; Brandner, S.; Griepentrog, H.W. 3D Imaging with a Sonar Sensor and an Automated 3-Axes Frame for Selective Spraying in Controlled Conditions. *J. Imaging* **2017**, *3*, 9. [CrossRef]
173. Vázquez-Arellano, M.; Reiser, D.; Paraforos, D.S.; Garrido-Izard, M.; Burce, M.E.C.; Griepentrog, H.W. 3-D Reconstruction of Maize Plants Using a Time-of-Flight Camera. *Comput. Electron. Agric.* **2018**, *145*, 235–247. [CrossRef]
174. Gerhards, R.; Kollenda, B.; Machleb, J.; Möller, K.; Butz, A.; Reiser, D.; Griegentrog, H.-W. Camera-Guided Weed Hoeing in Winter Cereals with Narrow Row Distance. *Gesunde Pflanz.* **2020**, *72*, 403–411. [CrossRef]
175. Munz, S.; Reiser, D. Approach for Image-Based Semantic Segmentation of Canopy Cover in Pea–Oat Intercropping. *Agriculture* **2020**, *10*, 354. [CrossRef]
176. Mahlein, A.-K. Plant Disease Detection by Imaging Sensors–Parallels and Specific Demands for Precision Agriculture and Plant Phenotyping. *Plant. Dis.* **2016**, *100*, 241–251. [CrossRef] [PubMed]
177. Thomas, S.; Kuska, M.T.; Bohnenkamp, D.; Brugger, A.; Alisaac, E.; Wahabzada, M.; Behmann, J.; Mahlein, A.-K. Benefits of Hyperspectral Imaging for Plant Disease Detection and Plant Protection: A Technical Perspective. *J. Plant. Dis. Prot.* **2018**, *125*, 5–20. [CrossRef]
178. Wahabzada, M.; Mahlein, A.-K.; Bauckhage, C.; Steiner, U.; Oerke, E.-C.; Kersting, K. Metro Maps of Plant Disease Dynamics— Automated Mining of Differences Using Hyperspectral Images. *PLoS ONE* **2015**, *10*, e0116902. [CrossRef]
179. Thomas, S.; Wahabzada, M.; Kuska, M.T.; Rascher, U.; Mahlein, A.-K.; Thomas, S.; Wahabzada, M.; Kuska, M.T.; Rascher, U.; Mahlein, A.-K. Observation of Plant–Pathogen Interaction by Simultaneous Hyperspectral Imaging Reflection and Transmission Measurements. *Funct. Plant. Biol.* **2016**, *44*, 23–34. [CrossRef]
180. Tsouros, D.C.; Bibi, S.; Sarigiannidis, P.G. A Review on UAV-Based Applications for Precision Agriculture. *Information* **2019**, *10*, 349. [CrossRef]
181. Watson, C.A.; Reckling, M.; Preissel, S.; Bachinger, J.; Bergkvist, G.; Kuhlman, T.; Lindström, K.; Nemecek, T.; Topp, C.F.E.; Vanhatalo, A.; et al. Chapter Four–Grain Legume Production and Use in European Agricultural Systems. In *Advances in Agronomy*; Sparks, D.L., Ed.; Academic Press: Cambridge, MA, USA, 2017; Volume 144, pp. 235–303.
182. Brankatschk, G.; Finkbeiner, M. Crop Rotations and Crop Residues Are Relevant Parameters for Agricultural Carbon Footprints. *Agron. Sustain. Dev.* **2017**, *37*, 58. [CrossRef]
183. Kolbe, H. Fruchtfolgegestaltung im ökologischen und extensiven Landbau: Bewertung von Vorfruchtwirkungen. *Pflanzenbauwissenschaften* **2006**, *10*, 82–89.
184. Bundessortenamt. *Beschreibende Sortenliste Getreide, Mais Öl- Und Faserpflanzen Leguminosen Rüben Zwischenfrüchte 2020*; Bundessortenamt: Hannover, Germany, 2020.
185. Miedaner, T.; Juroszek, P. Climate Change Will Influence Disease Resistance Breeding in Wheat in Northwestern Europe. *Theor. Appl. Genet.* **2021**, *134*, 1771–1785. [CrossRef] [PubMed]
186. Tian, J.; Xu, G.; Yuan, M. Towards Engineering Broad-Spectrum Disease-Resistant Crops. *Trends Plant. Sci.* **2020**, *25*, 424–427. [CrossRef]
187. Van Esse, H.P.; Reuber, T.L.; van der Does, D. Genetic Modification to Improve Disease Resistance in Crops. *New Phytol.* **2020**, *225*, 70–86. [CrossRef]
188. Langner, T.; Kamoun, S.; Belhaj, K. CRISPR Crops: Plant Genome Editing Toward Disease Resistance. *Annu. Rev. Phytopathol.* **2018**, *56*, 479–512. [CrossRef]
189. Eom, J.-S.; Luo, D.; Atienza-Grande, G.; Yang, J.; Ji, C.; Thi Luu, V.; Huguet-Tapia, J.C.; Char, S.N.; Liu, B.; Nguyen, H.; et al. Diagnostic Kit for Rice Blight Resistance. *Nat. Biotechnol.* **2019**, *37*, 1372–1379. [CrossRef]
190. Oliva, R.; Ji, C.; Atienza-Grande, G.; Huguet-Tapia, J.C.; Perez-Quintero, A.; Li, T.; Eom, J.-S.; Li, C.; Nguyen, H.; Liu, B.; et al. Broad-Spectrum Resistance to Bacterial Blight in Rice Using Genome Editing. *Nat. Biotechnol.* **2019**, *37*, 1344–1350. [CrossRef]
191. Griepentrog, H.W. Längsverteilung von Sämaschinen und ihre Wirkung auf Standfläche und Ertrag bei Raps. *Agrartech. Forsch.* **1995**, *1*. Available online: https://docplayer.org/70215040-Laengsverteilung-von-saemaschinen-und-ihre-wirkung-auf-standflaeche-und-ertrag-bei-raps.html (accessed on 25 August 2021).
192. Boulard, T.; Roy, J.-C.; Pouillard, J.-B.; Fatnassi, H.; Grisey, A. Modelling of Micrometeorology, Canopy Transpiration and Photosynthesis in a Closed Greenhouse Using Computational Fluid Dynamics. *Biosyst. Eng.* **2017**, *158*, 110–133. [CrossRef]
193. Sultan, S.E. Phenotypic Plasticity for Plant Development, Function and Life History. *Trends Plant. Sci.* **2000**, *5*, 537–542. [CrossRef]
194. Bongers, F.J.; Pierik, R.; Anten, N.P.R.; Evers, J.B. Subtle Variation in Shade Avoidance Responses May Have Profound Consequences for Plant Competitiveness. *Ann. Bot.* **2018**, *121*, 863–873. [CrossRef] [PubMed]

195. Zhu, J.; Vos, J.; van der Werf, W.; van der Putten, P.E.L.; Evers, J.B. Early Competition Shapes Maize Whole-Plant Development in Mixed Stands. *J. Exp. Bot.* **2014**, *65*, 641–653. [CrossRef] [PubMed]

196. Evers, J.B.; Bastiaans, L. Quantifying the Effect of Crop Spatial Arrangement on Weed Suppression Using Functional-Structural Plant Modelling. *J. Plant. Res.* **2016**, *129*, 339–351. [CrossRef]

197. Bradáčová, K.; Weber, N.F.; Morad-Talab, N.; Asim, M.; Imran, M.; Weinmann, M.; Neumann, G. Micronutrients (Zn/Mn), Seaweed Extracts, and Plant Growth-Promoting Bacteria as Cold-Stress Protectants in Maize. *Chem. Biol. Technol. Agric.* **2016**, *3*, 19. [CrossRef]

198. Datnoff, L.E.; Elmer, W.H.; Huber, D.M. Mineral Nutrition and Plant Disease. Amer Phytopathological Society: St. Paul, MN, USA, 2007.

199. Imran, M.; Mahmood, A.; Römheld, V.; Neumann, G. Nutrient Seed Priming Improves Seedling Development of Maize Exposed to Low Root Zone Temperatures during Early Growth. *Eur. J. Agron.* **2013**, *49*, 141–148. [CrossRef]

200. Huber, D.M.; Watson, R.D. Nitrogen Form and Plant Disease. *Annu. Rev. Phytopathol.* **1974**, *12*, 139–165. [CrossRef]

201. Sun, R.-C. Lignin Source and Structural Characterization. *ChemSusChem* **2020**, *13*, 4385–4393. [CrossRef]

202. Coskun, D.; Deshmukh, R.; Sonah, H.; Menzies, J.G.; Reynolds, O.; Ma, J.F.; Kronzucker, H.J.; Bélanger, R.R. The Controversies of Silicon's Role in Plant Biology. *New Phytol.* **2019**, *221*, 67–85. [CrossRef]

203. Datnoff, E.L.; Rodrigues, A.F. Highlights and Prospects for Using Silicon in the Future. In *Silicon and Plant Diseases*; Rodrigues, F.A., Datnoff, L.E., Eds.; Springer International Publishing: Cham, Switzerland, 2015; pp. 139–145. ISBN 978-3-319-22930-0.

204. Huber, D.M.; Wilhelm, N.S. The Role of Manganese in Resistance to Plant Diseases. In *Manganese in Soils and Plants, Proceedings of the International Symposium on 'Manganese in Soils and Plants' held at the Waite Agricultural Research Institute, The University of Adelaide, Glen Osmond, South Australia, 22–26 August 1988*; Graham, R.D., Hannam, R.J., Uren, N.C., Eds.; Developments in Plant and Soil Sciences; Springer: Dordrecht, The Netherlands, 1988; pp. 155–173. ISBN 978-94-009-2817-6.

205. Lutman, P.J.W.; Moss, S.R.; Cook, S.; Welham, S.J. A Review of the Effects of Crop Agronomy on the Management of Alopecurus Myosuroides. *Weed Res.* **2013**, *53*, 299–313. [CrossRef]

206. Hurle, K. Integrated Management of Grass Weeds in Arable Crops. In Proceedings of the International Conference, British Crop Protection Council, Brighton, UK, 22–25 November 1993; pp. 81–88.

207. Riemens, M.; Weide, R.V.D.; Bleeker, P.; Lotz, L. Effect of Stale Seedbed Preparations and Subsequent Weed Control in Lettuce (Cv. Iceboll) on Weed Densities. *Weed Res.* **2007**, *47*, 149–156. [CrossRef]

208. Melander, B.; Cirujeda, A.; Jørgensen, M.H. Effects of Inter-Row Hoeing and Fertilizer Placement on Weed Growth and Yield of Winter Wheat. *Weed Res.* **2003**, *43*, 428–438. [CrossRef]

209. Conrath, U.; Beckers, G.J.M.; Flors, V.; García-Agustín, P.; Jakab, G.; Mauch, F.; Newman, M.-A.; Pieterse, C.M.J.; Poinssot, B.; Pozo, M.J.; et al. Priming: Getting Ready for Battle. *MPMI* **2006**, *19*, 1062–1071. [CrossRef] [PubMed]

210. Kunz, J.; Löffler, G.; Bauhus, J. Minor European Broadleaved Tree Species Are More Drought-Tolerant than Fagus Sylvatica but Not More Tolerant than Quercus Petraea. *For. Ecol. Manag.* **2018**, *414*, 15–27. [CrossRef]

211. Fernández, F.G.; White, C. No-Till and Strip-Till Corn Production with Broadcast and Subsurface-Band Phosphorus and Potassium. *Agron. J.* **2012**, *104*, 996–1005. [CrossRef]

212. Kuska, M.; Wahabzada, M.; Leucker, M.; Dehne, H.-W.; Kersting, K.; Oerke, E.-C.; Steiner, U.; Mahlein, A.-K. Hyperspectral Phenotyping on the Microscopic Scale: Towards Automated Characterization of Plant-Pathogen Interactions. *Plant. Methods* **2015**, *11*, 28. [CrossRef]

213. Kuska, M.T.; Brugger, A.; Thomas, S.; Wahabzada, M.; Kersting, K.; Oerke, E.-C.; Steiner, U.; Mahlein, A.-K. Spectral Patterns Reveal Early Resistance Reactions of Barley against *Blumeria graminis* f. Sp. *Hordei*. *Phytopathology* **2017**, *107*, 1388–1398. [CrossRef]

214. Leucker, M.; Mahlein, A.-K.; Steiner, U.; Oerke, E.-C. Improvement of Lesion Phenotyping in Cercospora Beticola–Sugar Beet Interaction by Hyperspectral Imaging. *Phytopathology* **2015**, *106*, 177–184. [CrossRef] [PubMed]

215. Scheu, S. Effects of Earthworms on Plant Growth: Patterns and Perspectives: The 7th International Symposium on Earthworm Ecology Cardiff Wales 2002. *Pedobiologia* **2003**, *47*, 846–856. [CrossRef]

216. Stott, D.E.; Kennedy, A.C.; Cambardella, C.A. Impact of Soil Organisms and Organic Matter on Soil Structure: #. In *Soil Quality and Soil Erosion*; CRC Press: Boca Raton, FL, USA, 1999; ISBN 978-0-203-73926-6.

217. Kumar, A.; Verma, J.P. Does Plant—Microbe Interaction Confer Stress Tolerance in Plants: A Review? *Microbiol. Res.* **2018**, *207*, 41–52. [CrossRef]

218. Hallama, M.; Pekrun, C.; Lambers, H.; Kandeler, E. Hidden Miners–The Roles of Cover Crops and Soil Microorganisms in Phosphorus Cycling through Agroecosystems. *Plant. Soil* **2019**, *434*, 7–45. [CrossRef]

219. Hallama, M.; Pekrun, C.; Pilz, S.; Jarosch, K.A.; Frąc, M.; Uksa, M.; Marhan, S.; Kandeler, E. Interactions between Cover Crops and Soil Microorganisms Increase Phosphorus Availability in Conservation Agriculture. *Plant. Soil* **2021**, *463*, 307–328. [CrossRef]

220. Pelosi, C.; Barot, S.; Capowiez, Y.; Hedde, M.; Vandenbulcke, F. Pesticides and Earthworms. A Review. *Agron. Sustain. Dev.* **2014**, *34*, 199–228. [CrossRef]

221. Stukenbrock, E.H.; McDonald, B.A. The Origins of Plant Pathogens in Agro-Ecosystems. *Annu. Rev. Phytopathol.* **2008**, *46*, 75–100. [CrossRef]

222. Hubbard, A.; Lewis, C.M.; Yoshida, K.; Ramirez-Gonzalez, R.H.; de Vallavieille-Pope, C.; Thomas, J.; Kamoun, S.; Bayles, R.; Uauy, C.; Saunders, D.G. Field Pathogenomics Reveals the Emergence of a Diverse Wheat Yellow Rust Population. *Genome Biol.* **2015**, *16*, 23. [CrossRef] [PubMed]

223. Bueno-Sancho, V.; Persoons, A.; Hubbard, A.; Cabrera-Quio, L.E.; Lewis, C.M.; Corredor-Moreno, P.; Bunting, D.C.E.; Ali, S.; Chng, S.; Hodson, D.P.; et al. Pathogenomic Analysis of Wheat Yellow Rust Lineages Detects Seasonal Variation and Host Specificity. *Genome Biol. Evol.* **2017**, *9*, 3282–3296. [CrossRef]

224. Zebisch, M. Modellierung der Auswirkungen von Landnutzungsänderungen auf Landschaftsmuster und Biodiversität. Ph.D. Thesis, Technische Universität Berlin, Berlin, Germany, 2004. [CrossRef]

225. Lázaro, A.; Alomar, D. Landscape Heterogeneity Increases the Spatial Stability of Pollination Services to Almond Trees through the Stability of Pollinator Visits. *Agric. Ecosyst. Environ.* **2019**, *279*, 149–155. [CrossRef]

226. Thies, C.; Tscharntke, T. Landscape Structure and Biological Control in Agroecosystems. *Science* **1999**, *285*, 893–895. [CrossRef]

227. Sirami, C.; Gross, N.; Baillod, A.B.; Bertrand, C.; Carrié, R.; Hass, A.; Henckel, L.; Miguet, P.; Vuillot, C.; Alignier, A.; et al. Increasing Crop Heterogeneity Enhances Multitrophic Diversity across Agricultural Regions. *Proc. Natl. Acad. Sci. USA* **2019**, *116*, 16442–16447. [CrossRef] [PubMed]

228. Stark, A.; Wetzel, T. Fliegen Der Gattung Platypalpus (Diptera, Empididae)–Bisher Wenig Beachtete Prädatoren Im Getreidebestand. *J. Appl. Entomol.* **1987**, *103*, 1–14. [CrossRef]

229. Weber, G.; Franzen, J.; Büchs, W. Beneficial Diptera in Field Crops with Different Inputs of Pesticides and Fertilizers. *Biol. Agric. Hortic.* **1997**, *15*, 109–122. [CrossRef]

230. Stark, A. Zum Beutespektrum Und Jagdverhalten von Fliegen Der Gattung Platypalpus (Empidoidea, Hybotidae). *Studia Dipterol.* **1994**, *1*, 49–74.

231. Willocquet, L.; Meza, W.R.; Dumont, B.; Klocke, B.; Feike, T.; Kersebaum, K.C.; Meriggi, P.; Rossi, V.; Ficke, A.; Djurle, A.; et al. An Outlook on Wheat Health in Europe from a Network of Field Experiments. *Crop. Prot.* **2021**, *139*, 105335. [CrossRef]

232. Röll, G.; Batchelor, W.D.; Castro, A.C.; Simón, M.R.; Graeff-Hönninger, S. Development and Evaluation of a Leaf Disease Damage Extension in Cropsim-CERES Wheat. *Agronomy* **2019**, *9*, 120. [CrossRef]

233. Obata, T.; Fernie, A.R. The Use of Metabolomics to Dissect Plant Responses to Abiotic Stresses. *Cell. Mol. Life Sci.* **2012**, *69*, 3225–3243. [CrossRef]

234. Lechenet, M.; Dessaint, F.; Py, G.; Makowski, D.; Munier-Jolain, N. Reducing Pesticide Use While Preserving Crop Productivity and Profitability on Arable Farms. *Nat. Plants* **2017**, *3*, 17008. [CrossRef]

235. Dabbert, S.; Braun, J. *Landwirtschaftliche Betriebslehre. Grundwissen Bachelor*, 3rd ed.; Verlag Eugen Ulmer: Stuttgart, Germany, 2012; ISBN 978-3-8252-3819-3.

236. Hardaker, J.B.; Huirne, R.B.M.; Anderson, J.R.; Lien, G. (Eds.) *Coping with Risk in Agriculture*, 2nd ed.; CABI: Wallingford, UK, 2004; ISBN 978-0-85199-831-2.

237. Mußhoff, O.; Hirschhauer, N. *Modernes Agrarmanagement Betriebswirtschaftliche Analyse- Und Planungsverfahren*, 2nd ed.; Vahlen Franz GmbH: München, Germany, 2011; ISBN 978-3-8006-3827-7.

238. Bhattacharya, S. Deadly New Wheat Disease Threatens Europe's Crops. *Nature* **2017**, *542*, 145–146. [CrossRef]

239. Heap, I. International Survey of Herbicide Resistant Weeds. Available online: http://www.weedscience.org/Home.aspx (accessed on 1 October 2018).

240. North, D.C. *Institutions, Institutional Change and Economic Performance*; Cambridge University Press: Cambridge, UK, 1990.

241. Ostrom, E. *Governing the Commons: The Evolution of Institutions for Collective Action*; Cambridge University Press: Cambridge, UK, 1990.

242. European Commission. *Ernährung Und Landwirtschaft Der Zukunft*; European Commission: Brussels, Belgium, 2017.

243. Gutsche, V. Managementstrategien des Pflanzenschutzes der Zukunft im Focus von Umweltverträglichkeit und Effizienz. *J. Für Kult.* **2012**, *64*, 325–341.

244. Böl, G.-F.; Epp, A.; Michalski, B. Pflanzenschutzmittelrückstände in Lebensmitteln, Die Wahrnehmung der deutschen Bevölkerung. 57. Deutsche Pflanzenschutztagung: 6–9 September 2010; Humboldt-Universität zu Berlin; *Kurzfassungen der Beiträge* 2010. Available online: https://www.kulturkaufhaus.de/de/detail/ISBN-9783938163603/Epp-A./Pflanzenschutzmittel-R%C3%BCckst%C3%A4nde-in-Lebensmitteln (accessed on 25 August 2021).

245. BfR. *Pflanzenschutzmittel und -Rückstände in Lebensmitteln; Analyse der Medienberichterstattung PlantMedia*; Bundesinstitut für Risikobewertung: Berlin, Germany, 2013.

246. Bruhn, C.M.; Diaz-Knauf, K.; Feldman, N.; Harwood, J.; Ho, G.; Ivans, E.; Kubin, L.; Lamp, C.; Marshall, M.; Osaki, S.; et al. Consumer Food Safety Concerns and Interest in Pesticide-Related Information. *J. Food Saf.* **1991**, *12*, 253–262. [CrossRef]

247. Huang, C.L. Consumer Preferences and Attitudes towards Organically Grown Produce. *Eur. Rev. Agric. Econ.* **1996**, *23*, 331–342. [CrossRef]

248. Koch, S.; Epp, A.; Lohmann, M.; Böl, G.-F. Pesticide Residues in Food: Attitudes, Beliefs, and Misconceptions among Conventional and Organic Consumers. *J. Food Prot.* **2017**, *80*, 2083–2089. [CrossRef]

249. Misra, S.; Huang, C.L.; Ott, S.L. Georgia Consumers' Preference for Organically Grown Fresh Produce. *J. Agribus.* **1991**, *9*, 53–63.

250. Magnusson, M.K.; Arvola, A.; Koivisto Hursti, U.; Åberg, L.; Sjödén, P. Attitudes towards Organic Foods among Swedish Consumers. *Br. Food J.* **2001**, *103*, 209–227. [CrossRef]

251. Hemmerling, S.; Hamm, U.; Spiller, A. Consumption Behaviour Regarding Organic Food from a Marketing Perspective—A Literature Review. *Org. Agr.* **2015**, *5*, 277–313. [CrossRef]

252. Spiller, A.; Nitzko, S. Welchen Mehrwert Haben Agrarprodukte Ohne CPSM, Die Dennoch Nicht Öko Sind? *Landinfo* **2019**, *2*, 15–18.

253. Cranfield, J.A.L.; Magnusson, E. Canadian Consumer's Willingness-To-Pay for Pesticide Free Food Products: An Ordered Probit Analysis. *Int. Food Agribus. Manag. Rev.* **2003**, *6*, 1–18.
254. Weaver, R.D.; Evans, D.J.; Luloff, A.E. Pesticide Use in Tomato Production: Consumer Concerns and Willingness-to-Pay. *Agribusiness* **1992**, *8*, 131–142. [CrossRef]
255. Boccaletti, S.; Nardella, M. Consumer Willingness to Pay for Pesticide-Free Fresh Fruit and Vegetables in Italy. *Int. Food Agribus. Manag. Rev.* **2000**, *3*, 297–310. [CrossRef]
256. Román, S.; Sánchez-Siles, L.M.; Siegrist, M. The Importance of Food Naturalness for Consumers: Results of a Systematic Review. *Trends Food Sci. Technol.* **2017**, *67*, 44–57. [CrossRef]
257. Keller, R. *Diskursforschung: Eine Einführung für SozialwissenschaftlerInnen*, 4th ed.; Qualitative Sozialforschung; Springer: Berlin/Heidelberg, Germany, 2011; ISBN 978-3-531-17352-8.
258. Paltridge, B. *Discourse Analysis An. Introduction*, 2nd ed.; Bloomsbury Publishing Plc: New York, NY, USA, 2012; ISBN 978-1-4411-6762-0.
259. Dijkman, T.J.; Birkved, M.; Hauschild, M.Z. PestLCI 2.0: A Second Generation Model for Estimating Emissions of Pesticides from Arable Land in LCA. *Int. J. Life Cycle Assess.* **2012**, *17*, 973–986. [CrossRef]
260. Von Cossel, M. Renewable Energy from Wildflowers—Perennial Wild Plant Mixtures as a Social-Ecologically Sustainable Biomass Supply System. *Adv. Sustain. Syst.* **2020**, *4*, 2000037. [CrossRef]
261. Von Cossel, M.; Winkler, B.; Mangold, A.; Lask, J.; Wagner, M.; Lewandowski, I.; Elbersen, B.; van Eupen, M.; Mantel, S.; Kiesel, A. Bridging the Gap between Biofuels and Biodiversity through Monetizing Environmental Services of Miscanthus Cultivation. *Earth's Future* **2020**, *8*, e2020EF001478. [CrossRef]
262. Janusch, C.; Lewin, E.F.; Battaglia, M.L.; Rezaei-Chiyaneh, E.; Von Cossel, M. Flower-Power in the Bioenergy Sector—A Review on Second Generation Biofuel from Perennial Wild Plant Mixtures. *Renew. Sustain. Energy Rev.* **2021**, *147*, 111257. [CrossRef]
263. Nemecek, T.; Dubois, D.; Huguenin-Elie, O.; Gaillard, G. Life Cycle Assessment of Swiss Farming Systems: I. Integrated and Organic Farming. *Agric. Syst.* **2011**, *104*, 217–232. [CrossRef]
264. Meyer, F.; Wagner, M.; Lewandowski, I. Optimizing GHG Emission and Energy-Saving Performance of Miscanthus-Based Value Chains. *Biomass Conv. Bioref.* **2017**, *7*, 139–152. [CrossRef]
265. Pishgar-Komleh, S.H.; Sefeedpari, P.; Pelletier, N.; Brandão, M. Life cycle assessment methodology for agriculture: Some considerations for best practices. In *Assessing the Environmental Impact of Agriculture*; Weidema, B.P., Ed.; Burleigh Dodds Science Publishing Limited: Cambridge, UK, 2019; ISBN 978-1-78676-228-3.
266. Schmidt, T.; Fernando, A.L.; Monti, A.; Rettenmaier, N. Life Cycle Assessment of Bioenergy and Bio-Based Products from Perennial Grasses Cultivated on Marginal Land in the Mediterranean Region. *Bioenerg. Res.* **2015**, *8*, 1548–1561. [CrossRef]
267. Searchinger, T.D.; Beringer, T.; Holtsmark, B.; Kammen, D.M.; Lambin, E.F.; Lucht, W.; Raven, P.; Ypersele, J.-P. Europe's Renewable Energy Directive Poised to Harm Global Forests. *Nat. Commun.* **2018**, *9*, 3741. [CrossRef] [PubMed]
268. Ceballos, G.; Ehrlich, P.R.; Barnosky, A.D.; García, A.; Pringle, R.M.; Palmer, T.M. Accelerated Modern Human–Induced Species Losses: Entering the Sixth Mass Extinction. *Sci. Adv.* **2015**, *1*, e1400253. [CrossRef] [PubMed]
269. Von Cossel, M.; Wagner, M.; Lask, J.; Magenau, E.; Bauerle, A.; Von Cossel, V.; Warrach-Sagi, K.; Elbersen, B.; Staritsky, I.; van Eupen, M.; et al. Prospects of Bioenergy Cropping Systems for a More Social-Ecologically Sound Bioeconomy. *Agronomy* **2019**, *9*, 605. [CrossRef]
270. Goglio, P.; Brankatschk, G.; Knudsen, M.T.; Williams, A.G.; Nemecek, T. Addressing Crop Interactions within Cropping Systems in LCA. *Int. J. Life Cycle Assess.* **2018**, *23*, 1735–1743. [CrossRef]
271. Costa, M.P.; Chadwick, D.; Saget, S.; Rees, R.M.; Williams, M.; Styles, D. Representing Crop Rotations in Life Cycle Assessment: A Review of Legume LCA Studies. *Int. J. Life Cycle Assess.* **2020**, *25*, 1942–1956. [CrossRef]
272. Calvet, C.; Napoléone, C.; Salles, J.-M. The Biodiversity Offsetting Dilemma: Between Economic Rationales and Ecological Dynamics. *Sustainability* **2015**, *7*, 7357–7378. [CrossRef]
273. Wagner, M.; Lewandowski, I. Relevance of Environmental Impact Categories for Perennial Biomass Production. *Gcb Bioenergy* **2017**, *9*, 215–228. [CrossRef]
274. Alejandre, E.M.; van Bodegom, P.M.; Guinée, J.B. Towards an Optimal Coverage of Ecosystem Services in LCA. *J. Clean. Prod.* **2019**, *231*, 714–722. [CrossRef]
275. Bos, U.; Horn, R.; Beck, T.; Lindner, J.P.; Fischer, M. *LANCA®—Characterization Factors for Life Cycle Impact Assessment*; Version 2.0; Fraunhofer-Institut für Bauphysik: Stuttgart, Germany, 2016.
276. Chaudhary, A.; Brooks, T.M. Land Use Intensity-Specific Global Characterization Factors to Assess Product Biodiversity Footprints. *Environ. Sci. Technol.* **2018**, *52*, 5094–5104. [CrossRef] [PubMed]
277. Crenna, E.; Marques, A.; La Notte, A.; Sala, S. Biodiversity Assessment of Value Chains: State of the Art and Emerging Challenges. *Environ. Sci. Technol.* **2020**, *54*, 9715–9728. [CrossRef]

Article

Intercropping in Rice Farming under the System of Rice Intensification—An Agroecological Strategy for Weed Control, Better Yield, Increased Returns, and Social–Ecological Sustainability

Tavseef Mairaj Shah [1,*], Sumbal Tasawwar [1], M. Anwar Bhat [2] and Ralf Otterpohl [1]

[1] Rural Revival and Restoration Engineering, Institute of Wastewater Management and Water Protection, Hamburg University of Technology, 21073 Hamburg, Germany; sumbal.tasawwar@tuhh.de (S.T.); ro@tuhh.de (R.O.)

[2] Division of Agronomy, Faculty of Agriculture, Sher-e-Kashmir University of Agricultural Sciences and Technology of Kashmir, Sopore 193201, Jammu & Kashmir, India; hodagron_foa@skuastkashmir.ac.in

* Correspondence: tavseef.mairaj.shah@tuhh.de

Abstract: Rice is the staple food for more than half of the world's population. In South Asia, rice farming systems provide food to the majority of the population, and agriculture is a primary source of livelihood. With the demand for nutritious food increasing, introducing innovative strategies in farming systems is imperative. In this regard, intensification of rice farming is intricately linked with the challenges of water scarcity, soil degradation, and the vagaries of climate change. Agroecological farming systems like the System of Rice Intensification (SRI) have been proposed as water-saving and sustainable ways of food production. This study examines the effect of intercropping beans with rice under SRI management on the growth of weeds and on the different plant growth parameters. Intercropping led to a 65% decrease in weed infestation on average, which is important given that weed infestation is stated as a criticism of SRI in some circles and is a major factor in limiting yield in rice-producing regions. In addition to the water savings of about 40% due to the SRI methodology, the innovation led to an increase in rice yield by 33% and an increase in the net income of farmers by 57% compared to the conventional rice farming method. The results indicate that intercropping can be a positive addition to the rice farming system, hence contributing to social–ecological sustainability.

Keywords: agroecology; rice; intercropping; sustainable agriculture; sustainable intensification

Citation: Shah, T.M.; Tasawwar, S.; Bhat, M.A.; Otterpohl, R. Intercropping in Rice Farming under the System of Rice Intensification—An Agroecological Strategy for Weed Control, Better Yield, Increased Returns, and Social–Ecological Sustainability. *Agronomy* **2021**, *11*, 1010. https://doi.org/10.3390/agronomy11051010

Academic Editors: Moritz Von Cossel, Joaquín Castro-Montoya and Yasir Iqbal

Received: 2 May 2021
Accepted: 19 May 2021
Published: 20 May 2021

Publisher's Note: MDPI stays neutral with regard to jurisdictional claims in published maps and institutional affiliations.

1. Introduction

Agriculture is at the core of one of the greatest challenges of this century: the challenges of meeting the neglected and the growing nutritional needs of the world's population and remedying the environmental damage due to agricultural activities at the same time [1]. Currently, 800 million individuals worldwide do not have access to enough food, while more than two billion people experience key micronutrient deficiencies. This is a problem that is more serious in low-income countries where the percentage of food insecure individuals is around 60% [2]. On the other hand, agriculture is now counted as a major force contributing to planetary overshoot of natural resources, contributing dominantly towards climate change, biodiversity loss, and the degradation of land and freshwater [3–5].

In this regard, rice farming systems are particularly relevant, with their contribution to feeding the world as well as the large scope for improvements with respect to different environmental issues associated with rice farming systems, specifically flooded rice cultivation systems [6–11]. Rice is a staple food for more than half of the world's population and contributes about one-fourth of the global energy consumption [12]. The wide-ranging importance of rice is evident from the fact that it is grown in 112 different countries of the world in different climatic zones [12]. Irrigated rice cultivation covers about 60% of the total land under rice cultivation worldwide and contributes 75% to the total rice produce [13].

With a large water footprint, which is 2–3 times more than that of upland crops, irrigated flooded rice monocropping systems, dependent on agrochemical use, have been found to be detrimental to the environment and biodiversity [8]. For example, it has been reported that nitrogen losses due to ammonia volatilization account for up to 60% of nitrogen applied in flooded rice systems [8]. Flooding of rice paddies has also been found to have an association with health risks, particularly in tropical and sub-tropical regions, in terms of potentially aiding the proliferation of water-borne diseases [10]. Another health risk associated with flooded rice systems is the incidence of toxic residues of pesticides and metalloids/metals like arsenic, cadmium, and mercury in rice grains. This is particularly damaging for regions where rice forms a large percentage of the diet, for example, in South Asia and Southeast Asia. Arsenic pollution is a widespread problem in the Punjab and Bengal regions in South Asia, which straddle Pakistan, Bangladesh, and India [14–17]. In this regard, an aerobic water management system in rice, through the practice of alternate wetting and drying, has been reported to reduce the risk of uptake of metals like arsenic [18].

In this context, the increasing demand of nutritious food is putting an unprecedented pressure on the land and water resources. In this regard, 'another green revolution', which follows the methodology of the 'original' green revolution of the 1960s, is not a viable option, owing to its ecological and social costs [19]. Its positive effect on cereal production notwithstanding, the green revolution also contributed to the reduced availability of pulses and legumes (protein sources) in the rural diet [20]. Additionally, the per capita availability of land has halved from 1960 to 2010 and is expected to decrease by a further 30% by 2050 [21,22]. With respect to the livelihoods of farmers, stricter regulations on agrochemical residues in food grains are making agrochemical-based intensification unviable. It is in the backdrop of these socioeconomic challenges, highlighted and intensified during the COVID-19 pandemic, that sustainable intensification of agriculture is being proposed [23].

Agroecology-based farming strategies promise nutritious yields, with more consideration for the fast depleting natural resources vital for agriculture, i.e., water and soil. In a report by the FAO released in September 2020, the role of agroecology has been termed as vital for climate-resilient livelihoods and sustainable food systems [24]. The system of rice intensification (SRI) is an agroecological methodology of growing rice. It involves early plant establishment, wider spacing between plants transplanted individually, manual weeding, and alternate wetting and drying of the rice field instead of continuous flooding [25]. SRI-based rice farming systems have been associated with a range of environmental benefits, reportedly reducing water and energy use by 60% and 74%, respectively, and greenhouse gas emissions by 40% [11]. It has led to improved yields in different parts of the world and includes practices that enhance the soil health [26–29]. An oft-cited disincentive of this methodology, however, is the increased incidence of weeds due to the absence of continuous flooding of the fields, which creates aerobic conditions feasible for the growth of weeds [20,30,31]. This also leads to increased work-hour requirements for the weeding process [32]. Weeds are one of the single largest source of yield losses in rice farming, representing 6.6% of the yield gap in South Asia [33]. In the Mediterranean region as well, the occurrence of weeds has been described as the main reason behind yield variability and yield gap, specifically in organic rice farming systems [34–36]. Under SRI, without proper weed management, a reduction in rice yield by up to 70% has been reported due to weed infestation [37].

Hence, the incidence of weeds is reported as a challenge in SRI, along with the need for increased work-hours required during weeding, which occurs multiple times in a crop season [27,38]. In this regard, the availability of family labor has been found to influence the adoption of SRI practices by farm households [39]. Higher work-hour requirements may also have led to dis-adoption of SRI in some cases [40]. Intercropping has been reported as a natural way to control the growth of weeds in other cropping systems, including upland rice and maize farming systems [41–44]. The system of rice intensification is relevant in this regard, as it provides wider inter-row space to introduce intercropping [45]. In addition to comparing the effect of intercropping on weed infestation, the effect of intercropping on different plant growth characteristics under SRI management has also been studied. Improvements in different plant growth characteristics under SRI management in comparison to the conventional flooded rice (CFR) has already been widely documented in literature [26,46–50].

The results of this study recommend the same in the case of rice grown under alternate wetting and drying conditions of SRI. It was hypothesized that growing a leguminous crop, i.e., beans, as an intercrop in between rice rows grown under SRI would affect the incidence of weeds and the growth characteristics of rice plants. These parameters were recorded in field experiments. The effect of intercropping on the chlorophyll content of the rice crop was also determined in multiple experiments under controlled lab greenhouse conditions. Growing beans together with rice can also serve to diversify the diet of the rural population by restoring an important source of protein to it. A new terminology, denoted as SRIBI, has been used for the methodology of system of rice intensification with beans intercropping. The conventional flooded rice method is represented as CFR.

2. Materials and Methods

2.1. Experimental Research Design

Experiments were first conducted in pots at the laboratory level in a greenhouse chamber under controlled conditions in three batches in the years 2017 and 2018. The pot experiments followed a mirrored randomized complete block design to avoid any variations due to placement in the greenhouse. Sandy clay loam soil was used in the pot experiments, which contained 11.25% organic matter, 0.16% total nitrogen, 0.05% total phosphorous, and 0.47% total potassium. In the first batch of experiments, the conventional flooded rice cultivation (CFR) was compared with the system of rice intensification (SRI) and SRI with beans intercropping (SRIBI). The CFR treatment is referred to as 'flooded' treatment, while the SRIBI treatment is referred to SRI+I in the case of pot experiments. Pots of diameter 25 cm and height 25 cm were used in this study. In these experiments, 4 replications of each treatment were analysed. Once it was established that intercropping had a positive effect on the rice crop under SRI management, the next experiments were conducted to find the ideal time and space combination between the rice and the intercrop (beans). Hence, the subsequent batches involved treatments I9 (intercropping done between rice rows at 9 days after transplantation), I35 (intercropping done between rice rows at 35 days after transplantation), and IS (intercropping done at 9 days after transplantation as strip intercropping), in addition to the standard SRI treatment. In these experiments, 8 replications of each treatment were analysed. These experiments were conducted in mini-plots of size 60 cm (length) by 50 cm (width) by 25 cm (height). Based on the analysis of plant growth characteristics like nutrient uptake, chlorophyll content, and yield parameters, the proof of concept was established and field experiments were designed accordingly, which were conducted in 2019 and 2020 (in the local rice growing season May–October).

The field experiments followed a randomized complete block design with four replications and a plot size of 60 m^2 (10 m × 6 m) each. The experiments were carried out in two villages falling under the Sagam belt of the Islamabad region (District Anantnag) in Kashmir, a popular niche belt of the local heritage aromatic landraces of Kashmir, particularly the Mushkibudij (Mushk Budji, literally 'Aromatic Grain') variety that was also used in the studies. The experimental fields were located at 33°36′31″ N, 75°14′59″ E and 33°36′54″ N, 75°15′2″ E, in Jammu and Kashmir. The soil at the experimental site was characterised as silty clay loamy soil, with a neutral pH of 7.3. The soil was low in available nitrogen (140–280 Kg/ha) and medium in available phosphorous (11–22 Kg/ha) and potassium (110–280 Kg/ha). The elevation of the experimental site is at 1800 m amsl. The average maximum temperatures over the months (May–October) were 25.98 °C and 27.14 °C in 2019 and 2020, respectively, while the average minimum temperatures for the same time period were 12.01 °C and 12.19 °C, respectively. Total rainfall recorded in this period was 541.5 mm and 424.2 mm for 2019 and 2020, respectively.

For the CFR treatment, rice plants were transplanted according to the local conventions, at 5 weeks age. For field experiments, the control SRI treatment, in which weeds were allowed to grow for comparison, was referred to as SRI-w (Weedy control) to avoid any misunderstanding of the SRI method where weeding forms an integral part of the cultivation process. In the SRI-w and SRIBI treatments, rice plants were uniformly transplanted in a square pattern at a distance of 25 cm from each other, following the SRI method. Following the SRI method, seedlings were transplanted singly at the two-leaf stage (10 days after sowing). For the CFR treatment, NPK fertilizer was applied based on the standard application: 300 kg (150 N + 75 P_2O_5 + 75 K_2O) per ha. For the SRI and SRIBI treatments, only compost was applied at the rate of 6 t per ha. No pesticides were used in the SRI and SRIBI treatments. The details of individual practices followed in the three treatments are given in Table 1.

Table 1. A comparison of the different agricultural practices in standard SRI (System of Rice Intensification) and the two other SRI-based treatments in the current study.

Practice\System	**SRI**	**SRI-w (Weedy Control)**	**SRIBI**
Early Transplanting	Yes	Yes	Yes
Wide Spacing	Yes	Yes	Yes
One Plant per Hill	Yes	Yes	Yes
Compost Application	Yes	Yes	Yes
Alternate Wetting and Drying	Yes	Yes	Yes
Frequent Weeding	Yes	No	No
Intercropping	No	No	Yes

The SRI plots in this study were not weeded after the first weeding and were used as the weedy-control (SRI-w). Hence, this study is not an evaluation of the yield potential of SRI per se, but rather a study to determine how far intercropping can contribute to make the SRI method better.

2.2. Data Collection

The data on the reported parameters of weed density, plant height, tiller number, panicle length, and spikelet number per panicle were collected manually after 120 days after transplantation, which corresponds to the ripening stage of the rice crop. Work-hours were recorded during the course of the experiments. The occurrence of rice blast was recorded on the basis of on-farm experiences.

2.3. Laboratory Analysis

The chlorophyll content was measured following the method described by Arnon [51] used by Doni et al. [49,52]. The chlorophyll content in leaves was measured at different growth stages, i.e., at the seedling stage and then at different days after transplantation, using the following formulae:

$$C_{Chl-a} = 12.7\ A_{663} - 2.69\ A_{645};\ C_{Chl-b} = 22.9\ A_{645} - 4.68\ A_{663}$$

where A_{663} and A_{645} are the values of absorbance of the solution at wavelengths of 663 nm and 645 nm, respectively. The solution was prepared by cutting leaves into fine pieces and placing 0.1 g of the same pieces into a test-tube to which 20 mL of 80% acetone was added. This solution was kept in the dark for 48 h for incubation, and afterwards, it was analysed with a spectrophotometer.

Nutrient uptake in the plants was measured at the maturity stage (120 days). Rice plants were washed with water after harvesting and were allowed to dry, covered in paper bags at 65 °C for 7 days. A mixture of all plant parts was then ground up to pass through a 1 mm sieve. This ground mixture was then analysed for NPK content. Nitrogen was measured by the Kjeldahl method, phosphorus by cuvette test (Hach Lange LCK350), and potassium by the reflectometer method (Merck Reflectoquant RQflex 10 Reflectometer). Nitrogen content was also measured by a cuvette test (Hach Lange LCK138) and was found to be the same as the value determined using the Kjeldahl method.

2.4. Statistical Analysis

The statistical analysis for ANOVA (Analysis of Variance) was performed using Microsoft Excel 2013 (Microsoft Corporation, Redmond, DC, USA, https://www.microsoft.com/ Last accessed on: 19 May 2021). The statistical significance level in the ANOVA was set at $p \leq 0.05$. Post hoc analysis of ANOVA was also done using the Tukey–Kramer test for parameters involving comparisons of three treatments.

3. Results and Discussion

The experiments were carried out over a period of four years (2017–2020), with pot experiments in 2017–2018 and field experiments in 2019 and 2020. Certain parameters, like nutrient uptake, chlorophyll content, and tiller number, were measured in the lab experiments (pot experiments) to establish the proof of concept about any positive effects intercropping has on the growth of rice under the system of rice intensification (SRI). These parameters were measured for conventional flooded rice (CFR), SRI, and SRI with intercropping (SRIBI) in 2017, while in 2018, different intercropping configurations were compared with SRI. These configurations included I9/SRIBI-9 (intercropping at 9 days after transplantation, DAT), I35/SRIBI-35 (intercropping at 35 DAT), and IS (strip intercropping at 9 DAT). In the field experiments, the control SRI treatment was referred to as SRI-w (weedy control), which was compared to the intercropping treatment (SRIBI). In the pot experiments, the intercropping treatment (SRIBI) was referred to as SRI+I, while CFR was referred to as 'flooded' treatment. The detailed statistical data about the different measured parameters is included in Appendix A.

3.1. Field Experiments

3.1.1. Weed Incidence

The growth of weeds was compared in the treatments of SRI-w (weedy-control) and SRIBI (system of rice intensification with beans intercropping). The number of weed species found in the two treatments was the same, indicating similar growth conditions for the weeds. In total, six weed species were found to be present: Echinochloa colona, Cynodon dactylon, Ammania baccifera, Cyperus iria, Cyperus deformis, and Fimbristylis. However, the density of weeds was observed to be significantly less in the case of the plots with intercropping ($p \leq 0.05$; $p = 0.0048$) (Table 2). The mean values of weed density for plots with and without intercropping were 56 per m^2 and 196 per m^2, respectively, in 2019, as can be seen from Figure 1. In 2020, the mean values of weed density of plots with and without intercropping were 87 and 213, respectively, as shown in Figure 1. On an average, intercropping led to a decrease in weed incidence by 65% in the field experiments over two years. A reduction in the weed density under intercropping regimes has previously been observed in upland rice and dry seeded rice cropping systems [43]. Intercropping of Sesbania in between the crop rows has been found to increase soil fertility in addition to suppressing the growth of weeds [41]. Weeds have also been found to be more responsive to the nitrogen applied to the soil, as a result of which more yield losses and lower values of crop growth parameters are expected in the presence of weeds [42]. The following parameters measured in the current study exemplify this effect.

Table 2. Parameters from ANOVA statistical analysis for weed density comparison.

Weed Density	2019 SRI-w vs. SRIBI				2020 SRI-w vs. SRIBI			
	p-value	F-value	$F_{critical}$	DF	*p*-value	F-value	$F_{critical}$	DF
	0.0048	14.96	5.32	(1,8)	0.0001	51.81	4.74	(1,12)

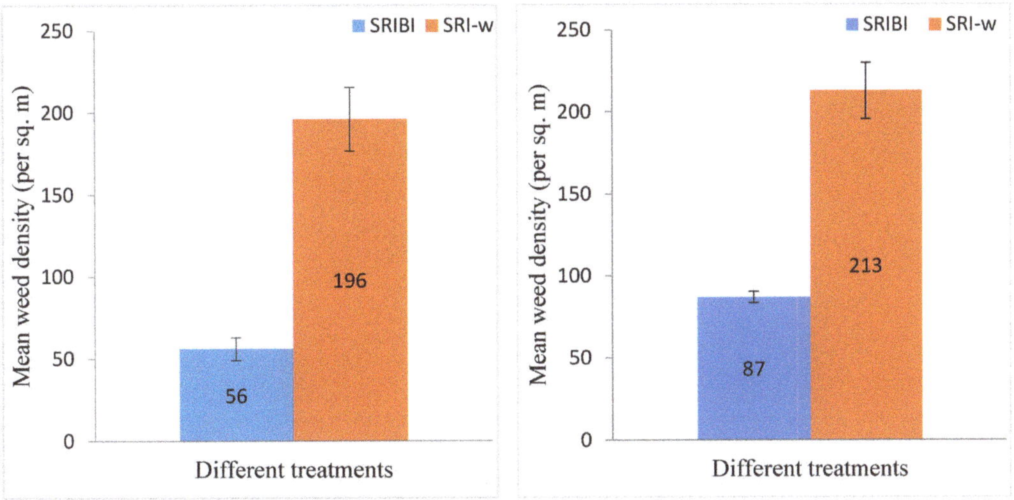

Figure 1. Weed density in the two SRI-based treatments (SRI-w: SRI weedy control; SRIBI: SRI with beans intercropping) (**Left**: 2019; **Right**: 2020).

The decreased weed incidence under the intercropping regime, as seen in Figure 2, makes the adoption of SRI as an agroecological methodology easier for the farmers. Weed infestation is otherwise a main criticism of the SRI methodology and has led to dis-adoption in some regions. The decreased weed incidence can also decrease the dependence of smallholder farmers on chemical solutions to tackle weeds. This can, in turn, lead to less contamination of soil and water which results from excessive use of agrochemicals. The lower weed incidence also leads to a lower labour requirement, which provides another incentive to those farmers for whom SRI is otherwise labour-intensive.

(a)   (b)

Figure 2. A comparison of weed infestation: (**a**) SRI weedy control, SRI-w; (**b**) SRI with intercropping, SRIBI (2019).

3.1.2. Plant Growth Parameters

Plant Height

A comparison of the plant height between rice with and without intercropping suggests that the lower weed density had visible effects on the growth of rice under the two dry rice systems. Rice plants under the intercropping regimen (SRIBI) showed higher plant height as compared to those without intercropping (SRI-w) (Tables 3 and 4. The mean values for plant height were 127.50 cm, 119.25 cm, and 125 cm for SRIBI, SRI-w, and CFR, respectively, at 120 days after transplantation, in 2019 (Table 3). In 2020, the observed mean values for plant height for CFR, SRI-w, and SRIBI were 101 cm, 104 cm, and 109 cm, respectively (Table 4). The effect on the plant height due to intercropping was observed to be significant ($p \leq 0.05$; $p = 0.05$) (Table 5). The maximum and mean heights observed in the three treatments are shown in Tables 3 and 4. The better plant height and earlier harvest-readiness in the intercropping treatment can be attributed to nitrogen fixation by the intercropped legumes as well as the higher availability of applied nitrogen to the rice crop with lower weed density [41,42].

Table 3. A comparison of plant height parameters observed for the three different treatments (2019).

Index	Crop Management System		
	CFR	**SRI-w**	**SRIBI**
Maximum height (cm)	136	125	148
Mean height (cm)	125	119	128

Table 4. A comparison of plant height parameters observed for the three different treatments (2020).

Index	Crop Management System		
	CFR	**SRI-w**	**SRIBI**
Maximum height (cm)	102	106	112
Mean height (cm)	101	104	109

Table 5. Parameters from ANOVA statistical analysis for plant height comparison.

Plant Height	2019 SRI-w vs. SRIBI				2020 CFR vs. SRI-w vs. SRIBI			
	p-value	F-value	$F_{critical}$	DF	*p*-value	F-value	$F_{critical}$	DF
	0.05	4.50	4.49	(1,14)	0.0001	35.73	3.88	(2,12)
					Post hoc analysis			
					$Q_{critical}$	$AMD_{CFR/SRI}$	$AMD_{SRI/SRIBI}$	$AMD_{CFR/SRIBI}$
					4.72	3	5.8	8.8

Post hoc tests were also done for plant growth parameters measured in the field experiments, using the Tukey–Kramer test. The absolute mean difference is used to test the significance of the difference of means. In case of the observed plant height, the increase in plant height in SRIBI was found to be statistically significant in comparisons with both SRI and CFR (Table 5). Here, AMD denotes the absolute mean difference, and its value should be higher than the critical Q value for the mean differences between two specific treatments to be deemed significant.

Number of Tillers

The number of tillers was found to be significantly higher in SRI-based management ($p \leq 0.05$; $p = 0.0001$) (Figure 3, Table 6). This is in line with the comparisons between SRI and flooded rice systems done in previous studies [27,53]. The higher number of tillers in the SRI method has been attributed to the synergetic development of roots and tillers [54]. Tillering ability of rice has been directly associated with the rice grain yield, as the panicle number per hill is directly proportional to the total number of tillers, irrespective of whether the tillers are productive or unproductive [55]. In 2019, the mean number of tillers in SRIBI was found to be 26, while it was 9.5 for conventional flooded rice (CFR), visualized in Figure 3A comparison of the number of tillers in SRI, SRI with intercropping, and CFR, as observed in the field experiments in 2020, is also shown. For the 2020 data, a comparison of the means of all the three treatments revealed that the increase in the tiller number in the case of SRIBI was more significant than that observed in case of SRI treatment (Table 6).

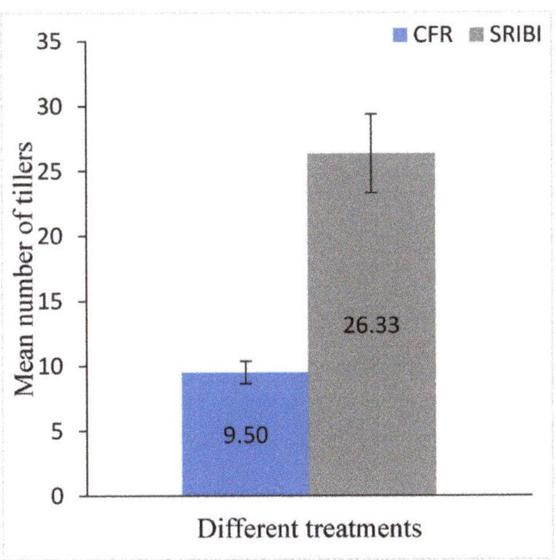

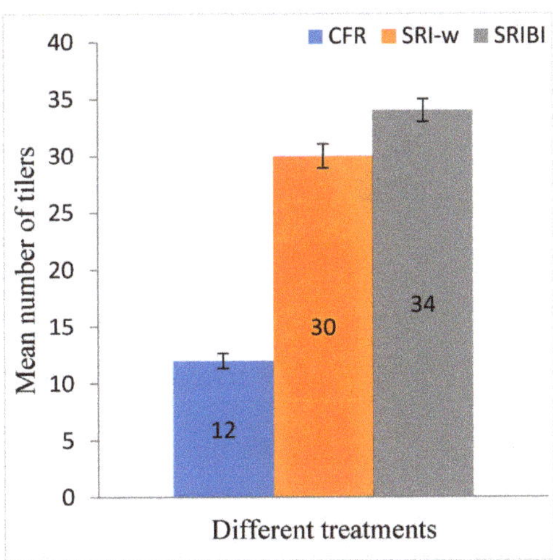

Figure 3. Mean number of tillers in the intercropping treatment and the conventional rice treatment (**Left**: 2019; **Right**: 2020).

Table 6. Parameters from ANOVA statistical analysis for tiller number comparison.

Tiller Number	2019 SRI-w vs. SRIBI				2020 CFR vs. SRI-w vs. SRIBI			
	p-value	F-value	$F_{critical}$	DF	*p*-value	F-value	$F_{critical}$	DF
	0.0001	32.49	4.75	(1,12)	0.0001	160	3.56	(2,18)
					Post hoc analysis			
					$Q_{critical}$	$AMD_{CFR/SRI}$	$AMD_{SRI/SRIBI}$	$AMD_{CFR/SRIBI}$
					8.12	17.86	4.14	22

Panicle Length

Notwithstanding a higher number of tillers and, by extension, a higher number of panicles [55], in order to quantify the effects of intercropping on the yield, a comparison of the panicle length was done for all three treatments, flooded rice, SRI-w, and SRI with beans intercropping (SRIBI) (Figure 4). The difference among the three treatments was found to be statistically insignificant ($p > 0.05$; $p = 0.065$), indicating that the yield potential of rice was not negatively affected by either SRI or intercropping (Table 7). However, since the number of tillers and panicles in SRI-based management was multiple times higher, it was expected to translate to higher yields, as reported in other studies [55].

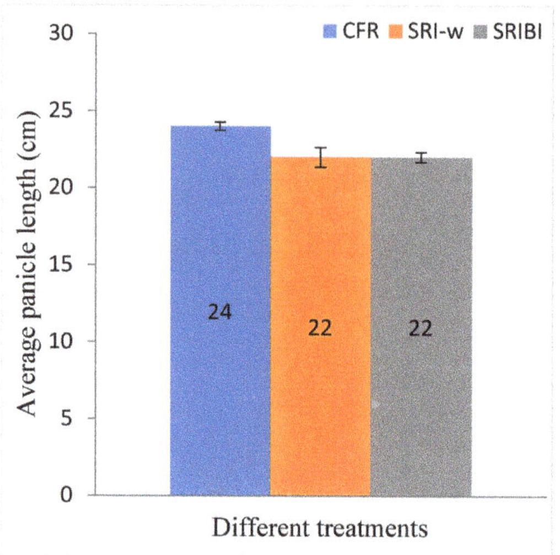
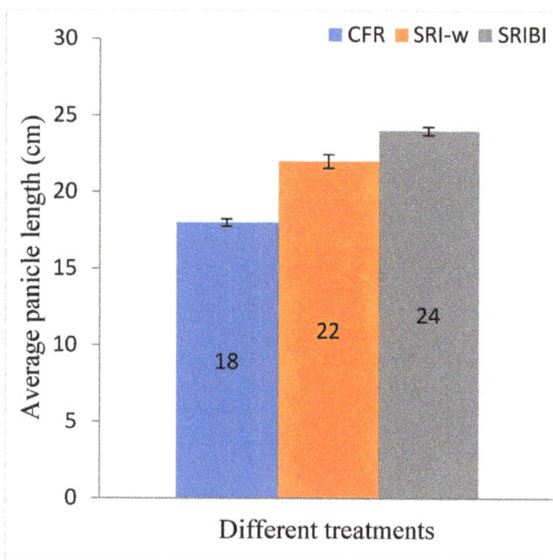

Figure 4. Averagepanicle length of the different treatments as observed in field experiments (**Left**: 2019; **Right**: 2020).

Table 7. Parameters from ANOVA statistical analysis for panicle length comparison.

Panicle Length	2019 SRI-w vs. SRIBI				2020 CFR vs. SRI-w vs. SRIBI			
	p-value	F-value	$F_{critical}$	DF	*p*-value	F-value	$F_{critical}$	DF
	0.0649	3.12	3.35	(2,7)	0.0001	85	3.56	(2,18)
					Post hoc analysis			
					$Q_{critical}$	$AMD_{CFR/SRI}$	$AMD_{SRI/SRIBI}$	$AMD_{CFR/SRIBI}$
					0.89	3.93	2	5.93

There was a significant increase in this yield parameter, observed in the field experiments in 2020. In the case of panicle length as well, the increase in the case of SRIBI was found to statistically more significant (with a higher absolute mean difference) as compared to the increase in the case of SRI (Table 7). The difference of the means between SRIBI and SRI was also found to be significant.

Spikelet Number Per Panicle

Intercropping was found to have a positive effect on another yield parameter that was observed in the studies, the spikelet number per panicle (SNPP). The SNPP was found to be significantly higher in the case of rice under intercropping ($p \leq 0.05$; $p = 0.0001$). The average SNPP of rice with intercropping and without intercropping was observed as 142 and 95 in 2019 and 161 and 140 in 2020, respectively (Figure 5). When compared with the CFR treatment, the statistical significance of the SNPP in the case of SRIBI was observed to be higher as compared to that in SRI, based on the Tukey–Kramer post hoc test (Table 8). Additionally, the rice plots under the intercropping regimen were observed to be harvest-ready earlier than the plots without intercropping. In this regard, it is pertinent to report the chlorophyll content of the rice plant leaves that was measured under controlled conditions in greenhouse experiments in the next section. The SNPP has been associated with the soil microbial composition [56], indicating that intercropping might have modified the soil microbial composition, leading to an increase in the SNPP. Changes in soil bacterial communities favourable to rice yield have also been confirmed with other changes in the rice cropping system, such as double-season rice cropping [57].

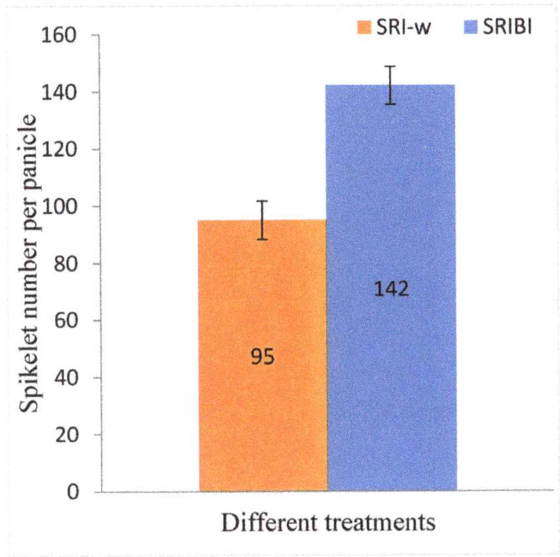

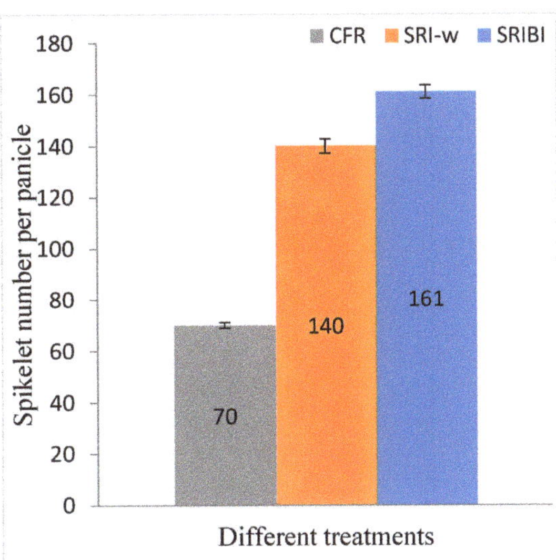

Figure 5. The spikelet number per panicle (SNPP) observed in the field experiments (**Left**: 2019; **Right**: 2020).

Table 8. Parameters from ANOVA statistical analysis for spikelet number (SNPP) comparison.

SNPP	2019 SRI-w vs. SRIBI				2020 CFR vs. SRI-w vs. SRIBI			
	p-value	F-value	$F_{critical}$	DF	*p*-value	F-value	$F_{critical}$	DF
	0.0001	25.31	4.49	(1,16)	0.0001	329.80	3.40	(2,24)
					Post hoc analysis			
					$Q_{critical}$	$AMD_{CFR/SRI}$	$AMD_{SRI/SRIBI}$	$AMD_{CFR/SRIBI}$
					70.46	70.55	18.11	88.66

Filled Grains Per Panicle

The number of filled grains per panicle was counted for the three different treatments, and intercropping was found to have a positive effect on this parameter, as can be seen in Figure 6. The differences in this parameter are statistically significant in both SRI and SRIBI, as compared to CFR. However, the level of significance was higher in case of SRIBI, as can be seen from the AMD (absolute mean difference) (Table 9).

Table 9. Parameters from ANOVA statistical analysis for filled grains per panicle (FGPP) comparison.

Grains Per Panicle	*CFR* vs. *SRI-w* vs. *SRIBI*			
	p-value	F-value	$F_{critical}$	DF
	0.0001	383	3.47	(2,21)
	Post hoc analysis			
	$Q_{critical}$	$AMD_{CFR/SRI}$	$AMD_{SRI/SRIBI}$	$AMD_{CFR/SRIBI}$
	65.43	73.5	21.625	95.125

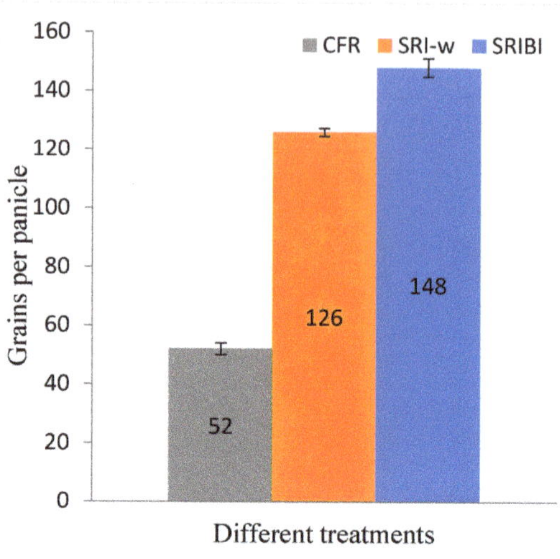

Figure 6. Grains per panicle observed in the different treatments (2020).

3.1.3. Rice Yield and Fodder Yield

Rice Yield

Based on the improvement in the plant growth parameters discussed in the previous sections, a higher grain yield was expected under SRIBI management. This was evident from the yields reported from the farmers' fields. The total yield measured in the field experiments of the year 2019 was higher under the SRIBI regime than the conventional flooded rice by 15–20%, and in 2020, it was found to be 33% higher under the SRI intercropping regimen as compared to the conventional flooded rice cultivation (Figure 7).

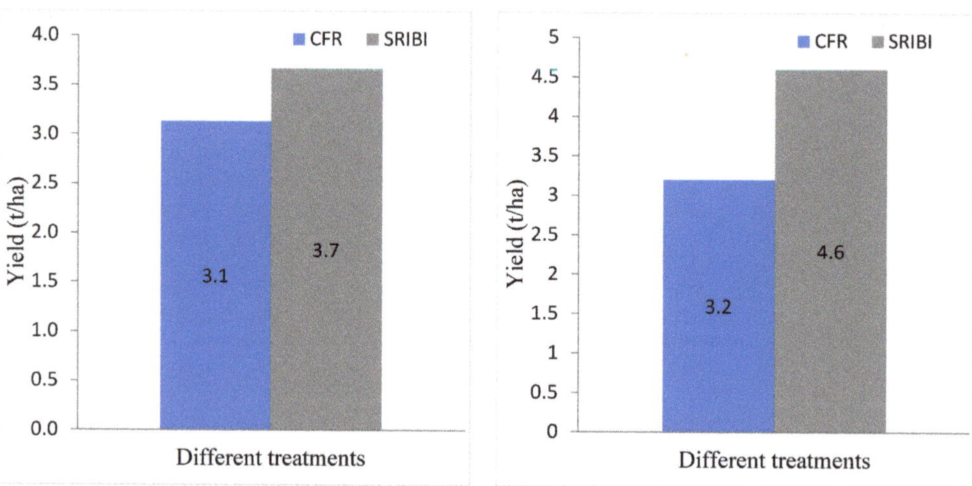

Figure 7. Yield observed in the different treatments (**Left**: 2019; **Right**: 2020).

Fodder Units

Rice husk is used either as fodder or as a filling material for horticultural practices and hence, has a significant economic value for the farmers. In 2019, the number of fodder units in CFR was found to be 90, while in case of SRI with intercropping, it was found to be 140. Similarly, in 2020, the number of fodder units under intercropping SRI was found to be 160 (Figure 8). Hence, the trend in the case of rice husk biomass was similar to the grain yield. This can be attributed to the higher number of tillers in the case of SRI management practices.

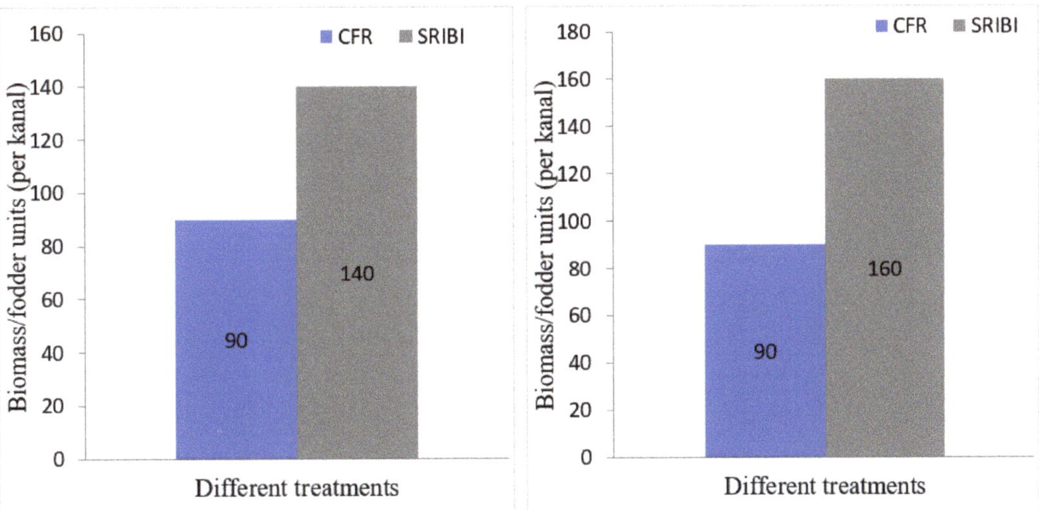

Figure 8. Number of fodder units observed in the different treatments (**Left**: 2019; **Right**: 2020).

A marked improvement was observed in different plant growth and yield parameters due to SRI and intercropping in comparison to the conventional flooded rice cultivation method. This indicates the potential of these methodologies in improving the yield of rice farming systems, which can result in improvements in the food security scenario and the socio-economic condition of rice farmers.

3.1.4. Economic Balance

Work-Hours

With respect to the other main criticism of the SRI methodology, which is that it leads to an increase in work-hour requirements due to the need of manually operated weeding, the intercropping regimen was beneficial. It was observed that for 1 kanal (0.05 ha) of land, 1 h of time was required for one round of weeding, which translates to 20 extra work-hours per hectare of land, per weeding. With the standard of 4 weedings per cycle of rice production in SRI, this translates to 80 work-hours per hectare. Assuming that intercropping is done after the first weeding, as was done in the current study, this can lead to savings of 75% of the extra work-hours required by SRI, which is 60 work-hours per hectare. This is a conservative estimate, given that higher work-hour requirements for weeding, of up to 160 work-hours per hectare per cycle of rice production under SRI-based rice farming techniques, have been reported. Thus, the work-hour savings could be even higher, depending on the stage of adoption of the SRI method [32].

Economic Balance Sheet

The economic balance of the new rice farming system with beans as the intercrop, as seen from the field experiments, is presented in Table 10. The increase in the net earnings of farmers was observed to be 57%. This was accompanied by an increase of 41% in the input costs with a corresponding increase of 51% in the output benefits. The cost of compost constituted 63% of the input costs, and this points to the potential of decreasing the inputs costs even further through local production of compost (Table 10). In this case, the net earnings would be more than double compared to the conventional rice farming system.

Table 10. Economic balance sheet of the innovations implemented during the course of research.

	Input	Quantity (per ha)	Cost (INR)	Output	Quantity (per ha)	Benefit (INR)	Earnings (INR)	Increase (%)
CFR	Seeds	400 kg	6000	Rice	5600 kg	84,000		
	Manure	0.33 trolley	20,000	Fodder (Rice Straw)	1800 units	72,000		
	Fertilizer	300 kg	6666					
	Pesticides	n.d.	4000					
	Labour for weeding	80 work hours	5000					
	Labour for irrigation	200 work hours	12,500					
	Total cost		**54,166**	**Total benefit**		**156,000**	**101,834**	
SRIBI	Seeds	40 kg	600	Rice	7200 kg	108,000		
	Manure	0.33 trolley	20,000	Fodder (Rice Straw)	3200 units	128,000		
	Compost	6000 kg	48,000					
	Intercrop Seeds	2 kg	400					
	Labour for weeding	10 work hours	660					
	Labour for watering	100 work hours	6260					
	Total cost		**75,920**	**Total benefit**		**236,000**	**160,080**	**57.20**

Disease Incidence in Mushkibudij (Mushk Budji) Rice Landrace

The study experiments were conducted with a local heritage aromatic rice landrace known as Mushkibudij. This landrace of rice was revived by the local agricultural university and has been reported to be susceptible to rice blast that is the most widely occurring disease in rice, which has led to huge losses in yield, of up to 70% [58–60]. This leads to huge losses for the farmers while also increasing the input costs in the form of insecticide sprays. However, in the case of current experiments, the incidence of rice blast was not observed in any of the plots grown under SRI and SRIBI. The wider spacing between rice plants under SRI management system could be one of the factors behind this decreased incidence of disease [61,62]. This could form a basis for further research in this direction.

3.2. Pot Experiments

Pot experiments were conducted to establish the proof of concept about the positive effect of intercropping under the system of rice intensification on rice crop. The following parameters were measured.

3.2.1. Nutrient Uptake

The rice plants were analysed for nutrient uptake in the pot experiments performed at the greenhouse level. The content of the three essential nutrients, nitrogen (Figure 9), potassium (Figure 10), and phosphorous (Figure 11), was found to be higher in the case of SRI with intercropping as compared to the treatment without intercropping.

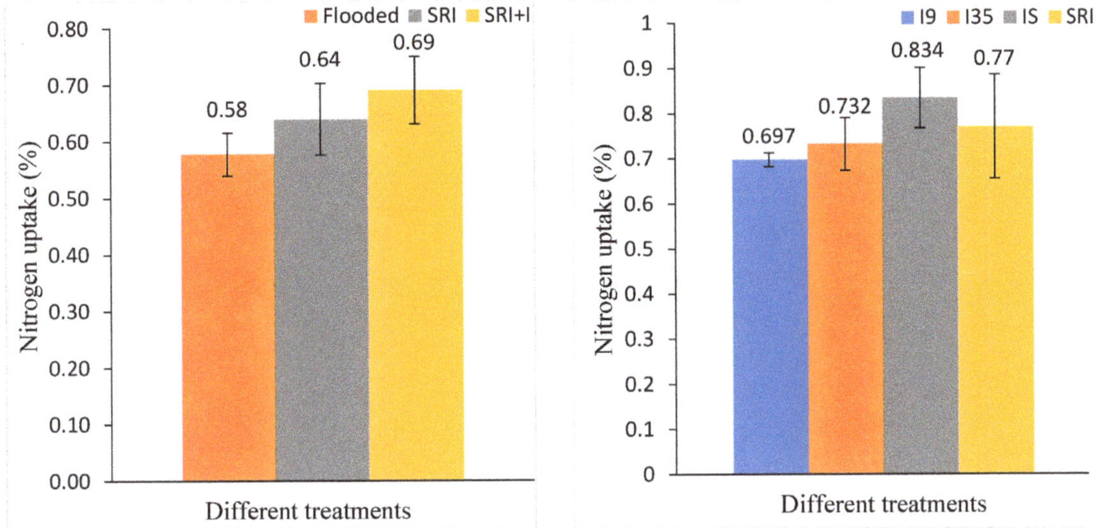

Figure 9. Nitrogen uptake as measured in the different treatments. Conventional flooded rice (CFR), SRI, and SRI with intercropping (SRI+I or SRIBI) in 2017 (**Left**). SRI and three different intercropping configurations in 2018 (**Right**).

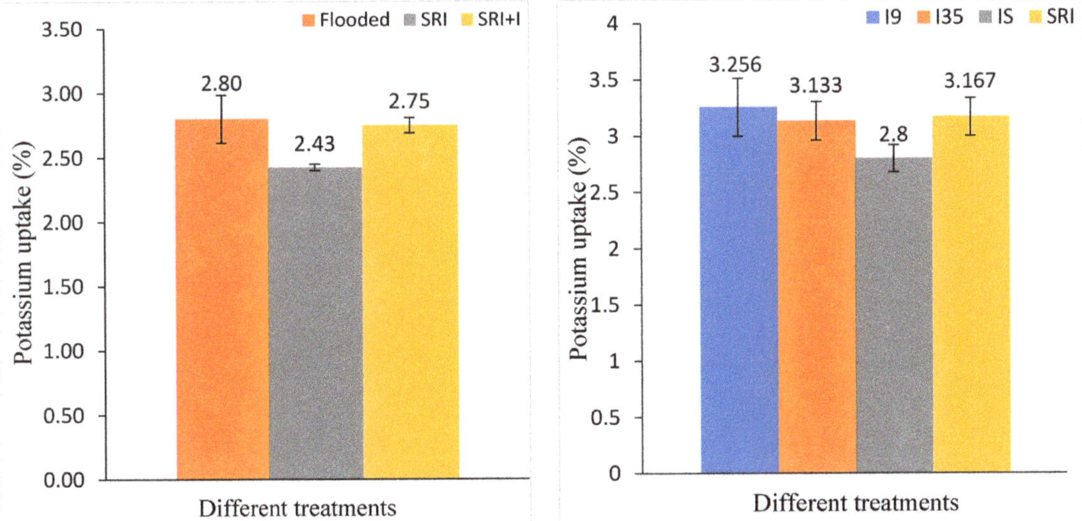

Figure 10. Potassium uptake as measured in the different treatments. Conventional flooded rice (CFR), SRI, and SRI with intercropping (SRI+I or SRIBI) in 2017 (**Left**). SRI and three different intercropping configurations in 2018 (**Right**).

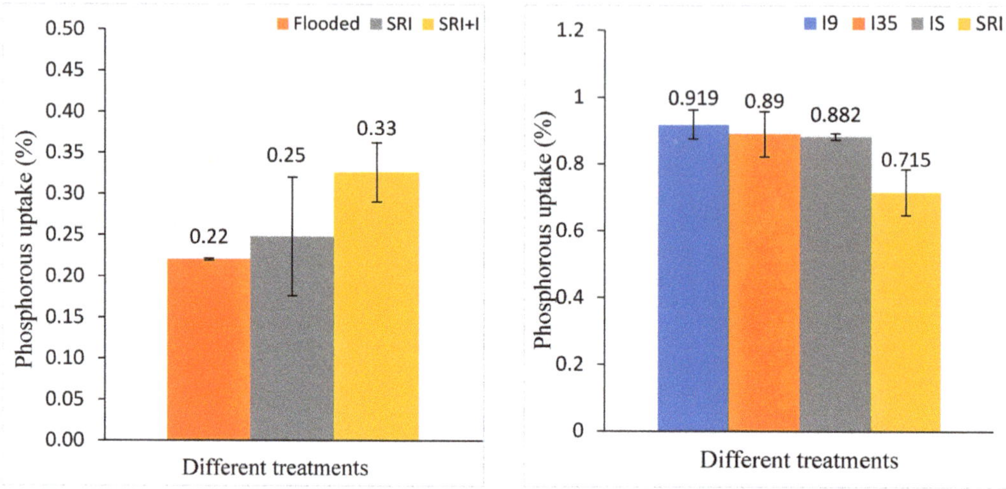

Figure 11. Phosphorous uptake as measured in the different treatments. Conventional flooded rice (CFR), SRI, and SRI with intercropping (SRI+I or SRIBI) in 2017 (**Left**). SRI and three different intercropping configurations in 2018 (**Right**).

As a general trend, the uptake of essential nutrients, nitrogen, phosphorous, and potassium, was found to be higher with intercropping as compared to the control SRI treatment.

3.2.2. Chlorophyll Content

A significant difference was observed in the Chlorophyll-a and Chlorophyll-b content of the rice plants grown under SRI and SRIBI ($p \leq 0.05$). A higher chlorophyll content in SRI as compared to the conventional flooded rice (CFR) has also been reported in previous studies [63]. While the mean value of Chlorophyll-a was 9.3 for CFR and SRI, it was 13.2 for SRIBI (SRI+I), which was significantly higher ($p = 0.05$) than that of SRI (9.2) (Figure 12). On the other hand, the mean value of Chlorophyll-b was 3.0 for CFR, while it was 4.1 for SRIBI, which was significantly higher ($p = 0.035$) than that of SRI (2.9) (Figure 13) (Table 11).

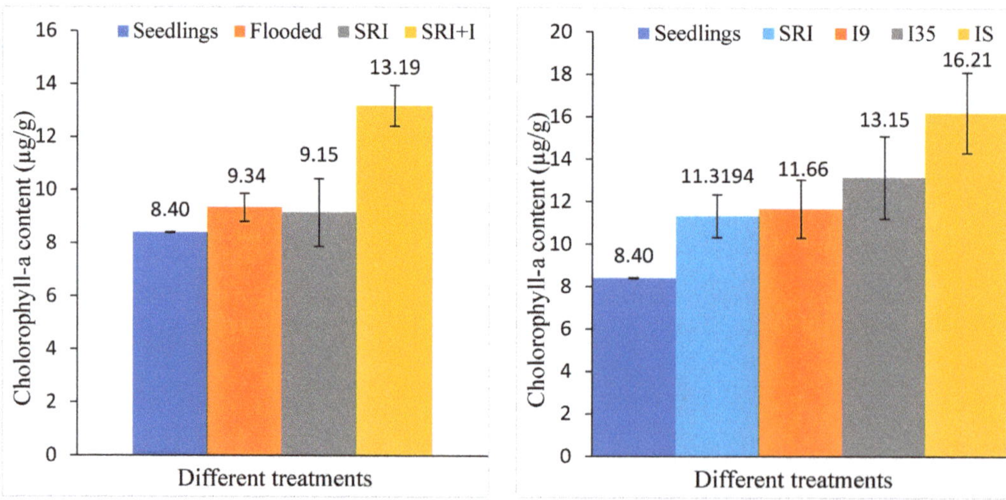

Figure 12. Comparison of chlorophyll-a content between the different treatments (**Left**: 2017; **Right**: 2018).

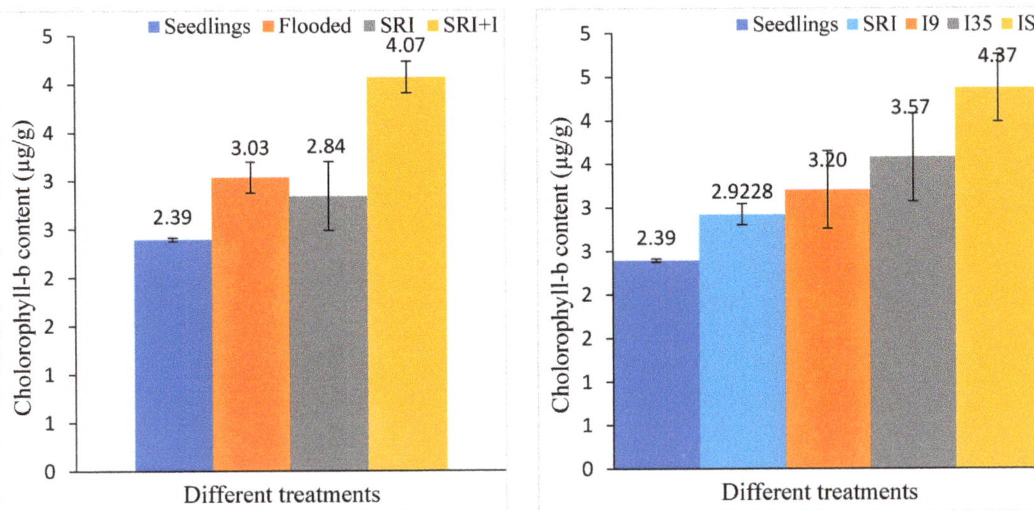

Figure 13. Comparison of chlorophyll-b content between the different treatments (**Left**: 2017; **Right**: 2018).

Table 11. Parameters from ANOVA statistical analysis for chlorophyll content comparison.

Chlorophyll Content	Chlorophyll-a SRI vs. SRIBI				Chlorophyll-b SRI vs. SRIBI			
	p-value	F-value	$F_{critical}$	DF	p-value	F-value	$F_{critical}$	DF
	0.05	7.39	7.70	(1,4)	0.035	9.87	7.71	(1,4)

Chlorophyll content was also measured in three different intercropping treatments, I9 (intercropping at 9 days after transplantation), I35 (intercropping at 35 days after transplantation), and IS (strip intercropping). The chlorophyll content in the three different intercropping treatments (I9, I35, IS) was not found to be statistically significantly different from each other ($p > 0.05$; $p = 0.1153$, $p = 0.1524$) (Figures 12 and 13). This indicates that while intercropping had an effect on the chlorophyll content of the rice plant, the time and spatial differences in intercropping did not have a significant effect.

A progressive decrease in the chlorophyll content in rice has been associated with the ripening of the rice plants [53,56]. This decrease in chlorophyll content was associated with senescence of the rice plants, which has been found to occur earlier in flooded rice than in the SRI-based techniques [53,56]. A higher chlorophyll content in the leaves has been linked to a higher photosynthesis rate and root activity [63].

3.2.3. Tiller Number

In pot experiments, the number of tillers was found to be the highest in the case of SRI with intercropping, followed by SRI. The conventional flooded rice (CFR) treatment had the lowest number of tillers (Figure 14).

The parameters measured in the pot experiments described above provided the required proof of concept to perform the intercropping experiments at the field level. The results of the field experiments conclusively indicated the potential of the intercropping innovation in making rice farming a socially and ecologically sustainable agricultural production system.

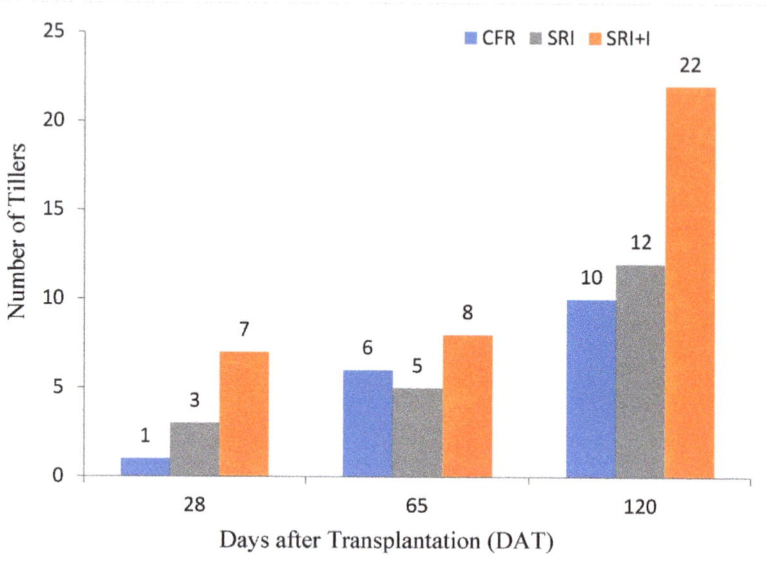

Figure 14. Comparison of the highest number of tillers observed in different phonological stages in the three different treatments: CFR, conventional flooded rice; SRI; SRI+I (SRIBI).

4. Conclusions

The results of experiments carried out at the lab scale in greenhouse (pot experiments) provided the proof of concept about the value addition that intercropping could provide in rice cultivation under the System of Rice Intensification (SRI). The results from the field experiments in 2019 and 2020 concurred with the findings of the pot experiments carried out from 2017 to 2018. The findings show that intercropping beans in the inter-row space in rice planted under the SRI management system provides value addition to the farmers, mainly visible in the form of reduced work hours and increased yields.

This study provided further clarity that SRI-based rice cultivation can be more productive and beneficial for farmers, while at the same time presenting SRIBI as an improved agroecological developmental stage of SRI. Intercropping serves to address two oft-cited criticisms of SRI, weed infestation and increased labour requirement, which have been linked to its dis-adoption. The improvements resulting from intercropping observed in this study include a clear reduction in weed infestation, higher nutrient uptake, improved plant growth characteristics, and better yield parameters. These improvements have the potential to control the overuse of pesticides in rice farming, which has been linked to human and planetary health concerns.

These improvements highlight the potential of such agroecological interventions to improve the socio-economic condition of rice farmers significantly, especially smallholder and subsistence farmers who otherwise depend on various agrochemical inputs. Intercropping legumes with rice can also lead to a diversification of farmers' incomes as well as their diets. This can have positive implications on the livelihood and food security scenario given that rice is a staple for more than half of the world's population.

Author Contributions: Conceptualization, T.M.S. and R.O.; methodology, T.M.S.; software, S.T.; validation, T.M.S., R.O. and M.A.B.; formal analysis, T.M.S.; investigation, T.M.S.; resources, T.M.S. and R.O.; data curation, S.T. and M.A.B.; writing—original draft preparation, T.M.S.; writing—review and editing, S.T.; visualization, T.M.S.; supervision, R.O. and M.A.B.; project administration, T.M.S. All authors have read and agreed to the published version of the manuscript.

Funding: This research received no external funding.

Institutional Review Board Statement: Not applicable.

Informed Consent Statement: Not applicable.

Data Availability Statement: Data are contained within the article. Further details about this study is available online here: https://doi.org/10.15480/882.3286 (Last accessed on: 19 May 2021).

Acknowledgments: We acknowledge support for Open Access publishing through APC funding by GFEU e.V. at TUHH. We acknowledge the financial support provided to the first author by the State of Hamburg (Free and Hanseatic City of Hamburg) through the HmbNFG (Hamburg State Graduate Funding Program) doctoral scholarship. We acknowledge the support provided to the first author by CCAFS-CGIAR through the CLIFF (Climate Food and Farming) fellowship. We acknowledge the feedback provided by Norman Uphoff during this research and the pioneering role played by SRI-RICE (SRI International Network and Resources Centre) in promoting socially, economically, and ecologically sustainable farming systems worldwide.

Conflicts of Interest: The authors declare no conflict of interest.

Appendix A

Table A1. Different Statistical Data about the Different Measured Parameters.

	Mean			Standard Deviation			Standard Error		
	CFR	**SRI**	**SRIBI**	**CFR**	**SRI**	**SRIBI**	**CFR**	**SRI**	**SRIBI**
Plant height (*Field 2020*)	101	104	109	1.14	1.95	1.82	0.51	0.87	0.81
Plant height (*Field 2019*)	125	119	128	4.88	4.59	9.56	1.47	1.38	3.38
Panicle length (*Field 2020*)	18	22	24	0.65	1.17	0.67	0.24	0.44	0.25
Panicle length (*Field 2019*)	24	22	22	0.81	2.03	1.05	0.25	0.64	0.34
Tiller Number (*Field 2020*)	12	30	34	1.72	2.82	2.64	0.65	1.07	0.99
Tiller Number (*Field 2019*)	9.5	n.a.	26.33	2.15	n.a.	7.45	0.81	n.a.	2.82
SNPP (*Field 2020*)	70	140	161	3.27	8.13	7.78	1.09	2.71	2.59
SNPP (*Field 2019*)	n.a.	95	142	n.a.	20.12	19.79	n.a.	6.70	6.59
FGPP (*Field 2020*)	52	126	148	6.04	4.10	9.55	2.01	1.36	3.18
Chlorophyll-A	9.3	9.2	13.2	0.9	2.2	1.3	0.5	1.3	0.8
Chlorophyll-B	3.0	2.8	4.1	0.3	0.6	0.3	0.2	0.4	0.2
N-content	0.58	0.64	0.69	0.10	0.10	0.08	0.06	0.04	0.03
P-content	0.22	0.25	0.33	0.01	0.04	0.05	0.003	0.02	0.02
K-content	2.83	2.43	2.75	0.15	0.08	0.07	0.09	0.04	0.03

References

1. Foley, J.A.; Ramankutty, N.; Brauman, K.A.; Cassidy, E.S.; Gerber, J.S.; Johnston, M.; Mueller, N.D.; O'Connell, C.; Ray, D.K.; West, P.C.; et al. Solutions for a cultivated planet. *Nature* **2011**, *478*, 337–342. [CrossRef]
2. Pérez-Escamilla, R. Food Security and the 2015–2030 Sustainable Development Goals: From Human to Planetary Health. *Curr. Dev. Nutr.* **2017**, *1*, e000513. [CrossRef] [PubMed]
3. Foley, J.A.; DeFries, R.; Asner, G.P.; Barford, C.; Bonan, G.; Carpenter, S.R.; Chapin, F.S.; Coe, M.T.; Daily, G.C.; Gibbs, H.K.; et al. Global consequences of land use. *Science* **2005**, *309*, 570–574. [CrossRef] [PubMed]

4. Power, A.G. Ecosystem services and agriculture: Tradeoffs and synergies. *Philos. Trans. R. Soc. B Biol. Sci.* **2010**, *365*, 2959–2971. [CrossRef] [PubMed]

5. Millennium Ecosystem Assessment Ecosystems and Human Well-Being: Synthesis. Available online: https://www. millenniumassessment.org/documents/document.356.aspx.pdf (accessed on 16 May 2021).

6. FAO. The State of Food and Agriculture—Climate Change, Agriculture and Food Security. Available online: http://www.fao. org/3/a-i6030e.pdf (accessed on 16 May 2021).

7. Chivenge, P.; Angeles, O.; Hadi, B.; Acuin, C.; Connor, M.; Stuart, A.; Puskur, R.; Johnson-Beebout, S. Ecosystem services in paddy rice systems. *Role Ecosyst. Serv. Sustain. Food Syst.* **2020**, *1*, 181–201. [CrossRef]

8. Buresh, R.J.; Reddy, K.R.; Van Kessel, C. Nitrogen transformations in submerged soils. *Nitrogen Agric. Syst.* **2008**, 401–436.

9. Sander, B.O.; Wassmann, R.; Palao, L.K.; Nelson, A. Climate-based suitability assessment for alternate wetting and drying water management in the Philippines: A novel approach for mapping methane mitigation potential in rice production. *Carbon Manag.* **2017**, *8*, 331–342. [CrossRef]

10. Bambaradeniya, C.; Amerasinghe, F. Biodiversity Associated with the Rice Field Agroecosystem in Asian Countries: A Brief Review. Available online: https://www.iwmi.cgiar.org/publications/iwmi-working-papers/iwmi-working-paper-63/ (accessed on 16 May 2021).

11. Gathorne-Hardy, A.; Reddy, D.N.; Venkatanarayana, M.; Harriss-White, B. System of Rice Intensification provides environmental and economic gains but at the expense of social sustainability—A multidisciplinary analysis in India. *Agric. Syst.* **2016**, *143*, 159–168. [CrossRef]

12. FAO. Save and Grow in Practice: Maize, Rice and Wheat. A Guide to Sustainable. Available online: http://www.fao.org/policy-support/tools-and-publications/resources-details/en/c/1263072/ (accessed on 16 May 2021).

13. Global Rice Science Partnership (GRiSP). *Rice Almanac*; International Rice Research Institute: Los Banos, CA, USA, 2013; ISBN 9789712203008.

14. Mishra, U. Indo-Pak Study Reveals Extensive Arsenic Problem in Punjab Groundwater. Available online: https://www. downtoearth.org.in/news/water/indo-pak-study-reveals-extensive-arsenic-problem-in-punjab-groundwater-62534 (accessed on 16 May 2021).

15. Rasool, A.; Farooqi, A.; Masood, S.; Hussain, K. Arsenic in groundwater and its health risk assessment in drinking water of Mailsi, Punjab, Pakistan. *Hum. Ecol. Risk Assess. Int. J.* **2016**, *22*, 187–202. [CrossRef]

16. Mukherjee, A.B.; Bhattacharya, P. Arsenic in groundwater in the Bengal Delta Plain: Slow poisoning in Bangladesh. *Environ. Rev.* **2001**, *9*, 189–220. [CrossRef]

17. Roychowdhury, T.; Uchino, T.; Tokunaga, H.; Ando, M. Arsenic and other heavy metals in soils from an arsenic-affected area of West Bengal, India. *Chemosphere* **2002**, *49*, 605–618. [CrossRef]

18. Arao, T.; Kawasaki, A.; Baba, K.; Mori, S.; Matsumoto, S. Effects of water management on cadmium and arsenic accumulation and dimethylarsinic acid concentrations in Japanese rice. *Environ. Sci. Technol.* **2009**, *43*, 9361–9367. [CrossRef]

19. Harwood, J. Could the adverse consequences of the green revolution have been foreseen? How experts responded to unwelcome evidence. *Agroecol. Sustain. Food Syst.* **2020**, *44*, 509–535. [CrossRef]

20. Bunch, R. Use of Green Manures and Cover Crops with SRI. In Proceedings of the Assessments of the System of Rice Intensification: Proceedings of an International Conference, Sanya, China, 1–4 April 2002; Cornell International Institute for Food, Agriculture and Development: Ithaca, NY, USA, 2002; pp. 184–185.

21. FAO. FAOSTAT. Statistical Database. Inputs. Available online: http://www.fao.org/faostat/en/ (accessed on 7 May 2020).

22. FAO. Investing in Smallholder Agriculture for Food Security. Available online: http://www.fao.org/family-farming/detail/en/ c/273868/ (accessed on 16 May 2021).

23. IPES-Food. COVID-19 and the Crisis in Food Systems: Symptoms, Causes, and Potential Solutions. Available online: http: //www.ipes-food.org/_img/upload/files/COVID-19_CommuniqueEN%282%29.pdf (accessed on 16 May 2021).

24. Leippert, F.; Darmaun, M.; Bernoux, M.; Mpheshea, M. The Potential of Agroecology to Build Climate-Resilient Livelihoods and Food Systems. Available online: http://www.fao.org/documents/card/en/c/cb0438en (accessed on 16 May 2021).

25. Uphoff, N. Higher Yields with Fewer External Inputs? The System of Rice Intensification and Potential Contributions to Agricultural Sustainability. *Int. J. Agric. Sustain.* **2003**, *1*, 38–50. [CrossRef]

26. Uphoff, N. SRI: An agroecological strategy to meet multiple objectives with reduced reliance on inputs. *Agroecol. Sustain. Food Syst.* **2017**, *41*, 825–854. [CrossRef]

27. Balamatti, A.; Uphoff, N. Experience with the system of rice intensification for sustainable rainfed paddy farming systems in India. *Agroecol. Sustain. Food Syst.* **2017**, *41*, 573–587. [CrossRef]

28. Kassam, A.; Stoop, W.; Uphoff, N. Review of SRI modifications in rice crop and water management and research issues for making further improvements in agricultural and water productivity. *Paddy Water Environ.* **2011**, *9*, 163–180. [CrossRef]

29. Stoop, W.A.; Adam, A.; Kassam, A. Comparing rice production systems: A challenge for agronomic research and for the dissemination of knowledge-intensive farming practices. *Agric. Water Manag.* **2009**, *96*, 1491–1501. [CrossRef]

30. Rao, K.C.; Satyanarayana, V.V.S. System of Rice Intensification Experiences of Farmers in India. Available online: http://www. sri-india.net/documents/Farmersexperiences.pdf (accessed on 16 May 2021).

31. Satyanarayana, A.; Thiyagarajan, T.M.; Uphoff, N. Opportunities for water saving with higher yield from the system of rice intensification. *Irrig. Sci.* **2007**, *25*, 99–115. [CrossRef]

32. Ly, P.; Jensen, L.S.; Bruun, T.B.; Rutz, D.; de Neergaard, A. The System of Rice Intensification: Adapted practices, reported outcomes and their relevance in Cambodia. *Agric. Syst.* **2012**, *113*, 16–27. [CrossRef]
33. John, A.; Fielding, M. Rice production constraints and "new" challenges for South Asian smallholders: Insights into de facto research priorities. *Agric. Food Secur.* **2014**, *3*, 1–16. [CrossRef]
34. Shennan, C.; Krupnik, T.J.; Baird, G.; Cohen, H.; Forbush, K.; Lovell, R.J.; Olimpi, E.M. Organic and Conventional Agriculture: A Useful Framing? *Annu. Rev. Environ. Resour.* **2017**, *42*, 317–346. [CrossRef]
35. Hazra, K.K.; Swain, D.K.; Bohra, A.; Singh, S.S.; Kumar, N.; Nath, C.P. Organic rice: Potential production strategies, challenges and prospects. *Org. Agric.* **2018**, *8*, 39–56. [CrossRef]
36. Delmotte, S.; Tittonell, P.; Mouret, J.-C.; Hammond, R.; Lopez-Ridaura, S. On farm assessment of rice yield variability and productivity gaps between organic and conventional cropping systems under Mediterranean climate. *Eur. J. Agron.* **2011**, *35*, 223–236. [CrossRef]
37. Wayayok, A.; Soom, M.A.M.; Abdan, K.; Mohammed, U. Impact of Mulch on Weed Infestation in System of Rice Intensification (SRI) Farming. *Agric. Agric. Sci. Procedia* **2014**, *2*, 353–360. [CrossRef]
38. Uphoff, N.; Kassam, A. Agricultural Technologies for Developing Countries: Case Study—"The System of Rice Intensification". Available online: https://www.europarl.europa.eu/RegData/etudes/etudes/stoa/2009/424734/DG-IPOL-STOA_ET%2820 09%29424734_EN%28PAR05%29.pdf (accessed on 16 May 2021).
39. Noltze, M.; Schwarze, S.; Qaim, M. Understanding the adoption of system technologies in smallholder agriculture: The system of rice intensification (SRI) in Timor Leste. *Agric. Syst.* **2012**, *108*, 64–73. [CrossRef]
40. Dobermann, A. A critical assessment of the system of rice intensification (SRI). *Agric. Syst.* **2004**, *79*, 261–281. [CrossRef]
41. Nawaz, A.; Farooq, M. Weed management in resource conservation production systems in Pakistan. *Crop. Prot.* **2016**, *85*, 89–103. [CrossRef]
42. Liebman, M.; Dyck, E. Crop Rotation and Intercropping Strategies for Weed Management. *Ecol. Appl.* **1993**, *3*, 92–122. [CrossRef]
43. Singh, S.; Ladha, J.K.; Gupta, R.K.; Bhushan, L.; Rao, A.N.; Sivaprasad, B.; Singh, P.P. Evaluation of mulching, intercropping with Sesbania and herbicide use for weed management in dry-seeded rice (*Oryza sativa* L.). *Crop. Prot.* **2007**, *26*, 518–524. [CrossRef]
44. Kermah, M.; Franke, A.C.; Adjei-Nsiah, S.; Ahiabor, B.D.K.; Abaidoo, R.C.; Giller, K.E. Maize-grain legume intercropping for enhanced resource use efficiency and crop productivity in the Guinea savanna of northern Ghana. *Field Crop. Res.* **2017**, *213*, 38–50. [CrossRef]
45. Stoop, W.A.; Sabarmatee; Pushpalatha, S.; Ravindra, A.; Sen, D.; Prasad, S.C.; Thakur, A.K. Opportunities for ecological intensification: Lessons and insights from the System of rice/crop intensification—Their implications for agricultural research and development approaches. *Cab Rev. Perspect. Agric. Vet. Sci. Nutr. Nat. Resour.* **2017**, *12*, 1–19. [CrossRef]
46. Stoop, W.A. The scientific case for system of rice intensification and its relevance for sustainable crop intensification. *Int. J. Agric. Sustain.* **2011**, *9*, 443–455. [CrossRef]
47. Adhikari, P.; Araya, H.; Aruna, G.; Balamatti, A.; Banerjee, S.; Baskaran, P.; Barah, B.C.; Behera, D.; Berhe, T.; Boruah, P.; et al. System of crop intensification for more productive, resource-conserving, climate-resilient, and sustainable agriculture: Experience with diverse crops in varying agroecologies. *Int. J. Agric. Sustain.* **2017**, *16*, 1–28. [CrossRef]
48. Uphoff, N.; Amod, K.T. An agroecological strategy for adapting to climate change: The system of rice intensification (sri). *Sustain. Solut. Food Secur. Combat. Clim. Chang. Adapt.* **2019**, 229–254. [CrossRef]
49. Doni, F.; Zain, C.R.C.M.; Isahak, A.; Fathurrahman, F.; Anhar, A.; Mohamad, W.N.W.; Yusoff, W.M.W.; Uphoff, N. A simple, efficient, and farmer-friendly Trichoderma-based biofertilizer evaluated with the SRI Rice Management System. *Org. Agric.* **2018**, *8*, 207–223. [CrossRef]
50. Thakur, A.K.; Mandal, K.G.; Raychaudhuri, S. Impact of crop and nutrient management on crop growth and yield, nutrient uptake and content in rice. *Paddy Water Environ.* **2020**, *18*, 139–151. [CrossRef]
51. Arnon, D.I. Copper Enzems in Isolated Chloroplasts. Polyphenoloxidase in Beta Vulgaris. *Plant. Physiol.* **1949**, *24*, 1–15. [CrossRef]
52. Doni, F.; Zain, C.R.C.M.; Isahak, A.; Fathurrahman, F.; Sulaiman, N.; Uphoff, N.; Yusoff, W.M.W. Relationships observed between Trichoderma inoculation and characteristics of rice grown under System of Rice Intensification (SRI) vs. conventional methods of cultivation. *Symbiosis* **2017**, *72*, 45–59. [CrossRef]
53. Mishra, A.; Salokhe, V.M. The effects of planting pattern and water regime on root morphology, physiology and grain yield of rice. *J. Agron. Crop. Sci.* **2010**, *196*, 368–378. [CrossRef]
54. Zhu, D.; Cheng, S.; Zhang, Y.; Lin, X. Tillering patterns and the contribution of tillers to grain yield with hybrid rice and wide spacing. In Proceedings of the Assessments of the System of Rice Intensification: Proceedings of an International Conference, Sanya, China, 1–4 April 2002; Cornell International Institute for Food, Agriculture and Development: Ithaca, NY, USA, 2002; pp. 125–131.
55. Ao, H.; Peng, S.; Zou, Y.; Tang, Q.; Visperas, R.M. Reduction of unproductive tillers did not increase the grain yield of irrigated rice. *Field Crop. Res.* **2010**, *116*, 108–115. [CrossRef]
56. Mishra, A.; Uphoff, N. Morphological and physiological responses of rice roots and shoots to varying water regimes and soil microbial densities. *Arch. Agron. Soil Sci.* **2013**, *59*, 705–731. [CrossRef]
57. Huang, M.; Tian, A.; Chen, J.; Cao, F.; Chen, Y.; Liu, L. Soil bacterial communities in three rice-based cropping systems differing in productivity. *Sci. Rep.* **2020**, *10*, 1–5. [CrossRef] [PubMed]

58. Khan, G.H.; Shikari, A.B.; Vaishnavi, R.; Najeeb, S.; Padder, B.A.; Bhat, Z.A.; Parray, G.A.; Bhat, M.A.; Kumar, R.; Singh, N.K. Marker-assisted introgression of three dominant blast resistance genes into an aromatic rice cultivar Mushk Budji. *Sci. Rep.* **2018**, *8*, 1–13. [CrossRef] [PubMed]

59. Najeeb, S.; Parray, G.A.; Shikari, A.B.; Bhat, Z.A.; Bhat, M.A.; Asif, I.M.; Bhat, S.J.; Shah, A.H.; Ahangar, M.A.; Shafiq, W.A. Revival of Endangered High Valued Mountain Rices of Kashmir Himalayas Through Genetic Purification and In-Situ Conservation. *Int. J. Agric. Sci.* **2016**, *8*, 2453–2458.

60. Dean, R.A.; Talbot, N.J.; Ebbole, D.J.; Farman, M.L.; Mitchell, T.K.; Orbach, M.J.; Thon, M.; Kulkarni, R.; Xu, J.R.; Pan, H.; et al. The genome sequence of the rice blast fungus Magnaporthe grisea. *Nature* **2005**, *434*, 980–986. [CrossRef]

61. Mahesh, P.; Shakywar, R.C.; Dinesh, S.; Singh, S. Prevalence of insect pests, natural enemies and diseases in SRI (System of Rice Intensification) of Rice cultivation in North East Region. *Ann. Plant. Prot. Sci.* **2012**, *20*, 375–379.

62. Tann, H.; Makhonpas, C.; Utthajadee, A.; Soytong, K. Effect of good agricultural practice and organic methods on rice cultivation under the system of rice intensification in Cambodia. *Int. J. Agric. Technol.* **2012**, *8*, 289–303.

63. Thakur, A.K.; Rath, S.; Kumar, A. Performance evaluation of rice varieties under the System of Rice Intensification compared with the conventional transplanting system. *Arch. Agron. Soil Sci.* **2011**, *57*, 223–238. [CrossRef]

agronomy

Article

Assessing Ecosystem Services of Rice–Fish Co-Culture and Rice Monoculture in Thailand

Noppol Arunrat and Sukanya Sereenonchai *

Faculty of Environment and Resource Studies, Mahidol University, Nakhon Pathom 73170, Thailand; noppol.aru@mahidol.ac.th
* Correspondence: sukanya.ser@mahidol.ac.th

Abstract: Increasing production costs for rice monoculture and concerns about farming households' food security have motivated farmers to adopt integrated rice–fish farming. To date, there has been little research that comparatively assesses the ecosystem services (ESVs) of both rice–fish co-culture and the rice monoculture system in Thailand. Therefore, this study aims to estimate the ESV values of these systems based on the Millennium Ecosystem Assessment. A total of 19 rice–fish co-culture farms were investigated, covering three regions of Thailand (northern, northeastern, and central regions) and consisting of 13 sub-districts, 13 districts, and 11 provinces. For a fair comparison, 19 conventional rice farms were selected as comparison sites. Rice–fish co-culture had a higher net ESV value of 48,450,968.4 THB ha^{-1} year^{-1} than rice monoculture with a net ESV value of 42,422,598.5 THB ha^{-1} year^{-1}. Rice–fish co-culture generated average economic values 25.40% higher than in rice monoculture farming. The most positive change in ESV was found in the regulation of temperature and humidity, with 3,160,862.9 THB ha^{-1} year^{-1}. Moreover, agrotourism can generate revenue and increase the ESV in rice–fish co-culture. Our findings showed that rice–fish co-culture gives more economic and ecological benefits compared to the rice monoculture system. Further studies are recommended to explore and analyze the potential advantages of the rice–fish system in more detail.

Keywords: ecosystem services; rice–fish co-culture; rice monoculture; Thailand

Citation: Arunrat, N.; Sereenonchai, S. Assessing Ecosystem Services of Rice–Fish Co-Culture and Rice Monoculture in Thailand. *Agronomy* **2022**, *12*, 1241. https://doi.org/10.3390/agronomy12051241

Academic Editors: Moritz Von Cossel, Joaquín Castro-Montoya and Yasir Iqbal

Received: 8 April 2022
Accepted: 20 May 2022
Published: 23 May 2022

Publisher's Note: MDPI stays neutral with regard to jurisdictional claims in published maps and institutional affiliations.

1. Introduction

Rice is the primary source of nutrition for approximately two-thirds of the world's population [1], which accounts for up to 75% of the daily calorie intake of people in some Asian countries [2]. It is projected that the world population will require 560 million tons of rice by 2035, which increased to around 120 million tons after 2010 [3]. With 11.17 million harvesting hectares, 21.3 million tons of rice were produced in the crop year 2020/2021, making Thailand the world's 6th largest rice producer after China, India, Bangladesh, Indonesia, and Vietnam [4]. However, future food security and the precarious livelihood of poor people are great challenges for rice farming.

The rice–fish co-culture system is a solution to improve the functioning of ecosystems and alleviate farmers' poverty in many locations [5]. Rice yields from modern monoculture rice are not realistically sustainable due to falling yields from reduced soil fertility and pest problems [6], and the detrimental environmental effects of intense fertilization and pesticide use have now been properly addressed. According to previous studies, the rice–fish co-culture system can efficiently reduce the use of pesticides and herbicides [7], as well as the amount of nitrogen consumed and absorbed by rice plants and fish [8–10]. Despite the environmental benefits of the rice–fish co-culture system, their adoption is extremely low. In Asian countries, e.g., Bangladesh [11], China [8], Malaysia [12], and Vietnam [13], the adoption rate is only marginally greater than 1% [5].

Integrated rice and fish farming has been conducted in Thailand for more than 200 years [13]. Capturing wild fish seed for stocking rice fields was necessary in the beginning. The Department of Fisheries (DOF) began to promote rice–fish production in the 1940s by providing fish seed and improving technology. The central plains saw a boom in rice–fish farming, with fish yields ranging from 137 to 304 kg ha^{-1} crop^{-1} [13]. Rice yields increased by 25 to 30% in fields that included fish. In the 1970s, however, the introduction of high-yielding rice varieties, as well as increasing fertilizer and pesticide applications, led to the near collapse of rice–fish farming in Thailand's central plains. Farmers had two options: separate their rice and fish operations or stop raising fish [13]. Currently, increasing production costs for rice cultivation (e.g., chemical fertilizers, insecticides, and herbicides) and concern for farming households' food security have motivated farmers to adopt integrated rice–fish farming due to its lower cost, higher economic returns, and additional food source. However, the number of rice–fish farms in Thailand remains low. Furthermore, integrated rice and fish farming is an organic agriculture system that the Thai government initially practiced in the 1980s. It has been promoted to persuade and subsidize farmers to adopt organic farming based on the philosophy of the late King Bhumibol Adulyadej as "sufficiency economy". There were only 2500 organic farmers in 2003, and this number increased to 44,418 organic farmers in 2019 [14,15], which accounted for only 0.003% of the total farmers in Thailand [16]. To increase the number of organic farmers, proactive policies need to be focused specifically on rice–fish co-culture farming; thus, comprehensive research is required.

To comprehensively understand the ecological and economic benefits, ecosystem services (ESVs) are widely considered appropriate quantitative and qualitative assessment methods. Following the Millennium Ecosystem Assessment, ESVs are defined as "the benefits people obtain from ecosystems" [17]. ESVs are classified into four types, namely cultural, provisioning, regulating, and supporting services [18], which connect ecological and sociological values for policy implications and decision making. ESVs are widely used and have achieved scientific results in rice–fish farming [8,19–22]. To date, there has been little research that comparatively assesses the rice–fish co-culture and rice monoculture systems in Thailand. Therefore, the objective of this study was to determine the ESV values of rice–fish co-culture and rice monoculture (conventional rice farming) in Thailand and to propose policy implications based on key findings to support government policy and decision making.

2. Materials and Methods

2.1. Study Sites and Description

The number of rice–fish co-culture farms in Thailand is very small, and there is no official record of the location and number of rice–fish co-culture farms. Thus, a purposive sampling method was used to select the farms. There were two criteria for rice–fish co-culture farm selection in this study: (1) the rice–fish co-culture farm must practice organic rice farming and feed fish in the paddy fields without using any chemical substances, and (2) the rice–fish co-culture farm must have practiced rice–fish co-culture for at least 2 years. Based on our survey in the crop years 2020 and 2021, 19 rice–fish co-culture farms were selected, and the data investigated. These farms covered three regions of Thailand (northern, northeastern, and central regions), consisting of 13 sub-districts, 13 districts, and 11 provinces (Table 1). For a fair comparison, 19 conventional rice farms were selected as comparison sites. These conventional rice farms were located near the rice–fish co-culture farms in each sub-district to avoid variations in soil texture, microclimate, and irrigation conditions (Table 1).

Table 1. Description of study areas.

Region	Province	District	Sub-District	Number of Rice-Fish Co-Culture (Farm)	Number of Rice Monoculture (Farm)	Climate *		
						T_{max} (°C)	T_{min} (°C)	Precipitation (mm year^{-1})
Northern	Mae Hong Son	Khun Yuam	Mueang Pon	1	1	33.0	20.0	1100.0
	Phichit	Dong Charoen	Samnak Khun Nen	1	1	32.9	23.3	1264.8
Northeastern	Amnat Charoen	Lue Amnat	Rai Khee	1	1	22.1	27.2	1581.7
	Sakon Nakhon	Phang Khon	Rae	1	1	22.0	31.7	1650.0
	Nakhon Phanom	Renu	Na Kham	1	1	21.8	31.8	1600.0
	Sisaket	Kantharalak	Phu Ngoen	1	1	22.3	33.6	1439.6
	Ubon Ratchathani	Khueang Nai	Ban Thai	3	3	22.1	33.0	1700.0
		Det Udom	Tha Pho Si	3	3			
	Surin	Prasat	Chok Na Sam	1	1	22.7	32.7	1432.2
		Rattanaburi	Rattanaburi	3	3			
	Buriram	Ban Kruat	Sai Ta Ku	1	1	22.2	33.0	1100.0
	Yasothon	Mueang	Nong Khu	1	1	22.5	32.3	1200.0
Central	Ang Thong	Pa Mok	Bang Sadet	1	1	22.0	34.0	1100.0

T_{max} = maximum temperature, T_{min} = minimum temperature. * Source: Thai Meteorological Department (TMD) in 2020.

2.2. Rice–Fish Co-Culture and Conventional Rice Systems

2.2.1. Rice–Fish Co-Culture System

Based on the 19 rice–fish co-culture farms in this study, two field types of rice–fish co-culture were identified, namely the canal refuge (Figure 1a) and pond refuge (Figure 1b). 'Khao Dawk Mali 105' (KDML 105), 'RD 6', and 'San Pah Tawng 1' varieties were found to be grown in paddy fields once a year. The transplanting method was used for planting, while harvesting was done by hand. The main species of farmed fish raised in the paddy fields were Nile tilapia (*Oreochromis niloticus*), Common snakehead (*Channa striata*), Common carp (*Cyprinus carpio*), Common silver barb (*Barbonymus gonionotus*), Mrigal carp (*Cirrhinus cirrhosus*), Seven-stripped carp (*Probarbus jullieni*), and Walking catfish (*Clarias batrachus* (Linnaeus)). Organic materials (rice husk, rice bran, pig manure, cattle manure, poultry manure, fruits and vegetables) were applied in the paddy fields to provide nutrients for rice and food for the fish.

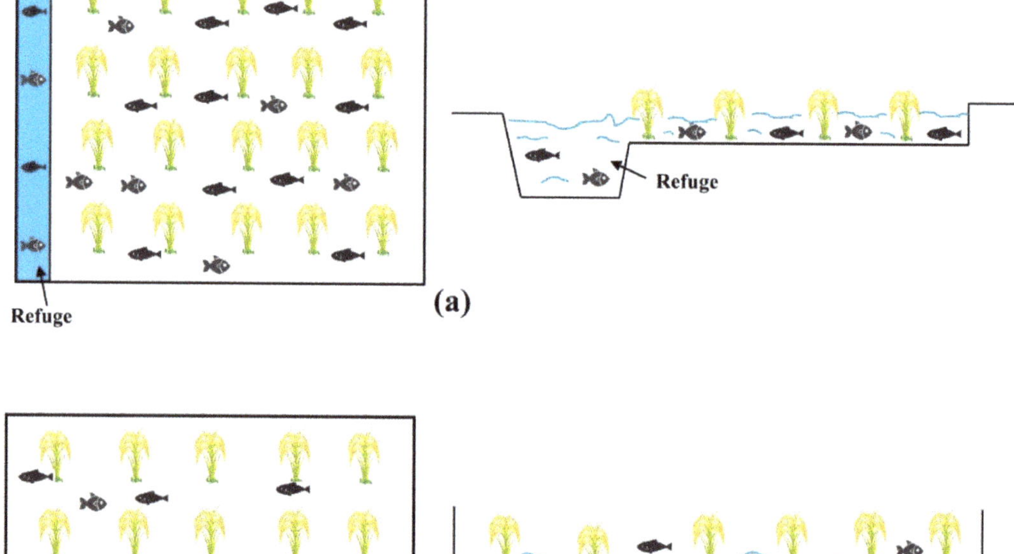

Figure 1. Rice–fish co-culture field type: (**a**) canal refuge and (**b**) pond refuge. Note: The refuges in our study sites were heterogeneous in size (depth and width).

2.2.2. Conventional Rice System

For a fair comparison, rice cultivation farms were chosen once a year. 'KDML 105', 'San Pah Tawng '1, 'RD 6', 'RD 41', 'RD 57', 'RD 79', and 'RD 85' varieties were grown on these farms. Chemical fertilizers (16-20-0, 46-0-0, 16-16-8, 16-8-8, and 15-15-15), insecticides, and herbicides were applied to enhance rice plant growth. Transplanting and broadcasting methods were found for conventional rice systems, depending on water availability. The broadcasting method is commonly used for areas subject to water shortages. A harvesting machine is usually used for harvesting.

2.3. Data Collection

Data on farm management practices in the two crop years (2019/2020 and 2020/2021) were collected from the owners of the rice–fish co-culture and conventional rice farms. The quantitative data were recorded from each farm, including rice field area, rice yield, fish yield, height of field ridge, volume of circular furrow, number of days of flooding in the field, annual irrigation volume, total number of tourists, and residence time. Moreover, the unit prices of rice, fish, and pesticides, reservoir engineering fee usage, water supply, and money received from tourism were recorded.

2.4. Ecosystem Service Value Evaluation Method

The Common International Classification for Ecosystem Services (CICES) version 5.1 (2018) was used [23] in this study. Based on the definition of ecosystem services in CICES version 5.1 (2018), Liu et al. [21] and Liu et al. [24] designed 23 ESV indicators and 3 sections (provisioning, regulation and maintenance, and cultural) (Table 2). Due to the lack of relevant studies in Thailand and limited data availability, 13 of the 23 indicators were applied in this study (Table 2).

Following The Economics of Ecosystems and Biodiversity (TEEB) method [25,26], three categories were widely used to express the ESVs in monetary units: the direct market method, equivalent factor method, and replacement costs method [27]. In this study, the direct market method was used to evaluate the "provisioning services", while the simulated market method was used to estimate the "development of tourism". Finally, the other ecosystem services were assessed based on the alternative market method. Based on Liu et al. [24], the formulas for calculating ESVs are presented below.

2.4.1. Provisioning Services

Rice and fish generate income for farmers depending on the yield and market prices.

$$V_1 = (Y_{rice} \times P_{rice}) + (Y_{fish} \times P_{fish})$$

where V_1 is the total income of primary products from paddy fields (THB ha^{-1} year^{-1}); Y_{rice} is rice yield (ton ha^{-1}); P_{rice} is the price of rice (THB ton^{-1} year^{-1}); Y_{fish} is the yield of fish (ton); and P_{fish} is the price of fish (THB ton^{-1} year^{-1}).

2.4.2. Gas Regulation

Rice farming regulates gases in the atmosphere by absorbing CO_2 and releasing O_2 through photosynthesis.

$$V_2 = E_{CO_2} + E_{O_2}$$

$$E_{CO_2} = Y_{Nrice} \times \alpha \times C_{CO_2} \times C_{STR}$$

$$E_{O_2} = Y_{Nrice} \times \varphi \times O_{cost}$$

$$Y_{Nrice} = Y_{rice} \times (1 - m)/\beta$$

where V_2 is the value of gas regulation from paddy fields (THB ha^{-1}), E_{CO_2} is the value of CO_2 fixed by rice (THB), Y_{Nrice} is the net rice yield (ton ha^{-1}), α is the amount of CO_2 fixed for 1 g of rice dry matter (1.63 g [24]), C_{CO_2} is the carbon content in CO_2 (27.27% [24]), C_{STR} is the Swedish carbon tax rate (133.26 USD ton^{-1} CO_2 on 1 November 2020 [28]), E_{O_2} is the value of rice-released O_2 (THB), φ is the amount of O_2 produced for 1 g of rice dry matter (1.19 g [24]), O_{cost} is the cost of industrial oxygen production (2092 THB ton^{-1} O_2, converted from Xu et al. [22]), Y_{rice} is rice yield (ton ha^{-1}), m is the moisture content of rice, and β is economic coefficient of rice (0.5 [24]).

2.4.3. Temperature and Humidity Regulation

Crop evapotranspiration and water evaporation in paddy fields can regulate heat and humidity in surrounding areas.

$$V_3 = W_{EV} \times H_{DS} \times \eta \times P_{Coal}$$

where V_3 is the value of temperature and humidity regulation from paddy fields (THB ha^{-1}), W_{EV} is the average daily water evaporation in the rice field (4.4 mm day^{-1}, generated using the CROPWAT 8.0 model), H_{DS} is the number of hot days in summer in the study area (days; obtained from the Thai Meteorological Department), η is the heat consumption for evaporating 50 mm of water in 1 ha of rice field (equal to burning 30.57 tons of coal) [24], and P_{Coal} is the price of standard coal (THB ton^{-1}).

2.4.4. Air Purification

Rice field ecosystems can purify the air by absorbing harmful gases (e.g., SO_2, NO_x, HF, and dust) in the atmosphere.

$$V_4 = (A_{SO_2} \times P_{SO_2}) + (A_{NO_x} \times P_{NO_x}) + (A_{HF} \times P_{HF}) + (A_D \times P_D)$$

where V_4 is the value of air purification from paddy fields, A_{SO2}, A_{NOX}, A_{HF}, and A_D are the average annual flux (kg) of SO_2, NO_x, HF, and dust absorbed by the paddy fields, respectively. Based on Ma et al. [29], the average annual flux of SO_2, NO_x, HF, and dust was 45.0, 33.3, 0.57, and 33,200 kg ha^{-1} year^{-1}, respectively. P_{SO_2}, P_{NO_x}, P_{HF}, and P_D are the costs of SO_2, NO_x, HF, and dust in the rice field, respectively (THB kg^{-1}). In this study, the costs of SO_2, NO_x, HF, and dust in the rice field were 7.53, 3.97, 4.34, and 0.94 THB kg^{-1}, respectively, which were converted from Ma et al. [29].

2.4.5. Pest Control

Fish can help reduce the weeds and pests in paddy fields by consuming them, resulting in a reduced demand for pesticides and herbicides.

$$V_5 = P_p \times R$$

where V_5 is the value of pest control from paddy fields (THB ha^{-1} year^{-1}); P_p is the average pesticide cost for the rice monoculture system (THB ha^{-1} year^{-1}); and R is the percentage of reduction in pesticide use for rice–fish co-culture.

2.4.6. Increase in Fauna Diversity and Microorganisms

Fish can control weeds and pests, which helps reduce the use of herbicides, pesticides, and chemical fertilizers, leading to increased species diversity.

$$V_6 = \tau \times V_P$$

where V_6 is the value of increase of fauna diversity and microorganisms from paddy fields (THB ha^{-1} year^{-1}), τ is the value-equivalent factor of the rice field ecosystem (0.21, [30]), and V_P is the equivalent product provisioning service (THB ha^{-1} year^{-1}).

2.4.7. Maintaining Soil Nutrients

Paddy fields are sources of GHG emissions, especially CO_2 and CH_4, whereas rice fields are sink pools of carbon through soil carbon sequestration.

$$V_7 = P_{OM} \times (IN_{OM} - OU_{OM})$$

$$IN_{OM} = (N_r \times C_r) + (N_s \times 11\% \times C_s)$$

$$OU_{OM} = (R_{CO_2} \times 0.27) + (R_{CH_4} \times 0.75)$$

where V_7 is the value of maintaining soil nutrient value from paddy fields (THB ha^{-1} year^{-1}); P_{OM} is the price of organic materials (7.69 THB kg^{-1} C, converted from Liu et al. [24]); IN_{OM} is the organic matter input from soil (kg C ha^{-1} year^{-1}); OU_{OM} is the output amount of soil organic matter (kg C ha^{-1} year^{-1}); R_{CO_2} is the amount of CO_2 emissions from rice fields (2123.63 kg ha^{-1} year^{-1}, [24]); R_{CH_4} is the amount of CH_4 emissions from rice fields (29.64 kg ha^{-1} year^{-1}, [24]); the constant values of 0.27 and 0.75 are the conversion coefficients of CO_2 and CH_4 into carbon, respectively; N_r and N_s are the biomass of the rice root system and straw (kg ha^{-1} year^{-1}), respectively; and C_r and C_s are the carbon content of the rice root system and straw (%), respectively.

In this study, a quadrat (1 m × 1 m) was used to randomly collect rice straw and rice roots with three replications from each field. Rice straw and rice roots were separated in the field and then put into plastic bags for laboratory analysis. The dry mass of rice straw and rice roots were determined after oven drying at 80 °C for 48 h. According to Ma et al. [31], the carbon content in rice straw and rice roots in this study was assumed to be 43.26% and 38.20%, respectively.

2.4.8. Water Conditions

Rice cultivation requires large amounts of water, mainly from rainfall, surface water, and groundwater. Moreover, paddy fields can provide water storage by storing rainwater on the surface and maintaining groundwater.

$$V_8 = E_{WS} + E_{GW}$$

$$E_{WS} = (H_R + V_{CF}/A) \times P_{RE}$$

$$E_{GW} = S_{WP} \times P_{WT} \times D_{FL}$$

where V_8 is the value of water conditions from paddy fields (THB ha^{-1} year^{-1}), E_{WS} is the value of the water storage function of the rice system (THB ha^{-1} year^{-1}), E_{GW} is the value of groundwater conservation (THB), H_R is the average height of the field ridge, V_{CF} is the volume of a circular furrow, A is the area of the rice field, P_{RE} is the unit price of the reservoir engineering fee usage (THB m^{-3}), S_{WP} is the soil water permeability in the rice field (6 mm, [24]), P_{WT} is the market price of water (THB m^{-3}, obtained from Provincial Waterworks Authority), and D_{FL} is the average days of flooding in the rice growing period (days).

2.4.9. Energy Losses for Irrigation

During the rice-growing period, maintaining the water level in the paddy field is very important, especially in rice–fish co-culture systems. However, water from rainfall may not be sufficient for rice cultivation throughout the growing period. Therefore, energy is required for pumping and lifting irrigation water from irrigation canals and groundwater.

$$V_9 = E_{IRR} \times P_{WS}$$

where V_9 is the value of energy losses from paddy fields (THB ha^{-1} year^{-1}), E_{IRR} is the average annual irrigation per area (m^3 ha^{-1} year^{-1}), and P_{WS} is the cost of the water supply in lifting irrigation (THB m^{-3}).

Table 2. Ecosystem services and goods and benefits valued from rice–fish co-cultures (following CICES V5.1; Liu et al. [21]; Liu et al. [24]).

CICES V5.1 Section	Division	Ecosystem Services of Rice–Fish Co-Cultures	Goods and Benefits Valued	Direction of Value
Provisioning	Biomass	1. Rice and fish provided food and nutrition	Provisioning service	Positive
Regulation and Maintenance	Transformation of biochemical or physical inputs to ecosystems	2. CO_2 fixation from photosynthesis	Gas regulation	Positive
		3. O_2 release from photosynthesis	Gas regulation	Positive
		4. SO_2, NOx, HF, and dust absorbed by the paddy field	Air purification	Positive
		5. Nutrient cycling and organic accumulation	Maintaining soil nutrients	Positive
		6. Reduction of GHG emissions	Maintaining soil nutrients	Positive
	Regulation of physical, chemical, biological conditions	7. Reducing land abandonment	X	X
		8. Improving soil salinization	X	X
		9. Pesticides and herbicides reduction	Pest control	Positive
		10. Regulation of temperature and humidity	Climate control	Positive
		11. Enhancing humidification and rain	X	X
		12. Increase of fauna diversity and micro-organisms	Biodiversity	Positive
		13. Increase water storage	Water storage and retention	Positive
		14. Groundwater conservation	Water storage and retention	Positive
		15. Energy losses in lifting irrigation	Energy losses for irrigation	Negative
	Other types of regulation and maintenance service	16. Securing the rural poor	X	X
Cultural	Direct, in situ, and outdoor interactions with living systems that depend on presence in the environmental setting	17. Development of tourism	Development of tourism	Positive
		18. Experiential use of plants, animals, and land	X	X
		19. Education opportunities	X	X
		20. Research subject	X	X
		21. Cultural value and heritage	X	X
		22. Artistic inspiration (theater, painting, sculpture)	X	X
		23. Willingness to preserve for future generations	X	X

Note: X represents ecological services that were not considered in this study due to no data available for calculation.

2.4.10. Development of Tourism

Paddy fields can serve as tourist attractions, enhancing the added value. The rice–fish co-culture system is a magnet used to attract visitors for relaxation and learning about rice–fish co-culture.

$$V_{10} = P_{TC} \times N_{TR} \times T$$

where V_{10} is the value of tourism development (THB ha^{-1} year^{-1}), P_{TC} is the amount of money from tourism consumption (THB person^{-1}), N_{TR} is the total number of tourists (person), and T is residence time.

3. Results

3.1. Farm Investigation

3.1.1. Rice–Fish Co-Culture Farms

The average height of the field ridge was 150 cm, and the average number of days of flooding during the rice–fish growing period was 110 days. The area of rice–fish fields was mostly around 0.15 ha, on average. The water level in rice–fish fields varied by 15–35 cm throughout the rice growing period. The amount of fish was approximately 1875–3125 fish ha^{-1}. One-month-old fish were released into the paddy field 30 days after rice planting. The water was drained out before rice harvesting at around 7–10 days; most of the fish escaped to the refuge pond and then were caught using nets after rice harvest.

The average rice yield was 3.6 ton ha^{-1} year^{-1}, with a range of 2.0 to 4.2 ton ha^{-1} year^{-1}. The average moisture content of rice was 14%. The average rice price ranged from 10 to 13 THB kg^{-1}. This is because the rice yield in rice–fish co-culture farms was organic rice, resulting in a higher price. The average yield of the fish products was 300 kg ha^{-1}, while the prices of the fish products ranged from 30 to 50 THB kg^{-1}. The average rice straw biomass was 12,008 kg ha^{-1}, while the average rice root biomass was 2401.6 kg ha^{-1}. The average rice root carbon content and rice straw carbon content were 37.2% and 42.1%, respectively. The average number of tourists who visited rice–fish co-culture farms was 53 persons year^{-1}.

3.1.2. Conventional Rice Farms

The average rice yield was 4.7 ton ha^{-1} year^{-1}, with a range of 3.8 to 5.6 ton ha^{-1} year^{-1}. The average moisture content of rice was 18%. The average rice price ranged from 7.5 to 8.0 THB kg^{-1}. The average biomass of the rice straw and rice roots was 1310 and 2135 kg ha^{-1}, respectively. The average rice root carbon content and rice straw carbon content were 32.1% and 37.4%, respectively.

3.2. Provisioning Services

The basic function of rice–fish co-culture is to provide rice and fish for food and nutrition, while the primary product of the rice monoculture system is rice. In 2020–2021, the revenue generated by rice–fish co-culture was approximately 50,400 THB ha^{-1} year^{-1}, on average. However, the average ecosystem service value of the rice monoculture system was estimated to be 37,600 THB ha^{-1} year^{-1}. Notably, ecosystem service values in this category increased 12,800 THB ha^{-1} year^{-1} annually, as rice–fish culture enhanced ecosystem services (25.40%) (Table 3). This is because the farmers received income from selling rice and fish.

Table 3. Ecosystem service values of rice–fish co-culture and rice monoculture systems during 2020–2021.

Ecosystem Services	Rice-Fish Co-Cultures (THB ha^{-1} year^{-1})		Rice Monoculture System (THB ha^{-1} year^{-1})		Changing of Ecological Service Values (THB ha^{-1} year^{-1})
	Mean	SD	Mean	SD	
Positive value					
Rice and fish provided food and nutrition	50,400.0	868.8	37,600.0	618.8	12,800.0
CO_2 fixation from photosynthesis	358,092.2	139,500.7	456,106.6	90,978.7	−98,014.4
O_2 release from photosynthesis	14,697.9	5725.8	18,720.9	3734.2	−4023.0
SO_2, NOx, HF, and dust absorbed by the paddy field	31,681.5	0	31,681.5	0	0
Nutrient cycling and organic accumulation	23,798,852.3	0	21,678,588.0	0	2,120,264.3
Reduction of GHG emissions					
Pesticides and herbicides reduction	937,500.0	0	187,500.0	0	750,000.0
Regulation of temperature and humidity	23,179,661.4	0	20,018,798.5	0	3,160,862.9
Increase of fauna diversity and micro-organisms	10,584.0	182.4	7896.0	129.9	2688.0
Increase water storage	16,682.3	6341.9	295.0	6341.9	16,387.3
Groundwater conservation	13,992.0	1017.6	9540.0	1017.6	4452.0
Development of tourism	53,000.0	7000.0	0	0	53,000.0
Sub total	*48,465,143.6*	-	*42,446,726.5*	-	*6,018,417.2*
Negative value					
Energy losses in lifting irrigation	14,175.2	1786	24,128.0	2115.0	−9952.8
Sub total	*14,175.2*	-	*24,128.0*	-	*−9952.8*
Net value	**48,450,968.4**	**-**	**42,422,598.5**	**-**	**6,028,370.0**

3.3. Regulation and Maintenance

3.3.1. Gas Regulation

The mean value of the regulation service for CO_2 fixation from photosynthesis was 358,092.2 THB ha^{-1} year^{-1} in the co-culture system, whereas the monoculture system earned 456,106.6 THB ha^{-1} year^{-1}. The annual decline in ESV can be seen in this regulation service. The O_2 released from photosynthesis in the two systems contributed to 14,697.9 and 18,720.9 THB ha^{-1} year^{-1}. A decrease of 4023.0 THB ha^{-1} year^{-1} per annum was observed when evaluating the ESV in this service. The paddy fields absorb SO_2, NOx, HF, and dust, and this regulation service generates revenue of 31,681.5 THB ha^{-1} year^{-1} in the co-culture system and 31,681.5 THB ha^{-1} year^{-1} in the monoculture system (Table 3).

3.3.2. Nutrient Cycling and Organic Accumulation, and Reduction of GHG Emissions

The calculation of ESV from the ecosystem service related to nutrient cycling, organic accumulation, and reduction of GHG emissions was 23,798,852.3 THB ha^{-1} year^{-1} in co-culture and 21,678,588.0 THB ha^{-1} year^{-1} in monoculture. Remarkably, approximately half of the total ESV comes from this service in both systems. A significant annual increase in ESV was also found in this service (Table 3). This is because the biomass and carbon content in rice straw and roots of the rice–fish co-culture farms were higher than in rice monoculture.

3.3.3. Pesticide and Herbicide Reduction

Reducing the use of pesticides and herbicides enhances ecosystem services in several ways. The rice–fish co-culture system obtained an ESV of 937,500.0 THB ha^{-1} year^{-1}, while an annual ESV of approximately 187,500.0 THB ha^{-1} year^{-1} was received in the rice monoculture system. Moreover, the increase in ESV in this category was estimated to be 750,000.0 THB ha^{-1} year^{-1} (Table 3). Organic rice farming practiced in rice–fish co-culture does not require the application of chemical substances, leading to lower production costs and a reduction in environmental pollution.

3.3.4. Regulation of Temperature and Humidity

The valuation of the ecosystem service related to the regulation of temperature and humidity of rice–fish co-culture and rice monoculture systems resulted in 23,179,661.4 THB ha^{-1} year^{-1} and 20,018,798.5 THB ha^{-1} year^{-1}, respectively. This service generates nearly half of the total ESV in both systems. Furthermore, a significant increase in the annual ESV was notable, as rice–fish culture developed in an area (Table 3).

3.3.5. Increase in Fauna Diversity and Microorganisms

Increasing fauna diversity and microorganisms can improve the performance of ecosystem services. An ESV of 10,584.0 THB ha^{-1} year^{-1} was received from the co-culture and 7896.0 THB ha^{-1} year^{-1} from the monoculture. The ESV has risen annually by 2688.0 THB ha^{-1} year^{-1} (Table 3). Due to the higher provisioning services in rice–fish co-culture than rice monoculture, the ESV of fauna diversity and microorganisms increased. This demonstrates that avoiding the use of pesticides and herbicides can increase biodiversity in paddy fields.

3.3.6. Increase in Water Storage

The rice–fish co-culture system gained 16,682.3 THB ha^{-1} year^{-1} from this service, while the monoculture system earned 6341.9 THB ha^{-1} year^{-1}. The yearly increase in ESV was noteworthy (Table 3). Under the rice–fish co-culture system, the value of the water storage function increased due to the high volume of water stored on the surface, as well as the long period of flooding during the rice–fish growing period.

3.3.7. Groundwater Conservation

Groundwater conservation in co-culture and monoculture contributes 13,992.0 and 1017.6 THB ha^{-1} year^{-1}, respectively, with an annual increase of ESV 4452.0 THB ha^{-1} year^{-1} (Table 3). A longer period of flooding in rice–fish co-culture fields means that more groundwater can be stored through percolation and infiltration.

3.3.8. Energy Losses in Lift Irrigation

As a negative ESV, the valuation of energy losses in lifting irrigation was 14,175.2 THB ha^{-1} year^{-1} in the co-culture and 2115.0 THB ha^{-1} year^{-1} in monoculture. In this category, an ESV decrease of 9952.8 THB ha^{-1} year^{-1} occurred due to the development of rice–fish culture. Based on the field survey, most farmers used fossil fuel (diesel) for pumping water into paddy fields, while a few farms installed solar panels and used solar energy for water management in their fields. Using solar energy can reduce 19.5% of the energy cost compared with diesel fuel.

3.4. Cultural Services

Development of Tourism

Agrotourism is becoming increasingly popular in rice–fish regions. According to the farmers who participated in the survey, approximately 53 tourists were attracted by rice–fish activities in 2020–2021. Each tourist spends one day, and their average expenditure is 1000 THB. Therefore, the tourism contribution value of the rice–fish system was 53,000.0 THB ha^{-1} year^{-1}

(Table 3). Most of the visitors came to see the rice–fish co-culture and gain knowledge and experiences, and the rest visited to buy organic rice and fish products.

4. Discussion

4.1. Ecosystem Service Value of the Rice–Fish Co-Culture System

Integrated rice and fish have been recommended as a sustainable strategy for improving soil nutrient status and water resources, which provide carbohydrates and proteins to humans and reduce environmental pollution [8,32]. Moreover, the rice–fish co-culture system can alleviate local farmers' poverty and enhance social welfare [11,33]. When comparing the two systems in the current study, the rice–fish co-culture system has a higher net ESV value of 48,450,968.4 THB ha^{-1} year^{-1} (Table 3). In addition, rice–fish co-culture generated average economic values 25.40% higher than rice monoculture farming (Table 3). The regulation services that occupied the largest portion of total ESV were nutrient cycling and organic accumulation, reduction of GHG emissions, and regulation of temperature and humidity (Figure 2). In contrast, the contributions of the remaining ESVs were not significant, and only a small portion of the net value was received from these services (Figure 2). Developing rice–fish co-culture has positive effects on provisioning services, as co-culture contributes to the increase of ESV in the area. Regarding gas regulation services, the benefits of the co-culture system cannot be seen in CO_2 fixation and O_2 release from photosynthesis. This is because rice yields from rice–fish co-culture farms were mostly lower than in the rice monoculture system. Furthermore, there was no significant change in the ESV of the two systems regarding SO_2, NOx, HF, and dust absorbed by the paddy field.

The most significant positive change in ESV can be seen in the regulation of temperature and humidity with 3,160,862.9 THB ha^{-1} year^{-1} (Table 3). The service related to nutrient cycling and organic accumulation, and reduction of GHG emissions, takes second place in contributing to the improvement of ecosystem services (Figure 2). Paddy fields have the potential to improve soil physical and chemical properties, increase soil organic carbon, and mitigate CO_2 emissions in the atmosphere [21,34,35]. Increasing fauna diversity, microorganisms (bacteria, protozoa, algae, and fungi), and water storage, as well as groundwater conservation, make minor contributions to the increase in net ESV. These are in line with the studies of Nayak et al. [36] and Ren et al. [37], who reported that rice–fish co-cultures maintain the genetic diversity of aquatic organisms in paddy fields due to the reduction in the use of pesticides, insecticides, and chemical fertilizers. Wan et al. [38] found that finless eel and loach rice–fish co-cultural practices in China can help reduce the abundance of pests, leading to lower use of pesticides and a reduction in labor costs. This is consistent with our study, which found that even though the rice–fish co-cultural farms in our study areas practiced organic rice farming, the yields of organic rice were high, and there were fewer pests and diseases as well as weeds. This is because fish excrement can improve soil nutrients, and fish consume insects in paddy fields, while the water level can control the abundance of weeds. This is similar to the study of Xie et al. [8], which found that the level of water in paddy fields can reduce the abundance of rice planthoppers. Wan et al. [38] also found that the abundance of herbivore insects decreased by 24.07%, weed abundance was reduced by 67.62%, and invertebrate predator abundance increased by 19.48%.

Although agrotourism can generate revenue and increase the ESV, its proportion in the total value is not significant. However, tourists are interested in visiting rice–fish farming areas but not traditional monocultures. This means that the co-culture system has the potential to receive a higher ESV from this cultural service. Tourism can have direct benefits for farmers by creating marketing opportunities to sell their products to tourists [39] and may provide additional income to farmers from other agricultural activities, such as developing creative tourism, which provides a true experience of connection for tourists.

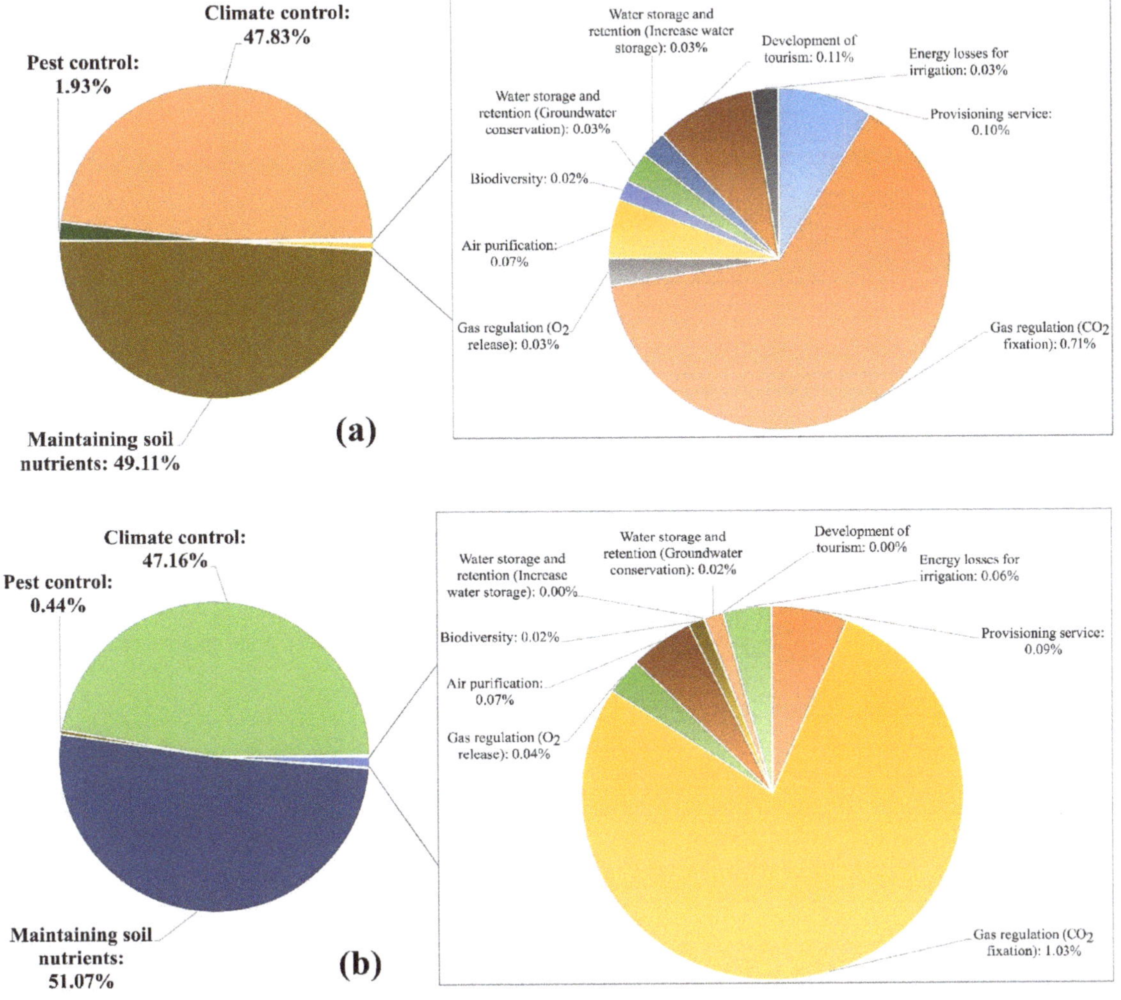

Figure 2. Proportion of service types: (**a**) rice–fish co-culture system, (**b**) rice monoculture system.

4.2. Policy Implications

Although integrated rice and fish farming has been practiced in Thailand for a long time, in recent years, the number of farms has been small, and the trend is declining. This is due to the intensification and modernization of rice cultivation focusing on maximizing yield, and urbanization involving converting paddy fields to commercial building and industrial factories. In addition, the impact of climate change is causing changes to the seasons and increasing the frequency and intensity of flood and drought events. Based on in-depth interviews, drought was the main cause of loss of rice and fish yields on farms in the northeastern region, while flooding caused damage in the northern and central regions. This indicates that rice–fish co-culture farming answers these challenges in Thailand.

The results demonstrated that rice–fish co-culture provides nutrient cycling and organic accumulation, reduction of GHG emissions, and regulation of temperature and humidity for the ecosystem (Table 3 and Figure 2). At the same time, rice–fish co-culture provides safe foods (rice and fish) and extra income for the farmers' households, implying that Thailand has great potential to be a rice–fish co-culture society because rice and fish are part of the ancestral food culture for Thai people. It is obvious that rice–fish co-culture could address more than one sustainable development goal (SDG), such as SDG 2 (zero hunger), SDG 12 (responsible consumption and production), SDG 13 (climate action), and SDG 14 (life below water). Therefore, policy implications should implement the following strategies to promote and support rice–fish co-culture: (1) develop innovation for better irrigation systems to reduce the impact from flood and drought events, (2) support the quantity of fish seed to increase the number of fish seed survival after release into paddy fields, (3) promote community learning centers for rice–fish co-culture to establish the farmer school, (4) strengthen the new innovative technology for pests and diseases control, (5) work as multi-stakeholders (farmer–officer–businessman–scholar), and (6) develop and promote the unique selling points of rice–fish co-culture, which are organic rice, organic fish, and destinations for travel. These strategies can help ensure the sustainability of the agricultural, environmental, and economic aspects of rice–fish co-culture in Thailand.

5. Conclusions

The rice–fish co-culture system has benefits for sustainability and ecology. At the same time, it must compete with commercial and advanced agricultural systems. Our findings showed that the rice–fish system provides more economic and ecological benefits than the rice monoculture system. The rice–fish co-culture system has a higher net ESV value of 48,450,968.4 THB ha^{-1} year^{-1} than rice monoculture (net ESV 42,422,598.5 THB ha^{-1} year^{-1}), which generated average economic values 25.40% higher than rice monoculture. The most positive change in ESV can be seen in the regulation of temperature and humidity, with 3,160,862.9 THB ha^{-1} year^{-1}. Services related to nutrient cycling and organic accumulation, and reduction of GHG emissions take second place in contributing to the improvement of ecosystem services. Further studies are recommended to explore and analyze the potential advantages of the rice–fish system in more detail.

Author Contributions: Conceptualization, N.A. and S.S.; methodology, N.A. and S.S.; investigation, N.A. and S.S.; writing—original draft preparation, N.A. and S.S.; writing—review and editing, N.A. and S.S. All authors have read and agreed to the published version of the manuscript.

Funding: This research project is supported by Mahidol University (Basic Research Fund: fiscal year 2021, contract number: BRF1-A8/2564). Moreover, this research project is supported by Mahidol University (contract number: A14/2564).

Institutional Review Board Statement: The study was conducted according to the guidelines of the Declaration of Helsinki and approved by the Institutional Review Board of Institute for Population and Social Research, Mahidol University (IPSR-IRB) (COA. No. 2021/01-001 and COA. No. 2021/01-003).

Informed Consent Statement: Not applicable.

Data Availability Statement: Not applicable.

Acknowledgments: The authors extend their appreciation to Mahidol University (contract number: A14/2564 and BRF1-A8/2564). Furthermore, we would also like to thank Thailand Science Research and Innovation (TSRI) for their funding (contract number: BRF1-A8/2564).

Conflicts of Interest: The authors declare no conflict of interest.

References

1. Wynn, T. *Rice Farming: How the Economic Crisis Affects the Rice Industry*; Rice Producers Forum USRPA: Houston, TX, USA, 2008.
2. Food and Agricultural Organization (FAO). *Rice in the World*; The Fifth External Programme and Management Review of the International Plant Genetic Resources Institute (IPGRI); FAO: Rome, Italy, 2001.
3. Global Rice Science Partnership (GRiSP). *Rice Almanac*, 4th ed.; International Rice Research Institute: Los Baños, Philippines, 2013.
4. Office of Agricultural Economics (OAE). *Agricultural Statistics of Thailand*; Ministry of Agriculture and Cooperatives: Bangkok, Thailand, 2021. Available online: https://www.oae.go.th/assets/portals/1/files/jounal/2565/yearbook2564.pdf (accessed on 20 December 2021).
5. Halwart, M.; Gupta, M.V. *Culture of Fish in Rice Fields*; FAO: Rome, Italy; The WorldFish Center: Penang, Malaysia, 2004.

6. Pingali, P.L.; Moya, P.F.; Velasco, L.E. *The Post-Green Revolution Blues in Asian Rice Production—The Diminished Gap between Experiment Station and Farmer Yields*; Social Science Division Paper No. 90-01; International Rice Researech Institute: Manila, Philippines, 1990.

7. Dwiyana, E.; Mendoza, T.C. Determinants of productivity and profitability of rice–fish farming systems. *Asia Life Sci.* **2008**, *17*, 21–42.

8. Xie, J.; Hu, L.L.; Tang, J.J.; Wu, X.; Li, N.; Yuan, Y.; Yang, H.; Zhang, J.; Luo, S.; Chen, X. Ecological mechanisms underlying the sustainability of the agriculture heritage rice–fish coculture system. *Proc. Natl. Acad. Sci. USA* **2011**, *108*, 1381–1387. [CrossRef] [PubMed]

9. Zhang, J.; Hu, L.L.; Ren, W.Z.; Guo, L.; Tang, J.J.; Shu, M.A.; Chen, X. Rice-soft shell turtle coculture effects on yield and its environment. *Agric. Ecosyst. Environ.* **2016**, *224*, 116–122. [CrossRef]

10. Frei, M.; Becker, K. Integrated rice–fish culture: Coupled production saves resources. *Nat. Resour. Forum.* **2005**, *29*, 135–143. [CrossRef]

11. Ahmed, N.; Garnett, S.T. Integrated rice-fish farming in Bangladesh: Meeting the challenges of food security. *Food Secur.* **2011**, *3*, 81–92. [CrossRef]

12. Ali, A.B. Seasonal dynamics of microcrustacean and rotifer communities in Malaysian rice fields used for rice-fish farming. *Hydrobiologia* **1990**, *206*, 139–148. [CrossRef]

13. Mackay, K.T.; Chapman, G.; Sollows, J.; Thongpan, N. Rice-fish culture in northeast Thailand: Stability and sustainability. In *Global Perspectives on Agroecology and Sustainable Agricultural System*; Allen, P., van Dusen, D., Eds.; University of California: Santa Cruz, CA, USA, 1987; pp. 355–372.

14. Willer, H.; Yussefi, M. The World of Organic Agriculture—Statistics and Emerging Trends 2006. International Federation of Organic Agriculture Movements (IFOAM), Research Institute of Organic Agriculture FiBL, 2006. Available online: https://orgprints.org/id/eprint/5161/2/willer-yussefi-2005-world-of-organic.pdf (accessed on 8 February 2022).

15. Office of Agricultural Economics (OAE). *Organic Agriculture Action Plan 2017–2022*; Ministry of Agriculture and Cooperatives: Bangkok, Thailand, 2020.

16. Lee, S. In the Era of Climate Change: Moving Beyond Conventional Agriculture in Thailand. *Asian J. Agric. Dev.* **2021**, *8*, 1–14. [CrossRef]

17. Millennium Ecosystem Assessment (MEA). *Ecosystems and Human Well-Beings: Synthesis*; Island Press: Washington, DC, USA, 2005; Available online: https://www.millenniumassessment.org/documents/document.356.aspx.pdf (accessed on 10 April 2021).

18. Costanza, R.; de Groot, R.; Braat, L.; Kubiszewski, I.; Fioramonti, L.; Sutton, P.; Farber, S.; Grasso, M. Twenty years of ecosystem services: How far have we come and how far do we still need to go? *Ecosyst. Serv.* **2017**, *28*, 1–16. [CrossRef]

19. Xie, J.; Liu, L.; Chen, X.; Chen, J.; Yang, X.X.; Tang, J.J. Control of Diseases, Pests and Weeds in Traditional Rice–fish Ecosystem in Zhejiang, China. *Bull. Sci. Technol.* **2009**, *25*, 802–810.

20. Berg, H.; Tam, N.T. Decreased use of pesticides for increased yields of rice and fish-options for sustainable food production in the Mekong Delta. *Sci. Total Environ.* **2018**, *619–620*, 319–327. [CrossRef]

21. Liu, D.; Tang, R.; Xie, J.; Tian, J.; Shi, R.; Zhang, K. Valuation of ecosystem services of rice-fish coculture systems in Ruyuan County. *China Ecosyst. Serv.* **2020**, *41*, 101054. [CrossRef]

22. Xu, Q.; Liu, T.; Guo, H.; Dou, Z.; Gao, H.; Zhang, H. Conversion from rice—Wheat rotation to rice—Crayfish coculture increases net ecosystem service values in Hung-tse Lake area, east China. *J. Clean. Prod.* **2021**, *319*, 128883. [CrossRef]

23. Haines-Young, R.; Potschin, M. *Common International Classification of Ecosystem Services (CICES) V5.1 and Guidance on the Application of the Revised Structure*; Fabis Consulting Ltd.: Nottingham, UK, 2018; Available online: https://cices.eu/content/uploads/sites/8/2018/01/Guidance-V51-01012018.pdf (accessed on 10 April 2021).

24. Liu, D.; Feng, Q.; Zhang, J.; Zhang, K.; Tian, J.; Xie, J. Ecosystem services analysis for sustainable agriculture expansion: Rice-fish co-culture system breaking through the Hu Line. *Ecol. Indic.* **2021**, *133*, 108385. [CrossRef]

25. Costanza, R.; D' Arge, R.; de Groot, R.; Farber, S.; Grasso, M.; Hannon, B.; Limburg, K.; Naeem, S.; O' Neill, R.V.; Paruelo, J.; et al. The value of the world 's ecosystem services and natural capital. *Nature* **1997**, *387*, 253–260. [CrossRef]

26. de Groot, R.; Brander, L.; van der Ploeg, S.; Costanza, R.; Bernard, F.; Braat, L.; Christie, M.; Crossman, N.; Ghermandi, A.; Hein, L.; et al. Global estimates of the value of ecosystems and their services in monetary units. *Ecosyst. Serv.* **2012**, *1*, 50–61. [CrossRef]

27. Christie, M.; Fazey, I.; Cooper, R.; Hyde, T.; Kenter, J.O. An evaluation of monetary and non-monetary techniques for assessing the importance of biodiversity and ecosystem services to people in countries with developing economies. *Ecol. Econ.* **2012**, *83*, 67–78. [CrossRef]

28. World Bank Group. Carbon Pricing Dashboard. Available online: https://carbonpricingdashboard.worldbank.org/ (accessed on 27 December 2020).

29. Ma, X.H.; Ren, Z.Y.; Sun, G.N. The calculation and assessment to the values of air purification by vegetation in Xi' An City. *Chin. J. Eco-Agric.* **2004**, *12*, 180–182. (In Chinese)

30. Xie, G.-D.; Zhang, C.-X.; Zhang, L.-M.; Chen, W.-H.; Li, S.-M. Improvement of the Evaluation Method for Ecosystem Service Value Based on Per Unit Area. *J. Nat. Resour.* **2015**, *30*, 1243–1254. (In Chinese)

31. Ma, S.; He, F.; Tian, D.; Zou, D.; Yan, Z.; Yang, Y.; Zhou, T.; Huang, K.; Shen, H.; Jingyun Fang, J. Variations and determinants of carbon content in plants: A global synthesis. *Biogeosciences* **2018**, *15*, 693–702. [CrossRef]

32. Hu, L.; Zhang, W.; Guo, L.; Cheng, Y.; Li, J.; Li, K.; Zhu, Z.; Zhang, J.; Luo, S.; Cheng, L.; et al. Can the co-cultivation of rice and fish help sustain rice production? *Sci. Rep.* **2016**, *6*, 28728. [CrossRef]

33. Berg, H.; Ekman Söderholm, A.; Söderström, A.-S.; Tam, N.T. Recognizing wetland ecosystem services for sustainable rice farming in the Mekong Delta, Vietnam. *Sustain. Sci.* **2017**, *12*, 137–154. [CrossRef] [PubMed]

34. Arunrat, N.; Kongsurakan, P.; Sereenonchai, S.; Hatano, R. Soil organic carbon in sandy paddy fields of Northeast Thailand: A Review. *Agronomy* **2020**, *10*, 1061. [CrossRef]

35. Arunrat, N.; Sereenonchai, S.; Kongsurakan, P.; Hatano, R. Assessing soil organic carbon, soil nutrients and soil erodibility under terraced paddy fields and upland rice in Northern Thailand. *Agronomy* **2022**, *12*, 537. [CrossRef]

36. Nayak, A.K.; Shahid, M.; Nayak, A.D.; Dhal, B.; Moharana, K.C.; Mondal, B.; Tripathi, R.; Mohapatra, S.D.; Bhattacharyya, P.; Jambhulkar, N.N.; et al. Assessment of ecosystem services of rice farms in eastern India. *Ecol. Process.* **2019**, *8*, 35. [CrossRef]

37. Ren, W.; Hu, L.; Guo, L.; Zhang, J.; Tang, L.; Zhang, E.; Zhang, J.; Luo, S.; Tang, J.; Chen, X. Preservation of the genetic diversity of a local common carp in the agricultural heritage rice-fish system. *Proc. Natl. Acad. Sci. USA* **2018**, *115*, E546–E554. [CrossRef] [PubMed]

38. Wan, N.-F.; Li, S.-X.; Li, T.; Cavalieri, A.; Weiner, J.; Zheng, X.-Q.; Ji, X.-Y.; Zhang, J.-Q.; Zhang, H.-L.; Zhang, H.; et al. Ecological intensification of rice production through rice-fish co-culture. *J. Clean. Prod.* **2019**, *234*, 1002–1012. [CrossRef]

39. Hjalager, A.M. Agricultural diversification into tourism: Evidence of a European Community development programme. *Tour. Manag.* **1996**, *7*, 103–111. [CrossRef]

Article

Techno-Economic Assessment of an Office-Based Indoor Farming Unit

Jedrzej Cichocki, Moritz von Cossel * and Bastian Winkler

Biobased Resources in the Bioeconomy (340b), Institute of Crop Science, University of Hohenheim, Fruwirthstr. 23, 70599 Stuttgart, Germany
* Correspondence: moritz.cossel@uni-hohenheim.de

Abstract: Decentralized, smart indoor cultivation systems can produce herbs and vegetables for fresh and healthy daily nutrition of the urban population. This study assesses technical and resource requirements, productivity, and economic viability of the "Smart Office Farm" (SOF), based on a 5-week production cycle of curled lettuce, lolo rosso, pak choi and basil at three photosynthetic photon flux density (PPFD) levels using a randomized block design. The total fresh matter yield of consumable biomass of all crops was 2.5 kg m^{-2} with operating expenses (without labor costs) of EUR 53.14 kg^{-1}; more than twice as expensive compared to large-scale vertical farm and open-field cultivation. However, there is no need to add trade margins and transportation costs. The electricity supply to SOF is 73%, by far the largest contributor to operational costs of office-based crop production. Energetic optimizations such as a more homogeneous PPFD distribution at the plant level, as well as adaptation of light quality and quantity to crop needs can increase the economic viability of such small indoor farms. With reduced production costs, urban indoor growing systems such as SOF can become a viable option for supporting fresh and healthy daily nutrition in urban environments.

Keywords: *Brassica rapa* L. ssp. Chinensis; *Lactuca sativa* L. var. cerbiata; *Lactuca sativa* L. var. Lollo Rosso; leafy greens; *Ocimum basilicum* L. var. Genovese; PPFD; sustainable intensification; techno-economic assessment; urban agriculture; indoor farm

Citation: Cichocki, J.; von Cossel, M.; Winkler, B. Techno-Economic Assessment of an Office-Based Indoor Farming Unit. *Agronomy* **2022**, *12*, 3182. https://doi.org/10.3390/agronomy12123182

Academic Editor: David W. Archer

Received: 12 October 2022
Accepted: 13 December 2022
Published: 15 December 2022

Publisher's Note: MDPI stays neutral with regard to jurisdictional claims in published maps and institutional affiliations.

1. Introduction

The number of people living on planet earth is expected to rise to 9.7 billion by 2050 [1] and the growing global population will require about 60% more food from 2007 to 2050 [2]. Additionally, urbanization is transforming our society, since more than half of the world's population now lives in cities. This proportion is expected to increase to 68% by 2050 [3].

These figures shape the agricultural sector and the way we feed our cities. To date, our cities are highly dependent on food imports and linear value chains with food being imported, consumed and waste moved out. However, sustainable cities need to close the open-loop system [4] and to design circular and decentralized value chains.

Urban agriculture is becoming increasingly recognized as a viable option to support global food security in times of climate change, resource constraints and growing food demand. Highly productive, resource-efficient urban cultivation systems can play a key role over the next decade in turning cities sustainable [5–7].

As defined by [8] *"Most broadly, urban agriculture refers to growing and raising food crops and animals in an urban setting to feed local populations"*. It includes indoor and outdoor agriculture and ranges from food grown on the balcony (Bio-Balkon [9], Geco-Gardens [10]), to large production facilities such as AeroFarms in the US [11].

Urban agriculture has multiple environmental, economic and social sustainability benefits to both developing and post-industrial cities [12]. *"Furthermore, it contributes to ten key societal challenges of urbanization: climate change, food security, biodiversity and ecosystem services, agricultural intensification, resource efficiency, urban renewal and regeneration, land management, public health, social cohesion, and economic growth"* [13].

According to the Worldwatch Institute, globally, about 15–20% of food is produced today in urban and peri-urban areas [13]. For example, urban cultivation accounts for 5–10% of total noncereal crop production, which has a total market volume of USD 1509 billion [6]. In addition to food, urban agriculture provides valuable ecosystem services through the creation of new green areas. These can, for example, reduce heat islands and mitigate storm water impacts, thus increasing cities' resilience to climate change impacts [14]. Globally, the ecosystem services provided by urban agriculture are worth USD 88 to 164 billion, increasing the well-being of urban inhabitants [6].

Urban agriculture can transform the typical structure of rural food production and urban food consumption towards more decentralized production systems [7] that increase local access to and availability of food, two crucial factors of food security [15]. For improving the supply of fresh fruits and vegetables, in particular, the crops that are deemed suitable for urban production are those consumed as fresh functional foods supporting fresh and healthy daily nutrition [16,17].

As a result of the change from traditional linear supply chains to short, local and circular supply chains based on urban production, a new food distribution infrastructure in urban areas can emerge with lower food miles and high resource recovery. For instance, Pirog and Benjamin [18] show that conventional US broccoli travels 92 times further than local broccoli and average vegetables travel 27 times longer. The food in US supermarkets has travelled on average about 2000 km between the production and consumption site, releasing between 0.8 and 1.9 kg of CO_2 Mg^{-1} km^{-1}. The production of Berlin's food is 72% from domestic land, 7% in the EU and 21% outside the EU [19,20]. Ackerman et al. [14] found that the decrease in food miles can reduce food waste through a reduction in spoilage during transport and storage, which is another important benefit of decentralized urban food systems. The decrease in food waste directly increases both resource-use and the energy-use efficiency of food production. In fact, resource circulation in cities through urban agriculture is an important enabler for the transition towards a sustainable urban bioeconomy [21].

Decentralized, smart and automated production systems can be important for producing vegetables and herbs at the place of consumption, e.g., at offices for providing fresh and healthy food for staff members. For the production of herbs and vegetables inside offices, artificial cultivation conditions need to be created. This requires several inputs, including lights, substrate, nutrients, water and energy as well as a careful monitoring and management of the cultivation conditions. In rainfed agriculture, light, temperature and humidity are controlled by nature and provided at no cost. Therefore, the questions arise how productive, resource efficient and economically viable is food production in small-scale urban indoor farming units?

To investigate this research question, a techno-economic assessment of vegetable and herb production in a small indoor farming unit, designed to automatically produce leafy greens in offices, was conducted. For the "Smart Office Farm" (SOF), the technical requirements, resource use, productivity and economic viability were analyzed, based on a 5-week production cycle of curled lettuce (*Lactuca sativa* L. var. Cerbiata), lolo rosso (*Lactuca sativa* L. var. Lollo Rosso), pak choi (*Brassica rapa* L. ssp. Chinensis) and basil (*Ocimum basilicum* L. var. Genovese) under different light conditions.

2. Materials and Methods

The leafy greens were cultivated from 8 July 2020 to 12 August 2020 at the "Smart Office Farm" (SOF{ XE "SOF" \t "Smart office farm" }), located at the office of the urban farming Start-Up Farmee GmbH, in Stuttgart.

2.1. Technical Setup of the Smart Office Farm

The SOF is designed for the automated production of herbs and leafy greens in offices (Figure 1). It has dimensions of 1.91 m × 1.38 m × 0.75 m, covering a total surface area of

0.79 m^2. The SOF has three production levels (PL{ XE "PL" \t "Production level" }) for the crops, which can be switched on and off individually.

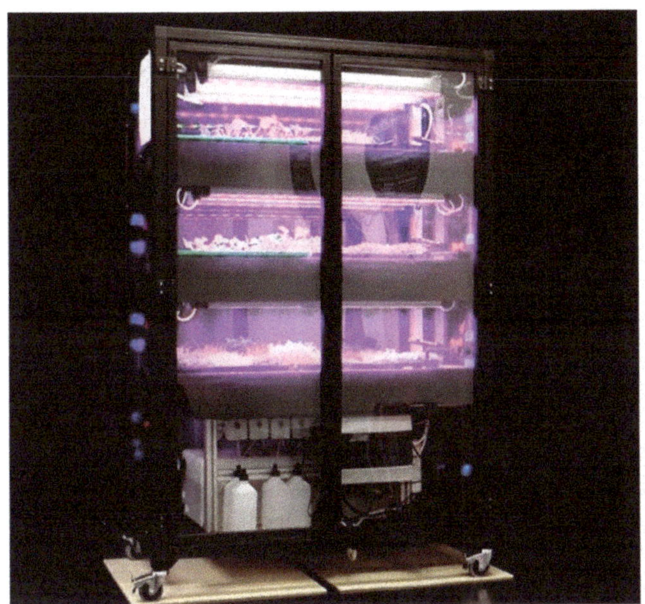 

Figure 1. Smart Office Farm (Photos: Farmee GmbH, 2020 (**left**) and Cichocki, 2020 (**right**)).

One PL (Figure 2) has a total cultivation area of 0.62 m^2 (0.98 m × 0.63 m), divided into four cultivation trays of 0.32 m × 0.49 m with 24 planting spaces each. This results in 96 planting spaces per PL. In total, 288 plants can be cultivated on 1.86 m^2, making efficient use of the often-limited surface areas in offices.

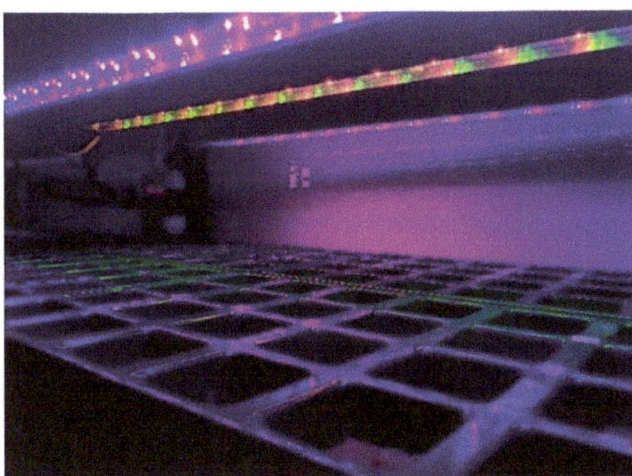 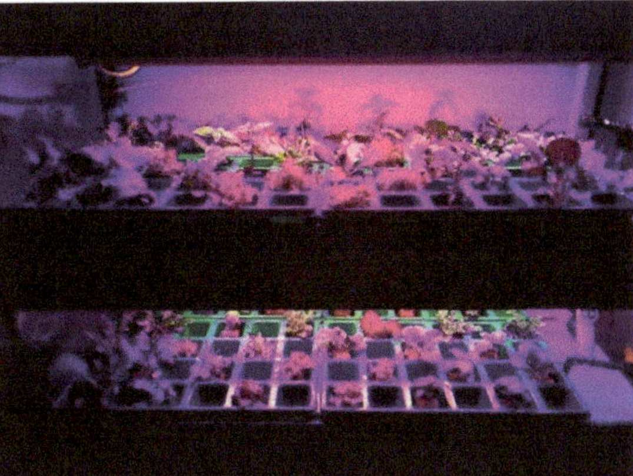

Figure 2. LED units and planting trays (**left**) and an overview of PL about two weeks after germination (**right**) (Photos: Cichocki, 2020).

For this experiment, the PLs were numbered starting from the bottom with PL 1. All production levels share one 50 L water tank. Therefore, the same nutrient solution is continuously recirculated between the three PLs and the tank. In between, it is sterilized with UV{ XE "UV" \t "Ultraviolet" } light. The irrigation of each level of the SOF can be set as an Ebb and Flow system or as Nutrient Film technique (NFT{ XE "NFT" \t "Nutrient film technique" }). For this study, the Ebb and Flow mode was chosen, because of a lower energy demand (compared to running the water pump 24/7 in NFT systems) and the higher resilience of this cultivation method against operational or technical problems. The nutrient solution was pumped every 6 h for 2 min to the PLs from where it drained within 10 min back into the water tank.

Each PL is equipped with three dimmable LED{ XE "LED" \t "Light emmiting diode" } units, with 32 blue light (450 nm) OSLON® Square, 32 red light (660 nm) OSLON® Square, 32 green light (520 nm) OSLON® SSL 80 and 32 far red (730 nm) OSLON® SSL 120 LEDs.

The cultivation conditions in the SOF are monitored with sensors on each production level and the water tank (Figure 3). At these four places, sensors measure temperature, relative humidity of the air and pH{ XE "pH" \t "Potential of hydrogen" }, electrical conductivity (EC{ XE "EC" \t "Electrical conductivity" }) and dissolved oxygen in the nutrient solution. All parameters, except dissolved oxygen, were measured and recorded throughout the experiment.

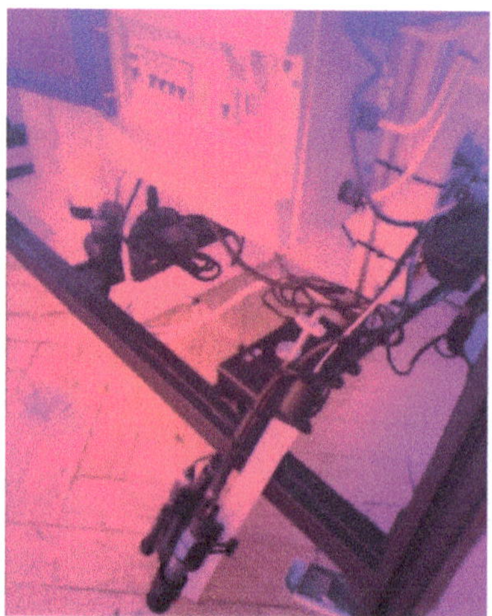

Figure 3. Monitoring unit, sensors for temperature, humidity, pH, EC and the irrigation control cabinet (**left**) and the user interface (**right**) (Photos: Cichocki 2020).

2.2. Experimental Design

In this study three leafy vegetables and one herb species were cultivated in polyculture, since it was assumed that the intended users in an office would prefer to have several crops at the same time, in order to account for different tastes.

For the experiment, only the two lower PLs were planted, because previous cultivation experiments showed that the water and nutrient distribution was more stable when only operating two PLs and therefore yielding more reliable results. However, to still be able to assess the whole SOF, the productivity of the third level was calculated, based on a linear extrapolation of the in- and outputs of the two cultivated levels. All data shown here account for three layers, to assess the techno-economic feasibility of the overall indoor farming unit.

The experiment in this study was conducted based on a randomized block design of the four crops at three different photosynthetic photon flux density (PPFD{ XE "PPFD" \t "*Photosynthetic photon flux density*" }) (μmol m^{-2} s^{-1}) levels. The artificial lighting of the SOF was analysed more closely, because energy is a major input and thus a main contributor to the operating expenses of indoor farming units [22]. The PPFD at crop surface level was determined in a preliminary experiment using a PPFD meter (Quantum meter MQ-200,

Apogee). This measurement revealed considerable differences in PPFD level at plant level as shown in Figure 4.

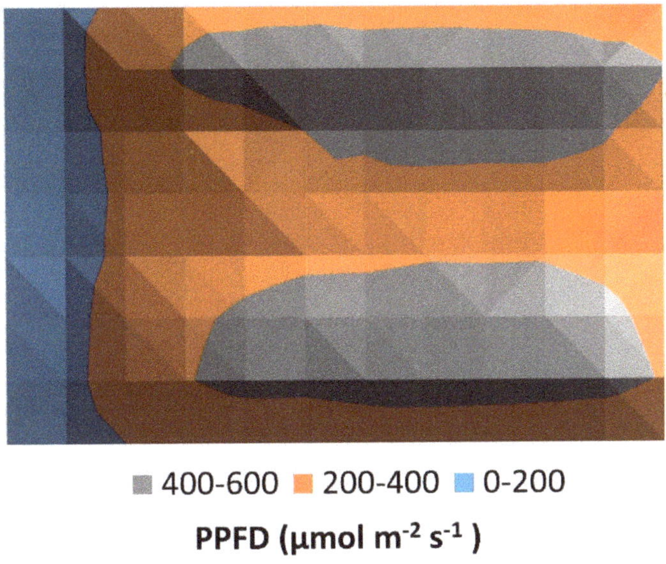

Figure 4. Photosynthetic photon flux density (PPFD) distribution at production level (μmol m^{-2} s^{-1}).

Taking this into account, three blocks with three different PPFD levels (0–200/200–400/400–600 μmol m^{-2} s^{-1}) were established to get a more adequate picture of light as a key factor for plant growth. Consequently, each of the four different crops was cultivated under each PPFD level in two repetitions. In order to maximize the light intensity for the crops, the dimmable LED units were set to 100% PPFD{ XE "PPFD" \t "Photosynthetic photon flux density" }.

The first block had a low PPFD with an average of 185.2 μmol m^{-2} s^{-1}. The second had a medium PPFD with an average of 355.7 μmol m^{-2} s^{-1} and the third had a high PPFD with an average of 466.4 μmol m^{-2} s^{-1}. The total average of PPFD was 335.8 μmol m^{-2} s^{-1} (Table 1).

Table 1. Three blocks with three different PPFD and DLI.

Mean	PPFD (μmol m^{-2} s^{-1})	DLI (mol m^{-2} d^{-1})
Block 1	185.19	10.6
Block 2	355.69	20.5
Block 3	466.44	26.6
Total	**335.77**	**19.3**

For this experiment, all LED lights were switched on for 16 h per day, which resulted in an average daily light integral (DLI{ XE "DLI" \t "Daily light integral" }) of 19.3 mol m^{-2} d^{-1}. The DLI is a useful unit when describing the light environment of plants, since it illustrates the rate of photosynthetic active radiation (PAR{ XE "PAR" \t "Photosynthetic active radiation" }) distributed to the plants over a 24 h period [23]. The first block had a DLI of 10.6 mol m^{-2} d^{-1}. The second block had a DLI of 20.5 mol m^{-2} d^{-1} and the third block had a DLI of 26.6 mol m^{-2} d^{-1}. All light conditions are summarized in Table 1.

All planting spaces were sorted by PPFD and divided into three blocks with the same number of planting spaces. The final distribution of plants per PL after randomisation and block construction is shown in Figure 5.

	A	B	C	D	E	F	G	H	I	J	K	L
1		Lolo		Curled		Basil		Basil		Basil		Lolo
2	Curled		Curled		Basil		Lolo		Pak		Pak	
3		Curled		Lolo		Basil		Pak		Pak		Pak
4	Pak		Basil		Lolo		Pak		Lolo		Curled	
5		Curled		Basil		Curled		Lolo		Curled		Curled
6	Curled		Lolo		Curled		Pak		Pak		Basil	
7		Pak		Curled		Lolo		Lolo		Basil		Pak
8	Pak		Basil		Lolo		Lolo		Basil		Basil	

Figure 5. Planting plan of the production levels based on a randomized block design with three photosynthetic photon flux density (PPFD) level: 185.2 µmol m^{-2} s^{-1} (blue); 355.7 µmol m^{-2} s^{-1} (orange); 466.4 µmol m^{-2} s^{-1} (grey). Every square represents one planting space (Curled: Curled lettuce, Lolo: lolo rosso, Pak: pak choi and Basil: basil).

Subsequently, four crops were cultivated on 36 mm Grodan rockwool blocks. For each PL, 48 substrate blocks were used, which resulted in a total usage of 144 planting spaces. The planting spaces marked in yellow in Figure 5 were not planted, in order to provide space for crop development.

Curled lettuce, lolo rosso and pak choi were cultivated as typical leafy greens and basil was the selected herb. Lettuce was chosen because of its fast growth and because it is cultivated worldwide, making it one of the most consumed leafy vegetables [24]. Basil was chosen because it can be found in most supermarkets and has been used as a spice and medicinal plant for ages [25]. A production cycle of five weeks was chosen from sowing till harvest, which is typical for lettuce and basil in indoor environments. In total, 12 plants of each crop were grown per production level at different PPFD levels. In order to minimize loses of productivity and because the germination rate in the faming unit was not known, 2–3 seeds of curled lettuce, lolo rosso and pak choi were sown per substrate block. For basil, 3–5 seeds per substrate block were sown. All seeds were sown by hand directly into wet rockwool plugs and subsequently transferred into the SOF according to the cultivation plan (Figure 5).

The crops were cultivated using municipal tap water. The pH and the electrical conductivity (EC) were adjusted at the beginning of the experiment and during the experiment about once a week. The pH was maintained between 5.5 and 6, because nutrients are optimally available [26,27]. During the experiment, however, the pH ranged between 4.8 and 6.8 due to manual adjustment using pH-up (nitric and phosphoric acid) and pH-down (potassium carbonate and silicate) (Terra Aquatica from General Hydroponics Europe). The EC value in the nutrient solution was steadily increased after the germination phase (about 5 days) to reach the crops' demands, allowing for a fast production cycle and the intended consumption as babyleaf lettuce. A three-component fertilizer was used with the following N-P-K concentration: Remo Nutrients 'Grow' (2-3-5), 'Magnifical' (3-0-0) and 'Micro' (3-0-3) [28]. Initially an EC of 800 µS cm^{-1} was chosen for the principal growth stage 1, which was terminated after 6–9 leaves per plant were developed. After 21 days, the EC was raised to 1300 µS cm^{-1}. It ranged from 1050–1300 µS cm^{-1} throughout the experiment. Regular measurements showed that the temperature always remained between 28–30 °C in PL two and 26–28 °C in PL one. Air humidity was 50–60% in PL two and 80–90% in PL one.

2.3. Techno-Economic Analysis

The techno-economic analysis performed here consists of: (i) a material flow analysis, following the approach of [29], and (ii) economic analysis following [30].

2.3.1. Material Flow Analysis

For the material flow analysis, all inputs that entered the SOF from planting stage to harvest of the crops were measured and recorded. These parameters included the amount of substrate, seeds, water, fertilizer, pH buffering solution and the energy for lights, pumps and monitoring devices. During the experiment, the water usage was measured by recoding the water level in the tank before and after filling. The energy consumption was measured with an electric meter (Energy Check 3000, Voltcraft, Hirschau, Germany). The weight and number of substrate blocks was measured with a scale (AMIR, DE-KA6). Fertilizer and pH buffer usage was documented over the whole period.

At harvest, the total fresh matter (FM{ XE "FM" \t "Fresh matter" }) yield of consumable biomass was considered as output. However, the FM{ XE "FM" \t "Fresh matter" } yield of all four crops produced in the two PLs with three PPFD levels was measured in order to analyze the yield per PL and PPFD level. This resulted in 24 different biomass samples at harvest. The biomass samples were dried in a drying oven (VTU 125/200, Weiss Technik GmbH, Reiskirchen, Germany) at 60 °C for 24 h to obtain the dry matter (DM) yield as well.

The productivity of PL 1 and PL 2 was analyzed and checked for significant differences due to different cultivation conditions with a paired t-test in Microsoft Office XP Excel.

Furthermore, the theoretical productivity of the farming unit was determined in four different scenarios assuming that one of the crops would have been cultivated in monoculture.

2.3.2. Economic Analysis

For the economic analysis, all prices for the input factors from the material flow analysis were determined and the total operating expenses (OPEX) of the SOF were calculated. This allowed for the determination of the production costs kg^{-1} consumable biomass.

In order to account for the possibility of operating an indoor farming unit on renewable energy two scenarios were made, distinguishing between conventional (EUR 0.2925 kWh^{-1}) and renewable (EUR 0.2768 kWh^{-1}) electricity prices. Subsequently, all cost factors were summarized and ranked based on their relative contribution to the total operating expenses.

3. Results

First, the results of the material flow assessment are presented and second the economical assessment.

3.1. Material Flow Assessment

First, the inputs of the SOF were determined in Section 3.1.1, followed by the outputs in Section 3.1.2 An overview of the results of techno-economic assessment of the SOF is provided in Figure 6.

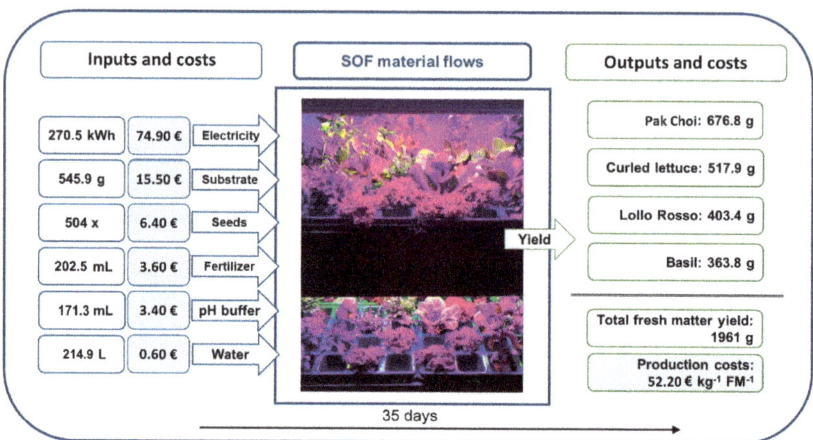

Figure 6. Overview of the material flow and production cost assessment, summarizing all inputs on the left side and all outputs on the right side (extrapolated values) (Photo: Winkler 2020).

3.1.1. Input

The inputs include seeds, substrate, fertilizer, pH-buffer, water and energy. The amounts used were documented along with the prices (Figure 6).

Seed Usage

As shown in Table 2, 108 seeds of curled lettuce, lolo rosso and pak choi were used. For basil 180 seeds were used. Collectively, seed costs were EUR 4.24 (Table 2).

Table 2. Overview of the SOF inputs (number of crops and seeds).

Plant	Seed Usage	Price (EUR)
Curled lettuce	72	1.35
Lollo rosso	72	1.35
Pak choi	72	0.60
Basil	120	0.95
Total		4.24

Fertilizer Usage

Over a period of 5 weeks, 202.5 mL fertilizer was used. The total fertilizer costs account for EUR 2.39 in the experiment (Table 3).

Table 3. SOF inputs: Overview of types, amounts and costs of the fertilizers used in this study.

Fertilizer	Costs (EUR L^{-1})	Usage (mL)	Price (EUR)
Remo 'grow'	13.30	45	0.60
Remo 'magnifical'	19.86	45	0.89
Remo 'Micro'	19.85	45	0.89
Total			2.39

The combined content of added nutrients is summarized in Table 4. The remaining nutrients after production were not assessed.

Table 4. SOF inputs: Overview of types, amounts and dosages of nutrients applied in this study [28].

Element	(mg Trial^{-1})	(g m^{-2})
Nitrogen (N)	2337.1	2.947
Phosphorus (P)	615.4	0.776
Potassium (K)	1680.3	2.119
Magnesium (Mg)	1552.5	1.958
Calcium (Ca)	3375.0	4.256
Boron (B)	27.0	0.034
Copper (Cu)	33.8	0.043
Iron (Fe)	681.8	0.860
Manganese (Mn)	67.5	0.085
Molybdenum (Mo)	0.3	0.0004
Zinc (Zn)	33.8	0.043

pH-Buffer Usage

Over the production cycle, 103.06 mL pH-down and 11.16 mL pH-up buffering solution were used. The costs of buffering solutions were EUR 2.25 for the whole production cycle (Table 5).

Table 5. SOF inputs: pH-buffer used in this study and their implied costs.

pH-Buffer	Usage (mL)	Costs (EUR L^{-1})	Price (EUR)
pH down	103.06	20.90	2.15
pH up	11.16	8.90	0.10
Total			2.25

Water Usage

In total, 143.28 L of water were consumed (Table 6). The water usage over the production cycle is shown as weekly cumulated values in Figure 7. In the beginning, the water usage was low and increased substantially later on (Figure 7). The total average water consumption was 4.1 L d^{-1} at a temperature of 28–30 °C maintained in PL two and 26–28 °C in PL one and air humidity of 50–60% in PL two and 80–90% in PL one.

Table 6. SOF inputs: application dates, amounts and costs of irrigation.

Date	Water Usage (L)	Costs (EUR L^{-1})
08.07.20	36	0.003
15.07.20	7.92	0.003
22.07.20	11.88	0.003
29.07.20	15.48	0.003
05.08.20	33.12	0.003
12.08.20	38.88	0.003
Total	143.28	0.29

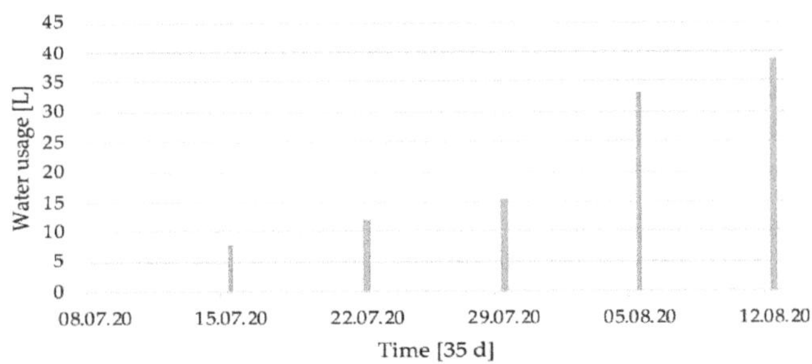

Figure 7. SOF inputs: weekly cumulated water consumption over 35 days.

Energy Consumption

The total energy consumption of the SOF (extrapolated to all three PL) for one production cycle of five weeks was 270.5 kWh or 973.8 MJ. This results in an energy input per produced kilogram biomass of 496.4 MJ kg^{-1} FM or 2977 MJ kg^{-1} DM (considering electricity inputs only, while omitting energy required for material and fertilizer production).

3.1.2. Output

First, the total consumable biomass yield (fresh matter and dry matter) was analysed for the entire SOF surface and then calculated on a square meter basis. Second, the fresh matter and dry matter yields of the different crops were analysed and compared for the three PPFD levels. Third, the biomass yields of the two PL were statistically compared. Additionally, a scenario analysis was performed to calculate the potential yields for the four individual crops of this experiment in the SOF.

The total fresh matter yield of the four crops, extrapolated to the entire SOF, was 1961.8 g per 0.79 m^2. This converts to 2473.9 g m^{-2}. Pak choi had the highest potential yield with 676.8 g and basil had the lowest potential yield with 363.8 g, which is 53.8% less (Table 7).

Table 7. Total yield of the crops investigated in this study.

Crop	Yield (g)
Curled lettuce	517.9
Lollo rosso	403.4
Pak choi	676.8
Basil	363.8
Total	1961.8

The individual yields of the four crops under the three PPFD blocks are summarized in Table 8. The lowest yield was measured for basil (24.1 g) on PL 1 at low PPFD level. The highest yield was obtained from pak choi with 105.9 g at intermediate PPFD level on PL two. Looking at PL two, curled lettuce and lolo rosso grew better with increasing PPFD level. On PL one, lolo rosso yield increased slightly the higher the PPFD was, while the yield of curled lettuce decreased with stronger PPFD. Pak choi had a peak at the intermediate PPFD again and basil yield was increasing with higher PPFD.

Table 8. Yield per PL and PPFD block.

Block	Crop	Yield PL 1	Yield PL 2
		[g FM]	[g FM]
	Curled lettuce	48.2	52.1
Block 1 (Low PPFD)	Lolo rosso	30.4	42.2
	Pak choi	52.5	76.8
	Basil	24.1	36.0
	Curled lettuce	45.3	73.2
Block 2 (Intermediate PPFD)	Lolo rosso	33.9	57.7
	Pak choi	69.8	105.9
	Basil	32.0	54.6
	Curled lettuce	43.1	83.3
Block 3 (High PPFD)	Lolo rosso	43.7	61.0
	Pak choi	65.3	80.8
	Basil	42.0	53.8
Total (2 PL)		530.3	777.5

Furthermore, dry matter (DM) yields have been analyzed (Figure 8). Pak choi produced the highest DM yield and lolo rosso the lowest. The total DM yield of the SOF was

172.3 g DM. All DM yields increased with higher PPFD, except pak choi, which had the highest DM at intermediate PPFD level.

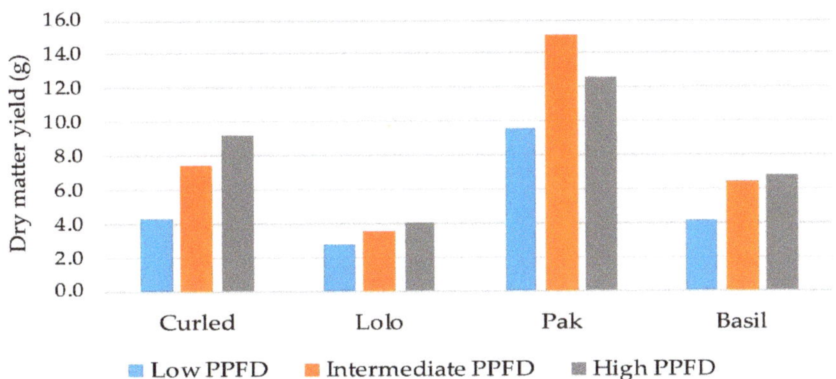

Figure 8. Dry matter (DM) yield of the crops under different photosynthetic photon flux density (PPFD) levels.

Subsequently, the productivity of both production levels was compared in order to assess whether the growth conditions were similar on the different PL. The total yield for the two PL was 1307.9 g FM per 0.79 m². However, PL one yielded 530.3 g FM, while PL two had a 46.6% higher yield with 777.5 g FM. The yield differences between the four crops on the two production levels at varying PPFD levels are displayed in Figure 9. It shows that PL 2 always produced higher yields at all PPFD levels.

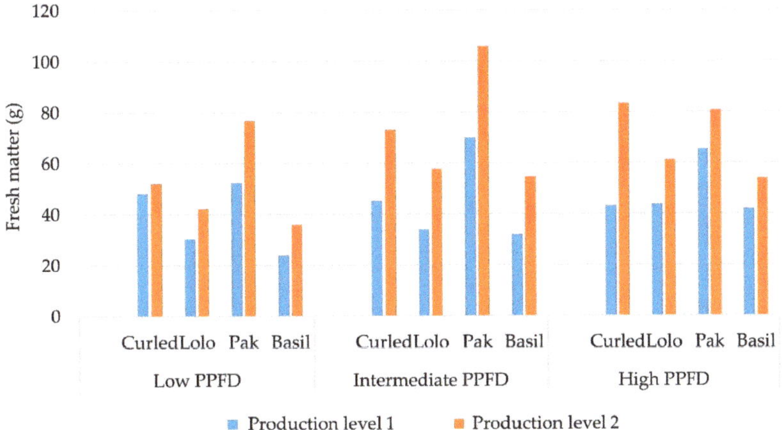

Figure 9. Fresh matter yield of the crops under different production levels and photosynthetic photon flux density (PPFD) levels.

A two-tailed paired *t*-test resulted in the rejection of the assumption that both production levels delivered the same output since P was higher than alpha (0.05). The results of the two-tailed paired t-test are shown in Table 9. The comparison revealed that the cultivation conditions varied significantly between the two layers.

Table 9. Results of two-tailed paired *t*-test ($p < 0.05$).

	PL 1	PL 2
Mean	44.19	64.80
Variance	186.05	390.27
Observations	12	12
Pearson Correlation	0.86	
df	11	
t Stat	−6.69	
P (T ≤ *t*) two-tail	3.42×10^{-5}	
t Critical two-tail	2.20	

In addition, the yield of the individual crops was extrapolated to the whole farm, assuming that only this crop would have been cultivated. This resulted in four scenarios revealing the crop with the highest potential biomass yield (Table 10). Scenario three (pak choi production only) would attain a FM yield of 2707 g, which is the highest potential yield. On the contrary, scenario four (basil production only) would result in 1455.2 g, the lowest yield of all crops.

Table 10. Scenario analysis: Calculated productivity of the individual crop.

Scenario	Crop	Yield (g)	Yield (g m^{-2})
1	Curled lettuce	2071.4	2612.1
2	Lolo rosso	1613.5	2034.7
3	Pak choi	2707.0	3413.7
4	Basil	1455.2	1835.1

For comparison of the biomass yield of the SOF with yield levels of other in- and outdoor biomass production systems, the productivity of the SOF was converted to Mg FM per hectare and year. The SOF has a total surface area of 0.79 m^2 (1.38 m × 0.7 m). The production cycle of five weeks would potentially allow for up to 10 harvests.

The yield levels on hectare basis are shown in Table 11. A total FM yield of 257 Mg FM ha^{-1} year^{-1} is potentially achievable with the four crops used in this study. Considering the cultivation of pak choi (scenario 3) only 355 Mg FM ha^{-1} year^{-1} would potentially be possible. The production of only basil (scenario 4) would lead to the lowest potential yield of 191 Mg FM ha^{-1} year^{-1}.

Table 11. Scenario analysis: Productivity in g m^{-2} 5 weeks^{-1} and Mg FM ha^{-1} year^{-1}.

Scenario	Productivity during Observation Period (g m^{-2} 5 Weeks^{-1})	Calculated Annual Productivity (Mg FM ha^{-1} Year^{-1})
1	2612.1	272
2	2034.7	212
3	3413.7	355
4	1835.0	191
Mixed	2473.9	257

3.2. Economic Assessment

The total input costs of the SOF were summed up to determine the price per kg consumable biomass produced in the SOF (Figure 6). The operating expenses (OPEX) for all production factors (without labor) and inputs are summarized and ranked based on their relative contribution to the total OPEX (Tables 12 and 13). Energy accounts for more than 70% of the OPEX. Considering this, the results are shown with the electricity price of electricity from renewable sources only (Table 12) and the conventional electricity mix of Germany (Table 13).

Table 12. Economical assessment (renewable energy mix): production factors, costs and their relative contribution to the total costs.

Production Factors	Costs (EUR)	Share of Total Costs (%)
Electricity	74.9	71.8
Substrate	15.5	14.8
Seeds	6.4	6.1
Fertilizer	3.6	3.4
pH-buffer	3.4	3.2
Water	0.3	0.6
Total	104.3	100

Table 13. Economical assessment (conventional energy mix): production factors, costs and their relative contribution to the total costs.

Production Factors	Costs (EUR)	Share of Total Costs (%)
Electricity	79.1	73.1
Substrate	15.5	14.3
Seeds	6.4	5.9
Fertilizer	3.6	3.3
pH-buffer	3.4	3.1
Water	0.3	0.3
Total	108.2	100

Fixed costs and labor costs were neglected because the latter can vary considerably and the intended users in offices would perform the maintenance as a hobby during their breaks. The fixed costs can hardly be estimated for an SOF at prototype stage.

When using renewable energy, total OPEX result in EUR 104.3 per five-week production cycle (Table 13). The main input was renewable energy (270.5 kWh) with EUR 74.9, representing 71.8% of the total costs. Rockwool as cultivation substrate accounts for EUR 15.5 and contributes the second highest share (14.8%) to the production costs, followed by seeds, fertilizer, and pH buffer. Water usage had the lowest impact on total costs, with EUR 0.3 and 0.6% of total costs.

With conventional energy, the total production costs were EUR 108.23 per five-week production cycle, with energy accounting for EUR 79.12 and 73.1% of the total production costs.

Overall, the cultivation of one kilogram of the four crops over five weeks results in total OPEX of EUR 53.14 when using renewable energy, while one kg of biomass grown with conventional energy costs EUR 55.17. In this case, the more sustainable renewable energy from the local power utility is also 3.7% cheaper than conventional electricity.

4. Discussion

The techno-economic assessment of the SOF is discussed with respect to (i) the technical setup and the design of the SOF (Section 4.1), (ii) the implications for cultivation management (Section 4.2) and (iii) the resource use and the productivity of the SOF compared to traditional field, greenhouse production and professional vertical farming units (Section 4.3).

4.1. Productivity

In the following section, the mixed scenario will be compared to the average lettuce production worldwide, to the average field and greenhouse lettuce production in Germany and the average field production in Baden Württemberg. Such a comparison with yields under field conditions is important to put the results of the indoor farming approach into perspective. A similar approach was also carried out by Wittmann et al. [31], who studied an indoor vertical farming method for marjoram production. Additionally, the lettuce yield will be compared with a greenhouse in Switzerland and a vertical farm feasibility study of the German Aerospace Centre (DLR). Finally, published yields of existing vertical farms such as Aerofarms, Plenty, Infarm and Skygreens were regarded.

The yield of the SOF in this study was 2473.9 g m^{-2} 5 weeks^{-1}. Since the SOF can produce all year long, 257 Mg FM ha^{-1} year^{-1} of mixed crops are potentially possible. In comparison, the mean yield for lettuce production worldwide in the year 2017 was 21.9 Mg FM ha^{-1} year^{-1} and the mean yield of lettuce produced in Germany in the year 2019, under field conditions, was 25.8 Mg ha^{-1} for lolo rosso, 25.5 Mg ha^{-1} for curled lettuce and 33.3 Mg ha^{-1} for cabbage lettuce [32]. The mean yield of romaine lettuce (*Lactuca sativa* var. longifolia) was 36.1 Mg ha^{-1}, while cabbage lettuce (*Lactuca sativa* var. capitata L.) yield was 42.6 Mg ha^{-1} under field conditions in Baden Württemberg (Germany) in 2019 [33,34].

In a greenhouse in Switzerland, Marton [35] found that cabbage lettuce yield was 48.1 Mg ha^{-1}. Zeidler et al. [36] calculated potential lettuce yields in a feasibility study of the vertical farm "EDEN". They found that a yield of 6436.2 g m^{-2} 5 weeks^{-1} is possible, which results in 669.4 Mg FM ha^{-1} year^{-1} with a minimum price of EUR 12.5 kg^{-1} of biomass.

Aerofarms in the US, one of the leading vertical farming companies, for example, says that they can produce 390-times more than conventional agriculture [37] and Skygreens in Singapore claim a 10-times higher yield in comparison to conventional agriculture [38]. Infarm, a Berlin based vertical farming company, mentioned that 400 times higher yields are possible [39] and Plenty, a San Francisco based vertical farming company, claims that 350 times higher yields are possible [34]. The mean world production of lettuce was used as a core value and multiplied with the claims of the vertical farming companies to give an overview. Toledano et al. [40] estimates that current market prices for one kg of leafy greens are around USD 33 for vertically-grown produce. When comparing the productivity, it is shown that the mean lettuce production has the lowest productivity and that Infarm has the highest claimed productivity with 8760 Mg FM ha^{-1} year^{-1} (Figure 10). Thus, field conditions are less productive than greenhouse conditions and vertical farming conditions are even more productive.

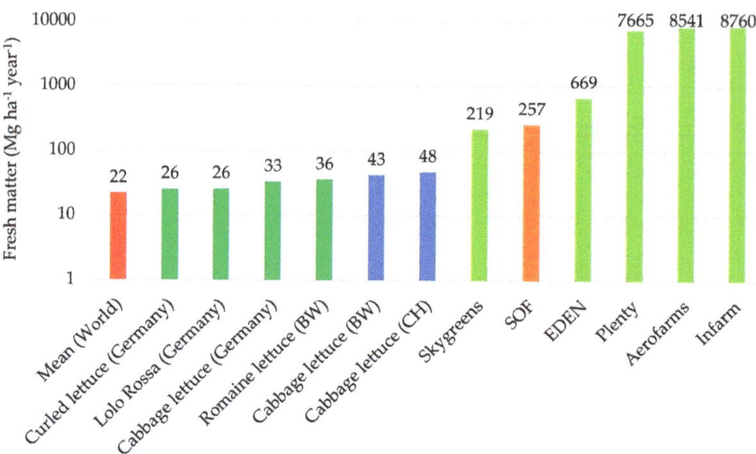

Figure 10. Comparison of yield; Mean World (red); field conditions (green); greenhouse (blue); SOF (orange); vertical farming companies (green).

The SOF has a smaller productivity than bigger farms since effects of scale are small, management is less efficient and resource use is high, making the production not economical viable at the moment. However, on the other hand, this trial implied only one production cycle and the cultivation measures and technical aspects of the farming unit can be improved with increasing experience with the SOF.

The agricultural sector could become more decentralized if similar farming units would be installed in many places. However, for the efficient use of such indoor farming units, they might need more plant production knowledge and the distribution of produced crops would need to be managed efficiently. This is a call for some kind of digital platform, which manages data, processes and resource flows and professionals, which can support and educate producers to make production economically viable.

From an economic point of view, one kg of biomass produced in the SOF costs EUR 53.1 and thus was more expensive than reported by Zeidler et al. [36] (EUR 12.5) and Toledano et al. [40] (USD 33). This is due to the rather low productivity compared to commercial indoor cultivation units (e.g., Infarm) and large-scale vertical farms (e.g., Plenty, Aerofarms). A closer look at the technical setup of the SOF and the cultivation conditions of this study can provide explanations for these yield differences and strategies for optimization of the SOF.

4.2. Technical Aspects

The technical design of the SOF revealed significant differences in growth conditions between the two production levels with respect to irrigation and fertilization as well as temperature and relative humidity. Based on this, future improvements of the SOF have been elaborated with respect to irrigation and fertilization (Section 4.2.1) as well as climate control (Section 4.2.2) in order to optimize SOF productivity.

4.2.1. Irrigation

The irrigation and fertilization of crops in soilless cultivation systems determine crop productivity to a large extent. The irrigation setup determines the supply of the plants with water and nutrients [41,42]. Keeping the EC and pH levels at crop specific optima is highly important for achieving high productivity in hydroponics and even small misconfigurations in irrigation, fertilization and pH level can lead to the total failure of the crops [16]). Automated monitoring and control of these cultivation factors are thus highly important, especially when untrained people, such as, e.g., in offices, operate the cultivation system.

The significant productivity differences between the two production levels can be attributed to differences in the irrigation regime. One reason might be the too small volume of the water tank in relation to the planting space, which does not allow for continuous irrigation of both production levels.

A larger water tank volume would additionally increase the pH buffering capacity, as small pH changes have direct impacts on nutrient availability [42]. With every irrigation cycle, the quantity of water, the nutrient composition and the pH change because plants take up water and nutrients and return metabolic products. The larger the volume of the tank, the smaller the impact of changes on EC, pH and water. Sufficiently large water reservoirs are important in hydroponic crop cultivation, reducing the dependency on the exact dosage of pH-buffering solution and fertilizer, making the amplitude of changes of EC, pH and water amount smaller and improving crop cultivation conditions [43]. This could improve plant health and increase biomass yields and decrease the maintenance requirements of indoor farming units [44].

Furthermore, it was observed that PL one received less water that PL two, which is located above PL one. The resulting differences in pump resistance due to different heights had to be carefully managed through manual valves in this study. A second improvement possibility would thus be a different configuration of water inflow and outflow. A solution would be a more precise dosage of water and nutrients to each PL, which can be conducted through modification of the valves for in and outflow. A larger water reservoir paired with flow sensor-based magnetic valves would allow for an automated control of the irrigation and fertilization regime of both production levels and can thus improve overall productivity of the SOF.

A third option would be a change in irrigation style, once a larger water reservoir would be installed. The current setup does only allow for very short ebb and flow cycles with pumping times of a maximum of two minutes. Maybe longer flow cycles or the shift towards nutrient film technique (NFT) would result in a higher and more even productivity of the two layers.

4.2.2. Climate Control

The differences in temperature and relative humidity may explain the varying productivity between the two production layers [45]. Plants are an integral part of the soil–plant–atmosphere continuum (SPAC{ XE "SPAC" \t "soil-plant-atmosphere continuum" }). The water gradient between the atmosphere of the office and the nutrient solution of the SOF is a main driver for plant transpiration and nutrient absorption [46]. Additionally, temperature in the office and thus inside the SOF (without climate control) plays a vital role in plant physiology [16]. Hence, the relative humidity of the SOF determines plant growth and may even cause phytopathogenic problems (see Section 4.3). For optimizing relative humidity and the microclimate at each production level of the SOF, a controlled airflow would be a solution. The additional installation of a sensor-based temperature and relative humidity control system with fans and defined temperature and humidity benchmarks adapted to crop requirements can improve productivity and resource-use efficiency.

Following the assumption that the use case for the SOF is an office with relatively low air humidity, an adjustable airflow from the inside of the farm to the outside of the building might decrease relative humidity inside the SOF (and the office room) and thus increase plant transpiration and nutrient uptake for faster growth. If this is not possible, e.g., due to high installation costs, sensor-controlled air circulation at each production level should be installed.

In addition, evaporation should be minimized to reduce water losses. Therefore, the planting trays should cover the production levels completely for as little direct contact with the air as possible.

The role of the CO_2 concentration has not been considered in this study, but should be investigated further, because beneficial outcomes for humans and plants are possible when CO_2 from the office (as a product of human transpiration) is provided as a resource to the crops [47].

Consequently, the productivity and resource-use efficiency of the SOF can be optimised by a larger water reservoir, enabling NFT cultivation, an improved water and nutrient distribution to the production levels and controlled temperature and relative humidity. The technical adaptations additionally facilitate crop production by untrained users, rendering the SOF a viable option for the automatized crop cultivation in offices.

4.3. Cultivation Measures

The applied cultivation measures in this experiment were determined by the technical setup of the SOF. Here, the implications of temperature and relative air humidity, pH and EC and the chosen production cycle are discussed in detail. Furthermore, adjustments in these parameters are discussed in order to optimize productivity.

4.3.1. Temperature and Relative Air Humidity

Following the discussion about the technical setup for climate control in Section 4.2.2, here, the actual values of the temperature and relative humidity of this study are examined and discussed in detail. Regular measurements of the installed sensors showed that the temperature inside the SOF was always between 28–30 °C in PL two and 26–28 °C in PL one. Relative air humidity was 50–60% in PL two and 80–90% in PL one.

Ahmed et al. [48] suggest a temperature of 22–25 °C during the light period and 70–80% relative air humidity to be optimal for lettuce cultivation. The differences in temperature and relative air humidity can partly explain the yield different between the two PLs [45], in addition to the difference in received water and nutrients, discussed in 4.2.1.

The suboptimal relative air humidity also led to symptoms of calcium deficiency, which were detected in young basil leaves of this study. Palzkill et al. [49] showed many years ago that high relative air humidity can cause calcium deficiency.

Overall, the suboptimal temperature and air humidity, which cannot be controlled so far, are subject of improvement through the installation of a sensor-controlled forced airflow technology. This measure can optimize productivity and thus decrease production costs of the SOF.

4.3.2. EC and pH

EC and pH of the nutrient solution are another two important factors for crop production. Essentially, the EC is a measure for concentration of plant nutrients (not about the composition), while the pH largely determines their plant availability.

Ding et al. [50] showed that the fresh and dry weight, and the leaf size of pak choi plants increased with higher EC values. Highest pak choi yields were achieved at an EC of 4800 μS cm^{-1} [50]. This reveals that the EC of this study (max. 1300 μS cm^{-1}) was too low for pak choi. This further indicates that pak choi is not suitable for polyculture with low-demanding lettuce and basil.

Walters and Currey [51] showed that for basil EC levels from 500 to 4000 μS cm^{-1} did not affect plant growth, but it was affected by increasing DLI from about 7 mol·m^{-2}·d^{-1} to about 15 mol·m^{-2}·d^{-1}. In this study, the DLI was even higher with 19.3 mol·m^{-2}·d^{-1}. Basil plants with low EC and high DLI showed no significant difference to plants with high EC and high DLI in terms of fresh matter [51]. Consequently, the EC and DLI set in the SOF were in the optimal range for basil cultivation.

The optimal pH range is generally assumed to be between 5.5 and 6 because nutrients are optimally available [26]. However, Gillespie et al. [27] showed that a pH of 4 in the nutrient solution can contribute to better plant health by reducing root rot severity in basil plants [27]. In the SOF, automated fertilizer and pH buffer dosing can keep pH and EC levels more stable and within the optimal ranges for particular crops. Despite the additional costs for this technical adaptation, a higher productivity of the SOF can be expected [30,52].

4.3.3. Production Cycle

A 'production cycle' refers to the time the crops are cultivated from seed or seedling to harvest and is determined by the producer and the desired product. It can range from a couple of days (e.g., for microgreens) to several months (e.g., herbs). The length of production cycle influences yield, morphology, content of nutrients, vitamins and other secondary metabolites of the crop parts and whether the plant is in its vegetative or generative phase. A multitude of social, ecological and economic factors determine the production cycle chosen by a producer.

For this experiment, it was assumed that a group of office workers has different food preferences, which can be met best with a polyculture and a rather short production cycle of five weeks till harvest. Thus, the harvest time of basil, curled lettuce, lolo rosso and pak choi were combined.

As a consequence, basil grew too tall, and some plants touched the LED unit resulting in burned tips of the uppermost leaves. In the trial, three to five basil seeds were sown per rockwool plug. Basil varieties with a more compact morphology can be used in the future if a polyculture will be cultivation in the SOF. Furthermore, a higher sowing density can probably reduce the size of the basil plants per substrate block. Lettuce plants were small in comparison to commercially grown lettuce. The average fresh matter of lolo rosso and curled lettuce was 71.5 g in the SOF, which is much less than the average selling weight of about 170 g in Germany [53]. For lettuce, a longer production cycle of about 10–14 days would have resulted in much higher yields, since lettuce plants were just at the beginning of their major growth phase at harvest time [54].

Consequently, the production cycle largely determines the productivity of indoor farming units, while the intended use of a crop (or certain parts or ingredients) in turn determines the production cycle and economic viability.

4.4. Resource Use

Urban areas have great potential for the increase in resource-use efficiency in terms of energy, materials (e.g., nutrients, water) and information and can at the same time contribute largely to a reduction in the impacts on environment and climate [55]. Decentralized food production through urban farming can increase the reuse and circulation of resources through the production of food, while providing improved access and availability to fresh and healthy food at low food miles [5,16,18]. High quality and nutritious food can be cultivated, e.g., in urban offices. Thereby, the consumer of food becomes the producer and values his own products potentially more than purchased food. This might also decrease the high share of food waste caused by consumers of up to 25% of the total food produced [56]. To turn cities sustainable through urban farming, resource flows need to circulate in the city [5]. In future, crop nutrients need to be derived from urban organic wastes. For example, Stoknes et al. [57] demonstrated that the vegetable and mushroom production from organic waste nutrient recovery is possible with the same or even higher yields. A shift from mineral-fertilizer-based hydroponics towards organic cultivation requires new approaches such as the terrabioponic cultivation currently used for outdoor urban gardening [58]. However, large scale food production based on nutrient and water recovery from urban organic wastes requires more research and development efforts in order to close the loops in resource flows [4].

4.4.1. Energy Consumption

Energy consumption determines the food production costs to the largest extent, with 73.1% of the total costs when conventional energy is used and 71.8% of the total costs when renewable energy is used. Therefore, energy can be seen as the main production factor in this experiment and in indoor urban farming generally [5,16]. The comparison of the renewable and conventional energy scenarios showed a small difference of 3.7% in this study. For improving the sustainability of urban indoor farming, the decrease in the energy consumption is of the highest importance. In this case, there is monetary incentive to utilize renewable instead of conventional energy. Energy self-production, e.g., on the roofs and facades of the building wherein the indoor farming unit is located, could improve the energy situation and render indoor food production more sustainable [59].

When considering the energy input (referring to electricity only), it becomes evident that 496.4 MJ kg^{-1} FM or 2977 MJ kg^{-1} DM are enormously high. Grain production through conventional agriculture shows energy inputs several orders of magnitude lower, with 5.3 MJ kg^{-1} grain for soybean, 3.3 MJ kg^{-1} grain for wheat and 2.6 MJ kg^{-1} grain for maize [60]. Modern greenhouse production cycles, comprising tomato, pepper and cucumber production in sequence (including tomato nursery), show average energy inputs for crop production (including all byproducts) ranging from 1.9 to 2.7 MJ kg^{-1} FM (total above ground biomass yield) [61]. These figures reveal the tremendous energy inefficiency of indoor crop production in small units such as the SOF.

At present, lettuce production in large-sale modern vertical farming requires about 10 kWh (36 MJ) kg^{-1} harvested fresh biomass on average [16]. The SOF, however, consumed 137.9 kWh kg^{-1} edible fresh biomass (Figure 11). Consequently, the SOF requires 13.8 times more electricity. If the energy usage of the SOF could be reduced by 13.8 times through technical optimizations and improved cultivation management, this would decrease total production costs and make the relative impact of energy on the costs substantially smaller. In the scenario where renewable energy was used, total energy costs were EUR 74.9 and accounted for 71.8% of costs. When decreased by 13.8 times, energy costs would be EUR 34.8 and account for only 15.6% of costs. Looking at the price per kg of biomass, it would decrease from EUR 53.1 kg^{-1} to EUR 17.7 kg^{-1}, thus cutting costs by 66.6%.

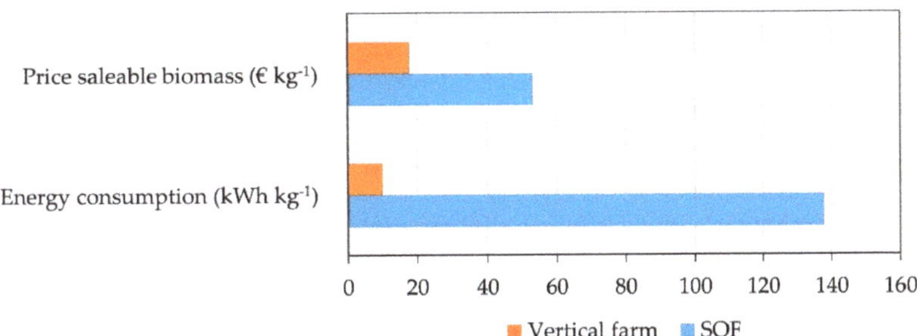

Figure 11. Energy consumption per kg of biomass SOF (blue), and vertical farm (orange) (adapted from [14]).

In particular, artificial lighting, which is not used in open-field agriculture and greenhouse production, is responsible for the high energy inputs in indoor farming units [5,16]. The use of or the self-production of renewable energy currently provides the most promising way to increase energy efficiency and render indoor farming a more viable and sustainable option for food production.

In addition, efficient usage of light through optimal PPFD values is of high importance for increasing the energy use efficiency and economic viability of indoor crop production. For the plants cultivated in this study, the optimum in PPFD is considered 200–250 μmol m^{-2} s^{-1} with a photoperiod of 16–18 h [22,48]. The arrangement of the LEDs should be adjusted to achieve an adequate and more equal PPFD distribution over the whole production area for optimizing crop productivity. In case of the SOF, the LED units should be longer to span over the whole production surface, while the PPFD output should be adjusted to the crop cultivated (which is already possible through dimming). PPFD levels above crop requirements, as was the case for pak choi in this study, increase production costs unnecessarily and reduce economic viability. Furthermore, crop arrangement in polyculture should be conducted according to the light demand of the crop. Crops with high light demand (e.g., lettuce in this study) should be placed directly under the light source, while crops with lower light demand can be placed in between the light sources (such as, e.g., pak choi in this study). Basil growth depends strongly on PPFD level and DLI [51], but grew too large in shoot length in this study resulting in tip burning, and, as such, should be placed between the LED stripes and the shoot may be cut at an early stage to trigger branching of the plant. The latter increases the cultivation cycle, but results in more shoots and harvestable leaves to be picked over a longer period of time. When basil is grown in polyculture, this could further help to adapt the harvest time to other crops with a longer production time.

The importance of LEDs, to decrease energy consumption and make production economically viable, increased since LEDs have a variety of advantages over traditional forms of horticultural lighting and a more efficient performance and longevity compared to any traditional lighting system [62]. Their small size, durability, long lifetime, cool emitting temperature and the option to select specific wavelengths for a targeted plant response make LEDs more suitable for plant-based uses than many other light sources. Furthermore spectral quality of LEDs can have dramatic effects on crop anatomy and morphology as well as nutrient uptake and pathogen development [63]. Global LED use has increased in recent years, since a market share of 5% in 2013 grew to nearly 50% of lighting sales in 2019 [64].

Whether the wavelengths of the LED units are optimal for this use has not been analyzed, but represents another promising option for optimizing resource use efficiency of indoor farming units such as the SOF.

Fixed costs have been neglected in this assessment because it is the prototype stage. However, it has to be noted that LED units are the part of the SOF with the highest individual costs [14].

Artificial lighting represents the driver for optimizing productivity in indoor farming units and reducing the production costs. Both energy inputs and production costs of small production units, such as the SOF, were found to be very high compared to large-scale modern vertical farming. From this assessment, it remains questionable whether office-based farming is a viable and sustainable option for urban production and more research and development efforts are necessary to substantially improve resource use efficiency.

4.4.2. Substrate

The cultivation substrate (rockwool) had a share of 14.3% (conventional energy) and 14.8% (renewable energy) of the total production costs of the SOF. When it comes to the choice of substrate, there are several options. Substrate fulfills three main functions in hydroponic cultivation systems, typically applied in modern indoor farming: (1) provide oxygen, nutrients and water to the roots, (2) allow for root growth and (3) support the stability of the crop [65]. Essentially, an optimal cultivation substrate must have a structure capable of providing a balance of oxygen and nutrient solution both during and between irrigation events to the roots [66,67].

Cultivation substrates can be categorized as organic, for example, peat and coconut coir, and inorganic components, for example, rockwool. Since peat and rockwool have large negative impacts on the environment, peat mining destroys wetlands, while rockwool can hardly be reused and recycled but production is very energy-intensive, current research and development focuses on organic residues from agriculture and organic waste products from biobased industries [68,69]. This can reduce environmental impacts and increase sustainability of indoor farming. In addition, organic substrates can be integrated into local material flows.

4.4.3. Seed Usage

The third biggest factor contributing to costs were the seeds, with shares of 5.9–6.1% on the total costs.

In this experiment, two to three seeds of curled lettuce, lolo rosso and pak choi were sown per substrate block. Whereas three to five seeds per substrate block were sown for basil. All seeds were sown directly into the wet rockwool plugs and then subsequently transferred into the SOF. Direct sowing into rockwool plugs showed fast and homogenous germination with minimum productivity losses and can thus be recommended for the operation of the SOF.

Additionally, a good choice of varieties is crucial in order to optimize yields. For example, basil had 53.8% less consumable biomass yield than pak choi, which is important to consider if maximal biomass yield is the objective. However, the aim of the SOF is to allow the consumer to cultivate according to her/his preference. Hence, direct sowing of crop combinations with similar growth requirements and automated cultivation in polyculture could be one of the use-cases of the SOF.

Since soil is not used in hydroponic production units, theoretically, diseases caused by soil-borne pathogens should not pose a problem [70]. Furthermore, under optimal indoor conditions, the whole environment can be controlled, and the production system can be closed. Hence, no pesticides are required under such conditions [16]. Pest management was not necessary in this study and was therefore not considered. Pesticide free production is a major asset of this type of cultivation system in terms of both consumer health and the environment.

4.4.4. Water Consumption

Vertical farming can be very water-efficient, with savings of up to 95% compared to open-field agriculture [71].

In this study, water had the lowest share on the total costs. One reason is the low water price in Stuttgart, Germany, another is the size of the unit and the recirculating ebb and flow cultivation system requiring only a small quantity of water.

In the future, efficient water usage will become more important since more than a quarter of the world's population lives in regions which will have to cope with water scarcity [72], while Europe experienced one of the most severe droughts ever recorded in 2018 [73].

Therefore, it can be expected that the importance and the competitiveness of water-efficient production will increase and support the implementation of water efficient food production systems in urban areas.

5. Conclusions

Urban farming is part of the solutions for the intensification of food production with the aim of meeting the growing demand of the urban population. The cultivation experiment shows and confirms the high yield potential of smart indoor cultivation systems, while the techno-economic analysis of the SOF revealed very high production costs, mainly caused by very low energy-use efficiency compared to large-scale vertical farming (10 MJ kg^{-1} FM), modern greenhouse production (1.9 to 2.7 MJ kg^{-1} FM) and conventional open-field wheat production (3.3 MJ kg^{-1} grain).

The polyculture in the SOF yielded 1961.8 g on 0.79 m^2 ground area, indicating a theoretical yield potential of 257 Mg FM ha^{-1} year^{-1} of edible and fresh biomass. However, as the economic assessment revealed, at production costs of EUR 53.10 kg^{-1} edible biomass in the renewable energy scenario (conventional energy EUR 55.17 kg^{-1}). The most costly production factor is electricity (73%) for operating the SOF, followed by substrate (14%) and seeds (6%). Since energy is the main cost driver, urban indoor production should use renewable energy and requires further research and development to reduce its high energy demand.

Light is a crucial factor for (indoor) crop production. The yields of curled lettuce, lolo rosso and basil increased proportionally with increasing PPFD, except pak choi with a peak yield at the intermediate PPFD level (355.7 µmol m^{-2} s^{-1}). Light distribution was very heterogeneous on plant level inside the SOF, revealing the need for optimizing PPFD at plant level to save energy and reduce production costs.

Small-scale cultivation systems, such as the SOF, for indoor food production are less productive and have lower energy-use efficiency than vertical farms. For the SOF, the effects of scale are small (e.g., smart, sensor-based control and operation unit for less than 2 m^2 productive area), the management is less efficient, and the resource-use is high, making food production not economically viable today.

However, this trial assessed a prototype SOF. The yield of this trial was low compared to vertical farming and modern greenhouse production and production conditions analyzed in the techno-economical assessment were found to be not optimal. Further trials are needed to optimize the SOF, focusing on technical improvement of the SOF and subsequently on developing optimal cultivation conditions and management with this particular indoor farm.

With improved resource-use efficiency and advanced cultivation management, the productivity of the SOF can be increased, rendering smart indoor cultivation systems such as the SOF a viable option to produce fresh and healthy food right at the place of consumption.

Author Contributions: Conceptualization, J.C., B.W. and M.v.C.; methodology, J.C. and B.W.; investigation, J.C. and B.W.; resources, J.C.; data curation, J.C. and B.W.; writing—original draft preparation, J.C., B.W. and M.v.C.; writing—review and editing, J.C., B.W. and M.v.C.; visualization, J.C. and B.W.; supervision, B.W.; project administration, J.C. and B.W.; funding acquisition, B.W. All authors have read and agreed to the published version of the manuscript.

Funding: This research received funding from the Ministry of Rural Affairs and Consumer Protection Baden Württemberg (14-(27)-8402.43/0405 E).

Institutional Review Board Statement: Not applicable.

Informed Consent Statement: Not applicable.

Data Availability Statement: Data sets of this research can be shared upon request.

Acknowledgments: The authors are very grateful to the Ministry of Rural Affairs and Consumer Protection Baden Württemberg, Jens, Florian and Steffen from farmee GmbH as well as Iris Lewandowski for making the SOF research work possible.

Conflicts of Interest: The authors declare no conflict of interest.

References

1. Christensen, P.; Gillingham, K.; Nordhaus, W. Uncertainty in Forecasts of Long-Run Economic Growth. *Proc. Natl. Acad. Sci. USA* **2018**, *115*, 5409–5414. [CrossRef] [PubMed]
2. Alexandratos, N.; Bruinsma, J. World Agriculture towards 2030/2050: The 2012 Revision. Available online: http://ageconsearch.umn.edu/record/288998 (accessed on 28 January 2021).
3. United Nations; Department of Economic and Social Affairs; Population Division. *World Urbanization Prospects: The 2018 Revision*; 2019. Available online: https://population.un.org/wup/publications/Files/WUP2018-Report.pdf (accessed on 10 December 2022).
4. Smit, J.; Nasr, J. Urban Agriculture for Sustainable Cities: Using Wastes and Idle Land and Water Bodies as Resources. *Environ. Urban.* **1992**, *4*, 141–152. [CrossRef]
5. Al-Kodmany, K. The Vertical Farm: A Review of Developments and Implications for the Vertical City. *Buildings* **2018**, *8*, 24. [CrossRef]
6. Clinton, N.; Stuhlmacher, M.; Miles, A.; Aragon, N.U.; Wagner, M.; Georgescu, M.; Herwig, C.; Gong, P. A Global Geospatial Ecosystem Services Estimate of Urban Agriculture. *Earths Future* **2018**, *6*, 40–60. [CrossRef]
7. Specht, K.; Zoll, F.; Schümann, H.; Bela, J.; Kachel, J.; Robischon, M. How Will We Eat and Produce in the Cities of the Future? From Edible Insects to Vertical Farming—A Study on the Perception and Acceptability of New Approaches. *Sustainability* **2019**, *11*, 4315. [CrossRef]
8. Goldstein, M.; Bellis, J.; Morse, S.; Myers, A.; Ura, E. *Urban Agriculture: A Sixteen City Survey of Urban Agriculture Practices across the Country*; Turner Environmental Law Clinic: Atlanta, GA, USA, 2011; pp. 1–94.
9. Bio-Balkon.de Dein Balkon Als Essbare Wohlfühloase Für Mensch Und Tier! (Transl.: Your Balcony as an Edible Oasis of Well-Being for Humans and Animals!). Available online: https://bio-balkon.de/ (accessed on 10 December 2022).
10. Geco-Gardens.de Ökologisch Gärtnern. Einfach & Überall. (Transl.: Garden ECOlogically Everywhere). Available online: https://geco-gardens.de/ (accessed on 10 December 2022).
11. AeroFarms—An Environmental Champion, AeroFarms Is Leading the Way to Address Our Global Food Crisis by Growing Flavorful, Healthy Leafy Greens in a Sustainable and Socially Responsible Way. Available online: https://aerofarms.com/ (accessed on 10 December 2022).
12. Zezza, A.; Tasciotti, L. Urban Agriculture, Poverty, and Food Security: Empirical Evidence from a Sample of Developing Countries. *Food Policy* **2010**, *35*, 265–273. [CrossRef]
13. Artmann, M.; Sartison, K. The Role of Urban Agriculture as a Nature-Based Solution: A Review for Developing a Systemic Assessment Framework. *Sustainability* **2018**, *10*, 1937. [CrossRef]
14. Ackerman, K.; Conard, M.; Culligan, P.; Plunz, R.; Sutto, M.-P.; Whittinghill, L. Sustainable Food Systems for Future Cities: The Potential of Urban Agriculture. *Econ. Soc. Rev.* **2014**, *45*, 189–206.
15. McKenzie, F.C.; Williams, J. Sustainable Food Production: Constraints, Challenges and Choices by 2050. *Food Secur.* **2015**, *7*, 221–233. [CrossRef]
16. Kozai, T.; Niu, G.; Takagaki, M. *Plant Factory: An Indoor Vertical Farming System for Efficient Quality Food Production*, 2nd ed.; Elsevier Inc.: Amsterdam, The Netherlands, 2020; ISBN 978-0-12-816691-8.
17. McCormack, L.A.; Laska, M.N.; Larson, N.I.; Story, M. Review of the Nutritional Implications of Farmers' Markets and Community Gardens: A Call for Evaluation and Research Efforts. *J. Am. Diet. Assoc.* **2010**, *110*, 399–408. [CrossRef]
18. Pirog, R.; Benjamin, A. *Checking the Food Odometer: Comparing Food Miles for Local versus Conventional Produce Sales to Iowa Institutions*; Leopold Center Pubs and Papers: Ames, IA, USA, 2003.
19. Hönle, S.E.; Meier, T.; Christen, O. Land Use and Regional Supply Capacities of Urban Food Patterns: Berlin as an Example. *Ernahr-Umsch* **2017**, *64*, 11–19. [CrossRef]
20. Ohyama, K.; Takagaki, M.; Kurasaka, H. Urban Horticulture: Its Significance to Environmental Conservation. *Sustain. Sci.* **2008**, *3*, 241–247. [CrossRef]
21. Lewandowski, I. Bioeconomy: Shaping the Transition to a Sustainable, Biobased Economy. p. 14. Available online: https://library.oapen.org/viewer/web/viewer.html?file=/bitstream/handle/20.500.12657/42905/2018_Book_Bioeconomy.pdf?sequence=1&isAllowed=y (accessed on 31 January 2021).
22. Pennisi, G.; Blasioli, S.; Cellini, A.; Maia, L.; Crepaldi, A.; Braschi, I.; Spinelli, F.; Nicola, S.; Fernandez, J.A.; Stanghellini, C.; et al. Unraveling the Role of Red:Blue LED Lights on Resource Use Efficiency and Nutritional Properties of Indoor Grown Sweet Basil. *Front. Plant Sci.* **2019**, *10*, 305. [CrossRef] [PubMed]
23. Faust, J.E.; Holcombe, V.; Rajapakse, N.C.; Layne, D.R. The Effect of Daily Light Integral on Bedding Plant Growth and Flowering. *HortScience* **2005**, *40*, 645–649. [CrossRef]

24. Wiggins, Z.; Akaeze, O.; Nandwani, D.; Witcher, A. Substrate Properties and Fertilizer Rates on Yield Responses of Lettuce in a Vertical Growth System. *Sustainability* **2020**, *12*, 6465. [CrossRef]
25. Dachler, M.; Pelzmann, H. *Arznei-Und Gewürzpflazen*; Agrarverlag Wien: Vienna, Austria, 1999.
26. Alam, S.M. Effects of Solution PH on the Growth and Chemical Composition of Rice Plants. *J. Plant Nutr.* **1981**, *4*, 247–260. [CrossRef]
27. Gillespie, D.P.; Kubota, C.; Miller, S.A. Effects of Low PH of Hydroponic Nutrient Solution on Plant Growth, Nutrient Uptake, and Root Rot Disease Incidence of Basil (*Ocimum basilicum* L.). *HortScience* **2020**, *55*, 1251–1258. [CrossRef]
28. Remonutrients Base Nutrient Formulas. Available online: https://www.remonutrients.com/product/remos-micro-grow-bloom/ (accessed on 10 December 2022).
29. Heck, P.; Bemmann, U. *Praxishandbuch Handbuch Stoffstrommanagement 2002/2003*; Deutscher Wirtschaftsdienst Köln: Koln, Germany, 2002.
30. Lauer, M. Methodology Guideline on Techno Economic Assessment (TEA). In *Workshop WP3B Economics, Methodology Guideline*; 2008. Available online: https://www.scribd.com/document/333629930/Thermalnet-Methodology-Guideline-on-Techno-Economic-Assessment (accessed on 10 December 2022).
31. Wittmann, S.; Jüttner, I.; Mempel, H. Indoor Farming Marjoram Production—Quality, Resource Efficiency, and Potential of Application. *Agronomy* **2020**, *10*, 1769. [CrossRef]
32. Ertragsmenge von Salatsorten in Deutschland 2019. Available online: https://de.statista.com/statistik/daten/studie/639108/umfrage/ertragsmenge-von-salatsorten-in-deutschland/ (accessed on 22 March 2021).
33. Zufriedenstellende Gemüseernte 2019—Statistisches Landesamt Baden-Württemberg. Available online: https://www.statistik-bw.de/Presse/Pressemitteilungen/2020033 (accessed on 18 February 2021).
34. Plenty, 2021 About Us. Available online: https://www.plenty.ag/about-us/ (accessed on 22 March 2021).
35. Marton, S. Ökobilanz von Gewächshausgurken Und Salaten. Stiftung myclimate—The Climate Protection Partnership, Zürich, Switzerland, 2010. Available online: https://www.researchgate.net/publication/228380499_Okobilanz_von_Gewachshausgurken_und_Salaten (accessed on 10 December 2022).
36. Zeidler, C.; Schubert, D.; Vrakking, V. Feasibility Study: Vertical Farm EDEN. Available online: https://elib.dlr.de/87264/ (accessed on 18 February 2021).
37. AeroFarms. 2021. Available online: https://aerofarms.com/ (accessed on 28 March 2021).
38. Sky Greens. Available online: https://www.skygreens.com/ (accessed on 28 March 2021).
39. Upshall, E. Infarm Unveils New Automated Growing Centre with Higher-Yield Capabilities. *FoodBev Media* **2021**. Available online: https://www.foodbev.com/news/infarm-unveils-new-automated-growing-centre-with-higher-yield-capabilities/ (accessed on 10 December 2022).
40. The Second Generation of Vertical Farming Is Approaching. Here's Why It's Important. Available online: https://agfundernews.com/the-second-generation-of-vertical-farming-is-approaching-heres-why-its-important.html (accessed on 27 March 2021).
41. Schröder, F.-G.; Lieth, J.H. *Irrigation Control in Hydroponics*; Embryo Publications: Athens, Greece, 2002.
42. Maucieri, C.; Nicoletto, C.; Van Os, E.; Anseeuw, D.; Van Havermaet, R.; Junge, R. Hydroponic Technologies. In *Aquaponics Food Production Systems*; Springer: Cham, Switzerland, 2019; pp. 77–110.
43. Higgins, C. Important Tips for Designing a Hydroponic Production Facility. *Urban Ag News* **2020**. Available online: https://urbanagnews.com/blog/exclusives/important-tips-for-designing-a-hydroponic-production-facility/ (accessed on 10 December 2022).
44. Automation Solutions for Quality Crops | Grow. Anywhere. Available online: https://autogrow.com (accessed on 26 March 2021).
45. Thompson, H.C.; Langhans, R.W.; Both, A.-J.; Albright, L.D. Shoot and Root Temperature Effects on Lettuce Growth in a Floating Hydroponic System. *J. Am. Soc. Hortic. Sci.* **1998**, *123*, 361–364. [CrossRef]
46. Goddek, S.; Joyce, A.; Kotzen, B.; Burnell, G. *Aquaponics Food Production Systems-Combined Aquaculture and Hydroponic Production Technologies for the Future*; Springer International Publishing: Cham, Switzerland, 2019; ISBN 978-3-030-15942-9.
47. Kimball, B.A.; Idso, S.B. Increasing Atmospheric CO2: Effects on Crop Yield, Water Use and Climate. *Agric. Water Manag.* **1983**, *7*, 55–72. [CrossRef]
48. Ahmed, H.A.; Yu-Xin, T.; Qi-Chang, Y. Optimal Control of Environmental Conditions Affecting Lettuce Plant Growth in a Controlled Environment with Artificial Lighting: A Review. *S. Afr. J. Bot.* **2020**, *130*, 75–89. [CrossRef]
49. Palzkill, D.A.; Tibbitts, T.W.; Struckmeyer, B.E. High Relative Humidity Promotes Tipburn on Young Cabbage Plants. *HortScience* **1980**, *15*, 659–660. [CrossRef]
50. Ding, X.; Jiang, Y.; Zhao, H.; Guo, D.; He, L.; Liu, F.; Zhou, Q.; Nandwani, D.; Hui, D.; Yu, J. Electrical Conductivity of Nutrient Solution Influenced Photosynthesis, Quality, and Antioxidant Enzyme Activity of Pakchoi (*Brassica campestris* L. Ssp. Chinensis) in a Hydroponic System. *PLoS ONE* **2018**, *13*, e0202090. [CrossRef] [PubMed]
51. Walters, K.J.; Currey, C.J. Effects of Nutrient Solution Concentration and Daily Light Integral on Growth and Nutrient Concentration of Several Basil Species in Hydroponic Production. *HortScience* **2018**, *53*, 1319–1325. [CrossRef]
52. Shamshiri, R.; Kalantari, F.; Ting, K.C.; Thorp, K.R.; Hameed, I.A.; Weltzien, C.; Ahmad, D.; Shad, Z.M. Advances in Greenhouse Automation and Controlled Environment Agriculture: A Transition to Plant Factories and Urban Agriculture. *Int. J. Agric. Biol. Eng.* **2018**, *11*, 1–22. [CrossRef]

53. Gemüse—Portionsgrößen—Lebensmittelwissen.De. Available online: https://www.lebensmittelwissen.de/tipps/haushalt/portionsgroessen/gemuese.php (accessed on 11 February 2021).
54. Shimizu, H.; Kushida, M.; Fujinuma, W. A Growth Model for Leaf Lettuce under Greenhouse Environments. *Environ. Control Biol.* **2008**, *46*, 211–219. [CrossRef]
55. Musango, J.K.; Currie, P.; Robinson, B. *Urban Metabolism for Resource Efficient Cities: From Theory to Implementation*; UN Environment: Paris, France, 2017.
56. Verma, M.; van den, B.; de Vreede, L.; Achterbosch, T.; Rutten, M.M. Consumers Discard a Lot More Food than Widely Believed: Estimates of Global Food Waste Using an Energy Gap Approach and Affluence Elasticity of Food Waste. *PLoS ONE* **2020**, *15*, e0228369. [CrossRef]
57. Stoknes, K.; Scholwin, F.; Krzesiński, W.; Wojciechowska, E.; Jasińska, A. Efficiency of a Novel "Food to Waste to Food" System Including Anaerobic Digestion of Food Waste and Cultivation of Vegetables on Digestate in a Bubble-Insulated Greenhouse. *Waste Manag.* **2016**, *56*, 466–476. [CrossRef]
58. Winkler, B.; Maier, A.; Lewandowski, I. Urban Gardening in Germany: Cultivating a Sustainable Lifestyle for the Societal Transition to a Bioeconomy. *Sustainability* **2019**, *11*, 801. [CrossRef]
59. Al-Chalabi, M. Vertical Farming: Skyscraper Sustainability? *Sustain. Cities Soc.* **2015**, *18*, 74–77. [CrossRef]
60. Alluvione, F.; Moretti, B.; Sacco, D.; Grignani, C. EUE (Energy Use Efficiency) of Cropping Systems for a Sustainable Agriculture. *Energy* **2011**, *36*, 4468–4481. [CrossRef]
61. Hedau, N.K.; Tuti, M.D.; Stanley, J.; Mina, B.L.; Agrawal, P.K.; Bisht, J.K.; Bhatt, J.C. Energy-Use Efficiency and Economic Analysis of Vegetable Cropping Sequences under Greenhouse Condition. *Energy Effic.* **2014**, *7*, 507–515. [CrossRef]
62. Bourget, C.M. An Introduction to Light-Emitting Diodes. *HortScience* **2008**, *43*, 1944–1946. [CrossRef]
63. Massa, G.D.; Kim, H.-H.; Wheeler, R.M.; Mitchell, C.A. Plant Productivity in Response to LED Lighting. *HortScience* **2008**, *43*, 1951–1956. [CrossRef]
64. IEA—International Energy Agency. Available online: https://www.iea.org (accessed on 28 March 2021).
65. Olle, M.; Ngouajio, M.; Siomos, A. Vegetable Quality and Productivity as Influenced by Growing Medium: A Review. *Zemdirbyste* **2012**, *99*, 399–408.
66. Caron, J.; Nkongolo, V.K.N. Aeration in growing media: Recent developments. *Acta Hortic.* **1999**, *481*, 545–552. [CrossRef]
67. Fonteno, W.C. Problems & considerations in determining physical properties of horticultural substrates. *Acta Hortic.* **1993**, *342*, 197–204. [CrossRef]
68. Barrett, G.E.; Alexander, P.D.; Robinson, J.S.; Bragg, N.C. Achieving Environmentally Sustainable Growing Media for Soilless Plant Cultivation Systems—A Review. *Sci. Hortic.* **2016**, *212*, 220–234. [CrossRef]
69. Gruda, N.S. Increasing Sustainability of Growing Media Constituents and Stand-Alone Substrates in Soilless Culture Systems. *Agronomy* **2019**, *9*, 298. [CrossRef]
70. Zlnnen, T.M. Assessment of Plant Diseases in Hydroponic Culture. *Plant Dis.* **1988**, *72*, 96. [CrossRef]
71. Podmirseg, D. *Vertical Farming—Vertical Farm Institute: Leading Research Network*; Vertical Farm Institute: Vienna, Austria. Available online: https://verticalfarminstitute.com/ (accessed on 10 December 2022).
72. Seckler, D.; Barker, R.; Amarasinghe, U. Water Scarcity in the Twenty-First Century. *Int. J. Water Resour. Dev.* **1999**, *15*, 29–42. [CrossRef]
73. Schuldt, B.; Buras, A.; Arend, M.; Vitasse, Y.; Beierkuhnlein, C.; Damm, A.; Gharun, M.; Grams, T.E.E.; Hauck, M.; Hajek, P.; et al. A First Assessment of the Impact of the Extreme 2018 Summer Drought on Central European Forests. *Basic Appl. Ecol.* **2020**, *45*, 86–103. [CrossRef]

Review

Vine and Wine Sustainability in a Cooperative Ecosystem—A Review

Agostinha Marques [1,2] and Carlos A. Teixeira [2,*]

1   Adega Cooperativa de Favaios, 5070-265 Vila Real, Portugal; agostinhamarques@adegadefavaios.com.pt
2   Centre for the Research and Technology of Agro-Environmental and Biological Sciences (CITAB), Institute for Innovation, Capacity Building and Sustainability of Agri-Food Production (Inov4Agro), University of Trás-os-Montes and Alto Douro (UTAD), 5000-801 Vila Real, Portugal
*   Correspondence: cafonso@utad.pt

Abstract: The world is changing, and climate change has become a serious issue. Organizations, governments, companies, and consumers are becoming more conscious of this impact and are combining their forces to minimize it. Cooperatives have a business model that differs from those in the private or public sector. They operate according to their own principles of cooperation, which makes it difficult to obtain results that are in harmony with the objectives of the organization and the cooperative members. However, they are also aware of climate change because their businesses are directly affected. Thus, in this review, we have tried to answer the following questions: What is necessary to meet the sustainability goals? Are wine cooperatives competitive in the context of the global market? How can we respond to the challenges of environmental sustainability while maintaining wine quality standards and economic profitability? What are the economic and social impacts of reducing the carbon footprint of cooperatives and their members?

Keywords: wine cooperative; sustainability; benchmarks

Citation: Marques, A.; Teixeira, C.A. Vine and Wine Sustainability in a Cooperative Ecosystem—A Review. *Agronomy* **2023**, *13*, 2644. https://doi.org/10.3390/agronomy13102644

Academic Editors: Moritz Von Cossel, Joaquín Castro-Montoya and Yasir Iqbal

Received: 8 September 2023
Revised: 6 October 2023
Accepted: 9 October 2023
Published: 19 October 2023

1. Introduction

In 2015, the United Nations established an agenda for 2030 to 2050 and defined the 17 Goals for Sustainable Development, which aim to improve living conditions; combat poverty, hunger, and social inequity; promote access to water, health, and education; combat climate change; and protect the environment [1–3]. In this respect, governments, companies, and organizations have been looking for ways to respond to the United Nations' challenge. Europe, for its part, has taken the leading role in combating climate change, particularly in the food sector, with the creation of the European Green Deal, whose goals include agriculture that is environmentally sustainable and a fair and healthy food system [4].

Wine production is one of the oldest economic activities, and environmental factors have always affected grape production, forcing people to select grape varieties according to the terroir and the soil in order for greater efficiency [5]. The cultivation of wine has transformed landscapes and has become one of the sectors that contributes most to the economic and social sustainability of communities. It is an integral part of culture, providing many experiences, encouraging tourism, and being a source of pride for communities [2,6].

One of the sectors that most contributes to greenhouse gas emissions is agriculture, with the wine sector accounting for 0.3% of global GHG emissions (considering a bottle of wine leaving a cellar), and promoting sustainable environmental behavior has consequently been the subject of certain policies [7]. Viticulture has a large impact on the environment, as the use of chemical products, soil tillage, irrigation, soil management, and mechanization are all responsible for GHG emissions [4,8].

In 2004, the OIV defined viticultural sustainability as a global strategy for grape and wine production which contributes to the economic sustainability of communities by producing quality products and practicing responsible viticulture. Sustainable viticulture is concerned with risks to the environment, product safety, and consumer health, as well as valuing local heritage, history, landscape, and culture [9–11]. There is a growing commitment in agriculture to more sustainable practices [12], not only because of the economy, but also for environmental reasons and the legacy for future generations. Sustainability and the efficient use of environmental, social, and economic resources are becoming increasingly important to wine consumers and winemakers. This is clear from the way that markets and consumers prefer products produced and labeled according to "sustainability indicators or terms", such as organic, sustainable, natural, free, ecological, etc. [9], because for the consumer, the term "sustainable" is associated with the environment and their carbon footprint. Governments, for their part, have been trying to impose measures that encourage consumers to choose products that are more sustainable, for example, by applying environmental taxes (on carbon) [7] or, in the case of monopolies, restricting products that do not meet sustainability standards.

Wine cooperatives are considered to be organizations with sustainable social and economic development as some of their multiple roles and objectives [13], and they feel pressure not only from consumers, but also from governments and monopoly markets. Ziegler [14] argues that wine cooperatives should have objectives and strategies to ensure circular social and ecological sustainability.

Growing pressure for political reasons and customers looking for sustainable products [5,15] have created the need for winegrowing organizations to develop indicators aimed at the efficient use of water, production methods, the use of phytopharmaceuticals, energy efficiency in the vineyard and winery, the promotion of clean energy rather than fossil fuels, waste management, community impact, and employee well-being [5,16]. This is because, for them, and in contrast to the consumer, sustainability is not only environmental, but also economic and social [9]. In addition, consumers are becoming increasingly aware of the need to be more sustainable, and wine producers need to implement sustainable practices in order to stand out in a market with growing competition [2,6,17]. This has led to the creation of various sustainable certification programs in winegrowing, which winegrowers have tried to adopt. Cooperatives, formed by small winegrowers, most of them with very limited literacy, cannot impose these rules; cooperatives will have to create tools to encourage their members to adopt sustainable practices in response to market demands.

In our review of the literature, we found that guidelines have been defined by the authors mentioned in Table 1. This was the first observation that research has only focused on one aspect of sustainability, and in the case of cooperatives, the focus is on economic and social sustainability.

Table 1. Literature review.

Ref.	Authors	Year	Country	Relevant Information
[1]	Chabin et al.	2023	France	Sustainability; 17 Sustainable Development Goals; Economy; Environment; Resources
[2]	Ferrer et al.	2022	Spain	Economy; Environment; Resources
[4]	Nazzaro et al.	2022	Italy	Innovation; Sustainability; Cooperatives; Governance; European Green Deal
[5]	Tsalidis et al.	2022	Greece	Organic; Viticulture; GHG emissions; Carbon Footprint
[6]	Martínez-Falcó et al.	2023	Spain	Sustainable Development Goals; Wine Industry
[7]	Soregaroli et al.	2021	Italy	Carbon Footprint; Climate Change; Wine Consumers
[9]	Lamastra et al.	2016	Italy	Vineyard Sustainability; Indicators; Environmental
[10]	Marras et al.	2015	Italy	Vineyard Management; Carbon Footprint; Agriculture; GHG Emissions
[11]	Casolani et al.	2022	Italy	LCA; Wine Sustainability; Environment Sustainability

2. Cooperative Ecosystem

In 1852, Great Britain declared the cooperative a business for the first time [18]. This shows the cooperative tradition in Europe [19]. According to the "Declaration of Cooperative Identity" defined by the International Cooperative Alliance in 1995, "a cooperative is an autonomous association of persons voluntarily united to meet their common economic, social and cultural needs and aspirations through a jointly owned and democratically controlled enterprise" [14,20,21]. Article 2°, paragraph 1 of the Portuguese Cooperative Code defines a cooperative as "collective and autonomous persons, free constituted, with variable capital and composition, which, through the cooperation and mutual help of their members, in compliance with the cooperative principles, aim, on a non-profit basis, to satisfy their economic, social or cultural needs and aspirations" [20,21]. Cooperatives are governed by seven main principles: voluntary and free membership, democratic management by members, economic participation by members, autonomy and independence, education, training and information, cooperation between cooperatives, and interest in the community [21,22]. In other words, cooperatives are socially based people's enterprises [22], and stand out for promoting social equality, community development, and the well-being of their members [20]. We can conclude that cooperatives are the best business model for local development, considering the cooperation between citizens and local, regional, and national organizations [20].

Lately, there has been growing interest in the cooperative model, as this business model has proved to be more resilient in times of prolonged economic crises than capitalist companies [13]. Cooperatives favor the maintenance of jobs, preferring to reduce salaries, and the distribution of surpluses is more balanced to meet needs in times of crisis [13]. Historically, there has been an increase in the creation of cooperatives in times of economic and social crisis [20], such as in the production of the liqueur muscatel in Portugal in the 1950s. Ziegler [14] has conducted a study showing that cooperatives are fundamental to the circular economy and its incorporation into regional economies, concerning revalorization, production, consumption, and lasting use.

These organizations are more sensitive to environmental, social, and economic issues due to their cooperative values [12]. Since equality, community development, the well-being of their members, and combating exclusion and poverty among the most disadvantaged classes are at the genesis of the creation of cooperatives, they are an alternative business model to capitalism [13]. This business model helps small producers to create scale, i.e., they are able to sell their products more easily as they gain the capacity to negotiate by volume [20]. However, there are also weaknesses, since a cooperative demands the acceptance of all the production of its cooperative members, without taking into consideration quality or production methods, and can only impose a few rules that benefit those who comply to the detriment of those who do not [19].

Figueiredo [20] defined cooperatives and cooperative members as "social entrepreneurs" who are orientated towards financial independence and sustainable entrepreneurship to create social value for the less privileged. We can therefore say that cooperatives enable the creation of stronger and more sustainable local economies because they reinvest profits, without forgetting social values and their mission [20].

As cooperatives are solutions for local development, agricultural cooperativism is very much in the spotlight, especially when we look at production. According to Figueiredo [13], 41% of the wine produced in Portugal is made by cooperatives, and the numbers are even more impressive when it comes to milk, which accounts for around 62%. This is why agricultural cooperatives are so important, given that they operate at a rural level and contribute to the conservation of these environments and the environment in general [20]. However, like other companies, they must be competitive and create value in order to become economically, socially, and environmentally sustainable [20].

Climate change has been challenging companies to take urgent action to maintain their competitive edge [19]. Some studies show that cooperatives are more proactive on environmental issues than private companies [12], but there is no evidence of their application in agricultural practices, such as in reducing their carbon footprint, water footprint, use of fossil fuels, etc., since there is a lack of documentation or sustainability reports by cooperatives; these reports could not only show their commitment and sustainability strategy, but could also be seen as an internal learning mechanism [14].

Figueiredo is one of the most widely published authors in the field of cooperatives and their dynamics. Analyzing the articles by Ritcher and Figueiredo has provided a better understanding of the fundamentals and the cooperative business model (Table 2).

Table 2. Cooperativism literature review.

Ref.	Authors	Year	Country	Relevant Information
[12]	Calle et al.	2020	Spain	Cooperatives; Environmental; Wine Sector
[13,20]	Figueiredo et al.	2018	Portugal	Cooperative; Sustainability; Social; Economic; Society Development
[14]	Ziegler et al.	2023	Canada	Cooperatives; Circular Economy; Business Model; Social Economy
[18,19]	Ritcher et al.	2021	Germany	Sustainable Management; Cooperatives; Cooperative Values; Social Capital
[21]	Ramos et al.	2023	Portugal	Cooperatives; Democracy; Governance

3. Difficulties in Respect to Responses from Cooperative Members

The cooperative model depends on the ability of cooperatives to satisfy the ambitions of their members, which sometimes do not meet the principles of cooperativism due to the external and internal pressures that management can face [20]. This disruption can lead to a loss of cooperative identity [20]. For this reason, when results are equal to or better than expected, satisfaction is high and fundamental to maintaining trust, cooperation, and commitment between everyone, cooperatives and cooperators, reducing disputes [20]. In addition, through the difficulties inherent in cooperativism, the wine sector suffers from the effects of demographics and land abandonment. According to Figueiredo's research [13], the average age of cooperative members is around 60, they are mostly men, and they have low literacy levels. They are also resistant to change, and issues of efficiency and performance are of lesser importance. The great challenge for cooperatives lies in their ability to attract younger members to maintain the sustainability of the organization [13].

Another difficulty is related to one of the cooperative principles, freedom, i.e., there is an "open door" policy, which enables the free entry but also the free exit of members, which leads to problems of opportunism and lack of commitment [20]. Differences between members, like quantity, grape production as a main or secondary activity, acceptance of risk, and organization, contribute to a high degree of heterogeneity between members, which slows down decision-making [18]. Due to this heterogeneity, the challenge is to persuade members to apply sustainability measures [19].

However, it is not only the cooperative members who create difficulties. One of the biggest problems is caused by the cooperative itself: the payment periods for cooperative members are long, never less than 90 days, and often more than two years, which is one of the main reasons why cooperative members leave, as they need immediate liquidity [13].

An advantage of the cooperative system is that when the governance model is oriented to innovation and development, this allows access to innovative technologies and techniques, such as precision agriculture [4]. As well as promoting knowledge, this can make investments in technology accessible to cooperative members, since individual investment would be economically unviable. However, this can be criticized due to differences in objectives between management and cooperative members; one of the most common situations is production vs. quality, with the cooperative looking for quality and the cooperative members seeking production [4].

It is difficult for farmers to measure all the indicators they need to take advantage of in a sustainability framework [23]. The lack of a clear standardization of indicators leaves winegrowers in doubt about which indicators are essential for understanding their company's level of sustainability, and in responding to market demands [24,25] and determining how to do so. The process is more complicated when applied to wine cooperatives. In a private company, the management board easily defines the objectives to be met by the organization, while in the case of cooperatives, the decision-making capacity of the management board is more limited not only because it is an elected position, but also because of the time limitation for implementing long-term objectives [19]. This difficulty is compounded by the fact that, in general, investments in sustainable measures have a long-term effect and the winegrower needs funding in the short-term, so money is more important [18]. Communication between the board and the members is essential; it is important that the members understand that consumers are now willing to pay more for sustainably produced wine [18].

Faced with the current situation and the analyses carried out in this study, it is necessary to provide cooperatives with tools that support them in materializing their values and responding to the markets [12,18], and that allow the cooperative to prevail in the long-term.

Understanding the dynamics of cooperatives requires an understanding of their strengths and weaknesses. Since cooperatives are created to help a large and heterogeneous number of individuals, this creates many challenges that are not found in private companies. In Table 3, the authors of this study have gathered some information, but many questions remain unanswered.

Table 3. Difficulties with cooperativism, literature review.

Ref.	Authors	Year	Country	Relevant Information
[23]	Withehead	2017	New Zealand	Sustainable Development; Sustainability Assessment; Wine
[24]	Borosato et al.	2020	Italy	Viticulture; Sustainability; Innovation
[25]	Merli et al.	2018	Italy	Indicators; Environmental Management Systems; Sustainability

4. Different Sustainability Benchmarks

Over the years, several sustainable certification benchmarks have emerged which differ from organic, biodynamic, and biological certification [15]. Although they have the same objectives, they are different in terms of methodology [15]. This diversity of benchmarks for certification in the wine sector [25] has led to some markets (export, national) feeling the need to create a set of rules in which the sustainability indicators fit in with greater or lesser importance, as is the case with SystemBolarget, created in Sweden [26], and Sonae's Producers Club in Portugal.

In the wine sector, there are various models of certification. These can involve the certification of vineyards, wineries, or both [15]. For example, although organic farming has a positive impact on the environment, it has little focus on sustainability [15]. ISO 14001 was designed in the 1990s and is an environmental management system with an auditing program. It is voluntary and includes all economic areas, including agriculture and more specifically the wine sector [15].

In order to regulate the sector in 2020, the OIV (International Organization of Vine and Wine) worked on a guide for implementing the principles of sustainability in viticulture [27]. The sustainable certifications that have subsequently emerged use the OIV's guidelines in this document as a basis [28]. However, while the key indicators are common across the different benchmarks for certification, the ways in which they are described vary; for example, in calculating the carbon footprint, energy consumption, impact of the carbon footprint on soil, GHG [28], water footprint [29], etc. However, the indicators usually tend to be more descriptive than analytical, making it difficult to determine the questions to be measured and their answers, which is a weakness of the system [23].

Sustainable certification benchmarks in the wine sector first started in New Zealand in 1997 with the "Sustainable Winegrowing New Zealand" program [30]; others have been emerging [15], most recently in Portugal with ViniPortugal's "National Reference for Sustainability Certification in the Wine Sector" in 2022 [31] and the IVDP's "Sustainability Manual for the Douro Wine Region" in 2023 [32] (Figures 1 and 2). Portugal currently has the Alentejo (PSVA), launched in 2015 and promoted by the Alentejo Regional Wine Commission [33].

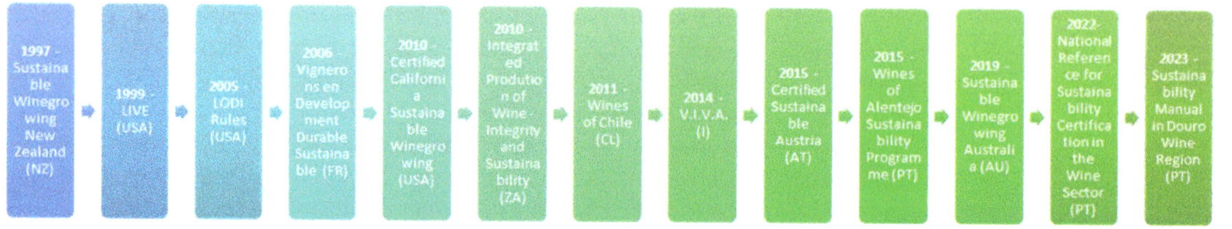

Figure 1. Timeline for the creation of the different sustainable certification models for the wine sector.

Figure 2. Labels associated with different sustainable certification models in the wine sector.

Some of the best-known sustainable certification benchmarks for the wine sector, created specifically for the vine and wine sector, are described below (Figure 1).

4.1. Sustainable Winegrowing New Zealand (1997)

In 1995, the New Zealand Winegrowers Association started the "Sustainable Winegrowing New Zealand" program, its success being such that in 1997 they began the process of certifying producers at a national level [15]. It is a national program with a sustainability label, financed by a tax on the sale of grapes and wine and by the cost of certification. The process has 62 chapters, based on various indicators such as biodiversity, soil, air, water, energy, chemicals, by-products, people, and the economy. To obtain certification, it is necessary to have an audit carried out by an independent auditor [15,30]. One of the main criteria for the label is that the grape and wine are produced 100% sustainably [30]. New Zealand's progress has given it a competitive advantage over other winegrowing regions in the world [15].

4.2. LIVE (1999)

LIVE is the first North American certification benchmark to originate in Oregon. A nonprofit organization, the LIVE program was created in 1999, based on the indicators of the International Organization of Biological Control (IOBC) for Integrated Pest Management (IPM). Today, the program is not so focused on chemical products; environmental, economic, and social indicators have been added [15,34].

4.3. LODI Rules (2005)

The LODI Rules certification came into being in 2005, but its basis was created in 1992 when the Lodi Winegrape commission and the University of California State Agricultural Extension created a document based on sustainable practices. For a producer to use the

LODI Rules label, 85% of the grapes used must be certified and the minimum score for each chapter is 50%, with a minimum total score of 70% for certification. The indicators are divided into six chapters: economy, human resources, ecosystems, soil, water, and pests [15,35].

4.4. Sustainable Development for Wine Growers (2006)

This is the first certification benchmark in Europe, which originated in France and was launched in 2006. It is based on four pillars: environmental preservation, wine quality, society factors, and a fair price for the consumer. To obtain certification, the producer must be a member of an association, fulfil 37 indicators, and obtain at least 50% [15,36].

4.5. Certified California Sustainable Winegrowing (2010)

In 2001, California developed a Sustainable Winegrowing program and in 2003 it was included in the California Sustainable Winegrowing Alliance (CSWA). In 2010, Certified California Sustainable Winegrowing was born with the aim of training and providing growers with tools to improve sustainable winegrowing practices. Today, it focusses on more transversal sustainability, with the main pillars focusing on the environment, economy, and society factors. To obtain certification in viticulture, 50 indicators must be met, and wine production must meet 32 indicators. Like other programs, after a self-assessment, an audit is carried out by an independent auditor [15,37].

4.6. Integrated Production of Wine—Integrity and Sustainability (2010)

This was perhaps the first sustainability program, as in 1998 the Integrated Production of Wine (IPW), run by the South African government, was established. However, it was only in 2010 that the sustainability label and certificate were created, which is why the start is attributed to 2010. This is one of the few programs that has no cost to the producer and therefore also has one of the highest uptakes. To obtain certification, producers must have a minimum of 162 points out of 270, in 27 indicators. In the case of wineries, 93 points out of 155 points are required in 31 indicators [15,38].

4.7. Wines of Chile (2011)

Chile's wine sustainability certification program was created in 2011 with the impetus of the wine industry and is managed by a non-profit organization. It focuses on the three pillars of sustainability, not just the environmental pillar, applied to the vineyard, winery, bottling, and human resources. It mainly seeks to reduce the risks of the production system and the vulnerability of the sector to environment and climate change [15,39].

4.8. V.I.V.A (2014)

The V.I.V.A. program appeared as a pilot project of the Italian government, Ministry of Environment, Land and Sea in 2011 and the first certification was made in 2014. Certification is financed by the government and aims at the sustainability of the sector and adding value to the certified product. It focuses on four chapters: water, vineyard, air, territory. It is also the first program to make publicly available the results of the audits made by an independent auditor, making it a transparent program [15,40].

4.9. Certified Sustainable Austria (2015)

Austria has taken existing programs and adapted them to its reality. Its program was created in 2015 by the Austrian Winegrowers Association and is national in scope. Austria is one of the European countries with the largest area of vineyard certified as organic, so its adaptation was easy and in the first year 23 wineries were certified. Certification works on a traffic-light scheme, with green being the most sustainable. Producers respond in an online tool that can be consulted by the consumer in a model of transparency, like the Italian program [15,41].

4.10. Wines of Alentejo Sustainability Program (2015)

This was the first sustainability program created in Portugal and adapted to the Alentejo region. The program was initiated in 2013 by the organization that controls the wines of Alentejo (Comissão Vitivinícola do Alentejo), and it was inspired by the California model, CSW, due to the similarities in production, climate, and terroir. It is divided into three sectors: vineyard, winery, and vineyard and winery. It has 18 chapters and 171 indicators, based on four global pillars. The first is supervision, management, and quality; the second is social, the third is environmental; and the fourth is exclusive requirements. For wine certification, 60% of the vineyard area must be registered in the PSVA. It has a scale of levels that starts at initial, where growers must achieve 60%, followed by intermediate, with the last being developed [33].

4.11. Fresh Australian Wine Industry Standard of Sustainable Practice (2020)

Launched in 2019, this program is based on the "Sustainable Australia Winegrowing" (2011) and Entwine Australia programs and was revised in 2020. In 2020, the benchmark was categorized into two parts, viticulture and winery, and was renamed. It is a national program aimed at winegrowers and winemakers. The main pillars are social, economic, and environmental, with landscape and soil, water, people, the economy, biodiversity, energy, and waste being the most prominent [15,42].

4.12. National Reference for Sustainability Certification in the Wine Sector (2022)

This program, launched in 2022, is one of the most recent sustainability programs in Portugal. It was developed by two public organizations, one for control and the other for promotion, i.e., Instituto da Vinha e do Vinho (IVV) and ViniPortugal, respectively, based on programs already implemented in other regions of the world, such as the Alentejo program (PSVA), California Sustainable Winegrowing (CSW), LODI Rules, Bodegas Argentinas, Sustainable Winegrowing Australia, etc. It is based on four pillars, which are management and continuous improvement, environmental, social, and economic, which are divided into 86 indicators spread over 17 chapters. To obtain certification, 50% of the grapes must meet the minimum requirements of the program. The classification corresponds to letters, the lowest being C (ranging from 50% to 65%) and the highest being A (more than 85%) [31].

4.13. Sustainability Manual for the Douro Wine Region (2022)

The more recent sustainability program in Portugal is the Sustainability Manual for the Douro Wine Region, developed by IVDP and the Faculdade de Ciências da Universidade do Porto, which is currently under public consultation. It is a program very similar to the Californian CSW, which works on a colored traffic-light system, like Austria's program. The scoring criteria take into account the size of the companies in terms of area, volume of liters, turnover, and number of employees. However, it has one of the lowest acceptance levels; from 33%, it already has a D classification, the remaining levels being similar to the ViniPortugal program. This program focuses only on one region, which is the Douro. Like the other programs, it addresses the SDGs and is based on the main pillars: economy, social, environmental, and quality [28,32].

5. Different Sustainability Benchmarks

It is possible to analyze which indicators are the most important or eliminatory for each certification model. Furthermore, there has been an evolution in the certification models, with the most recent ones not only being more demanding, but also having more indicators aimed at economic and social sustainability (Figure 3). The first certification models focused more on vineyard, water, and soil aspects [15]. Biodiversity and water management are indicators mentioned in all of the sustainable certification models [10] (Figure 4). In New Zealand, Whitehead [23] analyzed the priority indicators for sustainability analysis and concluded that the water indicator is the most valued. This may be due to the notion that it is a finite resource that is becoming increasingly scarce, with implications not only for agriculture, but also for everyone's day-to-day life.

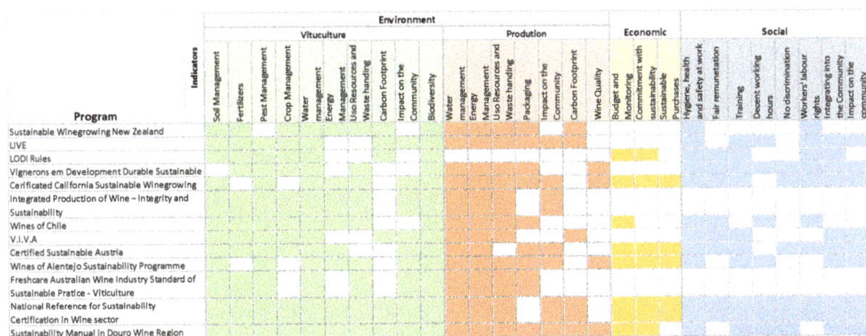

Figure 3. Sustainable certification benchmarks from the oldest to the most recent, showing the most important indicators for each benchmark (legend: green—indicators mentioned in the benchmarks for viticulture; orange—indicators mentioned in the benchmarks for wine production; yellow—economic sustainability indicators mentioned in the benchmarks; blue—social sustainability indicators mentioned in the benchmarks).

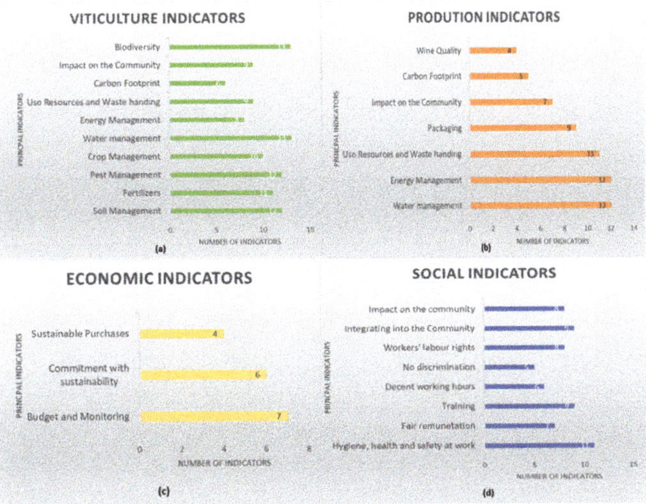

Figure 4. The most important sustainability indicators in the different sustainable certification benchmarks: (**a**) represents the environmental indicators for the vineyard (green color) and the number of benchmarks that measure them, and the most important are biodiversity and water management; (**b**) represents the environmental indicators for wine production (orange color) and the number of benchmarks that measure them, and again, water management is an important indicator, as is energy management; (**c**) represents the company's economic management indicators (yellow color) and the number of benchmarks that measure them, with budget and monitoring being the most relevant; and (**d**) represents social indicators (blue color) and the company's relationship with the community and the number of benchmarks that measure them, with hygiene, safety, and health at work being the most mentioned, but others also appear, such as training and integration into the community.

Biodiversity is approached in various ways. In older models, the focus was on maintaining the oldest and regional grape varieties, as well as the ecosystem. The most recent models focus on the vineyard's ecosystems, such as forest, riparian, small vegetation, and bird nesting sites, and the correct maintenance of these ecosystems. Mulch is becoming increasingly important [31–33,37]. In addition to increasing the soil's ability to retain water, it is a shelter for pest predators and a source of nutrients for the plant, as well as reducing the invasion of undesirable weeds. A good mulch helps to reduce tillage and the use of insecticides, herbicides, and fertilizers, creating greater water retention in the soil and the prevention of soil leaching.

While all the models give importance to the social aspect, it can be seen that "Hygiene, Health and Safety at Work" is present in most of them, as is training. However, these indicators are legal requirements in Europe and the USA, so this is more a way of checking legal compliance, although it can also be seen as an opportunity for improvement.

Below are some graphs (Figure 4) showing which indicators are most relevant to the different benchmarks. Only the most relevant were selected and/or were an eliminating factor in certification.

In this set of graphs (Figure 4), the importance of environmental indicators is clear, especially in the vineyard. Economic indicators are only evaluated in a macro way, which encourages analyses in the direction of economic sustainability. Social indicators are becoming increasingly important, especially on the part of consumers. Consumers prefer products whose production respects human rights, such as fair wages, non-discrimination, and social equity [9,26]. Interaction with the local community is also valued, in terms of the circular economy and minimizing the environmental impact of the activity [5,6,16].

The carbon footprint is an indicator that is not directly addressed in some of the certification benchmarks, but most organizations have online availability so that producers can calculate it [32,34,35,37,40]. However, this is the indicator that consumers recognize most easily, perhaps because it is applicable to all products and is valued more highly than the certification label [12].

There are other certification benchmarks that have not been mentioned, but which are also important for environmental sustainability, such as integrated production (management of natural resources, favoring natural regulation, control of agrochemicals used, and safety times), organic production (determining the type of agrochemicals used, favoring biodiversity, preservation of natural resources) [43] or the Global GAP (benchmark for good agricultural practices) [44]. These models only focus on agricultural practices, but they are also applicable to viticulture.

6. Method

In June 2023, we conducted a literature review on environmental sustainability in wine cooperatives and their difficulties in responding to the new demands of markets and governments. For this analysis, we used two databases, ScienceDirect and Scopus, employing keywords and various combinations of them, i.e., sustainability, environment, cooperative wineries, cooperativism, sustainability benchmarks, and indicators. Figure 5 shows the research strategy. First, the word "sustainability" was included; then "cooperative" was included, then "environment", and finally different variables were inserted. Some restrictions were imposed: years of publication between 2015 and 2013; only research and review articles; and environmental, agricultural, and social areas. The search resulted in 2628 articles (1850 articles in ScienceDirect, 778 articles in Scopus), from which 126 articles were extracted for analysis. The rest were rejected because they were not associated with the wine sector or cooperativism or environmental sustainability, and because of duplication. Finally, 27 articles were included in this study. No software was used to support the analysis.

Figure 5. Search strategy use. Different keyword combinations. For example: sustainability AND cooperative AND indicators.

7. Discussion

Cooperatives have an important role in agriculture, but also in the communities in which they are established. The origins of cooperative agricultural organizations are associated with moments of crisis, when small producers join forces to sell the farm products they produce [13,45].

As we discussed at the beginning of this paper, companies and organizations must respond to the United Nations' challenge by creating benchmarks to meet the SDGs. Cooperatives have some of these goals as a priority, such as SDG 1, No Poverty. Cooperatives have been created to improve community conditions, such as SDG 2, No Hunger. In the case of agricultural cooperatives, the promotion of sustainable practices contributes not only to SDG 2, but also to SDGs 12 (Responsible Consumption) and 15 (Life and Land). According to the indicators analyzed above, cooperatives should be able to respond to SDGs 4 (Quality Education), 5 (Gender Equality), 8 (Good Jobs and Economic Growth), and 10 (Reduce Inequalities) through training, improving working conditions, and promoting gender and pay equality. The implementation of measures to mitigate climate change, such as water management and the use of renewable energy sources, should respond not only to SDGs 12 and 15, discussed above, but also to SDGs 6 (Clean Water and Sanitation), 7 (Renewable Energy), and 13 (Climate Action). Cooperatives are always well integrated into the community, and often provide support, so they always create synergies with government, social, political, business, and educational/research institutions. Since their mindset involves overcoming difficulties, administrations are very receptive to innovation. After this brief analysis, we can say that they easily respond to SDGs 9 (Innovation and Infrastructure) and 17 (Partnership for the Goals). Figure 6 below represents the SDGs that can be met by cooperatives if they implement the indicators discussed in the previous point.

In a quick analysis of the indicators listed in the benchmarks studied, it is possible to see that they respond to practically all of the SDGs in a more or less exhaustive way, as shown in Figure 7.

At the beginning of this review, some questions were raised, and with the information that has been compiled, we will try to answer them.

7.1. Q1: What Is Necessary to Achieve Sustainability?

First, we need to define sustainability, which, according to Ferrer [2], is the adaptation of human activities to guarantee the future of the next generations. In other words, it means securing a future where climate change has little impact, but also economic and social stability.

Analyzing the different benchmarks for sustainable certification, we were able to suggest a broad range of indicators that are common to the various models. These indicators address not only environmental issues, but also economic and social ones. For example, the Swedish market, Systembolaget [26], not only values environmental indicators but also gives great importance to social indicators, such as fair remuneration, non-discrimination, and precarious labor.

7.2. Q2: Is the Wine Cooperative Competitive in the Global Market?

Like any other company or organization, the cooperative is equally exposed to market challenges. The business model has proved resilient in times of crisis [12,13]. The objective of cooperatives is to sell the products of their members, remunerate them as much as possible, and reinvest the profits. However, this depends on the governance model and the members' commitment to the cooperative, for which they must maintain a high level of satisfaction. However, we have not fully answered the question because the challenges of market sustainability are what is needed. In the area of social sustainability, cooperatives respond comfortably, since this is the genesis of their creation, as well as their own economic sustainability and that of the community in which they are inserted. If the question is asked to each producer individually, it is not possible to answer because there is a lack of documentation. In the case of environmental sustainability, the producer is more attentive, although they are not sensitive to some indicators and do not measure them. There are other factors that the producer monitors for legal reasons or economic interests; for example, to comply with the Integrated Production Mode or the Organic Production Mode [43].

Even if the cooperative is competitive at the moment, it must create tools to respond to sustainability criteria, because the market demands it and consumers are becoming increasingly aware of these issues.

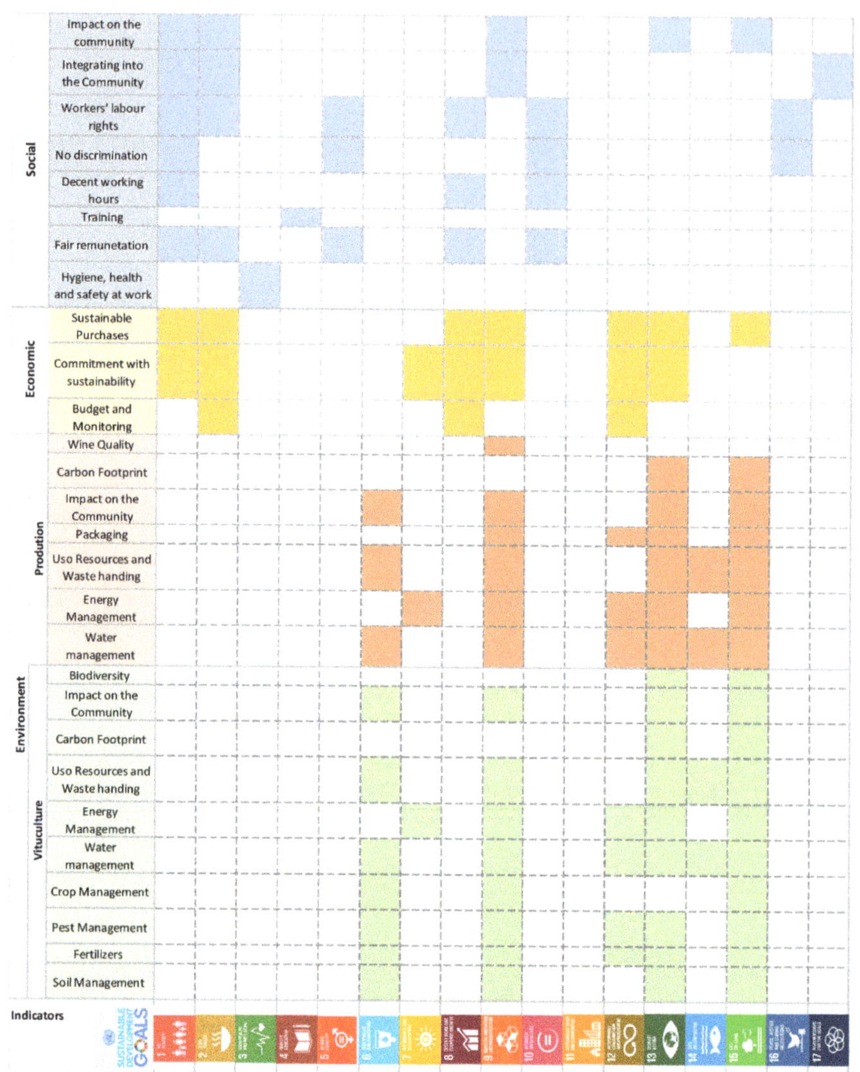

Figure 6. List of indicators identified in the benchmarks and their relationship with the SDGs (legend: green—indicators mentioned in the benchmarks for viticulture; orange—indicators mentioned in the benchmarks for wine production; yellow—economic sustainability indicators mentioned in the benchmarks; blue—social sustainability indicators mentioned in the benchmarks).

7.3. Q3: How Can We Respond to the Challenges of Environmental Sustainability While Maintaining Wine Quality Standards and Economic Profitability?

Although the literature explores environmental sustainability, it was not possible to find a relationship with quality and economic profitability in the cooperative model. The literature shows studies on the economic sustainability of cooperatives and their model, with its advantages and difficulties. However, when we tried to analyze whether the impact of environmental measures has a positive or negative economic impact, these data were not shown. We are unable to conclude whether some environmental measures have not been implemented for financial reasons, or if their implementation could present a cost reduction that would be attractive to the producer. To answer this question, there needs to be more research into the effect of measures to reduce the environmental impact on economic sustainability and also their economic viability, such as the effect on wine quality. Soregaroli [7], in a consumer survey, found that they valued the economic factor more than the carbon footprint. However, if the customer has the perception that wine with a low carbon footprint has higher quality, they will choose it [7]. Ferrer [2] analyzed the business model of 411 wineries in Spain and devised two types of business model: highly sustainable and low sustainability [2]. The major difference in the model was only related to the fact that the highly sustainable business model had a well-defined structure, favored the sale of bottled wine, and had knowledge of the entire process [2]. The model with a low level of sustainability sold mostly in bulk and did not know where the wine was going [2].

Figure 7. Sustainable development objectives that can be answered by wine cooperatives.

7.4. Q4: What Are the Economic and Social Impacts of Reducing the Carbon Footprint of the Winery and Its Members?

In the same way that the literature did not answer the previous question, we did not find any answers in the literature to this question either. Cooperatives, especially wine cooperatives, are made up of small producers, most of whom are older and have low literacy levels, which makes it difficult not only to communicate but also to obtain answers, as Figueiredo [13] mentions in his study of wine cooperatives in the Dão wine region. However, there are no works in the literature that answer this question in the case of other types of organizations.

It is possible to have a consistent and comprehensive group of sustainability indicators, already implemented and with a track record in the wine sector, but there is a lack of studies on the impact of these indicators on communities, organizations, and consumers. It is important for small producers to realize that they have a fundamental role to play on the road to sustainability, but they need to know what the economic advantage is. Their priority is to satisfy their needs, and selling their products to the cooperative will fulfil them.

8. Conclusions and Future Directions

To answer the questions raised, it is necessary to develop a methodology that allows wine cooperatives to calculate their level of sustainability in a credible way, as well as that of their members. This methodology should cover the most relevant indicators: water management, soil management, vine management (including crop practices, nutrition, and pest control), energy management, carbon footprint, and human resources (workers' rights, hygiene, health, safety at work). It should also respond via the cooperative organization to indicators on local biodiversity and the impact of activities on the community (not only environmentally, but also socially).

Through the analysis of the dynamics of cooperative wineries, we can transform weaknesses into added value, such as by giving members an active role in sustainability, creating integration tools to mitigate economic differences such as financial capacity or the area of land parcels. Providing the organization with tools with which they can integrate their members will enable them to respond to the current environmental, economic, and social challenges not only imposed by the wine markets, but also by the current socio-economic situation. In other words, this will create activities by which the environmental, social, and economic aspects of winegrowing members and wine production can be improved.

Author Contributions: Conceptualization, C.A.T. and A.M.; methodology, C.A.T.; software, A.M.; validation, C.A.T. and A.M.; formal analysis, A.M.; investigation, A.M.; resources, A.M.; data curation, C.A.T.; writing—original draft preparation, A.M.; writing—review and editing, C.A.T.; visualization, A.M.; supervision, C.A.T.; project administration, C.A.T.; funding acquisition, C.A.T. All authors have read and agreed to the published version of the manuscript.

Funding: This work was supported by National Funds from the FCT—Portuguese Foundation for Science and Technology, under the project UIDB/04033/2020.

Data Availability Statement: The data used to support the findings of this study are available from the corresponding author upon request.

Acknowledgments: We thank Adega Cooperativa de Favaios, CRL, for their support in the development of the project.

References

1. Chabin, Y.; Rochard, J. UN Sustainable du Development Goals (SDGs) and Corporate Social Responsibility in the wine sector: Concepts and applications. *BIO Web Conf.* **2023**, *56*, 03011.
2. Ferrer, J.R.; García-Cortijo, M.C.; Pinilla, V.; Castillo-Valero, J.S. The business model and sustainability in the Spanish wine sector. *J. Clean. Prod.* **2022**, *330*, 129810. [CrossRef]
3. United Nations. 2023. Available online: https://sdgs.un.org/goals (accessed on 3 August 2023).
4. Nazzaro, C.; Stanco, M.; Uliano, A.; Lerro, M.; Marotta, G. Collective smart innovations and corporate governance models in Italian wine cooperatives: The opportunities of the farm-to-fork strategy. *IFAMA* **2022**, *25*, 723–736. [CrossRef]
5. Tsalidis, G.A.; Kryona, Z.-P.; Tsirliganis, N. Selecting south European wine based on carbon footprint. *Resour. Environ. Sustain.* **2022**, *9*, 100066. [CrossRef]
6. Martínez-Falcó, J.; Martínez-Falcó, J.; Marco-Lajara, B.; Sánchez-García, E.; Visser, G. Aligning the Sustainable development Goals in the wine industry: A bibliometric analysis. *Sustainability* **2023**, *15*, 8172. [CrossRef]
7. Soregaroli, C.; Ricci, E.C.; Stranieri, S.; Nayga, R.M., Jr.; Capri, E.; Castellari, E. Carbon footprint information, prices, and restaurant wine choices by customers: A natural field experiment. *Ecol. Econ.* **2021**, *186*, 107061. [CrossRef]
8. Litskas, V.D.; Irakleous, T.; Tzortzakis, N.; Stavrinides, M.C. Determining the carbon footprint of indigenous and introduced grape varieties through Life Cycle Assessment using the island of Cyprus as a case study. *J. Clean. Prod.* **2017**, *156*, 418–425. [CrossRef]
9. Lamastra, L.; Balderacchi, M.; Di Guardom, A.; Monchiero, M.; Trevsisan, M. A novel fuzzy expert system to assess the sustainability of the viticulture at the wine-estate scale. *Sci. Total Environ.* **2016**, *572*, 724–733. [CrossRef]
10. Marras, S.; Masia, S.; Duce, P.; Spano, D.; Sirca, C. Carbon footprint assessment on a mature vineyard. *Agric. For. Meteorol.* **2015**, *214–215*, 350–356. [CrossRef]
11. Casolani, N.; D'Eusania, M.; Liberatore, L.; Raggi, A.; Petti, L. Life Cycle Assessment in the wine sector: A review on inventory phase. *J. Clean. Prod.* **2022**, *376*, 134404. [CrossRef]
12. Calle, F.; González-Moreno, A.; Carrasco, I.; Vargas-Vargas, M. Social economy, environmental proactivity eco-innovation and performance in the Spanish wine sector. *Sustainability* **2020**, *12*, 5908. [CrossRef]
13. Figueiredo, V.; Franco, M. Wine Cooperatives as a form of social entrepreneurship: Empirical evidence about their impact on society. *Land Use Policy* **2018**, *79*, 812–821. [CrossRef]
14. Ziegler, R.; Poirier, C.; Lacasse, M.; Murray, E. Circular Economy and Cooperatives—An Exploratory Survey. *Sustainability* **2023**, *15*, 2530. [CrossRef]
15. Moscovici, D.; Reed, A. Comparing wine sustainability certifications around the word: History, status and opportunity. *J. Wine Res.* **2018**, *29*, 1–25. [CrossRef]

16. D'Ammaro, D.; Capri, E.; Valentino, F.; Grillo, S.; Fiorini, E.; Lamastra, L. A multi-criteria approach to evaluate the sustainability performances of wines: The Italian red wine case study. *Sci. Total Environ.* **2021**, *799*, 149446. [CrossRef] [PubMed]
17. Dressler, M. Sustainable Business Model Design: A Multi-Case Approach Exploring Generic Strategies and Dynamic Capabilities on the Example of German Wine Estates. *Sustainability* **2023**, *15*, 3880. [CrossRef]
18. Ritcher, B.; Hanf, J.H. Cooperatives in wine Industry: Sustainable Management practices and digitalisation. *Sustainability* **2021**, *13*, 5543.
19. Ritcher, B.; Hanf, J.H. Sustainability as "Value of Cooperatives"—Can (wine) Cooperatives use sustainability as a driver for a brand concept? *Sustainability* **2021**, *13*, 12344.
20. Figueiredo, V.; Franco, M. Factors influencing cooperator satisfaction: A study applied to wine cooperatives in Portugal. *J. Clean. Prod.* **2018**, *191*, 15–25. [CrossRef]
21. Ramos, M.E.; Azevedo, A.; Meira, D.; Malta, M.C. Cooperatives and the Use of Artificial Intelligence: A Critical View. *Sustainability* **2023**, *15*, 329. [CrossRef]
22. CASES. *Código Cooperativo Lei nº119/2015 of 31 August Amended by Lei nº66/2017 of 9 August*; CASES: Lisboa, Portugal, 2018.
23. Whitehead, J. Prioritizing sustainability indicators: Using materiality analysis to guide sustainability assessment and strategy. *Bus. Strategy Environ.* **2017**, *26*, 399–412. [CrossRef]
24. Borsato, E.; Zucchinelli, M.; D'Ammaro, D.; Giubilato, E.; Zabeo, A.; Criscione, P.; Pizzol, L.; Cohen, Y.; Tarolli, P.; Lamastra, L.; et al. Use of multiple indicators to compare sustainability performance of organic vs conventional vineyard management. *Sci. Total Envorin.* **2020**, *711*, 135081. [CrossRef]
25. Merli, R.; Preziosi, M.; Acampora, A. Sustainability experiences in the wine sector: Toward the development of an international indicators system. *J. Clean. Prod.* **2018**, *172*, 3791–3805. [CrossRef]
26. Omsystmenbolaget. 2023. Available online: https://www.omsystembolaget.se/english/sustainability/labels/sustainable-choice/environmental-certifications (accessed on 4 August 2023).
27. OIV. 2020. Available online: https://www.oiv.int/public/medias/7601/oiv-viti-641-2020-en.pdf (accessed on 3 August 2023).
28. Da Silva, L.P.; Da Silva, J.C.G.E. Evaluation of the carbon footprint of the life cycle of wine production: A review. *Clean. Circ. Bioeconomy* **2022**, *2*, 100021.
29. Herath, I.; Green, S.; Singh, R.; Horne, D.; Zijpp, S.; Clothier, B. Water footprint of agricultural products: A hydrological assessment for the water footprint of New Zealand's wines. *J. Clean. Prod.* **2013**, *41*, 232–243. [CrossRef]
30. Discover Sustainable Wine. 2023. Available online: https://discoversustainablewine.com/new-zealand/ (accessed on 5 August 2023).
31. Referencial de Sustentabilidade Para o Setor Vitivinícola, ViniPortugal. 2023. Available online: https://www.viniportugal.pt/pt/sustentabilidade/ (accessed on 5 August 2023).
32. IVDP. 2023. Available online: https://digital.ivdp.pt/douro-sustentavel/manual-de-sustentabilidade-na-regiao-do-douro-vinhateiro/ (accessed on 6 August 2023).
33. PSA. 2023. Available online: https://sustentabilidade.vinhosdoalentejo.pt/pt/certificacao-psva (accessed on 5 August 2023).
34. LIVE. 2023. Available online: https://livecertified.org (accessed on 4 August 2023).
35. LODI WINE Growers. 2023. Available online: https://lodigrowers.com/certification/ (accessed on 4 August 2023).
36. VDD. 2023. Available online: https://www.icv.fr/en/viticulture-oenology-consulting/grower-sustainable-development-vdd (accessed on 4 August 2023).
37. California Sustainable Winegrowing Alliance (CSWA). 2023. Available online: https://www.sustainablewinegrowing.org/sustainable_winegrowing_program.php (accessed on 2 August 2023).
38. Integrated Protution Wine (IPW). 2023. Available online: https://www.ipw.co.za/ (accessed on 4 August 2023).
39. Sustainability Code of the Chilean Wine Industry (WOC). 2023. Available online: https://www.sustentavid.org/en/ (accessed on 5 August 2023).
40. VIVA. 2022. Available online: https://viticolturasostenibile.org/en/ (accessed on 4 August 2023).
41. Sustainable Austria. 2023. Available online: https://www.austrianwine.com/our-wine/environmental-consciousness-in-austrian-viticulture/sustainable-austria-certified-viticulture (accessed on 4 August 2023).
42. Fresh Australian Wine Industry Standard of Sustainable Practice. 2023. Available online: https://www.freshcare.com.au/our-standards/awissp-win1/ (accessed on 5 August 2023).
43. Direção-Geral de Agricultura e Desenvolvimento Rural. 2023. Available online: https://www.dgadr.gov.pt/producao-integrada2 (accessed on 8 August 2023).
44. Global GAP. 2023. Available online: https://www.globalgap.org/pt/ (accessed on 8 August 2023).
45. Deng, W.; Hendrikse, G.; Liang, Q. Internal social capital and the life cycle of agricultural cooperatives. *J. Evol. Econ.* **2021**, *31*, 301–323. [CrossRef]

Article

Farmers' Perceptions, Insight Behavior and Communication Strategies for Rice Straw and Stubble Management in Thailand

Sukanya Sereenonchai * and Noppol Arunrat

Faculty of Environment and Resource Studies, Mahidol University, Salaya, Nakhon Pathom 73170, Thailand; noppol.aru@mahidol.ac.th
* Correspondence: sukanya.ser@mahidol.ac.th

Abstract: The adoption of rice straw and stubble management approaches can be affected by various factors. To understand the psychological factors influencing Thai farmers' adoption of rice straw and stubble management approaches, three integrated behavioral theories were employed: the Theory of Planned Behavior (TPB), the Value-Belief-Norm (VBN) and the Health Belief Model (HBM). Then, a practical communication framework was synthesized and proposed to promote rice straw utilization for social-ecological benefits to achieve more sustainable agricultural production. Through a questionnaire survey and in-depth interviews with 240 local farmers, a statistical analysis was performed employing cross-tab, stepwise multiple linear regression, one-way ANOVA and descriptive content analysis using QDA lite miner software. The key results clearly showed that perceived pro-environmental personal norms, perceived cues to rice straw utilization, perceived behavioral control, perceived severity of rice straw burning, perceived ascription of responsibility, and the perceived benefits of rice straw utilization were significantly negatively influenced by burning, and that there was a significantly negative difference to non-burning approaches. Meanwhile, cost savings as perceived benefits of the current option of burning showed a significantly positive difference when compared with incorporation and free-duck grazing options. In communication strategies to promote rice straw utilization for achieving sustainable agriculture, key messages should highlight the clear steps of rice straw utilization, as well as the costs and benefits of each option in terms of economic, health, environmental and social perspectives. Moreover, messages designed to promote action knowledge and self-efficacy at the group level, to promote perceived responsibility via self-awareness and self-commitment, and convenient channels of communication to the farmers can help to achieve more effective non-burning rice straw and stubble management.

Keywords: farmers' perceptions; insight behavior; communication strategies; rice straw and stubble management; sustainable agricultural production; Thailand

Citation: Sereenonchai, S.; Arunrat, N. Farmers' Perceptions, Insight Behavior and Communication Strategies for Rice Straw and Stubble Management in Thailand. *Agronomy* **2022**, *12*, 200. https://doi.org/10.3390/agronomy12010200

Academic Editors: Moritz Von Cossel, Joaquín Castro-Montoya and Yasir Iqbal

Received: 3 December 2021
Accepted: 12 January 2022
Published: 14 January 2022

Publisher's Note: MDPI stays neutral with regard to jurisdictional claims in published maps and institutional affiliations.

1. Introduction

The open-field burning of rice straw and stubble is a common practice in many countries. Recently, burning has caused serious air pollution problems worldwide [1], which has resulted in calls for participation from stakeholders to deal with this issue. Air pollution from open burning has also been a serious environmental health risk impacting Thai people [2]. The air pollutants from crop residues due to open burning in Thailand were found to be primarily CO_2, CO, PM_{10}, $PM_{2.5}$ [3].

In Thailand, the Department of Agricultural Extension has recognized the importance of this problem. Therefore, a project to promote the cessation of burning in agricultural areas was initiated in 2014 and continues to this day, which aims to expand the results across the country [4]. Although some technologies have been introduced to Thai society as options to avoid burning and to benefit from the use of rice straw through compacting, soil covering, incorporation, growing mushrooms, etc., the problem of open-field burning remains a serious issue. In practice, however, many farmers still burn rice straw and stubble.

The relevant authorities have not yet been able to take legal action. Therefore, questions arise as to why farmers choose to burn rice straw and stubble, and which management options are the most cost-effective for farmers and society at large.

Based on a literature review, some previous studies employed behavioral theories to understand farmers' climate mitigation and adaptation behaviors (e.g., [5–8]). To the best of our knowledge, one study used the Theory of Planned Behavior (TPB) to investigate the factors and mechanisms driving the straw resource utilization behaviors of Chinese farmers [9]. Meanwhile, other relevant studies were more focused on using quantitative analysis to explore the factors influencing rice straw management practices [1,10] that were not directly based on behavioral theory. There was still a research gap in applying behavioural theory to understand farmers' behaviors especially regarding rice straw management, and the lack of practical communication strategies to promote non-burning approaches.

Consequently, the present study took both qualitative and quantitative approaches aiming to answer two main questions: (1) What are the key psychological factors influencing farmers' decisions to adopt each type of rice straw and stubble management; (2) What communication strategies should be planned and practiced further for promoting non-burning rice straw and stubble management approaches? The contribution of this study could promote rice straw utilization for social–ecological benefits for the achievement of more sustainable agricultural production.

2. Literature Review on Factors Influencing Farmers' Rice Straw Management Practice

2.1. Application of the TPB, VBN and HBM to Understand Farmers' Decision

This study utilized three integrated behavioral theories (Figure 1) consisting of the Theory of Planned Behavior (TPB), the Value-Belief-Norm (VBN) and the Health Belief Model (HBM), which were proposed by Zhang et al. [6], Abdollahzadeh and Sharifzadeh [7] and Ataei et al. [8], respectively, to understand the psychological factors influencing farmers' adoption of each rice straw and stubble management approach.

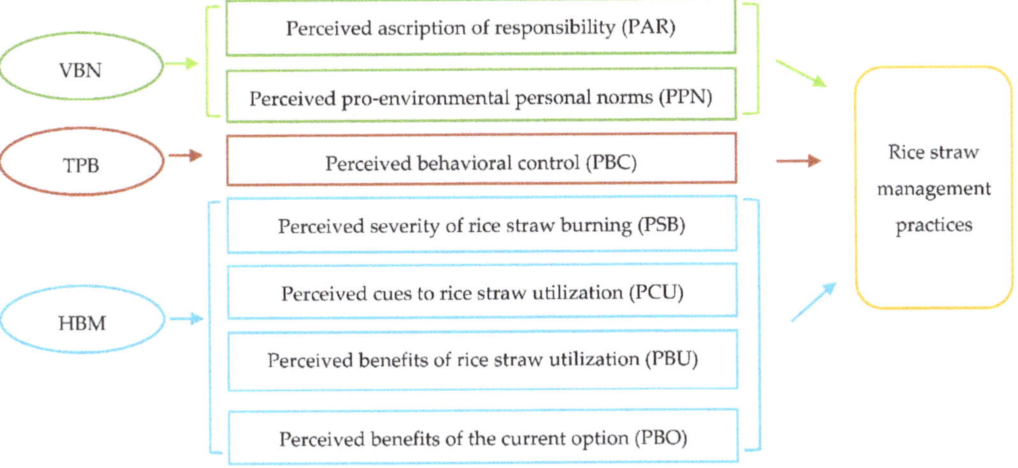

Figure 1. Integrated theories of TPB, VBN and HBM using in this study.

Based on the TPB, behavioral intention is influenced by attitude, subjective norms, and perceived behavioral control [11–13]. The Value-Belief-Norm (VBN) theory moves from personal values (biospheric, altruistic, and egoistic values) to the new environmental paradigm, which includes an awareness of consequences, the ascription of responsibility, and pro-environmental personal norms, and finally leads to pro-environmental behaviors [14].

The TPB and VBN models were employed to predict Chinese farmers' intentions in relation to farmers' climate mitigation and adaptation behaviors using partial least squares structural equation modelling (PLS-SEM) [6]. The TPB was found to be more successful for predicting self-interest-oriented behaviors like climate adaptation, while the VBN theory better explained altruistic behaviors like climate mitigation. The TPB and HBM models were also employed together to explore farmers' intentions to use green pesticides based on Structural equation modelling (SEM) for data analysis. Key results highlighted that both theories could predict the intention, while HBM was better than TPB [8]. Furthermore, the HBM was particularly applied with respect to health-issue relevance. One example employed the HBM to examine the factors affecting farmers' intentions to use personal protective equipment (PPE), three components were positively found: higher levels of perceived severity of pesticide adverse effects; cues to action; and perceived PPE benefits [7].

To understand health-related behaviors, the HBM has been claimed by some scholars to be the most appropriate and widely employed framework. The model consisted of perceived severity, perceived susceptibility, perceived barriers, perceived benefits, perceived self-efficacy, and cues to action [15]. The HBM was applied with some studies regarding farmers' behaviors such as farmers' intentions and safety behavior regarding pesticides [7,16,17]. For rice straw management options, there is also a health-related issue because the burning method can also generate air pollution. Therefore, this study also integrated the HBM to understand farmers' decisions with respect to their concern over people's health.

2.2. Factors Influencing Farmers' Decision to Choose Rice Straw Management Practices

2.2.1. Factors Influencing Burning Decision

Based on the logit regression model, which is used to understand why the Indian farmers choose each crop residue practice, social influence was found as a significant determinant of residue burning [18]. Additionally, weather (humidity and rain), disproportionate incentives, inefficient straw collection technology, inefficient management from agricultural agencies, lack of logistic facilities (baler machines, storage and transportation), lack of capital to manage straw, and a low level of skills and knowledge were found for Malaysian farmers, where farmers also realized the benefits of rice straw burning due to it having no serious impacts, and being the easier and cheaper option [10]. Although the farmers perceived high risks, few benefits, low acceptance for rice straw burning [10], and an awareness of its adverse environmental effects [18], they retained their burning practice [10,18].

2.2.2. Factors Influencing Non-Burning Decision

To analyze the factors influencing farmers' adoption of different rice straw management techniques, i.e., covering, burning, incorporation, or rice straw removal using multinomial logit models [1], farm type, location, number of household members, cow ownership, and distance from farm to house, were found to significantly influence farmers' use of alternative techniques, i.e., incorporation or removal instead of burning. Other factors that also influenced farmers to incorporate alternative techniques instead of burning were training attendance, perceptions of incorporation benefits, income from non-rice farming, cultivated area, tenure status, and provincial regulations of burning. Moreover, the significant perception variables for Vietnamese farmers to incorporate were the negative impacts of open-field burning, awareness of environmental regulations, and attitude towards incentives. They adopted and incorporated these perception variables for themselves rather than for the environment or society [10].

These attitudes were also mentioned by Kadam et al. [19] as the most important for changing straw management practices in the United States, particularly regulation with greater economic incentives for cooperation with rice straw collection advice, which might help to change farmers' attitudes. The economic or financial incentive could be a powerful driver for farmers to choose a non-burning approach [10,20,21]. In addition, to investigate the impact of policy measures on Chinese farmers' rice straw management using a regression model, key results highlighted that the burning ban has reduced rice straw burning dramatically and motivated farmers to retain straw in their soil. However, the straw retention subsidy seemed to have an insignificant effect because it was low and not directly provided to farmers [22].

In order to promote non-burning and more sustainable farming practices, acceptability, feasibility and benefit perception should be promoted [10,23].

3. Materials and Methods

3.1. Study Areas

The study areas were purposively selected covering 12 villages in the Taluk sub-district of Chainat Province in the middle region of Thailand (Figure 2). These areas were selected by Thailand's Department of Agricultural Extension based on published data on yields, costs and net income from rice plantations in 2005–2006, which were collected to avoid open-field burning in agricultural areas [24]. The main occupation of most people in the area is rice cultivation and it is an area that is suitable for growing rice according to the Agri-Map online platform that was jointly developed by the Ministry of Agriculture and Cooperatives, the Ministry of Science and Technology, and the National Electronics and Computer Technology Center as a member of the National Science and Technology Development Agency.

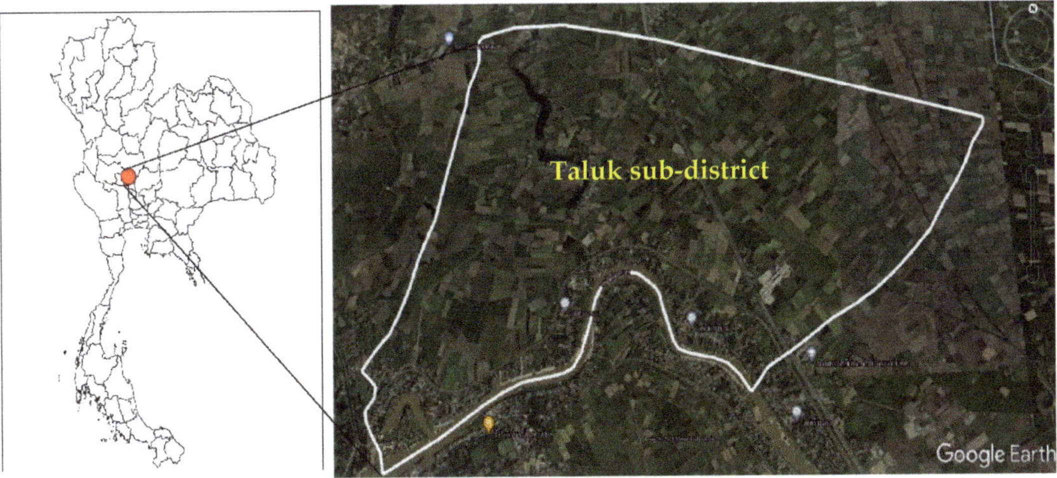

Figure 2. Study area.

3.2. Sampling Design and Data Collection

Farmers' household surveys were conducted by accidental or convenience sampling of farmers from 12 villages in the Taluk sub-district of Chainat Province, who were conveniently placed, willing to participate in answering the questionnaire and provide in-depth details. The researchers collected questionnaires from 240 local farmers via door-to-door visits and informal interviews to obtain details about their rice farming, and rice straw and stubble management.

The questionnaire was designed based on the TPB, VBN and HBM theories. The reliability of the questions regarding psychological factors was tested, with Cronbach's alpha being between 0.870 to 0.983 (Table 1). Consequently, the questions were deemed to be reliable and fit for the objectives of this study.

Table 1. Psychological factors, variables, reliability test and reference theories.

Psychological Factors	Cronbach's Alpha	Reference Theories	Sources
(1) Perceived pro-environmental personal norms (PPN)	0.983	Value-Belief-Norm (VBN): Pro-environmental personal norms	[6,8]
(2) Perceived cues to rice straw utilization (PCU)	0.915	Health Belief Model (HBM): Cues to action	[7,8]
(3) Perceived behavioral control of rice straw utilization (PBC)	0.962	Theory of planned behavior (TPB): Perceived behavioral control	[6,8]
(4) Perceived severity of rice straw burning (PSB)	0.941	Health Belief Model (HBM): Perceived severity	[7,8]
(5) Perceived ascription of responsibility (PAR)	0.929	Value-Belief-Norm (VBN): Ascription of responsibility	[6]
(6) Perceived benefits of rice straw utilization (PBU)	0.944	Health Belief Model (HBM): perceived benefits	[7,8]
(7) Perceived benefits of current option (PBO)	0.870	Health Belief Model (HBM): perceived benefits	[6,8]

To validate the questionnaire, three experts in different fields of agricultural management, environmental and health psychology, and environmental and health communication were invited individually in-person and online. The questionnaire was pre-tested with 10 farmers in the same province who were not be included in the final study. A few questions of the questionnaire were revised, alongside the language that was used based on the pilot farmers' suggestions that it be easier to understand by the farmers.

The questionnaire consisted of three main parts (Table 2): (1) demographic information; (2) rice straw and stubble management; and (3) psychological factors. A checklist, an open form and a five-point Likert scale were used to record their responses, where 1 = minimum and 5 = maximum scores.

3.3. Data Analysis

To understand farmers' perceptions regarding rice straw and stubble management including pro-environmental personal norms, the benefits of rice straw utilization, perceived behavioral control, ascription of responsibility, cues for rice straw utilization, severity of rice straw burning, and the benefits of the current option, cross-tabs and percentages were employed for data analysis. Moreover, stepwise multiple linear regression analysis was used to decode the psychological factors affecting farmers' rice straw and stubble management. The normal distribution test, a test of the homogeneity of variances, and a one-way ANOVA test using post hoc multiple comparisons (Scheffe) to compare the farmers' perceptions of rice straw and stubble management practices were also performed. These tests were analyzed using SPSS version 22.0.

The qualitative data from farmers' interviews were analyzed by thematic content analysis [25]. Firstly, recorded data was transcribed, read overall to familiarize the collected data, and classified into different issues based on the study objectives. Then, the QDA Miner Lite Program was employed to code, label the similar meaning, and group the data. The overall analysis from the Program was printed out and reviewed again. Then, the critical results were described by face-to-face discussions between two researchers according to the interview issues and study objectives.

Table 2. The questions used in the questionnaire to ask respondents.

Part	Questions
(1) Demographic information (a checklist and an open form)	1.1 Gender (1. male, 2. female) 1.2 Age (indicating years) 1.3 Schooling (indicating years) 1.4 Farmland owner (1. no, 2. yes) 1.5 Farm size (indicating areas: Rai) 1.6 Farming group attendance (1. no, 2. yes)
(2) Rice straw and stubble management (a checklist)	2.1 Burning 2.2 Compacting 2.3 Incorporation 2.4 Free-grazing ducks 2.5 Mixed method
(3) Psychological factors (a five-point Likert scale answer: 5 = most, 4 = more, 3 = moderate, 2 = low, 1 = very low)	3.1 Perceived pro-environmental personal norms (PPN) PPN1: The use of rice straw is consistent with your farming values. PPN2: You feel it is your responsibility to help reduce air pollution. PPN3: You feel guilty if you do not utilize the rice straw. 3.2 Perceived cues to rice straw utilization (PCU) PCU1: You have read/heard/received the information about using rice straw to reduce environmental problems. PCU2: Local authorities educate about rice straw utilization. PCU3: You can access facilities of rice straw utilization such as purchasing sites near the community. 3.3 Perceived behavioral control of rice straw utilization (PBC) PBC1: You have a good understanding of how to gain benefits from rice straw. PBC2: Rice straw utilization is easy and uncomplicated. PBC3: You feel confident that rice straw can be utilized. 3.4 Perceived severity of rice straw burning (PSB) PSB1: Rice straw burning is a threat to local people's health. PSB2: Rice straw burning is a serious threat to the environment and agriculture. PSB3: Rice straw burning is a serious threat for future generations. PSB4: People and the environment cannot cope with the effects of rice straw burning. 3.5 Perceived ascription of responsibility (PAR) PAR1: Everyone should be responsible for air pollution. PAR2: People in the community should avoid activities that cause air pollution. 3.6 Perceived benefits of rice straw utilization (PBU) PBU1: Rice straw utilization has a positive effect on agriculture and the environment around you. PBU2: Rice straw utilization helps increase income. PBU3: Rice straw utilization helps build good relationships in the community. PBU4: Rice straw utilization does not incur an extra cost. PBU5: Rice straw is more useful than burning, so it should be used for maximum benefit. 3.7 Perceived benefits of current option (PBO) PBO1: Your rice straw and stubble management option can help generate income for yourself/your family. PBO2: Your rice straw and stubble management option has low costs. PBO3: Your rice straw and stubble management option can help reduce air pollution. PBO4: Your rice straw and stubble management option does not cause trouble for others and allows people to coexist peacefully. PBO5: Your rice straw and stubble management option is appropriate for the available resources in your community.

4. Results and Discussion

4.1. Demographic Information

Most of the respondents were male (around 55.83%). Their average ages, farm sizes, and levels of education were 51 years old, 19 Rais (3.04 hectares) and secondary school, respectively. Most farmers owned their rice farming area (91.67%) but did not attend the farming group (94.17%). Five main rice straw and stubble management practices were found in this area consisting of (1) burning (43.75%), (2) compacting (40.83%), (3) incorporation (6.25%), (4) free-grazing ducks (1.25%) and (5) mixed methods (7.92%). The average costs and returns for each method of rice straw and stubble management are shown in Table 3.

Table 3. Average costs and returns for each method of rice straw management.

Rice Straw Management	Costs	Returns
(1) burning	Lighter 10 Baht/piece (can use more than one time)	Fast straw management
(2) compacting	-	120–180 Baht/rai or 750–1125 Baht/hectare
(3) incorporation	Wage for plowing rice straw (200 Baht/rai or 1250 Baht/hectare)	Soil nourishment
(4) free-grazing ducks	Water pumping (75 Baht/rai or 468.75 Baht/hectare)	Free duck eggs (around 30 duck eggs)
(5) mix method		
• compacting	-	120–180 Baht/rai or 750–1125 Baht/hectare
• incorporation	Water pumping (75 Baht/rai or 468.75 Baht/hectare)	Soil nourishment
• soil covering	-	Soil moisture

4.2. Farmers' Rice Straw and Stubble Management and Their Perceptions

Based on integrated quantitative and qualitative approaches, this study highlighted psychological factors that could reflect the farmers' demographic, social and economic factors as well. Five main rice straw and stubble management practices were found in the study area as described in detail below.

4.2.1. Rice Straw Burning

For the farmers who chose to burn rice straw, the lowest-rated perception was found for all sub-issues of PPN, PCU, PBC and PSB (Table S1).

For the perceived benefit of burning, cost-saving was found to be the highest-rated perception (PBO2, 69.52%), while generating income (PBO1, 85.71%) and reducing air pollution (PBO3, 80.95%) were the lowest-rated perceptions. The cost-saving from the burning method is in line with the findings of Ahmed et al. [26], who mentioned that rice straw utilization had higher costs than burning, so farmers in Pakistan adopted the burning method. This was also in line with the Malaysian farmers' perception of the ease and low cost of rice straw burning [27].

The lowest-rated perceptions regarding the utilization of rice straw were the fostering of good relationships (PBU3, 76.19%) and the lack of additional cost requirements (PBU4, 85.71%), which relate to the perceived benefits of rice straw utilization. This perception can hinder the adoption of rice straw utilization, as also supported by Launio et al. [1] who suggested that having a low income could prevent Filipino and Vietnamese farmers from adopting practices other than burning, as they were unwilling to pay more for rice straw and stubble management.

Moreover, a moderate-level perception rating (undecided) was found regarding the benefit for agriculture and the surrounding environment (PBU1, 80.95%). Some of the farmers added that they were also interested in methods of rice straw utilization instead of burning, especially compacting, which was the method that generated the most interest provided that facilities could be supplied for them as well.

In light of the perceptions of farmers who were not sure about the advantages of using rice straw to benefit their agricultural activities and surrounding environment, and given the interest in rice straw compacting, the comparative communication highlighted each rice straw utilization method, as well as the costs and benefits for the agricultural system, environment and people. This was also supported by the findings of Rosmiza et al. [27], who suggested that farmers would be more active in participating in and adopting the use of a new technology if they could increase their profits. Moreover, to promote the adoption of rice straw utilization, Launio et al. [1] raised the importance of accessing appropriate public and private support by the farmers. Therefore, continuous agricultural advice by local authorities and agricultural extensionists should be arranged to promote confidence and the improvement of abilities relating to rice straw utilization instead of burning. Additionally, knowledge on the benefits of rice straw, which is rich in various substances and can be used as fertilizer, fodder, for bioenergy, etc. should be highlighted as well.

4.2.2. Rice Straw Compacting

The farmers who chose to compact rice straw perceived the benefit of this option at the highest level across all sub-elements, especially for generating income (PBO1, 100%), reducing air pollution (PBO3, 100%), and helping to live with others in the community peacefully (PBO4, 100%), followed by saving costs for rice straw and stubble management (PBO2, 77.55%) and appropriateness for the available resources in the area (PBO5, 54.08%) (Table S1). This highest-level perception rating was consistent with the perceived benefits of rice straw utilization as follows: most farmers strongly agree that rice straw is more useful than burning, so it should be used for maximum benefit (PBU5, 98.98%) and the generation of income (PBU2, 92.86%); secondly, most agree with the use of rice straw to benefit agriculture and the surrounding environment (PBU1, 100%), that it helps to strengthen good relationships in the community (PBU3, 100%) and has no additional costs (PBU4, 88.78%).

Furthermore, the farmers gave high-level perception ratings regarding PPN and PBC. The utilization of rice straw has been a common practice for an average of 2–3 years. Rice straw compacting is convenient and generates income for the farmers because the private sector that buys straw directly contacts the farmers at their paddy fields, and offers a price for rice straw, meaning that farmers can decide whether to sell the straw or not. Therefore, the farmers considered that, compared to methods of rice straw and stubble management such as burning, compacting is a new way that they see has advantages without disadvantages, so other farmers in their village should also choose this method. In addition, during the past 2–3 years, there has been an issue of smog from open burning, therefore, compacting has been viewed as one solution that can help reduce air pollution. The farmers thought that everyone should take part in reducing air pollution as much as possible, and stated that they will continue to use this method if in the future the private sector still buys their rice straw.

It is interesting that the farmers perceived high (49.5%) and moderate (47.3%) levels of understanding for rice straw utilization, while perceived ease and confidence in rice straw utilization was at a high level (73.6%), and some responded at the highest level (26.4%). Consequently, the alternatives to rice straw utilization should be clearly communicated, both the methods/steps, as well as the costs and benefits to be earned by comparing each method. If these can be measured or compared in terms of money gained and lost, the clear alternatives to burning rice straw trend to be adopted by farmers.

The results of PAR showed that the farmers perceived at a high level the sub-issue that everyone should be responsible for air pollution (PAR1, 100%). Meanwhile, they perceived at a moderate level that people in the community should avoid activities that might cause air pollution (PAR2, 61.22%), because of an understanding of the necessity that some farmers need burning. They understood that some rice fields are not large enough for the private sector to directly contact and compact rice straw in their fields. In addition, most farmers rushed to maximize the amount of rice cultivation, so it is necessary to choose a rice straw and stubble management method that is fast, convenient, and does not cost a lot. The burning method fulfills these criteria compared with non-burning methods. Based on this reason, our survey resulted in a high-level perception rating (100%, not the highest level) regarding the issue that everyone should be responsible for air pollution.

PCU found that most farmers perceived the highest level of rice straw utilization facilities (PCU3, 57.1%). There was a high level of perception for rice straw utilization education by local authorities (PCU2, 67%) and a moderate level of perception for the reading/hearing/receiving of information expressing that utilizing rice straw reduces environmental problems (PCU1, 79.1%). For rice straw compacting, a private sector representative made contact and agreed on the price of purchasing and the date of compacting rice straw in the paddy fields, so it was convenient to manage the rice straw. There were also local authorities to promote knowledge about rice straw utilization. Some of the farmers gained knowledge from attending community agricultural groups, especially large-scale agriculture groups. However, most farmers do not have knowledge on how to gain the benefits of utilizing rice straw. Having agricultural agencies and academics promote knowledge and practice is, therefore, an important and effective way for farmers to learn how to gain benefits from rice straw. These findings reflect that farmers are more open to knowledge from local authorities than from other channels. Therefore, another viable approach may be to design communication streams through local authorities to directly promote rice straw utilization approaches and practical actions to farmers. Mechanisms to support rice straw utilization, which can be operated mainly by the farmers in the community should be also promoted.

Most farmers perceived the severity of rice straw burning at a moderate level. This was proven through the sub-issues that people and the environment were unable to cope with the effects of rice straw burning (PSB4, 91.0%), followed by perceptions that burning straw poses a serious threat to the environment and agriculture (PSB2, 86.8%), and that it poses a threat to community health (PSB1, 69.2%). Most disagreed that rice straw burning was a serious threat for future generations (PSB3, 100%).

The explanation provided by the farmers was that they did not feel rice straw burning was a very serious threat was because it did not cause a serious effect for a long time. Farmers only burnt the rice straw after harvesting and preparing for the next crop. In some years, rice could only be planted once or twice because of insufficient water for many crops. Each burning took a short period of time for the fire to be extinguished. Most farmers knew and understood their reasons for burning. At the same time, there were ways to protect themselves from the smog of burning, such as avoiding activities outside or in the open air, staying in the house, and keeping doors and windows closed. If they needed to be outside during the burning time, they also wore a mask to protect themselves from the burning smog.

4.2.3. Rice Straw Incorporation

Farmers who incorporated their rice straw perceived the highest benefits of rice straw utilization, i.e., that rice straw was more useful than burning, so it could be used for maximum benefit (PBU5, 100%) (Table S1), which supported the finding of Connor, et al. [10]. The farmers realized the benefits of soil nourishment, which could reflect their high-level perception of the benefits of rice straw for agriculture and the surrounding environment (PBU1, 100%). Furthermore, the farmers agreed that straw incorporation helped build good relationships in the community (PBU3, 100%) and did not require additional costs (PBU4, 100%), while the benefit of rice straw in terms of generating income (PBU2, 100%) was perceived at a moderate level. This is because rice straw incorporation does not directly result in monetary return, but this is rather a byproduct that results from incorporation for soil nourishment.

These results were consistent with PPN, which had a high-level perception rating regarding farmers feeling guilty if rice straw was not utilized (PPN3, 100%). Meanwhile, a moderate perception of rice straw utilization was consistent with farming values (PPN1, 100%) and their responsibility to reduce air pollution (PPN2, 100%). The crucial reasons for rice straw incorporation was lacking water resources to continue the next crop. So farmers let the rice straw dry naturally until enough water was available for the next crop of rice cultivation.

For farmers who had their own tractors, they incorporated rice straw and prepared for the next crop when water from the irrigation system was available. Most farmers did not think that rice straw incorporation would link to their responsibility for reducing air pollution, and they were mainly concerned about available resources in their area. Moreover, after gaining knowledge of the benefits of rice straw utilization from local authorities, agricultural academics and agricultural extensionists, they were more concerned about the benefits of rice straw in terms of soil nourishment, which would benefit rice growing in the future.

Regarding PBC, most of the incorporation farmers reported a high level of perception that rice straw utilization was simple and uncomplicated (PBC2, 84.62%), and that rice straw was believed to be usable (PBC3, 76.92%). Meanwhile, a moderate level of perception was expressed regarding how to use rice straw (PBC1, 61.54%). It is still an issue that should be communicated, and greater understanding should be promoted to increase their self-efficacy regarding rice straw utilization. This was because the farmers were initially confident in the benefits of rice straw, and its perceived ease of use. Promoting knowledge that is clearly understandable could increase the occurrence of rice straw utilization.

The fact that everyone should be responsible for air pollution (PAR1, 100%) was shown at a high level of perception for the ascription of responsibility, followed by a moderate perception that people in the community should avoid activities that cause air pollution (PAR2, 100%). This group of farmers explained similar reasons for understanding the necessity of burning, as mentioned by compacting farmers.

Cues to rice straw utilization were highly perceived by the incorporation farmers for its facilities (PCU3, 61.5%), while moderate perceptions were found for reducing environmental problems (PCU1, 100%) and gaining knowledge from local authorities (PCU2, 76.9%). Since possessing a tractor is an important aspect of rice straw incorporation, the farmers did not pay for the labor cost. They did not focus on acquiring knowledge by themselves or from local authorities, but they would attend training if it led to a direct benefit to themselves.

PSB was found at moderate to low perception levels. The ideas that people and the environment are unable to cope with the effects of rice straw burning (PSB4, 100%), and that rice straw burning poses a serious threat to the environment and agriculture (PSB2, 86.8%) were perceived at a moderate level, while rice straw burning being a serious threat for future generations (PSB3, 100%) and a threat to community health (PSB1, 76.9%) were perceived lower.

The incorporation farmers perceived the benefits of the current option at the highest levels for reducing air pollution (PBO3, 100%), helping people in the community live together peacefully (PBO4, 76.9%), and being suitable for the available resources in the area (PBO5, 53.85%), these were followed by cost-saving (PBO2, 76.92%), which was reported at a high level of perception.

On the contrary, a lower level of perception for generating income (PBO1, 100%) was found. The farmers focused more on environmental benefits, especially to the soil in their paddy fields than on monetary benefit. This could benefit the surrounding environment and cause no problems for others due to a lack of burning. Rice straw incorporation mainly benefited their soil quality, so the lowest level of perception for this point was shown.

4.2.4. Free-Grazing Ducks in the Paddy Field

In terms of PBU, farmers who practiced free-grazing ducks in the paddy field had the highest level of perception that rice straw was more useful than burning it, so it should be used for maximum benefit (PBU5, 100%). The high level of perception was shown for rice straw utilization having a positive effect on agriculture and the surrounding environment (PBU1, 100%), helping build good relationships in the community (PBU3, 100%) and not requiring additional costs (PBU4, 100%). Meanwhile, a moderate perception was mentioned for gaining income (PBU2, 100%) because the farmers did not earn a monetary return, but got duck eggs to eat that could reduce the cost of buying food (Table S1).

The results of perceived pro-environmental personal norms were found at a high level of perception for all sub-issues; rice straw utilization was in accordance with their farming values (PPN1, 100%), their responsibility to reduce air pollution (PPN2, 100%) and feeling guilty if rice straw was not utilized (PPN3, 100%).

Having free-grazing ducks in the paddy field has been practiced for about 1–2 years because duck farmers have offered to raise ducks in the paddy fields after harvesting, and offered to give duck eggs in return. Moreover, this method was selected as the farmers realized the benefit of duck manure in their paddy field as a good fertilizer to nourish their soil. These are the main motivating reasons for the perception that this approach is consistent with their farming values.

In addition, the farmers had a high level of self-efficacy in rice straw utilization (PBC) on the issue of rice straw utilization being easy and uncomplicated (PBC2, 100%), and they believed that rice straw could provide benefits (PBC3, 100%) while having a moderate understanding of how to gain benefits from rice straw (PBC1, 100%).

It is interesting that the results reflected the farmers' moderate knowledge and understanding of rice straw utilization, while the benefits of rice straw were the most recognized. Their recognition of its benefits should be employed as a key message of communication. Moreover, providing the clear steps of other methods on how to use rice straw for other benefits including mention of their costs and returns should be communicated, this could encourage farmers to consider alternative methods of rice straw utilization. This communication technique is called "action–knowledge" [28] and should be focused on relevance and usefulness, positive and negative examples, the suggestion of simple behaviors, the utilization of previous knowledge, fostering transferability, and providing information tailored for the specific context of action.

Ascription of responsibility of the incorporation farmers was reported at a high level of perception, i.e., that everyone should be responsible for air pollution (PAR1, 100%), while a moderate level of perception that people in the community should avoid activities that cause air pollution (PAR2, 57.14%) was noted. This is in line with the results concerning those farmers who chose other approaches to rice straw utilization (instead of burning), sharing similar reasons for understanding the necessity of burning by farmers. The farmers highlighted that everyone should help each other in any way they can to avoid activities that cause air pollution.

For the perception of cues to rice straw utilization, the majority of farmers employing free-grazing ducks in the paddy field expressed the highest perception of the straw utilization facilities (PCU3, 100%), followed by a high level of rice straw utilization education by local authorities (PCU2, 100%), and perceived reading/hearing/receiving knowledge that rice straw utilization reduces environmental problems (PCU1, 100%) moderately, which is similar to those farmers who use rice straw for other purposes. The farmers added that there is a free-grazing ducks operator who contacted the farmers at their field to determine whether they were interested in raising ducks in the field, and who expressed that duck eggs would be given in return. In addition, this method is used during resting periods for their paddy fields, so they were not in a hurry to continue planting rice. The findings on this issue are in good agreement with results from compacting rice straw farmers, although they were more open to knowledge from local authorities than from other channels. The approach can be also applied to design communication through local authorities to directly provide knowledge to farmers.

A moderate perception was proved for PSB in that people and the environment cannot cope with the effects of rice straw burning (PSB4, 100%), and that rice straw burning poses a serious threat to the environment and agriculture (PSB2, 100%). A low perception was shown for the opinions that rice straw burning poses a threat to community health (PSB1, 100%) and is a serious threat to future generations (PSB3, 100%). It was notable that the perception levels of farmers who employed rice straw compacting and free-grazing ducks in the paddy field were similar and quite close to those who employed other approaches of rice straw utilization. Overall, the severity of the rice straw burning was not highly perceived by non-burning farmers due to an understanding of the necessity of burning for some farmers.

A study on the PBO found that the farmers had the highest level of perception that using free-grazing ducks in the paddy field reduces air pollution (PBO3, 100%). It was clear that this method did not require rice straw burning, which could help to reduce air pollution. Additionally, this method helped people in the community coexist peacefully (PBO4, 100%) because of the lack of rice straw burning. Therefore, it did not affect others in the community and was suitable for the available resources in the area (PBO5, 100%), such as not having enough water for the next crop. Cost-saving in rice straw and stubble management (PBO2, 100%) was perceived at a moderate level for this approach because there is still a need to pump water into their paddy fields when the ducks are released. Meanwhile, a low perception of generating income (PBO1, 100%) was expressed, because raising ducks in the paddy field did not provide a monetary return, but duck eggs were provided instead.

4.2.5. Mixed Method

The selected mixed-method consisted of compacting, incorporation and soil covering. Farmers employing this mixed-method for rice straw and stubble management preferred not to burn rice straw as the first priority, so they tried to employ as many alternatives as they could.

Farmers' perceptions of the use of the mixed method were found to be consistent among PPN and PBC for all sub-issues at a high level as follows: rice straw utilization was in line with their farming values (PPN1, 100%); their responsibility to reduce air pollution (PPN2, 100%); and feeling guilty for not using rice straw (PPN3, 85.71%) (Table S1). The farmers stated that they realized the benefits of rice straw, so they tried to use it to benefit various purposes. Besides allowing the private sector to compact rice straw and incorporating rice straw into the soil, they owned orchards growing mango, banana and papaya, so they covered the soil with rice straw in order to increase moisture in the soil. They also realized that using rice straw reduced burning. Reducing rice straw burning as much as possible was considered to be part of a farmer's responsibility.

The farmers also believed that rice straw could be utilized (PBC3, 85.71%), they stated their understanding of how to use rice straw (PBC1, 66.67%) and expressed the opinion that rice straw utilization was simple and uncomplicated (PBC2, 66.67%). Most of the farmers participated in community farming groups, such as large-scale farming groups, which promoted more opportunities for training to enhance knowledge of rice straw utilization techniques. Some training also provided the opportunity to participate in practical activities and experiments, so they understood the process of implementing rice straw and felt that it was not too difficult to gain benefits from rice straw. Therefore, they have continued using rice straw to receive various benefits.

The results of PBU were in line with PBO, with the highest perception of rice straw being more useful than burning, so farmers thought it could be used for maximum benefit (PBU5, 90.48%). A high level of perception was shown that rice straw had a beneficial effect on agriculture and the surrounding environment (PBU1, 100%), increasing income (PBU2, 100%), promoting good relationships in the community (PBU3, 100%) and not requiring additional costs (PBU4, 95.24%), which are congruent with the perceived benefits of their methods of rice straw and stubble management. Most of them reflected a high perception of rice straw compacting being useful for reducing air pollution (PBO3, 100%), helping people in the community coexist peacefully (PBO4, 100%) and being appropriate for the available resources in the area (PBO5, 85.71%).

A high level of perception was mentioned for generating income (PBO1, 100%) and cost-saving for rice straw and stubble management (PBO2, 100%), with similar reasons being expressed as with compacting farmers. They realized the benefits of generating income first, then reducing air pollution was clearly a good result as well because rice straw was definitely not burned and resulted in peaceful coexistence in the community. The mixed-method chosen was suitable for their available resources because rice straw could also be used to cover the soil in their orchard to improve soil moisture. Overall, the mixed method did not incur expenses for rice straw utilization.

A study on the issue of the ascription of responsibility showed a high level of perception for all sub-issues, i.e., that everyone should be responsible for air pollution (PAR1, 100%) and that people in the community should avoid activities that cause air pollution (PAR2, 57.14%), which is in line with the perception of farmers employing other methods of rice straw utilization.

The results of the study on the issue of the perception of cues for rice straw utilization found that farmers who used rice straw to raise cows had a high perception of rice straw utilization facilities (PCU3, 80.95%), followed by a moderate perception of education on rice straw utilization by local authorities (PCU2, 85.71%), and of reading/hearing/receiving information that rice straw utilization reduced environmental problems (PCU1, 57.14%).

PSB also showed similar results to other methods of rice straw utilization, at low to moderate levels. Most farmers perceived that rice straw burning poses a serious threat to the environment and agriculture (PSB2, 85.71%) at a moderate level, alongside that it is a threat to community health (PSB1, 57.14%), and that people and the environment cannot cope with the effects of rice straw burning (PSB4, 100%). Meanwhile, there was a low perception of rice straw burning as being a serious threat for future generations (PSB3, 100%). The same level of perception of the PSB2, PSB3 and PSB4 issues were raised, which was slightly different only for PSB1, with similar reasons to those who were employing other methods of rice straw utilization being stated.

In conclusion (Table 4), the result of our study showed that the perceived benefits of current options for both burning and non-burning farmers played an important role in determining which practice was selected by the farmers, which was in line with the perceived benefits of PPE for influencing farmers' adoption of rice straw utilization techniques [7]. Similar to the results regarding perceived cues for practicing rice straw utilization techniques, access to facilities of rice straw utilization (PCU3) seemed to play an especially crucial role in the adoption of non-burning options.

However, the perceived severity of burning being at "disagree" to "undecided" levels did not really influence non-burning farmers' decisions, which was different from the finding of Abdollahzadeh and Sharifzadeh [7] who mentioned that perceived severity of the bad effects of pesticides influenced farmers' adoption of PPE.

The psychological factors based on HBM seemed to play a crucial role for non-burning farmers at the "strongly agree" and "agree" levels of perception. The role of HBM was consistent with Ataei et al. [8] who found it was a better predictor of intention than TPB.

PBC, as a part of TPB, was also mainly perceived by non-burning farmers as almost the highest level of perception, which was in quite good agreement with the results of Zhang et al. [6] regarding TPB prediction of self-interest-oriented behaviors.

Table 4. Summary of rice straw management options, influencing psychological factors and behavioural theories employed.

Options	Strongly Disagree (1)	Disagree (2)	Undecided (3)	Agree (4)	Strongly Agree (5)
Burning	PPN (VBN); PBC (TPB); PCU (HBM); PSB1,2,3 (HBM); PBU3,4 (HBM); PBO1,3 (HBM)	PAR2 (VBN); PBU2,5 (HBM); PSB4 (HBM)	PAR1 (VBN); PBU1 (HBM); PBO4,5 (HBM)	PBO2 (HBM)	
Non-burning	PBO1 (incorporation) (HBM)	PBO1 (Free-grazing ducks) (HBM); PSB3 (HBM) PSB1 (incorporation, Free-grazing ducks) (HBM)	PBO2 (Free-grazing ducks) (HBM) PBU2 (Incorporation, Free-grazing ducks) (HBM) PCU1 (HBM); PSB2,4 (HBM); PSB1 (compacting, mix) (HBM)	PBO4,5 (Free-grazing ducks) (HBM); PBU1,3,4 (HBM)	PBO3 (HBM); PBO4,5 (compacting, incorporation, mix) (HBM) PCU3 (HBM); PBU5 (HBM); PBU2 (compacting) (HBM)
			PPN (VBN); PAR (VBN); PBC (TPB); PCU2 (HBM)		

4.3. Decoding Rice Straw and Stubble Management Based on Farmers' Psychological Factors

Rice straw and stubble management practices could reflect farmers' psychological perceptions, which were analyzed by stepwise multiple linear regression models (Table S1).

(1) Perceived pro-environmental personal norms (PPN)

The results clearly show that PPN was significantly negatively influenced by rice straw burning (PN1: $t = -67.553$, $p = 0.000$; PN2: $t = -67.553$, $p = 0.000$; PN3: $t = -48.848$, $p = 0.000$). The results were also in line with the ANOVA of the burning approach with a significantly negative difference compared to non-burning approaches. These indicate that farmers adopting the burning method tended to have the lowest level of perception of PPN compared with non-burning approaches to rice straw and stubble management.

Farmers who chose to burn rice straw provided additional information that gaining benefits from rice straw was a method that they had never used before. The burning method has been employed for a long time, i.e., since the beginning of rice cultivation. Burning is therefore more in line with their farming values. Furthermore, they viewed that the responsibility to reduce air pollution would not be a matter of any one individual, but instead a communal responsibility.

Since not much damage is caused by burning, they did not feel that rice straw burning was a very serious threat. Burning is practiced for a short time of around 2 h and not frequently, only 1–3 times a year, it is practiced within the rice planting area only and does not spread to other areas. Surface fires were unlikely to cause much damage to the soil. The farmers also reflected that soil quality/properties have not yet been measured in some villages. Moreover, the negative effects of burning on the soil have not yet been shown. The farmers did not feel too guilty for choosing burning instead of rice straw utilization. Most people in the same community understood the reason for burning. When they noticed rice straw burning near their house, they stayed inside, closed windows and doors, and avoided being outside to protect their health.

Moreover, PPN1 and PPN2 were also significantly negatively influenced by farmers who employed rice straw incorporation (PPN1: $t = -7.165$, $p = 0.000$; PPN2: $t = -7.165$, $p = 0.000$). The results were consistent with the ANOVA of rice straw incorporation that showed a significantly negative difference when compared with other non-burning methods of rice straw compacting, free-grazing ducks and mixed methods. This result was supported by the farmers that do not rush for the next crop because of a lack of water from the irrigation system, so they chose to dry rice straw until enough water was available to start planting. In contrast, if there was enough water for growing rice, they would burn rice straw in order to continue planting the next crop. Consequently, the results of the stepwise regression found a significantly negative influence, but not a high value when compared to burning.

(2) Perceived cues to rice straw utilization (PCU)

Burning clearly negatively influenced PCU (PCU1: $t = -14.996$, $p = 0.000$; PCU2: $t = -19.450$, $p = 0.000$; PCU3: $t = -39.920$, $p = 0.000$). Compacting also negatively influenced PCU1 ($t = -2.059$, $p = 0.044$), indicating that farmers did not seek much rice straw utilization knowledge on their own, similar to those employing burning. Meanwhile, the mixed method positively influenced PCU2 ($t = 3.143$, $p = 0.002$). The results are also in line with the ANOVA of the burning approach showing a significantly negative difference compared to non-burning approaches.

The results showed that farmers who employed the burning method tended to have the lowest perception of all PCU components, while those who employed mixed methods more often realized the benefits of gaining knowledge about rice straw utilization from local authorities. This suggests that farmers who utilized mixed methods of rice straw and stubble management were part of an agricultural group, most likely a large-scale farming group. They could gain more knowledge on how to gain benefits from rice straw from local authorities and agricultural extensionists such as how incorporation could nourish their soil, and how covering the soil of orchards can maintain soil moisture.

In order to promote a higher PCU perception in farmers selecting the burning approach, which can then encourage a higher perception of rice straw utilization, the communication of knowledge and practices with clear inclusion of practical action knowledge is a crucial point of communication. The perception of knowledge of rice straw utilization, which was still at a low level, should be promoted through various channels that are convenient for farmers to gain knowledge, particularly through social media, group gathering, training and workshops for sharing experiences.

(3) Perceived behavioral control of rice straw utilization (PBC)

PBC was highly negatively influenced by burning practice (PBC1: $t = -36.035$, $p = 0.000$; PBC2: $t = -30.752$, $p = 0.000$; PBC3: $t = -61.495$, $p = 0.000$). The results were also essentially the same as the ANOVA of the burning approach, with a significantly negative difference compared with all non-burning approaches. This shows that burning farmers had the lowest perception of PBC, similar to the results of PPN and PCU, and the communication guidelines for burning farmers can be considered as discussed in the section on PPN and PCU above in order to promote all the factors together.

(4) Perceived severity of rice straw burning (PSB)

All sub-issues of PSB were highly negatively influenced by the burning practice (PSB1: $t = -28.227$, $p = 0.000$; PSB2: $t = -45.599$, $p = 0.000$; PSB3: $t = -28.342$, $p = 0.000$). The results also confirmed the ANOVA of the burning approach with a significantly negative difference compared with non-burning approaches. As expected, farmers who selected rice straw burning had the lowest perception of PSB, which was one reason for their burning practice. This result also supports the finding of Rosmiza et al. [27], who revealed farmers' perception that rice straw burning did not cause a serious effect on their rice fields. This was also in good agreement with Connor et al. [10], who showed that when higher risks were perceived, there was less acceptance of the practice.

The insight explanation provided by the farmers was that burning rice straw was unlikely to have a significant impact on the people's health in the community. No one in the village had been diagnosed with respiratory disease. Burning was infrequent, which made them think that people and the environment could cope with the effects of burning. Some farmers further reflected that their burning was unlikely to cause the high dust levels that have been reported in northern Thailand during the summer season.

Farmers who employed rice straw incorporation and chose to feed ducks on their rice farm area stated that they did not feel that burning was a very serious danger and understood the reasons for choosing this method. The farmers only burnt their rice straw after harvesting and preparing for the next crop. In some years, rice could be planted only once or twice because of insufficient water for many cycles of farming. Most of the people in the community knew and understood each other. They could protect themselves from burning smoke by avoiding doing activities outside/in the open air, staying in the house, and keeping the doors and windows closed. If they needed to be outside during this period, they wore a mask which was also done to prevent COVID-19. The smoke did not continue for a long time, so the feeling of a non-serious threat was perceived.

(5) Perceived ascription of responsibility (PAR)

PAR was highly negatively influenced by burning (PAR1: $t = -35.078$, $p = 0.000$; PAR2: $t = -27.237$, $p = 0.000$), while PAR2 ($t = -3.290$, $p = 0.001$) was also negatively influenced by incorporation. The results also agree with the ANOVA of the burning approach, showing a significantly negative difference compared with non-burning approaches.

The farmers who employed rice straw burning stated that local people in their community as a whole rarely did activities that clearly caused air pollution. Although the rice straw stubble was burned, it was not on a daily basis, and was considered unlikely to be a major contributor to air pollution. Overall, it was moderately agreed that everyone should be responsible for air pollution, not only the agricultural sector but also other sectors, such as the industrial and transportation sectors.

Those who employed rice straw incorporation reflected that their main reason to employ this method was due to a lack of water to continue rice growing. Therefore, they were not in a hurry to eliminate rice straw stubble. Some of them also realized the benefits of incorporation. That is why the stepwise regression of the incorporation method also negatively influenced PAR2.

(6) Perceived benefits of rice straw utilization (PBU)

The results clearly showed that the perceived benefits of rice straw utilization were significantly negatively influenced by rice straw burning for all elements (PBU1: $t = -35.078$, $p = 0.000$; PBU2: $t = -6.451$, $p = 0.000$, PBU3: $t = -75.030$, $p = 0.000$; PBU4: $t = -67.290$, $p = 0.000$; PBU5: $t = -70.586$, $p = 0.000$). Meanwhile, PBU2 was significantly positively influenced by rice straw compacting ($t = 14.585$, $p = 0.000$) and the mixed method ($t = 5.789$, $p = 0.000$).

Farmers who burned their rice straw reflected that, if rice straw was used, there would be an additional cost, but they were not clear on how much it costs. Some farmers stated that they had heard about rice straw for growing mushrooms, but this would incur a cost for preparing or cultivating. Moreover, they did not really know if it would be worth it or not.

In terms of utilizing rice straw to help strengthen community relationships, the farmers did not think that it would have much effect because burning did not cause conflict among the local people.

In terms of recognizing the benefits of rice straw, the farmers still did not know exactly how rice straw would be suitable for each purpose, and were not sure if it would be worth the cost of using rice straw or not. Conversely, the burning option was easy, convenient, and quickly provided a result as the next cycle of farming could be started immediately. Then, they would be able to earn money from rice production by selling to their customers or the market before other farmers. Therefore, a clear comparative message of rice straw utilization should be communicated by comparing the steps, costs and benefits of each method.

In addition, farmers who chose to burn were not convinced by rice straw utilization. They were not sure of the positive effect on agriculture and their surrounding environment. Therefore, it is important to communicate for better understanding and awareness, to visualize and realize the differences between rice straw burning and utilization. The benefits that could occur from rice straw utilization, in particular, need to be communicated and linked to benefits in terms of generating income for farmers. After that, benefits in terms of farmers' health, benefits for the environment, and for the society where they live should be highlighted.

The compacting farmers reflected that earning money from compacting was a key aspect motivating them to choose this option, and even more so as it generates income continuously. They realized the benefits of rice straw compacting as it required no investment, while it was also beneficial to agriculture and the environment because compacting rice straw could reduce burning. By obtaining knowledge from agricultural agencies and the village headman and publicizing this to the villagers through their local broadcasting tower and through the village meeting to avoid burning, farmers received information that burning in rice fields was a major cause of soil depletion causing their rice to grow poorly. The yield was lower than it should be and caused air pollution, which might cause a conflict between the farmers who burned and non-burning and non-farmer groups.

Farmers who employed mixed methods for rice straw and stubble management, including rice straw compacting, soil covering and incorporation, could also gain income from compacting. Moreover, using rice straw to cover their soil in orchards was another indirect way to generate income. Similar to the farmers employing rice straw incorporation, they thought that the practice was good for the soil in their rice fields. Additionally, they obtained knowledge from relevant agencies that promoted rice straw incorporation to increase soil nutrients. The farmers would feel guilty for not taking advantage of the rice straw to help nourish their soil. Most of the farmers who chose this method also owned a tractor for incorporation by themselves or by their family members, so they could save on the labor cost for incorporation.

(7) Perceived benefits of current option (PBO)

The current method of rice straw and stubble management did not cause trouble for others, and allowed farmers to coexist peacefully (PBO4), but was highly negatively influenced by burning ($t = -54.455$, $p = 0.000$), followed by free-grazing ducks ($t = -5.795$, $p = 0.000$). The results reflected that both groups of farmers perceived that their practices could make others feel uncomfortable due to the smog of burning, and annoyance from ducks in the area, which might cause some difficulties in terms of transportation and noise.

The perception of current options to generate income (PBO1) was highly positively influenced by compacting ($t = 34.396$, $p = 0.000$) and mixed methods ($t = 15.059$, $p = 0.000$), because both groups of farmers gained money from the private sector visiting their fields for rice straw compacting. The results also showed good congruence with PBU2 (rice straw utilization could generate income). These findings are also in accordance with Connor et al. [10], who mentioned that the benefit perceptions of rice straw management options can influence practice adoption.

Lower costs of the current practice (PBO2) were positively influenced by compacting ($t = 10.281$, $p = 0.000$) and burning ($t = 4.719$, $p = 0.000$), and negatively influenced by duck feeding ($t = -3.587$, $p = 0.000$). The negative result was in line with the reasons provided by the farmers employing rice straw for duck feed due to the cost of pumping water into the paddy field before feeding the ducks.

The current option to reduce air pollution (PBO3) was negatively influenced only by burning ($t = -93.247$, $p = 0.000$), proving that the farmers perceived that their practice could not help to reduce air pollution, but they maintained the practice because of being in a hurry to prepare for the next crop.

In terms of the current options being appropriate with resources in their area (PBO5), the results show that this was negatively influenced by burning ($t = -29.402$, $p = 0.000$) and free-grazing ducks ($t = -2.389$, $p = 0.018$), and positively influenced by mixed methods ($t = 2.606$, $p = 0.010$).

Based on the perceived benefits of the current rice straw and stubble management methods, it was observed that farmers who burned rice straw realized the most benefit of burning in terms of cost reduction, while they had the least benefit in terms of generating income and reducing air pollution. Burning did not generate income, which was different from rice straw compacting, but it helped to earn money later because of the chance to start the next crop faster. Noticeably, an important factor was water resources, i.e., not having enough water from the irrigation system or rainwater could slow down the next crop.

In terms of reducing air pollution, it was noted by the farmers who chose to burn in comparison with non-burning farmers, that they perceived the lowest severity of rice straw burning. At the same time, they frankly stated that compared to non-burning methods, burning was still the least effective method to reduce air pollution. This is an interesting point showing that indeed the farmers were already aware of the negative effects of burning. Therefore, this is an opportunity to communicate alternative rice straw utilization methods and make these options widely accessible to farmers. Encouraging farmers to understand and obtain the benefits of non-burning methods may be possible such that in the future they will choose rice straw utilization instead of burning.

In addition, the farmers considered the burning method as appropriate for the resources available in their area, especially during the rainy season, as wet rice straw cannot be used for any purpose. Therefore, they realized that burning was the most appropriate method. In addition, due to the urgency of continuous farming according to the availability of water from the irrigation system, the farmers had to speed their farming to catch up with the water release period. Importantly, straw burning is the least time-consuming method of straw management. In addition, straw burning also helped to prevent rice pests when cultivating rice.

Compacting farmers emphasized that they selected this approach mainly due to the benefits of generating income. To reduce air pollution, it was clearly a good result, because rice straw was definitely not burned and the by-product was peaceful coexistence in the community. In addition, most were aware of rice straw utilization facilities due to the fact that in the process of compacting, the private sector came to their rice field and made an agreement on the price beforehand. The result is also consistent with the stepwise analysis that found positive effects on PBO1 and PBO2.

Moreover, based on the results of the ANOVA on rice straw compacting, a significantly positive difference compared to other non-burning methods was shown in terms of causing no trouble to others (PBO4), earning income (PBO1) and cost-saving (PBO2), particularly when compared with the incorporation and free-grazing duck methods. This implies (from the farmers' viewpoint) that rice straw compacting seems to be the best option of the non-burning rice straw and stubble management techniques. Therefore, the relevant stakeholders should take part in promoting and facilitating this option in order to promote self-efficacy and the adoption of this practice by farmers. At the same time, in-depth research and knowledge translation of each rice straw and stubble management option should be employed continuously to make farmers active, and to expose them to new knowledge and innovations in the management of rice straw and stubble.

4.4. Communication Strategies to Promote Rice Straw Utilization for Achieving Sustainable Agriculture

In this part, the integrated frameworks, namely the "social–ecological model", "the six domains of the full-spectrum approach" [29], the "sustainable agricultural social system" [30], the "environmental health literacy (EHL)" [31], the "agricultural knowledge and information system (AKIS)" [32] and personal ecological norms [28] were employed to develop communication strategies for promoting non-burning rice straw and stubble management to achieve social–ecological benefits for more sustainable agricultural production (Figure 3).

Within the community, there were both farmers attending groups and those farming individually. They had their own knowledge, attitude and skills on rice straw and stubble management. Communication can be both internal and external to their community, in which two-way communication, and communication oriented towards networking, should be promoted to enhance more confidence in rice straw utilization. Regarding levels of communication, intrapersonal communication was added from Greiner [29] as a starting point to design how to communicate, then it was connected to interpersonal and group communication, to mass media, and new web-based media; the techniques and tools to be used also need to be considered.

Supporting mechanisms can be generated from key stakeholders, i.e., (1) agricultural extensionists/local authorities, (2) central authorities, (3) researchers/academic sector and (4) private sector NGOs and media practitioners. The forms of support are (1) policy and regulations on climate-friendly agriculture, technology and investment for rice straw utilization, and (2) communication strategies.

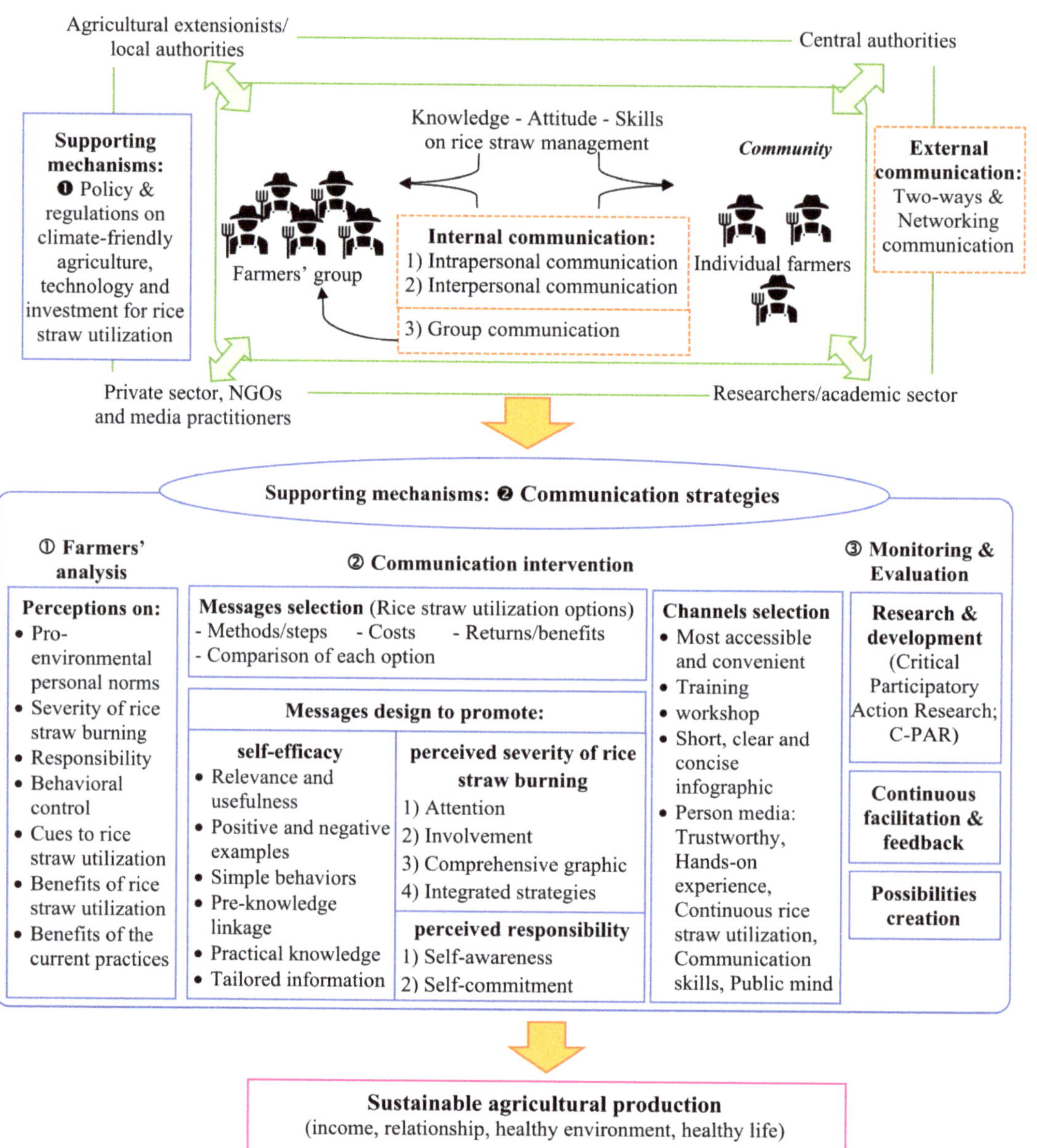

Figure 3. Communication strategies for promoting non-burning rice straw and stubble management to achieve social-ecological benefits for more sustainable agricultural production.

Regarding communication strategies, farmers' perceptions should be analyzed and understood first, particularly their pro-environmental personal norms, and their perceptions of the severity of rice straw burning, responsibility, perceived behavioral control, cues for rice straw utilization, the benefits of rice straw utilization and the benefits of the current practices.

Next, the processes of communication intervention should highlight the selection of messages, message design and channel selection to fit with farmers. Rice straw utilization options should be selected as the key messages to promote non-burning rice straw and stubble management practices. Clear steps or methods, costs and benefits, and a comparison of each rice straw utilization option should be highlighted. The benefits to farmers, family members, people in the community and consumers should be emphasized, together with key messages regarding the potential these approaches have for increasing income, being healthy, peacefully coexisting with people in the community, and improving the better environment should be promoted to change behavior.

In order to promote more self-efficacy, the messages should highlight the relevance and usefulness of these techniques [33], positive and negative examples [33], simple behaviors [34], pre-knowledge linkage and practical knowledge, and tailored information [35]. Additionally, promoting action-knowledge and self-efficacy at the group level can help to achieve more effective non-burning rice straw management [36]. Moreover, collective efficacy as a group should be also considered, which can help to promote more effective rice straw utilization than self-efficacy [37]. Individual farmers might feel that only one person changing from burning to rice straw utilization would not be very meaningful to society. Therefore, group action tends to see clearer benefits or good results for both individuals and as a group or a community, so collective efficacy should be given as the initial priority as well.

Message design to promote perceived responsibility can focus on self-awareness and self-commitment. To enhance effective self-awareness, practical options and clear steps of practicing should be highlighted as key messages [28] in order to make it possible for farmers to adapt their burning to utilizing rice straw. This kind of message for communication can motivate farmers to adapt their current practice (burning) to incorporate new knowledge (rice straw utilization). Otherwise, a burning farmer might easily feel contradicted inside their mind because new information on rice straw utilization is irrelevant compared with their previous perceptions, knowledge and current practices of rice straw burning. When denying that rice straw utilization occurs, it is possible that farmers have less awareness of environmental concerns and values than of their survival.

To promote self-commitment, written, public and voluntary commitment seems to be more effective than spoken, private and involuntary commitment [38]. Commitment should also come together with the reasons for rice straw utilization to achieve more sustainable behavior [35].

Messages designed to highlight the perceived severity of rice straw burning can employ attention, involvement, comprehensive graphics and integrated strategies [28]. In order to expose and understand the impact of rice straw burning, this kind of knowledge and information should be interesting enough for the farmers to pay attention to, such as updated references and well-known metaphors, and analogies that are relevant in their society [34]. The more involvement and direct impacts there are to the farmers, the more possibility there will be for the farmers to concern themselves with environmental values, particularly referring to the institutions inside their villages and places they are familiar with [28]. Comprehensive and easy-to-understand graphics should be selected to disseminate facts on both the positive and negative points of rice straw burning and rice straw utilization, and farmers should be allowed to decide for themselves whether they trust the information and knowledge that is disseminated [34]. Moreover, integrated strategies should be employed such as linkages between knowledge dissemination and self-commitment to create more possibilities to adopt rice straw utilization instead of burning; this strategy has succeeded before [39].

Communication channels that are the most accessible and convenient for farmers should be employed. Those using social media as a change agent should be trustworthy, should have taken action on rice straw utilization, should have seen the clear results of their practice, should have continued with methods of rice straw utilization, and should have communication skills to transfer knowledge and practices, as well as an outlook for passing on their practices as an example to fellow farmers. Thus, clear and concise online media and infographics focusing on how to gain benefits from rice straw should be also designed and distributed to farmers. Furthermore, farmers who still burn rice straw might not be interested in attending the training and workshops. Training and workshops should not take too much time (should not be more than three hours), and should try to provide a hands-on experience in order to demonstrate the practical success of rice straw utilization, which can influence farmers to change from burning to gain benefits from rice straw.

The last stage for communication strategies is monitoring and evaluation. Critical Participatory Action Research (C-PAR) should be applied as a practical framework for monitoring and evaluation, linking to continuous facilitation, feedback provision, and the creation of possibilities for farmers. C-PAR processes consist of practices and activities, including situation and context. The five main steps of C-PAR are as follows: (1) inspection of various operations, understanding and conditions under the intention of the participants; (2) critical questioning of participants' actions and outcomes; (3) applying communicative intervention to participants for an optional consensus conversation about the activities; (4) actions for behavior change, understanding the actions and conditions of the participants; and (5) evidence collection and monitoring of behavior change.

To ensure that farmers change their behavior from rice straw burning to rice straw utilization, support from local authorities should be also practiced, including monitoring and evaluating the perception, understanding the current practices of farmers. Any problems and limitations should be prioritized, brain-stormed, and discussed, and solutions proposed to support rice straw utilization should be easy and uncomplicated. After that, the farmers will be more aware of the benefits rather than feeling an additional burden, such as supporting the source of rice straw compacting service near the community, as well as contacting and coordinating the purchase of rice straw, which would be more convenient for farmers. Consultation from agricultural academics or extensionists and related parties should be performed to concretely and continuously facilitate rice straw utilization.

Furthermore, government agencies should take part in dealing with the problem as well. The overall communication management plan should be planned at the national level under the issue of air pollution control, starting from related ministries such as the Pollution Control Department together with the Department of Agricultural Extension (Ministry of Agriculture), Department of Industrial Works (Ministry of Industry) and Department of Land Transportation (Ministry of Transport) to see results from the collaboration of all sectors. At the same time, this communication plan should be passed on to relevant local authorities, such as the agricultural sector, through provincial, district and sub-district agricultural offices.

5. Conclusions and Recommendations

The results clearly show that PPN, PCU, PBC, PSB, PAR and PBU were significantly negatively influenced by rice straw burning for all sub-elements, and were significantly negative compared to non-burning approaches. Meanwhile, PBO of burning farmers showed a significantly positive difference for cost-saving to manage rice straw and stubble when compared with the approaches of incorporation and free duck grazing. This indicates that farmers adopting the burning method tended to have the lowest perception of PPN, PCU, PBC, PSB, PAR and PBU compared with other non-burning approaches of rice straw and stubble management. In contrast, cost-saving together with rapid management seemed to be the key points for motivating farmers to retain their burning practice. The results for incorporation farmers also found a significantly negative influence on PPN1 and PPN2 because the main reason for this practice was insufficient water for the next crop, rather than concern about the benefits of incorporation. Therefore, the benefits of incorporation should be also highlighted as a key message for knowledge enhancement by agricultural extensionists and relevant authorities. Furthermore, farmers employing mixed methods of rice straw and stubble management significantly positively influenced PCU2, PBO1, PBU2 and PBO5. This group of farmers should be supported as key change agents to convey their hands-on experience in rice straw and stubble management to motivate burning farmers to open their minds to other methods besides burning.

The integrated behavioral theories employed in this study proved that HBM, particularly the perceived benefits of current practice, seemed to play a crucial role for both burning and non-burning farmers. Moreover, perceived cues to practice, especially the accessing of facilities for rice straw utilization, seemed to play an important role in non-burning adoption. TPB, specifically PBC, appeared to be highly influenced by non-burning farmers.

To promote rice straw utilization and achieve sustainable agriculture, communication strategies should focus on message selection, message design and channel selection to fit with farmers. Key messages should focus on clear steps or methods, the costs and benefits of each rice straw utilization option, the benefits to farmers, their family members, people in the community and consumers, together with increasing income, being healthy, peacefully coexisting with people in the community, and an improved environment in order to promote the tendency to change behavior. Message design to promote action knowledge and self-efficacy at the group level, and to enhance perceived responsibility via self-awareness and self-commitment, can help to achieve more effective non-burning rice straw and stubble management.

Supplementary Materials: The following supporting information can be downloaded at: https://www.mdpi.com/article/10.3390/agronomy12010200/s1, Table S1: Cross-tab and Stepwise multiple linear regression analysis.

Author Contributions: Conceptualization, S.S.; data curation, S.S.; formal analysis, S.S.; funding acquisition, S.S.; investigation, S.S.; methodology, S.S.; supervision, N.A.; writing—original draft, S.S.; writing—review & editing, S.S and N.A. All authors have read and agreed to the published version of the manuscript.

Funding: This research project is supported by Mahidol University (Basic Research Fund: fiscal year 2021). We would also like to thank Thailand Science Research and Innovation (TSRI) for their funding to this project, contract number: BRF1-A8/2564.

Institutional Review Board Statement: The study was conducted according to the guidelines of the Declaration of Helsinki, and approved by the Institutional Review Board of Institute for Population and Social Research, Mahidol University (IPSR-IRB) (COA. No. 2021/01-003, date of approval: 28 January 2021).

Informed Consent Statement: Informed consent was obtained from all key informants involved in the study.

Data Availability Statement: Not applicable.

Acknowledgments: We express great thanks to all key informants for providing the information. We would also like to thank the anonymous reviewers for their comments and suggestions to improve this paper.

Conflicts of Interest: The authors declare no conflict of interest.

References

1. Launio, C.C.; Asis, C.A., Jr.; Manalili, R.G.; Javier, E.F.; Belizario, A.F. What factors influence choice of waste management practice? Evidence from rice straw management in the Philippines. *Waste Manag. Res.* **2014**, *32*, 140–148. [CrossRef]
2. Sereenonchai, S.; Arunrat, N.; Kamnoonwatana, D. Risk Perception on Haze Pollution and Willingness to Pay for Self-Protection and Haze Management in Chiang Mai Province, Northern Thailand. *Atmosphere* **2020**, *11*, 600. [CrossRef]
3. Arunrat, N.; Pumijumnong, N.; Sereenonchai, S. Air-Pollutant Emissions from Agricultural Burning in Mae Chaem Basin, Chiang Mai Province, Thailand. *Atmosphere* **2018**, *9*, 145. [CrossRef]
4. Office of Agricultural Economics. Agriculture Goes Forward with Campaign to Reduce Dust, Reduce Smoke, Inviting Farmers to Refrain from Burning in Agricultural Areas. 2020. Available online: https://www.oae.go.th/view/1/%E0%B8%A3%E0%B8%B2%E0%B8%A2%E0%B8%A5%E0%B8%B0%E0%B9%80%E0%B8%AD%E0%B8%B5%E0%B8%A2%E0%B8%94%E0%B8%82%E0%B9%88%E0%B8%B2%E0%B8%A7/%E0%B8%82%E0%B9%88%E0%B8%B2%E0%B8%A7%20%E0%B8%AA%E0%B8%A8%E0%B8%81./33626/TH-TH (accessed on 20 November 2021). (In Thai)
5. Arunrat, N.; Wang, C.; Pumijumnong, N.; Sereenonchai, S.; Cai, W. Farmers' intention and decision to adapt to climate change: A case study in the Yom and Nan basins, Phichit province of Thailand. *J. Clean. Prod.* **2017**, *143*, 672–685. [CrossRef]

6. Zhang, L.; Ruiz-Menjivar, J.; Luo, B.; Liang, Z.; Swisher, M.E. Predicting climate change mitigation and adaptation behaviors in agricultural production: A comparison of the theory of planned behavior and the Value-Belief-Norm Theory. *J. Environ. Psychol.* **2020**, *68*, 101408. [CrossRef]

7. Abdollahzadeh, G.; Sharifzadeh, M.S. Predicting farmers' intention to use PPE for prevent pesticide adverse effects: An examination of the Health Belief Model (HBM). *J. Saudi Soc. Agric. Sci.* **2021**, *20*, 40–47. [CrossRef]

8. Ataei, P.; Gholamrezai, S.; Movahedi, R.; Aliabadi, V. An analysis of farmers' intention to use green pesticides: The application of the extended theory of planned behavior and health belief model. *J. Rural Stud.* **2021**, *81*, 374–384. [CrossRef]

9. Guo, H.; Xu, S.; Wang, X.; Shu, W.; Chen, J.; Pan, C.; Guo, C. Driving Mechanism of Farmers' Utilization Behaviors of Straw Resources—An Empirical Study in Jilin Province, the Main Grain Producing Region in the Northeast Part of China. *Sustainability* **2021**, *13*, 2506. [CrossRef]

10. Connor, M.; Guia, A.H.D.; Quilloy, R.; Nguyen, H.V.; Gummert, M.; Sander, B.O. When climate change is not psychologically distant—Factors influencing the acceptance of sustainable farming practices in the Mekong river Delta of Vietnam. *World Dev. Perspect.* **2020**, *18*, 100204. [CrossRef]

11. Ajzen, I.; Fishbein, M. *Understanding Attitudes and Predicting Social Behavior*; Prentice-Hall: Englewood Cliffs, NJ, USA, 1980.

12. Fishbein, M.; Ajzen, I. *Belief, Attitude, Intention and Behavior: An Introduction to Theory and Research*; Addison-Wesley: Boston, MA, USA, 1975.

13. Ajzen, I. From intentions to actions: A theory of planned behavior. In *Action Control: From Cognition to Behavior*; Kuhl, J., Beckmann, J., Eds.; Springer: Berlin/Heidelberg, Germany, 1985.

14. Stern, P.C. New environmental theories: Toward a coherent theory of environmentally significant behavior. *J. Soc. Issues* **2000**, *56*, 407–424. [CrossRef]

15. Glanz, K.; Rimer, B.K.; Viswanath, K. *Health Behavior and Health Education: Theory, Research, and Practice*; Jossey-Bass, John Wiley and Sons: San Francisco, CA, USA, 2008.

16. Bhandari, G.; Atreya, K.; Yang, X.; Fan, L.; Geissen, V. Factors affecting pesticide safety behaviour: The perceptions of Nepalese farmers and retailers. *Sci. Total Environ.* **2018**, *631–632*, 1560–1571. [CrossRef] [PubMed]

17. Yazdanpanah, M.; Tavakoli, K.; Marzban, A. Investigating factors that influenced framers' intention regarding safe use of pesticides through health belief model. *Iran. Agric. Ext. Educ. J.* **2016**, *11*, 21–29.

18. Lopes, A.A.; Viriyavipart, A.; Tasneem, D. The role of social influence in crop residue management: Evidence from Northern India. *Ecol. Econ.* **2020**, *169*, 106563. [CrossRef]

19. Kadam, K.L.; Forrest, L.H.; Jacobson, W.A. Rice straw as a lignocellulosic resource: Collection, processing, transportation, and environmental aspects. *Biomass Bioenergy* **2000**, *18*, 369–389. [CrossRef]

20. Mendoza, T. Enhancing crop residues recycling in the Philippine landscape. In *Environmental Implications of Recycling and Recycled Products*; Springer: Berlin/Heidelberg, Germany, 2015; pp. 79–100. [CrossRef]

21. Thanh, T.N.T. *Comparative Assessment of Using Rice Straw for Rapid Composting and Straw Mushroom Production in Mitigating Greenhouse Gas Emissions in Mekong Delta Vietnam and Central Luzon, Philippines*; University of the Philippines Los Baños: Laguna, Philippines, 2011.

22. Sun, D.; Ge, Y.; Zhou, Y. Punishing and rewarding: How do policy measures affect crop straw use by farmers? An empirical analysis of Jiangsu Province of China. *Energy Policy* **2019**, *134*, 110882. [CrossRef]

23. United Nations ESCAP. Reducing straw residue burning and air pollution through sustainable agricultural mechanization. ESCAP/CED/2020/INF/4 2020. In Proceedings of the Economic and Social Commission for Asia and the Pacific Committee on Environment and Development, Bangkok, Thailand, 9–10 December 2020.

24. Department of Agriculture Extension. Stop Burning in Farmland. 2019. Available online: http://www.royalagro.doae.go.th/download/Bro_Burn.pdf (accessed on 20 November 2021). (In Thai)

25. Gray, D.E. *Doing Research in the Real World*, 4th ed.; SAGE Publications: Thousand Oaks, CA, USA, 2018.

26. Ahmed, T.; Ahmad, B.; Ahmad, W. Why do farmers burn rice residue? Examining farmers' choices in Punjab, Pakistan. *Land Use Policy* **2013**, *47*, 448–458. [CrossRef]

27. Rosmiza, M.Z.; Davies, W.P.; Rosniza, A.C.R.; Mazdi, M.; Jabil, M.J.; Wan, T.W.Y.; Che Rosmawati, C.M. Farmers' Participation in Rice Straw-Utilisation in the MADA Region of Kedah, Malaysia. *Mediterr. J. Soc. Sci.* **2014**, *5*, 23. [CrossRef]

28. Hamann, K.; Baumann, A.; Löschinger, D. *Psychology of Environmental Protection—Handbook for Encouraging Sustainable Actions*; Initiative Psychologie im Umweltschutz: Berlin, Germany, 2016. [CrossRef]

29. Greiner, K. The advantages and limits of a "full-spectrum" approach to communication for development (C4D). *J. Dev. Commun.* **2017**, *28*, 80–87.

30. Janker, J.; Mann, S.; Rist, S. Social sustainability in agriculture—A system-based framework. *J. Rural Stud.* **2019**, *65*, 32–42. [CrossRef]

31. Gray, K.M. From Content Knowledge to Community Change: A Review of Representations of Environmental Health Literacy. *Int. J. Environ. Res. Public Health* **2018**, *15*, 466. [CrossRef]

32. Demiryurek, K. Agricultural Knowledge and Innovation Systems and Social Communication Networks. In *Agricultural Extension and Consultacy: Volume II Agricultural Extension and Consulting Methodology*; Gaziosmanpaşa University Publication: Tokat, Turkey, 2014.

33. Salas, E.; Tannenbaum, S.I.; Kraiger, K.; Smith-Jentsch, K.A. The science of training and development in organizations. What matters in practice. *Psychol. Sci. Public Interest* **2012**, *13*, 74–101. [CrossRef] [PubMed]
34. Gardner, G.T.; Stern, P.C. *Environmental Problems and Human Behavior*; Pearson Custom Publishing: Boston, MA, USA, 2002.
35. Lehman, P.K.; Geller, E.S. Behavioral analysis and environmental protection accomplishments and potential for more. *Behav. Soc. Issues* **2004**, *13*, 13–32. [CrossRef]
36. Barth, M.; Jugert, P.; Fritsche, I. Still underdetected—Social norms and collective efficacy predict the acceptance of electric vehicles in Germany. *Transp. Res. Part F Traffic Psychol. Behav.* **2016**, *37*, 64–77. [CrossRef]
37. Homburg, A.; Stolberg, A. Explaining pro-environmental behavior with a cognitive theory of stress. *J. Environ. Psychol.* **2006**, *26*, 1–14. [CrossRef]
38. Clayton, S.; Myers, G. *Conservation Psychology: Understanding and Promoting Human Care for Nature*; Wiley-Blackwell: West Sussex, UK, 2009.
39. Kazdin, A. Psychological science's contributions to a sustainable environment. Extending our reach to a grand challenge of society. *Am. Psychol.* **2009**, *64*, 339–356. [CrossRef] [PubMed]

Communication

Why Can Green Social Responsibility Drive Agricultural Technology Manufacturing Company to Do Good Things? A Novel Adoption Model of Environmental Strategy

Stanley Y. B. Huang [1], Shih-Chin Lee [2],* and Yue-Shi Lee [3]

[1] Master Program of Financial Technology, School of Financial Technology, Ming Chuan University, Taipei 111, Taiwan; yanbin@mail.mcu.edu.tw
[2] Department of Finance, Chihlee University of Technology, New Taipei 220, Taiwan
[3] Department of Computer Science and Information Engineering, Ming Chuan University, Taoyuan City 333, Taiwan; leeys@mail.mcu.edu.tw
* Correspondence: icestorm@mail.chihlee.edu.tw

Abstract: The present research proposes the hierarchical linear modeling model (HLM) that describe how green social responsibility (GSR) predict the environmental strategy (ES) of agricultural technology manufacturing companies by the intermediary effects of the supervisor's green promise (GP) based on symbolic context theory. This study collected data with 150 supervisors from 50 different agricultural technology companies in Taiwan to analyze the HLM. The results suggest that vendors of agricultural technology companies should establish GSR to increase GP, which consequently can increase the companies' adoption of the ES. It is now the first to establish a milestone, propose a novel adoption model—GP and its antecedents through the HLM to predict the adoption of ES. These findings can upgrade the related literature of agriculture and can provide the procedure in implementing ES in agricultural technology companies.

Keywords: green social responsibility; green promise; environmental strategy; agricultural technology company

Citation: Huang, S.Y.B.; Lee, S.-C.; Lee, Y.-S. Why Can Green Social Responsibility Drive Agricultural Technology Manufacturing Company to Do Good Things? A Novel Adoption Model of Environmental Strategy. *Agronomy* **2021**, *11*, 1673. https://doi.org/10.3390/agronomy11081673

Academic Editors: Moritz Von Cossel, Joaquín Castro-Montoya, Yasir Iqbal and Gianni Bellocchi

Received: 6 July 2021
Accepted: 20 August 2021
Published: 23 August 2021

Publisher's Note: MDPI stays neutral with regard to jurisdictional claims in published maps and institutional affiliations.

1. Introduction

1.1. Background

Contemporary agricultural technology manufacturing companies should adopt a good strategy to optimize agricultural production and environmental strategy to handle environmental issues, which is also confirmed as a significant source of competitive advantage [1–4] because of external stakeholders [5–7]. Also, previous research has pointed out that agricultural production will cost huge resources and bring about pollution [8], which supports the emergency in studying the driving factor of environmental strategy (ES) [9–11]. ES is defined as the extent to which the company integrates environmental concerns into strategic planning, such as changing the production process to prevent pollution [8]. This study poses a novel perspective that using green social responsibility (GSR) predicts ES through an intermediary mechanism of green promise (GP) of supervisors based on symbolic context theory [11]. GSR denotes an environmentally responsible practices pol-icy that focuses on various stakeholders [12]. GP denotes the extent to which an employee's state of mind that is attachment and identity on environmental concerns [13]. Also, previous researcher [14] calls that little study to study corporate social responsibility at the organization level to yield a literature gap, so the present study poses how GSR and GP of supervisors s at cross-level can affect company's ES adoption at the same time by the multi-level growth curve model (HLM) [15] to respond this gap. Indeed, previous researchers of the agricultural field on ES implementation almost focus on technical aspects [16–18], and little study has examined the similar concept of GSR, GP, and ES on a HLM framework.

In sum, the present study uses HLM to explore GP and its antecedents to predict the adoption of ES and uses six-month longitudinal data to address the gaps discussed above.

1.2. Literature Reviewing

1.2.1. GSR and GP

According to the symbolic context theory [11], the GSR is a crucial symbol to guide the self-concept of supervisors to fit environmentally responsible, suggesting the antecedent role of GSR to GP. Indeed, past studies have suggested when the companies demonstrate responsibility and concern to the environment (GSR), the company's employee would reciprocate the company with GP [19,20]. Also, previous researchers found that socially and environmentally responsible activities can shape employees with similar attributes [21]. Thus:

Hypothesis 1 (H1). *GSR positively affects GP.*

1.2.2. GP and ES

In the same vein, GP of supervisors is also an important symbol to guide companies to select strategy according to the symbolic context theory [11], because supervisors have the power to allocate resources and manpower to perform companies' business activities, which are significant factors to determine what strategy the companies adopt. Thus:

Hypothesis 2 (H2). *GP positively affects ES.*

1.2.3. GSR and GP at the Organization Level

Previous studies [22–24] have examined corporate social responsibility and affective commitment at the organization level through the theory of the multilevel method [25], so GSR and GP should also have a similar context. For example, the organization-level GSR and GP are the atmosphere that is overspread within the group and are shared by people within the group [26]. In other words, individual-level GSR affects individual-level systems (e.g., individual-level GP and ES) when organization-level GSR affects organization-level systems (e.g., organization-level GP), which explains unique variations in different levels. Also, according to the theory of social learning [27], we pose that individual-level ES is affected by the organization-level and individual-level GSR and GP at the same time. Thus:

Hypothesis 3 (H3). *Organization-level GSR positively affects organization-level GP.*

Hypothesis 4 (H4). *Organization-level GP positively affects ES adoption.*

2. Material and Methods

Based on hypothesis 1 to hypothesis 4, the research model of this research is shown in Figure 1.

2.1. Sampling and Procedures

We investigated data at a three-phase time in six months from the agricultural technology manufacturing companies in Taiwan. The interval of each time point was three months to in line with past attitude changes studies [28–30]. We contacted these agricultural technology manufacturing companies to join the survey. These agricultural technology companies mainly use technology to produce upstream products related to agricultural products, such as rice seedlings, breeding chickens, fertilizers, etc. We collected 50 technology manufacturing companies, and each company was requested to recruit 3 supervisors to join this investigation. We used email to collect questionnaires. From the first phase time to the third phase time, we collected 150 supervisors' assessments toward the adoption of ES, GP and GSR.

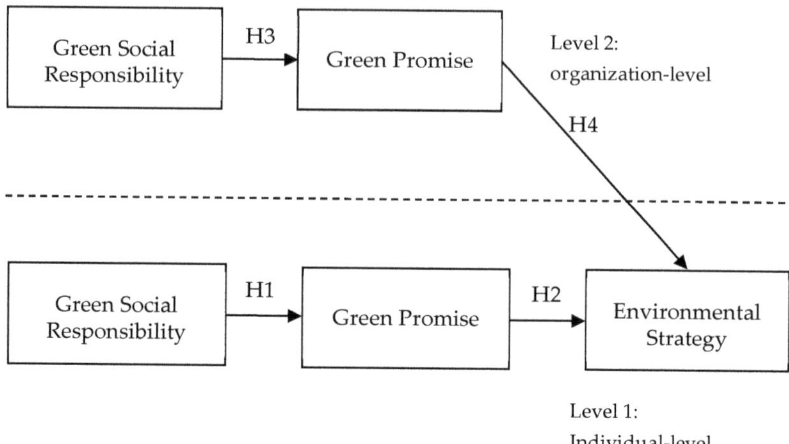

Figure 1. Research model of this research.

2.2. Measures

We adopted language conversion method to confirm quality [31], and James et al.'s [32] within-group consensus rwg(j) was adopted to confirm the variables aggregation. GSR, GP, and ES were assessed through past studies [8,12,33].

2.3. Model Validation

The minimum rwg(j) is 0.81 of GSR, GP, and ES, and it supports aggregating the individual-level GSR and GP into organization level variables. The minimum average variance extracted and the reliability respectively is 0.55 and 0.89. The model fit indexes of the research model are in line with the research of Fornell and Larcker [34].

3. Results

Analysis Results

Because the data framework of this research was nested within each workgroup (105 different companies), so this research employed HLM to analyze the cross-level frameworks [15]. The analysis results are shown in Table 1. First, the individual-level GSR significantly affected the individual-level GP ($\gamma = 0.32$, $p < 0.01$), and individual-level GP significantly affected the individual-level ES ($\gamma = 0.35$, $p < 0.01$).

Table 1. Results of HLM.

Hypothesis	Path	Coefficient	Results
H_1	Individual-level Green Social Responsibility → Individual-level Green Promise	0.32 **	Supported
H_2	Individual-level Green Promise → Individual-level Environmental Strategy	0.35 **	Supported
H_3	Organization-level Green Social Responsibility → Organization-level Green Promise	0.41 **	Supported
H_4	Organization-level Green Promise →Individual-level Environmental Strategy	0.37 **	Supported

** = $p < 0.01$; Second, the organization-level GSR significantly influenced organization-level GP ($\gamma = 0.41$, $p < 0.01$), and organization-level GP significantly influenced the individual-level ES ($\gamma = 0.37$, $p < 0.01$).

4. Discussion

4.1. Academic Contribution

This survey is the first to demonstrate the HLM that conceptualizes the ES adoption and its driving factors according to the theory of symbolic context in the agricultural field. According to the analysis results, individual-level and organization-level GSR would influence individual-level and organization-level GP, which consequently would influence the ES adoption, thereby indicating the validity of the HLM. Also, the HLM perspective is a novel mechanism to open the black box with ES and its antecedent at the multilevel framework that past study has not examined this pathway [9,22]. Therefore, this research has ex-tended GSR, GP, and ES literature into the agricultural field to guide these agricultural technology manufacturing companies to implement sustainable production through the ES.

4.2. Practice Contribution

In the past, research in the field of agriculture has almost adopted new agricultural technologies to implement ES [35,36], but this research proposes another way to implement ES. According to the empirical results, the vendors of agricultural technology manufacturing companies should keep in mind that investing resources in improving employees' attitudes is not the most effective investment and paying attention to the GSR and GP may be a more worthwhile investment. Indeed, GP of supervisors can transform GSR into the company's adoption of ES, and ES is a key source of sustainable production. Therefore, these vendors should learn how to increase GSR and GP by the management mechanism. For example, education training may be one of the effective management mechanisms.

4.3. Further Research and Limitations

The present study includes GSR and GP of supervisors to predict ES adoption, but there may be other key driving factors that could cause the company's ES adoption. Further researchers must explore key driving factors of ES in different contexts. For example, institutional theory has been examined as a key driving factor of ES [8]. Also, further re-searchers must employ more data in different countries to the proposed model in this research. Finally, a previous study proposed that information technology adoption behavior models can be used as the theoretical basis for strategy adoption of agricultural enterprises [37], and further research should test which models have better explanatory power in different contexts.

5. Conclusions

This survey proposes the novel HLM, that is, how GSR can predict the company's ES adoption through the mediation role of the GP in the organizational multi-level framework. This new type of HLM can significantly promote GSR, GP, and ES literature in the field of agriculture management. Indeed, previous studies in the field of agriculture lacked similar studies to the theoretical model of this research because these studies mainly explored how to use innovative agricultural technologies to increase yields. These results can offer references to firms to formulate ES and let these companies know that ES should be implemented by the GP of supervisors to achieve the goal of sustainable development.

Author Contributions: Conceptualization, S.Y.B.H.; Data curation, S.-C.L.; Formal analysis, S.-C.L.; Funding acquisition, Y.-S.L.; Methodology, S.-C.L.; Project administration, S.Y.B.H.; Resources, Y.-S.L.; Software, S.-C.L. and Y.-S.L.; Supervision, S.Y.B.H.; Visualization, Y.-S.L.; Writing—original draft, S.Y.B.H.; Writing—review & editing, S.Y.B.H. All authors have read and agreed to the published version of the manuscript.

Funding: This research received no external funding.

Institutional Review Board Statement: Not applicable.

Informed Consent Statement: Not applicable.

Data Availability Statement: Not applicable.

Conflicts of Interest: The authors declare no conflict of interest.

References

1. Calle, F.; González-Moreno, Á.; Carrasco, I.; Vargas-Vargas, M. Social Economy, Environmental Proactivity, Eco-Innovation and Performance in the Spanish Wine Sector. *Sustainability* **2020**, *12*, 5908. [CrossRef]
2. Ge, B.; Yang, Y.; Jiang, D.; Gao, Y.; Du, X.; Zhou, T. An Empirical Study on Green Innovation Strategy and Sustainable Competitive Advantages: Path and Boundary. *Sustainability* **2018**, *10*, 3631. [CrossRef]
3. Junquera, B.; Barba-Sánchez, V. Environmental Proactivity and Firms' Performance: Mediation Effect of Competitive Advantages in Spanish Wineries. *Sustainability* **2018**, *10*, 2155. [CrossRef]
4. Ryszko, A. Proactive Environmental Strategy, Technological Eco-Innovation and Firm Performance—Case of Poland. *Sustainability* **2016**, *8*, 156. [CrossRef]
5. Lehtonen, H.; Palosuo, T.; Korhonen, P.; Liu, X. Higher Crop Yield Levels in the North Savo Region—Means and Challenges Indicated by Farmers and Their Close Stakeholders. *Agriculture* **2018**, *8*, 93. [CrossRef]
6. Mantino, F.; Forcina, B. Market, Policies and Local Governance as Drivers of Environmental Public Benefits: The Case of the Localised Processed Tomato in Northern Italy. *Agriculture* **2018**, *8*, 34. [CrossRef]
7. Salvia, R.; Simone, R.; Salvati, L.; Quaranta, G. Soil Conservation Practices and Stakeholder's Participation in Research Projects—Empirical Evidence from Southern Italy. *Agriculture* **2018**, *8*, 85. [CrossRef]
8. Banerjee, S.B.; Iyer, E.S.; Kashyap, R.K. Corporate environmentalism: Antecedents and influence of industry type. *J. Mark.* **2003**, *67*, 106–122. [CrossRef]
9. Huang, S.Y.B.; Ting, C.-W.; Li, M.-W. The Effects of Green Transformational Leadership on Adoption of Environmentally Proactive Strategies: The Mediating Role of Green Engagement. *Sustainability* **2021**, *13*, 3366. [CrossRef]
10. Peng, B.H.; Tu, Y.; Elahi, E.; Wei, G. Extended producer responsibility and corporate performance: Effects of environmental regulation and environmental strategy. *J. Environ. Manag.* **2018**, *218*, 181–189. [CrossRef]
11. Hatch, M.J. The dynamics of organizational culture. *Acad. Manag. Rev.* **1993**, *18*, 657–693. [CrossRef]
12. Wei, Z.; Shen, H.; Zhou, K.Z.; Li, J.J. How does environmental corporate social responsibility matter in a dysfunctional institutional environment? Evidence from China. *J. Bus. Ethics* **2017**, *140*, 209–223. [CrossRef]
13. Yusliza, M.Y.; Amirudin, A.; Rahadi, R.A.; Nik Sarah Athirah, N.A.; Ramayah, T.; Muhammad, Z.; Dal Mas, F.; Massaro, M.; Saputra, J.; Mokhlis, S. An Investigation of Pro-Environmental Behaviour and Sustainable Development in Malaysia. *Sustainability* **2020**, *12*, 7083. [CrossRef]
14. Norton, T.A.; Zacher, H.; Ashkanasy, N.M. Organizational sustainability policies and employee green behavior: The mediating role of work climate perceptions. *J. Environ. Psychol.* **2014**, *38*, 49–54. [CrossRef]
15. Raudenbush, S.W.; Bryk, A.S. *Hierarchical Linear Models*; Sage: Thousand Oaks, CA, USA, 2002.
16. Connor, M.; de Guia, A.H.; Pustika, A.B.; Sudarmaji; Kobarsih, M.; Hellin, J. Rice Farming in Central Java, Indonesia—Adoption of Sustainable Farming Practices, Impacts and Implications. *Agronomy* **2021**, *11*, 881. [CrossRef]
17. Monjardino, M.; López-Ridaura, S.; Van Loon, J.; Mottaleb, K.A.; Kruseman, G.; Zepeda, A.; Hernández, E.O.; Burgueño, J.; Singh, R.G.; Govaerts, B.; et al. Disaggregating the Value of Conservation Agriculture to Inform Smallholder Transition to Sustainable Farming: A Mexican Case Study. *Agronomy* **2021**, *11*, 1214. [CrossRef]
18. Shah, T.M.; Tasawwar, S.; Bhat, M.A.; Otterpohl, R. Intercropping in Rice Farming under the System of Rice Intensification—An Agroecological Strategy for Weed Control, Better Yield, Increased Returns, and Social–Ecological Sustainability. *Agronomy* **2021**, *11*, 1010. [CrossRef]
19. Bingham, J.B.; Mitchell, B.W.; Bishop, D.G.; Allen, N.J. Working for a higher purpose: A theoretical framework for commitment to organization-sponsored causes. *Hum. Resour. Manag. Rev.* **2013**, *23*, 174–189. [CrossRef]
20. Cantor, D.E.; Morrow, P.C.; Montabon, F. Engagement in environmental behaviors among supply chain management employees: An organizational support theoretical perspective. *J. Supply Chain Manag.* **2012**, *3*, 3–51. [CrossRef]
21. Chou, C.J. Hotels' environmental policies and employeenpersonal environmental beliefs: Interactions and outcomes. *Tour. Manag.* **2014**, *40*, 436–446. [CrossRef]
22. Huang, S.Y.B.; Ting, C.-W.; Fei, Y.-M. A Multilevel Model of Environmentally Specific Social Identity in Predicting Environmental Strategies: Evidence from Technology Manufacturing Businesses. *Sustainability* **2021**, *13*, 4567. [CrossRef]
23. Kong, H.; Jeon, J.-E. Daily Emotional Labor, Negative Affect State, and Emotional Exhaustion: Cross-Level Moderators of Affective Commitment. *Sustainability* **2018**, *10*, 1967. [CrossRef]
24. Mahmood, F.; Qadeer, F.; Abbas, Z.; Muhammadi; Hussain, I.; Saleem, M.; Hussain, A.; Aman, J. Corporate Social Responsibility and Employees' Negative Behaviors under Abusive Supervision: A Multilevel Insight. *Sustainability* **2020**, *12*, 2647. [CrossRef]
25. Kozlowski, S.W.J.; Klein, K.J. A multilevel approach to theory and research in organizations: Contextual, temporal, and emergent processes. In *Multilevel Theory, Research, and Methods in Organizations: Foundations, Extensions, and New Directions*; Klein, K.J., Kozlowski, S.W.J., Eds.; Jossey-Bass: San Francisco, CA, USA, 2000; pp. 3–90.
26. Hackman, J.R. Group influences on individuals in organizations. In *Handbook of Industrial Organizational Psychology*; Dunnette, M.D., Hough, L.M., Eds.; Psychologists Press: Palo Alto, CA, USA, 1992; pp. 199–267.
27. Bandura, A. *Social Foundations of Thought and Action: A Social Cognitive Theory*; Prentice Hall: Upper Saddle River, NJ, USA, 1986.
28. Huang, S.Y.B.; Fei, Y.-M.; Lee, Y.-S. Predicting Job Burnout and Its Antecedents: Evidence from Financial Information Technology Firms. *Sustainability* **2021**, *13*, 4680. [CrossRef]

29. Huang, S.Y.B.; Li, M.-W.; Chang, T.-W. Transformational Leadership, Ethical Leadership, and Participative Leadership in Predicting Counterproductive Work Behaviors: Evidence From Financial Technology Firms. *Front. Psychol.* **2021**, *12*, 658727. [CrossRef]
30. Lee, C.-J.; Huang, S.Y.B. Double-edged effects of ethical leadership in the development of Greater China salespeople's emotional exhaustion and long-term customer relationships. *Chin. Manag. Stud.* **2020**, *14*, 29–49. [CrossRef]
31. Brislin, R.W. Translation and content analysis of oral and written materials. In *Handbook of Cross-Cultural Psychology*; Triandis, H.C., Berry, J.W., Eds.; Allyn and Bacon: Boston, MA, USA, 1980; Volume 2, pp. 389–444.
32. James, L.R.; Demaree, R.G.; Wolf, G. Estimating within group interrater reliability with and without response bias. *J. Appl. Psychol.* **1984**, *69*, 85–89. [CrossRef]
33. Raineri, N.; Paille, P. Linking corporate policy and supervisory support with environmental citizenship behaviors: The role of employee environmental beliefs and commitment. *J. Bus. Ethics* **2016**, *137*, 129–148. [CrossRef]
34. Fornell, C.; Lacker, D.F. Evaluating structural equation models with unobservable variables and measurement error. *J. Mark. Res.* **1981**, *18*, 39–50. [CrossRef]
35. Chukwudi, U.P.; Kutu, F.R.; Mavengahama, S. Influence of Heat Stress, Variations in Soil Type, and Soil Amendment on the Growth of Three Drought–Tolerant Maize Varieties. *Agronomy* **2021**, *11*, 1485. [CrossRef]
36. Martínez-Gómez, P.; Rahimi Devin, S.; Salazar, J.A.; López-Alcolea, J.; Rubio, M.; Martínez-García, P.J. Principles and Prospects of Prunus Cultivation in Greenhouse. *Agronomy* **2021**, *11*, 474. [CrossRef]
37. Montes de Oca Munguia, O.; Pannell, D.J.; Llewellyn, R. Understanding the Adoption of Innovations in Agriculture: A Review of Selected Conceptual Models. *Agronomy* **2021**, *11*, 139. [CrossRef]

Article

Chinese Residents' Perceived Ecosystem Services and Disservices Impacts Behavioral Intention for Urban Community Garden: An Extension of the Theory of Planned Behavior

Can Wu [1,2,3], Xiaoma Li [1,3], Yuqing Tian [4], Ziniu Deng [2,5], Xiaoying Yu [2,6], Shenglan Wu [7], Di Shu [1,3], Yulin Peng [2], Feipeng Sheng [8] and Dexin Gan [1,3,*]

[1] College of Landscape Architecture and Art Design, Hunan Agricultural University, Changsha 410128, China; Wcan@hunau.edu.cn (C.W.); lixiaoma@hunau.edu.cn (X.L.); Shudi@hunau.edu.cn (D.S.)

[2] College of Horticulture, Hunan Agricultural University, Changsha 410128, China; dengzn@hunau.net (Z.D.); yxy1578@hunau.edu.cn (X.Y.); pengyulin@stu.hunau.edu.cn (Y.P.)

[3] Hunan Provincial Key Laboratory of Landscape Ecology and Planning & Design in Regular Higher Ed-ucational Institutions, Changsha 410128, China

[4] School of Environment, Tsinghua University, Beijing 100084, China; tyq21@mails.tsinghua.edu.cn

[5] Research Center for Horticultural Crop Germplasm Creation and New Variety Breeding, Ministry of Education, Changsha 410128, China

[6] Mid-Subtropical Quality Plant Breeding and Utilization Engineering Technology Research Center, Changsha 410128, China

[7] Orient Science & Technology College, Hunan Agricultural University, Changsha 410128, China; wushenglan@hunau.edu.cn

[8] Department of Architecture, Hunan Urban Construction College, Xiangtan 411104, China; shenfeipeng@youfangkj.com

* Correspondence: Gandexin@hunau.edu.cn

Citation: Wu, C.; Li, X.; Tian, Y.; Deng, Z.; Yu, X.; Wu, S.; Shu, D.; Peng, Y.; Sheng, F.; Gan, D. Chinese Residents' Perceived Ecosystem Services and Disservices Impacts Behavioral Intention for Urban Community Garden: An Extension of the Theory of Planned Behavior. *Agronomy* **2022**, *12*, 193. https://doi.org/10.3390/agronomy12010193

Academic Editors: Moritz Von Cossel, Joaquín Castro-Montoya and Yasir Iqbal

Received: 13 December 2021
Accepted: 3 January 2022
Published: 13 January 2022

Publisher's Note: MDPI stays neutral with regard to jurisdictional claims in published maps and institutional affiliations.

Abstract: Urban community gardens (UCGs), greenspace cultivated and managed for vegetables by local communities, provide substantial ecosystem services (ES) and are warmly welcomed by residents. However, they also have many ecosystem disservices (EDS) and are almost always refused by the decision-makers of the government, especially in China. Better understanding the residents' perceived ES and EDS and the impact on the behavioral intention (BI) toward UCGs is of great value to solve the conflicts between residents and the government concerning UCGs and to develop sustainable UCGs. Following the theory of planned behavior (TPB), we measured perceived ES/EDS, attitudes (ATT), perceived behavioral control (PBC), subjective norm (SN), and BI of 1142 residents in Changsha, China, and investigated their direct and indirect causal relationships using structural equation modeling (SEM). The results showed that: (1) ATT, PBC, and SN significantly and positively impact the BI of UCGs and together explained 54% of the variation of BI. (2) The extended TPB model with additional components of perceived ED/EDS improved the explanatory ability of the model, explaining 65% of the variance of BI. Perceived ES and perceived EDS showed significant direct positive and negative impacts on UCGs, respectively. They also indirectly impacted BI by influencing ATT, PBC, and SN. The findings of this study can extend our understanding of residents' attitudes, behavior, and driving mechanism toward UCGs, and can help decision makers to design better policies for UCG planning and management.

Keywords: urban community garden; ecosystem service and disservice; behavioral intention; theory of planned behavior; structural equation model

1. Introduction

According to the American Community Garden Association (ACGA), an urban community garden (UCG) is any piece of land managed and cultivated by local communities to grow vegetables or flowers [1,2]. UCGs are hybrid parts of the city belonging to both the built environment and the green infrastructure, the public and the private, and the planned and the unplanned [3]. They are productive landscapes where people and place, mind and body, social and physical, and past and present intermingle [4]. UCGs offer many ecosystem services (ES) [5] such as the provision of food and medicinal plants [6], local climate regulation [7], biodiversity [8], habitat for species [9], the facilitation of active and healthy lifestyles [10], neighborhood relationships [11], opportunities for relaxation and recreation [12], increased social cohesion [13], and environmental education [14]. However, UCGs are also associated with various ecosystem disservices (EDS) [15] such as sheltering harmful animals and vectors of diseases, contaminating soil, destroying landscape aesthetics, and increasing interpersonal tension [3,16]. The current theoretical research mainly focuses on the direct or indirect social and environmental impacts of urban community gardens [1,5], where the understanding of ES and EDS focuses on assessing its level of delivery benefits and value [2,6]. However, to our knowledge, few studies investigate whether these services and impairments affect personal and social cognitive factors such as perception and attitudes, which affect their behavioral intentions or practices.

In the 19th century, the UCG emerged in the European and American urban areas in response to the lack of fresh food during the Industrial Revolution, and became an effective measure to prevent social unrest during war periods and the Great Depression [17]. The UCG is currently expanding worldwide and is seen as one strategy for responding to the dietary, social, and environmental problems caused by rapid urbanization in developed and developing countries. UCGs have attracted increasing attention from scholars and practitioners [18], with examples of UCGs including the Verge Garden in Australia [19], the Community in Bloom in Singapore [20], the P-Patch program in Seattle [21], and the Empty-Spaces Plan in Barcelona [22]. At the same time, many countries see UCGs as illegal and informal spaces; gardeners grow land in community public spaces without permission and consultation almost around the world [23], such as in Zimbabwe, Kenya, and South Africa [24].

Before 2010, there were no formal UCGs in Chinese cities [25]; residents spontaneously occupied open public space, usually green space, by creating self-claimed vegetable lots in many corners of cities [26]. Like many illegal UCG practices abroad, this spontaneously formed open gardening space [15] constitutes an infringement of land use rights without consultation with stakeholders such as non-gardeners and community officials. These spontaneous practices lead to disputes and conflicts, posing challenges to urban community governance in China [25,26]. In recent years, the Chinese government has increasingly supported the intensive development and ecological utilization of idle urban land. Within this context, the UCG is also seen as an effective measure for many urban regeneration policies, such as "Stock Regeneration" and "City Betterment and Ecological Restoration" [25]. Furthermore, since the UCG is regarded as a beneficial way to meet the well-being and needs of contemporary urban people, many local government agencies began to legally establish and maintain UCGs [27]. As of 2020, the number of legal UCGs has reached more than 300 in Shanghai and 30 in Changsha [28]. Most UCGs in China are initiated and created by residents or community grassroots organizations; in contrast, UCGs in developed countries are more often planned, developed, and managed by governmental or commercial agencies [25]. Therefore, to facilitate the establishment of comprehensive UCG planning, construction, and management policies in China, and to ensure that UCG projects remain vital and relevant, a deep understanding of urban residents' community gardening behavior is necessary.

To develop more effective community gardening interventions for urban residents, it is important to conduct research based on theoretical models that adequately explain and predict gardening behavioral intention. A primary framework for predicting and explaining behavioral intention is the theory of planned behavior (TPB) [29,30]. The theory emphasizes that the behavioral intention (BI) of individuals is influenced by their attitude (ATT), subjective norms (SN), and perceived behavioral control (PBC) [29]. Due to different research focuses, previous TPB-based research did not examine the interaction of residents' perceived ecosystem services and disservices associated with the formation of UCG behavioral intention. Therefore, assessing the perceptions of ecosystem services and disservices is pivotal to developing guidelines and policies that improve UCG planning and management [31].

Given the lack of a comprehensive theoretical framework, the present study aimed to create a structural equation modeling (SEM) of the extended TPB model, including residents' perceived UCG ecosystem services and disservices, as a means of predicting the community gardening behavioral intention of Chinese urban residents. Our main objectives were to (i) investigate the perception of UCGs' ecosystem services and disservices by urban residents in inland China and (ii) examine the possible causal relationships among residents' perception of UCG ecosystem services and disservices and individual and social cognitive factors. SEM was applied to examine the hypothetical causal relationships driving the behavioral intention. The analysis in this paper focuses on the impact of perceived UCG ecosystem services/disservices on residents' individual cognitive factors (attitudes and perceived behavioral control) and social cognitive factors (subjective norms), as well as the significance and magnitude of impact on the gardening behavior intention. Therefore, the study findings contribute to the literature by revealing the mechanism for residents' perceived ecosystem services/disservices, attitude, self-efficacy, social stress, and UCG behavioral intention.

This study's first original contribution was novel evidence for the cognition of ecosystem services associated with UCG. The second original contribution was that we adopted an integrated view to jointly consider UCGs' abilities to provide ecosystem services and disservices. The third original contribution was that by adding the structure of perceived ecosystem services and disservices, we constructed an extended theoretical framework based on classical TPB theory. Our findings could help urban decision-makers and managers to better understand the ways UCGs are perceived and predict the possible impacts of changes in UCG ecosystem services and disservices on urban residents' gardening participation. Increasing residents' awareness of benefits that the UCG could provide (perception of ecosystem services) can motivate their continued behavioral intention. Reducing the various obstacles associated with UCGs and thus reducing residents' perceived ecosystem disservices will help provide useful information for future scientific regulations, planning concepts, construction models, and management policies related to UCG practice, giving community gardens sustained long-term vitality.

The rest of this article is organized as follows. In Section 2, we review the relevant literature and develop hypotheses. Then, the research design and methodologies are described in Section 3. Furthermore, in Section 4, we analyzed the results and examined the influential mechanism for residents' perceived ES/EDS, individual/social cognition factors, and behavioral intention. Finally, the discussion and conclusions are given in Section 5.

2. Research Background and Hypotheses Development

To develop more effective community gardening interventions for general urban residents, it is important to conduct research based on theoretical models that adequately explain and predict gardening behavioral intention. The theory of planned behavior (TPB) is a socio-psychological model for predicting and explaining individual motivation and behaviors, especially pro-environmental behaviors [29,30]. From the psychological perspective, TPB theory is a useful theoretical framework that can shed light on the complex psychological processes underlying individual behavior, which makes it an important tool to apply while clarifying the role of human volitional behaviors [32,33]. The theory emphasizes that the behavioral intention (BI) of individuals is influenced by their attitude (ATT), subjective norms (SN), and perceived behavioral control (PBC) [29]. Attitude is the first determinant and refers to the degree of a person's positive/negative [34,35], support/non-support [36], and/or favorable/unfavorable [37] evaluation or appraisal for performing a behavior [38]. Perceived behavioral control (PBC) is the second determinant and refers to the personal perceptions that make it difficult or easy to conduct a certain action and the extent to which performing a particular act is under the individual's volitional control [39]. Subjective norm as a social predictor is the third determinant of behavioral intention and refers to an individual's perception of the degree to which social pressure prompts them to perform or avoid a particular behavior and whether important others support or do not support the activity [40]. Current studies indicate that an individual's intention to participate in a behavior is stronger when they have positive attitudes towards that behavior, a greater sense of self-efficacy, or strong social support [35,41]. Behavioral intention is at the core of TPB and is the key link between psychological perception and behavior. PBC, ATT, and SN all have an impact on BI, but in different ways and to various extents [41]. The stronger an individual's intention to undertake a behavior, the more likely the behavior will be performed [42,43]. If a person has a favorable appraisal and a better perceived behavior control ability of the UCG, he may also expect his important referent individuals (or groups) to endorse his behavior. The higher a person's evaluation and confidence in his participation in UCG behavior the more likely he is to perceive less pressure from important representatives. TPB has been successfully applied to many domains of pro-environmental behavior, including sustainable housing purchase intention [44], energy-saving and emission reduction [45], recycling behaviors [46], environmental protection [47], and participatory natural resource management [48].

Due to different priorities, previous researchers have investigated the determinants of a range of pro-environmental behaviors among citizens. Nevertheless, in the context of the UCG, few studies have used the TPB to investigate the relationship between residents' individual or social psychological factors, and their impact and extent on behavioral intention [39,40]. These lead to the following hypotheses:

Hypothesis 1 (H1). *Attitudes are positively associated with UCG behavioral intention.*

Hypothesis 2 (H2). *Subjective norms are positively associated with UCG behavioral intention.*

Hypothesis 3 (H3). *Perceived behavioral control is positively associated with UCG.*

These lead to the following hypotheses:

Hypothesis 4 (H4). *Attitudes are positively associated with subjective norms.*

Hypothesis 5 (H5). *Perceived behavioral control is positively associated with subjective norms.*

The TPB is a very parsimonious model that allows researchers to include additional predictors associated with a particular behavior [42]. Recent interpretations of the theory suggest that the original TPB model is extended through new factors if behavioral intention or practice achieve a significant amount of the variance by including these predictors [49]. Researchers have explained that UCG ecosystem services (or disservices) are a series of beneficial (or harmful) consequences of promoting (or hindering) social and ecological sustainability produced through gardening behavior. Therefore, the perception of ecosystem services (or disservices) is rooted in an awareness of beneficial (or harmful) consequences [50,51]. Although it is rare to intervene in perceived ecosystem services as an extension factor in the TPB theory to explain behavioral intention and actual action, some researchers have studied residents' perceptions of ecosystem services and disservices for urban green spaces as predictors of behavioral intention and suggest that the decision-making process is related to protective behavior [52]. Moreover, in a meta-analysis of psychological–social determinants of pro-environmental behavior, it was found that the awareness of consequences has important and direct effects on individuals, social elements, and intentions in the TPB and indirectly affects behavior [53]. These results suggest that perceived detrimental consequences affect behavioral intention [33,34,54,55] and that a high level of awareness of positive consequences will help foster a more favorable attitude [32]. In addition to the relationship between awareness of consequences and attitude, the fact that individuals recognizing specific behavioral consequences will better understand the impact of the behavior on others and the environment [56] implies that people who are highly aware of consequences have a strong sense of social expectation [57]. Consequently, some studies have provided empirical evidence for the direct impact of awareness of consequences on subjective norms [53,57,58]. Moreover, research has shown that an awareness of consequences precedes attitudes, perceived behavioral control, and subjective norms [58]. These results suggest that an awareness of beneficial (or harmful) consequences may influence behavioral intention and real action through individual and social variables.

However, few studies have included residents' perception of ecosystem services and disservices in the UCG as additional factors in TPB models and explored the driving effects of perception on residents' attitudes, self-efficacy, and social stress. These lead to the following hypotheses:

Hypothesis 6 (H6). *People who perceive higher ES will have more positive attitudes toward UCGs.*

Hypothesis 7 (H7). *People who perceive higher ES will have more positive SN toward UCGs.*

Hypothesis 8 (H8). *People who perceive higher ES will have more positive PBC toward UCGs.*

Hypothesis 9 (H9). *People who perceive lower EDS will have more positive attitudes toward UCGs.*

Hypothesis 10 (H10). *People who perceive lower EDS will have more positive toward UCGs.*

Hypothesis 11 (H11). *People who perceive lower EDS will have more positive PB toward UCGs.*

Moreover, ecosystem services and disservices are often formed based on the same set of ecosystem characteristics, ecological functions, or species groups [59]. This can easily make people conscious of the connection between the perceived ecosystem services and disservices. These lead to the following hypothesis:

Hypothesis 12 (H12). *Perceived ecosystem services are negatively correlated with perceived ecosystem disservices.*

Recently, increasing attention has been given to the importance of ecosystem services and the application of ecological models to analyze the multi-level effects of pro-environmental behavior [19,39,40]. The concept of ES has been mainstreamed as an interdisciplinary guiding framework in urban sustainability and resilience agendas [60]. The psychological theory of motivational functionalism states that motivation is the process that initiates, guides, and maintains goal-oriented behaviors and involves biological, emotional, social, and cognitive forces [61]. Motivation is the need to drive people in a specific direction to a certain purpose and to conduct a certain behavior. However, as ecosystem services help shed light on the relationship between the ecosystem and human behavior [62], ES can not only motivate the development of personal and social processes that initiate, direct, and sustain human action [38,63], but also stimulate changes in attitudes and behaviors regarding some symbol or object or aspect. Thus, the psychological theory of motivational functionalism defines ecosystem services as motivation [38,62]. The ecosystem service motivations of UCGs include those related to food supply, biodiversity preservation, soil fertility maintenance, air purification, climate regulation, physical health, recreation, interpersonal communication, and cultural education [39,51,61,64–67]. While fewer existing studies have focused on the impact of residents' perceived UCG ecosystem services on their gardening behavior, many studies confirm a significant association between resident or gardener motivation and urban gardening intention or behavior, focusing on the causes of motivation and how motivation affects behavioral intention or practice. The literature relevant to motivation-related urban gardening behavioral intention or practice studies is summarized in Table 1.

Table 1. Motivations related to urban gardening behavioral intention or behavior studies.

Reference	Garden Type	Sample Size	Data Collection	Interviewee Type	Main Finding
[39]	UCG/UAG	180	Questionnaire survey	Gardener	Used TPB theory; the gardener's behavioral intention to participate is influenced by functional motivation, emotional motivation, and conditional motivation.
[67]	UCG/UAG	300	Questionnaire survey	Resident	Discussed the associations between personal characteristics and perceptions, needs, and motivations of potential future gardeners.
[15]	UCG	300	Questionnaire survey	Resident Gardener	Discussed respondents' corresponding motivation, the existing barriers and challenges of their behavior, and the behavioral intention to participate in paid gardening.
[65]	UAG	873	Online survey	Resident	Identified the main characteristics and motivation of potential urban gardeners and determined the mechanism of the influence of these features on motivation.
[61]	UAG	141	Questionnaire survey	Resident	Explored the correspondence between the six resident motivations and the eight gardening needs.
[68]	UHG	126	Questionnaire survey	Gardener	Divided the motivation for home gardening into two categories, one involving social interests and function and the other inherent in nature.
[69]	UCG	23	Interview	Gardener	Explored six functional motivations and three conditional motivations that influence community garden behavior.
[51]	UAG/UHG	23	Interview	Gardener	Indicates gardeners with different cultural backgrounds have different motivations for participating in gardening.
[18]	UG	60	Questionnaire survey	Gardener	Gardener's gardening motivation is divided into two types: a clear preference for gardening as a means of physical and mental health and learning new skills, and another mainly a yield-based motivation.
[64]	UCG/UAG	40	Interview	Gardener	Gardener functional motivation has direct relevance to their gardening practice; the connection between gardeners and garden occurs in gardening behavior.

Note: UG stands for urban gardens, UCG stands for urban community gardens, UAG stands for urban allotment gardens, URG stands for urban roof gardens, and UHG stands for urban home gardens.

Furthermore, motivation and barriers always appear simultaneously in many research articles. There is a range of motivations to promote gardening behavior and various barriers to the behavioral development have also been studied. Barriers regarding UCGs are related to finances, external damage, space, water, soil, organizational structure, communication, interpersonal relationships, and participation [70–72]. In many articles, ecosystem disservices have been described as harmful consequences of ecological change or as deficient ecosystem services [73,74]. Ecosystem disservices belong mainly to four fields [73]. To begin with, ecosystem structure, processes, and services provided are negatively affected by population and development systems [75]. Then, (socio-) economic structures and processes are negatively affected by ecosystem disservices [76,77]. Furthermore, human health is negatively affected by ecosystem disservices [52]. Finally, people's quality of life is reduced by ecosystem disservices, which in turn leads to negative emotions such as anxiety and depression [52]. Studies have pointed out that many of the barriers to the occurrence and sustainability of behavior are due to ecosystem disservices [74]. Recent studies have reported that UCGs' ecosystem disservices obstruct the development of UCGs and negatively impact gardening behavior. Studies have also found that residents' motivation to grow organic or healthier food, maintain personal health, and improve quality of life [39] were positively associated with their behavioral intention and actual gardening behavior, whereas environmental impacts and health threats were negatively associated with horticultural participation and behavior [68]. Notably, people of different social backgrounds have different motivations or obstructions, but all factors have positive or negative effects on gardening behavior [66]. However, few articles have talked about the perception of UCG ecosystem services and services affecting horticultural behavioral intention or practice. The more residents believe in high ecosystem services, the greater their motivation, and the more forward their behavioral intention. In contrast, the more they think that UCGs produce various ecosystem disservices and hinder the development of gardening behavior, the more cautious and even less intent they have to carry out this behavior [70]. In addition, studies that rely on survey responses while examining the perception of motivation or obstruction have focused on gardeners in Western countries and have rarely centered on urban residents in Asian environments [15]. These lead to the following hypotheses:

Hypothesis 13 (H13). *Perceived ecosystem services are positively associated with UCG behavioral intention.*

Hypothesis 14 (H14). *Perceived ecosystem disservices are negatively associated with UCG behavioral intention.*

In conclusion, all previous studies have pointed to a relationship between perceived ecosystem services or disservices and gardening behavior. However, a deeper interpretation of the interactions between individual, social, and perceived ecosystems should explore a theoretical framework. Given the absence of an integrated theoretical framework to predict the gardening behavioral intention of Chinese urban residents, the present study adopted an extended TPB model. Perceived ecosystem services and disservices variables were hypothesized as common precursors of attitude, perceived behavioral control, and subjective norms as determinants of gardening behavioral intention among urban residents (Figure 1).

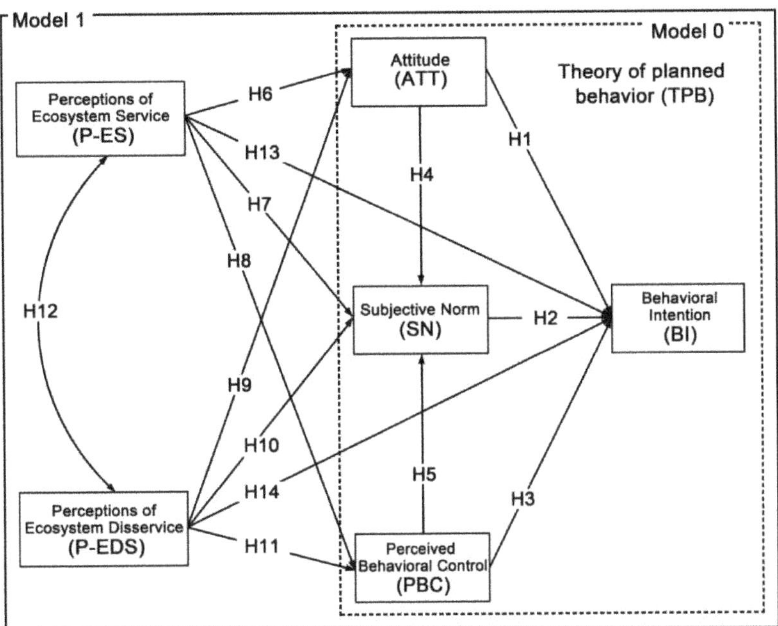

Figure 1. The initial conceptual diagram of urban residents' behavioral intention for UCGs.

3. Methodology

3.1. Participants and Data Collection

This paper gathered data in Changsha, China. The city is situated in the subtropical monsoon climate and has an urbanization rate of 79.6%, absorbing 2 million permanent migrants (23.8% of the permanent population). Since 2017, the government has launched the UCG project in 30 communities within the city. This survey went through three phases over a period of two years—two pre-surveys and a formal survey stage. First, between January and May 2019, we conducted the first pre-survey. This consisted of semi-structured face-to-face interviews with 84 gardeners living in six adjacent communities in the Furong District of Changsha to identify and characterize several of the most important ecosystem services in the area. Respondents could choose among eighteen ecosystem services available for urban gardens with content derived from previous research [22]. All respondents were audio recorded with their permission and each session lasted between 30 and 75 min. Twelve ecosystem service categories that were easily understood and perceived by gardeners were finalized. Subsequently, we conducted a second pre-survey. Between May and June 2019, we conducted a household survey of residents living in the same community as the previously interviewed gardeners, with 778 residents completing a structured questionnaire. We examined which of the twelve ecosystem services proposed by gardeners are more easily perceived by residents. At the same time, residents' perceptions of ten ecosystem disservices from UCGs were investigated. Variable preparation for EDS was derived from previous studies [70,73]. We placed the six ecosystem service functions that residents can easily perceive under the four major categories covered in established ES classifications: provisioning, regulating, habitat/supporting, and cultural services [78]; four EDS categories that were easily understood and perceived by residents were finalized.

Through the above two pre-surveys, the focused content of the more easily perceived ES and EDS by the urban residents of the surveyed area was identified. Furthermore, the individual/social cognitive factors and behavioral intention of UCGs were investigated. To obtain more representative data, with the help of the Changsha City Administration, the community managers of the five main urban areas became key informants. Using a "snowball technique," interviewed community managers were asked to provide connections to new informants. An online survey was conducted in July 2020. In cooperation with Changsha Ranxing Information Technology Co., Ltd., a professional online survey service provider, we designed and collected the questionnaire. The company's users have covered more than 90% of the universities and research institutes in China; their questionnaire survey is a well-known brand trusted by leading enterprises in various industries. The company's strict privacy regulations for personal information and incentives to stimulate participation help to reduce potential bias. This survey received 1692 replies with a valid response rate of 67.5%. Statistical analysis was performed on the data of 1142 valid collected questionnaires.

The respondents' socioeconomic status is shown in Table 2. Respondents consisted of 465 men (40.72%) and 677 women (59.28%). The demographic and social attributes of the respondents were surveyed in the research, including their gender, age, education level, and family monthly income. There were 326 individuals aged between 18 and 35 (28.55%), 342 aged between 36 and 44 (29.95%), 265 aged between 45 and 64 (23.2%), and 209 people aged over 65 (18.32%). A total of 12% of the respondents had a master's degree and above, and 39.32% had a college diploma. People with senior middle school education accounted for 25.31% of the total population, and 23.38% population had completed junior middle school and below. With respect to income level, 17.08% of the respondents had a monthly income below CNY 1000, 10.25% had a monthly income over CNY 10,000, and 23.82% had a monthly income of CNY 1000–2999.

Table 2. Distribution of the socioeconomic status characteristics of the respondents.

Socioeconomic Status	Items	Frequency (*n*)	Proportion (%)
Gender	Male	465	40.72
	Female	677	59.28
Age	18–35	326	28.55
	36–44	342	29.95
	45–64	265	23.20
	>65	209	18.32
Education	Junior middle school and below	267	23.38
	Senior middle school	289	25.31
	College	449	39.32
	Master's degree and above	137	12.00
Income (CNY per month)	<1000	195	17.08
	1000–2999	272	23.82
	3000–4999	225	19.70
	5000–6999	207	18.13
	7000–9999	126	11.03
	>10,000	117	10.25

3.2. Measures

To compile a reliable set of items to measure the constructs and test the proposed model, we conducted a comprehensive literature review. Table 3 provides a summary of the items with their corresponding constructs.

Table 3. Measurement items for the personal, social, and perceived ES/EDS constructs.

Constructs	Measurement Items
Personal and Social Factors	
Behavioral Intention	Do you intend to do the following behavior in the community garden? (BI1) Gardening activities, such as planting, watering, digging, and weeding; (BI2) mental recovery activities, such as meditation and relaxation; (BI3) and recreational communicative activities, such as chatting and gathering.
Attitude	What is your attitude towards UCG as follows? (ATT1) health, education, communication, and other functions; (ATT2) community gardening space and fields; (ATT3) a variety of planting, leisure, and communicative behaviors; and (ATT4) fruit and vegetable products.
Perceived behavioral control	How do you view your behavioral control capabilities associated with UCG? (PBC1) Gardening experience, (PBC2) gardening skills, (PBC3) time available for gardening, and (PBC4) gardening information possession.
Subjective Norms	To what extent are you willing to obey the attitudes and expectations of these people? (SN1) Family members, (SN2) friends, (SN3) neighbors, and (SN4) municipal authorities.
Perceived ecosystem services and disservices	
perceived ecosystem services	To what extent do you think UCGs could bring the following benefits? (P-ES1) Vegetable food supply, (P-ES2) physical recreation/leisure, (P-ES3) social cohesion/integration, (P-ES4), maintained soil fertility/purified air/regulated climate, (P-ES5) biodiversity conservation/habitat maintenance, and (P-ES6) maintenance of agricultural or horticultural cultural heritage/education.
perceived ecosystem disservices	To what extent do you think UCGs could bring the following four damages? (P-EDS1) Volatile organic compounds and greenhouse gas emissions pollute the community environment, (P-EDS2) urban infrastructure is damaged by growing plants and microbes, (P-EDS3) vector-borne disease lurking in urban wetlands affecting residents' health, and (P-EDS4) the messy and dense growth of vegetation or crops makes people irritable.

Note: P-ES = perceptions of ecosystem service; BI = behavioral intention; P-EDS = perceptions of ecosystem disservice; ATT = attitude; PBC = perceived behavioral control; SN = subjective norm.

A validated TPB questionnaire [49] was translated into Chinese. The questionnaire was reworded so the questions and answers were more in line with our national conditions, making it more easily understood by Chinese people. The revised questionnaire, used to measure individual and social structure, was hypothesized to reflect gardening behavioral intention in UCGs. We used the questionnaire to conduct face-to-face interviews with 84 gardeners and 778 residents.

In the theory of planned behavior, behavioral intention (BI) is the most direct factor affecting behavior, which is the individual volitional intention to perform (or not to perform) a behavior [49]. This paper defines BI as the volitional intention of residents to carry out or not carry out community gardening behavior. Residents' BI toward UCG was measured with three items: "Do you intend to do the following behavior in the community garden? Gardening activities, such as planting/watering/digging/weeding [79], mental recovery activities, such as meditation and relaxation [48]; and recreational communicative activities, such as chatting and gathering [44]." Participants were asked to rate each item described above on a 5-point scale ranging from 1 (never) to 5 (always). The BI scale internal consistency (Cronbach's alpha) was 0.913.

Attitude (ATT) is defined as a psychological tendency that is expressed by evaluating a particular entity with some degree of favor or disfavor. This study defines attitude as the evaluative response of residents to the community gardening behavior itself and its target products. Residents' attitude toward UCG was measured with four items: "What is your attitude towards UCG as follows? Health, education, communication, and other functions [80]; community gardening space and fields [77]; a variety of planting/leisure and communicative behaviors [81]; and fruit/vegetable products [82]." Participants were asked to rate each item described above on a 6-point scale ranging from 1 (strongly dislike) to 6 (strongly like). The ATT scale internal consistency (Cronbach's alpha) was 0.963.

Perceived behavioral control (PBC) is the degree of ease or difficulty with which an individual thinks he can control and perform a behavior [49]. PBC is influenced by two factors: controlled belief and perceived power [43]. This study defines the perceived level of four important factors: information, skills, experience, and abilities related to the development of community gardening behavior [44]. PBC was measured by confidence under four conditions in response to the question: "How do you view your behavioral control capabilities associated with UCG? Gardening experience [43], gardening skills [46], time available for gardening [51], and gardening information possession [44]". Participants were asked to rate each item described above on a 6-point scale ranging from 1 (not confident at all) to 6 (very confident). The internal consistency (Cronbach's alpha) was 0.940.

Subjective norm (SN) is a person's perception of social pressures that enables them to perform or not perform the behavior [49]. SN is influenced by normative beliefs and submissive motivations. This paper defines SN as the degree to which residents follow the expectations of a "significant other" for community gardening behavior. Subjective norms were measured by asking about the following four "significant others": To what extent are you willing to obey the attitudes and expectations of these people? Family members, friends, neighbors and municipal authorities [47,54]." Participants were asked to rate each item described above on a 6-point scale ranging from 1 (strongly unwilling) to 6 (strongly willing). The internal consistency (Cronbach's alpha) was 0.903.

Before measuring the perceived UCG's ecosystem service, we first evaluated the ecosystem services and disservices for gardeners. Eighteen ES items (food supply, biodiversity, maintenance of soil fertility, local climate regulation, air purification, global climate regulation, pollination, entertainment and leisure, quality of food, nature and spiritual experiences, exercise and physical recreation, relaxation and stress reduction, aesthetic information, place-making, social cohesion and integration, biophilia, learning and education, and maintenance of cultural heritage) were extracted using items from the studies of Marta Camps-Calvet et al. [22]. Six EDS items (volatile organic compounds and greenhouse gas emissions, excessive use of pesticides and chemical fertilizers, urban infrastructure damaged by growing plants and microbes, vector-borne disease lurking in urban wetlands, disgust about animal feces or plant waste, and irritability about the messy and dense growth of vegetation or crops) were also extracted from the studies of Von Dohren [73] and Becker [70]. We visited the gardeners' UCGs and 178 gardeners were asked to score the above 18 ES items and 6 EDS items on a scale from 1 (strongly disagree) to 6 (strongly agree). The reliability and validity of each item was reported. This step shows us that in selected regions the UCG is relatively significant in the provision of nine ES and four EDS; this result is similar to the previous article [15]. However, several recent research articles have identified differences between the actual delivery of ecosystem services and human perception [83]. Therefore, before the formal survey we surveyed 778 residents, targeting the nine perceived ES and four perceived EDS indicators described above, to examine whether their perceived ES and EDS were significant for measured items.

Finally, in the formal questionnaire, perceived ES was measured by the following six items: "To what extent do you think UCGs could bring the following benefits? Vegetable food supply [84], physical recreation/leisure [85], social cohesion/integration [86], maintained soil fertility/purified air/regulated climate [87], biodiversity conservation/habitat maintenance [88], and maintenance of agricultural or horticultural cultural heritage/education [89]". Additionally, perceived EDS was measured by the following four items: "To what extent do you think UCGs could bring the following four damages? Volatile organic compounds and greenhouse gas emissions pollute the community environment [90], urban infrastructure is damaged by growing plants and microbes [91], vector-borne disease lurking in urban wetlands affects residents' health [73], and the messy and dense growth of vegetation or crops makes people irritable [92]". The internal consistency (Cronbach's alpha) was as follows: perceived ES scale was 0.928 and perceived EDS was 0.848.

3.3. Statistical Analysis

SEM is a combination of two statistical methods: confirmatory factor analysis and path analysis. Confirmatory factor analysis (CFA), which originated in psychometrics, estimates latent psychological traits [93,94]. Path analysis, on the other hand, had its beginning in biometrics and finds the causal relationship among variables by creating a path diagram [95]. This article follows the two-step approach for the practical structural equation model (SEM) recommended by Anderson and Gerbing [96].

First, the CFA was conducted with IBM SPSS AMOS to test data reliability by calculating the Cronbach's alpha of all observed variables [97]. This step tested whether measurement models of residents' perceptions of ecosystem service/disservice, attitude, subjective norms, perceived behavioral control, and behavioral intention towards the UCG were consistent with our hypothesis. Before data analyses were carried out, we completed data descriptive analysis for socioeconomic status (SES), which was measured as a construct of education and income.

Second, SEM was performed. SEM is a special form of multivariate analysis and was used to examine the hypothetical causality among multiple variables and how their inter-relationships may play a role in determining a particular outcome in this study. Hypothetical causal relationships were illustrated using a path diagram and analyzed for the standardized partial regression coefficients, which can be interpreted as the magnitude of direct causal influence [98]. The maximum likelihood method was used during both steps to estimate parameters. The tests involved the two models: the TPB-only model (M0) and the TPB plus perceived ES/EDS model (M1), which integrated the perceived ecosystem service and disservice constructs. Model fit was assessed using the chi-square test (X^2), likelihood ratio (X^2/df), comparative fit index (CFI), and root mean square error of approximation (RMSEA) following the best practices in SEM as suggested by Mueller and Hancock [99]. All statistical analyses were performed using the Statistical Package for the Social Sciences (SPSS) 22.0 for Windows and Analysis of Moment Structures (AMOS) 23.0.

4. Results
4.1. Descriptive Statistics

The descriptive statistics for the measured variables are shown in Table 4. Overall, participants had a positive attitude (M = 4.077, SD = 1.526), and a moderate level of confidence in perceived behavioral control associated regarding UCG (M = 3.197, SD = 1.557). Furthermore, residents' behavioral intention to participate in the UCG, which was measured by using a five-point scale, had a moderately high score (M = 2.881, SD = 1.451) In addition, means of 4.048 (SD = 1.403) for the perceived ecosystem services indicated that the participants had a high level of evaluation with the UCG ecosystem services. The mean score of 3.452 (SD = 1.232) for perceived ecosystem disservices indicated that residents favored a certain amount of the possible damages caused by UCGs. However, social support for community gardening from "important others" around them is inadequate (M = 2.921, SD = 1.347).

Table 4. Standardized estimates and psychometric properties of the parameters of the measurement model.

Measurement Item	Mean (S.D.)	Standard Deviation	Standard Error	Factor loadings	Squared Multiple Correlations	Average Variance Extracted	Composite Reliability
BI	2.881	1.451				0.781	0.914
BI1	3.045	1.42	0.042	0.686	0.712		
BI2	2.824	1.463	0.043	0.772	0.757		
BI3	2.774	1.469	0.043	0.773	0.699		
P-ES	4.048	1.403				0.682	0.928
P-ES1	4.305	1.503	0.044	0.675	0.634		
P-ES2	4.566	1.406	0.042	0.714	0.677		
P-ES3	3.884	1.366	0.04	0.854	0.719		
P-ES4	3.665	1.357	0.04	0.849	0.703		
P-ES5	3.848	1.385	0.041	0.798	0.687		
P-ES6	4.022	1.401	0.041	0.782	0.714		
P-EDS	3.452	1.232				0.598	0.853
P-EDS1	4.147	1.119	0.033	−0.705	0.538		
P-EDS2	3.165	1.275	0.038	−0.827	0.59		
P-EDS3	3.835	1.311	0.039	−0.798	0.652		
P-EDS4	2.659	1.222	0.036	−0.735	0.53		
ATT	4.077	1.526				0.869	0.964
ATT1	4.306	1.426	0.042	0.773	0.828		
ATT2	3.887	1.614	0.048	0.818	0.844		
ATT3	3.963	1.517	0.045	0.806	0.855		
ATT4	4.153	1.545	0.046	0.803	0.847		
PBC	3.197	1.557				0.806	0.943
PBC1	3.173	1.614	0.048	0.883	0.829		
PBC2	3.131	1.598	0.047	0.901	0.861		
PBC3	3.071	1.539	0.046	0.883	0.782		
PBC4	3.411	1.478	0.044	0.786	0.643		
SN	2.921	1.347				0.701	0.903
SN1	2.984	1.362	0.04	0.842	0.683		
SN2	2.757	1.307	0.039	0.82	0.694		
SN3	2.934	1.329	0.039	0.83	0.685		
SN4	3.007	1.388	0.041	0.789	0.583		

Note: S.D.: standard deviation; P-ES = perceptions of ecosystem service; BI = behavioral intention; P-EDS = perceptions of ecosystem disservice; ATT = attitude; PBC = perceived behavioral control; SN = subjective norm.

4.1.1. Behavioral Intention for UCG

Respondents reported a high degree of behavioral intention for UCGs. Residents were asked to choose based on the five-point scale, with higher scores representing stronger behavioral intention. Residents scored the highest enthusiasm for PI1-Gardening activities, such as planting, watering, digging, and weeding (3.05). They also participated in community gardening to undergo spiritual relaxation and recovery; therefore, the score for PI2-Mental recovery activities, such as meditation and relaxation (2.82), was also high. In addition, residents were interested in the social interactions that took place in community gardening, which was reflected in the score for PI3-Recreational communicative activities (e.g., chatting and gathering) (2.77).

4.1.2. Attitude, Perceived Behavioral Control and Subjective Norm

The study investigated the respondents' attitudes towards four types of the behavioral objects for the UCG, which all received high support. ATT1-Attitude towards health, education, communication, and other functions had the highest score (4.31), followed by ATT4-Attitude towards fruit and vegetable products (4.15) and ATT3-Attitude towards a variety of planting, leisure, and communicative behaviors (3.96). The lowest score was for ATT2-Attitude towards community gardening space and fields (3.89).

Participants had a certain high level of perceived behavioral control for UCG. The highest score was PBC4-Perceived level of gardening information possession (3.41), which may stem from the fact that we live in an informational age and people have easy access to knowledge about urban gardening and agriculture. PBC1-Perceived level of gardening experience (3.17) and PBC2-Percieved level of gardening skills (3.13) had similar scores, which indicates a strong association between these two factors. Respondents with long-term agricultural experience also had higher levels of perceived skills in the areas of agriculture and gardening; however, respondents who feel weaker behavioral control over PBC3 (perceived level of time available for gardening) reported lower scores (3.07), which indicates that they were uncertain about whether they could dedicate the time necessary to maintain an urban community garden or participate in those types of activities.

To a certain extent, residents were willing to follow the expectations of "significant others" or be influenced by pressure from individuals or institutions while making a decision about whether to support or oppose community gardens. Under the current urban management system, respondents had the highest compliance with urban and community managers, hence the high score for SN4-Municipal authorities (3.00). Their level of compliance towards family and neighbors was relatively neutral, so SN1-Family members (2.98) and SN3-Neighbors (2.93) received similar scores. Finally, they showed the lowest compliance toward their friends, which is reflected in the score for SN2-Friends (2.76).

4.1.3. Perceptions of Ecosystem Service and Disservice for UCG

The survey listed six types of common ecosystem services and allowed respondents to score them based on their perceptions of UCGs. Using statistical analysis, the mean value of each term was obtained. Residents had diverse sensitivity to the different types of ES provided by the UCG, which showed high perception in general. Cultural services, such as PES2-Physical recreation and Leisure services, were the most easily perceived by the respondents and had the highest score (4.57) followed by the product offering functions of Community gardening, such as PES1-Vegetable food supply, which were also highly recognized (4.31). Factors of cultural services related to human well-being, such as PES6-Maintenance of agricultural or horticultural cultural heritage/education (4.02) and PES3-Social cohesion/integration (3.88), were also favored by the respondents. Compared to several other ecosystem services, the scores for the two regulatory services related to ecological environment declined dramatically; PES5-Biodiversity conservation/habitat maintenance received a score of 3.85, and PES4-Maintain soil fertility/purified air/regulate climate received a score of 3.67.

Generally, in these negative statements on ecological, economic, and physical health impacts, more respondents agreed or strongly agreed that urban community gardens led to these ecosystem disservices. Respondents believed that UCGs contaminated the community environment due to sewage, toxic or irritating gas, and rubbish; therefore, the P-EDS1-Volatile organic compounds and greenhouse gas emissions pollute the community environment had the most negative impact with the highest score (4.15), followed by P-EDS3-Vector-borne disease lurking in urban wetlands affecting residents health (3.84), which includes plant allergies, animal attacks, and vector-borne diseases hidden in urban wetlands. The residents expressed a neutral attitude towards the factors related to economic damage, such as P-EDS2-Urban infrastructure is damaged by growing plants and microbes (3.17), and they were not sensitive to those damages. Overall, respondents did not believe that urban community gardening would have a significant negative impact on human psychology; thus, the opinion about P-EDS4-The messy and dense growth of vegetation or crops makes people irritable, has the lowest score (2.66).

4.2. Measurement Model

This study first conducted a confirmatory factor analysis of the measurement model. The results showed that the measurement model was significant, and all the other measurement loads on the latent variables were also significant, as shown in Table 4. The collected data and measurement models reflected a good fit because all model fits exceeded the acceptable levels proposed by previous researchers. The fit indices for the TPB model were as follows: CMIN = 813.88, df = 245, CMIN/df = 3.322, GFI = 0.942, CFI = 0.978, AGFI = 0.924, NFI = 0.969, RFI = 0.962, IFI = 0.978, TLI = 0.973 and RMSEA = 0.045. The initial model M0 is a simple TPB model, and model M1 is an extended TPB plus perceived ES/EDS model. The composite reliability values for the constructs all exceeded the 0.80 guideline. Previous studies state that the average variance extracted (AVE) should not be below the recommended 0.50 level, and the AVEs in this study were between 0.64 to 0.97, fulfilling the requirements [100]. Additionally, the CFA detected that the factor composite score was greater than 0.50, which indicates that the convergent validity is good [101]. In this study, the AVE of an individual construct was compared with the squared shared variances between constructs and was used to examine discriminant validity. Additionally, as shown in Table 5, the AVE of each structure of the factor correlation matrix is greater than the square correlation coefficients between constructs and thus has discriminant validity [100]. Overall, the model performs well in terms of reliability, convergent validity, and discriminant validity. The result showed that the Cronbach's alpha coefficient is 0.902. The KMO value is 0.937, and the P value of Bartlett's test of sphericity is significant [102].

Table 5. Discriminant validity: average variance extracted (AVE) and shared variances.

Variance	1	2	3	4	5	6
1 BI	**0.884**					
2 P-ES	0.545	**0.827**				
3 P-EDS	−0.485	−0.411	**0.773**			
4 ATT	0.599	0.640	−0.546	**0.932**		
5 PBC	0.521	0.429	−0.413	0.435	**0.898**	
6 SN	0.565	0.466	−0.322	0.493	0.352	**0.837**

(1) Off-diagonals are shared variance, and diagonals are AVEs (bolded). (2) All were significant at 0.001 levels; P-ES = perceptions of ecosystem service; BI = behavioral intention; P-EDS = perceptions of ecosystem disservice; ATT = attitude; PBC = perceived behavioral control; SN = subjective norm.

4.3. Possible Causal Relations Influencing Behavioral Intention

4.3.1. The Impact of ATT, PBC and SN

After fitting the measurement model, this study tested the M0 structural model and assessed the fitted data. Furthermore, to explore whether the extended TPB model (M1 model) increases the interpretation strength of the variance of behavioral intention, Figure 2 presents the standardized path coefficients and explanatory variance (R^2) for the M0 structural model. The M0 model shows good goodness-of-fit measures (GOFs) to the data in Table 6. The three variables of TPB had significant positive effects on behavioral intention, indicating that residents had a more positive attitude towards the UCG ($\beta = 0.38$, $p < 0.001$) and greater subjective norms ($\beta = 0.38$, $p < 0.001$); stronger perceived behavioral control ($\beta = 0.30$, $p < 0.001$) was also associated with greater behavioral intention. Respectively, attitudes ($\beta = 0.47$, $p < 0.001$) and perceived behavioral control ($\beta = 0.17$, $p < 0.001$) showed significant correlations for subjective norms. In the M0 model, the TPB variables accounted for 54% of the variance in behavioral intention. Therefore, the results of the M0 model tested the theoretical principles of TPB and supported the study's first hypothesis.

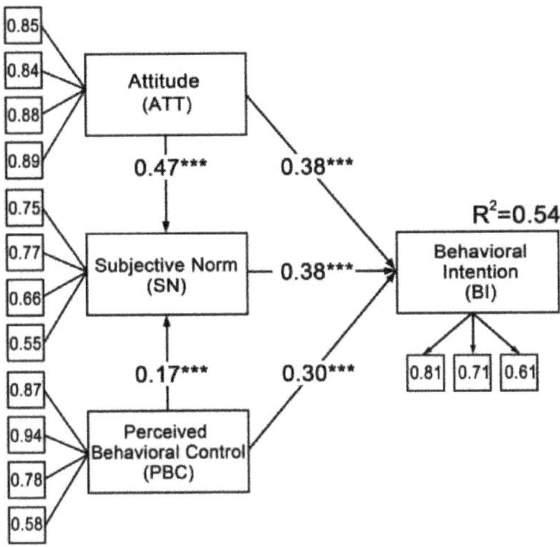

Figure 2. Possible causal relationships between the attitude, subjective norm, perceived behavioral control, and behavioral intention; *** $p < 0.001$.

Table 6. Goodness-of-fit measures (GOFs) for the structural equation models.

Goodness-of-Fit Measures (GOFs)	Adequate Level	Recommended Level	M0	M1
Root mean sq. error of approx. (RMSEA)	<0.1	<0.05	0.062	0.045
Goodness-of-fit index (GFI)	>0.9	>0.95	0.955	0.942
Adj. goodness-of-fit index (AGFI)	>0.8	>0.9	0.932	0.924
Comparative fit index (CFI)	>0.9	>0.95	0.980	0.978
Tucker–Lewis index (TLI)	>0.9	>0.95	0.973	0.973
Normal fit index (NFI)	>0.9	>0.95	0.975	0.969
Incremental fit index (IFI)	>0.9	>0.95	0.980	0.978
Relative fit index (RFI)	>0.9	>0.95	0.967	0.962

4.3.2. The Effects of Perceived Ecosystem Services and Disservices

The M1 model is an extended TPB model that includes factors of perceived ecosystem services and disservices, and it shows a good GOF in Table 5. The M1 structure model explained 65.0% of the variance in behavioral intention. By comparing the results of the M1 model to the M0 model, the research found an 11% increase in the explained variances of behavioral intention. Of all the 14 path coefficients, 13 were statistically significant ($p < 0.01$). Most of the results are the same as the ones that we initially hypothesized, with the perception of ecosystem service having significant and direct positive effects on attitudes ($\beta = 0.56$, $p < 0.001$), perceived behavioral control ($\beta = 0.33$, $p < 0.001$), and subjective norms ($\beta = 0.26$, $p < 0.001$). Additionally, the perception of ecosystem disservice had a significant direct negative effect on attitudes ($\beta = -0.34$, $p < 0.001$) and perceived behavioral control ($\beta = -0.29$, $p < 0.001$), but the direct effect on the subjective norm was not statistically significant. Standardized parameter estimates indicated that perceived ecosystem services and disservices were associated the most strongly with attitudes. Subsequently, attitudes ($\beta = 0.13$, $p < 0.001$), subjective norms ($\beta = 0.31$, $p < 0.001$), and perceived behavioral control ($\beta = 0.20$, $p < 0.01$) significantly predicted behavioral intention. Subjective norms are the most predictive of behavioral intention, as they are influenced by attitudes and perceived behavioral control. As shown in Figure 3, the path diagram has completely standardized parameter estimates. In support of the second hypothesis, the inclusion of the perceived ES and EDS variables based on the M0 model led to an increase in the explained variances of behavioral intention.

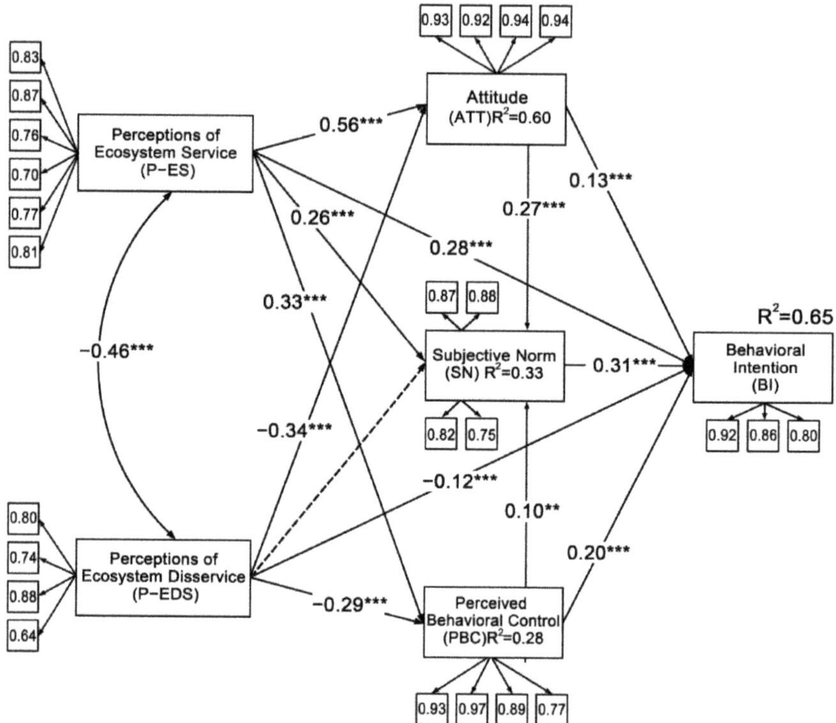

Figure 3. Possible causal relationships between perception of ecosystem service/disservice, attitude, subjective norm, perceived behavioral control, and behavioral intention; The dashed paths indicate removed hypothetical relationships leading to failed GOF measures; *** $p < 0.001$, ** $p < 0.01$.

5. Discussion

UCG development not only conforms to China's national concept of ecological construction, promoting the intensive development and ecological utilization of idle urban land, but is also an effective way to meet the needs of urban people, such as allowing close access to nature, improving lifestyle quality, and advancing well-being. This study used SEM to incorporate ecosystem services and disservices into the TPB model to explore the relationship between perceived ecosystem services by ordinary urban residents and community gardening behavioral intentions. It added novel evidence for cognition of the ecosystem services associated with UCG and adopted an integrated view to jointly incorporate UCGs' abilities to provide ecosystem services and deliver ecosystem disservices into an extended theoretical framework based on TPB theory. Overall, this study provides a better understanding of the relationships between individual, social, and perceived ecosystems by revealing that the perception of ecosystem services and disservices has important effects on predicting the behavioral intention of ordinary residents through individual and social factors. Furthermore, this study provides some support for the effectiveness and utility of the extended theory of planned behavior model in predicting the gardening behavioral intention of Chinese urban residents.

We reveal the mechanism of how residents' perceptions jointly influence their UCG behavioral intention. The first hypothesis of this study holds that behavioral intention is directly and positively influenced by attitude, perceived behavioral control, and subjective norms, which are positively influenced by attitude and perceived behavioral control. Consistent with the TPB theory, the study concluded that residents' attitudes towards UCGs have the greatest influence on their behavioral intention. Attitude or emotional motivation is an important determinant of all horticulture behavioral intention [15,19,67]. A study of gardeners' participation in community (allocation) gardens found that the positive emotional connection they established with the garden strengthened their intention to participate [39].

Second, our study showed that residents' subjective norms of UCG had almost the same effect as attitudes on behavioral intention; however, subjective norms are largely influenced by attitudes; therefore, attitudes indirectly influence behavioral intention through subjective norms. Residents' subjective norms of UCGs affect behavioral intention to almost the same degree as attitudes. As subjective norms are largely influenced by attitudes, and attitudes are a greater influence on behavioral intent than subjective norms. This study pointed out that subjective norms are the most important cognitive structure that affects residents' gardening behavior. Compared with gardeners who have participated in a UCG, residents who were not involved in urban gardening thought that important others or institutions around them, especially municipal authorities, would not approve of the existence of community gardening in the city. Existing research has already proposed that municipal authorities could promote gardening participation by readjusting policies to change people's beliefs in subjective norms, which is a suggestion that our study confirms [19]. Additionally, this study found that the effect of perceived behavioral control or the self-efficacy of overcoming barriers on behavioral intention was the weakest of the TPB structures, which aligns with previous findings [3,71]. We thus confirmed that in order to consider the effects of attitude, perceived behavioral control, and subjective norms on behavioral intention, future interventions should adopt a multi-level research approach.

We simultaneously considered two perception constructs, i.e., perceived ecosystem services and disservices. We found that the two perceptions are negatively correlated with each other. These perceptions can explain and predict people's personal and social cognition towards UCG, thus playing a major role in influencing gardening participation. Among them, perceived ecosystem services become a non-negligible factor that can help them for residents to develop continuous gardening intentions by forming more positive attitudes, greater confidence, and less social pressure towards UCG. As we initially predicted, the inclusion of perceived ecosystem services and disservices in the TPB model led to an increase in the explanatory variance of residents' behavioral intention toward the UCG. In the initial M0 model, the explanatory variance of behavioral intention influenced by individual and social factors of TPB was 54%; however, with the extended M1 model, comparing the initial model led to an increase in the explanatory variance of behavioral intention from 54% to 65%, which is higher than the results from previous studies [39]. This finding provides support for the predictive validity of the extended TPB model. Our study did not incorporate residents' actual gardening behavior into the TPB model because many studies have shown that behavioral intention, though the closest and most direct prediction factor of behavior, does not necessarily explain real behavioral changes [42]. To narrow the gap between behavioral intention and real action and increase the explanatory variance of actual gardening behavior, future studies could delve deeper into the different roles of perceived ecosystem services and disservices.

The results of this study show that perceived ecosystem services (or disservices) not only have an indirect impact on gardening behavioral intention through the ATT, PBC, and SN factors in the TPB but also have positive (or negative) direct effects on behavioral intention. This suggests that the three variables in the TPB model do not fully mediate the relationship between perceived ecosystem and behavioral intention and that other important individual and social variables may exist. The present study examined other perceptions associated with ecosystem service functions, with individual and social factors, such as the neighborhood attachment [103], perceptions of environmental impact [68], perceived risks [104], life satisfaction [40], and identity perception [105], and found that they all have an impact on gardening behavioral intention or actions. Numerous studies have shown that a strong sense of community attachment linked to communal gardening encourages people to form connections between physical and social environments that promote interaction with others and spark changes in sustainable gardening behavior [61,103,106,107]. Studies have also shown that perceptions of environmental impact from urban gardening ecosystems affect the promotion of more sustainable horticultural behaviors and practices [68,108]. The literature suggests that perception and identifying risk factors for urban gardening behavior that may have detrimental effects on human well-being are important considerations for achieving the various ecosystem service functions brought about by urban gardens [82,109,110]. In some studies on urban gardening promoting better life satisfaction and eating habits, these good perceptions have a positive effect on gardening behavior [40,68]. Some research highlights that urban gardening can reshape urban citizenship through residents' identity perception with their "environmental gardening identity" [105]. This identity perception, brought about by the social and cultural functions of ecosystem services, offers far-reaching potential for changes in gardening behavior and practice [111–114]. Based on our findings, we believe that further research should be conducted that examines the effects of various perceptual factors linked to urban gardening on behavioral intention and action.

Although aspects of urban horticultural ecosystem services and disservices have been explored in this study, gaining deeper insight into other service functions, such as community resilience [115], urban sustainability [116], landscaping [117], treatment function [118], and urban health service potential [69], would help shed light on gardening behavioral intention and practices. Community gardens have been proven to play an important role in the social-ecological resilience of urban ecosystems and offer residents a place where they can reduce stress, share experiences, and gain community support [119,120]. Previous research has shown that community gardens provide a model that can promote sustainable urban living by linking individuals and communities to the food system and promoting the development of a deep reconnection and long-term commitment to sustainable living practices [84,85,121,122]. It has been concluded that the aesthetic value of urban gardens helps enhance the attractiveness of urban areas and that a scientific garden distribution is more likely to contribute to a larger network of ecosystem services across a broader urban landscape [123–125]. Current research also indicates that site spatial and design characteristics affect people's gardening participation [126,127]. For nearly a decade, community gardens, especially at the local level, have been increasingly praised for revealing the relationship processes that connect people, ecology, and health [128] [129–132]. Notably, the latest research indicates that urban gardens have functioned as a potential health resource during the COVID-19 pandemic by giving gardeners a source of freedom and joy while facing difficult and potentially isolating situations [133,134]. On the other hand, while examining ecosystem disservices, researchers have found that vegetables grown in community gardens pose a health threat arising from their potential exposure to pollutants, such as heavy metals in the environment [104,135]. In conclusion, the metrics of six ES and four EDS are relatively common in this paper, but deepening the understanding of the function and damage of their ecosystem services is meaningful to illustrate community gardening behavioral intentions and practices in different cultural and social contexts.

Our findings provide a strategy adopted on the premise of adapting to the intensive development and ecological utilization of idle urban land and meeting the well-being and needs of contemporary urban people. That is: (1) to improve the level of delivery of the UCG's ecosystem services by developing and promoting new gardening technologies; (2) to construct and publicize a series of high-quality UCG exemplary cases, to spread its various ecosystem service functions to urban residents, and to improve their perception of UCG benefits; (3) to consider and address the various gardening technical problems and social obstacles, to reduce the UCG's ecosystem disservices; (4) to publicize effective technical measures and methods to reduce the UCG's ecosystem disservices, address potential participant concerns about possible obstacles and reduce their perception of UCG damage; and (5) to find and cultivate institutions or individuals as important reference groups, such as management agencies good at organizing and operating UCGs and enthusiastic gardeners who have successfully developed UCGs; their demonstration and drive will help improve the subjective norms of residents and thus promote their gardening behavioral intention. The proposed strategy implies that the roadmap for UCG sustainable development can be implemented by early practitioners to later participants, from megacities to middle-and-small cities.

This study had several limitations that should be considered. First, potential confounding factors, such as demographic factors and UCG location (i.e., whether it was on a roof of a public building, on a communal inland surface, or in a vacant area outside of a given community), were not strongly distinguished. Although this study is based on TPB theory as well as previous studies, variables of the SEM model containing more confounding factors would be beneficial to elucidate our understanding of the perceived ecosystem correlates of gardening behavioral intention. Second, the survey of perceived ecosystem services/disservices, ATT, PBC, and SN measures in this study were self-managed by the authors, and although the reliability and validity of the measurements of these items were derived from previous conclusions, there was not a widely tested maturity scale, and this may have led to potential errors. Using objective techniques would help to reduce biases. Third, because research on UCG perceptual ecosystem services and disservices is in its infancy, further measurement tools need to be developed. Identifying the ecosystem supply characteristics of the different types of UCGs associated with resident gardening behavior will help develop more effective policies and incentives. Finally, the structural equation model that this study proposes is based on previous research and the TPB theories of environmental psychology, but it would also be productive to explore other plausible models to help explain community gardening behavioral intention and the practices of urban residents.

6. Conclusions

The results showed that perceived ES and EDS had significant effects on gardening behavioral intention and that factors of attitude, perceived behavioral control, and subjective norms in the TPB model partly mediated this association. These results have important practical implications for both policymakers and managers. For policymakers, the benefits and positive impact of UCG can be gradually advanced by improving its capacity to deliver ecosystem services and reducing the level of disservices to their ecosystems. Therefore, the findings of this study are valuable for developing guidelines and intervention policies for the planning and management of UCGs, which can promote more scientific and sustainable gardening behavioral intention among Chinese urban residents. Furthermore, the findings of this study also have important implications for community managers or local institutions following the legalization of UCGs. Since perceived ecosystem disservices were important for new participants who focused on innocuous practice, UCG managers should pay more attention to reducing these damages by advancing gardening techniques and optimizing management policies. Future studies could distinguish between the correlation between personal characteristics of potential gardeners in residents and their perceived ecosystem services and disservices. It is necessary to further explore the impact of the perception of ecosystem services and disservices in UCG, and of different spatial and governance types, on gardening behavioral intention and behaviors. To conclude, future studies are necessary to develop evidence-based theoretical models of changes in community gardening behavior among urban residents.

Author Contributions: C.W.: conceptualization, data curation, formal analysis, investigation, methodology, writing—original draft; Y.T.: conceptualization, methodology; Z.D.: conceptualization; X.Y.: conceptualization, writing—review and editing; S.W.: investigation; D.S.: investigation; Y.P.: investigation; F.S.: conceptualization; D.G.: conceptualization, data curation, methodology, resources, writing—review and editing; X.L.: conceptualization, methodology, project administration, resources, writing—review and editing. All authors have read and agreed to the published version of the manuscript.

Funding: This study was financially supported by the State Key Laboratory of Urban and Regional Ecology (SKLURE 2018-2-2), and was supported by the project of research on urban gardening planning based on the demand of national central city construction (Changsha City Administration and Integrated Law enforcement Bureau 17020).

Institutional Review Board Statement: Not applicable.

Informed Consent Statement: Not applicable.

Data Availability Statement: Not applicable.

Acknowledgments: We would like to acknowledge the Changsha City Administration and Integrated Law enforcement Bureau—since 2017, with their strong support and help, the urban garden research team of Hunan Agricultural University, represented by the corresponding author and the first author, has successfully carried out a series of scientific research and practical explorations on urban gardens in Changsha. In addition, the first author would like to thank Shuai Xue for his assistance in restructuring and improving this manuscript.

Conflicts of Interest: The authors declare no conflict of interest.

References

1. Guitart, D.; Pickering, C.; Byrne, J. Past results and future directions in urban community gardens research. *Urban For. Urban Green.* **2012**, *11*, 364–373. [CrossRef]
2. Filkobski, I.; Rofè, Y.; Tal, A. Community gardens in Israel: Characteristics and perceived functions. *Urban For. Urban Green.* **2016**, *17*, 148–157. [CrossRef]
3. Fox-Kämper, R.; Wesener, A.; Münderlein, D.; Sondermann, M.; McWilliam, W.; Kirk, N. Urban community gardens: An evaluation of governance approaches and related enablers and barriers at different development stages. *Landsc. Urban Plan.* **2018**, *170*, 59–68. [CrossRef]
4. Dennis, M.; Armitage, R.P.; James, P. Appraisal of social-ecological innovation as an adaptive response by stakeholders to local conditions: Mapping stakeholder involvement in horticulture orientated green space management. *Urban For. Urban Green.* **2016**, *18*, 86–94. [CrossRef]
5. Caneva, G.; Cicinelli, E.; Scolastri, A.; Bartoli, F. Guidelines for urban community gardening: Proposal of preliminary indicators for several ecosystem services (Rome, Italy). *Urban For. Urban Green.* **2020**, *56*, 126866. [CrossRef]
6. Glavan, M.; Schmutz, U.; Williams, S.; Corsi, S.; Monaco, F.; Kneafsey, M.; Rodriguez, P.A.G.; Istenič, M.Č.; Pintar, M. The economic performance of urban gardening in three European cities—Examples from Ljubljana, Milan and London. *Urban For. Urban Green.* **2018**, *36*, 100–122. [CrossRef]
7. Cabral, I.; Keim, J.; Engelmann, R.; Kraemer, R.; Siebert, J.; Bonn, A. Ecosystem services of allotment and community gardens: A Leipzig, Germany case study. *Urban For. Urban Green.* **2017**, *23*, 44–53. [CrossRef]
8. Beumer, C. Show me your garden and I will tell you how sustainable you are: Dutch citizens' perspectives on conserving biodiversity and promoting a sustainable urban living environment through domestic gardening. *Urban For. Urban Green.* **2018**, *30*, 260–279. [CrossRef]
9. Niemelä, J.; Saarela, S.-R.; Söderman, T.; Kopperoinen, L.; Yli-Pelkonen, V.; Väre, S.; Kotze, D.J. Using the ecosystem services approach for better planning and conservation of urban green spaces: A Finland case study. *Biodivers. Conserv.* **2010**, *19*, 3225–3243. [CrossRef]
10. Veldheer, S.; Winkels, R.M.; Cooper, J.; Groff, C.; Lepley, J.; Bordner, C.; Wagner, A.; George, D.R.; Sciamanna, C. Growing Healthy Hearts: Gardening Program Feasibility in a Hospital-Based Community Garden. *J. Nutr. Educ. Behav.* **2020**, *52*, 958–963. [CrossRef]
11. Torres, A.C.; Prévot, A.-C.; Nadot, S. Small but powerful: The importance of French community gardens for residents. *Landsc. Urban Plan.* **2018**, *180*, 5–14. [CrossRef]
12. Exner, A.; Schützenberger, I. Creative Natures. Community gardening, social class and city development in Vienna. *Geoforum* **2018**, *92*, 181–195. [CrossRef]
13. Schram-Bijkerk, D.; Otte, P.; Dirven, L.; Breure, A.M. Indicators to support healthy urban gardening in urban management. *Sci. Total Environ.* **2018**, *621*, 863–871. [CrossRef] [PubMed]
14. Gregory, M.M.; Leslie, T.W.; Drinkwater, L.E. Agroecological and social characteristics of New York city community gardens: Contributions to urban food security, ecosystem services, and environmental education. *Urban Ecosyst.* **2015**, *19*, 763–794. [CrossRef]
15. He, B.-J.; Zhu, J. Constructing community gardens? Residents' attitude and behaviour towards edible landscapes in emerging urban communities of China. *Urban For. Urban Green.* **2018**, *34*, 154–165. [CrossRef]
16. Egendorf, S.P.; Cheng, Z.; Deeb, M.; Flores, V.; Paltseva, A.; Walsh, D.; Groffman, P.; Mielke, H.W. Constructed soils for mitigating lead (Pb) exposure and promoting urban community gardening: The New York City Clean Soil Bank pilot study. *Landsc. Urban Plan.* **2018**, *175*, 184–194. [CrossRef]

17. Lawson, L.J. *City Bountiful: A Century of Community Gardening in America*; University of California Press: Berkeley, CA, USA, 2005; pp. 18–112.
18. Ruggeri, G.; Mazzocchi, C.; Corsi, S. Urban Gardeners' Motivations in a Metropolitan City: The Case of Milan. *Sustainability* **2016**, *8*, 1099. [CrossRef]
19. Marshall, A.J.; Grose, M.J.; Williams, N.S. Of mowers and growers: Perceived social norms strongly influence verge gardening, a distinctive civic greening practice. *Landsc. Urban Plan.* **2020**, *198*, 103795. [CrossRef]
20. Tan, L.H.; Neo, H. "Community in Bloom": Local participation of community gardens in urban Singapore. *Local Environ.* **2009**, *14*, 529–539. [CrossRef]
21. Hou, J.; Grohmann, D. Integrating community gardens into urban parks: Lessons in planning, design and partnership from Seattle. *Urban For. Urban Green.* **2018**, *33*, 46–55. [CrossRef]
22. Camps-Calvet, M.; Langemeyer, J.; Calvet-Mir, L.; Gómez-Baggethun, E. Ecosystem services provided by urban gardens in Barcelona, Spain: Insights for policy and planning. *Environ. Sci. Policy* **2016**, *62*, 14–23. [CrossRef]
23. Adams, D.; Hardman, M.; Larkham, P. Exploring guerrilla gardening: Gauging public views on the grassroots activity. *Local Environ.* **2014**, *20*, 1231–1246. [CrossRef]
24. Mwakiwa, E.; Maparara, T.; Tatsvarei, S.; Muzamhindo, N. Is community management of resources by urban households, feasible? Lessons from community gardens in Gweru, Zimbabwe. *Urban For. Urban Green.* **2018**, *34*, 97–104. [CrossRef]
25. Ding, X.; Zhang, Y.; Zheng, J.; Yue, X. Design and Social Factors Affecting the Formation of Social Capital in Chinese Community Garden. *Sustainability* **2020**, *12*, 10644. [CrossRef]
26. Zhu, J.; He, B.-J.; Tang, W.; Thompson, S. Community blemish or new dawn for the public realm? Governance challenges for self-claimed gardens in urban China. *Cities* **2020**, *102*, 102750. [CrossRef]
27. Yu, B. Ecological effects of new-type urbanization in China. *Renew. Sustain. Energy Rev.* **2021**, *135*, 110239. [CrossRef]
28. Kou, H.; Zhang, S.; Liu, Y. Community-Engaged Research for the Promotion of Healthy Urban Environments: A Case Study of Community Garden Initiative in Shanghai, China. *Int. J. Environ. Res. Public Health* **2019**, *16*, 4145. [CrossRef] [PubMed]
29. Yuriev, A.; Dahmen, M.; Paillé, P.; Boiral, O.; Guillaumie, L. Pro-environmental behaviors through the lens of the theory of planned behavior: A scoping review. *Resour. Conserv. Recycl.* **2020**, *155*, 104660. [CrossRef]
30. Ajzen, I. The theory of planned behavior. *Organ. Behav. Hum. Decis. Processes* **1991**, *50*, 179–211. [CrossRef]
31. Miller, S.M.; Montalto, F.A. Stakeholder perceptions of the ecosystem services provided by Green Infrastructure in New York City. *Ecosyst. Serv.* **2019**, *37*, 100928. [CrossRef]
32. Rezaei, R.; Safa, L.; Damalas, C.A.; Ganjkhanloo, M.M. Drivers of farmers' intention to use integrated pest management: Integrating theory of planned behavior and norm activation model. *J. Environ. Manag.* **2019**, *236*, 328–339. [CrossRef]
33. Shahangian, S.A.; Tabesh, M.; Yazdanpanah, M. How can socio-psychological factors be related to water-efficiency intention and behaviors among Iranian residential water consumers? *J. Environ. Manag.* **2021**, *288*, 112466. [CrossRef] [PubMed]
34. Zhang, X.; Bai, X.; Shang, J. Is subsidized electric vehicles adoption sustainable: Consumers' perceptions and motivation toward incentive policies, environmental benefits, and risks. *J. Clean. Prod.* **2018**, *192*, 71–79. [CrossRef]
35. Lam, T.W.L.; Tsui, Y.C.J.; Fok, L.; Cheung, L.T.O.; Tsang, E.P.K.; Lee, J.C.K. The influences of emotional factors on householders' decarbonizing cooling behaviour in a subtropical Metropolitan City: An application of the extended theory of planned behaviour. *Sci. Total Environ.* **2021**, *807*, 150826. [CrossRef] [PubMed]
36. Tsantopoulos, G.; Varras, G.; Chiotelli, E.; Fotia, K.; Batou, M. Public perceptions and attitudes toward green infrastructure on buildings: The case of the met-ropolitan area of Athens, Greece. *Urban For. Urban Green.* **2018**, *34*, 181–195. [CrossRef]
37. Ertz, M.; Huang, R.; Jo, M.-S.; Karakas, F.; Sarigöllü, E. From single-use to multi-use: Study of consumers' behavior toward consumption of reusable containers. *J. Environ. Manag.* **2017**, *193*, 334–344. [CrossRef]
38. Katz, D. The Functional Approach to the Study of Attitudes. *Public Opin. Q.* **1960**, *24*, 163–204. [CrossRef]
39. Lee, J.H.; Matarrita-Cascante, D. The influence of emotional and conditional motivations on gardeners' participation in community (allotment) gardens. *Urban For. Urban Green.* **2019**, *42*, 21–30. [CrossRef]
40. Lautenschlager, L.; Smith, C. Understanding gardening and dietary habits among youth garden program participants using the Theory of Planned Behavior. *Appetite* **2007**, *49*, 122–130. [CrossRef]
41. Armitage, C.J.; Conner, M. Efficacy of the theory of planned behaviour: A meta-analytic review. *Br. J. Soc. Psychol.* **2001**, *40*, 471–499. [CrossRef]
42. Lee, H.-S. Examining neighborhood influences on leisure-time walking in older Korean adults using an extended theory of planned behavior. *Landsc. Urban Plan.* **2016**, *148*, 51–60. [CrossRef]
43. Wang, S.; Wang, J.; Yang, S.; Li, J.; Zhou, K. From intention to behavior: Comprehending residents' waste sorting intention and behavior formation process. *Waste Manag.* **2020**, *113*, 41–50. [CrossRef] [PubMed]
44. Judge, M.; Warren-Myers, G.; Paladino, A. Using the theory of planned behaviour to predict intentions to purchase sustainable housing. *J. Clean. Prod.* **2019**, *215*, 259–267. [CrossRef]
45. Chen, M.F. Extending the theory of planned behavior model to explain people's energy savings and carbon reduction be-havioral intentions to mitigate climate change in Taiwan—moral obligation matters. *J. Clean. Prod.* **2016**, *112*, 1746–1753. [CrossRef]
46. Chen, M.-F.; Tung, P.-J. The Moderating Effect of Perceived Lack of Facilities on Consumers' Recycling Intentions. *Environ. Behav.* **2010**, *42*, 824–844. [CrossRef]

47. Gkargkavouzi, A.; Halkos, G.; Matsiori, S. Environmental behavior in a private-sphere context: Integrating theories of planned behavior and value belief norm, self-identity and habit. *Resour. Conserv. Recycl.* **2019**, *148*, 145–156. [CrossRef]

48. Grilli, G.; Notaro, S. Exploring the influence of an extended theory of planned behaviour on preferences and willingness to pay for participatory natural resources management. *J. Environ. Manag.* **2019**, *232*, 902–909. [CrossRef] [PubMed]

49. Ajzen, I. *Constructing a Theory of Planned Behavior Questionnare*; University of Massachusetts Amherst: Amherst Center, MA, USA, 2006.

50. Alaimo, K.; Beavers, A.W.; Crawford, C.; Snyder, E.H.; Litt, J.S. Amplifying Health Through Community Gardens: A Framework for Advancing Multicomponent, Behaviorally Based Neighborhood Interventions. *Curr. Environ. Health Rep.* **2016**, *3*, 302–312. [CrossRef]

51. Lewis, O.; Home, R.; Kizos, T. Digging for the roots of urban gardening behaviours. *Urban For. Urban Green.* **2018**, *34*, 105–113. [CrossRef]

52. Tian, Y.; Wu, H.; Zhang, G.; Wang, L.; Zheng, D.; Li, S. Perceptions of ecosystem services, disservices and willingness-to-pay for urban green space conservation. *J. Environ. Manag.* **2020**, *260*, 110140. [CrossRef]

53. Bamberg, S.; Möser, G. Twenty years after Hines, Hungerford, and Tomera: A new meta-analysis of psycho-social determinants of pro-environmental behaviour. *J. Environ. Psychol.* **2007**, *27*, 14–25. [CrossRef]

54. Han, H. Travelers' pro-environmental behavior in a green lodging context: Converging value-belief-norm theory and the theory of planned behavior. *Tour. Manag.* **2015**, *47*, 164–177. [CrossRef]

55. Obeng, E.A.; Aguilar, F.X. Value orientation and payment for ecosystem services: Perceived detrimental consequences lead to willingness-to-pay for ecosystem services. *J. Environ. Manag.* **2018**, *206*, 458–471. [CrossRef]

56. Schwartz, S.H. Normative influences on altruism. In *Advances in Experimental Social Psychology*; Academic Press: Cambridge, MA, USA, 1977; Volume 10, pp. 221–279.

57. Park, J.; Ha, S. Understanding consumer recycling behavior: Combining the theory of planned behavior and the norm activation model. *Fam. Consum. Sci. Res. J.* **2014**, *42*, 278–291. [CrossRef]

58. Zhang, X.; Geng, G.; Sun, P. Determinants and implications of citizens' environmental complaint in China: Integrating theory of planned behavior and norm activation model. *J. Clean. Prod.* **2017**, *166*, 148–156. [CrossRef]

59. Campagne, C.S.; Roche, P.K.; Salles, J.-M. Looking into Pandora's Box: Ecosystem disservices assessment and correlations with ecosystem services. *Ecosyst. Serv.* **2018**, *30*, 126–136. [CrossRef]

60. Calderón-Argelich, A.; Benetti, S.; Anguelovski, I.; Connolly, J.J.; Langemeyer, J.; Baró, F. Tracing and building up environmental justice considerations in the urban ecosystem service literature: A systematic review. *Landsc. Urban Plan.* **2021**, *214*, 104130. [CrossRef]

61. Partalidou, M.; Anthopoulou, T. Urban Allotment Gardens During Precarious Times: From Motives to Lived Experiences. *Sociol. Rural.* **2017**, *57*, 211–228. [CrossRef]

62. Asah, S.T.; Guerry, A.D.; Blahna, D.J.; Lawler, J.J. Perception, acquisition and use of ecosystem services: Human behavior, and ecosystem management and policy implications. *Ecosyst. Serv.* **2014**, *10*, 180–186. [CrossRef]

63. Snyder, M. Basic Research and Practical Problems: The Promise of a "Functional" Personality and Social Psychology. *Pers. Soc. Psychol. Bull.* **1993**, *19*, 251–264. [CrossRef]

64. Scheromm, P. Motivations and practices of gardeners in urban collective gardens: The case of Montpellier. *Urban For. Urban Green.* **2015**, *14*, 735–742. [CrossRef]

65. Da Silva, I.M.; Fernandes, C.; Castiglione, B.; Costa, L. Characteristics and motivations of potential users of urban allotment gardens: The case of Vila Nova de Gaia municipal network of urban allotment gardens. *Urban For. Urban Green.* **2016**, *20*, 56–64. [CrossRef]

66. Trendov, N.M. Comparative study on the motivations that drive urban community gardens in Central Eastern Europe. *Ann. Agrar. Sci.* **2018**, *16*, 85–89. [CrossRef]

67. Cepic, S.; Tomicevic-Dubljevic, J.; Zivojinovic, I. Is there a demand for collective urban gardens? Needs and motivations of potential gardeners in Belgrade. *Urban For. Urban Green.* **2020**, *53*, 126716. [CrossRef]

68. Clayton, S. Domesticated nature: Motivations for gardening and perceptions of environmental impact. *J. Environ. Psychol.* **2007**, *27*, 215–224. [CrossRef]

69. Kingsley, J.; Foenander, E. "You feel like you're part of something bigger": Exploring motivations for community garden par-ticipation in Melbourne, Australia. *BMC Public Health* **2019**, *19*, 745. [CrossRef] [PubMed]

70. Becker, S.L.; Von Der Wall, G. Tracing regime influence on urban community gardening: How resource dependence causes barriers to garden longer term sustainability. *Urban For. Urban Green.* **2018**, *35*, 82–90. [CrossRef]

71. Wesener, A.; Fox-Kämper, R.; Sondermann, M.; Münderlein, D. Placemaking in action: Factors that support or obstruct the development of urban community gardens. *Sustainability* **2020**, *12*, 657. [CrossRef]

72. Shackleton, C.M.; Njwaxu, A. Does the absence of community involvement underpin the demise of urban neighbourhood parks in the Eastern Cape, South Africa? *Landsc. Urban Plan.* **2021**, *207*, 104006. [CrossRef]

73. Von Döhren, P.; Haase, D. Ecosystem disservices research: A review of the state of the art with a focus on cities. *Ecol. Indic.* **2015**, *52*, 490–497. [CrossRef]

74. Kremer, P.; Hamstead, Z.A.; McPhearson, T. A social–ecological assessment of vacant lots in New York City. *Landsc. Urban Plan.* **2013**, *120*, 218–233. [CrossRef]

75. Yadav, P.; Duckworth, K.; Grewal, P.S. Habitat structure influences below ground biocontrol services: A comparison between urban gardens and vacant lots. *Landsc. Urban Plan.* **2012**, *104*, 238–244. [CrossRef]
76. Blanco, J.; Sourdril, A.; Deconchat, M.; Barnaud, C.; Cristobal, M.S.; Andrieu, E. How farmers feel about trees: Perceptions of ecosystem services and disservices associated with rural forests in southwestern France. *Ecosyst. Serv.* **2020**, *42*, 101066. [CrossRef]
77. Drake, L.; Lawson, L.J. Validating verdancy or vacancy? The relationship of community gardens and vacant lands in the U.S. *Cities* **2014**, *40*, 133–142. [CrossRef]
78. Millennium Ecosystem Assessment. Ecosystems and human well-being. In *Millennium Ecosystem Assessment*; Island Press: Washington, DC, USA, 2003.
79. Schniederjans, D.G.; Starkey, C.M. Intention and willingness to pay for green freight transportation: An empirical examination. *Transp. Res. Part D Transp. Environ.* **2014**, *31*, 116–125. [CrossRef]
80. Bauske, E.; Cruickshank, J.; Hutcheson, B. Healthy life community garden: Food and neighborhood transformation. *Acta Hortic.* **2017**, 395–398. [CrossRef]
81. Villalobos, A.; Alaimo, K.; Erickson, C.; Harrall, K.; Glueck, D.; Buchenau, H.; Coringrato, E.; Decker, E.; Fahnestock, L.; Hamman, R.; et al. CAPS on the move: Crafting an approach to recruitment for a randomized controlled trial of community gardening. *Contemp. Clin. Trials Commun.* **2019**, *16*, 100482. [CrossRef]
82. Alaimo, K.; Packnett, E.; Miles, R.A.; Kruger, D.J. Fruit and Vegetable Intake among Urban Community Gardeners. *J. Nutr. Educ. Behav.* **2008**, *40*, 94–101. [CrossRef]
83. Chen, S.; Wang, Y.; Ni, Z.; Zhang, X.; Xia, B. Benefits of the ecosystem services provided by urban green infrastructures: Differences between perception and measurements. *Urban For. Urban Green.* **2020**, *54*, 126774. [CrossRef]
84. Turner, B. Embodied connections: Sustainability, food systems and community gardens. *Local Environ.* **2011**, *16*, 509–522. [CrossRef]
85. Middle, I.; Dzidic, P.; Buckley, A.; Bennett, D.; Tye, M.; Jones, R. Integrating community gardens into public parks: An innovative approach for providing ecosystem services in urban areas. *Urban For. Urban Green.* **2014**, *13*, 638–645. [CrossRef]
86. Opitz, I.; Berges, R.; Piorr, A.; Krikser, T. Contributing to food security in urban areas: Differences between urban agriculture and peri-urban agriculture in the Global North. *Agric. Hum. Values* **2016**, *33*, 341–358. [CrossRef]
87. Bowler, D.E.; Buyung-Ali, L.; Knight, T.M.; Pullin, A.S. Urban greening to cool towns and cities: A systematic review of the empirical evidence. *Landsc. Urban Plan.* **2010**, *97*, 147–155. [CrossRef]
88. Vogl-Lukasser, B.; Vogl, C.R.; Gütler, M.; Heckler, S. Plant Species with Spontaneous Reproduction in Homegardens in Eastern Tyrol (Austria): Perception and management by women farmers. *Ethnobot. Res. Appl.* **2010**, *8*, 001–015. [CrossRef]
89. Speak, A.; Mizgajski, A.; Borysiak, J. Allotment gardens and parks: Provision of ecosystem services with an emphasis on biodiversity. *Urban For. Urban Green.* **2015**, *14*, 772–781. [CrossRef]
90. Celata, F.; Coletti, R. The policing of community gardening in Rome. *Environ. Innov. Soc. Transit.* **2018**, *29*, 17–24. [CrossRef]
91. Lyytimäki, J.; Petersen, L.K.; Normander, B.; Bezák, P. Nature as a nuisance ecosystem services and disservices to urban lifestyle. *Environ. Sci.* **2008**, *5*, 161–172. [CrossRef]
92. Perera, L.N.; Mafiz, A.I. Antimicrobial-resistant *E. coli* and *Enterococcus* spp. Recovered from urban community gardens. *Food Control* **2020**, *108*, 106857. [CrossRef]
93. Galton, F. Personal identification and description. *Nature* **1888**, *38*, 173–177. [CrossRef]
94. Spearman, C. "General Intelligence," objectively determined and measured. *Am. J. Psychol.* **1904**, *15*, 201–292. [CrossRef]
95. Wright, S. Correlation and causation. *J. Agric. Res.* **1921**, *20*, 557–585.
96. Ketchen, D.J. A Primer on Partial Least Squares Structural Equation Modeling. *Long Range Planning.* **2013**, *46*, 184–185. [CrossRef]
97. Tenenhaus, M.; Vinzi, V.E.; Chatelin, Y.M.; Lauro, C. PLS path modeling. *Comput. Stat. Data Anal.* **2005**, *48*, 159–205. [CrossRef]
98. Arbuckle, J. *Amos 17.0 User's Guide*; SPSS Inc.: Chicago, IL, USA, 2008.
99. Mueller, R.O.; Hancock, G.R. Best practices in structural equation modeling. In *Best Practices in Quantitative Methods*; Osborne, J.W., Ed.; SAGE Publications, Inc.: Louisville, KY, USA, 2008; pp. 488–508.
100. Fornell, C.; Larcker, D.F. Evaluating Structural Equation Models with Unobservable Variables and Measurement Error. *J. Mark. Res.* **1981**, *18*, 39–50. [CrossRef]
101. Anderson, J.C.; Gerbing, D.W. Structural equation modeling in practice: A review and recommended two-step approach. *Psychol. Bull.* **1988**, *103*, 411–423. [CrossRef]
102. Yunus, F. Statistics Using SPSS: An Integrative Approach, second edition. *JAPS* **2010**, *37*, 2119–2120. [CrossRef]
103. Comstock, N.; Dickinson, L.M.; Marshall, J.A.; Soobader, M.-J.; Turbin, M.S.; Buchenau, M.; Litt, J.S. Neighborhood attachment and its correlates: Exploring neighborhood conditions, collective efficacy, and gardening. *J. Environ. Psychol.* **2010**, *30*, 435–442. [CrossRef]
104. Rouillon, M.; Harvey, P.; Kristensen, L.J.; George, S.G.; Taylor, M.P. VegeSafe: A community science program measuring soil-metal contamination, evaluating risk and providing advice for safe gardening. *Environ. Pollut.* **2017**, *222*, 557–566. [CrossRef]
105. Kiesling, F.M.; Manning, C.M. How green is your thumb? Environmental gardening identity and ecological gardening practices. *J. Environ. Psychol.* **2010**, *30*, 315–327. [CrossRef]
106. Farrier, A.; Dooris, M.; Morley, A. Catalysing change? A critical exploration of the impacts of a community food initiative on people, place and prosperity. *Landsc. Urban Plan.* **2019**, *192*, 103663. [CrossRef]

107. Petrovic, N.; Simpson, T.; Orlove, B.; Dowd-Uribe, B. Environmental and social dimensions of community gardens in East Harlem. *Landsc. Urban Plan.* **2019**, *183*, 36–49. [CrossRef]

108. Caputo, S.; Rumble, H. "I like to get my hands stuck in the soil": A pilot study in the acceptance of soil-less methods of culti-vation in community gardens. *J. Clean. Prod.* **2020**, *258*, 120585. [CrossRef]

109. Amato-Lourenco, L.F.; Lobo, D.J.; Guimarães, E.T.; Moreira, T.C.; Carvalho-Oliveira, R.; Saiki, M.; Saldiva, P.H.; Mauad, T. Biomonitoring of genotoxic effects and elemental accumulation derived from air pollution in community urban gardens. *Sci. Total. Environ.* **2017**, *575*, 1438–1444. [CrossRef] [PubMed]

110. Laidlaw, M.A.; Alankarage, D.H.; Reichman, S.; Taylor, M.P.; Ball, A. Assessment of soil metal concentrations in residential and community vegetable gardens in Melbourne, Australia. *Chemosphere* **2018**, *199*, 303–311. [CrossRef]

111. Pudup, M.B. It takes a garden: Cultivating citizen-subjects in organized garden projects. *Geoforum* **2008**, *39*, 1228–1240. [CrossRef]

112. Ghose, R.; Pettygrove, M. Actors and networks in urban community garden development. *Geoforum* **2014**, *53*, 93–103. [CrossRef]

113. Litt, J.S.; Schmiege, S.J.; Hale, J.W.; Buchenau, M.; Sancar, F. Exploring ecological, emotional and social levers of self-rated health for urban gardeners and non-gardeners: A path analysis. *Soc. Sci. Med.* **2015**, *144*, 1–8. [CrossRef]

114. Passidomo, C. Community gardening and governance over urban nature in New Orleans's Lower Ninth Ward. *Urban For. Urban Green.* **2016**, *19*, 271–277. [CrossRef]

115. Shimpo, N.; Wesener, A.; McWilliam, W. How community gardens may contribute to community resilience following an earthquake. *Urban For. Urban Green.* **2019**, *38*, 124–132. [CrossRef]

116. Nova, P.; Pinto, E.; Chaves, B.; Silva, M. Urban organic community gardening to promote environmental sustainability practices and increase fruit, vegetables and organic food consumption. *Gac. Sanit.* **2020**, *34*, 4–9. [CrossRef]

117. Hale, J.; Knapp, C.; Bardwell, L.; Buchenau, M.; Marshall, J.; Sancar, F.; Litt, J.S. Connecting food environments and health through the relational nature of aesthetics: Gaining insight through the community gardening experience. *Soc. Sci. Med.* **2011**, *72*, 1853–1863. [CrossRef] [PubMed]

118. Marsh, P. D09-C Walking Each Other Home: Weaving Informal Palliative Supports into a Community Garden. *J. Pain Symptom Manag.* **2016**, *52*, e35–e36. [CrossRef]

119. Chan, J.; DuBois, B.; Tidball, K.G. Refuges of local resilience: Community gardens in post-Sandy New York City. *Urban For. Urban Green.* **2015**, *14*, 625–635. [CrossRef]

120. Spilková, J. Producing space, cultivating community: The story of Prague's new community gardens. *Agric. Hum. Values* **2017**, *34*, 887–897. [CrossRef]

121. Tappert, S.; Klöti, T.; Drilling, M. Contested urban green spaces in the compact city: The (re-)negotiation of urban gardening in Swiss cities. *Landsc. Urban Plan.* **2018**, *170*, 69–78. [CrossRef]

122. Tharrey, M.; Sachs, A.; Perignon, M.; Simon, C.; Mejean, C.; Litt, J.; Darmon, N. Improving lifestyles sustainability through community gardening: Results and lessons learnt from the JArDinS quasi-experimental study. *BMC Public Health* **2020**, *20*, 1798. [CrossRef]

123. Niinemets, Ü.; Peñuelas, J. Gardening and urban landscaping: Significant players in global change. *Trends Plant Sci.* **2008**, *13*, 60–65. [CrossRef]

124. Dennis, M.; James, P. User participation in urban green commons: Exploring the links between access, voluntarism, biodiversity and well being. *Urban For. Urban Green.* **2016**, *15*, 22–31. [CrossRef]

125. Lindemann-Matthies, P.; Brieger, H. Does urban gardening increase aesthetic quality of urban areas? A case study from Ger-many. *Urban For. Urban Green.* **2016**, *17*, 33–41. [CrossRef]

126. Danford, R.S.; Strohbach, M.; Warren, P.S.; Ryan, R.L. Active Greening or Rewilding the city: How does the intention behind small pockets of urban green affect use? *Urban For. Urban Green.* **2018**, *29*, 377–383. [CrossRef]

127. Anderson, E.C.; Egerer, M.H.; Fouch, N.; Clarke, M.; Davidson, M.J. Comparing community garden typologies of Baltimore, Chicago, and New York City (USA) to understand potential implications for socio-ecological services. *Urban Ecosyst.* **2019**, *22*, 671–681. [CrossRef]

128. Milliron, B.-J.; Tooze, J.; Vitolins, M. Process Evaluation of a Community Garden at an Urban Medical Center. *J. Nutr. Educ. Behav.* **2014**, *46*, S167. [CrossRef]

129. Marsh, P.; Gartrell, G.; Egg, G.; Nolan, A.; Cross, M. End-of-Life care in a community garden: Findings from a Participatory Action Research project in regional Australia. *Health Place* **2017**, *45*, 110–116. [CrossRef] [PubMed]

130. Litt, J.S.; Lambert, J.R.; Glueck, D.H. Gardening and age-related weight gain: Results from a cross-sectional survey of Denver residents. *Prev. Med. Rep.* **2017**, *8*, 221–225. [CrossRef]

131. Morris, A.; Campbell, K.; Britton, R.; Shabel, A.; Pacumbaba, R.; Taylor, D.; Wood, S. Urban Supplemental Nutrition Assistance Program -Education (USNAP-Ed) Community Garden Project. *J. Nutr. Educ. Behav.* **2017**, *49*, S130. [CrossRef]

132. Chalmin-Pui, L.S.; Roe, J.; Griffiths, A.; Smyth, N.; Heaton, T.; Clayden, A.; Cameron, R. "It made me feel brighter in myself"-The health and well-being impacts of a residential front garden horticultural intervention. *Landsc. Urban Plan.* **2021**, *205*, 103958. [CrossRef] [PubMed]

133. Corley, J.; Okely, J.A.; Taylor, A.M.; Page, D.; Welstead, M.; Skarabela, B.; Redmond, P.; Cox, S.R.; Russ, T.C. Home garden use during COVID-19: Associations with physical and mental wellbeing in older adults. *J. Environ. Psychol.* **2021**, *73*, 101545. [CrossRef]

134. Lehberger, M.; Kleih, A.-K.; Sparke, K. Self-reported well-being and the importance of green spaces—A comparison of garden owners and non-garden owners in times of COVID-19. *Landsc. Urban Plan.* **2021**, *212*, 104108. [CrossRef]
135. Corrigan, M.P. Growing what you eat: Developing community gardens in Baltimore, Maryland. *Appl. Geogr.* **2011**, *31*, 1232–1241. [CrossRef]

MDPI

St. Alban-Anlage 66

4052 Basel

Switzerland

www.mdpi.com

Agronomy Editorial Office

E-mail: agronomy@mdpi.com

www.mdpi.com/journal/agronomy